Fascinating Life Sciences

This interdisciplinary series brings together the most essential and captivating topics in the life sciences. They range from the plant sciences to zoology, from the microbiome to macrobiome, and from basic biology to biotechnology. The series not only highlights fascinating research; it also discusses major challenges associated with the life sciences and related disciplines and outlines future research directions. Individual volumes provide in-depth information, are richly illustrated with photographs, illustrations, and maps, and feature suggestions for further reading or glossaries where appropriate.

Interested researchers in all areas of the life sciences, as well as biology enthusiasts, will find the series' interdisciplinary focus and highly readable volumes especially appealing.

More information about this series at http://www.springer.com/series/15408

Manuel Soler
Editor

Avian Brood Parasitism

Behaviour, Ecology, Evolution and Coevolution

Springer

Editor
Manuel Soler
Departamento de Zoología, Facultad de Ciencias
Universidad de Granada
Granada, Spain

Grupo Coevolución, Unidad Asociada al CSIC
Universidad de Granada
Granada, Spain

ISSN 2509-6745 ISSN 2509-6753 (electronic)
Fascinating Life Sciences
ISBN 978-3-030-10321-7 ISBN 978-3-319-73138-4 (eBook)
https://doi.org/10.1007/978-3-319-73138-4

Softcover re-print of the Hardcover 1st edition 2017

Cover illustration: Common cuckoo, the best studied brood parasite, fed by a robin. Photo by Oldřich Mikulica

Printed on acid-free paper

This Springer imprint is published by the registered company Springer International Publishing AG part of Springer Nature.
The registered company address is: Gewerbestrasse 11, 6330 Cham, Switzerland

Foreword

Among the many ways in which individuals of each species exploit other species, brood parasitism has provided some of the richest opportunities for probing the many ways in which species can coevolve. This insightful and broad-ranging volume grapples with many of the difficult questions common to all coevolutionary studies, showing why it takes approaches from multiple scientific disciplines to answer how any form of coevolution develops in nature. How do these interactions begin? What traits drive coevolutionary selection? How and why do coevolutionary interactions diversify into geographic mosaics of traits and ecological outcomes? To what extent has coevolution contributed to the diversification of lineages? How, and under what conditions, may coevolution shape interactions among webs of interacting species rather than pairs of species? Individual chapters by major researchers studying brood parasitism probe these questions and others from multiple perspectives; collectively they illustrate—through elegant experiments, field observations, models, and summaries of past work—the many ways in which natural selection can shape interactions between species.

The long history of the study of brood parasitism has provided answers to some of these questions and, inevitably, produced yet more questions. These chapters show us that brood parasites can adapt to the local populations of their hosts, responding to host defenses with fine-tuned counterdefenses. But they simultaneously make us ask why some forms of defense and counterdefense are common, whereas others are rare, and why some behaviours important to defense or counterdefense seem genetically constrained while others seem so plastic. The remarkably wide range of morphological, physiological, behavioural, and life history adaptations found among brood parasites and their hosts makes us marvel at how natural selection hones combinations of coevolving interactions, but it makes us wonder why some trait combinations are more common than others. These are problems that must be confronted in all coevolutionary studies, but those studying brood parasitism have been adept at bringing them into sharp focus.

The worldwide effort by so many scientists to understand this form of coevolving interactions has become an exemplar of what it takes to understand the coevolutionary process. The breadth of studies of brood parasites and their hosts found in these chapters shows us that no field of science and no single approach to science can provide us with all the answers we need to understand the diversity of ways by which the coevolutionary process shapes traits, populations, species, and coevolving lineages. The deep insights found among these studies have been won through painstaking work that has included long days and years observing parasites and hosts in nature, experimentally manipulating traits in the wild and in the laboratory, evaluating morphological and physiological traits within the mathematical constraints known from physics and chemistry, assessing life histories within the boundaries known from life-history theory, comparing behaviours both within and among species, and determining through molecular and related methods how species and traits have evolved and diversified from ancestors to descendants over millions of years.

The questions and their answers addressed here span the full range of biological hierarchy, requiring approaches and perspectives at the molecular, organ, individual, population, species, lineage, community, regional, and global levels. We need all those levels to get the answers. Collectively, these studies tell us that we probably are still underestimating the role of coevolution in all forms of interaction in nature. By often focusing on a few traits, a few populations, and a few species, we probably often miss the real foci of reciprocal selection as it molds pairs or groups of interacting species.

Because the distribution of brood parasitism among species is so well known, these studies also show us that there are not inherent relationships between adaptation, coadaptation, speciation, and adaptive radiation. Brood parasitism has arisen multiple times, but each origin has produced a different combination of adaptive traits, coadaptive traits, and speciation patterns. At least that appears to be so from what we can tell from the study of extant species. Lacking a full fossil record, it is difficult to infer reliably how coevolving lineages wax and wane over time at any fine-scale of phylogenetic resolution. Even if we did have a full fossil record, however, it would be a matter of viewpoint in deciding the extent to which brood parasitism is successful. Does success mean the ability of a brood parasitic species to continue to coadapt with a host species over millions of years, or does success mean that a species has managed to coevolve with different populations coevolving in different ways? Or is success not just about adaptation but also about divergence into many descendent species with divergent life histories? Regardless of our definition of success, brood parasitism has arisen time and again during the history of life, and it has persisted, and sometimes diversified, over long periods of time in earth's history.

The studies and approaches summarized here also show the interconnectedness of the major questions in ecology, evolution, and coevolution as we try to understand the web of life. Studies of brood parasitism show us the effects of interactions on habitat and host selection, social communication and mating systems, correlations among traits, and population dynamics. In turn, these studies show us how environmental context, behaviour and physiology, cognitive systems and mating systems, genetic architecture of traits, and population structure affect the evolution of interactions and the wide range of potential evolutionary solutions to similar ecological and social problems. And they show us the interplay between intraspecific and interspecific interactions. The diversity of studies on brood parasitism therefore has provided us with an important lens through which to assess the interconnectedness of approaches to coevolutionary biology and, more generally, to our understanding of the network structure of biodiversity.

Even as they help us understand general principles in ecological and evolutionary processes, however, these studies caution us against overextending general from individual case studies. Parasitism by cuckoos and cowbirds provides as many contrasts as similarities, as does parasitism within each lineage of brood parasites. The differences in conclusions among studies are as important as the similarities. Each coevolving interaction represents only a subset of the potential range of coevolutionary trajectories. The collective body of work worldwide on how brood parasites interact with their hosts, summarized with such great care and insight in this volume, shows us the remarkable ability of natural selection to shape the evolution of interactions and the coevolution of species.

Department of Ecology and Evolutionary Biology
University of California, Santa Cruz
Santa Cruz, CA
USA

John N. Thompson

Preface

A female laying eggs in the nests of other females of its own or a different species avoids laborious and costly parental care. This behaviour has frequently been reported in conspecific nests, although there are only about one hundred obligate avian brood parasites. Why? Probably because many host species have evolved efficient defences. They are able to recognize an adult brood parasite as an enemy and attack it, and they can also recognize and reject parasitic eggs, chicks or fledglings. This means that to be a brood parasite is not easy. However, in the coevolutionary arms race involving brood parasites and their hosts, brood parasites in turn can evolve ingenious and often efficient counter-defences such as rapid and secretive laying behaviour, nearly perfect egg mimicry, destruction or removal of host eggs, revenge killing of host young, manipulative begging behaviour or chick and fledgling mimicry.

During the present century, hundreds of empirical and experimental studies have been published on brood parasitism, prompting an explosion of compelling discoveries on topics such as the evolution of sequences of adaptations and counter-adaptations and the factors that affect the evolution of both host defences and parasite counter-defences. Brood parasitism has become one of the most flourishing areas of research in evolutionary ecology and one of the most influential topics in the coevolutionary theory.

The idea for the present book arose from Lars Koerner, publishing editor at Springer, who, after noting that brood parasitism had undergone extraordinary advances over the last two decades but has not been covered by any book during the present century, suggested to me the possibility of publishing such a book. I agreed that an edited volume would be both valuable and timely. Then, I presented a preliminary detailed plan for the book including the title of 20 chapters specifying the name of expert researchers suggested as authors to write them. Fortunately, following a review by Springer editors and four expert reviewers, the publishers committed to the project.

The idea was to include about 30 moderate-length chapters of general interest and up-to-date reviews of high quality that could each be treated as a self-contained unit to be read independently while preventing thematic duplications and avoiding overlapping. For this objective, each corresponding author was provided with the outlines of other chapters potentially related to the one that author was going to write. To improve the quality of the chapters, I suggested to the authors the possibility that each chapter would be reviewed in a constructive way by at least two non-anonymous reviewers (mainly other authors of the book). The idea was well received, and, undoubtedly, the constructive and valuable comments and suggestions provided by these expert reviewers have greatly enhanced the quality of the overall work.

This book covers the major areas of brood parasitism in 30 chapters written by an authority or authorities in each area, thereby offering a broad survey of a diverse array of topics and perspectives concerning the behaviour and ecology of brood parasites, but, especially, of the coevolutionary arms race between brood parasites and their hosts. Each chapter starts with an abstract and ends with a section containing concluding remarks and recommendations for future research, which identify crucial gaps in our knowledge on the subject and the major questions that need to be addressed in the near future.

One of my concerns during the preparation of this book has been the use of several terms, which are not standardized in the brood parasitism literature. This is not a trivial point because inconsistency in the use of existing terms can add confusion, which could slow down scientific advance. Therefore, regarding some terms, whose use is confusing, the involved authors and I together decided upon the more appropriate term. The most significant case is the one concerning the term used to designate egg laying in the nests of conspecific females. Five different terms have been used in the brood parasitism literature: "conspecific brood parasitism" (CBP), "conspecific nest parasitism", "intraspecific brood parasitism" (IBP), "intraspecific nest parasitism" and, more rarely, "egg dumping". Regarding the four terms used frequently, we decided that the two using "nest parasitism" were less appropriate because there is confusion with species that steal nests without laying eggs. With respect to the other two, we decided that "conspecific brood parasitism" was the most appropriate because of two reasons: first, this term is used more in papers on egg recognition behaviour (the start of brood parasitism studies), and second, "intraspecific brood parasitism" (IBP) is broadly accepted; given that "intraspecific brood parasitism" would also be IBP, the use of conspecific brood parasitism (CBP) would be less confusing. Thus, we decided to use it along this book and suggest it to the scientific community in order to standardize criteria.

Other conflicting terms are "nestmate tolerance" vs. "nestmate acceptance". In this case, although the former has been more frequently used than the latter, we have decided upon the use of the latter because "nestmate tolerance" runs counter to already well-established terminology in animal host–parasite coevolution (tolerance is a kind of host defence, different from resistance, that consists of accepting parasitism while developing strategies that minimize the harm caused by the parasite).

Sometimes, the terms "evictors" and "ejectors" ("eviction" and "ejection") are used as synonymous in the brood parasitism literature. This should be avoided. Our suggestion is, in agreement with the most frequent use in the brood parasitism literature, to restrict the use of "evictors" when referring to parasitic nestlings that throw out of the nest host nestlings and "ejectors" when referring to adult hosts that throw out of the nest the parasitic egg. Other two terms whose use has been standardized are "multiparasitism" instead of "multiple parasitism" and "nestmate-killing" behaviour instead of "nesmate-cide" behaviour. I hope that the use of these standardized terminologies will improve efficiency and communication in brood parasitism research.

The chapters are organized into eight parts that represent major themes in brood parasitism. Following an introductory chapter, the first part is devoted to the "Coevolution and Diversification of Interspecific Brood Parasitism". It includes three loosely related chapters. The first examines the rates of coevolution between brood parasites and their hosts, the second explores the different ideas concerning the evolution and phylogenetic history of avian brood parasitism, and the third presents a comprehensive taxonomic review of the brood parasite species corresponding to the four orders which contain such parasites (only Cuculiformes had previously been reviewed). The second part includes three chapters dealing with "Conspecific Brood Parasitism". The five chapters in the third part are concerned with different areas of growth in research that use "Brood Parasites as Study Models". The fourth part presents two chapters that examine "Host Use by Brood Parasites" in two different geographic areas. The last four parts analyse the coevolutionary interactions between brood parasites and their hosts at the different stages of the breeding cycle. The sixth part includes two chapters that consider "Coevolutionary Interactions at Any Stage", and the last three parts include chapters that examine coevolutionary interactions in the subsequent phases of the nestling cycle at the prelaying stage (Part VI), at the laying stage (Part VII) and at the nestlings and fledgling stages (Part VIII).

These eight parts present a comprehensive survey of the latest research in brood parasitism, including topics such as the evolution of conspecific brood parasitism in birds, evolution and phylogenetic history of avian brood parasites, parasitic behaviour used by brood parasites, adaptations and counter-adaptations of brood parasites and their hosts at every stage of the breeding cycle (before laying, egg, chick and fledgling stages), factors that affect the evolution of host defences and parasite counter-defences, the role of phenotypic plasticity in host defences, mechanisms driving egg recognition and rejection, the evolution of nest sharing or chick killing by brood parasite chicks, begging behaviour in parasitized nests vs. food delivery by host adults and the recognition of conspecifics by juvenile brood parasites.

The book provides up-to-date reviews concerning theoretical, empirical and methodological advances made in each of these and other topics, presenting hypotheses, offering predictions and pointing the way for further research, which will offer state-of-the-art knowledge on this topic. Thus, this book will constitute an invaluable resource for new or established researchers in brood parasitism as well as a comprehensive reference for ornithologists, ecologists, ethologists and researchers in general interested in the evolution of ecology, behaviour and coevolution of brood parasites and their hosts. In addition, the book may also appeal to academics and students in other related disciplines.

Edited volumes are usually extensive collaborative projects in which the merit is attributable mainly to the authors of the chapters. In the present brood parasitism book, this is especially true. A total of 38 authors have participated in writing the 30 chapters making up the book (most of them, 29, have participated in the writing of only one chapter). I would like to thank all of the authors (many of them leading authorities in this area of research), not only for their enthusiastic response to my invitation to participate by writing one chapter (usually suggested by me) and for their excellent contributions but also for accepting my editorial suggestions, mainly two: to use a broad focus in their chapters taking into account as many brood parasites as possible and to present most of the descriptive information in tables in order to devote most of the main text to providing a real synthesis of the biological or evolutionary relevance of published discoveries, an approach that makes the chapter a practical scientific contribution to readers. In addition, I would like to thank them for their generosity in accepting to be the reviewers of other chapters (credits for constructive reviews are provided in the acknowledgements section of each chapter).

I would also like to extend my gratitude to Lars Koerner and Andrea Schlitzberger of Springer who provided invaluable help and advice throughout the whole process. Many thanks go also to Oldřich Mikulica for the cover image. And, finally, but not less importantly, I wish to thank my partner, Mayte Molina, for her patience and support during the time dedicated to this project.

Granada, Spain Manuel Soler

Contents

Brood Parasitism in Birds: A Coevolutionary Point of View

1

Manuel Soler

Abstract

Interspecific brood parasitism is a breeding strategy in which the brood parasite female evades all parental care by laying its eggs in host nests. Fitness costs imposed on hosts by brood parasitic adults or nestlings select for host defences setting the stage for a coevolutionary arms race in which brood parasites evolve counter-adaptations, which subsequently select for better host defences and so on. These so-called coevolutionary interactions have proven to be an excellent system for understanding coevolution. Here I review brood parasite–host interactions from the point of view of coevolutionary theory. After emphasizing the relevance of the costs provoked by brood parasitism, I present the traditional model of coevolution between brood parasites and their hosts, but I also incorporate new discoveries reported during the last two decades, which, frequently, do not support important predictions of coevolutionary theory. Next I describe the current situation of coevolution in brood parasite–host systems emphasizing three points that should be taken into account in studies centred on this subject. Later, I suggest three potentially important topics that have been almost neglected until now: (1) it would be superior to study as many defences as possible at all stages of the nesting cycle, (2) to examine the role of tolerance because this type of host defence contrary to what happen with resistance does not induce coevolutionary arms races and (3) to study the role of hormones in regulating host defensive responses to brood parasitism. Finally, a list of eleven future directions of research is provided.

M. Soler (✉)
Departamento de Zoología, Facultad de Ciencias, Universidad de Granada, Granada, Spain

Grupo Coevolución, Unidad Asociada al CSIC, Universidad de Granada, Granada, Spain
e-mail: msoler@ugr.es

M. Soler (ed.), *Avian Brood Parasitism*, Fascinating Life Sciences,
https://doi.org/10.1007/978-3-319-73138-4_1

1.1 Introduction

Brood parasitism is a type of parental-care parasitism in which brood parasitic females lay their eggs in the nests of other females of the same (conspecific brood parasitism, CBP) or different (interspecific brood parasitism, IBP) species (Roldán and Soler 2011). This implies that brood parasites fool hosts into raising their parasitic young and that hosts have to rear parasitic offspring, diverting resources away from their own young.

Two types of brood parasitism can be differentiated, one in which parasitic females do not provide parental care (obligate brood parasites) and another in which parasitic females rear their own offspring but, in addition, place some of their eggs in the care of other females (facultative brood parasites). Although facultative brood parasitic females can lay their parasitic eggs in the nests of other species (IBP), or in the nests of both conspecific and different species, they often parasitize only conspecific females, CBP being the most common form of facultative brood parasitism (see Chap. 4).

CBP is widespread in birds, having been documented in 236 avian species (Chap. 3), being much more frequent among precocial than altricial birds. While in the latter, the highest level of CBP does not reach 15%, in the former, most species (i.e. 60% among Anatidae or 100% in some small families) are engaged in CBP (Chap. 3). Although some cases of CBP have been described in insects (Müller et al. 1990; Roldán and Soler 2011), this reproductive strategy has been studied in detail mainly in birds. Similarly, although IBP is widespread throughout the animal kingdom having been described in spiders, insects, fish, frogs and birds (Sato 1986; Davies et al. 1989; Baba et al. 1990; Boulton and Polis 2002; Als et al. 2004; Brown et al. 2009; Kilner and Langmore 2011; Roldán and Soler 2011), it has been studied in depth mainly in birds [almost 3000 published papers listed on the Web of Science (Møller and Soler 2012)]. About 1% of bird species (109 species of 27 genera; Chap. 4) are known to be obligate brood parasites. This reproductive strategy has evolved independently seven times in birds (Chap. 3), being present in five families belonging to four orders: Anseriformes (family Anatidae), Cuculiformes (three lineages in the family Cuculidae), Piciformes (family Indicatoridae) and Passeriformes (families Viduidae and Icteridae) (see Chap. 4).

1.2 Costs Provoked by Brood Parasitism

By laying eggs in host nests, brood parasites impose fitness costs on hosts, which are mainly determined by the behaviour of the parasitic nestlings. Evictor and nestmate-killer brood parasite species are those that trigger the highest costs for hosts because parasitic nestlings evict all hosts' eggs or nestlings reducing host reproductive success to zero (Chap. 26). At the other extreme are conspecific brood parasites, in which brood parasitic costs involve only rearing one extra nestling or incubating one extra egg in altricial and precocial species, respectively. In between there are nonevictor brood parasitic species in which a great deal of variation in costs provoked to hosts can be found. Loss of host nest productivity due to parasitism by these nonevictor species extends throughout the nesting period. First, females frequently remove or destroy host eggs when they lay their eggs (e.g. Soler et al. 1997; Cossa et al. 2017). Second, given that brood parasitic nestlings usually hatch earlier, are larger and beg more intensively than host nestlings, competition for parental care usually decreases host brood size because parasitic nestlings outcompete host nestlings (Chap. 27). The effect of this competition depends on the relative size of host and parasite; in larger hosts both host and parasite nestlings can fledge, while in smaller hosts, which is the more frequent situation, parasitic nestlings provoke complete loss of the host brood (Chap. 27).

This negative effect of brood parasitism on host production of fledglings can be extended to the post-fledging period as has been shown in some host species (Payne and Payne 1998; Hoover and Reetz 2006; Mark and Rubenstein 2013), although in other studies, it has been reported that parasitism has no effect on survival during the post-fledging period (Smith 1981; Jenkins and Faaborg 2016).

In addition to these direct costs provoked by brood parasitism, other indirect costs need to be mentioned: first, clutch size decreases due to egg removal or destruction increasing the likelihood of nest abandonment by hosts (Tuero et al. 2007); second, exaggerated begging of brood parasitic nestlings can increase nest predation rate (Hannon et al. 2009; Ibáñez-Álamo et al. 2012a); and, third, brood parasites themselves might also act as predators (Chap. 15).

Brood parasitic nestlings are usually very efficient eliciting high provisioning effort by host parents both during nestling and fledging periods. This likely can lead to increased predation risk and decreased future fitness of host foster parents. This important potential cost of brood parasitism on host future survival and reproductive capacity has been studied in only a handful of papers with mixed results. Some studies have reported no differences between host foster parents after care for parasitized vs. nonparasitized broods [see Hauber (2006) and references therein]. However, in other studies different negative effects have been found. For example, parasitized parents laid a smaller clutch size (Hauber 2002; Hoover and Reetz 2006), delayed renesting in subsequent clutches during the same breeding season and were less likely to renest the following year (Mark and Rubenstein 2013). Then, logic dictates that there should be a major cost of brood parasitism to future reproduction, although more work is needed to demonstrate this prediction.

1.3 Coevolution in Brood Parasite–Host Systems

The severe costs imposed by brood parasitism on hosts (see Sect. 1.2) select for the evolution of host defences, which set the stage for a coevolutionary arms race, in which parasites evolve adaptive counter-defences that select for improved host defences, further parasitic adaptations and so on (Rothstein 1990; Davies 2000; Soler and Soler 2000; Stokke et al. 2005; Feeney et al. 2014; Soler 2014). This type of reciprocal evolution among interacting species is known as coevolution (Thompson 1994).

Steve Rothstein entitled his influential review of brood parasitism published nearly 30 years ago "A model system for coevolution: avian brood parasitism" (Rothstein 1990). This statement was justified at that time given that a great deal of adaptations and counter-adaptations had been reported and this system had provided some of the clearest experimental demonstrations of a coevolutionary process (e.g. de Brooke and Davies 1988; Soler and Møller 1990). Today, we can confirm that Rothstein's statement is even more justified at present, because a multitude of experimental studies have supported new and old predictions based on coevolutionary theory and several intriguing long-standing points have been solved driving new paradigm shifts (see below).

1.3.1 The Traditional Model Revised: Incorporating New Discoveries

The sequence of coevolutionary events based on the idea of a continuous arms race proposed by Davies and Brooke, developed on the basis of their studies on the common cuckoo (*Cuculus canorus*) and its hosts (Davies and de Brooke 1988, 1989a, b), and later slightly modified by Rothstein on the basis of his studies on the brown-headed cowbird (*Molothrus ater*) and its hosts (Rothstein 1990), was the model that drove brood parasitism studies during two decades, and, of course, it is a valuable model still today. However, an update of brood parasitism theory is needed because (1) since then many other brood parasite–host systems have been studied showing that host–parasite interactions can vary greatly in other species; (2) some highly influential topics of the traditional brood parasitism theory have been found to be less generalizable or useful than previously assumed; and (3) since 2003, new and previously even unsuspected adaptations and counter-adaptations have been discovered.

1.3.1.1 Some Classic Topics that Need to Be Reconsidered

1. The presumed costs of misimprinting are not generalizable. The evolution of egg rejection defence implies recognition of the parasitic egg, which has to be differentiated from the host's own eggs. It was assumed that egg recognition was based on imprinting-like processes. This implies, first, that learning the aspect of the female's own eggs during her first breeding attempt would allow her to reject the foreign egg differing in appearance from the learned morphology of the eggs (Rothstein 1974, 1978) and, second, that this learning process could be at risk of misimprinting costs (i.e. to imprint on the parasite egg if parasitism occurs during the first breeding attempt). Misimprinting costs were suggested to be the main explanation for the lack of nestling recognition given that at this stage misimprinting costs are enormous. If a host is parasitized at its first breeding attempt by an evictor brood parasite, it would imprint only on the parasite nestling and so will reject its own nestlings in all subsequent unparasitized breeding attempts (Lotem 1993). However, recent experimental studies have demonstrated that egg rejection is not affected by egg appearance in the first breeding attempt

(Moskát et al. 2014) and, more importantly, that misimprinting costs do not occur (Soler et al. 2013). With respect to other potential mechanisms for egg learning, by parasitizing first-time breeding females, Soler et al. (2013) found that the possibility of recognition templates being inherited was not supported, opening up the possibility of the existence of a memory-based mechanism in which recognition templates are learned in each new breeding attempt.

The existence of an imprinting-like process applied to chick recognition is not supported either because contrary to predictions of the misimprinting hypothesis, chick rejection has been demonstrated in some hosts of Australian cuckoos (Langmore et al. 2003).

In addition, the misimprinting hypothesis does not work at the fledgling stage either given that fledgling discrimination by hosts has been demonstrated in two host species of nonevictor brood parasites. On one hand, the baywing (*Agelaioides badius*) feeds mimetic fledglings of the specialist screaming cowbird (*Molothrus rufoaxillaris*) but do not feed nonmimetic fledglings from the generalist shiny cowbird (*Molothrus bonariensis*) (De Mársico et al. 2012). On the other hand, magpies (*Pica pica*) feed great spotted cuckoo (*Clamator glandarius*) fledglings when they are reared alone in the nest but are reluctant to feed fledgling cuckoos reared in mixed broods (Soler et al. 2014a). In this system, great spotted cuckoo nestlings have not counteracted magpie discrimination by evolving fledgling mimicry but by evolving a fascinating behaviour: undernourished cuckoo fledglings reared in mixed broods abandon their foster parents and join other cuckoo fledglings that are being cared for by magpies that reared only-cuckoo broods (Soler et al. 2014b). Taking into account that magpies are long-lived birds (Birkhead 1991), we can assume that at least some of those parasitized magpies that reared only-cuckoo broods were not parasitized during their first breeding attempt; however, all of them accepted to feed cuckoo fledglings, even those that were not reared in their nests (Soler et al. 2014b). This means, first, that the only relevant point in the case of magpies is the presence of the host's own nestlings, which provides them with the opportunity of comparison between own and parasitic fledglings, and, second, the existence of a memory-based mechanism in which recognition templates are learned in each new breeding attempt.

Thus, current available information does not support the misimprinting hypothesis. It seems that hosts learn neither the appearance of their eggs nor their chicks from their earlier breeding experiences, and, consequently, the cost of misimprinting on foreign eggs or nestlings does not have any effect on future nesting attempts.

2. The response to egg mimicry is not always an increase in discrimination ability. Another counter-adaptation in hosts as a response to egg mimicry different from an increase in discrimination ability is to generate variation in egg appearance among individuals, i.e. polymorphism, which would make effective mimicry more difficult to evolve and will select in turn for polymorphism in brood parasites (Yang et al. 2010; Spottiswoode and Stevens 2011). Egg polymorphism was discovered through the study of a greater range of host species. Notably, it appeared not only in newly studied brood parasite-host systems but also in some

Chinese hosts of the common cuckoo. In Europe, common cuckoo hosts only respond to egg mimicry by increasing egg-recognition ability, while some cuckoo hosts in China have evolved egg polymorphism and cuckoos have also evolved egg polymorphism in turn as a consequence of frequency-dependent selection (Yang et al. 2010). This important difference between European and Chinese host populations implies that the coevolutionary arms race between one brood parasite and its hosts can vary greatly geographically, even provoking different evolutionary trajectories.

3. Recognition does not imply rejection. Traditionally it was assumed that rejection implies recognition, whereas acceptance implies absence of recognition (Rothstein 1990; Davies 2000). However, nowadays it is broadly accepted that egg rejection is a much more complicated three-stage process (Fig. 1.1). Indeed, recognition without rejection has been experimentally demonstrated in four host species (Antonov et al. 2009; Guigueno and Sealy 2012; Soler et al. 2012; Ruiz-Raya et al. 2015), which confirms that rejection is not always the result of recognition. Nowadays it is clear that accepting parasitic eggs is not always the consequence of recognition failure (Soler 2014). In fact, it has been shown that

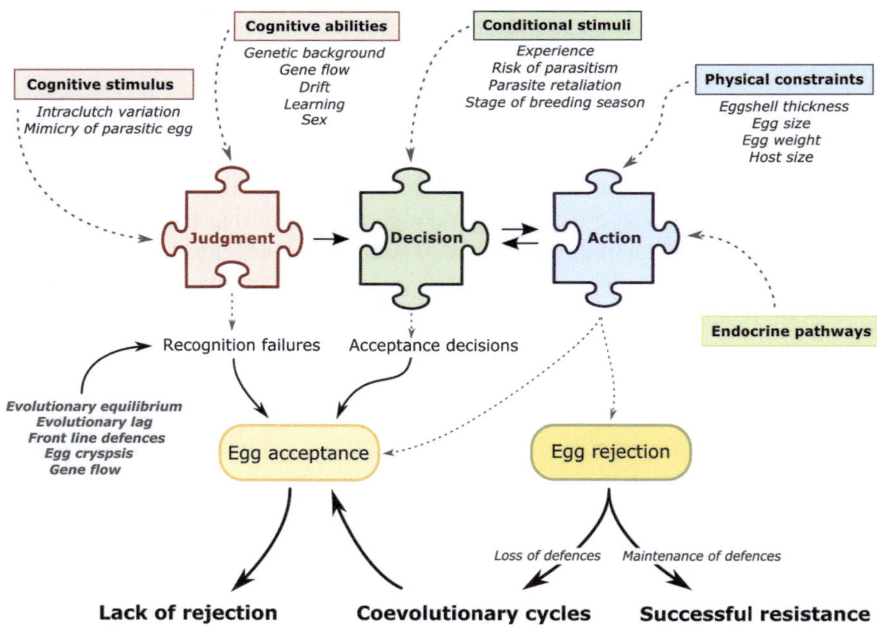

Fig. 1.1 Egg rejection is a three-stage process (puzzle pieces) that can be affected separately by different factors (*coloured squares*). Proximate mechanisms affecting the egg-rejection process (*dashed lines*). Figure based on Stokke et al. (2005), Soler et al. (2012) and Chap. 25. Although the term "judgment" is usually related to humans only, it is included here following the new approach of adopting a unified conceptual framework for the study of animal decisions ("judgment and decision-making"; JDM) both in behavioural ecology studies in general (Mendelson et al. 2016) and in brood parasitism studies in particular [Ruíz-Raya and Soler (2018). After Ruiz-Raya (2017)]

sometimes mimetic and larger experimental eggs are recognized but accepted (Ruiz-Raya et al. 2015). In addition, accepting parasitic eggs is not always the consequence of recognition failure but of acceptance decisions (Soler et al. 2017). The fact that sometimes mimetic and larger experimental eggs are recognized but accepted because of recognition or mechanical constraints implies that there are two crucial points regarding egg ejection: motivation and a flexible rejection threshold based on phenotypic plasticity and risk of parasitism (Soler et al. 2017).

4. The arms race is not confined to the egg stage. Conventional theory considered that the arms race between brood parasites and their hosts was confined to the egg stage of the breeding cycle and, even, that a lack of recognition implies a lack of defences (Rothstein 1990; Davies 2000). On the contrary, nowadays it is clear that hosts have evolved many different lines of defence that may operate at any stage of the breeding cycle and that brood parasites have evolved counter-adaptations against most of them (Fig. 1.2). Two clear examples can be provided. First, the "frontline" defences. Direct nest defence (i.e. mobbing or attacking) against brood parasites that approach host nests was known to be a potentially efficient host defence evolved as a specific adaptation against brood parasites (Moksnes et al. 1991; Welbergen and Davies 2008, 2009). However, during recent years the existence of several lines of defence prior to the deposition of the parasitic egg has been demonstrated (the "frontline" of the arms race; Feeney et al. 2012; Chap. 17). Second, nestling and fledgling recognition. Until the beginning of the present century, it was assumed that nestling rejection was highly unlikely to evolve; however, today it has clearly been demonstrated both that some host species have evolved defensive mechanisms against parasitic nestlings and some brood parasites have evolved adaptations to counteract this host defence (Grim 2006; Soler 2009; Chaps. 29 and 30). In addition, even more surprisingly, both fledgling discrimination by hosts and counter-adaptations by fledgling parasites have been demonstrated in two brood parasite–host systems in which brood parasites are nonevictors and the presence of the host's own nestlings provides hosts with the opportunity of comparison between own and parasitic fledglings (De Mársico et al. 2012; Soler et al. 2014a, b; Chap. 30).

Nowadays it is broadly accepted that the arms race can operate at each stage and that adaptations at one stage of the nesting cycle can influence evolution at other stages (Davies 2011; Feeney et al. 2014; Soler 2014). In addition, it is well known that different lines of defence are closely linked. An efficient defensive strategy at an earlier stage of the breeding cycle will be more beneficial than another at a later stage because costs of brood parasitism accumulate as the breeding cycle advances (Davies 2011; Feeney et al. 2014). One efficient line of defence (or two or more lines of defence acting additively) could prevent the evolution of later lines of defence ("strategy blocking"; Britton et al. 2007). This implies, first, that later-stage defences are less common than early stage defences and, second, that a host defence at a later stage of the breeding cycle would be more likely to appear if defences have not evolved at earlier stages (Britton et al. 2007; Feeney et al. 2012, 2014).

	Better host defences	More sophisticated parasite counter-defences
FRONTLINE ADAPTATIONS	Nests further from parasite perches Secretive nesting Concealed nests	Host´s nest monitoring
	Nest architecture	Nest choice Small size
	Active nest defence	Male lure behaviour to enable female to lay Hawk mimicry
	Nest at higher densities, in large groups or colonies	
	Evaluation of risk of parasitism	Secretive behaviours Rapid laying
	Female remaining sitting on the nest	Parasitic event while hosts are incubating
EGG STAGE	Egg rejection	Egg mimicry Egg crypsis Supernormal eggs Stronger eggshells Mafia tactics Predator-like calls
	Low intra-clutch variation Higher inter-clutch variation	
	Polymorphism in host eggs	Polymorphism in parasite eggs
CHICK STAGE	Chick rejection Chick signatures	Chick mimicry Mimic signatures
	Host reluctance to feed chicks	Manipulative signals
FLEDGLING STAGE	Fledgling discrimination	Fledgling mimicry Mimicry in begging calls
	Host reluctance to feed fledglings	

Fig. 1.2 Coevolution of host defences and parasite counter-defences at different stages of the breeding cycle [based on Davies (2011) and Soler (2014). After Ruiz-Raya (2017)]

5. The coevolutionary process is much more dynamic than previously assumed. Three points deserve to be emphasized. First, some features of host-breeding biology such as inaccessible nest sites, unsuitable diet for parasite chicks, colonial nesting, nest cup design, well-hidden nests, secretive behaviour of hosts near their

nests, inappropriate breeding habitat for the brood parasite and host–egg or host-nestling size considerably larger than parasitic eggs or nestlings have sometimes been used to classify potential host species as suitable or unsuitable. The two former features clearly impede parasitism, but the latter ones only make parasitism more difficult. A model that represents the long-term trajectories and outcomes of coevolutionary interactions between brood parasites and their hosts with respect to the evolution of egg-rejection defence shows that as more host species acquire successful resistance, other previously unused host species become more profitable and will be parasitized (Soler 2014). Thus, parasitism of these species and the costs imposed by brood parasitism on them at the population level will increase, selecting for the evolution of host defences. This means that such species considered to be unsuitable hosts because of specific features that make parasitism more difficult actually later can become suitable hosts (Soler 2014).

Second, the implications of the evolutionary equilibrium hypothesis, which states that the evolution of egg rejection has not evolved in some host species because parasitism rate and/or the costs of parasitism are not sufficiently high for this defence to be adaptive (Rothstein 1990; Davies et al. 1996; Davies 2000), deserve to be nuanced. Although lack of egg-recognition ability is adaptive when the parasitism rate and the costs of parasitism are very low for the reasons stated under the first point, this situation cannot be considered an evolutionary equilibrium because it is unstable (Soler 2014).

And third, given that adaptations and counter-adaptations occurring at different stages of the nesting cycle can influence the evolution of adaptations and counter-adaptations at other stages (see above), the link between what happens at different stages can lead to different coevolutionary trajectories in different brood parasite–host systems (Britton et al. 2007; Feeney et al. 2014).

6. Coevolutionary cycles are very scarce. The theoretical framework of coevolutionary theory assumes that antiparasite adaptations are lost in the absence of parasitism. Therefore, fluctuating selection will favour the existence of coevolutionary cycles that could potentially continue indefinitely, at least in complex coevolutionary interactions in which even mutualism and parasitism can exist in different populations depending on the interacting species (Thompson and Cunningham 2002; Thompson 2005). This is the theoretical basis for most published studies on egg discrimination by hosts (Rothstein 1990; Davies 2000, 2011), and it implies that coevolutionary cycles should be the most common long-term outcome of brood parasite–host coevolution. However, it has been shown that coevolutionary cycles are very scarce (only 3.3% of host species), while other long-term outcomes such as lack of rejection and successful resistance, which were considered evolutionary enigmas, are much more frequent (Soler 2014). This scarcity of coevolutionary cycles is due to the frequently reported fact that rejection behaviour persists in many host species in the absence of brood parasitism (Rothstein 2001; Peer et al. 2011; Martín-Vivaldi et al. 2012), which implies that maintenance of rejection behaviour imposes a minimal cost in the absence of brood parasitism. In fact, evidence for the absence of the two types

of costs involved in egg rejection, recognition errors (when hosts mistakenly reject one or more of their own oddly coloured eggs) and rejection costs (when hosts accidentally damage their own eggs while trying to eject the parasitic egg; Davies 2000) has been accumulating rapidly in many studies and species (Martín-Vivaldi et al. 2002; Røskaft et al. 2002; Stokke et al. 2002; Hale and Briskie 2007; Vikan et al. 2010; Soler et al. 2011a). The persistence of rejection behaviour in the absence of brood parasitism also contradicts the prediction that host defences should have to decline in the absence of selection pressures imposed by brood parasites. The scarcity of coevolutionary cycles contrary to traditional theoretical expectations has two nonmutually exclusive explanations. First, that plastic responses, which are frequent in many host species presenting intermediate rejection rate and/or coevolutionary cycles (Soler 2014), given that they are costly (DeWitt et al. 1998; Ghalambor et al. 2007), will be lost if the risk of parasitism is frequently high and plasticity does not provide any benefits. As suggested by Pigliucci et al. (2006), the loss of phenotypic plasticity would select for canalization (i.e. the reduction in variability of a trait; Gibson and Wagner 2000) of rejection behaviour, and this trait will be fixed stopping coevolutionary cycles and provoking the outcome of successful resistance. And second, the scarcity of coevolutionary cycles could be a consequence of the fact that coevolutionary interactions involving brood parasites and their hosts are purely antagonistic, and then cycles would occur only during an intermediate phase of the interactions. But the arms race in most cases will end with the extinction of the host or with the host acquiring successful resistance, while the brood parasite can switch to other host species or go extinct (Soler 2014).

1.4 Coevolutionary Interactions

The study of coevolutionary processes is often complicated since they imply the evolution of reciprocal adaptations between interacting species (Janzen 1980), which are driven by geographic and temporal differences among populations and migration of individuals between them (i.e. gene flow; the geographic mosaic theory of coevolution; Thompson 2005). However, most studies usually consider a single host population over a short time span, which prevents to reach well-supported conclusions about coevolutionary interactions given that not only parasitism rate and host resistance to parasitism but even the outcome of the interaction can vary considerably among geographic areas (Thompson and Cunningham 2002).

Of course, studies made in a single location are important because adaptations and counter-adaptations acting at the local level are important in themselves, and, in addition, their results could be used in future comparative studies. However, it should be taken into account that, first, although selection pressures at the local level are the main determinants of both parasite and antiparasite strategies, they also depend on immigration of both hosts and parasites from other populations and, second, that differences between populations can be enormous implying that the coevolutionary process is driven by gene flow acting on the geographic mosaic of the interaction (Thompson 2005).

Geographic differences are inherent to the coevolutionary process because different selective pressures at the local level provoke different evolutionary trajectories. Indeed, there can exist populations with few or no coevolutionary interactions (cold spots), and others characterized by intense interactions between parasites and hosts with reciprocal adaptations and counter-adaptations (hot spots) that result in a higher reproductive success of well-adapted hosts (or parasites), which will provoke migration of individuals (i.e. gene flow) from these populations to others (Thompson 1994, 2005; Gomulkiewicz et al. 2000).

1.5 Coevolution in Brood Parasite–Host Systems: The Current Situation

A lot of remarkable adaptations and counter-adaptations have been reported in brood parasites and their hosts, which confirms that this system is one of the best for studying coevolution. However, as it occurs in other coevolutionary systems, appropriate studies covering several populations of the same host species are rare. In most cases only potential adaptations and counter-adaptations, often with only partial evidence of coevolution, have been described in studies performed in just one single location. Indeed, less than twenty papers address the study of dynamic coevolutionary interactions in the same brood parasite–host system in several locations, and/or at different years (Møller and Soler 2012). This means that the spatial and temporal ecology of brood parasite–host interactions, which is crucial for a good understanding of coevolutionary interactions given that even host selection can vary both spatially and temporarily (Baglione et al. 2017), remains poorly known.

Three points about coevolution in brood parasite–host systems deserve to be emphasized:

1. Most potential host species (68%) are currently not involved in coevolutionary interactions with brood parasites, at least regarding egg-rejection defence (39% present lack of rejection, and 30% present successful resistance outcomes; Soler 2014). This point is closely linked to the scarcity of coevolutionary cycles in brood parasite–host systems (see above).
2. Considering the brood parasite–host systems involved in coevolutionary interactions, good experimental demonstrations of dynamic coevolutionary interactions using geographically separated populations and sometimes taking gene flow into account have been reported in some of these systems (reviewed in Møller and Soler 2012), especially in the great spotted cuckoo–magpie (Soler and Møller 1990; Martínez et al. 1999; Soler et al. 1999, 2001; Martín-Gálvez et al. 2007) and the common cuckoo–reed warbler (*Acrocephalus scirpaceus*) (Lindholm 1999; Moskát et al. 2008; Stokke et al. 2008; Avilés et al. 2012) systems, but also, though more rarely, in other systems (Cruz and Wiley 1989;

Briskie et al. 1992; Kuehn et al. 2014). Temporal variation in parasitism rate and ejection rate has also been reported in some studies (Soler et al. 1994; Brooke et al. 1998; Lahty 2005; Møller and Soler 2012). All these studies provide convincing evidence of coevolutionary interactions regarding the egg-rejection defence, but coevolutionary interactions have also been reported considering other factors. For example, cuckoo eggs improve their mimicry with host eggs over time (Avilés et al. 2006, 2012; Antonov et al. 2010), hosts are more aggressive against brood parasites at sites with higher density of brood parasites (Langmore et al. 2012), and polymorphism evolves in hosts and brood parasites as a consequence of frequency-dependent selection (Yang et al. 2010). However, there are very few brood parasite–host systems. Thus, this means that appropriate demonstrations of dynamic coevolutionary interactions are lacking in most systems.

3. Coevolution between brood parasites and hosts is a powerful force that has been shown to affect to varying degrees ecological and life-history traits, including some related to sexual selection, patterns in brood parasite morphology and social systems (Krüger et al. 2007; Møller and Soler 2012; Feeney et al. 2013). However, this does not mean that everything that can be observed in brood parasite–host relationships is due to coevolution. As Møller and Soler (2012) emphasized in their review, both abiotic (environmental conditions, temperature and climatic change) and biotic (those affecting life history, sexual selection and evolution of signals) factors may affect the coevolutionary process by affecting some characteristics of hosts and/or brood parasites such as clutch size and egg size of hosts and population density and egg colour of both hosts and parasites, which is crucial for egg mimicry.

1.6 Topics that Should Be Considered in Future Studies of Coevolution of Brood Parasitism

1. Most studies on parasite–host coevolution have dealt with a single defence, even though in nature, coevolutionary interactions among species are affected by multiple traits (Vermeij 1982; Gilman et al. 2012). That is also the case in brood parasite–host studies, which usually are focused on a single defence, mainly egg rejection. Obviously, the best approach would be to study all defences at all stages of the nesting cycle, mainly considering that it is well known that the number of traits may affect the outcome of coevolution (Doebeli and Ispolatov 2010; Gilman et al. 2012).
2. The assumed arms race driving the coevolutionary relationships between brood parasites and their hosts is a consequence of the fact that host defences based on resistance reduce the breeding success of brood parasites. However, there exists another component of host defence different from resistance: tolerance, which consists of responses that minimize the negative impact of parasitism on host fitness without triggering negative effects for parasites (Svensson and Råberg 2010; Medzhitov et al. 2012). Therefore, tolerance does not induce

coevolutionary arms races, which reduces escalation in counter-defences facilitating continuous coevolutionary cycles and preventing parasite specialization (Krüger and Kolss 2013; Medina and Langmore 2015; Soler and Soler 2017). Reported evidence of tolerance in brood parasite–host systems includes reduction in clutch size while increasing reproductive attempts (Medina and Langmore 2016) and a positive association between levels of tolerance and prevalence of parasitism (Soler et al. 2011b). Future studies involving coevolutionary interactions between brood parasites and their hosts should take tolerance into account.

3. Very little is known about the proximate mechanisms underlying host responses to brood parasitism. Hormones are known to be important in regulating both parental care and individual responses to stressful situations. Specifically, corticosterone has an important role in the individual response to environmental perturbations leading to physiological and behavioural changes that promote individual survival over other activities (Wingfield et al. 1998; Breuner et al. 2008) and prolactin in the maintenance of parental behaviour (Angelier and Chastel 2009; Ouyang et al. 2011).

   There is clear evidence that brood parasitism is a stressful factor for hosts, given that corticosterone levels of host nestlings are higher when sharing the nest with a brood parasite nestling than when reared alone (Ibáñez-Álamo et al. 2012b), and that parasitized adult hosts suffer physiological costs during the fledgling stage (Mark and Rubenstein 2013). However, nothing is known about how hosts adjust their hormonal response as a consequence of brood parasitism nor about how hormonal mechanisms affect adult-host responses against the parasitic egg.

Concluding Remarks and Future Directions

New findings during the last 20 years about the relationships between brood parasites and their hosts have incorporated fascinating improvements in our previous knowledge on coevolutionary theory, but many crucial points remain poorly understood. Lots of work need to be done, because some longstanding questions and many new ones remain unanswered.

Some of the main future directions of research are as follow:

1. The characteristics and outcomes are highly variables in different brood parasite–host systems, and, thus, it is very important to collect information about those brood parasite–host systems that remain poorly studied.
2. It would be great to collect information on interactions considering all potential host species in the same study area.
3. Egg-recognition experiments need to be done in many potential host species that still never have been tested for this frequently evolved host defence.

(continued)

4. Nest defence and other potential frontline host defences remain poorly studied in most potential host species, including those that lack rejection ability. In these ones, it is especially important to study frontline defences because their existence would explain the absence of ejection.
5. Studies covering more than one defence (ideally all potential ones during all stages of the nesting cycle), in both parasitized and currently unparasitized potential host species, would provide important insights into how the arms race between brood parasites and their hosts drives the outcomes of the long-term coevolution of both counterparts.
6. New findings related to egg-rejection defence, such as absence of rejection after recognition, suggest the necessity of video monitoring the host recognition and ejection behaviours. Its use will facilitate to fully understand the proximate mechanisms driving egg-recognition and egg-rejection processes (Fig. 1.1).
7. Additional experimental tests of the misimprinting cost hypothesis are needed.
8. The possibility of both nestling and fledgling discrimination (e.g. reluctance to feed the parasite) deserves to be studied in as more brood parasite–host systems as possible. In those in which hosts are unable to reject parasitic eggs and/or brood parasites are nonevictors, because in these cases there is a higher probability of occurrence of nestling or fledgling recognition, and in the others, these studies would allow to determine the level of generalization concerning the blocking of the later defence hypothesis.
9. The study of proximate mechanisms such as the endocrine pathways underlying egg rejection would provide a new perspective on understanding host defences.
10. Special attention should be paid to the metapopulation approach. Given that characteristics of both parasitism and resistance can vary considerably among geographic areas due to adaptations at the local level and gene flow, studies covering several populations of the same host species are crucial. Detailed information on the spatial ecology of coevolutionary interactions between brood parasites and hosts is essential for an adequate understanding of the coevolutionary process.
11. Studies covering the same host population during different breeding seasons are also very important because temporal variation would provide direct evidence of the existence of coevolutionary cycles.

Acknowledgements I thank Anders Møller and Francisco Ruiz-Raya for their constructive and helpful review.

References

Als TD, Vila R, Kandul NP, Nash DR, Yen SH, Hsu YF, Mignault AA, Boomsma JJ, Pierce NE (2004) The evolution of alternative parasitic life histories in large blue butterflies. Nature 432:386–390

Angelier F, Chastel O (2009) Stress, prolactin and parental investment in birds: a review. Gen Comp Endocrinol 163:142–148

Antonov A, Stokke BG, Moksnes A, Røskaft E (2009) Evidence for egg discrimination preceding failed rejection attempts in a small cuckoo host. Biol Lett 5:169–171

Antonov A, Stokke BG, Vikan JR, Fossøy F, Ranke PS, Røskaft E, Moksnes A, Møller AP, Shykoff JA (2010) Egg phenotype differentiation in sympatric cuckoo *Cuculus canorus* gentes. J Evol Biol 23:1170–1182

Avilés JM, Stokke BG, Moksnes A, Røskaft E, Asmul M, Møller AP (2006) Rapid increase in cuckoo egg matching in a recently parasitized reed warbler population. J Evol Biol 19:1901–1910

Avilés JM, Vikan JR, Fossøy F, Antonov A, Moksnes A, Røskaft E, Shykoff JA, Møller AP, Stokke BG (2012) Egg phenotype matching by cuckoos in relation to discrimination by hosts and climatic conditions. Proc R Soc Lond B 279:1967–1976

Baba R, Nagata Y, Yamagishi S (1990) Brood parasitism and egg robbing among three freshwater fish. Anim Behav 40(776):778

Baglione V, Bolopo D, Canestrari D, Martínez JG, Roldán M, Vila M, Soler M (2017) Spatiotemporal variation of host use in a brood parasite: the role of the environment. Behav Ecol 28:49–58

Birkhead T (1991) The Magpies. The ecology and behaviour of Black-billed and Yellow-billed Magpies. Poyser, London

Boulton AM, Polis GA (2002) Brood parasitism among spiders: interaction between salticids and *Diguetia mojavea*. Ecology 83:282–287

Breuner CW, Patterson SH, Hahn TP (2008) In search of relationships between the acute adrenocortical response and fitness. Gen Comp Endocrinol 157:288–295

Briskie JV, Sealy SG, Hobson KA (1992) Behavioral defenses against avian brood parasitism in sympatric and allopatric host populations. Evolution 46:334–340

Britton NF, Planque R, Franks NR (2007) Evolution of defence portfolios in exploiter-victim systems. Bull Math Biol 69:957–988

Brooke MD, Davies NB, Noble DG (1998) Rapid decline of host defences in response to reduced cuckoo parasitism: behavioural flexibility of reed warblers in a changing world. Proc R Soc Lond B 265:1277–1282

Brown JL, Morales V, Summers K (2009) Tactical reproductive parasitism via larval cannibalism in Peruvian poison frogs. Biol Lett 5:148–151

Cossa NA, Tuero DT, Reboreda JC, Fiorini VD (2017) Egg pecking and puncturing behaviors in shiny and screaming cowbirds: effects of eggshell strength and degree of clutch completion. Behav Ecol Sociobiol 71:60. https://doi.org/10.1007/s00265-017-2289-1

Cruz A, Wiley JW (1989) The decline of an adaptation in the absence of a presumed selection pressure. Evolution 43:55–62

Davies NB (2000) Cuckoos, cowbirds and other cheats. T&AD Poyser, London

Davies NB (2011) Cuckoo adaptations: trickery and tuning. J Zool 284:1–14

Davies NB, de Brooke ML (1988) Cuckoos versus reed warblers: adaptations and counteradaptations. Anim Behav 36:262–284

Davies NB, de Brooke ML (1989a) An experimental study of co-evolution between the cuckoo, *Cuculus canorus*, and its hosts. I. Host egg discrimination. J Anim Ecol 58:207–224

Davies NB, de Brooke ML (1989b) An experimental study of co-evolution between the cuckoo, *Cuculus canorus*, and its hosts. II. Host egg markings, chick discrimination and general discussion. J Anim Ecol 58:225–236

Davies NB, Bourke AFG, de Brooke ML (1989) Cuckoos and parasitic ants: interspecific brood parasitism as an evolutionary arms race. Trends Ecol Evol 4:274–278

Davies NB, de Brooke ML, Kacelnik A (1996) Recognition errors and probability of parasitism determine whether reed warblers should accept or reject mimetic cuckoo eggs. Proc R Soc Lond B 263:925–931

de Brooke ML, Davies NB (1988) Egg mimicry by cuckoos *Cuculus canorus* in relation to discrimination by hosts. Nature 335:630–632

De Mársico MC, Gantchoff MG, Reboreda JC (2012) Host–parasite coevolution beyond the nestling stage? Mimicry of host fledglings by the specialist screaming cowbird. Proc R Soc Lond B 279:3401–3408

DeWitt TJ, Sih A, Wilson DS (1998) Costs and limits of phenotypic plasticity. Trends Ecol Evol 13:77–81

Doebeli M, Ispolatov I (2010) Complexity and diversity. Science 328:494–497

Feeney WE, Welbergen JA, Langmore NE (2012) The frontline of avian brood parasite-host coevolution. Anim Behav 84:3–12

Feeney WE, Medina I, Somveille M, Heinsohn R, Hall ML, Mulder RA, Stein JA, Kilner RM, Langmore NE (2013) Brood parasitism and the evolution of cooperative breeding in birds. Science 342:1506–1508

Feeney WE, Welbergen JA, Langmore NE (2014) Brood parasites and their hosts. Annu Rev Ecol Evol Syst 45:227–246

Ghalambor CK, McKay JK, Carroll SP, Reznick DN (2007) Adaptive versus non-adaptive phenotypic plasticity and the potential for contemporary adaptation in new environments. Funct Ecol 21:394–407

Gibson G, Wagner G (2000) Canalization in evolutionary genetics: a stabilizing theory? Bioessays 22:372–380

Gilman RT, Nuismer SL, Jhwueng DC (2012) Coevolution in multidimensional trait space favours escape from parasites and pathogens. Nature 483:328–330

Gomulkiewicz R, Thompson JN, Holt RD, Nuismer SL, Hochberg ME (2000) Hot spots, cold spots, and the geographic mosaic theory of coevolution. Am Nat 156:156–174

Grim T (2006) The evolution of nestling discrimination by hosts of parasitic birds: why is rejection so rare? Evol Ecol Res 8:785–802

Guigueno MF, Sealy SG (2012) Increased investigation of manipulated clutches suggests egg recognition without rejection in a brown-headed cowbird (*Molothrus ater*) host, the yellow warbler. Auk 129:17–25

Hale K, Briskie JV (2007) Response of introduced European birds in New Zealand to experimental brood parasitism. J Avian Biol 38:198–204

Hannon SJ, Wilson S, McCallum CA (2009) Does cowbird parasitism increase predation risk to American Redstart nests? Oikos 118:1035–1043

Hauber ME (2002) Is reduced clutch size a cost of parental care in Eastern Phoebes (*Sayornis phoebe*)? Behav Ecol Sociobiol 51:503–509

Hauber ME (2006) A future cost of misdirected parental care for brood parasitic young? Folia Zool 55:367–374

Hoover JP, Reetz MJ (2006) Brood parasitism increases provisioning rate and reduces offspring recruitment and adult return rates in a cowbird host. Oecologia 149:165–173

Ibáñez-Álamo JD, Arco L, Soler M (2012a) Experimental evidence for a predation cost of begging using active nests and real chicks. J Ornithol 153:801–807

Ibáñez-Álamo JD, de Neve L, Roldán M, Rodríguez J, Trouvé C, Chastel O, Soler M (2012b) Corticosterone levels in host and parasite nestlings: is brood parasitism a hormonal stressor? Horm Behav 61:590–597

Janzen DH (1980) When is it coevolution? Evolution 34:611–612

Jenkins JMA, Faaborg J (2016) Potential effects of brown-headed cowbirds (*Molothrus ater*) on host postfledging dispersal and survival. Wilson J Ornithol 128:404–411

Kilner RM, Langmore NE (2011) Cuckoos versus hosts in insects and birds: adaptations, counter-adaptations and outcomes. Biol Rev Camb Philos Soc 86:836–852

Krüger O, Kolss M (2013) Modelling the evolution of common cuckoo host-races: speciation or genetic swamping? J Evol Biol 26:2447–2457

Krüger O, Davies NB, Sorenson MD (2007) The evolution of sexual dimorphism in parasitic cuckoos: sexual selection or coevolution? Proc R Soc Lond B 274:1553–1560

Kuehn MJ, Peer BD, Rothstein SI (2014) Variation in host response to brood parasitism reflects evolutionary differences and not phenotypic plasticity. Anim Behav 88:21–28

Lahty DC (2005) Evolution of bird eggs in the absence of cuckoo parasitism. Proc Natl Acad Sci USA 102:18057–18062

Langmore NE, Hunt S, Kilner RM (2003) Escalation of a co-evolutionary arms race through host rejection of brood parasitic young. Nature 422:157–160

Langmore NE, Feeney WE, Crowe-Riddell J, Luan H, Louwrens KM, Cockburn A (2012) Learned recognition of brood parasitic cuckoos in the superb fairy-wren *Malurus cyaneus*. Behav Ecol 23:798–805

Lindholm AK (1999) Brood parasitism by the cuckoo on patchy reed warbler populations in Britain. J Anim Ecol 68:293–309

Lotem A (1993) Learning to recognize nestlings is maladaptive for cuckoo *Cuculus canorus* hosts. Nature 362:743–744

Mark MM, Rubenstein DR (2013) Physiological costs and carry-over effects of avian interspecific brood parasitism influence reproductive tradeoffs. Horm Behav 63:717–722

Martínez JG, Soler JJ, Soler M, Møller AP, Burke T (1999) Comparative population structure and gene flow of a brood parasite, the great spotted cuckoo (*Clamator glandarius*), and its primary host, the magpie (*Pica pica*). Evolution 53:269–278

Martín-Gálvez D, Soler JJ, Martínez JG, Krupa AP, Soler M, Burke T (2007) Cuckoo parasitism and productivity in different magpie subpopulations predict frequencies of the 457bp allele: A mosaic of coevolution at a small geographic scale. Evolution 61:2340–2348

Martín-Vivaldi M, Soler M, Møller AP (2002) Unrealistically high costs of rejecting artificial models for cuckoo *Cuculus canorus* host. J Avian Biol 33:295–301

Martín-Vivaldi M, Soler JJ, Møller AP, Pérez-Contreras T, Soler M (2012) Importance of nest site and habitat in egg recognition ability of potential hosts of the European Cuckoo *Cuculus canorus*. Ibis 155:140–155

Medina I, Langmore NE (2015) Coevolution is linked with phenotypic diversification but not speciation in avian brood parasites. Proc R Soc B 282:20152056

Medina I, Langmore NE (2016) The evolution of acceptance and tolerance in hosts of avian brood parasites. Biol Rev 91:569–577

Medzhitov R, Schneider DS, Soares MP (2012) Disease tolerance as a defense strategy. Science 335:936–941

Mendelson TC, Fitzpatrick CL, Hauber ME, Pence CH, Rodríguez RL, Safran RJ, Stern CA, Stevens JR (2016) Cognitive phenotypes and the evolution of animal decisions. Trends Ecol Evol 31:850–859

Moksnes A, Røskaft E, Braa AT (1991) Rejection behavior by common cuckoo hosts towards artificial brood parasite eggs. Auk 108:348–354

Møller AP, Soler JJ (2012) A coevolutionary framework based on temporal and spatial ecology of host-parasite interactions: a missing link in studies of brood parasitism. Chinese Birds 3:259–273

Moskát C, Hansson B, Barabas L, Bartol I, Karcza Z (2008) Common cuckoo *Cuculus canorus* parasitism, antiparasite defence and gene flow in closely located populations of great reed warblers *Acrocephalus arundinaceus*. J Avian Biol 39:663–671

Moskát C, Bán M, Hauber ME (2014) Naïve hosts of avian brood parasites accept foreign eggs, whereas older hosts fine-tune foreign egg discrimination during laying. Front Zool 11:45

Müller JK, Eggert AK, Dressel J (1990) Intraspecific brood parasitism in the burying beetle, *Necrophorus vespilloides* (Coleoptera, Silphidae). Anim Behav 40:491–499

Ouyang JQ, Sharp PJ, Dawson A, Quetting M, Hau M (2011) Hormone levels predict individual differences in reproductive success in a passerine bird. Proc R Soc B Biol Sci 278:2537–2545

Payne RB, Payne LL (1998) Brood parasitism by cowbirds: risks and effects on reproductive success and survival in Indigo Buntings. Behav Ecol 9:64–73

Peer BD, Kuehn MJ, Rothstein SI, Fleischer RC (2011) Persistence of host defence behaviour in the absence of brood parasitism. Biol Lett 7:670–673

Pigliucci M, Murren CJ, Schlichting CD (2006) Phenotypic plasticity and evolution by genetic assimilation. J Exp Biol 209:2362–2367

Roldán M, Soler M (2011) Parental care parasitism: how unrelated offspring attain acceptance by foster parents? Behav Ecol 22:679–691

Røskaft E, Moksnes A, Meilvang D, Bicík V, Jemelíková J, Honza M (2002) No evidence for recognition errors in *Acrocephalus* warblers. J Avian Biol 33:31–38

Rothstein SI (1974) Mechanisms of avian egg recognition: possible learned and innate factors. Auk 91:796–807

Rothstein SI (1978) Mechanisms of avian egg-recognition: additional evidence for learned components. Anim Behav 26:671–677

Rothstein SI (1990) A model system for coevolution: avian brood parasitism. Annu Rev Ecol Syst 21:481–508

Rothstein SI (2001) Relic behaviours, coevolution and the retention versus loss of host defences after episodes of avian brood parasitism. Anim Behav 61:95–107

Ruiz-Raya F (2017) Defences against brood parasitism in the common blackbird (*Turdus merula*): plasticity, physiology and evolution. PhD thesis, University of Granada, Spain

Ruiz-Raya F, Soler M (2018) Rejection of parasitic eggs: an updated terminology for a complex process. J Avian Biol (in press)

Ruiz-Raya F, Soler M, Sánchez-Pérez LL, Ibáñez-Álamo JD (2015) Could a factor that does not affect egg recognition influence the decision of rejection? PLoS One 10(8):e0135624

Sato T (1986) A brood parasitic catfish of mouth-brooding cichlid fishes in Lake Tanganyika. Nature 323:58–59

Smith JNM (1981) Cowbird parasitism, host fitness, and age of the host female in an island Song Sparrow population. Condor 83:152–161

Soler M (2009) Co-evolutionary arms race between brood parasites and their hosts at the nestling stage. J Avian Biol 40:237–240

Soler M (2014) Long-term coevolution between avian brood parasites and their hosts. Biol Rev 89:688–704

Soler M, Møller AP (1990) Duration of sympatry and coevolution between the great spotted cuckoo and its magpie host. Nature 343:748–750

Soler JJ, Soler M (2000) Brood-parasite interactions between great spotted cuckoos and magpies: a model system for studying coevolutionary relationships. Oecologia 125:309–320

Soler JJ, Soler M (2017) Evolutionary change: facultative virulence by brood parasites and tolerance and plastic resistance by hosts. Anim Behav 125:101–107

Soler M, Soler JJ, Martínez JG, Møller AP (1994) Micro-evolutionary change in host response to a brood parasite. Behav Ecol Sociobiol 35:295–301

Soler M, Soler JJ, Martínez JG (1997) Great spotted cuckoos improve their reproductive success by damaging magpie host eggs. Anim Behav 54:1227–1233

Soler JJ, Martínez JG, Soler M, Møller AP (1999) Genetic and geographic variation in rejection behavior of cuckoo eggs by European magpie populations: an experimental test of rejecter-gene flow. Evolution 53:947–956

Soler JJ, Martínez JG, Soler M, Møller AP (2001) Coevolutionary interactions in a host-parasite meta-population. Ecol Lett 4:470–476

Soler M, Ruiz-Castellano C, Fernández-Pinos MC, Rösler A, Ontanilla J, Pérez-Contreras T (2011a) House sparrows selectively eject parasitic conspecific eggs and incur very low rejection costs. Behav Ecol Sociobiol 65:1997–2005

Soler JJ, Martín-Gálvez D, Martínez JG, Soler M, Canestrari D, Abad-Gómez M, Møller AP (2011b) Evolution of tolerance by magpies to brood parasitism by great spotted cuckoos. Proc R Soc B 278:2047–2052

Soler M, Fernández-Morante J, Espinosa F, Martín-Vivaldi M (2012) Pecking but accepting the parasitic eggs may not reflect ejection failure: the role of motivation. Ethology 118:662–672
Soler M, Ruiz-Castellano C, Carra LG, Ontanilla J, Martín-Galvez D (2013) Do first-time breeding females imprint on their own eggs? Proc R Soc B 280:20122518
Soler M, de Neve L, Roncalli G, Macías-Sánchez E, Ibáñez-Álamo JD, Pérez-Contreras T (2014a) Great spotted cuckoo fledglings are disadvantaged by magpie host parents when reared together with magpie nestlings. Behav Ecol Sociobiol 68:333–342
Soler M, Pérez-Contreras T, Ibáñez-Álamo JD, Roncalli G, Macías-Sánchez E, de Neve L (2014b) Great spotted cuckoo fledglings often receive feedings from other magpie adults than their foster parents: which magpies accept to feed foreign cuckoo fledglings? PLoS One 9(10):e107412
Soler M, Ruiz-Raya F, Roncalli G, Ibáñez-Álamo JD (2017) Relationships between egg-recognition and egg-ejection in a grasp-ejector species. PLoS One 12(2):e0166283
Spottiswoode CN, Stevens M (2011) How to evade a coevolving brood parasite: egg discrimination versus egg variability as host defences. Proc R Soc B 278:3566–3573
Stokke BG, Honza M, Moksnes A, Røskaft E, Rudolfsen G (2002) Costs associated with recognition and rejection of parasitic eggs in two European passerines. Behaviour 139:629–644
Stokke BG, Moksnes A, Røskaft E (2005) The enigma of imperfect adaptations in hosts of avian brood parasites. Ornithol Sci 4:17–29
Stokke BG, Hafstad I, Rudolfsen G, Moksnes A, Møller AP, Røskaft E, Soler M (2008) Predictors of resistance to brood parasitism within and among reed warbler populations. Behav Ecol 19:612–620
Svensson EI, Råberg L (2010) Resistance and tolerance in animal enemy-victim coevolution. Trends Ecol Evol 25:267–274
Thompson JN (1994) The coevolutionary process. University of Chicago Press, Chicago
Thompson JN (2005) The geographic mosaic of coevolution. University of Chicago Press, Chicago
Thompson JN, Cunningham BM (2002) Geographic structure and dynamics of coevolutionary selection. Nature 417:735–738
Tuero DT, Fiorini VD, Reboreda JC (2007) Effects of shiny cowbird *Molothrus bonariensis* parasitism on different components of house wren *Troglodytes aedon* reproductive success. Ibis 149:521–529
Vermeij GJ (1982) Unsuccessful predation and evolution. Am Nat 120:701–720
Vikan JR, Stokke BG, Rutila J, Huhta E, Moksnes A, Røskaft E (2010) Evolution of defences against cuckoo (*Cuculus canorus*) parasitism in bramblings (*Fringilla montifringilla*): a comparison of four populations in Fennoscandia. Evol Ecol 24:1141–1157
Welbergen JA, Davies NB (2008) Reed warblers discriminate cuckoos from sparrowhawks with graded alarm signals that attract mates and neighbours. Anim Behav 76:811–822
Welbergen JA, Davies NB (2009) Strategic variation in mobbing as a front line of defense against brood parasitism. Curr Biol 19:235–240
Wingfield JC, Manney DL, Breuner CW, Jacobs JD, Lynn S, Ramenofsky M, Richardson RD (1998) Ecological bases of hormone—behavior interactions: the "Emergency Life History Stage". Am Zool 38:191–206
Yang C, Liang W, Cai Y, Shi S, Takasu F, Møller AP, Antonov A, Fossøy F, Moksnes A, Røskaft E, Stokke BG (2010) Coevolution in action: disruptive selection on egg colour in an avian brood parasite and its host. PLoS One 5:e10816

Part I

Coevolution and Diversification of Interspecific Brood Parasites

Factors Affecting the Rates of Coevolution Between Obligate Avian Brood Parasites and Their Hosts

2

Virginia E. Abernathy and Naomi E. Langmore

Abstract

Coevolution is a process in which two species, populations, or groups of individuals evolve reciprocal adaptations through their interactions with one another. Obligate avian brood parasitism is a model example of coevolution, and several reviews have discussed the different types of adaptations and counter-adaptations hosts and brood parasites evolve. However, there has been less focus on the rate at which this process proceeds. Here we review factors influencing the rate of coevolution between avian obligate brood parasites and their hosts. We also suggest that evidence from brood parasite coevolution studies concurs with other developments in evolutionary biology more broadly, which indicates that evolution can be a more rapid process than previously recognized and can proceed on a timescale similar to that of ecological dynamics. Finally, we discuss the difficulties of studying rates of coevolution in bird populations empirically and mention current studies that are resolving this problem by focusing on recently parasitized host populations. Understanding how rapidly hosts can evolve defenses to circumvent brood parasitism is an important step in uncovering aspects of speciation, determining which traits are actually indicative of true genetic change, and can aid in conservation decisions of endangered potential hosts, especially as brood parasites expand their breeding ranges with rising global temperatures and other environmental changes.

V. E. Abernathy (✉) · N. E. Langmore
Ecology and Evolution, Research School of Biology, Australian National University, Acton, ACT, Australia
e-mail: vabernathy123@gmail.com

M. Soler (ed.), *Avian Brood Parasitism*, Fascinating Life Sciences,
https://doi.org/10.1007/978-3-319-73138-4_2

2.1 Introduction

Coevolution occurs when two species evolve reciprocal adaptations through their interactions with one another and can occur in a range of different systems, including predator-prey, plant-pollinator, parasite-host, and pathogen-host interactions. One of the most well-known coevolved systems is the evolutionary arms race between avian obligate brood parasites and their hosts. A plethora of studies have revealed the adaptations that have evolved in hosts to prevent parasitism and the ingenious counter-adaptations in brood parasites that trick hosts into rearing their young (Rothstein 1990; Davies 2011; Soler 2014). However, we are only just beginning to understand how rapidly these arms races can proceed in birds (Sorenson and Payne 2001; Spottiswoode and Stevens 2012; Abernathy 2017). Are newly annexed hosts completely defenseless against brood parasitism (Abernathy and Langmore 2017), and how rapidly can they acquire the defenses needed to circumvent the high costs of parasitism? Moreover, do all host defenses require genetic change, representing true evolution, or are some acquired through learning (Langmore et al. 2012)? Understanding rates of coevolution may have significant implications for fields as diverse as conservation (Rothstein 1975a; Dinets et al. 2015) and speciation (Krüger et al. 2009; Medina and Langmore 2015a), yet no reviews have focused on the many factors affecting coevolutionary rate in brood parasites, such as generation time, mutation rate, fecundity, genetic variation, heritability, and selection pressure (but see Soler and Soler 2017). The aim of this review is to discuss these factors, specifically as they relate to avian interspecific brood parasites and their hosts, in order to better understand how quickly hosts and brood parasites can evolve adaptations and counter-adaptations.

2.2 Genetic Factors Affecting the Rate of Coevolution

2.2.1 Generation Time and Mutation Rate

Faster generation time can result in a higher mutation rate, because the germline DNA is copied more frequently within a given period of time increasing the probability of copying errors (Bromham 2009). Thus, the asymmetry of generation time and mutation rate between two coevolving species can affect how quickly coevolution will occur (Thompson 1994; Brandt et al. 2005). A bacterium with a short generation time and fast mutation rate is likely to evolve more rapidly than its vertebrate host, allowing the bacterium to adapt quickly to any host defenses that evolve (Thompson 1994). Where there is less asymmetry among these factors, as between an avian brood parasite and its host, the host may be able to catch up to the parasite, and so coevolution is more likely to occur in a stepwise fashion (Brandt et al. 2005).

A theoretical model on the spread of egg rejection in brood parasite hosts demonstrated that when mutation rate was slower, egg rejecter alleles spread more slowly (Kelly 1987). This effect was amplified as the allele became rarer in the population. While higher mutation rate can lead to faster evolution, antagonistic coevolution can also select for higher mutation rates (Brockhurst and Koskella 2013). In brood parasitic systems, there is evidence that coevolution has led to higher rates of subspeciation and extinction in parasites (Krüger et al. 2009), that egg size and plumage color and pattern evolve faster in parasitic cuckoos than nonparasitic cuckoos (Medina and Langmore 2015a), and that parasitism has led to increased diversity of egg patterns in host species (Medina et al. 2016). Sorenson and Payne (2001) found higher rates of DNA sequence evolution of the parasitic *Vidua* finches, an ancient brood parasitic lineage, compared to the cowbird (*Molothrus* spp.) lineage, a much younger parasitic group. One explanation for this increased rate of divergence is that the evolution of host defenses selects for counter-adaptations in brood parasites, which eventually leads to specialization on a few hosts (Medina and Langmore 2016a). Further, cuckoo lineages that are sympatric and use similar host genera showed higher rates of phenotypic evolution (Medina and Langmore 2015a), which could indicate that competition for hosts among brood parasites can also lead to higher mutation rates.

2.2.2 Fecundity

For two competing species occupying the same niche, even the slightest difference in fecundity can confer an advantage and allow one species to outcompete the other (Hardin 1960). Micro- and endoparasites tend to produce many offspring in a short amount of time compared to their larger vertebrate hosts, allowing parasites to adapt quickly, while host defenses lag behind (Thompson 1994; Brandt et al. 2005). In contrast, coevolution is likely to be more stable and may proceed more quickly between species with similar fecundity rates, such as a brood parasite and its host. The brood parasitic brown-headed cowbird (*Molothrus ater*) appears to have higher fecundity than its hosts because it has the potential to produce an average of 16–40 eggs per season, whereas a host might lay two to three clutches of three to four eggs (e.g., Scott and Ankney 1980; Fleischer et al. 1987; Jackson and Roby 1992). This could allow cowbirds to adapt more quickly than their hosts, slowing down coevolution.

However, the number of offspring that survive to reproduce in the subsequent generation should be considered when measuring fecundity. Some cowbird eggs will be rejected by hosts, laid in inactive nests, laid in multiply parasitized nests, laid in nests that become abandoned, or laid at the wrong time relative to the host's laying period and so will not hatch. Moreover, some cowbird nestlings may not survive with larger host nestlings (Kilner 2003) or unsuitable hosts (Middleton 1991; Peer and Bollinger 1998). When all these limitations are combined, the higher egg-laying potential of cowbirds may ultimately result in similar recruitment rates to their hosts. Indeed, several molecular genetic field studies have revealed that the annual fecundity of brood parasites is similar to that of other similarly sized bird species. In free-ranging brown-headed cowbirds, the mean realized fecundity (i.e., the number of eggs laid in nests that could potentially fledge a cowbird chick) was between one to nine eggs each season, much lower than estimates based on captive cowbirds (Hahn et al. 1999; Woolfenden et al. 2003). Similarly, individual Horsfield's bronze-cuckoos (*Chalcites basalis*) laid on average seven eggs per breeding site, but only 33% of cuckoo eggs fledged (Langmore et al. 2007), yielding around two fledglings/female as is typical for their hosts during a breeding season.

2.2.3 Genetic Variation

The effect of gene flow between populations on the rate of coevolution is quite complex and is dependent on numerous variables (Thompson 1994). For instance, increasing migration rates between populations will initially increase genetic variation in the entire species and, therefore, the rate of coevolution (e.g., Lenormand 2002; Brockhurst et al. 2003, 2007; Vogwill et al. 2008). Indeed, many studies indicate that parasite populations with higher gene flow are able to adapt more quickly to hosts because of higher genetic variation (Greischar and Koskella 2007; Hoeksema and Forde 2008). Likewise, low genetic variation may be responsible for a lack of defenses in hosts against avian brood parasitism, which could result in evolutionary lag (Rothstein 1990; Hosoi and Rothstein 2000). However, too much gene flow can result in homogenization between populations, which can reduce the rate of coevolution if different populations experience different selection pressures (Thompson 1994; Lenormand 2002; Vogwill et al. 2008; Fig. 2.1). Barabás et al. (2004) suggest that if there is high gene flow from a source population of naïve brood parasite hosts to a sink population of parasitized individuals, coevolution could be slowed (but see Brockhurst et al. 2007; Fig. 2.1). Several studies have revealed very low levels of natal philopatry in brood parasites (0–5% of banded fledglings returned, Nice 1937; Smith and Arcese 1994; Langmore et al. 2007; Moskát et al. 2008), which are comparable to those of many other migratory, nonparasitic landbirds (Schlossberg 2009; Hauber et al. 2012). This may facilitate high rates of evolution but constrain local adaptation for both the parasite and host.

Zölei et al. (2015) found that common cuckoo (*Cuculus canorus*) parasitism rate had remained high (52–64%) and relatively constant at their study site for seven decades, as had the rate of rejection of cuckoo eggs (34–39%) by their great reed warbler (*Acrocephalus arundinaceus*) hosts (see also Moskát et al. 2008). Because this host is unable to raise both the cuckoo and its own young, there should be strong selection for hosts to evolve higher rejection rates. This consistency of parasitism and rejection rate over such a long period may be maintained if naïve genotypes migrate into the site and prevent faster evolution of host defenses (Barabás et al. 2004; Zölei et al. 2015; Fig. 2.1). Indeed, an earlier study on the same system found three great reed warbler populations were genetically similar, even though each population was exposed to very different parasitism rates (41–68%, 11%, and less than 1%; Moskát et al. 2008). Moskát et al. (2008) also found very low return rates of fledglings and adults to the same site in subsequent years. Moreover, two other studies found that neighboring great reed warbler populations were genetically similar (Bensch and Hasselquist 1999; Hansson 2003). Thus, the high rates of

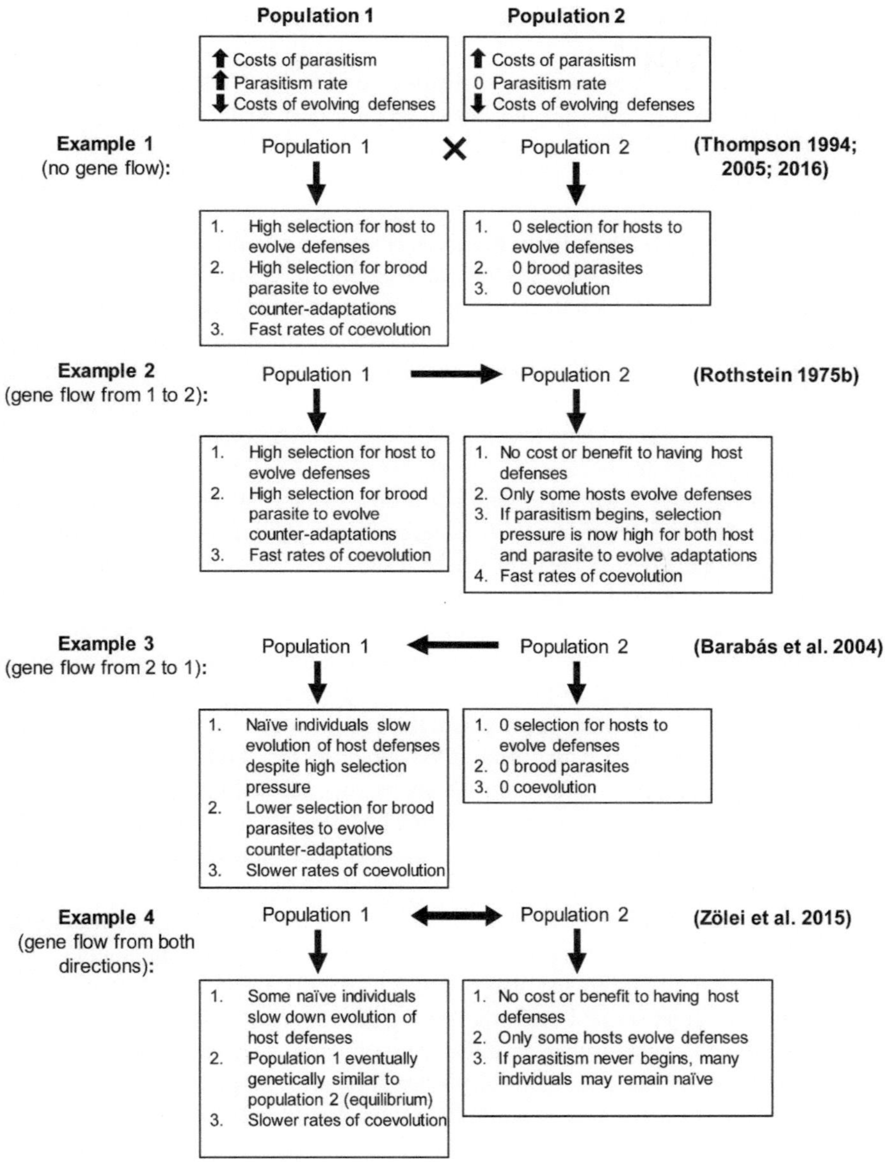

Fig. 2.1 Examples of how gene flow of host individuals between two host populations can influence rates of coevolution. In this scenario, both populations experience high costs when parasitized and there is low cost to evolving defenses, but "population 1" experiences high rates of parasitism, while "Population 2" is completely naïve to parasitism

gene flow between populations experiencing different parasitism rates may impede coevolution in this system.

If host and parasite populations are distributed in isolated patches, local adaptation can occur, where parasites and hosts become best adapted to the genotypes within their own population (e.g., Lively and Dybdahl 2000; Brockhurst et al. 2007). If gene flow is highly reduced, coevolution may occur at high rates in some populations and at low rates or not at all in others ("geographic mosaic of coevolution" Thompson 1994, 2005, 2016; Fig. 2.1). Because bird populations are often not completely isolated from one another (Martín-Gálvez et al. 2006), once a host population has evolved defenses against a brood parasite, defense genes could spread rapidly to other populations also exposed to brood parasitism (Rothstein 1975b; Fig. 2.1). Additionally, many host species maintain egg rejection in the absence of brood parasitism (e.g., Peer et al. 2005; Soler 2014; Yang et al. 2014) even through speciation events (Peer et al. 2013), which could prevent successful exploitation by any parasite.

Still, responses to parasitism among a few host species have been shown to vary between populations (e.g., Sealy 1996; Haas and Haas 1998; Lindholm and Thomas 2000; Liang et al. 2016), especially where some populations are allopatric with a brood parasite, while others are sympatric (e.g., Cruz and Wiley 1989; Soler and Møller 1990; Langmore et al. 2012). However, variation in host defense does not prove that genetic differences exist between populations. There is ample evidence that brood parasite hosts can adjust their level of rejection according to the probability of parasitism through phenotypic plasticity, which would not require any genetic changes in the population (Cruz et al. 2008; Soler 2014). Additionally, hosts that require an imprinting process to learn the appearance of their eggs may misimprint on a parasite egg if parasitized during their first breeding attempt (Lotem et al. 1995), which could lead to higher acceptance rates in areas where parasitism rates are higher (Strausberger and Rothstein 2009).

However, genetic differences were found between populations of both the great spotted cuckoo (*Clamator glandarius*) and its main host, the Eurasian magpie (*Pica pica*) (Martinez et al. 1999; see also Soler et al. 1999). Gene flow was higher in magpies than in great spotted cuckoos and higher in magpie populations that were sympatric with the cuckoo than in allopatric populations. Even though sympatric populations were closer geographically than allopatric populations, which might have accounted for the higher gene flow, the genetic differentiation between sympatric populations was lower than expected when considering their geographic distances (Martinez et al. 1999). This might indicate that increased levels of genetic variation can help maintain host defenses, as allopatric populations accepted more eggs than sympatric populations (Soler and Møller 1990; Soler et al. 1999). Additionally, the frequency of a particular allele covaried with magpie rejection rate among several subpopulations experiencing different levels of parasitism, suggesting that variation in magpie defenses was at least partially related to a genetic component and not just phenotypic plasticity (Martín-Gálvez et al. 2006, 2007).

There is evidence that common cuckoos have diverged genetically as they specialized on particular host species, forming host races (Gibbs et al. 2000; Fossøy et al. 2011). Møller et al. (2011) showed that cuckoo host races match the breeding times of their particular host species. If cuckoo host races are temporally isolated due to different breeding phenology of host species, this could reduce gene flow among cuckoo host races and lead to local adaptation, speeding up the process of specialization in the cuckoo (Møller et al. 2011). Additionally, large areas of homogenous habitat could limit the number of host species breeding in an area, reducing gene flow among cuckoo host races and increasing the rate of local adaptation, while areas of heterogeneous habitat may allow more host species to breed in close proximity, thereby increasing gene flow among cuckoo host races (Southern 1954; Moksnes and Røskaft 1987, 1995).

2.2.4 Heritability

The importance of heritability in determining the rate at which coevolution progresses may override the importance of selection pressure in certain situations (Brodie and Brodie 1999). Various factors determine whether one trait is more heritable than another (Thompson 1999), such as how many genes control a trait (Futuyma 2010), the inheritance pattern, and which chromosome contains the trait. If egg rejection is dominant, it will spread more quickly in a population than if it is recessive (Kelly 1987). If the genes controlling egg mimicry in brood parasites are found on the W-sex chromosome (only passed from mothers to daughters), mimicry will spread faster than if the genes are on an autosomal chromosome (Kelly 1987). While studies on the genetic inheritance of host defenses (Martín-Gálvez et al. 2006, 2007; Procházka et al. 2014) and egg mimicry have been performed (Gibbs et al. 2000; Gosler et al. 2000; Mahler et al. 2008; Fossøy et al. 2011), results among the studies were inconsistent. However, Fossøy et al. (2016) recently demonstrated that, for the common cuckoo, genes controlling blue-egg color are maternally inherited. This supports the well-accepted hypothesis that cuckoo host races must be maintained genetically by the female alone because males mate indiscriminately with females raised by different host species (Marchetti et al. 1998; but see Fossøy et al. 2011).

2.3 Selection Pressure

Selection pressure is one of the most important factors affecting the rate of coevolution between brood parasites and their hosts and can be influenced by many variables, including the cost of parasitism on the host, parasitism rate, and the relative costs and benefits of evolving certain adaptations (Futuyma 2010; Fig. 2.1). In the case of obligate brood parasitism, selection should generally be stronger on the brood parasite, because all parasites must interact with hosts in order to reproduce, whereas only a portion of hosts may ever encounter a parasite (the "rare enemy effect" Dawkins 1982; see also Davies 2000). This could mean brood parasites that impose a minimal cost on their hosts may evolve adaptations more quickly than the host (see Sect. 2.3.1).

2.3.1 The Cost of Parasitism and Life-History Traits

When parasitism results in the death of all host young, there should be strong selection pressure on the host to evolve defenses, which would then select for counter-defenses in the brood parasite (Soler and Soler 2017). This is because the costs of losing the battle for the nest are high for both parties and sometimes, for evicting parasites, only one can reproduce successfully at an individual nest (Molnár 1939; Grim et al. 2011). However, not all parasites are highly virulent, and selection pressure on the host may be minimal if the costs to the host are minimal (Lyon and Eadie 2004). A recent comparative analysis revealed that egg rejection is more likely to evolve in hosts of evictor brood parasites because costs of parasitism are much higher for these hosts (Medina and Langmore 2015b). Moreover, for hosts of evicting brood parasites only, body size is an important predictor of egg rejection; hosts that are relatively small compared to the parasite, and therefore experience higher energetic costs in rearing the parasite, are more likely to evolve egg rejection (Medina and Langmore 2015b). Similarly, Hosoi and Rothstein (2000) found that non-forest-nesting hosts of the brown-headed cowbird are more likely to abandon parasitized nests if costs of parasitism are higher (they are unable to raise their young with the cowbird young).

Specific life-history traits of a host, such as incubation time, migration, and breeding season duration, may influence the cost of parasitism and the strength of selection on host defenses (Abernathy and Peer 2016). Nonmigratory hosts often have longer breeding seasons than their brood parasites (Scott 1963; Scott and Ankney 1980), allowing a brood to be reared before or after the breeding season of the parasite. For some hosts, there appears to be directional selection for earlier breeding in order to avoid parasitism (Medina and Langmore 2016b; Boves et al. 2014). A new host of the pacific koel (*Eudynamys orientalis*), the red wattlebird (*Anthochaera carunculata*), showed little difference in reproductive success (measured as the total number of nests that fledged at least one red wattlebird out of the total number of monitored nests) at sites with high and low rates of koel parasitism, despite a lack of egg rejection in this naïve host (Abernathy and Langmore 2017; Abernathy 2017). The low impact of koel parasitism on the population may be explained by three life-history traits: the earlier breeding season of the host than the parasite; the small clutch size of the host (usually two eggs), which could make it difficult for the parasite to time egg laying accurately; and wattlebirds' abandonment of koel eggs laid before the host's laying period (Abernathy and Langmore 2017).

2.3.2 Species Density and Parasitism Rate

Several studies have demonstrated the importance of host density on the likelihood of parasitism (Álvarez 2003; Jensen and Cully 2005; Stokke et al. 2007; Jewell and Arcese 2008) and parasitism rate can directly influence selection pressure on a host (Soler et al. 2001). Models by Takasu et al. (1993) and Davies et al. (1996) showed that when parasite density and parasitism rates are low and egg rejection carries some cost, individuals that accept parasitism will have higher reproductive success than rejecters, so the rejecter gene will not spread. As parasite density increases, rejecter genes will spread more quickly. If the brood parasite is common, but the host is extremely rare, rejecter alleles, once evolved, would likely spread more rapidly than in a common and widespread host, as selection pressure would be high and the host population size would be small (Rothstein 1975a). Likewise, a rare or unevenly distributed host may slow down specialization in parasites (Soler et al. 2009). Strausberger and Rothstein (2009) found, counterintuitively, that high density of a parasite can actually cause high rates of acceptance in hosts if first-time breeders misimprint on parasite eggs.

Selection pressures on specific host traits could vary from year to year depending on the relative abundance of hosts to parasites. In years when great spotted cuckoo parasitism rate was lower and more Eurasian magpie nests were available, cuckoos mainly parasitized magpies with larger nests (Molina-Morales et al. 2016), as this is a sign of parental quality (Soler et al. 1995a; de Neve et al. 2004). When parasitism rate was higher and host nests more scarce, cuckoos parasitized both large and small nests at equal rates (Molina-Morales et al. 2016). This means, whenever host-nest availability is high in this population (there are more hosts than parasites), some individuals will be gaining experience with cuckoos, while others will not, and this could slow down coevolution in the population (see Sect. 2.3.4).

2.3.3 Costs of Coevolved Traits

Even when selection pressure is high for a species, coevolution could be prevented if the costs of evolving an adaptation are higher than the benefits. Costs of brood parasite host defenses include recognition and rejection errors; abandoning parasitized nests, which requires extra energy and time to renest (Hosoi and Rothstein 2000; Peer et al. 2005; Antonov et al. 2009; Guigueno and Sealy 2010); and mafia behavior, where the adult brood parasite punishes a rejecter host by destroying the host nest (Soler et al. 1995b; Hoover and Robinson 2007; Chap. 15). In areas where multiple parasitism is frequent, removing parasitic eggs from the nest could increase the chances of host eggs being removed or destroyed by parasites that visit the nest later (Sato et al. 2010; Gloag et al. 2012). These costs can be so high that it may be more profitable for hosts to accept a suspicious egg under some circumstances, such as when the cost of parasitism is low and the parasite's eggs appear similar to the host eggs (Begum et al. 2011a; Abernathy and Peer 2016; Abernathy 2017; Abernathy et al. 2017); when parasitism rates are low (e.g., Lotem et al. 1995; Davies et al. 1996; Peer and Bollinger 1997; Brooke et al. 1998); when detection of parasitic eggs is physically constrained, as is the case for some hosts that nest in cavities or build dark, dome-shaped nests (Langmore et al. 2005; Avilés 2008; see also Avilés et al. 2006a); or when the parasitic egg is unlikely to hatch (Boves et al. 2014). In these circumstances, hosts may instead evolve tolerance, whereby they accept the parasite, but adjust life history or other traits to minimize the costs of parasitism (Boves et al. 2014; Medina and Langmore 2016c; Soler and Soler 2017). This type of defensive strategy (defensive tolerance) would slow down specialization in the parasite, and coevolution in general, compared to a host-brood parasite system where egg rejection had evolved, as defensive tolerance would not necessarily reduce the parasite's reproductive fitness, unlike egg rejection (Soler and Soler 2017).

2.3.4 The Effects of Host Choice

When brood parasites choose between host individuals based on particular traits, such as nest size (Soler et al. 1995a; Molina-Morales et al. 2016), nest site or concealment (e.g., Hahn and Hatfield 2000; Hauber 2001; Underwood et al. 2004), egg color (Avilés et al. 2006b; Cherry et al. 2007), or the proximity of the nest to perches (vantage points for brood parasites; e.g., Røskaft et al. 2002; Antonov et al. 2007; Begum et al. 2011b), certain host individuals will become experienced with parasitism, while others may remain naïve. This could slow down coevolution within the entire population. As a contrasting example, *Vidua* finches imprint on their foster parents' song, and females choose to mate with males singing the same song as the species that reared them (Payne et al. 2000). This explains the rapid speciation that has occurred in this group because each host switch by a *Vidua* finch generates a new host-specific race, which eventually leads to a new parasitic species (Sorenson et al. 2003). As mentioned previously, this lineage has significantly faster rates of DNA sequence evolution than its hosts and than the younger cowbird lineage (Sorenson and Payne 2001; see Sect. 2.2.1), indicating that the speed of coevolution can also be affected by the parasite's life-history traits.

2.4 Coevolution in Action

Evolution has been considered by most to be slow and gradual, taking thousands or millions of years to observe. However, recent studies in evolutionary biology have found that evolutionary processes often proceed at rapid rates, similar to that of ecological processes (Thompson 1998, 2005, 2016). Thompson (1998) defines rapid evolution as occurring within the time span of about 100 years. Several theoretical models of avian brood parasitic systems predict evolution of host defenses could occur within this timeframe, depending on the surrounding circumstances (Rothstein 1975a; Takasu et al. 1993; Takasu 1998; Robert and Sorci 1999; Barabás et al. 2004). According to the model by Takasu et al. (1993), rejecter alleles could reach levels of 60% after about 30 years of parasitism, providing 40% of the population already has rejecter alleles before parasitism. If rejecter alleles are not initially present at a high rate, rejecter alleles could reach 60% after about 60 years of parasitism. In Rothstein's (1975a) model, an accepter with low selection pressure could reach a level of 80% rejection between 73 and 146 years, while an accepter with a higher selection pressure could reach 80% rejection within 30–60 years. According to Robert and Sorci (1999), if rejection behavior is never costly and parasitism rate is consistently high, a level of 80% rejection could be achieved within 40 or so years. If costs are associated with rejection behavior, it may take over 50 years for a host population to reach 80% rejection.

However, it is difficult to study empirically how quickly coevolution can proceed between brood parasites and their hosts. Bird populations cannot be readily isolated in a controlled environment, and the genetic underpinnings of most avian traits are poorly understood. Most brood parasites and their hosts have already been involved in coevolutionary relationships for thousands, if not millions, of years (Davies 2000; Sorenson and Payne 2001; Rothstein et al. 2002; Spottiswoode et al. 2011). Moreover, brood parasites rarely switch to completely naïve hosts, and lack of documentation on historical parasitism rates can further reduce our ability to know how many years are required for a naïve population to evolve an adaptation. Nevertheless, it is still important to understand how quickly avian hosts, especially those that may be completely naïve to brood parasitism, can acquire defenses, as this has conservation implications (Dinets et al. 2015), and understanding how rapidly brood parasites evolve counter-adaptations can aid in our understanding of the process of speciation (Sorenson et al. 2003; Medina and Langmore 2015a).

A number of studies have investigated how rapidly coevolution proceeds between brood parasites and hosts that have a history of parasitism (Robert and Sorci 1999; Spottiswoode and Stevens 2011, 2012; Medina et al. 2016), and at least three studies have demonstrated how quickly host defenses may evolve in newly parasitized populations (Soler et al. 1994; Nakamura et al. 1998; Abernathy 2017). Soler et al. (1994) showed that ejection rate of great spotted cuckoo eggs by Eurasian magpies increased from 0 to 10% over a 10-year period in Guadix, Spain, an area of recent sympatry with the cuckoo (Soler 1990). In addition, ejection rate of mimetic and nonmimetic model eggs increased from 14 to 33% over a 5-year period and 61–89% over an 11-year period, respectively (Soler et al. 1994). This increase in egg ejection was not due to the host using the presence of cuckoos in the area to determine risk of parasitism (Soler et al. 1994). Gene flow could be responsible for this rapid change in host defense, as the Guadix population is about 60 km away from another population in Santa Fe, Spain, which is thought to have ancient sympatry with cuckoos and showed higher rates of rejection than the population in Guadix (Soler 1990; Soler and Møller 1990; Martinez et al. 1999; see Sect. 2.2.3; Fig. 2.1).

In Japan, geographic elevation levels previously separated the common cuckoo and the azure-winged magpie (*Cyanopica cyanus*), but around the late 1950s–1960s, both species expanded their breeding ranges and became sympatric (Nakamura 1990; Nakamura et al. 1998). Parasitism of magpies increased steadily from 30 to 80% from 1981 to 1988 in the Nobeyama region. Magpies showed very low levels of rejection when parasitism first began, but after about 15–20 years of parasitism, some magpie populations exhibited aggression towards the cuckoo, nest desertion of parasitized nests, and higher rates of egg ejection (Nakamura 1990; Nakamura et al. 1998). However, there is some evidence that azure-winged magpies and cuckoos were sympatric in the Kanto Plain-Tokyo area before magpies expanded their breeding range, so the ability to recognize and eject eggs may have already been a part of this population's genetic makeup (Nakamura et al. 1998; see also Davies 2000; Avilés 2004).

While both of these studies show rapid changes in host defenses, a more recent study demonstrated egg ejection may sometimes require more than 38 years to evolve (Abernathy 2017). The red wattlebird in SE Australia was adopted as a host of the pacific koel, first in Sydney, NSW (38–86 years ago), and later in Canberra, ACT (8–33 years ago), after both species expanded their breeding ranges towards one another (Abernathy and Langmore 2017). Prior to this contact, wattlebirds in the SE were naïve to koel parasitism and had only rarely been parasitized by another brood parasite, the pallid cuckoo (*Cacomantis pallidus*) (Brooker and Brooker 1989, 2005; Abernathy and Langmore 2017). Despite the high costs of parasitism by koels, wattlebirds had not yet evolved egg rejection at the time of the study (3% rejection rate, $N = 60$), whereas two hosts that have been parasitized by the koel for over 100 years ejected model eggs at high rates (61%, $N = 41$ and 88%, $N = 43$; Abernathy 2017). Thus, the rate of evolution of host defenses appears to be occurring more slowly in this system than in the systems described by Soler et al. (1994) and Nakamura et al. (1998). This could be due to several factors. First, the parasitism rates in Sydney (24%) and Canberra (4%; Abernathy and Langmore 2017) are much lower than in the other two studies (e.g., parasitism rate of Eurasian magpies by great spotted cuckoos was 61% from 1990 to 1992, Soler et al. 1994; see also Nakamura et al. 1998), which could result in reduced selection pressure to evolve egg ejection for wattlebirds (see Sect. 2.3.2). Second, koel parasitism was often unsuccessful, which could mean the costs of parasitism for wattlebirds is lower (Abernathy and Langmore 2017; see Sect. 2.3.1). Third, rejection costs could be high for wattlebirds, as their eggs appear very similar to koel eggs, and, thus, wattlebirds may be more likely to make rejection errors (Abernathy et al. 2017; see Sect. 2.3.3). These studies demonstrate how variable rates of evolution can be between different populations and species and how important it is to investigate the many factors that could be influencing evolutionary rate (Thompson 1994, 2005, 2016; Soler and Soler 2017).

Concluding Remarks and Future Directions

The rate of coevolution between two species is the outcome of multiple, interacting variables, making generalizations difficult. One recurring pattern, however, is that when two interacting species have less asymmetry in mutation rate, generation time, fecundity, genetic variation, density, and selection pressure, coevolution should proceed more quickly than in highly

(continued)

asymmetrical species. Avian brood parasitism is a model example of coevolution mainly because both parties (the parasite and host) must be similar in most of the abovementioned characteristics in order for the host to successfully raise the brood parasitic chick. The selection pressures on parasites and hosts can vary greatly depending on their respective habits or traits, but selection is generally stronger on the parasite than the host and may lead to increasing specialization and, ultimately, speciation of parasites. While speciation may typically occur over millennia, there is increasing evidence that the reciprocal behavioral and morphological adaptations of brood parasites and their hosts can evolve within decades.

Discovering how quickly hosts can acquire defenses against brood parasites is important on several levels. First, it provides a rare opportunity to observe the process of coevolution in action in the natural world. Second, the rapidity with which defenses appear in a naïve host may indicate whether evolution is actually occurring. If a newly parasitized host species expresses defenses within a short time period, this may indicate that no genetic change is required for those defenses, and instead they are acquired within an individual's lifetime through learning. Third, with increasing environmental change, the distributions of many brood parasites are expanding, leading to annexation of new hosts. Understanding how rapidly a host can acquire defenses will allow predictions about the impact of brood parasites and whether control measures are required. Therefore, future studies should focus efforts on hosts that are naïve to brood parasitism to determine empirically the rate at which host defenses evolve in a new avian brood parasite host.

Acknowledgments We thank A. Edworthy and W. Feeney for their helpful editorial comments on the first version of this manuscript, as well as S. Rothstein and M. Soler whose comments on later versions greatly improved the quality of this review. We are also grateful for funding from the Australian National University and the Australian Research Council (DP110101966).

References

Abernathy VE (2017) Investigating the first stages of coevolution between the pacific koel and its newest host, the red wattlebird. PhD Thesis, Australian National University

Abernathy VE, Langmore NE (2017) The first stages of coevolution between a brood parasite and its new host: are naïve hosts defenseless? Emu 117:114–129. https://doi.org/10.1080/01584197.2017.1296330

Abernathy VE, Peer BD (2016) Reduced ultraviolet reflectance does not affect egg rejection by Northern Cardinals (*Cardinalis cardinalis*). Wilson J Ornithol 128:334–342

Abernathy VE, Troscianko J, Langmore NE (2017) Egg mimicry by the pacific koel: mimicry of one host facilitates exploitation of other hosts with similar egg types. J Avian Biol 48:1414–1424

Álvarez F (2003) Parasitism rate by the common cuckoo *Cuculus canorus* increases with high density of host's breeding pairs. Ornis Fenn 80:193–196

Antonov A, Stokke BG, Moksnes A, Røskaft E (2007) Factors influencing the risk of common cuckoo *Cuculus canorus* parasitism on marsh warblers *Acrocephalus palustris*. J Avian Biol 38:390–393

Antonov A, Stokke BG, Moksnes A, Røskaft E (2009) Evidence for egg discrimination preceding failed rejection attempts in a small cuckoo host. Biol Lett 5:169–171

Avilés JM (2004) Egg rejection by Iberian azure-winged magpies *Cyanopica cyanus* in the absence of brood parasitism. J Avian Biol 35:295–299

Avilés JM (2008) Egg colour mimicry in the common cuckoo *Cuculus canorus* as revealed by modelling host retinal function. Proc R Soc Lond B 275:2345–2352

Avilés JM, Soler JJ, Pérez-Contreras T (2006a) Dark nests and egg colour in birds: a possible functional role of ultraviolet reflectance in egg detectability. Proc R Soc Lond B 273:2821–2829

Avilés JM, Stokke BG, Moksnes A, Røskaft E, Åsmul M, Møller AP (2006b) Rapid increase in cuckoo egg matching in a recently parasitized reed warbler population. J Evol Biol 19:1901–1910

Barabás L, Gilicze B, Takasu F, Moskát C (2004) Survival and anti-parasite defense in a host metapopulation under heavy brood parasitism: a source-sink dynamic model. J Ethol 22:143–151

Begum S, Moksnes A, Røskaft E, Stokke BG (2011a) Interactions between the Asian koel (*Eudynamys scolopacea*) and its hosts. Behaviour 148:325–340

Begum S, Moksnes A, Røskaft E, Stokke BG (2011b) Factors influencing host nest use by the brood parasitic Asian Koel (*Eudynamys scolopacea*). J Ornithol 152:793–800

Bensch S, Hasselquist D (1999) Phylogeographic population structure of great reed warblers: an analysis of mtDNA control region sequences. Biol J Linn Soc 66:171–185

Boves TJ, Sperry JH, Comolli K, Weatherhead PJ (2014) Brood parasitism of black-capped vireos: frontline and post-laying behavioral responses and effects on productivity. J Field Ornithol 85:364–378

Brandt M, Foitzik S, Fischer-Blass B, Heinze J (2005) The coevolutionary dynamics of obligate ant social parasite systems—between prudence and antagonism. Biol Rev 80:251–267

Brockhurst MA, Koskella B (2013) Experimental coevolution of species interactions. Trends Ecol Evol 28:367–375

Brockhurst MA, Morgan AD, Rainey PB, Buckling A (2003) Population mixing accelerates coevolution. Ecol Lett 6:975–979

Brockhurst MA, Buckling A, Poullain V, Hochberg ME (2007) The impact of migration from parasite-free patches on antagonistic host-parasite coevolution. Evolution 61:1238–1243

Brodie ED, Brodie ED (1999) Predator-prey arms races. BioScience 49:557–568

Bromham L (2009) Why do species vary in their molecular evolution? Biol Lett 5:401–404

Brooke ML, Davies NB, Noble DG (1998) Rapid decline of host defences in response to reduced cuckoo parasitism: behavioural flexibility of reed warblers in a changing world. Proc R Soc Lond B 265:1277–1282

Brooker MG, Brooker LC (1989) Cuckoo hosts in Australia. Aust Zool Rev 2:1–67

Brooker L, Brooker M (2005) Records of parasitism by the common koel *Eudynamys scolopacea*. http://lesmikebrooker.com/KOELDB.HTM. Accessed 9 Apr 2013

Cherry MI, Bennett ATD, Moskát C (2007) Do cuckoos choose nests of great reed warblers on the basis of host egg appearance? J Evol Biol 20:1218–1222

Cruz A, Wiley JW (1989) The decline of an adaptation in the absence of a presumed selection pressure. Evolution 43:55–62

Cruz A, Prather JW, Wiley JW, Weaver PF (2008) Egg rejection behaviour in a population exposed to parasitism: village weavers on Hispaniola. Behav Ecol 19:398–403

Davies NB (2000) Cuckoos, cowbirds and other cheats. Academic Press, London

Davies NB (2011) Cuckoo adaptations: trickery and tuning. J Zool 284:1–14

Davies NB, Brooke ML, Kacelnik A (1996) Recognition errors and probability of parasitism determine whether reed warblers should accept or reject mimetic cuckoo eggs. Proc R Soc Lond B 263:925–931

Dawkins R (1982) The extended phenotype. Oxford University Press, Oxford

de Neve L, Soler JJ, Soler M, Pérez-Contreras T (2004) Nest size predicts the effect of food supplementation to magpie nestlings on their immunocompetence: an experimental test of nest size indicating parental ability. Behav Ecol 15:1031–1036

Dinets V, Samaš P, Croston R, Grim T, Hauber ME (2015) Predicting the responses of native birds to transoceanic invasions by avian brood parasites. J Field Ornithol 86:244–251

Fleischer RC, Smyth AP, Rothstein SI (1987) Temporal and age-related variation in the laying rate of the parasitic brown-headed cowbird in the eastern Sierra Nevada, California. Can J Zool 65:2724–2730

Fossøy F, Antonov A, Moksnes A, Røskaft E, Vikan JR, Møller AP, Shykoff JA, Stokke BG (2011) Genetic differentiation among sympatric cuckoo host races: males matter. Proc R Soc Lond B 278:1639–1645

Fossøy F, Sorenson MD, Liang W, Ekrem T, Moksnes A, Møller AP, Rutila J, Røskaft E, Takasu F, Yang C, Stokke BG (2016) Ancient origin and maternal inheritance of blue cuckoo eggs. Nat Commun 7:10272

Futuyma DJ (2010) Evolutionary constraint and ecological consequences. Evolution 64:1865–1884

Gibbs HL, Sorenson MD, Marchetti K, Brooke ML, Davies NB, Nakamura H (2000) Genetic evidence for female host-specific races of the common cuckoo. Nature 407:183–186

Gloag R, Fiorini VD, Reboreda JC, Kacelnik A (2012) Brood parasite eggs enhance egg survivorship in a multiply parasitized host. Proc R Soc Lond B 279:1831–1839

Gosler AG, Barnett PR, Reynolds SJ (2000) Inheritance and variation in eggshell patterning in the great tit *Parus major*. Proc R Soc Lond B 267:2469–2473

Greischar MA, Koskella B (2007) A synthesis of experimental work on parasite local adaptation. Ecol Lett 10:418–434

Grim T, Samaš P, Moskát C, Kleven O, Honza M, Moksnes A, Røskaft E, Stokke BG (2011) Constraints on host choice: why do parasitic birds rarely exploit some common potential hosts? J Anim Ecol 80:508–518

Guigueno MF, Sealy SG (2010) Clutch abandonment by parasitized yellow warblers: egg burial or nest desertion? Condor 112:399–406

Haas CA, Haas KH (1998) Brood parasitism by brown-headed cowbirds on brown thrashers: frequency and rates of rejection. Condor 100:535–540

Hahn DC, Hatfield JS (2000) Host selection in the forest interior: cowbirds target ground-nesting species. In: Smith JNM, Cook TL, Rothstein SI, Robinson SK, Sealy SG (eds) Ecology and management of cowbirds and their hosts: studies in the conservation of North American passerine birds. University of Texas Press, Austin, pp 120–127

Hahn DC, Sedgwick JA, Painter IS, Casna NJ (1999) A spatial and genetic analysis of cowbird host selection. Stud Avian Biol 18:204–217

Hansson B (2003) Dispersal, inbreeding and fitness in natural populations. PhD Thesis, Lund University

Hardin G (1960) The competitive exclusion principle. Science 131:1292–1297

Hauber ME (2001) Site selection and repeatability in brown-headed cowbird (*Molothrus ater*) parasitism of eastern phoebe (*Sayornis phoebe*) nests. Can J Zool 79:1518–1523

Hauber ME, Strausberger BM, Feldheim KA, Lock J, Cassey P (2012) Indirect estimates of breeding and natal philopatry in an obligate avian brood parasite. J Ornithol 153:467–475

Hoeksema JD, Forde SE (2008) A meta-analysis of factors affecting local adaptation between interacting species. Am Nat 171:275–290

Hoover JP, Robinson SK (2007) Retaliatory mafia behavior by a parasitic cowbird favors host acceptance of parasitic eggs. Proc Natl Acad Sci 104:4479–4483

Hosoi SA, Rothstein SI (2000) Nest desertion and cowbird parasitism: evidence for evolved responses and evolutionary lag. Anim Behav 59:823–840

Jackson NH, Roby DD (1992) Fecundity and egg-laying patterns of captive yearling brown-headed cowbirds. Condor 94:585–589

Jensen WE, Cully JF (2005) Density-dependent habitat selection by brown-headed cowbirds (*Molothrus ater*) in tallgrass prairie. Oecologia 142:136–149

Jewell KJ, Arcese P (2008) Consequences of parasite invasion and land use on the spatial dynamics of host populations. J Appl Ecol 45:1180–1188

Kelly C (1987) A model to explore the rate of spread of mimicry and rejection in hypothetical populations of cuckoos and their hosts. J Theor Biol 125:283–299

Kilner RM (2003) How selfish is a cowbird nestling? Anim Behav 66:569–576

Krüger O, Sorenson MD, Davies NB (2009) Does coevolution promote species richness in parasitic cuckoos? Proc R Soc Lond B 276:3871–3879

Langmore NE, Kilner RM, Butchart SHM, Maurer G, Davies NB, Cockburn A, Macgregor NA, Peters A, Magrath MJL, Dowling DK (2005) The evolution of egg rejection by cuckoo hosts in Australia and Europe. Behav Ecol 16:686–692

Langmore NE, Adcock GJ, Kilner RM (2007) The spatial organization and mating system of Horsfield's bronze-cuckoos, *Chalcites basalis*. Anim Behav 74:403–412

Langmore NE, Feeney WE, Crowe-Riddell J, Luan H, Louwrens KM, Cockburn A (2012) Learned recognition of brood parasitic cuckoos in the superb fairy-wren *Malurus cyaneus*. Behav Ecol 23:798–805

Lenormand T (2002) Gene flow and the limits to natural selection. Trends Ecol Evol 17:183–189

Liang W, Møller AP, Stokke BG, Yang C, Kovařík P, Wang H, Yao C, Ding P, Lu X, Moksnes A, Røskaft E, Grim T (2016) Geographic variation in egg ejection rate by great tits across 2 continents. Behav Ecol 27:1405–1412

Lindholm AK, Thomas RJ (2000) Differences between populations of reed warblers in defences against brood parasitism. Behaviour 137:25–42

Lively CM, Dybdahl MF (2000) Parasite adaptation to locally common host genotypes. Nature 405:679–681

Lotem A, Nakamura H, Zahavi A (1995) Constraints on egg discrimination and cuckoo-host co-evolution. Anim Behav 49:1185–1209

Lyon BE, Eadie JM (2004) An obligate brood parasite trapped in the intraspecific arms race of its hosts. Nature 432:390–393

Mahler B, Confalonieri VA, Lovette IJ, Reboreda JC (2008) Eggshell spotting in brood parasitic shiny cowbirds (*Molothrus bonariensis*) is not linked to the female sex chromosome. Behav Ecol Sociobiol 62:1193–1199

Marchetti K, Nakamura H, Gibbs HL (1998) Host-race formation in the common cuckoo. Science 282:471–472

Martinez JG, Soler JJ, Soler M, Møller AP, Burke T (1999) Comparative population structure and gene flow of a brood parasite, the great spotted cuckoo (*Clamator glandarius*), and its primary host, the magpie (*Pica pica*). Evolution 53:269–278

Martín-Gálvez D, Soler JJ, Martínez JG, Krupa AP, Richard M, Soler M, Møller AP, Burke T (2006) A quantitative trait locus for recognition of foreign eggs in the host of a brood parasite. J Evol Biol 19:543–550

Martín-Gálvez D, Soler JJ, Martínez JG, Krupa AP, Soler M, Burke T (2007) Cuckoo parasitism and productivity in different magpie subpopulations predict frequencies of the 457bp allele: a mosaic of coevolution at a small geographic scale. Evolution 61:2340–2348

Medina I, Langmore NE (2015a) Coevolution is linked with phenotypic diversification but not speciation in avian brood parasites. Proc R Soc Lond B 282:20152056

Medina I, Langmore NE (2015b) The costs of avian brood parasitism explain variation in egg rejection behaviour in hosts. Biol Lett 11:20150296

Medina I, Langmore NE (2016a) The evolution of host specialization in avian brood parasites. Ecol Lett 19:1110–1118

Medina I, Langmore NE (2016b) Batten down the thatches: front-line defences in an apparently defenceless cuckoo host. Anim Behav 112:195–201

Medina I, Langmore NE (2016c) The evolution of acceptance and tolerance in hosts of avian brood parasites. Biol Rev 91:569–577

Medina I, Troscianko J, Stevens M, Langmore NE (2016) Brood parasitism is linked to egg pattern diversity within and among species of Australian passerines. Am Nat 187:351–362

Middleton ALA (1991) Failure of brown-headed cowbird parasitism in nests of the American goldfinch. J Field Ornithol 62:200–203

Moksnes A, Røskaft E (1987) Cuckoo host interactions in Norwegian mountain areas. Ornis Scand 18:168–172

Moksnes A, Røskaft E (1995) Egg-morphs and host preference in the common cuckoo (*Cuculus canorus*): an analysis of cuckoo and host eggs from European museum collections. J Zool 236:625–648

Molina-Morales M, Martínez JG, Avilés JM (2016) Criteria for host selection in a brood parasite vary depending on parasitism rate. Behav Ecol 27:1441–1448

Møller AP, Antonov A, Stokke BG, Fossøy F, Moksnes A, Røskaft E, Takasu F (2011) Isolation by time and habitat and coexistence of distinct host races of the common cuckoo. J Evol Biol 24:676–684

Molnár B (1939) About the cuckoo. Observations and experiments on the eviction instinct of the cuckoo chick. Published by the author, Szarvas

Moskát C, Hansson B, Barabás L, Bártol I, Karcza Z (2008) Common cuckoo *Cuculus canorus* parasitism, antiparasite defence and gene flow in closely located populations of great reed warblers *Acrocephalus arundinaceus*. J Avian Biol 39:663–671

Nakamura H (1990) Brood parasitism by the cuckoo *Cuculus canorus* in Japan and the start of new parasitism on the azure-winged magpie *Cyanopica cyana*. Jpn J Ornithol 39:1–18

Nakamura H, Kubota S, Suzuki R (1998) Coevolution between the common cuckoo and its major hosts in Japan. In: Rothstein SI, Robinson SK (eds) Parasitic birds and their hosts: studies in coevolution. Oxford University Press, Oxford, pp 94–112

Nice MM (1937) Studies in the life history of the song sparrow. Trans Linn Soc Lond 4:1–246

Payne RB, Payne LL, Woods JL, Sorenson MD (2000) Imprinting and the origin of parasite-host species associations in brood-parasitic indigobirds, *Vidua chalybeata*. Anim Behav 59:69–81

Peer BD, Bollinger EK (1997) Explanations for the infrequent cowbird parasitism on common grackles. Condor 99:151–161

Peer BD, Bollinger EK (1998) Rejection of cowbird eggs by mourning doves: a manifestation of nest usurpation? Auk 115:1057–1062

Peer BD, Rothstein SI, Kuehn MJ, Fleischer RC (2005) Host defenses against cowbird (*Molothrus* spp.) parasitism: implications for cowbird management. Ornithol Monogr 57:84–97

Peer BD, Rivers JW, Rothstein SI (2013) Cowbirds, conservation, and coevolution: potential misconceptions and directions for future research. Chin Birds 4:15–30

Procházka P, Konvičková-Patzenhauerová H, Požgayová M, Trnka A, Jelínek V, Honza M (2014) Host genotype and age have no effect on rejection of parasitic eggs. Naturwissenschaften 101:417–426

Robert M, Sorci G (1999) Rapid increase of host defence against brood parasites in a recently parasitized area: the case of village weavers in Hispaniola. Proc R Soc Lond B 266:941–946

Røskaft E, Moksnes A, Stokke BG, Moskát C, Honza M (2002) The spatial habitat structure of host populations explains the pattern of rejection behavior in hosts and parasitic adaptations in cuckoos. Behav Ecol 13:163–168

Rothstein SI (1975a) Evolutionary rates and host defenses against avian brood parasitism. Am Nat 109:161–176

Rothstein SI (1975b) An experimental and teleonomic investigation of avian brood parasitism. Condor 77:250–271

Rothstein SI (1990) A model system for coevolution: avian brood parasitism. Annu Rev Ecol Syst 21:481–508

Rothstein SI, Patten MA, Fleischer RC (2002) Phylogeny, specialization, and brood parasite-host coevolution: some possible pitfalls of parsimony. Behav Ecol 13:1–10

Sato NJ, Mikamf OK, Ueda K (2010) The egg dilution effect hypothesis: a condition under which parasitic nestling ejection behaviour will evolve. Ornithol Sci 9:115–121

Schlossberg S (2009) Site fidelity of shrubland and forest birds. Condor 111:238–246

Scott DM (1963) Changes in the reproductive activity of the brown-headed cowbird within the breeding season. Wilson Bull 75:123–129

Scott DM, Ankney CD (1980) Fecundity of the brown-headed cowbird in southern Ontario. Auk 97:677–683

Sealy SG (1996) Evolution of host defenses against brood parasitism: implications of puncture-ejection by a small passerine. Auk 113:346–355

Smith JNM, Arcese P (1994) Brown-headed cowbirds and an island population of song sparrows: a 16-year study. Condor 96:916–934

Soler M (1990) Relationships between the great spotted cuckoo *Clamator glandarius* and its corvid hosts in a recently colonized area. Ornis Scand 21:212–223

Soler M (2014) Long-term coevolution between avian brood parasites and their hosts. Biol Rev 89:688–704

Soler M, Møller AP (1990) Duration of sympatry and coevolution between the great spotted cuckoo and its magpie host. Nature 343:748–750

Soler JJ, Soler M (2017) Evolutionary change: facultative virulence by brood parasites and tolerance and plastic resistance by hosts. Anim Behav 125:101–107

Soler M, Soler JJ, Martinez JG, Møller AP (1994) Micro-evolutionary change in host response to a brood parasite. Behav Ecol Sociobiol 35:295–301

Soler JJ, Soler M, Møller AP, Martínez JG (1995a) Does the great spotted cuckoo choose magpie hosts according to their parenting ability? Behav Ecol Sociobiol 36:201–206

Soler M, Soler JJ, Martinez JG, Møller AP (1995b) Magpie host manipulation by great spotted cuckoos: evidence for an avian mafia? Evolution 49:770–775

Soler JJ, Martínez JG, Soler M, Møller AP (1999) Genetic and geographic variation in rejection behavior of cuckoo eggs by European magpie populations: an experimental test of rejecter-gene flow. Evolution 53:947–956

Soler JJ, Martínez JG, Soler M, Møller AP (2001) Coevolutionary interactions in a host-parasite system. Ecol Lett 4:470–476

Soler JJ, Martín Vivaldi M, Møller AP (2009) Geographic distribution of suitable hosts explains the evolution of specialized gentes in the European cuckoo *Cuculus canorus*. BMC Evol Biol 9:1

Sorenson MD, Payne RB (2001) A single ancient origin of brood parasitism in African finches: implications for host-parasite coevolution. Evolution 55:2550–2567

Sorenson MD, Sefc KM, Payne RB (2003) Speciation by host switch in brood parasitic indigobirds. Nature 424:928–931

Southern HN (1954) Mimicry in the cuckoo eggs. In: Huxley JS, Hardy AC, Ford EB (eds) Evolution as a process. George Allen and Unwin, London, pp 219–232

Spottiswoode CN, Stevens M (2011) How to evade a coevolving brood parasite: egg discrimination versus egg variability as host defences. Proc R Soc Lond B 278:3566–3573

Spottiswoode CN, Stevens M (2012) Host-parasite arms races and rapid changes in bird egg appearance. Am Nat 179:633–648

Spottiswoode CN, Stryjewski KF, Quader S, Colebrook-Robjent JFR, Sorenson MD (2011) Ancient host specificity within a single species of brood parasitic bird. Proc Natl Acad Sci 108:17738–17742

Stokke BG, Hafstad I, Rudolfsen G, Bargain B, Beier J, Bigas Campàs D, Dyrcz A, Honza M, Leisler B, Pap PL, Patapavičius R, Procházka P, Schulze-Hagen K, Thomas R, Moksnes A, Møller AP, Røskaft E, Soler M (2007) Host density predicts presence of cuckoo parasitism in reed warblers. Oikos 116:913–922

Strausberger BM, Rothstein SI (2009) Parasitic cowbirds may defeat host defense by causing rejecters to misimprint on cowbird eggs. Behav Ecol 20:691–699

Takasu F (1998) Why do all host species not show defense against avian brood parasitism: evolutionary lag or equilibrium? Am Nat 151:193–205

Takasu F, Kawasaki K, Nakamura H, Cohen JE, Shigesada N (1993) Modeling the population dynamics of a cuckoo-host association and the evolution of host defenses. Am Nat 142:819–839
Thompson JN (1994) The coevolutionary process. University of Chicago Press, Chicago
Thompson JN (1998) Rapid evolution as an ecological process. Trends Ecol Evol 13:329–332
Thompson JN (1999) The raw material for coevolution. Oikos 84:5–16
Thompson JN (2005) The geographic mosaic of coevolution. University of Chicago Press, Chicago
Thompson JN (2016) Relentless evolution. University of Chicago Press, Chicago
Underwood TJ, Sealy SG, McLaren CM (2004) Eastern wood-pewees as brown-headed cowbird hosts: accepters but infrequently parasitized. J Field Ornithol 75:165–171
Vogwill T, Fenton A, Brockhurst MA (2008) The impact of parasite dispersal on antagonistic host-parasite coevolution. J Evol Biol 21:1252–1258
Woolfenden BE, Gibbs HL, Sealy SG, McMaster DG (2003) Host use and fecundity of individual female brown-headed cowbirds. Anim Behav 66:95–106
Yang C, Liu Y, Zeng L, Liang W (2014) Egg color variation, but not egg rejection behavior, changes in a cuckoo host breeding in the absence of brood parasitism. Ecol Evol 4:2239–2246
Zölei A, Bán M, Moskát C (2015) No change in common cuckoo *Cuculus canorus* parasitism and great reed warblers' *Acrocephalus arundinaceus* egg rejection after seven decades. J Avian Biol 46:570–576

Evolution of Avian Brood Parasitism and Phylogenetic History of Brood Parasites

3

Oliver Krüger and Martina Pauli

Abstract

The fascination with avian brood parasitism has spawned a wealth of research into breeding strategies and their evolutionary consequences. Yet until today, there is no clear consensus how brood parasitism has evolved. On current evidence, it is more likely that interspecific brood parasitism evolved directly from an ancestral species with parental care, rather than via the stepping stone of conspecific brood parasitism. Interspecific brood parasitism has evolved independently seven times in birds, but only about 1% of all bird species are interspecific brood parasites. The evolution of this breeding strategy was accompanied by profound changes in both ecology and life history. In cuckoos, it is more likely to be a later adaptation, possibly reducing the cost of raising chicks, whereas the reduction of egg size is a direct adaptation in the coevolutionary interaction with host species. The study of the evolution of brood parasitism and its phylogenetic history is likely to further advance rapidly in the near future as genomic approaches become feasible in non-model species and phylogenetic analyses techniques are developed at an equally breath-taking rate.

3.1 Introduction

Recent decades have witnessed an explosion of interest in the subject of brood parasitism, an explosion that has, we suggest, received crucial catalysis from three entirely different directions. Firstly, some 40 years ago, Stephen Rothstein (Rothstein 1975) realised that the behaviour of birds that were potential hosts of cowbirds could be explored by placing model eggs into their nests. This proved a

O. Krüger (✉) · M. Pauli
Department of Animal Behaviour, Bielefeld University, Bielefeld, Germany
e-mail: oliver.krueger@uni-bielefeld.de

M. Soler (ed.), *Avian Brood Parasitism*, Fascinating Life Sciences,
https://doi.org/10.1007/978-3-319-73138-4_3

powerful technique for exploring many of the subtle coevolutionary interactions between parasites and hosts and has since been widely adopted by other workers (see Chaps. 19, 22, 24 and 25). As the complexity of these interactions has become appreciated, evolutionary modellers have entered and contributed to the field. Secondly, refined DNA techniques have allowed the more reliable determination of the occurrence of conspecific brood parasitism and hence the asking of more precise questions about its occurrence and evolution (McRae and Burke 1996). These genetic techniques have also revolutionised our understanding of phylogenetic relationships among taxa (Prum et al. 2015) and gave rise to comparative genetics and genomics, even in non-model species (Krüger et al. 2017). This in turn, thirdly, has led to a flurry of comparative analyses that compare and contrast species while explicitly taking phylogeny into account, a development which was initiated by Joe Felsenstein over 30 years ago (Felsenstein 1985). Considering the evolution of conspecific brood parasitism briefly as a potential stepping stone for interspecific brood parasitism, our chapter focuses on the evolutionary history of brood parasitism and brood parasites.

3.1.1 Conspecific Brood Parasitism as a Stepping Stone?

Assuming that conspecific brood parasitism (CBP) is a precursor to interspecific parasitism is tempting. One of the earliest studies to analyse the origin of parasitism in an evolutionary, ultimate approach were by Hamilton and Orians (1965). They argued that parasitism was most likely to evolve in colonial nesters, where nesting pairs are more tolerant of the proximity of conspecifics and females would occasionally lay eggs into other nests after having lost their own. Once females benefited from this strategy, there would be selection on decoupling egg formation from nest building, leading females to seek out more nests, including in other species. This selection to shift to other species was assumed to be extremely strong by Payne (1977), as he predicted populations containing conspecific parasites to go extinct quickly. In a modelling approach, Cichon (1996) took up this assumption but instead found obligate, i.e. interspecific parasites highly likely to go extinct, unless hosts showed some level of defence against parasitism. Arguing that birds will most likely gain defences if they experienced some facultative, conspecific parasitism, he supported the idea of CBP as a stepping stone to interspecific parasitism, which had already been advocated by Lyon and Eadie a few years earlier (1991). In a modelling approach, Robert and Sorci (2001) further strengthened the stepping-stone hypothesis, as they found two evolutionary stable strategies for conspecific parasites: either stay conspecific or turn into obligate, and hence interspecific, parasites.

The major change from seeing CBP as a precursor to interspecific parasitism came with the advent of more sophisticated phylogenetic analyses. Yom-Tov and Geffen (2006) used both parsimony and maximum likelihood approaches to identify the ancestral states of avian clades containing brood parasites. Contrary to even their own expectations, they found parental behaviour to be ancestral in all but one group (the exception being the *Hirundinidae*, where conspecific parasitism is highly common). Subsequent analyses (Krüger and Davies 2002; Boerner and Krüger 2008) have supported this result of parental care as the ancestral state.

3.2 Evolution of Conspecific Brood Parasitism

While the recent focus in studying brood parasitism has often been on its intricate, finely tuned adaptations, earlier observers perceived the females' behaviour as defunct. Gilbert White (1789) described it as "violence" on maternal instincts, "one of the first great dictates on nature", capitulating before this puzzle "that the methods of providence are not subjected to any mode or rule". Despite a few authors' doubts (Darwin being the most prominent), this view of parasitism as a "degenerative" loss of nesting and parental behaviour persisted well into the twentieth century. Only in 1965, Hamilton and Orians (1965) dismissed most previous hypotheses on the origin of parasitism which focused on proximate causes. Arguing that degenerative behaviour will neither occur simultaneously in all females nor be able to spread throughout the population, they suggested an ultimate approach, asking for the ecological conditions under which these genes can result in more offspring and hence spread through the population.

An early suggestion for the benefits of parasitism was that of spreading risk (Payne 1977), quite literally to not "put all your eggs in one basket". This effect has been hotly debated, as models failed to show strong effects of risk spreading (Bulmer 1984; Sorenson 1991). Instead, parasites may increase their mean fitness by not simply spreading risk among random nests but choosing low-risk nests, thereby reducing the predation risk for their offspring (Pöysä and Pesonen 2007). Similarly, overall chick survival can be higher because parasitic eggs spend on average less time in a nest, i.e. at risk of nest predation (Andersson and Åhlund 2012). This is true especially in the case of large clutches and when incubation will only begin once a clutch is complete. Eggs laid first therefore spend more time in the nest than eggs added later.

More broadly speaking, when asking "When does it benefit female birds to lay their eggs into the nest of other females?", three main hypotheses have emerged (reviewed in Eadie et al. 1988; Lyon and Eadie 2008):

1. Best of a bad job: Females which find themselves unable to nest or to continue nesting gain at least some reproductive success. The reasons can be either intrinsic or extrinsic, for example, insufficient condition of the female to care for a brood of chicks (Sorenson 1991), destruction of the nest (Andersson 1984) or an insufficient number of nest sites (Yom-Tov 1980).
2. Mixed ESS strategy: Some females in a population only nest; others only parasitise. In this scenario, females consistently use either one or the other breeding strategy throughout their entire lives. The balance between these two strategies is maintained through negative frequency-dependent selection (reviewed in Eadie et al. 1988). While models suggest this strategy may be viable

(Pöysä and Pesonen 2007), no such population with lifelong specialisation of females has been found to date.

3. Mixed reproductive strategy: Females build their own nest but also place additional eggs into nests of other females to increase their reproductive success. Nesting and parasitizing by one female may either overlap in time, or she may have separate clutches, one of which she incubates, while the eggs of the other clutch(es) are being laid parasitically (Sorenson 1991). How many eggs a female should allocate to either strategy has been modelled several times, with a unified approach offered by Broom and Ruxton (2004) and more recently by Andersson and Åhlund (2012).

The mixed reproductive strategy can be difficult to distinguish from doing the best of a bad job, as females may start nesting, parasitise others and then lose their nest. A comprehensive approach to modelling such situations has been suggested by Michael Sorenson (1991). The degree to which a female is parasitic is treated as a continuous variable rather than as a set of alternative options. Sorenson argued that individuals' strategies will depend on other continuous variables, namely, environmental and individual conditions. Females thus take into account the costs of breeding to their own survival and can (1) either not nest at all if conditions are very poor, (2) they can only lay parasitic eggs if they can afford to form eggs but not to care for the young, (3) nest and care for their own young if conditions are good or (4) nest, rear their own young and lay additional eggs parasitically which they can form but not care for. The latter will only be possible if conditions are exceptionally good. The importance of individual condition for parasitism has also been supported by Jaatinen et al. (2011b) and Baran and Reeve (2015).

All these models stress the fact that parasitism can only occur if it is the best available strategy for the individual female, given her individual circumstances. Lyon and Eadie (2008) broke this down to understand which aspect females are maximising: their current fecundity, offspring survival or own survival, i.e. future fecundity. Females have to trade off these factors against one another, with decisive factors for these decisions being life history, ecology, physiology and social environment (Lyon and Eadie 2008). We will look at some of these factors in the following section.

3.2.1 The Role of Ecology, Social Group and Life History

Ecological circumstances favouring conspecific brood parasitism are far from well understood. Hamilton and Orians (1965), although focussing on the evolution of interspecific brood parasitism, argued that high predation pressure would result in many females having their nests destroyed before finishing their clutch, potentially making them seek out alternative nests for their eggs. Recent evidence, however, seems to challenge the idea that nest loss is of major importance (see Chap. 6). Additionally, they speculated that in areas of unpredictable food supply, females may show extended periods of egg laying, giving a parasite a longer window of opportunity to deposit her egg. If food supply is high, however, a female may also obtain opportunities for parasitism, as she can afford to form more eggs than she and a mate can care for. In this situation, breeding systems can evolve where females lay several clutches for several males, predisposing females to lay in nests they have not built and will not attend (Davies 2000).

Once females start laying eggs into each other's nests, having ample opportunity to use this strategy may fuel its spread. High fecundity and a large number of available host nests have been suggested to do just that, as comparative analyses have shown large clutch sizes and nesting in cavities and/or colonies to correlate with CBP (Beauchamp 1997; Yom-Tov 2001). The most striking correlation with CBP, however, comes from its high occurrence in waterfowl and in grebes and their allies (see Chaps. 5 and 6). The phylogenetic distribution of CBP among waterfowl strongly suggests multiple independent evolutionary origins and most likely what followed these events was an eco-evolutionary feedback. These species predominantly raise precocial young who feed themselves shortly after hatching. These low costs to hosts and therefore low selection for defences (Rohwer and Freeman 1989) and/or the high chances of parasitic chick survival (Sorenson 1992) seem to predispose this group to frequent parasitism. In fact, as host parents have to invest so little into additional chicks, it has been suggested that these costs may be outweighed by some benefits that come with additional chicks. In larger groups, hosts' own offspring may be less vulnerable to predators, due to dilution effects, increased vigilance or a "selfish herd" effect in which parasitic chicks get effectively sacrificed in the event of a predator attack (reviewed by Eadie et al. 1988). Empirical evidence and the fitness effects for these hypotheses are still lacking, though. However, the idea of brood parasitism being not as clear-cut detrimental to hosts as initially assumed merits a closer look.

3.2.2 The Role of Relatedness

A parasite gains fitness benefits from placing eggs into other females' nests, severely impacting the hosts' fitness. This clear-cut statement has recently been challenged, even in an interspecific brood parasitic relationship (Canestrari et al. 2014, but see Soler et al. 2017). In conspecific brood parasitism, evidence against the purely antagonistic view of many "parasitic" relationships has been mounting for much longer. In one of the earliest models of brood parasitism, Andersson (1984) suggested that kin selection may play a key role in the evolution of CBP. If parasites, or rather egg donors, and hosts (egg recipients) were related, recipients would gain indirect fitness through incubating their relatives' eggs. Empirical evidence for relatedness closer than that expected by chance has been found in several species since (McRae and Burke 1996; Andersson and Åhlund 2000; Nielsen et al. 2006; Andersson et al. 2015), but spatial population structure can sometimes be the driving factor rather than targeted parasitism (McRae and Burke 1996). Modern modelling studies have nonetheless explored cost-benefit analyses (Zink 2000; Jaatinen et al. 2011b; Fulmer and Hauber 2016; Andersson 2017) and the roles of kin recognition

(Pöysä 2004), egg recognition (Jaatinen et al. 2011b) and mixed ESS versus mixed strategy for closely related donors and recipients (Jaatinen et al. 2011a).

Relatedness between parasites and their conspecific hosts has also led to the suggestion that the evolutionary road to parasitism started at cooperative breeding (see Chap. 12). Helpers at a nest are usually related to the main breeding pair. They will sometimes also lay eggs, making it a small step to laying and then abandoning the nest (Zink 2000; Andersson 2001). Similar to the sliding scale of environmental and individual condition influencing the degree of (non-)nesting and parasitic behaviour (see above), the transition from cooperation to parasitism is now also seen on a continuous scale of relatedness (Lyon and Eadie 2008; Zink 2000). When females lay eggs into nests of close kin, this may benefit the inclusive fitness of both but only if the donor female cannot otherwise breed (Zink and Lyon 2016). With smaller degrees of relatedness, this relationship becomes less mutualistic, turning into commensalism in case of very low costs to the receiving female and finally into parasitism (Zink and Lyon 2016).

The evolution of conspecific brood parasitism is thus a more complex and nuanced breeding strategy than previously assumed. Its occurrence can be facilitated by both harsh and beneficial environmental conditions, and the cost-benefit balances for parasite and host strongly depend on their relatedness, potentially even turning seemingly parasitic relationships into cooperative ones.

3.3 Evolutionary History and Evolution of Interspecific Brood Parasitism

On current evidence, interspecific brood parasitism has evolved independently seven times in birds (Sorenson and Payne 2002, 2005): three times among cuckoos (order *Cuculiformes*); twice among *Passeriformes*, namely, in the cowbirds (genus *Molothrus*, family *Icteridae*) and African brood parasitic finches (family *Viduidae*); once among the honeyguides (family *Indicatoridae*, order *Piciformes*); and once among waterfowl (black-headed duck *Heteronetta atricapilla*, order *Anseriformes*). Only about a 100 species are obligate interspecific brood parasites (see Chap. 4), out of the currently recognised 10,800 or so bird species worldwide (Davies 2000; Krüger 2007; Feeney et al. 2014; Gill and Donsker 2017). Interestingly, interspecific brood parasitism evolved only once in precocial birds, although they show a much higher occurrence of conspecific brood parasitism (see Chaps. 5 and 6). This may imply that the benefits of interspecific brood parasitism are much higher in altricial birds where the costs of raising chicks are much higher (Lyon and Eadie 1991; Feeney et al. 2014). In five instances, the evolution of interspecific brood parasitism is considered to be relatively old (the three instances in the *Cuculiformes*, the *Viduidae* and the *Indicatoridae*: >10 million years, Davies 2000; Sorenson and Payne 2002) whereas the evolution of brood parasitism in the *Icteridae* and *H. atricapilla* is considered to be rather recent (<5 million years, Sorenson and Payne 2002). This is not to be confused with subsequent radiation of brood parasitic taxa, as Sorenson et al. (2003, 2004) found that the extant *Viduidae* species are less than 5 million years old, the radiation being most likely the result of colonization of new hosts rather than co-speciation.

Obligate interspecific brood parasitism is hence confined to 4 out of 40 orders in birds (Davies 2000; Prum et al. 2015; Chap. 4). Four of the seven brood parasitic taxa (the three cuculiform taxa and *Molothrus*) have received considerably more research effort than the other three. Because there are excellent reviews on *Cuculidae* and *Molothrus* (Ortega 1998; Rothstein and Robinson 1998; Payne 2005) and the other taxa are discussed in detail in Davies (2000), the taxonomic overview in this chapter will be very concise (see Chap. 4 for more details).

The most widely accepted taxonomy of the cuckoos now recognises 149 species (Payne 2005; Gill and Donsker 2017), of which 64 are brood parasites and 85 have parental care. They are found on all continents with the exception of Antarctica. They are an evolutionarily ancient taxon of birds, going back at least 60 million years (Rothstein and Robinson 1998). Among the brood parasitic species, a few parasitise their own species, some sometimes parasitise their own or other species and many are obligate brood parasites that rely entirely on host species for parental care (see Chap. 4). Obligate brood parasites are found in three monophyletic taxa: the Old World brood parasitic *Cuculinae* (57 species), the Old-World genus *Clamator* (four species) and the New-World brood parasitic genera *Tapera* and *Dromococcyx* (three species) (see Chap. 4). All brood parasitic cuckoos either evict or kill the host young, with the exception of *Clamator* cuckoos, where the young cuckoo is reared, at least initially, with the host young (Soler 1990).

The brood parasitic species of the *Icteridae* of North and South America comprise a monophyletic taxon with six species of the genus *Molothrus* (Johnson and Lanyon 1999; Gill and Donsker 2017) nested within the family *Icteridae* (grackles and allies). They are much younger in evolutionary terms, at most 3–5 million years old (Rothstein et al. 2002). The parasitic *Molothrus* species do not evict host young from the nest, and hence host and parasitic young are raised together (Ortega 1998).

The second group of passerines to evolve interspecific brood parasitism are the African finches in the genera *Vidua* (ten species of indigo birds and nine species of whydah) and *Anomalospiza* (one species) (see Chap. 4). Molecular phylogenies suggest that the origin of brood parasitism in this taxon might be 20 million years old (Sorenson and Payne 2001). However, the extant species of *Viduidae* are most likely comparatively recent (<5 million years, Sorenson et al. 2004). Most species only parasitise a single host species and hence are the most specialised brood parasites (Davies 2000).

The 17 species of *Indicatoridae* from sub-Saharan Africa (15 species) and Asia (2 species) are included as a family with other families in the order *Piciformes* (Gill and Donsker 2017). Knowledge about the *Indicatoridae* used to be very poor compared to the aforementioned taxa, but they do seem to be a relatively old evolutionary group (Prum et al. 2015), and duration of sympatry with hosts has been estimated to be around 3 million years (Spottiswoode et al. 2011). Detailed studies dealing with the evolution of brood parasitism in species of this group have recently been performed by Claire Spottiswoode and have shown that the coevolutionary interactions are as diversified and as fascinating as in the cuckoos (Spottiswoode and Colebrook-Robjent 2007; Spottiswoode and Koorevaar 2011).

The final interspecific brood parasite is the black-headed duck (*Heteronetta atricapilla*) from South America. This species was only confirmed as an obligate brood parasite in 1968 (Weller 1968) and parasitises mainly coots and more rarely other duck species (Lyon and Eadie 2013). The species and its brood parasitism are considered relatively recent (McCracken et al. 1999).

One of the most intriguing questions is how brood parasitism evolved. As so often Darwin (1859) proposed a useful hypothesis, namely, that interspecific brood parasites evolved from an ancestor with parental care. Hamilton and Orians (1965), Davies (2000) and Payne (2005) consider three potential avenues towards the evolution of interspecific brood parasitism.

Interspecific brood parasitism might have evolved from conspecific brood parasitism (Payne 1977; Lyon and Eadie 1991; Yamauchi 1995; Robert and Sorci 2001). Females that augment their fitness by laying eggs in nests of conspecifics will be selected for, and indeed some cuckoo species are supposed to be facultative con- and interspecific brood parasites (Fleischer et al. 1985, but see Dearborn et al. 2009). Under strong nestling competition, interspecific brood parasitism will be selected over conspecific brood parasitism (Robert and Sorci 2001). The majority of facultative interspecific brood parasites are also conspecific brood parasites (Lyon and Eadie 2008; Feeney et al. 2014). However, there is no close correlation between the occurrence of conspecific brood parasitism and interspecific brood parasitism (compare from Chaps. 4 and 5), and it is therefore unlikely that conspecific brood parasitism is always a necessary precursor for the evolution of interspecific brood parasitism (Payne 2005; Yom-Tov and Geffen 2006).

A variant of this evolutionary scenario emphasises the fact that cooperative breeding occurs in some cuckoo species. In cooperatively breeding species, there are ample opportunities for helpers to lay eggs parasitically into the nest where they help, and this might be a precursor of laying eggs first in nests of conspecifics and subsequently also in nests of other species. The potential importance of this precursor was emphasised by Hamilton (1964), but cooperative breeding is either rare or absent in the other taxa where interspecific brood parasitism evolved.

Finally, interspecific brood parasitism might have evolved directly from parental care without the stepping stone of conspecific brood parasitism (Cichon 1996; Yom-Tov and Geffen 2006). Potential evolutionary pathways might include nest takeover (Viduidae, Payne 2005), competition for nest sites leading to parasitism (Indicatoridae, Davies 2000) or parasitism of closely related species (Davies 2000; Sorenson et al. 2004). There is experimental evidence that parasitism on closely related species can produce healthy parasitic fledglings (Slagsvold 1998), but these were later shown to have lower recruitment and reproductive success, possibly due to false imprinting on their host species (Slagsvold and Hansen 2001). In conclusion, none of these pathways is entirely satisfactory in explaining the evolution of brood parasitism across all taxa, and it might be that a unique combination of these pathways is required in each taxon.

Feeney et al. (2014) have argued and reiterated that regardless of the ancestral breeding strategy, obligate interspecific brood parasitism most likely arose from facultative brood parasitism, possibly because of either nest predation (Shaw et al. 2014), nest usurpation (Davies 2000) or unusually favourable ecological conditions that allow females to increase their reproductive success by dumping their eggs into the nests of other species (Lyon and Eadie 1991; Robert and Sorci 2001). In contrast, Fulmer and Hauber (2016) have recently argued that the costs of siblicide under stressful conditions might be sufficiently high to favour the reduction of competition between nest mates via either conspecific or interspecific brood parasitism, thereby emphasising cost constraints rather than opportunities.

What are the changes that occurred in ecology and life history when brood parasites evolved brood parasitism from an ancestor with parental care? Did these changes precede the evolution of brood parasitism, or were they consequences? These questions have been tackled in comparative analyses. Krüger and Davies (2002) and Boerner and Krüger (2008) explained large amounts of variation of cuckoo reproductive strategies. Using a maximum-likelihood approach (Pagel 1994), it was possible to construct the most likely evolutionary pathway between the presumed ancestral state and that displayed by modern brood parasitic species (Fig. 3.1). Except for egg size, migration and preferred habitat, most changes in ecology were more likely to precede the evolution of brood parasitism than to result from it (Fig. 3.2). Hence, the evolution of brood parasitism in the *Cuculidae* is more likely to be a later adaptation, reducing the cost of raising chicks and thereby possibly enhancing fecundity, whereas the reduction of egg size is a direct adaptation in the coevolutionary interaction with the host species. Similar to this, Mermoz and Ornelas (2004) found that parasitic *Molothrus* had increased egg thickness compared to nonparasitic species but did not differ in any other life-history trait they examined.

Changes in relative brain size were also correlated with the evolution of brood parasitism (Boerner and Krüger 2008). A reduction in relative brain size most likely evolved before brood parasitism (Fig. 3.1a). Parental cuckoos first evolved smaller brains (transition 1 $\rightarrow$ 3), then parasitism (3 $\rightarrow$ 4). Parasitic cuckoos could then re-evolve large brains (4 $\rightarrow$ 2). There was no significant pathway for the direct evolution of parasitism in large-brained species. This seems to point towards selection pressures to reduce the energetic costs of living.

Ancestral cuckoos were resident, but no evolutionary pathways between the ancestral and the recent state (migratory, small brained) could be found. To reconstruct a linear sequence of evolutionary events, Boerner and Krüger (2008) therefore tested for the evolutionary position of migration relative to breeding strategy (which changed after the reduction in brain size). This showed that the evolution of parasitism preceded the increase in migration (Fig 3.1b: transitions 1 $\rightarrow$ 3 to parasitism and 3 $\rightarrow$ 4 to migration). To further reconstruct the evolutionary sequence, Boerner and Krüger (2008) built a model for habitat productivity and breeding strategy. This showed that the usage of less-productive habitats only evolved after the species had become parasitic (Fig. 3.1c, transitions 1 $\rightarrow$ 3 to parasitism and 3 $\rightarrow$ 4 to poor habitats). Since habitat productivity and migration pattern both changed after the evolution of brood parasitism, a test was needed as to which of the two traits evolved

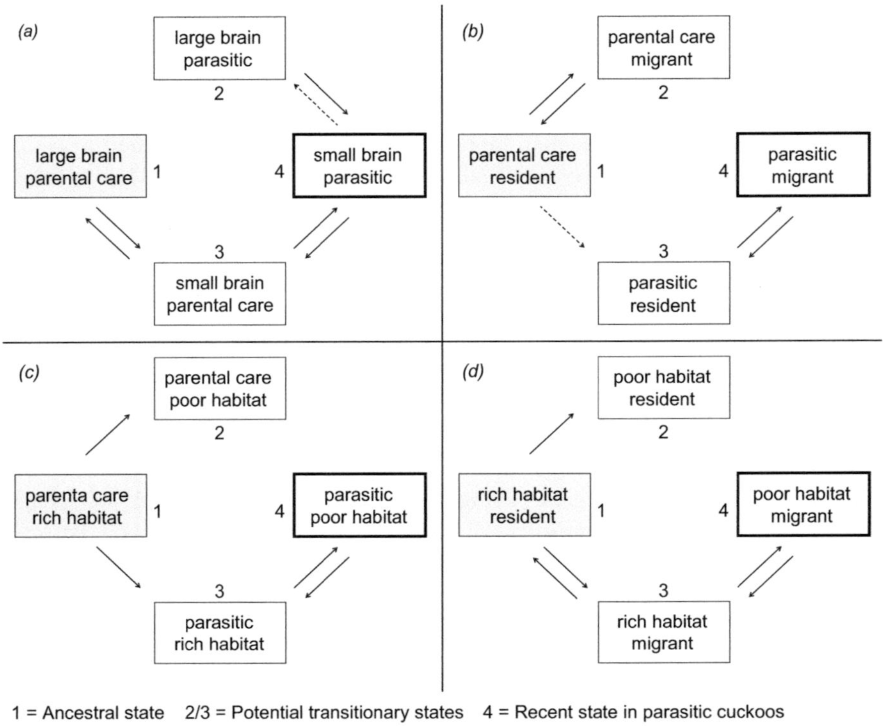

Fig. 3.1 Flow diagrams of the most likely evolutionary pathways between (**a**) relative brain size and breeding strategy, (**b**) breeding strategy and migration pattern, (**c**) breeding strategy and habitat productivity and (**d**) habitat productivity and migration pattern. The ancestral state is shown in a grey box, the most derived state in a bold one. Solid arrows indicate a significant transition ($P < 0.05$); dashed arrows indicate a trend ($P < 0.1$). Modified after Boerner and Krüger (2008)

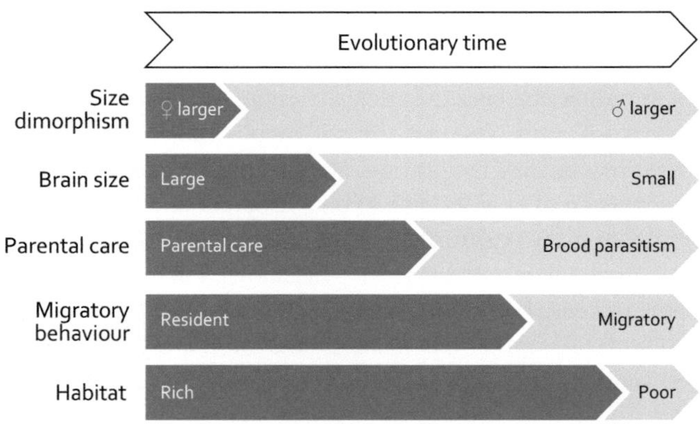

Fig. 3.2 Scheme showing the most likely sequence of evolutionary events going from the ancestral state to the one found in most extant brood parasitic cuckoo species. The sequence is based on the pathway analyses of Fig. 3.1. Modified after Boerner and Krüger (2008)

before the other (Fig. 3.1d). The model was highly significant and shows that the change from resident to migratory behaviour (1 → 3) occurred before the species changed from rich to poor habitats (3 → 4) (Boerner and Krüger 2008).

All the models were then combined to find the most likely sequence of these changes during the evolution of cuckoo brood parasitism (Fig. 3.2; Boerner and Krüger 2008). If a model shows that trait A evolved before trait B and a second model shows trait B evolved before trait C, then their most likely sequence would be A → B → C (Fig. 3.2). The sequence shows that the first most likely trait change was a change in size dimorphism, from females being larger to males being the larger sex. Afterwards, a reduction of brain size was followed by the evolution of brood parasitism. Next, cuckoos became migratory, then shifted towards habitats of lower productivity.

These evolutionary sequence analyses suggest possible hypotheses worth further investigating and testing. In the case of cuckoos, such an alternative hypothesis is the exact reversal of the energy trade-off hypothesis: the colonisation of new niches through increased migration and life in poorer habitats may only have been possible as a consequence of the reduction of energetic requirements earlier in evolution. This reduction in energy demands is not only seen in the change of brain size but also in the evolution of brood parasitism. Payne (1974) calculated that the omission of chick provisioning saves more energy than is invested into the higher number of eggs laid by female parasites. Therefore, parasitic cuckoos are likely to have reduced metabolic rates due to their smaller brain size (see also Armstrong and Bergeron 1985) and to be under less energy stress during the breeding season. This may enable them to live in poorer habitats, to gain sufficient condition for migration quickly after the breeding season and to require less energy during migration than heavier species with higher metabolic rates.

While the preceding evidence strongly suggests that the evolution of brood parasitism was associated with many profound changes in ecological and life-history traits, this notion can be extended to traits classically associated with sexual selection, such as sexual dimorphism. In fact, comparative analyses have shown that the evolution of sexual size dimorphism in cuckoos is more likely explained by coevolution than by sexual selection (Krüger et al. 2007). Instead of selection on males to become larger when freed from paternal care, it was found that both sexes became slightly smaller with the evolution of brood parasitism. While Owens and Hartley (1998) pointed out that size and plumage dimorphism is not necessarily explained by the same factors, there seems to be support for the coevolution hypothesis explaining not only size dimorphism but also plumage dimorphism (Krüger et al. 2007). There was also no evidence for male plumage becoming showier in parasitic cuckoos, but female plumage became more cryptic with the evolution of brood parasitism (Fig. 3.3).

Evolutionary pathway analyses suggested that the transition from size dimorphism with larger females to dimorphism with larger males most likely occurred before the evolution of brood parasitism. This suggests that smaller female size was a preadaptation for brood parasitism. Plumage dimorphism, however, most likely evolved after parasitism. This suggests that more cryptic female plumage, perhaps

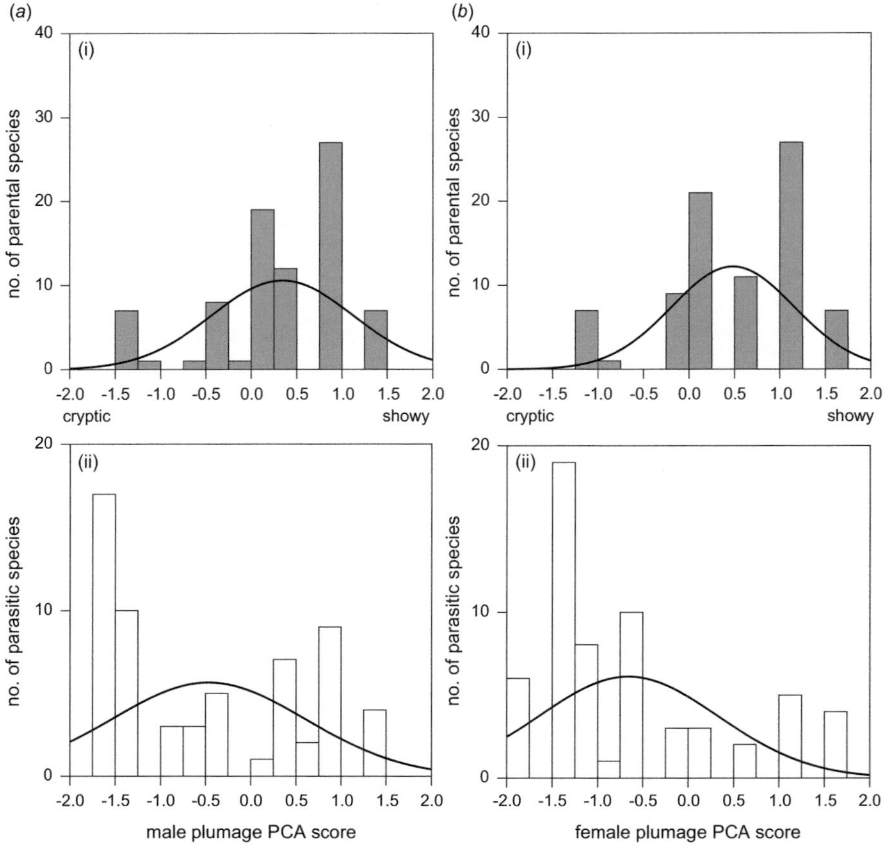

Fig. 3.3 Distribution of plumage scores (first principal component axis for (**a**) males and (**b**) females). Histograms are shown separately for species with (i) parental care and (ii) brood parasitic species. After Krüger et al. (2007)

functioning to reduce detection by hosts before egg laying, evolved in the context of host–parasite coevolution. Along those lines, the evolution of plumage barring in parasitic cuckoos seems to have evolved in response to the evolution of brood parasitism (Krüger et al. 2007; Gluckman and Mundy 2013).

Common to all three origins of cuckoo brood parasitism is a decrease in egg size, with small egg size most likely being an adaptation for successful parasitism of relatively small hosts (Payne 1974; Krüger and Davies 2002, 2004). This might help to explain male-biased size dimorphism in parasitic lineages as decreased female size could enable them to lay smaller eggs with shorter developmental periods. However, birds show considerable variation in egg size for a given body size and hence; it is probably more likely that a small egg was favoured, and this allowed females to decrease in body size as size constraints due to egg laying were removed.

With regard to plumage, it seems that while more cryptic females have evolved in both the New- and Old-World parasitic cuckoos, the *Clamator* cuckoos have followed a different path, evolving more showy plumage. This could reflect their different laying tactics: while Old-World *Cuculinae* cuckoos tend to rely on secretive behaviour to reduce detection by hosts, some *Clamator* cuckoos adopt conspicuous displays to distract hosts away from their nests (Davies 2000). However, Soler et al. (2014) have recently shown in the majority of cases (71%) that great spotted cuckoo females laid their eggs while their magpie hosts were incubating.

From all of these results, it should be obvious that coevolution can have a profound influence on traits that traditionally have been explained by sexual selection, both in hosts (Garamszegi and Avilés 2005) and parasites (Krüger and Davies 2004). Second, selection on females can be as important in generating sexual dimorphism (Irwin 1994).

These examples serve to highlight the fact that interspecific brood parasites often force evolutionary biologists to critically evaluate initial ideas about trait evolution (Krüger 2007). Likewise, it is now clear that the reciprocal coevolution with hosts is such a powerful force that it is likely that not only trait evolution (Soler et al. 1998) but also evolutionary rates (Medina and Langmore 2015), niche dynamics (Medina and Langmore 2016) and speciation and extinction rates are affected (Krüger et al. 2009; Krüger and Kolss 2013; Soler and Soler 2017). These aspects, however, are dealt with in much more detail in Chap. 2. The notion that such coevolutionary interactions will always escalate has recently been challenged, and once again a more refined picture emerges where different evolutionary equilibria are also possible outcomes (Krüger 2011; Soler and Soler 2017).

Concluding Remarks and Future Directions

We hope we have made clear why we believe that the study of the evolution of brood parasitism has lost none of its appeal to evolutionary ecologists. In fact, we believe that the coming years will witness a plethora of key results that will continue to challenge existing views and refine our understanding of the coevolutionary process. These future milestones will, in our opinion, be based on two major developments taking place right now. Before we explore those, we would like to highlight that phylogenetic details will be further elucidated by comparative genomics and will resolve relationships to even finer detail, but the overall picture is probably already relatively clear.

However, individual-based genomics in non-model species will allow the evolution of brood parasitism to be tackled from an entirely new perspective, and the genomic basis of adaptations will be comparable in unprecedented detail. The rapid development of genomic techniques has opened up the possibility of obtaining genome-wide sequence data, not only in traditional model species (e.g. humans, mice, fruit flies and nematodes) but essentially in any organism. Comparative genomics uses this growing availability of genomic data to compare sequences among species in order to infer evolutionary histories and investigate genetic footprints of past selection and adaptation.

(continued)

The evolution of brood parasitism will have left genetic and genomic marks, and these will be elucidated in comparison to closely related nonparasitic species. In parallel, facultative brood parasites could be a fruitful study object for transcriptomic approaches to test how and under which circumstances facultative parasitism arises.

Likewise, new comparative approaches are being developed at an equally breath-taking rate and will allow tackling questions that seem to be completely out of reach right now. Combined with accurate molecular phylogenies, rates of evolution, macroecological patterns and trait evolution will continue to be rewarding research topics.

Acknowledgements We are very grateful for the constructive and very helpful comments made by Bruce Lyon and Manuel Soler which significantly improved this chapter. O. K. and M. P. were both funded by the Marie Curie programme of the EU. O.K. also received funding via a Royal Society University Research Fellowship, a Churchill College Junior Research Fellowship and a Heisenberg-Professorship of the German Research Foundation (DFG, KR 2089/2-1).

References

Andersson M (1984) Brood parasitism within species. In: Barnard CJ (ed) Producers and scroungers: strategies of exploitation and parasitism. Croom Helm, London, pp 195–228

Andersson M (2001) Relatedness and the evolution of conspecific brood parasitism. Am Nat 158:599–614

Andersson M (2017) Helping relatives survive and reproduce: inclusive fitness and reproductive value in brood parasitism. Am Nat 189:138–152

Andersson M, Åhlund M (2000) Host-parasite relatedness shown by protein fingerprinting in a brood parasitic bird. Proc Natl Acad Sci U S A 97:13188–13193

Andersson M, Åhlund M (2012) Don't put all your eggs in one nest: spread them and cut time at risk. Am Nat 180:354–363

Andersson M, Waldeck P, Hanssen SA, Moe B (2015) Female sociality and kin discrimination in brood parasitism: unrelated females fight over egg laying. Behav Ecol 26:755–762

Armstrong E, Bergeron R (1985) Relative brain size and metabolism in birds. Brain Behav Evol 53:271–276

Baran NM, Reeve HK (2015) Coevolution of parental care, parasitic, and resistance efforts in facultative parasitism. Am Nat 186:594–609

Beauchamp G (1997) Determinants of intraspecific brood amalgamation in waterfowl. Auk 114:11–21

Boerner M, Krüger O (2008) Why do parasitic cuckoos have small brains? Insights from evolutionary sequence analyses. Evolution 62:3157–3169

Broom M, Ruxton GD (2004) A framework for modelling and analysing conspecific brood parasitism. J Math Biol 48:529–544

Bulmer MG (1984) Risk avoidance and nesting strategies. J Theor Biol 106:529–535

Canestrari D, Bolopo D, Turlings TC, Röder G, Marcos JM, Baglione V (2014) From parasitism to mutualism: unexpected interactions between a cuckoo and its host. Science 343:1350–1352

Cichon M (1996) The evolution of brood parasitism: the role of facultative parasitism. Behav Ecol 7:137–139

Darwin C (1859) The origin of species by means of natural selection. John Murray, London
Davies NB (2000) Cuckoos, cowbirds and other cheats. T and AD Poyser, London
Dearborn DC, MacDade LS, Robinson S, Fink ADD, Fink ML (2009) Offspring development mode and the evolution of brood parasitism. Behav Ecol 20:517–524
Eadie JMA, Kehoe FP, Nudds TD (1988) Pre-hatch and post-hatch brood amalgamation in North American Anatidae: a review of hypotheses. Can J Zool 66:1709–1721
Feeney WE, Welbergen JA, Langmore NE (2014) Advances in the study of coevolution between avian brood parasites and their hosts. Annu Rev Ecol Evol Syst 45:227–246
Felsenstein J (1985) Phylogenies and the comparative method. Am Nat 125:1–15
Fleischer RC, Murphy MT, Hunt LE (1985) Clutch size increase and intraspecific brood parasitism in the yellow-billed cuckoo. Wilson Bull 97:125–127
Fulmer AG, Hauber ME (2016) A siblicidal origin for avian brood parasitism? J Ornithol 157:219–227
Garamszegi LZ, Avilés JM (2005) Brood parasitism by brown-headed cowbirds and the expression of sexual characters in their hosts. Oecologia 143:167–177
Gill F, Donsker D (2017) IOC World bird list (v 7.1). https://doi.org/10.14344/IOC.ML.7.1
Gluckman TL, Mundy NI (2013) Cuckoos in raptors' clothing: barred plumage illuminates a fundamental principle of Batesian mimicry. Anim Behav 86:1165–1181
Hamilton WD (1964) The genetical evolution of social behaviour, II. J Theor Biol 7:17–52
Hamilton WJ, Orians GH (1965) Evolution of brood parasitism in altricial birds. Condor 67:361–382
Irwin RE (1994) The evolution of plumage dichromatism in the new-world blackbirds: social selection on female brightness. Am Nat 144:890–907
Jaatinen K, Öst M, Gienapp P, Merilä J (2011a) Differential responses to related hosts by nesting and non-nesting parasites in a brood-parasitic duck. Mol Ecol 20:5328–5336
Jaatinen K, Lehtonen J, Kokko H (2011b) Strategy selection under conspecific brood parasitism: an integrative modeling approach. Behav Ecol 22:144–155
Johnson KP, Lanyon SM (1999) Molecular systematics of the grackles and allies, and the effect of additional sequence (cyt b and ND2). Auk 116:759–768
Krüger O (2007) Cuckoos, cowbirds and hosts: adaptations, trade-offs and constraints. Philos Trans R Soc Lond B 362:1873–1886
Krüger O (2011) Brood parasitism selects for no defence in a cuckoo host. Proc R Soc Lond B 278:2777–2783
Krüger O, Davies NB (2002) The evolution of cuckoo parasitism: a comparative analysis. Proc R Soc Lond B 269:375–381
Krüger O, Davies NB (2004) The evolution of egg size in the brood parasitic cuckoos. Behav Ecol 15:210–218
Krüger O, Kolss M (2013) Modelling the evolution of common cuckoo host-races: speciation or genetic swamping? J Evol Biol 26:2447–2457
Krüger O, Davies NB, Sorenson MD (2007) The evolution of sexual dimorphism in parasitic cuckoos: sexual selection or coevolution? Proc R Soc Lond B 274:1553–1560
Krüger O, Sorenson MD, Davies NB (2009) Does coevolution promote species richness in parasitic cuckoos? Proc R Soc Lond B 276:3871–3879
Krüger O, Korsten P, Hoffman JI (2017) The rise of behavioral genetics and the transition to behavioral genomics and beyond. In: Call J (ed) APA handbook of comparative psychology. Academic Press, Washington, pp 365–379
Lyon BE, Eadie JM (1991) Mode of development and interspecific avian brood parasitism. Behav Ecol 2:309–318
Lyon BE, Eadie JM (2008) Conspecific brood parasitism in birds: a life-history perspective. Annu Rev Ecol Evol Syst 39:343–363
Lyon BE, Eadie JM (2013) Patterns of host use by a precocial obligate brood parasite, the Black-headed Duck: ecological and evolutionary considerations. Chin Birds 4:71–85

McCracken KG, Harshmann J, McClellan DA, Afton AD (1999) Data set incongruence and correlated character evolution: an example of functional convergence in the hind-limbs of stifftail diving ducks. Syst Biol 48:683–714

McRae SB, Burke TA (1996) Intraspecific brood parasitism in the moorhen: parentage and parasite-host relationships determined by DNA fingerprinting. Behav Ecol Sociobiol 38:115–129

Medina I, Langmore NE (2015) Coevolution is linked with phenotypic diversification but not speciation in avian brood parasites. Proc R Soc B 282:20152056

Medina I, Langmore NE (2016) The evolution of host specialisation in avian brood parasites. Ecol Lett 19:1110–1118

Mermoz ME, Ornelas JF (2004) Phylogenetic analysis of life-history adaptations in parasitic cowbirds. Behav Ecol 15:109–119

Nielsen CR, Semel B, Sherman PW, Westneat DF, Parker PG (2006) Host-parasite relatedness in wood ducks: patterns of kinship and parasite success. Behav Ecol 17:491–496

Ortega CP (1998) Cowbirds and other brood parasites. University of Arizona Press, Tucson

Owens IPF, Hartley IR (1998) Sexual dimorphism in birds: why are there so many different forms of dimorphism? Proc R Soc Lond B 265:397–407

Pagel MD (1994) Detecting correlated evolution on phylogenies: a general method for the comparative analysis of discrete characters. Proc R Soc Lond B 255:37–45

Payne RB (1974) The evolution of clutch size and reproductive rates in parasitic cuckoos. Evolution 28:169–181

Payne RB (1977) The ecology of brood parasitism in birds. Annu Rev Ecol Syst 8:1–28

Payne RB (2005) The cuckoos. Oxford University Press, Oxford

Pöysä H (2004) Relatedness and the evolution of conspecific brood parasitism: parameterizing a model with data for a precocial species. Anim Behav 67:673–679

Pöysä H, Pesonen M (2007) Nest predation and the evolution of conspecific brood parasitism: from risk spreading to risk assessment. Am Nat 169:94–104

Prum RO, Berv JS, Dornburg A, Field DJ, Townsend JP, Moriarty Lemmon E, Lemmon AR (2015) A comprehensive phylogeny of birds (Aves) using targeted next-generation DNA sequencing. Nature 526:569–573

Robert M, Sorci G (2001) The evolution of obligate interspecific brood parasitism in birds. Behav Ecol 12:128–133

Rohwer FC, Freeman S (1989) The distribution of conspecific nest parasitism in birds. Can J Zool Rev 67:239–253

Rothstein SI (1975) An experimental and teleonomic investigation of avian brood parasitism. Condor 77:250–271

Rothstein SI, Robinson SK (1998) Parasitic birds and their hosts: studies in coevolution. Oxford University Press, Oxford

Rothstein SI, Patten MA, Fleischer RC (2002) Phylogeny, specialization, and brood parasite-host coevolution: some possible pitfalls of parsimony. Behav Ecol 13:1–10

Shaw RC, Feeney WE, Hauber ME (2014) Nest destruction elicits indiscriminate con- versus heterospecific brood parasitism in a captive bird. Ecol Evol 4:4500–4504

Slagsvold T (1998) On the origin and rarity of interspecific nest parasitism in birds. Am Nat 152:264–272

Slagsvold T, Hansen BT (2001) Sexual imprinting and the origin of obligate brood parasitism in birds. Am Nat 158:354–367

Soler M (1990) Relationship between the great spotted cuckoo *Clamator glandarius* and its corvid hosts in a recently colonized area. Ornis Scand 21:212–223

Soler JJ, Soler M (2017) Evolutionary change: facultative virulence by brood parasites and tolerance and plastic resistance by hosts. Anim Behav 125:101–107

Soler M, Perez-Contreras T, de Neve L (2014) Great spotted cuckoos frequently lay their eggs while their magpie host is incubating. Ethology 120:965–972

Soler M, de Neve L, Roldan M, Perez-Contreras T, Soler JJ (2017) Great spotted cuckoo nestlings have no antipredatory effect on magpie or carrion crow host nests in southern Spain. PLoS One 12:e0173080

Soler M, Soler JJ, Martinez JG, Perez-Contreras T, Møller AP (1998) Micro-evolutionary change and population dynamics of a brood parasite and its primary host: the intermittent arms race hypothesis. Oecologia 117:381–390

Sorenson MD (1991) The functional-significance of parasitic egg-laying and typical nesting in redhead ducks – an analysis of individual behavior. Anim Behav 42:771–796

Sorenson MD (1992) Comment: why is conspecific nest parasitism more common in waterfowl than in other birds? Can J Zool Rev 70:1856–1858

Sorenson MD, Payne RB (2001) A single, ancient origin of obligate brood parasitism in African finches: implications for host-parasite coevolution. Evolution 55:2550–2567

Sorenson MD, Payne RB (2002) Molecular genetic perspectives on avian brood parasitism. Integr Comp Biol 42:388–400

Sorenson MD, Payne RB (2005) A molecular genetic analysis of cuckoo phylogeny. In: Payne RB (ed) The cuckoos. Oxford University Press, Oxford, pp 68–94

Sorenson MD, Sefc KM, Payne RB (2003) Speciation by host switch in brood parasitic indigobirds. Nature 424:928–931

Sorenson MD, Balakrishnan CN, Payne RB (2004) Clade-limited colonization in brood parasitic finches (*Vidua* spp.). Syst Biol 53:140–153

Spottiswoode CN, Colebrook-Robjent JFR (2007) Egg puncturing by the brood parasitic greater Honeyguide and potential host counteradaptations. Behav Ecol 18:792–799

Spottiswoode CN, Koorevaar J (2011) A stab in the dark: chick killing by brood parasitic honeyguides. Biol Lett 8:241–244

Spottiswoode CN, Stryjewski KF, Quader S, Colebrook-Robjent JFR, Sorenson MD (2011) Ancient host specificity within a single species of brood parasitic bird. Proc Natl Acad Sci U S A 108:17738–17742

Weller MW (1968) The breeding biology of the parasitic black-headed duck. Living Bird 7:169–208

White G (1789) The natural history of selborne. Project Gutenberg Etext 1998. www.gutenberg.org/cache/epub/1408/pg1408.txt

Yamauchi A (1995) Theory of evolution of nest parasitism in birds. Am Nat 145:434–456

Yom-Tov Y (1980) Intraspecific nest parasitism in birds. Biol Rev 55:93–108

Yom-Tov Y (2001) An updated list and some comments on the occurrence of intraspecific nest parasitism in birds. Ibis 143:133–143

Yom-Tov Y, Geffen E (2006) On the origin of brood parasitism in altricial birds. Behav Ecol 17:196–205

Zink AG (2000) The evolution of intraspecific brood parasitism in birds and insects. Am Nat 155:395–405

Zink AG, Lyon BE (2016) Evolution of conspecific brood parasitism versus cooperative breeding as alternative reproductive tactics. Am Nat 187:35–47

A Taxonomic Review of Obligate and Facultative Interspecific Avian Brood Parasitism

4

Clive F. Mann

Abstract

This chapter discusses the taxonomic relationships of brood parasites and their distribution through tribes, families and higher taxa, providing information about their breeding biology and host species when available. Distinction is made between obligate and facultative brood parasitism and, in the latter case, between interspecific and conspecific brood parasitism. It is shown that obligate brood parasites, with one exception, are all species that have altricial young, whereas facultative brood parasites, with two exceptions, all have precocial young. Reasons for this are discussed. Where data are available on success or failure of such parasitism, this is commented on. All known and suspected hosts are listed in the appendices. The possible evolutionary pathways between the three forms of brood parasitism (conspecific, facultative interspecific and obligate interspecific) are briefly discussed. Since obligate brood parasitism is shown to be a very effective way of increasing the individual's fecundity, the reasons for only about 1% of bird species evolving this strategy are discussed.

4.1 Introduction

Brood parasitism is found in other taxa of animals as well as birds, e.g. fish and insects. This chapter investigates the avian examples and the different forms that it takes. Distinction must first be made between those species that are *obligate* brood parasites versus those that are *facultative* brood parasites. One hundred and nine

---

Electronic Supplementary Material The online version of this chapter (https://doi.org/10.1007/978-3-319-73138-4_4) contains supplementary material, which is available to authorized users.

C. F. Mann (✉)
London, UK
e-mail: clivefmann@gmail.com

M. Soler (ed.), *Avian Brood Parasitism*, Fascinating Life Sciences,
https://doi.org/10.1007/978-3-319-73138-4_4

species (about 1% of all bird species) of 27 genera are the former, and such species never brood their own eggs or raise their own young. Facultative brood parasitism is known in some species where the birds normally care for their own eggs and young but occasionally place their eggs in the nest of another species—interspecific parasites, or another individual of the same species—conspecific parasites. The latter is by far the more common form of facultative brood parasitism and will not be dealt with further in this chapter (see Chaps. 5 and 6 for detailed information on this type of brood parasitism). Some species are involved in both interspecific and conspecific parasitism. It is generally assumed that normal brood behaviour through mistakes gave rise to conspecific brood parasitism, which in turn, again probably through error, became facultative interspecific brood parasitism (Hamilton and Orians 1965; Payne 1977; Yamauchi 1995), rather than the unlikely and less parsimonious route of normal nesting behaviour re-evolving from species that had previously evolved a parasitic system suggested by Siebel (1988) and Hughes (2000) (see Chap. 3 for detailed information on the origin and evolution of interspecific avian brood parasitism). This is most likely the scenario in precocial species, but in altricial birds, it appears that obligate brood parasitism, and probably facultative, may arise directly from nonparasitic behaviour, except perhaps for Hirundinidae where there is much conspecific parasitism (Slagsvold 1998; Sorenson and Payne 2001, 2002; Yom-Tov and Geffen 2006).

The majority of facultative interspecific brood parasites are also conspecific parasites, and those that are not may be proven to be so in future. Amongst altricial birds facultative interspecific parasitism is found in the same families as obligate interspecific brood parasitism (Lyon and Eadie 1991, 2013; Feeney et al. 2014). Evidence suggests that interspecific brood parasitism in the precocial black-headed duck *Heteronetta atricapilla* evolved from conspecific brood parasitism (Lyon and Eadie 2013).

Lyon and Eadie (1991) propose that altricial brood parasites 'obtain such a large increase in realized fecundity by avoiding the costs of parental care that obligate parasitism is favoured over facultative parasitism', in contrast 'precocial brood parasites gain relatively little in terms of increased fecundity', 'and much of this increase could potentially be gained by facultative parasitism'. Thus, obligate interspecific brood parasitism will not be favoured in precocial birds. They also suggest that facultative interspecific parasitism in birds may originate simply through a carry-over of conspecific parasitism; the majority of facultative interspecific parasites also parasitize conspecifics.

4.2 Method and Aims

Data on host relationships have been mostly gleaned from material in peer-reviewed publications, and where conclusions are controversial, this is mentioned.

The aims of this chapter are to review what is known, and what is suggested but yet to be discovered or proven, of avian brood parasitism, how it may have evolved and what advantages or disadvantages it may bestow upon the parasite. The incidences in the various taxonomic groups are shown. Full lists of known and suspected hosts for each parasitic species are shown in the appendices. The taxonomy of orders and higher levels of classification follow Cracraft (2013, 2014); families and lower levels follow Dickinson and Remsen (2013) and Dickinson and Christidis (2014), except where stated. As much of the systematics is based on genetic studies, quite different taxonomic groupings can be expected in the future as a greater proportion of the avian genome is studied.

The numbers in brackets after a species of parasite, e.g. (4, 3, 2), indicate number of species, number of genera and number of families it parasitizes, respectively.

+ indicates that the parasite has had some success, at least to the point of having its egg(s) incubated, but proof of complete success in many cases is lacking or is only circumstantial.

? indicates that the host has been suggested, but the relationship has yet to be proven.

4.3 Obligate Brood Parasites

Avian obligate brood parasites are found in only five families belonging to four orders: Anseriformes, containing one species, *Heteronetta atricapilla*, in the family Anatidae and the only obligate parasite to have precocial chicks; Cuculiformes containing 65 brood parasites in three lineages in Cuculidae, indicating that here obligate brood parasitism evolved on three occasions; Piciformes with one family entirely brood parasitic, Indicatoridae, of 17 species; and Passeriformes, with two families, one family, Viduidae of 20 species, entirely brood parasitic and, another, Icteridae, with six such species, all in the genus *Molothrus* (Table 4.1). Only in the Cuculidae has there been much taxonomic dissent.

4.3.1 Family Anatidae

Although facultative and conspecific parasitism is widespread in this family, the only obligate parasite is the **black-headed duck** *Heteronetta atricapilla* (19, 17, 9) of southern South America, which typically lays in nests of other waterfowl (usually 1–2 eggs in each nest). Hosts include species from a variety of families, but other anatids are particularly important. Members of Anatidae recorded as hosts are **black-necked swan** *Cygnus melancoryphus*, **coscoroba swan** *Coscoroba coscoroba*, **rosybill** *Netta peposaca*+, **fulvous whistling duck** *Dendrocygna bicolor* and **silver** and **red shovelers** *Spatula platalea* and S. *versicolor* (Höhn 1975; de la Peña 1986). It also lays in the nests of another Anseriformes family, Anhimidae, parasitizing the **southern screamer** *Chauna torquata* (Kear 2005). Amongst the **Rallidae** it utilizes two species of *Fulica*, the **red-gartered coot** *F. armillata*+ and **red-fronted coot** *F. rufifrons*+ and much less frequently **spotted rail** *Pardirallus maculatus* and **common gallinule** *Gallinula galeata* (Weller 1968; Powell 1979). A species of

Table 4.1 Obligate parasites

Taxon	Number of genera	Number of species
Order Anseriformes	1	1
Family Anatidae	1	1
Order Cuculiformes	19	65
Family Cuculidae	19	65
Sub-family Crotophaginae	2	3
Tribe Neomorphini	2	3
Sub-family Cuculinae	17	62
Tribe Phaenicophaeini	1	4
Tribe Cuculini	16	58
Order Piciformes	4	17
Family Indicatoridae	4	17
Sub-family Prodotiscinae	1	3
Sub-family Indicatorinae	3	14
Order Passeriformes	3	26
Family Viduidae	2	20
Family Icteridae	1	6
Totals	27	109

Laridae is also parasitized, the **brown-hooded gull** *Chroicocephalus maculipennis*+ (Weller 1968).

In Argentina it is particularly dependent on the two coots and the gull; hatching success is fairly low in its three main hosts mainly due to host rejection or neglect (Lyon and Eadie 2013). It is the only brood parasite species with precocial hatchlings, and logically it could only have a successful outcome by parasitizing hosts with precocial young, but it has been recorded laying in nests of two species of ibis, a spoonbill, **limpkin**, a hawk and a caracara, all with altricial young (Weller 1968; de la Peña 1986).

4.3.2 Family Cuculidae

Obligate brood parasitism is found in 65 species, whereas facultative parasitism is found only in two species in one genus, *Coccyzus.* Obligate brood parasitism is geographically widespread and found in family members on five continents.

4.3.2.1 Subfamily Crotophaginae

This subfamily contains three brood parasitic species all of them included in the tribe Neomorphini distributed mostly in the Neotropic region.

Tribe Neomorphini

The **striped cuckoo** *Tapera naevia* (28+, 16, 4) parasitizes small insectivorous birds, mostly furnariids, with closed nests by making a hole in wall or accessing nest through entrance tunnel. Usually one egg, but may be more, is laid per nest. Host nestlings are killed by the cuckoo nestling by biting (Payne 2016b). The **pheasant cuckoo** *Dromococcyx phasianellus* (5+, 5, 3) parasitizes birds with closed nests with or without entrance tunnel or open nests (Schönwetter 1967; Wilson 1992; Payne 2005). In the **pavonine cuckoo** *D. pavoninus* (4, 4, 2), hosts include species with domed or bag-shaped nests. Host young disappears after the cuckoo hatches (Payne 2016c). There is no information on success of hosts in raising parasites in any of these three species.

4.3.2.2 Subfamily Cuculinae

Two tribes contain obligate brood parasites. They are widespread in Eurasia, Africa, Australasia and Pacific, parasitizing a vast range of passerines and a few species of nonpasserines (numerous references in Erritzøe et al. 2012). Tribe Phaenicophaeini holds four species of obligate brood parasite in one genus, *Clamator*. All 58 species here accepted of tribe Cuculini are thought to be obligate brood parasites, but there are a few species where the breeding biology is little known and no hosts have yet been found.

Tribe Phaenicophaeini

The Oriental **chestnut-winged cuckoo** *Clamator coromandus* (25, 14, 7) parasitizes a wide variety of hosts. The **lesser necklaced laughingthrush** *Garrulax monileger* is suspected to account for up to 45% of hosts (Erritzøe et al. 2012). There is no information on success of hosts in raising parasite. The **great spotted cuckoo** *C. glandarius* (26, 7, 2) is more selective in the families it parasitizes (Mundy and Cook 1977; Cramp 1985; Irwin 1988; Shirihai 1996; Erritzøe et al. 2012); it has parasitized **hoopoe** *Upupa epops* in Africa (Jensen and Jensen 1969) and **jackdaw** *Corvus monedula* in Israel (Charles et al. 2005). In Europe the **common magpie** Pica pica is the usual host and in Africa crows *Corvus* spp. and starlings (Mundy and Cook 1977; Cramp 1985; Irwin 1988; Shirihai 1996; Erritzøe et al. 2012). Multiparasitism is very frequent in this species either by different females or by the same female laying several eggs in the same nest, but when there is more than one parasitic nestling, some of them may starve (Soler 1990; Martínez et al. 1998; Soler et al. 1996). The host young are not evicted and in some cases may be raised with nestling cuckoos, but more usually no magpie nestlings remain (Payne and Garcia 2016; Soler et al. 1996). The Afrotropical **Levaillant's cuckoo** *C. levaillantii* (10, 4, 4) parasitizes various *Turdoides* babblers, amongst other species. As other *Clamator* species, the female may puncture or destroy some host eggs, but young (up to four in one nest) do not eject host young and are often raised together (Erritzøe et al. 2012). The **Jacobin cuckoo** *C. jacobinus* (60, 35, 19) parasitizes a wide variety of hosts, and there are numerous others suspected, but are lacking in hard evidence. In India hosts are chiefly *Argya* and *Turdoides* babblers; in South Africa *Pycnonotus* bulbuls, **sombre greenbul** *Andropadus importunus* and **fiscal shrike** *Lanius collaris*; in Mauritania, Ethiopia and Kenya *Argya* babblers (Payne and de Juana 2016a).

Tribe Cuculini

The **thick-billed cuckoo** *Pachycoccyx audeberti* (>4, >1, 1?) utilizes nests of three or four species of helmet shrikes *Prionops* in mainland Africa. As *Prionops* does not occur in Madagascar, at least one other species must be a host there, perhaps a vangid, **Chabert's vanga** *Leptopterus chabert* being suspected. After hatching young cuckoo evicts host's eggs and nestlings (Payne 2016d).

Crows *Corvus* spp. are the most popular hosts in India and Pakistan for the **common koel** *Eudynamys scolopaceus* (16, 11, 6), but the cuckoo is more catholic in its choice of hosts in other areas. Usually one but up to 11 eggs may be laid in a single host nest, perhaps by more than one female; the nestling cuckoo does not evict host's eggs or nestlings, but some of the latter may starve (Hellebrekkers and Hoogerwerf 1967; Wells 2007). Of limited distribution, the **black-billed koel** *Eudynamys melanorhynchus* (1, 1, 1) is only known to utilize nests of the **Sulawesi myna** *Basilornis celebensis* (Payne 2005). The **Pacific koel** *Eudynamys orientalis* (20, 18, 9) is quite catholic in its choice of host, e.g. meliphagids, artamids and monarchids, but only one egg is laid per nest; of 22 eggs studied 11 hatched and 8 produced fledglings (Gosper 1962), and of nests that failed, most were either deserted by the hosts or the cuckoo egg disappeared (Higgins 1999; Payne 2005).

The **long-tailed koel** *Urodynamis taitensis* (14, 13, 10) nestling ejects the eggs and young of its host (Payne and de Juana 2016b); it has been reported to parasitize ten host species of eight different genera. Unsurprisingly, the huge **channel-billed cuckoo** *Scythrops novaehollandiae* (14, 8, 6) chooses large hosts. The cuckoo lays one to five eggs per nest (2+ in 52% of parasitized nests) and often removes or damages host eggs when laying; the host young sometimes disappear, but it is not known if eviction takes place (Payne 2016e).

The **black-eared cuckoo** *Misocalius osculans* (in *Chalcites* in Dickinson and Remsen 2013) (23, 15, 7), parasitizing chiefly acanthizids and malurids, evicts host's eggs and young (Brooker and Brooker 1989b; Higgins 1999; Lowther 2015). **Horsfield's bronze cuckoo** *Chalcites basalis* (genus *Chalcites* was subsumed in *Chrysococcyx* in Erritzøe et al. 2012) (104, 71, 24), utilizing the nests, amongst others, of numerous species of malurids, acanthizids and meliphagids, evicts host's eggs and chicks (Brooker and Brooker 1989a; Higgins 1999) as does the **shining bronze cuckoo** *C. lucidus* (99, 52, 23), with a similar catholic choice of hosts. There is some confusion between the **little bronze cuckoo** *C. minutillus* (10, 3, 2) and **Gould's bronze cuckoo** *C. poecilurus* (19, 13, 8) in the field, and the latter is frequently lumped with the former (e.g. Dickinson and Remsen 2013). There is an overlap in hosts, although *C. minutillus* relies more heavily on gerygones *Gerygone* spp., but *C. poecilurus* also uses the genus, but less so, and both utilize meliphagids. Cuckoo nestlings of both species evict host's chicks and perhaps also eggs (Brooker and Brooker 1989b; Higgins 1999; Payne et al. 2016).

The hosts of the **dwarf koel** *Microdynamis parva*, **rufous-throated bronze cuckoo** *Chalcites ruficollis*, **white-eared bronze cuckoo** *C. meyerii* and **pied**

bronze cuckoo *C. crassirostris* are unknown (Payne 2005; Erritzøe et al. 2012), but the **rufous-sided gerygone** *Gerygone dorsalis* suggested for last (Parker 1981).

Only one host is known for the Afrotropical **yellow-throated cuckoo** *Chrysococcyx flavigularis* (1, 1, 1). This cuckoo is very poorly known in the field. One of its two Oriental congeners, **violet cuckoo** *C. xanthorhynchus* (13, 12, 6), is commoner and much better known with a number of hosts recorded. Its other Oriental congener **Asian emerald cuckoo** *C. maculatus* (14, 10, 6) has been recorded as parasiting a number of species, some of which are thought to be erroneous (Friedmann 1968; Erritzøe et al. 2012). At the time of writing, Becking (1981) thought that all hosts claimed except for sunbirds were erroneous. However, the **chestnut-crowned warbler** *Seicercus castaniceps* has recently been confirmed as a host (Yang et al. 2012a, b). Their other three Afrotropical congeners are well studied and numerous hosts are known. **Klaas's cuckoo** *C. klaas* (68, 35, 17) is quite catholic in its choice of hosts, but sunbirds and insectivores predominate; the nestling evicts host's eggs and chicks (Payne 2005; Erritzøe et al. 2012). The **African emerald cuckoo** *C. cupreus* (41, 27, 14) parasitizes mostly hosts that feed their young on insects, although in two cases the hosts, **yellow-whiskered greenbul** *Eurillas latirostris* and **common bulbul** *Pycnonotus barbatus* feed young on fruit (Friedmann 1948; Moreau 1949). The nestling evicts host's eggs and chicks (Gatter 1997; Cheke and Mann 2001). The **diederik cuckoo** *C. caprius* (90, 39, 18) chiefly parasitizes seed-eating hosts that feed their young on insects, e.g. weavers *Ploceus* spp., but also a number of insectivorous species; the nestling evicts host's eggs and chicks (Payne 2005; Erritzøe et al. 2012). The **long-billed cuckoo** *Rhamphomantis megarhynchus* (in *Chalcites* in Dickinson and Remsen 2013) has as yet no known host (Payne 1997). The **pallid cuckoo** *Heteroscenes pallidus* (115, 65, 27) has been placed in *Cuculus* (Peters 1940) or *Cacomantis* (Payne 2005). It is quite catholic in its selection of hosts but seems to prefer honeyeaters with open nests. The nestling evicts host's eggs and nestlings (Brooker and Brooker 1989b; Higgins 1999).

The **fan-tailed cuckoo** *Cacomantis flabelliformis* (87, 52, 21) mostly utilizes hosts with domed nests, particularly malurids, acanthizids, meliphagids and petroicids. As with other members of the genus, the nestling cuckoo evicts host's eggs and chicks (Brooker and Brooker 1989b; Higgins 1999). There may have been some confusion of hosts with those of *C. castaneiventris* and *C. variolosus*. The **plaintive cuckoo** *C. merulinus* (14, 6, 3) usually chooses hosts with domed nests. Known hosts are mostly cisticolids throughout its range. The **grey-bellied cuckoo** *C. passerinus* (13, 7, 3) prefers hosts constructing closed nests with narrow entrance, particularly cisticolids. The **rusty-breasted cuckoo** *C. sepulcralis* (18, 14, 11) uses a greater variety of hosts. The **brush cuckoo** *C. variolosus* (66, 39, 14) has been recorded parasitizing a large variety of species, but there may have been confusion of hosts with *C. flabelliformis* and *C. castaneiventris* (Brooker and Brooker 1989b; Higgins 1999). The **Halmahera cuckoo** *C. heinrichi* (1, 1, 1) was previously considered a 'variant' of *C. variolosus infaustus* (Payne 2005;

Dickinson and Remsen 2013), but recent field evidence suggests it is best separated, and the **island leaf warbler** *Seicercus maforensis* has been found to be its host on Obi, Moluccas (Reeve et al. 2015). The **chestnut-breasted cuckoo** *C. castaneiventris* (2, 2, 2) has been claimed to parasitize four species of three different genera and families, but confusion with *C. flabelliformis* and *C. variolosus* is possible (Higgins 1999). Although the **banded bay cuckoo** *C. sonneratii* (>9, >9, >9) is not uncommon, its hosts are incompletely known (Erritzøe et al. 2012).

The little-known **white-crowned cuckoo** *Caliechthrus leucolophus* of New Guinea has no known hosts (Payne 2005).

The three members of the genus *Cercococcyx* are poorly known Afrotropical species. There is good evidence that, in the **dusky long-tailed cuckoo** *C. mechowi* (3, 3, 3), the hatchling evicts host's young (Chapin 1939; Brosset and Érard 1986). Two possible but not confirmed host species have been suggested for the **olive long-tailed cuckoo** *C. olivinus*, whereas the **barred long-tailed cuckoo** *C. montanus* (1, 1, 1) has just one confirmed host.

The Oriental genus *Surniculus* is here considered to contain four species, whereas these are lumped as one (*S. lugubris*) by Dickinson and Remsen (2013). The **Philippine drongo-cuckoo** *S. velutinus* has a juvenile plumage quite unlike its congeners, which gives strong reason for specific separation; as with the **Moluccan drongo-cuckoo** *S. musschenbroeki*, hosts have yet to be discovered (Erritzøe et al. 2012). There is some confusion over hosts between the **fork-tailed drongo-cuckoo** *S. dicruroides* (15, 12, 9) and the **square-tailed drongo-cuckoo** *S. lugubris* (10, 5, 4) in areas of sympatry; nestlings of both species eject hosts' eggs and chicks (Becking 1981; Payne 2005).

The Oriental genus *Hierococcyx* contains eight species, although Dickinson and Remsen (2013) accept only seven, retaining *bocki* in *sparverioides.* Few hosts are known for a number of species in this genus, e.g. **large hawk-cuckoo** *H. sparverioides* (8, 6, 4) has been recorded parasitizing eight species and the **dark hawk-cuckoo** *H. bocki* (2?, 2?, 2?) only one or two. There is some controversy over which species host the **common hawk-cuckoo** *H. varius* (8, 6, 3) whose nestlings evict host's chicks. There are only two hosts known for the uncommon **moustached hawk-cuckoo** *H. vagans* (2, 2, 2) and just a few for **Hodgson's hawk-cuckoo** *H. fugax* (2 or 3, 2 or 3, 2 or 3). The **Philippine hawk-cuckoo** *H. pectoralis* has no known hosts (Payne 2005). Both the **whistling hawk-cuckoo** *H. nisicolor* (17, 15, 8) and the **northern hawk-cuckoo** *H. hyperythrus* (13, 10, 5) have been recorded utilizing a number of hosts, the former particularly pellorneids and muscicapids, the latter muscicapids.

The poorly known **Sulawesi cuckoo** *Cuculus crassirostris* has no known hosts (Payne 2005). The widespread Afrotropical **red-chested cuckoo** *C. solitarius* (45, 32, 11) nestlings evict hosts' eggs and chicks and are recorded parasitizing a wide variety of species, particularly muscicapids, turdids and motacillids, whereas the less well-known Afrotropical **black cuckoo** *C. clamosus* (14, 9, 6) also evicts host's eggs but appears to parasitize a much smaller variety of species, particularly malaconotids. The **Indian cuckoo** *C. micropterus* (19, 18, 16) of the Eastern Palearctic and Oriental Regions is quite catholic in its choice of hosts. The Palearctic and Oriental **common cuckoo** *C. canorus* (276, 133, 42), due to its huge range and the amount of field work focussing upon it, is known to parasitize a large number and great variety of hosts, with confirmation of success in over 25% of the host species. As well as confirmed hosts, many species are suspected or claimed but with no confirmation; some, such as those species with precocial chicks, could not lead to a positive outcome for the parasite. It is a breeding summer migrant to the Palearctic and Oriental Regions (Erritzøe et al. 2012; Lowther 2015). The **African cuckoo** *C. gularis* (14, 12, 9) is widespread but uncommon in the Afrotropical Region and has a number of recorded hosts including muscicapids. The next three species in this genus are sometimes lumped, or as in Dickinson and Remsen (2013), *lepidus* is lumped in *saturatus*, but *optatus* is separated. The **Oriental cuckoo** *C. optatus* (29, 18, 11) is a breeding summer migrant in the Palearctic Region (Payne 1997; Lowther 2015), whereas the **Himalayan cuckoo** *C. saturatus* (11, 6, 4) migrates within the Oriental Region, both parasitizing particularly phylloscopids. The eponymous **Sunda cuckoo** *C. lepidus* (5, 2, 2) is uncommon and not well-known and also specializes in phylloscopids. The Oriental and Eastern Palearctic **lesser cuckoo** *C. poliocephalus* (25, 16, 10), parasitizing mostly scotocercids and phylloscopids, is a migrant, some wintering in Africa (Higuchi and Payne 1986; Brazil 1991), as do some of the eponymous **Madagascar cuckoo** *C. rochii* (5, 5, 4), which so far has few hosts attributed to it, including two cisticolids. As it is very widespread and common within Madagascar, this may be more an indication of specialization rather than lack of knowledge (Safford and Hawkins 2013).

4.3.3 Family Indicatoridae

This family may be separated into two subfamilies on anatomical grounds – Indicatorinae containing all genera except for *Prodotiscus*, which forms Prodotiscinae (Short and Horne 2016). The entire family of 17 species is thought to be obligate brood parasitic, although the hosts for some are as yet unknown. They are confined to the Afrotropical Region, with the exception of *Indicator xanthanotus* and *I. archipelagicus*, of the Oriental Region.

4.3.3.1 Subfamily Prodotiscinae

The **green-backed honeybird** *Prodotiscus zambesiae* (7, 5, 5), **Wahlberg's honeybird** *P. regulus* (12, 5, 3) and **Cassin's honeybird** *P. insignis* (3+, 3+, 3+) parasitize small insectivorous passerines or as in the host **yellow-throated bush sparrow** *Gymnoris superciliaris*, seedeaters that feed their young on insects; none of the hosts is a hole or cavity nester.

4.3.3.2 Subfamily Indicatorinae

All members of this subfamily are thought to parasitize only species that use holes or cavities or tunnels for nesting, with the exception of some hosts of *Indicator minor* and *I. indicator*. All hosts are nonpasserines. The nestlings of some species, e.g. *I. minor*, have hooks on their bills, which they use to puncture eggs or kill host's nestlings (Spottiswoode and Koorevaar 2011); these are perhaps present on all species.

No hosts are known for **Zenker's honeyguide** *Melignomon zenkeri* or **yellow-footed honeyguide** *M. eisentrauti* (Short and Horne 1985, 1988; Lowther 2012). Similarly, for the **dwarf honeyguide** *Indicator pumilio* and **Willcocks's honeyguide** *I. willcocksi*, the hosts are yet to be discovered (Short and Horne 2001). The **pallid honeyguide** *I. meliphilus* (1, 1, 1) is only recorded parasitizing one host species, but few other species are suggested. The **least honeyguide** *I. exilis* and the **spotted honeyguide** *I. maculatus* have no confirmed hosts (Friedmann 1955, 1978; Short and Horne 2001) and similarly the two Oriental species **yellow-rumped honeyguide** *I. xanthonotus* and **Malaysian honeyguide** *I. archipelagicus* (Ali 1983; Wells 1999). The **thick-billed honeyguide** *I. conirostris* (1, 1, 1), lumped in *I. minor* by Dickinson and Remsen (2013), is only known to parasitize one species of the genus *Gymnobucco*, but other members of the genus may also act as hosts. The next two species have been more thoroughly studied. The **scaly-throated honeyguide** *I. variegatus* (10, 6, 2) has been recorded as a successful host for a number of ramphastids and picids, and the **lesser honeyguide** *I. minor* (30, 18, 8) has a greater variety of hosts, almost all being successful.

The greater honeyguide *I. indicator* (43, 25, 14) is the best known of the subfamily, with a wide variety of hosts recognized. Nestling behaviour has been described in detail (Spottiswoode and Koorevaar 2011). Spottiswoode et al. (2012) have shown ancient divergent mtDNA lineages within this species associated with two groups of host species: honeyguides parasitizing hosts breeding in terrestrial burrows and those parasitizing hosts breeding in tree cavities.

The **lyre-tailed honeyguide** *Melichneutes robustus* is poorly known, with no host species yet identified. It has an aerial display and possibly a lek system (Dowsett-Lemaire 2008).

4.3.4 Family Viduidae

The family, confined to the Afrotropical Region, consists of two genera thought to be entirely obligate brood parasitic. Some species of *Vidua* are highly host-specific, but in a number of species, hosts are yet to be discovered. As far as is known, all except two species of this genus only parasitize members of **Estrildidae**, the family to which **Viduidae** is most closely related.

Relationships between *Vidua* species and their hosts show close ties in several aspects of their biology: male *Vidua* sing the song of the host that raised the male, female *Vidua* mate with those male *Vidua* that sing the song of the host species that raised the female and nestling ***Vidua*** mimic mouth patterns of their hosts' nestlings, as do their begging calls (Sorenson et al. 2003).

Anomalospiza has in the past been placed in Fringillidae, Estrildidae or Ploceidae, but anatomical features indicate its closer relationship with *Vidua* (Lahti and Payne 2003).

Payne (2016a) found genetic comparisons of *Vidua*, and their hosts indicate that speciation of host species and *Vidua* did not coevolve. Speciation in *Vidua* occurred with shifts of host species. However, where different individuals of a species mimic different hosts, there is no distinction in genetic markers. This might be expected in the future, ultimately resulting in speciation.

The **eastern paradise whydah** *Vidua paradisaea* (2, 2, 1) has two known hosts, but other seedeaters are suggested. The next four species, the **broad-tailed paradise whydah** *V. obtusa* (1, 1, 1), the **Sahel paradise whydah** *V. orientalis* (1, 1, 1), the **exclamatory whydah** *V. interjecta* (2, 1, 1) and the **Togo paradise whydah** *V. togoensis* (1?, 1?, 1?) are known or believed to only parasitize one or a few species of *Pytilia*. The **pin-tailed whydah** *V. macroura* (15, 8, 4), however, utilizes a larger number of hosts, all seedeaters, but perhaps also some cisticolids. The **steel-blue whydah** *V. hypocherina* (2, 1, 1) is much more restricted, parasitizing only two host species, but the **shaft-tailed whydah** *V. regia* (3, 3, 3) is less so, parasitizing an insectivorous cisticolid as well as two seedeaters, one of which is a ploceid. One host of the **straw-tailed whydah** *V. fischeri* is known, the **purple grenadier** *Granatina ianthinogaster*. **Wilson's indigobird** *V. wilsoni* (2, 1, 1) has two known hosts. The **Barka indigobird** *V. larvaticola* (2, 1, 1) parasitizes two species of *Lagonosticta*, whereas the **Jos Plateau indigobird** *V. maryae* probably lays in nests of another species of *Lagonosticta* (*sanguinodorsalis*?) as it mimics song of this species (Payne 2004), but this has yet to be confirmed. Hosts vary geographically in the **Fonio indigobird** *V. camerunensis* (4, 2, 1). Different song mimicry populations may be sympatric, but no morphological or genetic differences found (Payne et al. 2002; Sorenson et al. 2003; Payne 2004). The remaining six congeners, **quailfinch indigobird** *V. nigeriae* (1, 1, 1), **Jambandu indigobird** *V. raricola* (1, 1, 1), **Peter's twinspot indigobird** *V. codringtoni* (1 or 2, 1, 1), **dusky indigobird** *V. funerea* (1, 1, 1), **village indigobird** *V. chalybeata* (2, 1, 1) and **purple indigobird** *V. purpurascens* (1 or 2, 1, 1), are all extreme specialists as regards host selection, all parasitizing estrildids.

The **parasitic weaver** or **cuckoo-finch** *Anomalospiza imberbis* (15, 2, 1) has 15 known hosts, all members of Cisticolidae.

4.3.5 Family Icteridae

This family contains six species of obligate brood parasite, all in one genus, *Molothrus*, confined to the Americas. The somewhat different characters of *M. oryzivorus* are thought to be due to allometric changes related to its large size, evolutionary convergence of icterid hosts and parasite and mimicry and that it is correctly placed in *Molothrus* (Fraga and Bonan 2016). Well-known and widespread species affect a great range of hosts, mostly passerines. Some of those recorded could not be successful hosts, particularly those with precocial young, and may result from egg-dumping or observer error; those obviously unsuitable are not included in the table.

The **screaming cowbird** *Molothrus rufoaxillaris* (6, 6, 2) and the **giant cowbird** *M. oryzivorus* (11, 5, 2) are known to parasitize only a small number of hosts, mainly icterids, but also two species of Corvidae that are thought to be utilized only if its normal icterid hosts are absent (Di Giacomo 2010; Lowther 2016a, b). The former utilizes three main hosts which are all co-operative breeders, usually laying one egg per nest after puncturing host's eggs; up to 19 eggs have been found in one nest, usually as a result of more than one female utilizing a single host nest (Fraga and Garcia 2016a). The latter normally lays one or two eggs per host nest but up to six recorded, perhaps from more than one female (Fraga 2016c).

The **bronzed cowbird** *M. aeneus* (99, 57, 16) has considerably more known hosts, particularly passerellids, parulids, icterids and cardinalids, usually laying only one egg per host nest; the cowbird often punctures or removes host's eggs (Fraga 2016d; Lowther 2016a, b). The **bronze-brown cowbird** *M. armenti* has been recently separated from *M. aeneus* (Fraga 2016a). No hosts have been specifically recorded for this form.

The **shiny cowbird** *M. bonariensis* (269, 133, 30) and **brown-headed cowbird** *M. ater* (284, 116, 34) are both generalized parasites utilizing a large number of taxonomically diverse species, some of them, e.g. those with precocial nestlings such as wildfowl and waders, could not possibly lead to a satisfactory outcome for the parasite and presumably are the results of 'egg dumping' (Fraga and Garcia 2016b; Lowther 2016a, b). The latter is rarely successful in **common grackle** *Quiscalus quiscula* nests because host ejects eggs, and in about one-third of recorded hosts, the parasite is not successfully raised (Fraga 2016b).

4.4 Facultative Brood Parasites

Many such species are only recorded as conspecific nest parasites, others only as interspecific and some as both. It is likely that those in the second category will also be found to be conspecific parasites. There are many more incidents of conspecific than interspecific parasitism. Wildfowl and gamebirds are well-represented, all birds with precocial young. In cases of single, or few, records of associations, egg dumping, or a mistake on the part of the parasite that is frequently a conspecific parasite, are thought to be the explanations. In some cases adoption of a young 'parasite' by the apparent host may be the explanation where there is no evidence of actual egg laying in the 'host' nest. Another situation could give rise to apparent nest parasitism, i.e. interspecific egg retrieval. The **Canada goose** *Branta canadensis* has been recorded rolling eggs from a nearby **northern pintail** *Anas acuta* nest into its own nest and incubating them; **snow geese** *Anser caerulescens* have been shown experimentally to roll nearby eggs into their nests (Duncan 1984).

Facultative brood parasitism has been reported in 64–68 species of 45–46 genera included in 16 or 17 families (see Table 4.2) corresponding to nine orders. Although in most cases citations of facultative brood parasitism for each species are very scarce (frequently just one), in three orders they are frequent (Anseriformes, Galliformes and Cuculiformes), with observations in many species and, sometimes, many observations in some species.

4.4.1 Anseriformes

In the family **Anatidae** interspecific brood parasitism is more frequent and has been described in 35–39 species corresponding to 18 genera. In some cases there are a number of records indicating that it is probably not just the result of errors on the part of the laying bird. However, these cases are less frequent than reported cases of conspecific brood parasitism in this family (256 species, Chap. 5). In most of these species exhibiting facultative interspecific brood parasitism, conspecific brood parasitism can usually be found (Joyner 1976). There are many situations where only a few cases are known, usually involving one or two host species, usually other anatids, sometimes congeners. These could be interpreted as mistakes on the part of the laying bird.

As to be expected, the hosts are all birds that nest on or near water and are mostly other anatids but some rallids may be chosen. There is one case each of an anatid choosing a crane, a gull and a wader that are quite likely to have been mistakes on the part of the laying bird.

Some species have been recorded more frequently as parasites and the practice could be viewed as a definite strategy being developed. All three species of *Bucephala* lay in each other's nests and one of them also in the nests of two other species of duck of separate genera. The **redhead** *Aythya americana* utilizes ten species of five genera of anatids, including three other species of *Aythya* and two rallids. Other ducks have parasitized between two and five anatid hosts, sometimes congeners, but usually not, and a few rallids. More observations of such parasitism in most species are required to claim that this is a definite strategy, rather than 'egg dumping'.

4.4.2 Galliformes

In the order **Galliformes**, interspecific brood parasitism is less frequent than in Anseriformes and has been described in 11 species corresponding to 10 genera. In

Table 4.2 Facultative interspecific parasites

Taxon	Number of genera	Number of species
Order Anseriformes	18	35/39
Family Anatidae	18	35/39
Subfamily Dendrocygninae	1	2
Subfamily Oxyurinae	1	3
Subfamily Anserinae	2	2/3
Subfamily Anatinae	12	26/29
Tribe Mergini	6	10/12
Tribe Tadornini	1	1
Tribe Aythyini	2	8/9
Tribe Anatini	3	7
Genera incertae sedis	2	2
Order Galliformes	10	11
Family Megapodidae	1	1
Family Numididae	1	1
Family Odontophoridae	2	2
Family Phasianidae	6	7
Subfamily Phasianinae	5	6
Tribe Coturnicini	1	2
Tribe Phasianini	2	2
Tribe Tetraonini	2	2
Genus incertae sedis	1	1
Order Phoenopteriformes	3	3
Family Podicipedidae	3	3
Order Columbiformes	1	1
Family Columbidae	1	1
Order Cuculiformes	3	4
Family Cuculidae	3	4
Subfamily Crotophaginae	2	2
Tribe Crotophagini	1	1
Tribe Neomorphini	1	1
Subfamily Cuculinae	1	2
Order Gruiformes	1	1
Family Rallidae	1	1
Order Pelicaniformes	2	2
Family Ardeidae	2	2
Order Charadriiformes	3	3
Family Charadriidae	1	1
Family Scolopacidae	1	1
Family Laridae	1	1
Order Passeriformes	3/4	3/4
Family Nectariniidae	1?	1?
Family Passeridae	1	1
Family Estrildidae	1	1
Family Sturnidae	1	1
Total	45/46	63/68

some cases the 'host' has altricial nestlings, and such cases are best viewed as mistakes on the part of the laying bird, or 'egg dumping'. Only one species is known to parasitize more than three hosts. The **ring-necked pheasant** *Phasianus colchicus* (family **Phasianidae**) has parasitized nine species of phasianids, six successfully, one odontophorid, six species of anatids, one successfully, as well as three rallids and one scolopacid. The other ten species mostly lay in galliformes' nests but two also in anatids' nests. Only in one species is the 'host' congeneric (see Appendix 5).

It appears that in only one species, i.e. *P. colchicus*, could nest parasitism be seen as developing into a viable strategy. Without more information of success or otherwise, it is impossible to judge if facultative brood parasitism in the other species is anything more than egg dumping.

4.4.3 Cuculiformes

In the family Cuculidae, interspecific brood parasitism has only been described in five species corresponding to four genera. In the subfamily Crotophaginae the two species involved, the **guira cuckoo** *Guira guira* and **greater roadrunner** *Geococcyx californianus*, are probably 'egg dumpers', particularly as the former has laid in the nest of a species with precocial young, i.e. the **southern lapwing** *Vanellus chilensis* (see Appendix 5).

However in the genus *Coccyzus* of the subfamily **Cuculinae**, there are two species, the **yellow-billed cuckoo** *C. americanus* and the **black-billed cuckoo** *C. erythropthalmus*, which are conspecific nest parasites and lay in each other's nests. Each of the two species parasitizes ten species of ten genera of eight families, with four species in each case being successful and four species being unique to each cuckoo. There is also a successful attempt on a species not specifically recorded for either cuckoo, but the cuckoo species was not differentiated between *C. americanus* and *C. erythropthalmus* (Darwin 1859).

In the cases of these two species, a case could be made that a variable mixed strategy is successful. Whether this would evolve to obligate brood parasitism is conjecture. Little is known about the breeding biology of their congeners, but it would be interesting to discover.

The suggestion that *C. americanus* was a brood parasite that later regained normal brooding behaviour is rejected on the grounds that it is a much less parsimonious evolutionary route and that there are so many species where this appears not to be the case (see Sect. 4.1).

4.4.4 Other Orders

In four other orders, Phoenicopteriformes, Pelecaniformes, Charadriiformes and Passeriformes, there are various instances of laying in other species nests, but in no case does it suggest that it is other than a mistake or case of egg dumping. In the first order, there is a single record of a grebe being imprinted on an anatid (Rogers 1992).

In the second order, there are cases of two species of ardeid laying in the nests of a third (González-Martin and Ruiz 1996; Safford and Hawkins 2013). In the Charadriiformes, a charadiid has been known to lay in nests of three species of wader from two other families (Hampshire and Russell 1993; Niemcynowicz et al. 2015). In the same order, there is a case of gull apparently imprinting on a falcon (Baker 2002).

Four families of Passeriformes are recorded or suspected as laying in other species nests. One species of **Nectariniidae** apparently imprinted on a malkoha (**Cuculidae**) (Rozendaal and Dekker 1989); a species of sparrow occasionally lays in nests of three species of ploceid (Summers-Smith 1988, 2016). An estrildid species has been recorded appropriating nests of **chestnut weaver** *Ploceus rubiginosus* and laying in them; apparently some weavers deserted, but others did not, but no weavers hatched, and the fate of the estrildid eggs is not recorded (Craig 2016). A species of starling occasionally lays in the nests of another, non-congeneric starling (Craig and Feare 2016).

In the last three examples, a case could be made for an evolving strategy, although mistakes are perhaps the more likely conclusion.

Concluding Remarks and Future Directions

Fieldwork is of the essence here. The hosts of a number of species of cuckoos, honeyguides and indigobirds are still unknown, or in some cases only a few hosts have been identified. The relative success of different hosts is unknown for a significant number of obligate parasites. The nesting of many Neotropical members of the Tribe **Phaenicophaeini** is incompletely known or even unknown. Possibly other examples of brood parasitism await discovery there.

Much more work on DNA, particularly chromosomal, will lead to a more thorough understanding of the systematics of some groups of brood parasites.

Acknowledgements I am very grateful to J. Eadie, M. Soler and Y. Yom-Tov for constructive criticism of the manuscript and to Frederik P. Brammer for assistance in tracking recalcitrant references.

References

Abbott GA (1902) A grebe colony. Bird Lore 4:86–87

Ali S (1983) Some notes on the ecology and the status of the Orange-rumped Honeyguide *Indicator xanthonotus* in the Himalayas. J Bombay Nat Hist Soc 80:564–574

Ali S, Ripley SD (1968–1974) Handbook of the birds of India and Pakistan. Oxford University Press, Oxford

Allen RP (1932) Eggs of royal tern in laughing gull's nest. Auk 49:219–220

Allen DG (2002) First record of striped cuckoo *Clamator levaillantii* parasitizing Pied Babbler *Turdoides bicolor*. Bird Numb 11:30

Álvarez F (1994) A gens of cuckoo *Cuculus canorus* parasitizing Rufous Bush Chat *Cercotrichas galactotes*. J Avian Biol 25:239–243

Amat JA (1991) Effects of red-crested Pochard parasitism on Mallards. Wilson Bull 103:501–503

Anderegg K (2005) Eiderente *Somateria mollissima* führt junge Kolbenente *Netta rufina* [Common Eider *Somateria mollissima* leading young red-crested Pochard *Netta rufina*]. Ornithol Beob 102:33–52

Andersson M, Eriksson MOG (1982) Nest parasitism in goldeneyes *Bucephala clangula*: some evolutionary aspects. Am Nat 120:1–16

Antonov A, Stokke BG, Moksnes A, Røskaft E (2007) First evidence of regular common cuckoo, *Cuculus canorus*, parasitism on Eastern Olivaceous Warblers, *Hippolais pallida elaeica*. Naturwissenschaften 94:307–312

Archer AM (2012) First record: Levaillant's Cuckoo fed by Kurrichane Thrush. Ornithol Obs 3: 107–108

Arnold E (1894) My '94 outing trip in northwest Canada. Nidiologist 1:168, 2:11–13, 23–25

Artyukin YB (1997) Lanceolated Warbler *Locustella lanceolata*—a new host of oriental cuckoo *Cuculus saturatus*. Russ Ornithol J 6: 9–20 [in Russian]

Bagg E Jr (1877) A cuckoo's egg in a Cedar-bird's nest. Bull Nuttall Ornithol Club 2:110

Baker ECS (1927) The fauna of British India including Ceylon and Burma, Birds, vol 4. Taylor and Francis, London

Baker ECS (1934) The nidification of birds of the Indian Empire, vol 3. Taylor and Francis, London

Baker ECS (1942) Cuckoo problems. Witherby, London

Baker R (2002) Peregrine Falcon fostering Herring Gull chicks. Br Birds 95:190

Bakewell DN, Lim KC, Sah SAM (2011) Brood-host record for oriental (Sunda) Cuckoo *Cuculus* (*saturatus*) *lepidus* in the Cameron highlands, Peninsular Malaysia. BirdingASIA 15:48

Balatskiy NN (1988) The ecological races of *Cuculus canorus* L. parasitizing *Acrocephalus dumetorum* Blyth and *Saxicola torquata* L. near Novosibirsk. Nauch Dohl Vissh Shk Biol Nauki 1988(4):57–61 [in Russian]

Balatskiy NN (1989) The ecological race of the Cuckoo parasite of the Little Bunting in north Urals. Nauch Dohl Vissh Shk Biol Nauki 1989 (10):46–49 [in Russian]

Balatskiy NN (1990) The ecological race of the Cuckoo parasite of the Siberian Rubythroat in Siberia. Nauch Dohl Vissh Shk Biol Nauki 1990 (4): 37–44 [in Russian]

Balatskiy NN (2008) Cuckoo *Cuculus canorus* nest parasite of the Common Redstart near Novosibirsk. Russian Ornithol J 17 (Express issue 455):1835–1837 [in Russian]

Ballard BM (2001) Parasitism of a Laughing Gull nest by Black-bellied Whistling Ducks. Wilson Bull 11:339–340

Barker BW (1938) American eider and herring gull—mixed sets. Oologist 55:136–137

Barret R (1989) Deux hôtes rarissimes du Coucou gris (*Cuculus canorus*): la Bauscarle de Cetti (*Cettia cetti*) et la Locustelle luscinioïde (*Locustella luscinioides*) [Two rare hosts of the Cuckoo *Cuculus canorus*: Cetti's Warbler *Cettia cetti* and Savi's Warbler *Locustella luscinioides*]. Alauda 57:132

Barrientos R, Bueno-Enciso J, Serrano-Davies E, Sanz JJ (2015) Facultative interspecific brood parasitism in tits: a last resort to coping with nest-hole shortage. Behav Ecol Sociobiol 69:1603

Batchelor DM (1958) A checklist of the birds seen at Asahan, Malacca, part 1. Malay Nat J 12: 164–182

Becking JH (1975) New evidence of the specific affinity of *Cuculus lepidus* Müller. Ibis 117: 275–284

Becking JH (1981) Notes on the breeding of Indian cuckoos. J Bombay Nat Hist Soc 78:201–231

Beisenhertz W (1998) Mountain Tailorbird as host of Rusty-breasted Cuckoo. Kukila 10:159

Belcher C, Smooker GD (1936) Birds of the colony of Trinidad and Tobago, part 3. Ibis (13th Series) 6:1–35

Bellrose FC (1976) Ducks, geese and swans of North America. Stackpole, Harrisburg

Bendire C (1895) Life histories of North American birds from the parrots to the grackles, with special reference to their breeding habits and eggs. US Nat Mus Special Bull 3. Govt Printing Office, Washington

Bennett LJ (1936) The ring-necked Pheasant as a nesting parasite of other game birds. Iowa State Coll J Sci 10:373–375

Bennett LJ (1938) The blue-winged teal. Collegiate Press, Ames, pp 1–144

Benson CW, Benson FM (1977) The birds of Malawi. Mountford Press, Limbe

Bent AC (1902) Nesting habits of the Anatidae in North Dakota. Auk 19(1–12):165–174

Bent AC (1923) Life histories of North American wild fowl, part 1. US Nat Mus Bull 126

Bent AC (1940) Life histories of North American cuckoos, goatsuckers, hummingbirds, and their allies. US Nat Mus Bull 176

Beresford RP (1922) Curlews' eggs in wild ducks' nest. Irish Nat 13:72

Berruti A, McIntosh B, Walter R (1995) Parasitism of the Blue Swallow *Hirundo atrocaerulea* by the Greater Honeyguide *Indicator indicator*. Ostrich 66:94

Bharucha EK (1982) Sunbirds fostering fledglings of the Plaintive Cuckoo. J Bombay Nat Hist Soc 79:670–671

Birds Korea (2009) http://www.birdskorea.org. Accessed 10 Jun 2016

Bishop DR, Bishop D, Bishop MS (1995) Occurrence of Coppery Sunbird *Nectarinia cuprea* being parasitised by Klaas's Cuckoo *Chrysococcyx klaas*. Babbler 29/30:36

Bizeau EG (1951) Factors affecting waterfowl production at Grays' Lake, Idaho. MS thesis, Univ Idaho

Bodrati A, Salvador SA (2015) Curutié Blanco (*Cranioleuca pyrrhophia*), Pijuí Corona Rojiza (*Synallaxis ruficapilla*) y Picochato Grande (*Tolmomyias sulphurescens*): hospedadores de cría del Crespín (*Tapera naevia*). Nuestras Aves 60:15–18

Bouvier JM (1974) Breeding biology of the hooded merganser in Quebec, including interactions with common goldeneyes and wood ducks. Can Field Nat 88:323–330

Brazil MA (1991) The birds of Japan. Christopher Helm, A and C Black, London

Bregulla HL (1992) Birds of Vanuatu. Anthony Nelson, Oswestry

Brooker MG, Brooker LC (1989a) The comparative breeding behaviour of two sympatric cuckoos Horsfield's Bronze-Cuckoo *Chrysococcyx basalis* and the Shining Bronze-Cuckoo *C. lucidus* in Western Australia: a new model for the evolution of egg morphology and host specificity in avian brood parasites. Ibis 131:528–547

Brooker MG, Brooker LC (1989b) Cuckoo hosts in Australia. Aust Zool Rev, R Zool Soc New South Wales 2:1–67

Brooker LC, Brooker MG (2005) Australian cuckoo-host database. http://www.users.bigpond.com/LesMikeBrooker/database.htm. Accessed 21 Jul 2016

Brosset A, Érard C (1977) New faunistic records from Gabon. Bull Br Orn Club 97:125–132

Brosset A, Érard C (1986) Les oiseaux des régions forestières du nord-est du Gabon. Écologies et comportement des especes, vol 1. Société National du Protection de la Nature, Paris

Brown LH, Urban EK, Newman K (eds) (1982) Birds of Africa, vol 1. Academic Press, London

Bryant HC (1914) A survey of the breeding grounds of ducks in California in 1914. Condor 16:217–239

Cannell PF, Harrington BA (1984) Interspecific egg dumping by a Great Egret and Black-crowned Night Herons. Auk 101:889–891

Carboneras C, Kirwan GM (2016) Black-bellied Whistling-duck (*Dendrocygna autumnalis*). In: del Hoyo J, Elliott A, Sargatal J, Christie DA, de Juana E (eds) Handbook of the birds of the world alive. Lynx Edicions, Barcelona. http://www.hbw.com/node/52801. Accessed 17 Nov 2016

Carboneras C, Kirwan GM (2017) Rosy-billed Pochard (*Netta peposaca*). In: del Hoyo J, Elliott A, Sargatal J, Christie DA, de Juana E (eds) Handbook of the birds of the world alive. Lynx Edicions, Barcelona. http://www.hbw.com/node/52901. Accessed 17 Apr 2017

Carlson CE (1943) Unusual pheasant nests from Minnesota. Flicker 15:29–31

Carriker MA Jr (1910) An annotated list of the birds of Costa Rica including Cocos Island. Ann Carnegie Mus 6:314–915

Casa F, Mougeot F, Viñuela J (2010) Occurrence of Common Quail *Coturnix coturnix* eggs in Red-legged Partridge *Alectoris rufa* nests. Bird Study 57:560–562

Chalton DO (1976) Weight loss in Spectacled Weavers raising a cuckoo chick. Ostrich 47(1):69

Chapin JP (1939) The birds of the Belgian Congo, part 2. Bull Am Mus Nat Hist 75

Charter M, Bouskila A, Aviel S, Leshem Y (2005) First record of European Jackdaw (*Corvus monedula*) parasitism by the Great Spotted Cuckoo (*Clamator glandarius*) in Israel. Wilson Bull 117:201–204

Cheke RA, Mann CF (2001) Sunbirds: a guide to the sunbirds, flowerpeckers, spiderhunters and sugarbirds of the world. Christopher Helm, London

Cheke RA, Mann CF (2008) Family Nectariniidae (sunbirds). In: del Hoyo J, Elliottt A, Christie DA (eds) Handbook of the birds of the world, vol 13. Lynx Edicions, Barcelona, pp 196–320

Clement P (2016) American Goldfinch (*Carduelis tristis*). In: del Hoyo J, Elliott A, Sargatal J, Christie DA, de Juana E (eds) Handbook of the birds of the world alive. Lynx Edicions, Barcelona. http://www.hbw.com/node/61356. Accessed 12 Nov 2016

Clunie F (1973) Fan-tailed Cuckoo parasitises Fiji Warbler. Notornis 20(2):168

Coates BJ (1985) The birds of Papua New Guinea, vol 1. Dove Publications, Alderly

Coates BJ, Bishop KD (1997) A guide to the birds of Wallacea. Sulawesi, the Moluccas and Lesser Sunda Islands, Indonesia. Dove Publications, Alderly

Colebrook-Robjent JFR (1984a) A nest of the double-toothed Barbet *Lybius bidentatus* parasitized by a honeyguide in Uganda. Bull Br Ornithol Club 104:123–124

Colebrook-Robjent JFR (1984b) The breeding of the didric cuckoo *Chrysococcyx caprius* in Zambia. Proc V Pan-Afr Congr:763–778

Colebrook-Robjent JFR, Stjernstedt R (1976) Chaplin's Barbet *Lybius chaplini*: first description of eggs, a new host record for the Lesser Honeyguide *Indicator minor*. Bull Br Ornithol Club 96:109–111

Collar N (2016a) American Robin (*Turdus migratorius*). In: del Hoyo J, Elliott A, Sargatal J, Christie DA, de Juana E (eds) Handbook of the birds of the world alive. Lynx Edicions, Barcelona. http://www.hbw.com/node/58287. Accessed 17 Nov 2016

Collar N (2016b) Rufous-backed Robin (*Turdus rufopalliatus*). In: del Hoyo J, Elliott A, Sargatal J, Christie DA, de Juana E (eds) Handbook of the birds of the world alive. Lynx Edicions, Barcelona. http://www.hbw.com/node/58289. Accessed 17 Nov 2016

Collar N (2016c) Rufous-bellied Thrush (*Turdus rufiventris*). In: del Hoyo J, Elliott A, Sargatal J, Christie DA, de Juana E (eds) Handbook of the birds of the world alive. Lynx Edicions, Barcelona. http://www.hbw.com/node/58319. Accessed 17 Nov 2016

Collar N (2016d) Groundscraper Thrush (*Psophocichla litsitsirupa*). In: del Hoyo J, Elliott A, Sargatal J, Christie DA, de Juana E (eds) Handbook of the birds of the world alive. Lynx Edicions, Barcelona. http://www.hbw.com/node/58321. Accessed 18 Nov 2016

Collar N, de Juana E (2016a) Kurrichane Thrush (*Turdus libonyanus*). In del Hoyo J, Elliott A, Sargatal J, Christie DA, de Juana E (eds). Handbook of the birds of the world alive. Lynx Edicions, Barcelona. http://www.hbw.com/node/58283. Accessed 18 Nov 2016

Collar N, de Juana E (2016b) Creamy-bellied Thrush (*Turdus amaurochalinus*). In: del Hoyo J, Elliott A, Sargatal J, Christie DA, de Juana E (eds) Handbook of the birds of the world alive. Lynx Edicions, Barcelona. http://www.hbw.com/node/58290. Accessed 18 Nov 2016

Cracraft J (2013) Higher-level relationships and classification: Nonpasseriforms. In: Dickinson EC, Remsen JV Jr (eds) The Howard & Moore complete checklist of the birds of the world, 4th edn, vol 1. Aves Press, Eastbourne, pp xxi–xli

Cracraft J (2014) Avian higher-level relationships and classification: Passeriformes. In: Dickinson EC, Christidis L (eds) The Howard & Moore complete checklist of the birds of the world, vol 2, 4th edn. Aves Press, Eastbourne, pp xvii–xxlv

Craig A (2016) Chestnut Weaver (*Ploceus rubiginosus*). In: del Hoyo J, Elliott A, Sargatal J, Christie DA, de Juana E (eds) Handbook of the birds of the world alive. Lynx Edicions, Barcelona. http://www.hbw.com/node/61008. Accessed 10 Nov 2016

Craig A, Feare C (2016) White-cheeked Starling (*Poliopsar cineraceus*). In: del Hoyo J, Elliott A, Sargatal J, Christie DA, de Juana E (eds) Handbook of the birds of the world alive. Lynx Edicions, Barcelona. http://www.hbw.com/node/60854. Accessed 14 Nov 2016

Cramp S (ed) (1985) Handbook of the birds of Europe, the Middle East and North Africa. The birds of the Western Palearctic, vol IV. Oxford University Press, Oxford

Cranbrook E, Wells DR (1981) Observations of fledgling cuckoos and their fosters in Gunung Mulu National Park. Sarawak Mus J 29:147–149

Cunningham JM, Welch EO (1955) The grey teal in New Zealand: some nesting and plumage notes. Emu 55:303–308

Cunningham-van Someren GR (1970) On *Prodotiscus insignis* (Cassin) parasitising *Zosterops abyssinica* Guérin. Bull Br Ornithol Club 90:129–131

Cunningham-van Someren GR (1973) *Vidua fischeri* (Reichenow) parasitic on *Granatina ianthinogaster* (Reichenow). J East Afr Nat Hist Soc Natl Mus 139:1–4

Curson J (2016) Yellow Warbler (*Dendroica petechia*). In: del Hoyo J, Elliott A, Sargatal J, Christie DA, de Juana E (eds) Handbook of the birds of the world alive. Lynx Edicions, Barcelona. http://www.hbw.com/node/61466. Accessed 12 Nov 2016

Darwin C (1859) On the origin of species. Murray, London

Davey P (1994) Woodpecker feeding immature honeyguide. Scopus 18:61

Davidson JL (1887) Birds laying their eggs in the nests of other birds. Auk 4:263–264

Davis DE (1940) A suggestion concerning territorialism in *Tapera naevia*. Wilson Bull 52:208

de Balsac H (1926) Un eas de parasitisme chez le Rongequeue Ph. ph. phoenicurus (L.). Rev Fran d' Ornithol 10:137–138

de la Peña MR (1986) Guía de Aves Argentinas, vol 1. Literature of Latin America, Buenos Aires

de la Peña MR (1993) Parasitismo en el crespín, Tapera naevia chochi (Vieillot, 1817) en la provincial de Santa Fé, república Argentina. Nótulas Faunísticas, Corrientes 37:1–4 [in Spanish]

de la Peña MR (2005) Reprodución de las aves argentinas (con descripción de pichones). Literature of Latin America, Buenos Aires

de Naurois R (1979) The Emerald Cuckoo of São Thomé and Príncipe Islands (Gulf of Guinea). Ostrich 50:88–93

del Hoyo J, Collar N, Kirwan GM (2017a) Eastern Koel (*Eudynamys orientalis*). In: del Hoyo J, Elliott A, Sargatal J, Christie DA, de Juana E (eds) Handbook of the birds of the world alive. Lynx Edicions, Barcelona. http://www.hbw.com/node/467235. Accessed 13 Jun 2017

del Hoyo J, Collar N, Kirwan GM (2017b) Moluccan Cuckoo (*Cacomantis aeruginosus*). In: del Hoyo J, Elliott A, Sargatal J, Christie DA, de Juana E (eds) Handbook of the birds of the world alive. Lynx Edicions, Barcelona. http://www.hbw.com/node/467238. Accessed 14 Jun 2017

Dement'ev GP, Gladkov NA, Isakov YA, Kartashev NN, Kirikov AV, Mikheev AV, Ptushenko ES (1967) Birds of the Soviet Union, vol 4 [Translated from the 1952 Russian book by Israel Program for Scientific Translations, Jerusalem]

Deng W (2013) Brood parasitism on the Yellow-rumped Flycatcher (*Ficedula zanthopygia*) by the Oriental Cuckoo (*Cuculus optatus*) in an artificial nestbox in Beijing. Chin Birds 4:187–188

Di Giacomo A (2005) Aves de la Reserva El Bagual. In: Di Giacomo AG, Krapovickas SF (eds) Historia natural y paisaje de la Reserva El Bagual, provincia de Formosa, Argentina. Inventario de los vertebrados y de la flora vascular de un área protegida del Chaco Húmedo. Temas de Naturaleza y Conservación, vol 4. Aves Argentinas/Asociación Ornitológica del Plata, Buenos Aires, pp 201–465

Di Giacomo A (2010) Aves de la Reserva El Bagual. In: Di Giacomo AG, Krapovickas SF (eds) Historia natural y paisaje de la Reserva El Bagual, provincial de Formosa, Argentina. Inventario de los vertebrados y de la flora vascular de un área protegida del Chaco Húmedo, Temas de Naturaleza y Conservación, vol 4. Aves Argentinas/Asociación Ornitológica del Plata, Buenos Aires, pp 201–465

Di Giacomo AG, Mahler B, Reboreda JC (2010) Screaming Cowbird parasitism of nests of Solitary Caciques and Cattle Tyrants. Wilson J Ornithol 122:795–799

Dickinson EC, Christidis L (eds) (2014) The Howard & Moore complete checklist of the birds of the world, vol 2, 4th edn. Aves Press, Eastbourne

Dickinson EC, Remsen JV Jr (eds) (2013) The Howard & Moore complete checklist of the birds of the world, vol 1, 4th edn. Aves Press, Eastbourne

Dickinson EC, Kennedy R, Parkes KC (1991) The birds of the Philippines. An annotated checklist. BOU checklist 12. British Ornithologists' Union, Tring

Dowsett RJ (2005) Does the Didric Cuckoo *Chrysococcyx caprius* parasitize Cricket Warbler *Spiloptila clamans* in Nigeria? Malimbus 27:41–42

Dowsett RJ, Dowsett-Lemaire F (2011) The avifauna of Benin: additions and corrections. Bull Afr Bird Club 18:148–167

Dowsett-Lemaire F (1990) Ecoethology, distribution and status of Nyungwe Forest birds (Ruanda). Tauraco Res Rep 3:310–385

Dowsett-Lemaire F (2008) On the singing habits of forest honeyguides of the Guineo-Congolian region, with a request for further information. Bull Afr Bird Club 15:24–35

Dowsett-Lemaire F, Dowsett RJ (2006) The birds of Malawi: an Atlas and handbook. Tauraco Press and Aves, Liège

Dugger BD, Blums P (2001) Effect of conspecific brood parasitism on host fitness for tufted duck and common pochard. Auk 118:717–726

Dugger BD, Bollman LC, Fredrickson LH (1999) Response of female Hooded Mergansers to eggs of an interspecific brood parasite. Auk 116:269–273

Duncan DC (1984) Egg retrieval by Canada Geese: apparent interspecific retrieval and tests of egg displacement. Auk 101:886–887

Dwyer PD (1980) Two species of megapodes laying in the same mound. Emu 81:172–173

Earlé RA, Herholdt JJ (1987) Notes on a Greater Honeyguide *Indicator indicator* chick raised by Anteater Chats *Myrmecocichla formicivora*. Bull Br Ornithol Club 107:70–73

Eisentraut M (1935) Biologische Studien im bolivianischen Chaco. VI. Beitragzur Biologie der Vogelfauna. Mitt Zool Mus Berlin 20:367–443

Elgood JH (1982) The birds of Nigeria. British Ornithologists' Union Checklist Series, 4. BOU, London

Erritzøe J, Mann CF, Brammer FP, Fuller RA (2012) Cuckoos of the world. Christopher Helm, London

Erskine AJ (1990) Joint laying in Bucephala ducks "parasitism" or nest-site competition? Ornithol Scand 21:52–56

Feeney WE, Welbergen JA, Langmore NE (2014) Advances in the study of coevolution between avian brood parasites and their hosts. Annu Rev Ecol Evol Syst 45:227–246

Finley WL (1896) N.O.A. at work. Oreg Nat 3:102–103

Follett B (1990) Lesser double-collared sunbirds rearing Klaas's cuckoo. Promerops 196:13–14

Fraga R (2016a) Bronze-brown Cowbird (*Molothrus armenti*). In: del Hoyo J, Elliott A, Sargatal J, Christie DA, de Juana E (eds) Handbook of the birds of the world alive. Lynx Edicions, Barcelona. http://www.hbw.com/node/62294. Accessed 13 Aug 2016

Fraga R (2016b) Common Grackle (*Quiscalus quiscula*). In: del Hoyo J, Elliott A, Sargatal J, Christie DA, de Juana E (eds) Handbook of the birds of the world alive. Lynx Edicions, Barcelona. http://www.hbw.com/node/62283. Accessed 12 Nov 2016

Fraga R (2016c) Giant Cowbird (*Molothrus oryzivorus*). In: del Hoyo J, Elliott A, Sargatal J, Christie DA, de Juana E (eds) Handbook of the birds of the world alive. Lynx Edicions, Barcelona. http://www.hbw.com/node/62292. Accessed 25 Nov 2016

Fraga R (2016d) Bronzed cowbird (*Molothrus aeneus*). In: del Hoyo J, Elliott A, Sargatal J, Christie DA, de Juana E (eds) Handbook of the birds of the world alive. Lynx Edicions, Barcelona. http://www.hbw.com/node/62293. Accessed 25 Nov 2016

Fraga R, Bonan A (2016) New world blackbirds (Icteridae). In: del Hoyo J, Elliott A, Sargatal J, Christie DA, de Juana E (eds) Handbook of the birds of the world alive. Lynx Edicions, Barcelona. http://www.hbw.com/node/52383. Accessed 13 Aug 2016

Fraga R, Garcia EFJ (2016a) Screaming cowbird (*Molothrus rufoaxillaris*). In: del Hoyo J, Elliott A, Sargatal J, Christie DA, de Juana E (eds) Handbook of the birds of the world alive. Lynx Edicions, Barcelona. http://www.hbw.com/node/62291. Accessed 13 Aug 2016

Fraga R, Garcia EFJ (2016b) Brown-headed cowbird (*Molothrus ater*). In: del Hoyo J, Elliott A, Sargatal J, Christie DA, de Juana E (eds) Handbook of the birds of the world alive. Lynx Edicions, Barcelona. http://www.hbw.com/node/62295. 25 Nov 2016

Friedmann H (1927) Notes on some Argentina birds. Bull Mus Comp Zool 68:139–236

Friedmann H (1948) The parasitic cuckoos of Africa. Monograph 1. Washington Academy Sciences, Washington, DC

Friedmann H (1955) The honeyguides. Bull US Nat Mus 208

Friedmann H (1956) Further data on African parasitic cuckoos. Proc US Natl Mus 106:377–408

Friedmann H (1960) The parasitic weaverbirds. US Natl Mus Bull 223

Friedmann H (1963) Host relations of the parasitic cowbirds. US Natl Mus Bull 233

Friedmann H (1968) The evolutionary history of the avian genus *Chrysococcyx*. US Natl Mus Bull 265

Friedmann H (1970) Further information on the breeding biology of the honey-guides. Los Angeles Co Mus Contrib Sci 205

Friedmann H (1978) Current knowledge of the Lyre-tailed Honey-guide, *Melichneutes robustus* (Bates), and its implications. Rev Zool Afr 92:644–656

Friedmann H, Kiff LF (1985) The parasitic cowbirds and their hosts. Proc West Found Vert Zool 2:225–304

Fuller RW (1953) Studies in the life history and ecology of the American pintail (*Anas acuta tzitzihoa*) in Utah. Unpublished MS thesis, Utah State Agriculture College, Logan

Gaston AJ, Zacharias VJ (2000) Hosts of the common Hawk cuckoo *Hierococcyx varius* in India. Forktail 16:182

Gatter W (1997) Birds of Liberia. Pica Press, East Sussex

Gill BJ (1982) The Grey Warbler's care of nestlings: a comparison between unparasitized broods and those comprising a Shining Bronze-cuckoo. Emu 82:177–181

González-Martin M, Ruiz X (1996) Brood parasitism in herons. Colon Waterbirds 19:31–38

Gore MEJ (1981) The birds of the Gambia. British Ornithologists' Union Check List 3. BOU, London

Gosper DG (1962) Breeding records of the Koel. Austr Bird Watcher 17:226–228

Gosper DG (1997) Aspects of breeding of the Common Koel *Eudynamys scolopacea* and one of its biological hosts, the Magpie-lark *Grallina cyanoleuca*. Austr Bird Watcher 17(1):11–19

Grimes LG (1979) The yellow-billed shrike *Corvinella corvina*: an abnormal host of the yellow-billed cuckoo Cuculus gularis. Bull Br Ornithol Club 99:36–38

Grinnell J, Bryant HC, Storer TI (1918) The game birds of California. University of California Press, Berkeley

Grobler JH, Steyn P (1980) Breeding habits of the Boulder Chat and its parasitism by the red-chested cuckoo. Ostrich 51:253–254

Gross AO (1938) Eider ducks of Kent's Island. Auk 55:387–400

Hagen CA, Jamieson BE, Robel RJ, Applegate RD (2002) Ring-necked Pheasant parasitism of Lesser Prairie-Chicken nest in Kansas. Wilson Bull 114:522–542

Hamerstrom FN Jr (1936) A study of the nesting habits of the Ring-necked Pheasant in northwest Iowa. Iowa State Coll J Sci 10:173–199

Hamilton WJ, Orians GH (1965) Evolution of brood parasitism in altricial birds. Condor 67: 361–382

Hampshire JS, Russell FJ (1993) Oystercatchers rearing Northern Lapwing chick. Br Birds 86: 17–19

Hanmer DB (1982) Klaas's Cuckoo parasitizing Whitebellied Sunbird. Ostrich 53:58

Hellebrekkers WPJ, Hoogerwerf A (1967) A further contribution to our oological knowledge of the island of Java (Indonesia). Zool Verh Leiden 88:1–164

Henry CJ (1940) Northern great plains region. Bird Lore 42:469–470

Henry GM (1955) A guide to the birds of Ceylon. Geoffrey Cumberlege and OUP, Oxford

Herrick EH (1910) Life and behaviour of the Cuckoo. J Exp Zool 9:169–233

Higgins PJ (ed) (1999) Handbook of Australian, New Zealand and Antarctic birds, vol 4. OUP, Melbourne

Higuchi H, Payne RB (1986) Nestling and fledgling plumages *Cuculus saturatus horsfieldi* and *C. poliocephalus poliocephalus* in Japan. Jpn J Ornithol 35:61–65

Hitoshi N, Shigeru A, Biun A (1996) A photographic guide to the birds of Mt. Kinabalu, Borneo. Bun-ichi Sogoshuppan, Tokyo

Hochbaum HA (1944) The canvasback on a prairie marsh. American Wildlife Institute, Washington, pp 1–201

Hoffmann A (1950) Der Indische Kuckuck (*Cuculus micropterus* Gould). Bonn Zool Beitr 1:21–30

Höhn EO (1975) Notes on Black-headed Ducks, Painted Snipe and Spotted Tinamous. Auk 92: 566–575

Holm W (2006) Star Sturnus vulgaris als Kuckuckswirt. Ornithol Mitt 58:324–325

Hopwood JC, Mackenzie JMD (1917) A list of the birds from the North Chin Hills. J Bombay Nat Hist Soc 25:72–91

Hosken JH (1966) Sharp-billed Honeyguide *Prodotiscus regulus* being fed by a pair of Cisticolas. Ostrich 37:235. https://www.fieldmuseum.org/sites/default/files/OW-cuckoos-26apr2013.pdf. Accessed 10 Jun 2016

Hu Y-B, Wang X-P, Chang H-Z, Sun Y-H (2013) Brood parasitism on Elliot's Laughingthrush by Large Hawk Cuckoo. Chin J Zool 48(2):292–293

Hughes JM (1997) Taxonomic significance of host-egg mimicry by facultative brood parasites of the genus *Coccyzus* (Cuculidae). Can J Zool 75:1380–1386

Hughes JM (1999) Yellow-billed Cuckoo *Coccyzus americanus*. In: Poole A, Gill F (eds) The birds of North America, no. 418. The Birds of North America, Philadelphia

Hughes JM (2000) Monophyly and phylogeny of cuckoos (Aves, Cuculidae) inferred from osteological characters. Zool J Linn Soc Lond 130:263–307

Hunter C (1961) Parasitism of the Masked Weaver *Ploceus velatus arundinaceus*. Ostrich 32:55–63

Iqbal M, Handoko M (2014) First record of Bold-striped Tit Babbler *Macronus* (*gularis*) *bornensis* brood-hosting Drongo Cuckoo *Surniculus lugubris*. BirdingASIA 1:102–103

Irwin MPS (1988) Family Cuculidae. In: Fry CH, Keith S, Urban E (eds) The birds of Africa, vol III. Academic Press, London

Jackson WM (1992) Relative importance of parasitism by *Chrysococcyx* cuckoos versus conspecific nest parasitism in the Northern Masked Weaver *Ploceus taeniopterus*. Ornithol Scand 23:203–206

Jenny J (1997) Unusual breeding behaviour of the Guira Cuckoo *Guira guira*. Cotinga 7:18

Jensen RAC, Jensen MK (1969) On the breeding biology of Southern African cuckoos. Ostrich 40:163–181

Jessop JP (1960) Red-chested Cuckoo egg in Cape Paradise Flycatcher nest. Ostrich 31:179

Jiang Y-x, Liang W, Yang C-c, Sun Y-h (2007) *Hierococcyx sparverioides* parasitized and reproduced in nest of *Garrulax sannio*. Sichuan J Zool 26:509

Jiang A, Zhou F, Liu N (2014) Significant recent ornithological records from the limestone area of south-west Guangxi, south China, 2004–2012. Forktail 30:122–129

Job HK (1902) Among the waterfowl. Doubleday, Page and Co, New York

Johnsgard PA (1997) Avian brood parasites—deception at the nest. Oxford University Press, Oxford and New York

Johnsingh AJT, Paramanandham K (1982) Group care of White-headed Babblers *Turdoides affinis* for a Pied Crested Cuckoo *Clamator jacobinus* chick. Ibis 124:179–183

Jones JMB (1985) Stripe Crested Cuckoo parasitizing Arrow-marked Babbler. Honeyguide 31: 170–171

Joyner DE (1976) Effects of inter-specific nest parasitism by redheads and ruddy ducks. J Wildl Manage 40:33–38

Kear J (ed) (2005) Ducks, Geese and Swans. Oxford University Press, Oxford

Kennamer JE (1989) Oddities in Wild Turkey biology. Turkey Call March/April:26–27

Kiff LF, Williams A (1978) Host records of the Striped Cuckoo from Costa Rica. Wilson Bull 90:138–139

Kim C-H (1996) Behavioural characteristics between the parasite and host: crow Tits *Paradoxornis webbianus* and Common Cuckoos *Cuculus canorus*. Korean J Ornithol 3:51–57 [in Korean]

Knystautas AJV, Sibnev JB (1987) Die Vogelwelt Ussuriens. A Ziemens Verlag, Lutherstadt

Koons DN, Rotella JJ (2003) Comparative nesting success of sympatric Lesser Scaup and Ring-necked Ducks. J Field Ornithol 74:222–229

Krakauer AH (2003) California Quail lays egg in Wild Turkey nest. West Birds 34:169–170

Kroodsma D, Brewer D, Christie DA, Bonan A (2016) Northern Wren (*Troglodytes troglodytes*). In: del Hoyo J, Elliott A, Sargatal J, Christie DA, de Juana E (eds) Handbook of the birds of the world alive. Lynx Edicions, Barcelona. http://www.hbw.com/node/58150. Accessed 13 Nov 2016

La Touche JDD (1931) A handbook of the birds of eastern China, vol 2, part 1. Taylor and Francis, London

Lachenaud O (2006) Les oiseaux du Parc National du Banco et de la Forêt Classée de l'Anguédédou, Côte d'Ivoire. Malimbus 28:107–133

Lahti DC, Payne RB (2003) Morphological and behavioural evidence of relationships of the Cuckoo Finch *Anomalospiza imberbis*. Bull Br Ornithol Club 123:113–125

Lamarche B (1980) List commentée du oiseaux du Mali. Malimbus 2:121–158

Lansverk AL, Dogmo J-B, Schuetz JG, Balakrishnan CN (2015) Parasitism of the Blackcrowned Waxbill (*Estrilda nonnula*) by the Pin-tailed Whydah (*Vidua macroura*): implications for host-specific adaptation by a generalist brood-parasite. Wilson J Ornithol 127:733–739

Lasley GW, Sexton C (1985) The nesting season June 1-July 31, 1985. South Texas region. Am Birds 39:933–936

Lebedeva NV, Markitan LV (2001) Problems of population dynamics of the White-eyed Pochard (*Aythya nyroca* Guld.,1770) in the eastern Sea of Azov region. Russian J Ecol 32:425–431

Lee J-W, Noh H-J, Lee Y, Kwon Y-S, Kim C-H, Yoo J-C (2014) Spatial patterns, ecological niches, and interspecific competition of avian brood parasites: inferring from a case study of Korea. Ecol Evol 4:3689–3702

Leffingwell DJ (1928) The Ring-Necked Pheasant—its history and habits. Occasional papers of the Charles R. Conner Museum. 1. Pullman, State College of Washington

Lethwaite R (1995) Blue Magpie as a foster parent of Koel. Hong Kong Bird Rep 1995:222

Lewis TE, Garrettson PR (2010) Parasitism of a Blue-winged Teal by a Northern Shoveler in S Dakota. Wilson J Ornithol 122:612–614

Lim KS (2011) Notes on a new brood-host of Drongo Cuckoo *Surniculus lugubris* in the Thai-Malay Peninsula. BirdingASIA 15:49

Liu Y, Pyzhjanow S (2012) Apparent inter- and intraspecific brood-parasitism in a nest of Tufted Duck *Aythya fuligula*. Forktail 28:147–148

Liversidge R (1955) Observations on a Piet-my-vrou (*Cuculus solitarius*) and its host the Cape Robin (Cossypha caffra). Ostrich 26:18–27

Livesey TR (1938) Egg-laying of the Khasia Hills Cuckoo (*C. c. bakeri*) in the nest of the Burmese Stone Chat (*Saxicola caprata burmanica*). J Bombay Nat Hist Soc 40:125–127

Loetscher FW (1952) Striped Cuckoo fed by Rufous-and-white Wren in Panama. Condor 54:169

Lokemoen JT (1991) Brood parasitism among waterfowl nesting on islands and peninsulas in North Dakota. Condor 93:340–345

López F, Morujo S (2012) Eurasian oystercatcher feeding Common Shelduck chick. Dutch Birding 34:313–314

Louette M (1981) Contribution to the ornithology of Liberia, part 5. Rev Zool Afr 95:342–355

Low JB (1941) Nesting of the ruddy duck in Iowa. Auk 58:506–517

Low A (2008) Bird report October 2008. Singapore Avi 22(10):13
Lowther PE (2012) Host list of avian brood parasites—4—Piciformes; Indicatoridae. https://www.fieldmuseum.org/sites/default/files/HoneyguideHosts.pdf. Accessed 10 Jan 2016
Lowther PE (2015) Host list of avian brood parasites—2—Cuculiformes—Old World cuckoos. https://www.fieldmuseum.org/sites/default/files/OW-cuckoos-26apr2013.pdf. Accessed 10 Jan 2016
Lowther PE (2016a) Host list of avian brood parasites—5—Passeriformes; Viduidae. https://www.fieldmuseum.org/sites/default/files/plowther/2016/01/21/vidua-hosts-19jan2016.pdf. Accessed 10 Jun 2016
Lowther PE (2016b) Host list of victims and hosts of the parasitic cowbirds (*Molothrus*). https://www.fieldmuseum.org/sites/default/files/plowther/2017/01/27/molothrus-hosts-03may2016.pdf. Accessed 10 Jan 2016
Lowther PE (2017) Host list of avian brood parasites—5—Passeriformes; Viduidae. https://www.fieldmuseum.org/sites/default/files/plowther/2017/04/18/vidua-hosts-18apr2017.pdf. Accessed 10 Jan 2016
Lyon BE, Eadie JM (1991) Mode of development and interspecific avian brood parasitism. Behav Ecol 2:309–318
Lyon BE, Eadie JM (2013) Patterns of host use by a precocial obligate brood parasite, the Black-headed Duck: ecological and evolutionary considerations. Chin Birds 4:71–85
Maccun J, Maccun JM (1909) Catalogue of Canadian birds. Publ 973. Can Dept Mines, Geol Surveys Branch, Ottawa
Maclean GL, Maclean C (1993) Roberts's "Birds of Southern Africa", 6th edn. Trustees of the John Voelcker Bird Book Fund, Cape Town
Madoc GC (1956) Observations on two Malayan cuckoos. Malay Nat J 10:97–103
Mallory ML, Lamsden HG (1994) Notes on egg-laying and incubation in the Common Merganser. Wilson Bull 106:757–759
Masterson HB (1953) Klaas' Cuckoo, *Chrysococcyx klaas*, parasitizing Red-headed Weavers, *Anaplectes rubriceps*. Ostrich 24:51
McAllen IAW (1995) Young Channel-billed Cuckoo being fed by Little Ravens. Aust Bird Watcher 16:79–81
McLean IG (1988) Breeding behaviour of the Long-tailed Cuckoo on Little Barrie Island. Notornis 35:89–98
Medway L, Wells DR (1976) The birds of the Malay Peninsula, vol V. Witherby, London
Meneghetti F (1944) Further notes on the nestling of Paradise Flycatchersand on parasitism by Klaas' Cuckoo. J E Afr Nat Hist Soc 18:96
Michel SM (2013) Surprise in the nest box: Great Spotted Cuckoo. Ornithol Obs 4:76–80
Millais JG (1909) The natural history of British Game Birds. Longmans, Green and Co, London
Miller AW, Collins BD (1954) A nesting study of ducks and coots on Tule Lake and lower Klamath national wildlife refuges. Calif Fish Game 43:17–37
Mills MSL (2010a) Rock Firefinch *Lagonosticta sanguinodorsalis* and its brood parasite, Jos Plateau Indigobird *Vidua maryae*, in northern Cameroon. Bull Afr Bird Club 17:86–89
Mills MSL (2010b) A record of brood parasitism of Barratt's Warbler by African Emerald Cuckoo. Bull Afr Bird Club 17:212
Mills MSL (2014) Dusky Twinspot *Euschistospiza cinereovinacea*, a new host species for indigobirds Vidua. Bull Afr Bird Club 21:193–199
Mills MSL, Caddick J, Coetzer I, Hoff R, Myers D, Reynolds D, Walton A, Walton R (2007) First host species record for Yellow-throated Cuckoo *Chrysococcy flavigularis*. Bull Afr Bird Club 14:68–69
Mines BM (1999) Hosts of the Pintailed Whydah (*Vidua macroura*) in southern Africa. Bird Numbers 8:15–21
Mitrus C, Socko B (2005) Pierwsze stwierdzenie pasozytnictwa legowego kukulki *Cuculus canorus* na mucholówce małej *Ficedula parva* [First record of nest parasitism by Cuckoo *Cuculus*

canorus on the Red-breasted Flycatcher *Ficedula parva*]. Notatki Ornitologiczne 46:247–249 [in Polish]

Mogensen J (1927) Nota sobre el parasitismo del "crespín" (*Tapera naevia*). Hornero 4:68–70

Mogensen J (1930) Argentinas dyr. Græbes Bogtrykkeri København (Copenhagen) [in Danish]

Moreau RE (1949) Special review. Friedmann on African cuckoos. Ibis 91:529–537

Moreau RF, Moreau WM (1939) Observations on some East African birds. Ibis (14th series) 3:296–323

Mori S, Kondo Y, Higuchi H (2012) An Eastern Crowned Leaf Warbler *Phylloscopus coronatus* nest parasitized by an Oriental Cuckoo *Cuculus saturatus* with a reddish egg in Hokkaido. Jpn Ornithol Sci 11:109–112

Morton ES, Farabough SM (1979) Infanticide and other adaptations of the nestling Striped Cuckoo *Tapera naevia*. Ibis 121:212–213

Moyle RG, Biun A, Butit B, Sumpongol D (2001) Brood hosts of Oriental Cuckoo *Cuculus saturatus* in Sabah, Malaysia. Bull Br Ornithol Club 121:107–110

Mulder RA, Ramiarison R (2003) *Terpsiphone mutata*, Madagascar Paradise Flycatcher, Remaly, Tsingetra, Rengetra, Singetraka, Dady, Adetry. In: Goodman SM, Benstead JP (eds) The natural history of Madagascar. University Chicago Press, Chicago, pp 1134–1138

Mulyawati D, Hua F, Zetra B (2009) First Indonesian record of probable brood parasitism of Chestnut-winged Babbler *Stachyris erythroptera* by Asian Drongo-cuckoo *Surniculus lugubris*. Kukila 14:51–53

Mundy PJ, Cook AW (1977) Observations on the breeding of the Pied Crow and Great Spotted Cuckoo in northern Nigeria. Ostrich 48:72–84

Nahid MI, Begum S, Feeroz MM (2016) Brood parasitic cuckoos and their hosts in Jahangirnagar University Campus. Ind Birds 12:64–69

Naumberg EMB (1930) The birds of Mato Grosso, Brazil. A report on the birds secured by the Roosevelt-Rondon Expedition. With field notes by George K. Cherrie. Bull Am Mus Nat Hist 60:1–432

Neunteufel A (1951) Observaciones sobre el *Dromococcyx pavoninus* Pelzeln y el parasitismo de los cucúlidos. Hornero 9:288–290

Nicéforo H, Olivares A (1966) Adiciones a la avifauna Columbiana—III (Columbidae – Caprimulgidae). Bol Soc Venez Cienc Nat 26:370–393

Nicholson FG (1897) Eggs of South African golden cuckoo in nest of Cape Wagtail. Zoologist (Series 4) 1:142–143

Nickell WP (1954) Redwings hatch and raise a Yellow-billed Cuckoo. Wilson Bull 60:137–138

Nicolai J (1969) Beobachtungen an Paradieswitwen (*Steganura paradisaea* L., *Steganura obtusa* Chapin) und der Strohwitwe (*Tetraenura fischeri* Reichenow) in Ostafrika. J Ornithol 110: 421–447

Nicolai J (1974) Mimicry in parasitic birds. Sci Am 231:92–98

Nicolai J (1977) Der Rotmaskenastrid (*Pytilia hypogrammica*) als Wirt der Togo-Paradiswitwe (*Steganura togoensis*). J Ornithol 118:175–188

Nicolai J (1987) Die Rachenzeichnungen von Großem und Kleinem Pünktchenamarant (*Lagonosticta nitidula* und *L. rufopicta*). Trochilus 8:116–120

Nicolai J (1991) Vom Buntastrild (*Pytilia melba*) und seinen Brutparasiten. Gefiederte Welt 115: 369–372

Niemcynowicz A, Świetochowski P, Zalewski A, Chętnicki W (2015) Facultative interspecific brood parasitism in colonial breeding waterbirds in Biebrza National Park, Poland. Waterbirds 38:282–289

Nolan V, Thompson CF (1975) The occurrence and significance of anomalous reproductive activities in two North American non-parasitic cuckoos. Ibis 117:496–503

Noske RA (2001) The breeding biology of the Mangrove Gerygone, *Gerygone laevigaster*, in the Darwin region, with notes on brood parasitism by the Little Bronze—cuckoo, *Chrysococcyx minutillus*. Emu 101:129–135

Oatley TB (1980) Eggs of two cuckoo genera in one nest, and a new host for Emerald Cuckoo. Ostrich 51:126–127
Oberholser HC (1974) Birdlife of Texas, vol 1. University of Texas Press, Austin
Odin CR (1957) California gull predation on water-fowl. Auk 74:185–202
Oliver WRB (1955) New Zealand birds. AH and AW Reed, Wellington
Ottow J, Verheijen JAJ (1969) Zur Lebensweise der Kuckucke von Flores. J Ornithol 110:27–29
Palmer RS (1976) Handbook of North American birds, vol 2: Waterfowl, part 1. Yale University Press, New Haven, p i-xi, 1-521
Parker SA (1966) Coracias abyssinica: an unrecorded host of *Indicator indicator*. Bull Br Ornithol Club 86:81
Parker SA (1981) Prolegomon to further studies in the *Chrysococcyx* "*malayanus*" group (Aves, Cuculidae). Zool Verh Leiden 187:1–56
Payne RB (1973) Behavior, mimetic songs and song dialects, and relationships of the parasitic indigobirds (Vidua) of Africa. Ornitho Monogr 11
Payne RB (1977) The ecology of brood parasitism in birds. Annu Rev Ecol Syst 8:1–28
Payne RB (1982) Species limits in the indigobirds (Ploceidae, Vidua) of West Africa: mouth mimicry, song mimicry, and description of new species. Misc Publ Mus Zool Univ Michigan USA 162:1–96
Payne RB (1996) Field identification of the indigobirds. Bull Afr Bird Club 3:14–25
Payne RB (1997) Family Cuculidae. In: del Hoyo J, Elliott A, Sargatal J (eds) Handbook of the birds of the world, vol 4. Barcelona, Lynx Edicions, pp 508–607
Payne RB (1998) A new species of firefinch *Lagonosticta* from northern Nigeria and its association with the Jos Plateau Indigobird *Vidua maryae*. Ibis 140:368–381
Payne RB (2004) Family Viduidae: whydahs, indigobirds, Cuckoo-Finch. In: Fry CH, Keith S (eds) The birds of Africa, vol 7. Princeton University Press, Princeton, pp 416–451
Payne RB (2005) The cuckoos. Bird families of the world. Oxford University Press, Oxford
Payne RB (2010) Family Viduidae. In: del Hoyo J, Elliott A, Christie DA (eds) Handbook of the birds of the world, vol 4. Barcelona, Lynx Edicions, pp 198–232
Payne R (2016a) Cuckoo Finch (*Anomalospiza imberbis*). In: del Hoyo J, Elliott A, Sargatal J, Christie DA, de Juana E (eds) Handbook of the birds of the world alive. Lynx Edicions, Barcelona. http://www.hbw.com/node/61214. Accessed 13 Aug 2016
Payne R (2016b) Striped Cuckoo (*Tapera naevia*). In: del Hoyo J, Elliott A, Sargatal J, Christie DA, de Juana E (eds) Handbook of the birds of the world alive. Lynx Edicions, Barcelona. http://www.hbw.com/node/54910. Accessed 24 Sept 2016
Payne R (2016c) Pavonine Cuckoo (*Dromococcyx pavoninus*). In: del Hoyo J, Elliott A, Sargatal J, Christie DA, de Juana E (eds) Handbook of the birds of the world alive. Lynx Edicions, Barcelona. http://www.hbw.com/node/54912. Accessed 24 Sept 2016
Payne R (2016d) Thick-billed Cuckoo (*Pachycoccyx audeberti*). In: del Hoyo J, Elliott A, Sargatal J, Christie DA, de Juana E (eds) Handbook of the birds of the world alive. Lynx Edicions, Barcelona. http://www.hbw.com/node/54789. Accessed 24 Sept 2016
Payne R (2016e) Channel-billed Cuckoo (*Scythrops novaehollandiae*). In: del Hoyo J, Elliott A, Sargatal J, Christie DA, de Juana E (eds) Handbook of the birds of the world alive. Lynx Edicions, Barcelona. http://www.hbw.com/node/54834. Accessed 24 Sept 2016
Payne R (2017a) Cuckoo-finch (*Anomalospiza imberbis*). In: del Hoyo J, Elliott A, Sargatal J, Christie DA, de Juana E (eds) Handbook of the birds of the world alive. Lynx Edicions, Barcelona. http://www.hbw.com/node/61214. Accessed 29 May 2017
Payne R (2017b) Long-tailed Paradise-whydah (*Vidua paradisaea*). In: del Hoyo J, Elliott A, Sargatal J, Christie DA, de Juana E (eds) Handbook of the birds of the world alive. Lynx Edicions, Barcelona. http://www.hbw.com/node/61217. Accessed 28 May 2017
Payne R (2017c) Broad-tailed Paradise-whydah (*Vidua obtusa*). In: del Hoyo J, Elliott A, Sargatal J, Christie DA, de Juana E (eds) Handbook of the birds of the world alive. Lynx Edicions, Barcelona. http://www.hbw.com/node/61218. Accessed 28 May 2017

Payne R (2017d) Pin-tailed Whydah (*Vidua macroura*). In: del Hoyo J, Elliott A, Sargatal J, Christie DA, de Juana E (eds) Handbook of the birds of the world alive. Lynx Edicions, Barcelona. http://www.hbw.com/node/61215. Accessed 28 May 2017

Payne R (2017e) Shaft-tailed Whydah (*Vidua regia*). In: del Hoyo J, Elliott A, Sargatal J, Christie DA, de Juana E (eds) Handbook of the birds of the world alive. Lynx Edicions, Barcelona. http://www.hbw.com/node/61223. Accessed 28 May 2017

Payne R (2017f) Straw-tailed Whydah (*Vidua fischeri*). In: del Hoyo J, Elliott A, Sargatal J, Christie DA, de Juana E (eds) Handbook of the birds of the world alive. Lynx Edicions, Barcelona. http://www.hbw.com/node/61222. Accessed 28 May 2017

Payne R (2017g) Green Indigobird (*Vidua codringtoni*). In: del Hoyo J, Elliott A, Sargatal J, Christie DA, de Juana E (eds) Handbook of the birds of the world alive. Lynx Edicions, Barcelona. http://www.hbw.com/node/61233. Accessed 29 May 2017

Payne R (2017h) Village Indigobird (*Vidua chalybeata*). In: del Hoyo J, Elliott A, Sargatal J, Christie DA, de Juana E (eds) Handbook of the birds of the world alive. Lynx Edicions, Barcelona. http://www.hbw.com/node/61224. Accessed 29 May 2017

Payne R, Bonan A (2016) Whydahs and Indigobirds (Viduidae). In: del Hoyo J, Elliott A, Sargatal J, Christie DA, de Juana E (eds) Handbook of the birds of the world alive. Lynx Edicions, Barcelona. http://www.hbw.com/node/52374. Accessed 13 Aug 2016

Payne R, de Juana E (2016a) Jacobin Cuckoo (*Clamator jacobinus*). In: del Hoyo J, Elliott A, Sargatal J, Christie DA, de Juana E (eds) Handbook of the birds of the world alive. Lynx Edicions, Barcelona. http://www.hbw.com/node/54785. Accessed 24 Sept 2016

Payne R, de Juana E (2016b) Long-tailed Koel (*Urodynamis taitensis*). In: del Hoyo J, Elliott A, Sargatal J, Christie DA, de Juana E (eds) Handbook of the birds of the world alive. Lynx Edicions, Barcelona. http://www.hbw.com/node/54833. Accessed 24 Sept 2016

Payne R, Garcia EFJ (2016) Great Spotted Cuckoo (*Clamator glandarius*). In: del Hoyo J, Elliott A, Sargatal J, Christie DA, de Juana E (eds) Handbook of the birds of the world alive. Lynx Edicions, Barcelona. http://www.hbw.com/node/54788. Accessed 24 Sept 2016

Payne R, Kirwan GM (2016) Large Hawk-cuckoo (*Hierococcyx sparverioides*). In: del Hoyo J, Elliott A, Sargatal J, Christie DA, de Juana E (eds) Handbook of the birds of the world alive. Lynx Edicions, Barcelona. http://www.hbw.com/node/54791. Accessed 28 Sept 2016

Payne RB, Hustler K, Stjernstedt R, Sefc KM, Sorenson MD (2002) Behavioural and genetic evidence of a recent population switch to a novel host species in brood-parasitic indigobirds *Vidua chalybeata*. Ibis 144:373–383

Payne RB, Barlow CR, Balakrishnan CN, Sorenson MD (2005) Song mimicry of Blackbellied Firefinch *Lagonosticta rara* and other finches by the brood-parasitic Cameroon Indigobird *Vidua camerunensis* in West Africa. Ibis 147:130–143

Payne R, de Juana E, Kirwan GM (2016) Little Bronze-cuckoo (*Chalcites minutillus*). In: del Hoyo J, Elliott A, Sargatal J, Christie DA, de Juana E (eds) Handbook of the birds of the world alive. Lynx Edicions, Barcelona. http://www.hbw.com/node/54817. Accessed 26 Sept 2016

Pemberton JR (1925) Parasitism in the Road-runner. Condor 27:35

Pennard FP, Pennard AP (1910) De Vogels van Guyana (Suriname, Cayenne en Demerara), vol 2. FP Pennard, Paramaribo

Peters JL (1940) Check-list of birds of the world, vol 4. Harvard University Press, Cambridge, MA, pp i–xii, 1–291

Phillips JC (1923–1926) A natural history of the ducks. Houghton Mifflin, Boston

Phillipps Q (1970) Some important nesting notes from Sabah. Sabah Soc J 8:153–154

Pitman CRS (1961) The Kurrichane Thrush *Turdus libonyanus tropicalis* Peters a host of the Red-chested Cuckoo *Cuculus solitarius* Stephens in Southern Rhodesia. Bull Br Ornithol Club 81:48–49

Plumb WJ (1976) Observations on Hinde's Babbler *Turdoides hindei*. Scopus 3:61–67

Powell A (1979) Cuckoo in the nest. Wildfowl News 81:15–16

Prevett JP, Lieff BC, Maclnnes CD (1972) Nest parasitism at McConnell River. N.W.T. Can Field Nat. 86:369–372

Pryce E (1989) A Black Cuckoo raised by Swamp Boubous. Babbler 18:38
Raney EC (1939) Robin and mourning dove use same nest. Auk 56:337–338
Rasamison AS, Woog F (2012) First record of Souimanga Sunbird *Cinnyris souimanga* as a host of Madagascar Lesser Cuckoo *Cuculus rochii*, at Ankamenabe, Madagascar. Bull Afr Bird Club 19:56–58
Reeve AH, Mittermeir JC, Fabre P-H, Rosyadi I, Kennedy JD, Haryoko T (2015) New additions to the avifauna of Obi Island, Indonesia, with comments on migration and breeding seasonality of Moluccan birds. Forktail 31:98–102
Richardson D (1983) Didric Cuckoo: possible parasitizing of Malachite Sunbird. Promerops 157:12
Ridgely RS, Gwynne JA (1989) A guide to the birds of Panama, with Costa Rica, Nicaragua and Honduras, 2nd edn. Princeton University Press, Princeton
Robson C (2000) A field guide to the birds of Southeast Asia. Princeton University Press, Princeton
Robson C (2008) A field guide to the birds of Southeast Asia, 2nd edn. New Holland, London
Rogers A (1992) Little Grebe apparently imprinted on female Mallard. Br Birds 85:466
Rowan MK (1983) Doves, parrots, louries and cuckoos of southern Africa. Croom Helm, London
Rowley I, Russell E (1989) Splendid Fairy-wren. In: Newton I (ed) Lifetime reproduction in birds. Academic Press, London, pp 233–252
Rozendaal FG, Dekker RWRJ (1989) Annotated checklist of the birds of the Dumoga-Bone National Park, North Sulawesi. Kukila 4:85–109
Safford R, Hawkins F (eds) (2013) The birds of Africa, vol VIII. The Malagasy region, Helm
Salvador SA (1982) Estudio de paratismo del Crespin Tapera naevia chochi (Vieillot) (Aves Cuculidae). Hist Nat 2:65–70
Sargent J (1886) Grey Hen and Partridge laying in the same nest. Zoologist 10:298
Schellekens M (2009) The Timor Leaf-Warbler Phylloscopus presbytes is a probable host of the Sunda Cuckoo Cuculus lepidus. Kukila 14:49–50
Schmutz JA (1988) Ring-necked Pheasant parasitism of Wild Turkey nests. Wilson Bull 100: 508–509
Schönwetter M (1967) Handbuch der Oologie, band 1. Nonpasseres. Akademie-Verlag, Berlin
Seaton C (1962) The Yellow-breasted Sunbird as a host to the Rufous-breasted Bronze-Cuckoo. Emu 62:174–176
Sénéchal H, Gauthier G, Savard JP-L (2008) Nesting ecology of Common Goldeneyes and Hooded Mergansers in a boreal river system. Wilson J Ornithol 120:732–742
Serié P (1923a) Un huevo de pirincho *Guira guira* en un nido de *Phytotoma rutila*. Hornero 3:100
Serié P (1923b) Huevos de pirincho en nido de chimango. Hornero 3:189
Serventy DL, Whittell HM (1962) Birds of Western Australia. Paterson Brokensha Ltd, Perth
Sethi VK, Saxena V, Bhatt D (2006) An instance of the Asian Koel *Eudynamys scolopacea* destroying the nest of a Black-headed Oriole *Oriolus xanthornus*. Ind Birds 2:175–176
Shao ST (2017) Two photographs. https://www.facebook.com/photo.php?fbid=10212882965647867&set=a.2043396283480.2131449.1201547363&type=3&theater. Accessed 12 Jun 2017
Sharp WM, McClure HE (1945) The pheasant in the sandhill region of Nebraska. In: McAtee WL (ed) The ring-necked pheasant and its management in North America. American Wildlife Institute, Washington, pp 203–233
Sharpe RB (1889) On the ornithology of Northern Borneo. Ibis (6th series) 1:63–85, 185–205, 265–283, 409–443
Shields AM (1899) Nesting of the fulvous tree duck. Bull Cooper Ornithol Club 1:9–11
Shirihai H (1996) The birds of Israel. Academic Press, Unipress, London
Short LM, Horne JFM (1983) The relationship of the male Lesser Honeyguides *Indicator minor* with duetting barbet pairs. Bull Br Ornithol Club 103:25–32
Short LM, Horne JFM (1985) Behavioral notes on the nest-parasitic Afrotropical honeyguides (Aves: Indicatoridae). Am Mus Nov 2825
Short LM, Horne JFM (1988) Indicatoridae, honeyguides. In: Fry CH, Keith S, Urban EK (eds) The birds of Africa, vol III. Academic Press, London, pp 486–511

Short LM, Horne JFM (2001) Toucans, barbets and honeyguides. Ramphastidae, Capitonidae and Indicatoridae. Oxford University Press, Oxford

Short LM, Horne JFM (2016) Honeyguides (Indicatoridae). In: del Hoyo J, Elliott A, Sargatal J, Christie DA, de Juana E (eds) Handbook of the birds of the world alive. Lynx Edicions, Barcelona. http://www.hbw.com/node/52285. Accessed 13 Aug 2016

Short LL, Horne JFM (2017a) Zenker's Honeyguide (*Melignomon zenkeri*). In: del Hoyo J, Elliott A, Sargatal J, Christie DA, de Juana E (eds) Handbook of the birds of the world alive. Lynx Edicions, Barcelona. http://www.hbw.com/node/56107. Accessed 29 May 2017

Short LL, Horne JFM (2017b) Least Honeyguide (*Indicator exilis*). In: del Hoyo J, Elliott A, Sargatal J, Christie DA, de Juana E (eds) Handbook of the birds of the world alive. Lynx Edicions, Barcelona. http://www.hbw.com/node/56112. Accessed 29 May 2017

Sick H (1993) Birds in Brazil. A natural history. Princeton University Press, Princeton, p i–xvii, 1–703

Siebel DE (1988) A phylogenetic analysis of the Cuculiformes and Opisthocomus, based on postcranial skeletal characters. University of Kansas, Lawrence

Siegfried WR (1985) Nest-site fidelity of Malachite Sunbird and parasitism by Klaas's Cuckoo. Ostrich 56:277

Skead CJ (1950) A study of the Black-collared Barbet, *Lybius torquatus*. With notes on its parasitism by the Lesser Honeyguide. Ostrich 21:84–96

Slagsvold T (1998) On the origin and rarity of interspecific nest parasitism in birds. Am Nat 152: 264–272

Smith J, Surtees A (2001) Mute Swans raising young Canada Goose. Br Birds 94:438

Smythies BE (1953) The birds of Burma, 2nd edn. Oliver and Boyd, Edinburgh

Smythies BE (1957) An annotated checklist of the birds of Borneo. Sarawak Mus J 7:523–818

Smythies BE (1981) The birds of Borneo, 3rd edn (Earl of Cranbrook, ed). Sabah Society and Mal Nat Society, Kota Kinabalu and Kuala Lumpur

Smythies BE, Davison GWH (1999) The birds of Borneo, 4th edn. Natural History Publications, Kota Kinabalu

Snethlage E (1913) Über die Verbreitung der Vogelarten in Unteramazonien. J Ornithol 61:469–539

Snethlage E (1935a) Beiträge zur Brutbiologie brasiklianischer Vögel. J Ornithol 83:1–24

Snethlage E (1935b) Beiträge zur Fortpflanzungsbiologie Vögel. J Ornithol 83:532–562

Söderholm S (2005) Blandkull mellan grågås *Anser anser* och kanadagås *Branta canadensis* – boparasitism ell er kullsammanslagning? [Mixed brood of Greylag Goose *Anser anser* and Canada Goose *Branta canadensis* – nest parasitism or brood amalgamation?]. Ornithol Svecica 15:48–51 [in Swedish]

Soler M (1990) Relationships between the Great Spotted Cuckoo *Clamator glandarius* and its corvid hosts in a recently colonized area. Ornithol Scand 21:212–223

Soler M, Martínez JG, Soler JJ (1996) Effects of brood parasitism by the great spotted cuckoo on the breeding success of the magpie host: an experimental study. Ardeola 43:87–96

Solti B (2010) A Mátra Múzeum madártani gyüutej.nye III. Németh Márton toj.sgyüjtem.ny. Folia Historico Naturalia Musei Matraensis, Suppl 5 [in Hungarian]

Sorenson MD (1993) Parasitic egg laying in Canvasbacks—frequency, success, and individual behaviour. Auk 110:57–69

Sorenson MD, Payne RB (2001) A single ancient origin of brood parasitism in African finches: implications for host-parasite coevolution. Evolution 55:2550–2567

Sorenson MD, Payne RB (2002) Molecular genetic perspectives on avian brood parasitism. Integr Comp Biol 42:388–400

Sorenson MD, Sefc KM, Payne RB (2003) Speciation by host switch in brood parasitic indigobirds. Nature 424:928–931

Spencer HE Jr (1953) The cinnamon teal (*Anas cyanoptera* Vieillot) its life history, ecology, and management. Unpublished MS thesis, Utah State Agric Coll, Logan

Spottiswoode CN (1994) Lesser Honeyguide parasitizing Cardinal Woodpecker. Promerops 213:9

Spottiswoode CN, Colebrook-Robjent JFR (2007) Egg puncturing by the brood parasitic Greater Honeyguide and potential host counter adaptations. Behav Ecol 18:792–799

Spottiswoode CN, Koorevaar J (2011) A stab in the dark: chick killing by brood parasitic honeyguides. Biol Lett. https://doi.org/10.1098/rsbl.20110739

Spottiswoode CN, Stevens M (2011) How to evade a coevolving brood parasite: egg discrimination versus egg variability as host defences. Proc R Soc Lond B 278:3566–3573

Spottiswoode CN, Stryjewski KF, Quader S, Colebrook-Robjent JFR, Sorenson MD (2012) Ancient host specificity within a single species of brood parasitic bird. Proc Natl Acad Sci 108:17738–17742

St Paul R (1976) A bushman's seventeen years of noting birds, part D. Shining cuckoo and long-tailed cuckoo. Notornis 23:289–298

Steyn P, Myburgh N (1980) Red-chested Cuckoo parasitising Cape Siskin. Ostrich 51:53

Stoddard HL (1931) The bobwhite quail. Scribner's Sons, New York

Storrs HT (1944) Indian Cuckoo (*Cuculus micropterus micropterus*) laying in the nest of the Himalayan Black Drongo. J Bengal Nat Hist Soc 18:116–120

Štovičeк O, Kreisinger J, Javůrková V, Albrecht T (2013) High rate of conspecific brood parasitism by microsatellite analysis in a diving duck, the common pochard. J Av Biol 44:369–375

Strong RM (1912) Some observations on the life history of the red-breasted merganser. Auk 29: 479–488

Summers-Smith JD (1988) The sparrows. Poyser, Calton

Summers-Smith D (2016) Chestnut Sparrow (*Passer eminibey*). In: del Hoyo J, Elliott A, Sargatal J, Christie DA, de Juana E (eds) Handbook of the birds of the world alive. Lynx Edicions, Barcelona. http://www.hbw.com/node/60949. Accessed 11 Nov 2016

Sweetman JC (1979) Klaas's Cuckoo laying an egg in a Yellow-bellied Eremomela nest. Bull Zamb Ornithol Soc 11:47

Tanner WD, Hendrickson GO (1954) Ecology of the Virginia rail in Clay County, Iowa. Iowa Bird Life 24:65–70

Tarboton W (1975) African Cuckoo parasitizing Forktailed Drongo. Ostrich 46:186–188

Taylor B, Kirwan GM, Christie DA (2016) American Coot (*Fulica americana*). In del Hoyo J, Elliott A, Sargatal J, Christie DA, de Juana E (eds). Handbook of the birds of the world alive. Lynx Edicions, Barcelona. http://www.hbw.com/node/53697. Accesssed 8 Nov 2016

Tegala A, Stubbings EM (2013) Egg-dumping by Grey Partridge. Br Bird 106:285

Tegetmeier WB (1911) Pheasants, their natural history and practical management. Horace Cox, London

Temba EM, Cabeza M, Andrianarimisa A (2014) Breeding biology and reproductive success of the Spectacled Tetraka *Xanthomixis zosterops* (Bernieridae) in a rainforest of Madagascar. Ostrich 85:119–123

Thomas BT (1995) Black-billed Cuckoo parasitizes the nest of a Yellow-breasted Chat. Raven 66: 3–5

Uchida H (2011) Breeding biology of the Little Cuckoo *Cuculus poliocephalus*: a brood parasite specializing in one host species, the Japanese Bush Warbler *Cettia diphone*, in Honshu, Japan. Jpn J Ornithol 60:78–87

Valkenburg T (2015) Personal communication to CFM

Vernon CJ, Macdonald IA, Dean WRJ (1990) The birds of the Haroni-Lusitu. Honeyguide 36: 14–35

Vincent F (1965) Le Coucou émeraude (*Chrysococcyx cupreus intermedius* Hartlaub 1857) parasite d'un nid de Souimanga (Nectariniidae) au Congo (Brazzaville). Oiseau et RFO 35:81–86

von Ihering H (1914) Biologia e classificao das cuculidas brazilieras. Rev Mus Paulista 9:371–390

Waltert M, Faber K (2000) Olive-bellied Sunbird *Nectarinia chloropygia* host to Cassin's Honeyguide *Prodotiscus insignis*. Malimbus 22:86

Watling D (1982) The birds of Fiji, Tonga and Samoa. Millwood Press, Wellington, NZ

Weller MW (1959) Parasitic egg laying in the redhead (*Aythya americana*) and other North American Anatidae. Ecol Monogr 29:333–365

Weller MW (1967) Notes on some marsh birds of Cape San Antonio, Argentina. Ibis 109:391–416

Weller MW (1968) The breeding biology of the parasitic black-headed duck. Living Bird 7: 169–208

Wells DR (1999) The birds of the Thai-Malay Peninsula, vol 1. Non-passerines. London, Academic Press

Wells DR (2007) The birds of the Thai-Malay Peninsula, vol 2. Passerines. Christopher Helm, London

Wetmore A (1968) Birds of the Republic of Panama, part 2. Smithson Misc Coll 150:1–605

Widman O (1882) Cuckoos laying in the nests of other birds. Bull Nuttall Ornithol Club 7:56–57

Wiens JA (1965) Nest parasitism of the Dickcissel by the Yellow-billed Cuckoo in Marshall County, Oklahoma. Southw Nat 10:142

Wilkinson R, Dutson G, Sheldon B (1991) The avifauna of Barito Ulu, Central Borneo. ICBP Study Report 48

Williams GS, Nelson MC (1943) Management of the redhead duck in Utah. Unpublished report, US Fish and Wildlife Service, Washington, DC

Wilmé L (1994) Note sur le paratisme du Coucou de Madagascar Cuculus rochi. Working group on birds in the Madagascar Region Newsletter 3(2):4–5

Wilson RG (1992) Parasitism of Yellow-olive Flycatcher by the Pheasant Cuckoo. Euphonia 1: 34–36

Wilson JM, Sallinen P (2003) First records of Didric Cuckoo *Chrysococcyx caprius* parasitizing Cricket Warbler *Spiloptila clamans*. Malimbus 25:95–96

Wingfield BH (1951) A waterfowl productivity study in Knudson Marsh, Salt Lake Valley, Utah. MS thesis, Utah State Agric. Coll

Wolf KE (1953) Grebe-duck nesting parasitism. Auk 70:84

Wolfe DH (1994) Yellow-billed Cuckoo hatched in Mourning Dove nest. Bull Okla Ornithol Soc 27:29–30

Wolff S (1976) Egg dumping by the Helmeted Guineafowl. Bokmakierie 28:97–98

Wyllie I (1981) The Cuckoo. Universe Books, New York

Yamagishi S, Fujioka M (1986) Heavy brood parasitism by the Common Cuckoo *Cuculus canorus* on the Azure-winged Magpie *Cyanopica cyana*. Tori 34:91–96

Yamauchi A (1995) Theory of evolution of nest parasitism in birds. Am Nat 145:434–456

Yang C, Liang W, Antonov A, Cai Y, Stokke BG, Fossøy F, Moksnes A, Røskaft E (2012a) Diversity of parasitic cuckoos and their hosts in China. Chinese Birds 3:9–32

Yang C, Antonov A, Cai Y, Stokke BG, Moksnes A, Røskaft E, Liang W (2012b) Large Hawk-Cuckoo Hierococcyx sparverioides parasitism on the Chinese Babax Babax lanceolatus may be an evolutionarily recent host-parasite system. Ibis 154:200–204

Yang C, Su T, Liang W, Møller AP (2015) Coevolution between the large hawk-cuckoo (*Cuculus sparverioides*) and its two sympatric Leiothrichidae hosts: evidence for recent expansion and switch in host use? Biol J Lin Soc 115:919–926

Yasukawa R (2010) Yellow-billed Cuckoo hatched and fed by a Red-winged Blackbird. Wilson J Ornithol 122:402–405

Yom-Tov Y, Geffen E (2006) On the origins of brood parasitism in altricial birds. Behav Ecol 17: 196–205

Zheng-wang Z (1995) Studies on habitat and management of Brown-eared Pheasants in China. Ann Rev WPA 2:28–31

Zicus MC (1990) Nesting biology of Hooded Mergansers using nest boxes. J Wildl Man 1990: 637–643

Part II

Conspecific Brood Parasitism

Conspecific Brood Parasitism Among Birds: The Effects of Phylogeny, Mode of Reproduction and Geographic Distribution

5

Yoram Yom-Tov and Eli Geffen

Abstract

We updated the list of species for which conspecific brood parasitism (CBP) has been reported, and it contained 256 species. We used this list to examine the effects of four factors on the rate of CBP (defined as the percentage of species where CBP was observed in a family), namely, phylogeny, geographic distribution, mode of reproduction (i.e. altricial/precocial) and diversification rate (mean number of species per genus in a family). CBP is influenced by phylogeny, appears to have multiple origins and has evolved many times in various branches of the avian phylogeny tree. As already well known, it is much more prominent among precocial than altricial birds, in which in some families of the former, most species (around 60% among *Anatidae*) and in some small families, even all species engage in CBP. In contrast, among altricial birds the highest levels of CBP are much lower and are found in the *Ploceidae* (14.5% of species) and *Turdidae* (12.8%). Of the three other examined parameters, only the mode of reproduction was significantly related to the rate of CBP, but accounted for only ~2% of the variation. Diversification rate and geographic distribution area were not correlated with CBP. These results further confirm that CBP is influenced by phylogeny and mode of reproduction. Literature searches revealed that the rate of appearance of articles on CBP typically increased for several years, starting in the late 1970s and early 1980s, levelled off during the first decade of the twenty-first century and then declined. This would seem to represent a recent decline in interest in the CBP phenomenon.

Electronic Supplementary Material The online version of this chapter (https://doi.org/10.1007/978-3-319-73138-4_5) contains supplementary material, which is available to authorized users.

Y. Yom-Tov (✉) · E. Geffen
Department of Zoology, Tel Aviv University, Tel Aviv, Israel
e-mail: yomtov@post.tau.ac.il

M. Soler (ed.), *Avian Brood Parasitism*, Fascinating Life Sciences,
https://doi.org/10.1007/978-3-319-73138-4_5

5.1 Introduction

Conspecific brood parasitism (CBP) in birds was previously considered to be rare compared to interspecific parasitism (Payne 1977). This view, however, has since been reversed, and CBP has been shown to be a more common phenomenon than obligate interspecific brood parasitism (IBP) (Yom-Tov 2001). Despite this, it seems that much more is known about IBP than about CBP.

CBP species tend to be precocial (Yom-Tov 2001), to have a large clutch size (a covariant of being precocial) and to be hole nesters (Geffen and Yom-Tov 2001). Additional factors, which can be defined as temporal, such as the proportion of unmated females in the population, females that have lost their nests and the availability of nesting sites, may also affect the rate of CBP (Yom-Tov 1980; Sorenson 1992), as well as other factors (i.e. relatedness; Andersson 2001; Lyon and Eadie 2000), that have been discussed in a thorough review by Lyon and Eadie (2008).

The various factors that affect the rate of CBP have often been studied according to the proportion of bird species that were reported to engage in CBP, but rarely by taking into consideration their phylogeny (but see Yom-Tov and Geffen 2006). In this chapter we examined the effects of four factors on the rate of CBP, namely, phylogeny, geographic distribution, mode of reproduction (i.e. altricial/precocial) and diversification rate.

In addition to the above, we examined the question of the historical development of the study of CBP in birds. The CBP phenomenon was only occasionally studied prior to the 1980s, but since then it has attracted the attention of many ornithologists.

Regarding CBP, our hypotheses were: (1) since precocial species tend to concentrate on certain branches of the avian phylogeny tree, the rate of CBP will be related to phylogeny; (2) since precocial species tend to form a higher proportion of the tropical (and probably southern) avifauna than of the northern avifauna (Ar and Yom-Tov 1978), and precocials tend to have a high proportion of CBP, the rate of CBP will be related to geographical distribution; and (3) since geographical diversification rates are hemispheric rather than latitudinal, with bird assemblages in Asia, North America and southern South America containing a disproportionate number of species from recent rapid radiations (Jetz et al. 2012), this factor will not be related to CBP rates. This is logical, because related species tend to have similar breeding habits.

5.2 Materials and Methods

Our analysis is based on the list of species reported for CBP in Yom-Tov (2001), which we updated by searching the literature for new reports on the occurrence of CBP among new species. CBP rate differs greatly between species and even between populations of the same species, depending on various parameters such as population density and environmental factors such as climate (e.g. high rainfall may cause flooding of nests, thus forcing laying females to lay in other females' nests). Hence, we decided to include in our analyses all species for which CBP was reported, even those where CBP might be accidental.

Data on the number of genera and species per family were gathered from the series *Birds of the World* (del Hoyo et al. 1992–2011). The rate of diversification was calculated by dividing the number of extant species by the number of extant genera (Jetz et al. 2012). To analyze the effect of geographical distribution, species were sorted into three categories: southern (including the tropics), northern or both (cosmopolitan). Each CBP species was also sorted into one of two categories of mode of reproduction: precocial or altricial. This later division followed Nice's (1962) classification of the mode of reproduction into eight categories. Accordingly, we sorted precocial and semi-precocial species as precocial and altricial and semi-altricial as altricial.

We used a family-level phylogenetic tree for the birds published in Jetz et al. (2012). The level of CBP for each node on the tree was reconstructed using the parsimony reconstruction method (i.e. selecting the ancestral CBP level that minimizes the linear cost of change from one state to another given the tree and observed CBP distribution), implemented using the program Mesquite (version 3.10; Maddison and Maddison 2016). To correlate between CBP level (dependent variable) and the other independent variables (geographical distribution, mode of reproduction, rate of diversification), we used the independent contrasts method (Felsenstein 1985) and the PDAP package (Midford et al. 2002) in Mesquite.

The historical development of the study of CBP in birds was examined by searching the Web of Science (WoS) for annual appearance of the terms "intraspecific nest parasitism", "intraspecific brood parasitism", "conspecific nest parasitism" and "conspecific brood parasitism" in the titles and abstracts of articles that appeared in the zoological literature between 1965 and August 2016. Theoretically, an abstract could contain two or more of the above four phrases, but this is unlikely, and we assume that the total number of articles engaging with the above phenomenon is the sum of these four phrases. The search output comprised mostly birds, but also other animals.

Finally, we used our updated list in order to trace the number of new species for which CBP was reported.

5.3 Results and Discussion

Our data enabled us to reconstruct the ancestral state of every node of the phylogenetic tree. CBP appears to have multiple origins (Fig. 5.1). It has evolved many times in various branches of the bird phylogeny tree, but, as already shown by Yom-Tov (1980), it is much more prominent among precocial than altricial birds. Among precocial birds, some families demonstrate a very high level of CBP (Fig. 5.1; around 60% among *Anatidae*) and in some small families, all species are engaged in CBP. In contrast, among altricial birds the highest levels of CBP are found in the *Turdidae* (12.8%) and *Ploceidae* (14.5% of species). Due to this large difference in CBP, occurrence between precocial and altricial birds and the fact that most

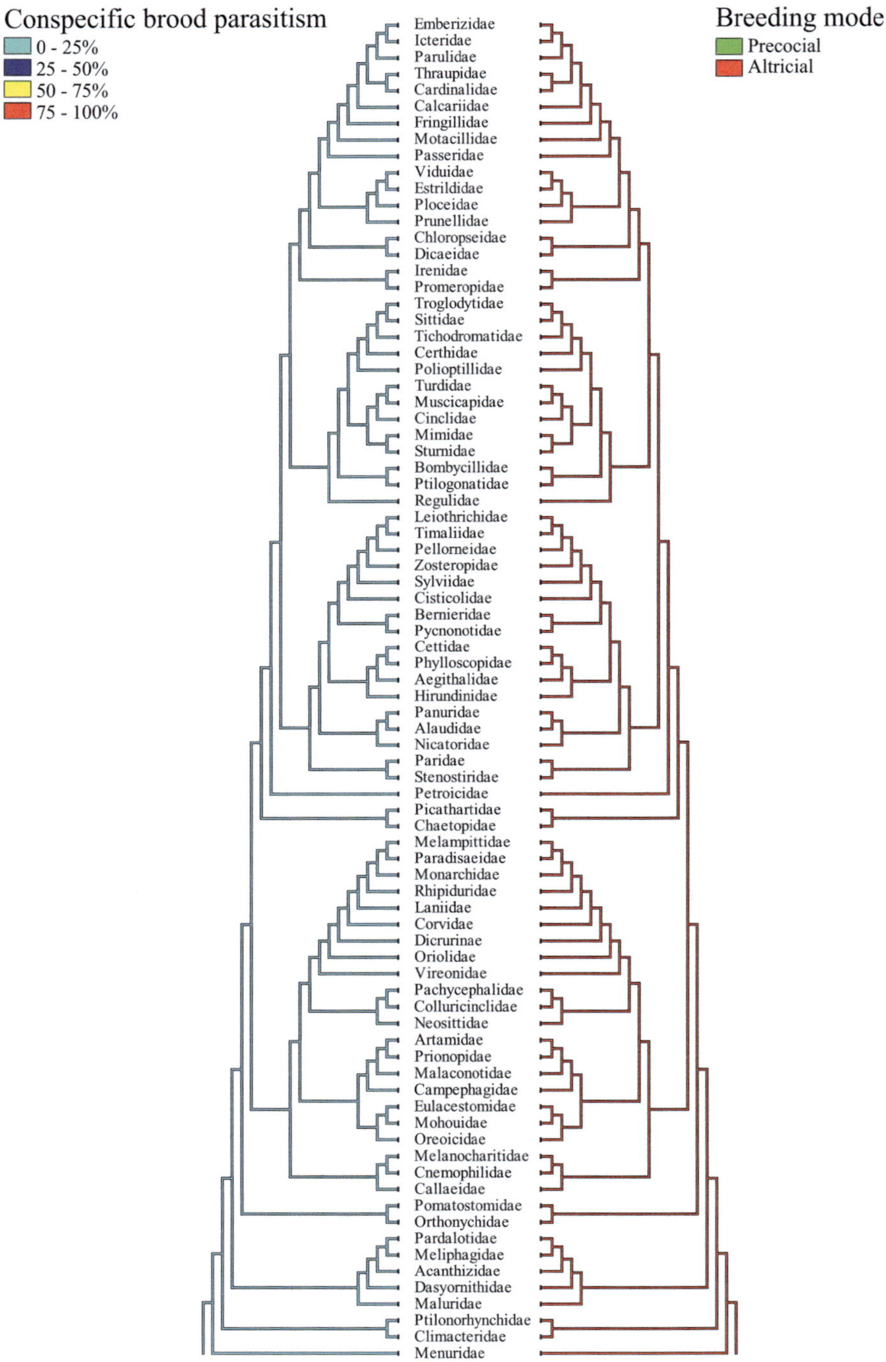

Fig. 5.1 Mirrored bird phylogeny showing the ancestral state reconstruction for the rate of CBP (Conspecific Brood parasitism, left) and the mode of reproduction (right). The rate of CBP is expressed as a percentage of the species engaging in CBP from all species in each family

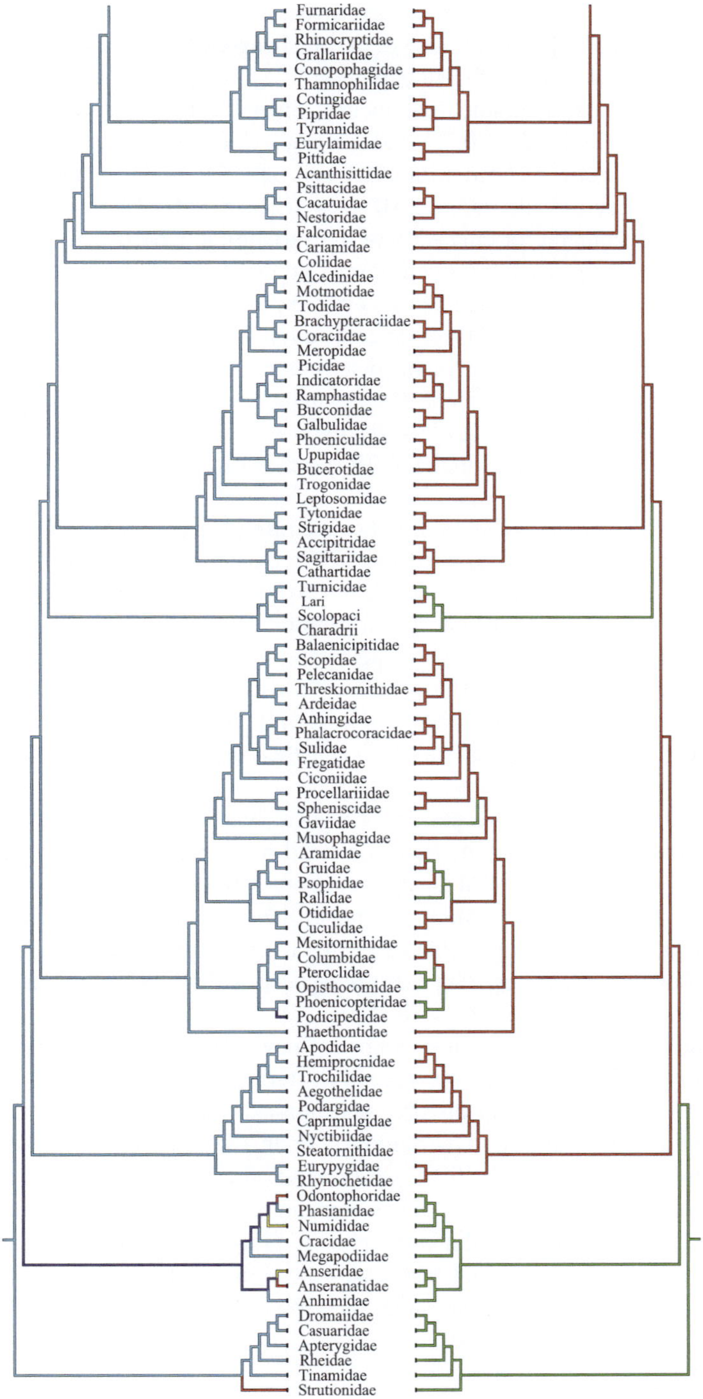

Fig. 5.1 (continued)

precocial bird families are clustered near the root of the phylogeny (Fig. 5.1), a more detailed division of CBP occurrence does not change the inner node reconstruction or show a different pattern of CBP evolution.

Of the three other examined parameters, diversification rate (177 independent contrasts, $r = -0.028$, $P = 0.72$) and geographic distribution area ($r = -0.036$, $P = 0.64$) were not correlated with CBP, and only the mode of reproduction was significantly related to the rate of CBP ($r = 0.16$, $P = 0.04$). However, mode of reproduction accounted for only ~2% of the variation in CBP.

To evaluate the possible bias contributed by families with few species and high level of CBP, we repeated the analysis after excluding all families with one genus and fewer than five species. Both diversification rate (145 independent contrasts, $r = -0.034$, $P = 0.66$) and geographic distribution area ($r = -0.026$, $P = 0.76$) were not correlated with CBP. The mode of reproduction was related to the rate of CBP as before ($r = 0.15$, $P = 0.07$), but this trend was no longer significant.

These results further confirm that CBP is influenced by phylogeny (Fig. 5.1) and mode of reproduction.

The results of the above WoS searches are provided in Fig. 5.2. Articles of any incorporating the four phrases used for CBP typically increased for several years, levelled off and then declined. The first two phrases above ("intraspecific nest parasitism" [n = 414] and "intraspecific brood parasitism" [n = 424]) showed increased use from 1974 to 1985, respectively, and levelled off around 2002 and 2007, respectively, while the last two terms ("conspecific nest parasitism" [n = 312] and "conspecific brood parasitism" [n = 302]) increased from 1985 to 1987, respectively, until levelling off in 2012 and then declining. The total number of articles found was 1473.

We suggest that the above pattern of a sharp increase, levelling off and then a decline represents what may be termed a trend in interest. It would seem that the publication of the first review of the subject (Yom-Tov 1980) generated an interest in CBP among ornithologists. Many scientists then re-examined at their past and present data, searching for possible cases of CBP and published them, followed by studying and experimenting with various aspects of CBP, until the trend declined and abated.

It is interesting to note that the number of new species in which CBP had been found and published was 82, 40, 12 and 9, during the 1980s, 1990s, 2000s and 2010–2016. This has several possible and non-exclusive reasons. First, since the number of species within a given family is finite, the more one searches, the less likely it is to find additional CBP cases because all possibilities have already been exhausted. Second, most ornithologists work in the northern hemisphere, whereas the majority of bird species inhabit the tropics. Thus, the likelihood of finding new cases diminishes the more that new cases are found. Third, it is possible that a decline in interest in this research subject is responsible for at least part of the decline in discovery of new CBP species.

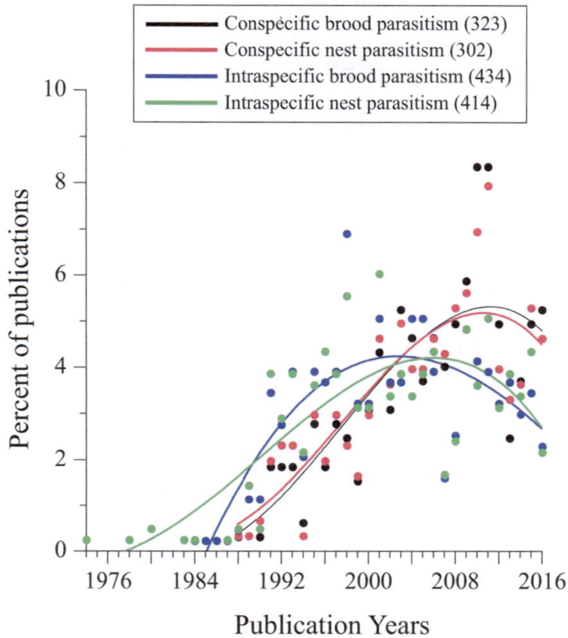

Fig. 5.2 The annual percentage of studies published between 1974 and 2016 sorted according to the following four search phrases: "intraspecific nest parasitism", "intraspecific brood parasitism", "conspecific nest parasitism" and "conspecific brood parasitism". The curves were fitted to the data using quadratic equations, because they explained more of the variation than linear ones (for the 4 relationships average $R^2 = 0.62$ and $R^2 = 0.48$, respectively)

Concluding Remarks and Future Directions

We found that conspecific brood parasitism evolved many times in various branches of the bird phylogeny tree, appears to have multiple origins, and at present it is known to occur in 256 bird species. However, its rate is highly variable, and, as already shown, it is much more prominent among precocial than altricial birds. CBP is influenced by phylogeny. The mode of reproduction was significantly related to the rate of CBP, but accounted for only ~2% of the variation. Diversification rate and geographic distribution area were not correlated with CBP. Literature searches revealed that the rate of appearance of articles on CBP typically increased for several years, starting in the late 1970s and early 1980s, levelled off during the first decade of the twenty-first century and then declined. The rate of finding and publishing the existence of CBP also declined since the 1990s. This would seem to represent a recent decline in interest in the CBP phenomenon.

We strongly suggest that future work on this subject will be experimental, system specific and will examine the effects of various environmental factors and the availability of resources on the rate of CBP.

Acknowledgements We are grateful to Manolo Soler for inviting us to write a chapter for this book and for his constructive comments. We thank Fugo Takasu and especially Bruce Lyon for their helpful criticism and suggestions for the improvement of the article. We thank Naomi Paz for editing the article.

References[1]

Andersson M (2001) Relatedness and the evolution of conspecific nest parasitism. Am Nat 158:599–614

Ar A, Yom-Tov Y (1978) The evolution of parental care among birds. Evolution 32:655–669

Bent AC (1932) Life histories of North American Gallinaceous birds. US Nat. Mus. Bull. no. 162

Bird KL, Aldridge CL, Carpenter JE, Paszkowski CA, Boyce MS, Coltman DW (2013) The secret sex lives of sage-grouse: multiple paternity and intraspecific nest parasitism revealed through genetic analysis. Behav Ecol 24:29–38

Craig A, Feare C (2009) Family Sturnidae (starlings). In: del Hoyo J, Elliott A, Christie DA (eds) Handbook of the birds of the world – Vol 14, Bush-shrikes to old world sparrows. Lynx ediciones, Barcelona

Davanço PV, Souza LMS, Oliveira LS, Francisco MR (2012) Intraspecific brood parasitism of the pale-breasted thrush (*Turdus leucomelas*). Wilson J Ornithol 124:611–614

del Hoyo J, Elliott A, Christie DA (eds) (1992–2011) Handbook of the birds of the world. Lynx Edicions, Barcelona

Duda N, Chetnicki W, Jadwiszczak P (2003) Intra-specific nest parasitism in black-headed gull *Larus ridibundus*. Ardea 91:122–124

Felsenstein J (1985) Phylogenies and the comparative method. Am Nat 125:1–15

Geffen E, Yom-Tov Y (2001) Factors affecting the rate of intraspecific nest parasitism among Anseriformes and Galliformes. Anim Behav 62:1027–1038

Jamieson IG, McRae SB, Simmons RE, Trewby M (2000) High rates of conspecific brood parasitism and egg rejection in Coots and Moorhens in ephemeral wetlands in Namibia. Auk 117:250–255

Jetz W, Thomas GH, Joy JB, Hartmann K, Mooers AO (2012) The global diversity of birds in space and time. Nature 491:444–448

Lesobre, L, Lacroix F, Nuz E, Hingrat Y, Chalah T, Saint Jaime M (2010) Absence of male reproductive skew, along with high frequency of polyandry and conspecific brood parasitism in the lekking Houbara bustard Chlamydotis undulata undulata. J Avian Biol 41:117-127.

Ležalová-Piálková R (2011) Molecular evidence for extra-pair paternity and intraspecific brood parasitism in the black-headed gull. Intraspecific brood parasitism. J Ornithol 152:291–295

Ležalová-Piálková R, Honza M (2008) Responses of black-headed gulls *Larus ridibundus* to conspecific brood parasitism. J Ornithol 149:415–421

Lyon BE, Eadie JMA (2000) Family matters: kin selection and the evolution of conspecific brood parasitism. Proc Natl Acad Sci USA 97:12942–12944

Lyon BE, Eadie JMA (2004) An obligate brood parasite trapped in the intraspecific arms race of its hosts. Nature 432:390–393

Lyon BE, Eadie JMA (2008) Conspecific brood parasitism in birds: a life-history perspective. Annu Rev Ecol Evol Syst 39:343–363

Maddison WP, Maddison DR (2016) Mesquite: a modular system for evolutionary analysis. Version 3.10. http://mesquiteproject.org

Midford PET, Garland T Jr, Maddison W (2002) PDAP: PDTREE package for Mesquite, Version 1.00

Minias P, Minias A, Jarosław A, Dziadek J (2014) Occurrence of extra-pair paternity and intraspecific brood parasitism in the whiskered tern *Chlidonias hybrida*. Bird Study 61:130–134

Nice MM (1962) Development and behavior in precocial birds. Trans Linn Soc NY 8:1–211

Paillisson J-M, Latraube F, Marion L, Bretagnolle V (2008) Indirect evidence of conspecific nest parasitism in the colonial whiskered tern (*Chlidonias hybrida*). C R Biol 331:559–567

Payne RB (1977) The ecology of brood parasitism in birds. Annu Rev Ecol Syst 8:1–28

---

[1]This references list includes references mentioned in the text since the publication of Yom-Tov (2001).

Peer BD (2010) Conspecific brood parasitism by the Dickcissel. Wilson J Ornithol 122:186–187

Petersen MR, Grand JB, Dau CP (2000) Spectacled eider (*Somateria fischeri*). In: Poole A, Gill F (eds) The birds of North America, no. 547. USGS, Philadelphia, PA

Quillfeldt P, Masello J, Segelbacher G (2012) Extra-pair paternity in seabirds: a review and case study of thin-billed prions *Pachyptila belcheri*. J Ornithol 153:367–373

Riehl C (2010) A simple rule reduces costs of extragroup parasitism in a communally breeding bird. Curr Biol 20:1830–1833

Ryan PG, Dyer BM, Martin AP, Ward VL, Whittington PA, Williams AJ (2013) Supernormal clutches in southern African kelp gulls *Larus dominicanus vetula*. Ostrich 84:157–160

Scholl T, Winkel W, Lubjuhn T (2008) Molecular genetic evidence for mixed maternity in broods of the coal tit *Parus ater*. Vogelwarte 46:223–227

Sorenson MD (1992) Comment: why is conspecific nest parasitism more frequent in waterfowl than in other birds? Can J Zool 70:1856–1858

Sorenson MD (1995) Evidence of conspecific nest parasitism and egg discrimination in the Sora. Condor 97:819–821

Tucker AM, Dyer RJ, Huber SK, Bulluck LP (2016) Opportunistic conspecific brood parasitism in a box-nesting population of prothonotary warblers (*Protonotaria citrea*). Auk 133:298–307

Vedder O, Komdeur J, van der Velde M, Magrath MJL (2010) Conclusive evidence for conspecific brood parasitism in the blue tit *Cyanistes caeruleus*: a reply to Griffith et al. J Avian Biol 41:348–349

Weaver HB, Brown CR, Sealy SG (2004) Brood parasitism and egg transfer in cave swallows *(Petrochelidon fulva)* and cliff swallows (*P. pyrrhonota*) in South Texas. Auk 121:1122–1129

Wiebe KL, Kempenaers B (2009) The social and genetic mating system in flickers linked to partially reversed sex roles. Behav Ecol 20:453–458

Yom-Tov Y (1980) Intraspecific nest parasitism in birds. Biol Rev 55:93–108

Yom-Tov Y (2001) An updated list and some comments on the occurrence of intraspecific nest parasitism in birds. Ibis 143:133–143

Yom-Tov Y, Geffen E (2006) On the origin of brood parasitism in altricial birds. Behav Ecol 17:196–205

Why *Do* Birds Lay Eggs in Conspecifics' Nests?

6

Bruce E. Lyon and John M. Eadie

Abstract

Conspecific brood parasitism (CBP) resembles interspecific brood parasitism, except that parasitic females lay eggs in the nests of conspecifics. CBP is a female alternative reproductive behavior, and understanding its evolution requires a life history approach. We review studies that investigate life history aspects of CBP. One or more life history components have been examined for 56 species, 49 of which are relevant to the adaptive basis of CBP. Various approaches have been used to detect CBP, but there is increasing reliance on molecular methods. Molecular methods are not always foolproof, and false exclusions are incorrectly interpreted as CBP; future studies should employ multiple lines of evidence. Two types of parasites have been documented, often in the same species: non-nesting females without their own nest in a given year and nesting females that lay some of their eggs parasitically. There is no evidence for lifelong professional brood parasites; non-nesting females apparently adopt parasitism to make the best of a bad job, although we lack a clear understanding of the specific constraints that prevent these females from nesting or the costs that make restraint a better option. Parasitism by nesting females is common, surprisingly so in the precocial waterfowl, and is only rarely associated with breeding interruption or nest loss. Nesting females appear to use parasitism to enhance their reproductive output, but for

B. E. Lyon (✉)
Department of Ecology and Evolutionary Biology, University of California, Santa Cruz, CA, USA
e-mail: belyon@ucsc.edu

J. M. Eadie
Department of Wildlife, Fish and Conservation Biology, Animal Behavior and Ecology Graduate Groups, University of California, Davis, CA, USA
e-mail: jmeadie@ucdavis.edu

M. Soler (ed.), *Avian Brood Parasitism*, Fascinating Life Sciences,
https://doi.org/10.1007/978-3-319-73138-4_6

most species it is not clear why females don't lay the additional eggs in their own nest. A full understanding of this form of brood parasitism will require a better understanding of clutch size constraints.

6.1 Introduction

Conspecific brood parasitism (CBP), where parasitism occurs among females of the same species, might seem to be a rather simple analog of interspecific brood parasitism (IBP). There are indeed the same potential benefits to the parasites through emancipation from parental care, similar potential costs to the host of providing parental care to foreign offspring, and possibilities of diverse parasite and host tactics to promote or avoid successful parasitism, respectively. Yet, on closer examination, it quickly becomes evident that CBP represents a unique and intriguing breeding system on its own. The critical distinction between CBP and IBP is that CBP is a facultative form of brood parasitism—since CBP involves hosts of the same species—a population of pure parasites cannot logically exist. This simple fact underpins every aspect of CBP—how the behavior is adaptive and when it is expressed in populations, the unique coevolutionary dynamics that arise because a given individual can be both parasite and host, the possibility that relatedness between host and parasite might make the behavior a form of cooperation rather than parasitism, and the potential for the behavior to influence population dynamics through density-dependent and frequency-dependent influences.

These attributes of CBP make for a rich and challenging system of considerable potential for both theoretical and empirical research. Moreover, CBP is not a rare phenomenon—it occurs most frequently in birds and has now been documented in almost 250 species (Chap. 5). However, CBP has received only a fraction of interest afforded its ecological cousin IBP. Despite its widespread occurrence, detailed demographic studies of CBP have been conducted in only a few well-studied species. In most species where CBP has been documented, it is unclear even which females are pursuing this behavior or why. The dearth of information is somewhat surprising given that CBP occurs in more species than IBP (245 vs 100 species, Davies 2000, Chap. 5) and is not that much less common than cooperative breeding (358 species of birds; Ligon and Burt 2004)—yet the past several decades have been dominated by interest in cooperative breeding and interspecific brood parasitism and a deep understanding of those breeding systems has developed. Studies of CBP offer a similar but yet untapped opportunity.

The key question, and the focus of this chapter, is why do females engage in CBP in the first place? Why would natural selection favor a female who lays some or all of her eggs (in a given year) in the nests of other conspecific females? We have argued previously that CBP should be viewed through the lens of life history theory, and indeed in many cases, we believe that CBP can be considered as a classic clutch size problem—how many eggs to lay and when and where to lay them to maximize lifetime fitness (Lyon and Eadie 2008). At its essence, CBP represents a fundamental life history decision that allows females to diversify their reproductive options under varying ecological and physiological conditions (Lyon and Eadie 2008). For birds without CBP, there are only two options available to females that are physiologically capable of breeding: nest or refrain from nesting if the costs of nesting are high or the resources needed for nesting are limited. CBP increases the range of allocation options for females (Sorenson 1991; Lyon and Eadie 2008): rather than having to choose between all (nesting) or nothing (not breeding), CBP allows the intermediate option of laying some eggs without paying the full costs of establishing a nest, incubating the clutch, and providing care for the offspring. Similarly, nesting also comes with constraints: females in many species are capable of laying more eggs than they can raise in their own nests, and CBP allows nesting females to circumvent this constraint and increase their total production of offspring.

CBP is an intriguing life history phenomenon because it is an example of an alternative reproductive tactic (ART), and this opens the study of CBP to the well-developed conceptual framework for the evolution and ecology of alternative life histories (Field 1992; Brockmann 2001; Oliveira et al. 2008). Virtually all of the ART literature focuses on males, usually in a mating context. Female ARTs have received a small fraction of the attention lavished on their male counterparts (but see Henson and Warner 1997), possibly because female ARTs fall outside the purview of sexual selection (Lyon and Montgomerie 2012). Most female ARTs, and specifically CBP, involve allocation of eggs rather than matings and thus fall within the scope of classic life history theory (Andersson 1984; Eadie et al. 1988; Lyon 1993, 1998; Lyon and Eadie 2008; Andersson and Åhlund 2012). A balanced and richer understanding of ART evolution should include both a male and female perspective (Lyon and Montgomerie 2012).

Our goal in this chapter is to provide a comprehensive review of the empirical work conducted over the past several decades to describe and characterize CBP in birds and to provide some guidance on future research needs. In a previous review, we summarized a conceptual framework for the study of CBP (Lyon and Eadie 2008) that emphasized CBP as a life history problem. Our intent here is to summarize what we have learned in the past three decades about the adaptive basis of CBP from field studies and to determine where the gaps remain in our understanding of CBP as part of a flexible alternative female life history tactic. We focus specifically on the parasites, rather than the hosts, to gain a general understanding of why some females, sometimes, act as conspecific brood parasites and to determine what fitness gains they might obtain by doing so. This focus on parasites means that we do not cover quasi-parasitism, where host males sire the parasitic eggs in their nests, because quasi-parasitism is more of an explanation for why a host would accept parasitic eggs than why the female lays parasitically in the first place. Similarly, we do not consider the costs of parasitism, nor do we review parasite or host tactics (we consider these elsewhere). However, to fully address the question of why females lay eggs in conspecifics' nests, we do consider briefly the patterns of parasitism through females' entire lives and the role that kinship might play.

Fig. 6.1 Examples of some of the species where life history aspects of CBP have been particularly well studied. (**a**) Common moorhen. (**b**) American coot nest with two eggs (darker eggs) that have been laid by a conspecific brood parasite. (**c**) Colonial nesting cliff swallows, a species where parasites transfer their eggs to a host nest in their beak. (**d**) A study of redhead ducks, a species with both CBP and IBP, led to an influential life history framework for investigating CBP. (**e**) Two species of cavity-nesting ducks, Barrow's (shown here) and common goldeneyes, have been particularly well studied and motivated much of the early theory on CBP. (**f**) European starling, a cavity-nesting species where parasites appear to be non-nesting females unable to establish a nest of their own. All photos by Bruce Lyon

Our list of species was compiled based on our general familiarity of the field and from Web of Science searches with the following terms: conspecific brood parasitism (331 hits), intraspecific nest parasitism (422 hits), and egg dumping (175 hits). Because our focus in this chapter is on the adaptive basis of CBP for parasitic females, we included in our survey only studies that provide information relevant to life history aspects (i.e., we did not include studies that only report the frequency of parasitism). Based on our survey, information on one or more life history components of CBP is available for 56 species from 19 different families of birds. Figure 6.1 illustrates a few of the model species that have been particularly well studied. As is true of the distribution of CBP generally (Yom-Tov 2001), waterfowl are particularly well represented. Our survey is available as an online Appendix (see Lyon and Eadie 2017; posted at Figshare; DOI https://doi.org/10.6084/m9.figshare.4787872); here we provide summaries of the main patterns from the data compilation. We attempted to complete a thorough survey, but we acknowledge that some papers may have been overlooked. Further, our determination of whether a particular pattern or result was found in any given paper was based as closely as possible on the authors' conclusions, but occasionally our interpretation of the data differed, and we report our assessments in the online Appendix. Finally, in most cases we decided to report the results of our compilation in terms of the frequency of studies that reported a particular feature, rather than the frequency of species. We did this because in most

Table 6.1 Methods used to detect conspecific brood parasitism

Method	Count	% of methods ($N = 197$)	% of studies ($N = 106$)
Observational methods			
Laying patterns (L)	64	32	60
Clutch size (CS)	31	16	25
Direct observation (DO)	24	12	23
Egg features (EF)	21	11	20
Camera (C)	7	4	7
Genetic methods			
Microsatellite DNA (M)	19	10	18
Minisatellite DNA fingerprinting (DF)	8	4	8
Protein fingerprinting (PF)	11	6	10
Allozymes (AZ)	4	2	4
Egg white proteins (EWP)	2	1	2
Genetic polymorphism chicks (GP)	1	<1	1
Captive populations			
Semi-captive population (SCV)	1	<1	1
Captive population (CV)	4	2	4
Total	197		

Data are the frequency (number of studies) in which each method was used ($N = 106$ studies). Multiple methods may have been used in the same study. Abbreviations in parentheses refer to the codes used in the online Appendix

cases, results varied among studies, and there was no simple way to accurately represent the findings at the species level.

6.2 Methods for Detecting CBP and Why They Matter

Conspecific brood parasitism often requires different methods of detection than interspecific brood parasitism because CBP can be much more difficult to detect—distinguishing between eggs of different females of the same species can be far more challenging than distinguishing eggs of different species (Andersson 1984). Accordingly, a variety of methods have been used to detect CBP (Yom-Tov 1980; Eadie et al. 2010), including egg laying rates, eggs laid well after clutch size is complete, unusually large clutch size, egg features, and several genetic methods (Table 6.1).

Egg laying rates can be potentially powerful for detecting CBP. Females cannot lay more than one egg per day due to the physiology of egg production and laying, so the addition of two or more new eggs to a nest in 24 h is typically strong evidence that more than one female has laid eggs in a nest (Yom-Tov 1980; Brown 1984; Gibbons 1986). Egg features like egg shape and the pattern and color of markings can be reliable in systems with extreme variation among females, particularly when combined with other information like egg laying rates (Jackson 1992; Lyon 1993; Eadie et al. 2010), and in some cases it also makes it possible to identify the parasitic females that lay the eggs (Lyon 1993). The accuracy of these field methods was confirmed by McRae and Burke's (1996) comparison of field and genetic methods for determining parasitic eggs in moorhens (*Gallinula chloropus*)—they found perfect correspondence between demographic and genetic methods.

The advent of increasingly powerful genetic techniques has revolutionized the study of CBP, and our survey shows that researchers have been quick to adopt each new method as it became available (44% of studies in the online Appendix, Table 6.1). The earliest approach—enzyme polymorphisms—allowed researchers to exclude parasitic offspring as legitimate offspring from the social parents, but not to identify the female brood parasites themselves (Gowaty and Karlin 1984) at least without additional information such as direct observation of egg laying by parasitic females (Wrege and Emlen 1987). However, as the power of the methods increased, it became possible to reliably identify the parasitic females based on genetic inference alone: e.g., minisatellite DNA (McRae and Burke 1996) and microsatellite DNA (Nielsen et al. 2006). Relatively few of the studies we reviewed used microsatellite DNA to detect CBP (18%, Table 6.1) perhaps because this method has only recently become relative easy and inexpensive. The newest addition to the genetic toolbox, protein fingerprinting based on egg albumin samples, is particularly powerful because it focuses on the maternal rather than the offspring genotype (Andersson and Åhlund 2001). Moreover, samples can be obtained from fresh eggs, and no embryonic development is needed, an important issue because parasitic eggs often fail to develop in some species, and eggs may be destroyed, rejected, or depredated before hatching, precluding typical DNA sampling methods.

Researchers working on CBP need to be vigilant not only to reduce the risk of false assignments of maternity but also the risk of false exclusions—false exclusions are interpreted as examples of CBP. It is notable that of 106 studies where one or more methods were used to detect CBP, only 18% used *both* genetic and observational methods, whereas 26% used only genetic techniques, and 56% used only observational methods. Moreover, while microsatellite DNA markers are generally reliable, given a sufficient number of variable loci (e.g., Queller et al. 1993), our own simulations suggest that risk of false exclusion and assignment may be elevated for species with a combination of high levels of female relatedness and a lack of information about sires (Thow et al., unpublished information). Unfortunately this is a situation that often applies to waterfowl, a group that accounts for a large portion of the detailed studies of CBP (26 of our 56 species). Errors in parentage assignment are a concern not only because they can provide false evidence for the occurrence of CPB but also because such errors could generate spurious life history patterns.

6.3 Frequency of CBP: Variation Between and Within Species

The frequency of parasitism reported in the studies reviewed ranged from extremely rare (e.g., 3% of pied avocet (*Recurvirostra avosetta*) nests, 1% chicks; Hötker 2000) to extremely common (100% of ostrich (*Struthio camelus*) nests; Kimwele and Graves 2003). It has long been recognized that parasitism is particularly widespread in the waterfowl in terms of occurring in many species (Yom-Tov 1980; Andersson and Eriksson 1982); our survey now indicates that parasitism also occurs at a higher frequency within species in the waterfowl than in non-waterfowl taxa (Fig. 6.2). Reviews of CBP show that it is particularly common in precocial birds (Chap. 5); in our review this pattern is dominated almost entirely by the waterfowl (only a single precocial species does not belong to Anseriformes), and so we refrain from drawing conclusions about precocial birds generally.

Many studies report parasitism only in terms of the percentage of nests that are parasitized (online Appendix, Fig. 6.2). Reporting the frequency of parasitic eggs or chicks would also be useful because it gives a more accurate representation of the frequency of the tactic and it is also useful in determining the fitness costs (to hosts) or benefits (to parasites) of the behavior (e.g., Lyon 2003). Our survey indicates clearly that the percentage of nests parasitized considerably exceeds the percentage of eggs or chicks that are actually parasitic (Fig. 6.2).

One other measure that is only rarely obtained is the frequency of females in a population that engage in parasitism. For non-nesting females, this would be virtually impossible to determine, but it has been determined for nesting females in few different species: 25% of American coots (*Fulica americana*) in a given year (Lyon 1993) and 27% of moorhens over a 3-year period (McRae 1998).

6.4 The Adaptive Basis of Parasitism: Why Do Females Lay Parasitically?

6.4.1 Review of the Hypotheses

The key life history issue for CBP is how and why laying eggs in the nests of other conspecifics enhances the fitness of the brood parasite. To answer this, we must first determine which females in the population are the brood parasites. Do they have a nest of their own or are they non-nesting females? What alternative options are available to these females, and what specific life history trade-offs favor laying eggs parasitically over the alternatives that are possible for those eggs? We provided a review of existing hypotheses previously (Lyon and Eadie 2008) and so only briefly summarize the hypotheses here.

It is important to first consider the possibility that "apparent" CBP may not be adaptive brood parasitism at all. Behaviors other than CBP can result in a pattern identical to brood parasitism—nests that contain eggs laid by more than one female but where only one female incubates. There are two possibilities: nest site competition (Semel and Sherman 2001) and nest take-over (Robertson 1998).

Two hypotheses can explain why *non-nesting females* engage in parasitism:

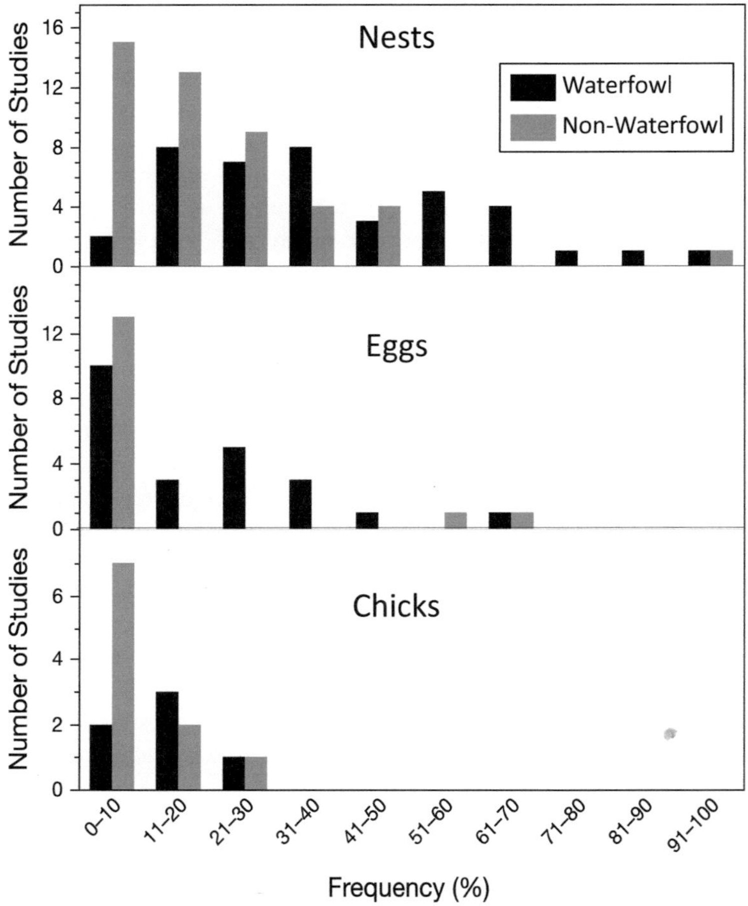

Fig. 6.2 The frequency of conspecific brood parasitism (CBP) in birds. Each bar is the number of studies that report a given frequency of CBP (arranged in ten percentiles) as measured by the percentage of nests parasitized (top), percentage of parasitic eggs detected (middle), or percentage of parasitic chicks detected (bottom). Black bars represent waterfowl; gray bars are all other bird species. In studies that report a range of frequencies, we used the minimum frequency

1. *Lifelong Specialist Parasite*—parasites never nest, and complete emancipation from nesting increases fecundity; the parasites depend completely on nonparasites for success, and frequency-dependent trade-offs lead to an equilibrium balance of professional parasites and nonparasites (Yom-Tov 1980; Eadie and Fryxell 1992).
2. *Best of a Bad Job* (BOBJ)—females lay parasitically because they are unable to nest due to limited nest sites or territories (constraint) or because their poor environmental or phenotypic situation makes parasitism a better option than nesting (restraint). Distinguishing between restraint and constraint aspects of BOBJ has proven exceptionally difficult (Lyon and Eadie 2008).

Two hypotheses can explain why *nesting females* lay some of their eggs parasitically:

1. *Breeding Interruption*—females resort to parasitism when forced to by nest predation or interference from conspecifics and salvage some fitness from reproductively committed eggs that would otherwise be wasted.
2. *Reproductive Enhancement* (or *Side Payment)*—parasitism allows a nesting female to bypass some constraint of investing parental care in offspring and thereby increase her reproductive success. The specific constraint being circumvented by parasitism could include any of the key life history variables: fecundity, offspring survival, or adult survival.

Two other hypotheses deserve mention. Pöysä (1999) suggested parasites gain by choosing nest sites safe from predation (*nest predation hypothesis*). However, because parasites are always expected to maximize fitness gained from parasitism, it is important to determine whether this factor drives parasitism or is simply an aspect of host choice by parasites. The *risk-spreading hypothesis* proposes that parasitism allows females to spread the risk of predation and increase the probability that a female will fledge some offspring (Payne 1977; Rubenstein 1982). Simulations show that the probability of fledgling at least one offspring is not important to natural selection: the mean fitness of the strategy matters, and in all but the tiniest population size, the average fitness of the two strategies is virtually identical (Bulmer 1984; Hopper et al. 2003). Andersson and Åhlund (2012) recently proposed a new, viable strategy that resembles risk spreading—parasitism reduces the time that eggs are exposed to predation and increases mean reproductive success.

6.4.2 Which Females Are the Parasites?

The identity of parasites is known for just over a third of the species in our survey (22 of 56 species, online Appendix). Most of these species ($n = 16$; 73%) have both nesting and non-nesting parasites, indicating that there will be more than one context and explanation for the CBP. For the species that have both types of females, nesting females often account for over half of the parasitic females. In five species (23%), only nesting parasites were recorded, and only one species (5%) had *only* non-nesting parasites. As we show below, in virtually all of the species with nesting parasites, nest loss is not a factor, so these females are not being forced to be parasites.

6.4.3 Why Do Female Birds Engage in CBP?

"Apparent" brood parasitism that results from competition over nest sites has only been invoked as an explanation in four studies and for only three cavity-nesting species (Table 6.1): bufflehead (*Bucephala albeola*), wood duck (*Aix sponsa*), and

Table 6.2 Support for alternative hypotheses on the adaptive basis of conspecific brood parasitism

Hypothesis	Yes	No	Total	% Yes	% of studies ($N = 43$)
Breeding interruption (BI)	10	7	17	59	13
Best of a bad job (BOBJ)	22	5	27	81	51
Nest competition (NC)	4	1	5	80	9
Reproductive enhancement (RE)	15	3	18	83	35
Risk spreading (RS)	2	2	4	50	5
Total	53	18	71		

Data are the frequency (number of studies) for which each hypothesis was claimed to be supported (yes) or refuted (no). Multiple hypotheses may have been tested in a single study. Abbreviations in parentheses refer to the codes used in the online Appendix

European starling (*Sturnus vulgaris*). Two of these studies presented limited evidence, but Semel and Sherman (2001), in particular, provided convincing evidence of nest competition in wood ducks by testing and rejecting alternative hypotheses. However, two lines of evidence can be used to reject the *nest competition hypothesis* as the primary or general explanation for CBP in a population (these tests have been applied to goldeneye ducks (*Bucephala* spp.) but not wood ducks). First, behavioral observations show clear differences in the behaviors of the females that lay in the same box, some of which indicate that only one female intends to nest (Åhlund 2005). Second, parasitic females can often be induced to lay in unoccupied nests baited with eggs, but they do not stay and care for the eggs they lay, counter to a key prediction of the *nest competition hypothesis* (Pöysä 2003; Odell and Eadie 2010).

In terms of adaptive hypotheses for CBP, at least one hypothesis has been claimed as supported for 28 species. For many of these species, more than one explanation for CBP seems to apply (Table 6.2, online Appendix). In species with non-nesting parasites, there is no evidence for specialist lifelong parasites. Twelve studies were able to follow parasites across breeding attempts, and in all cases the parasites had nests of their own at some point. In contrast, there is considerable evidence that non-nesting females resort to parasitism to make the best of a bad job: BOBJ was reported in 22 studies of 19 species, and the cases were evenly spread between waterfowl and non-waterfowl species. For most of these, it is unclear whether restraint or constraint explains the lack of nesting, but the aggregate results of our survey suggest that constraint may be more important than restraint. Nest limitation (i.e., *constraint hypothesis*) was invoked as an explanation in ten studies (Table 6.3). Experiments with cavity-nesting birds provide particularly compelling evidence that lack of a nest site forces some birds to resort to parasitism. In four species, nest boxes were added or removed to alter nest availability; all four studies altered the frequency of parasitism (Eadie 1991 (two species); Gowaty and Bridges 1991; Saitou 2001). More commonly, indirect evidence is used to infer that nest limitation is important; density is the most commonly reported correlate of parasitism (Table 6.3). However, parasitism could correlate with population density for other reasons, such as increased host availability (Rohwer and Freeman 1989). For example, parasitism correlates with colony size in cliff swallows (*Petrochelidon pyrrhonota*) (Brown 1984), but nest limitation is not an issue because they build their own nests. In the

Table 6.3 Ecological correlates associated with conspecific brood parasitism

Correlated variable	Yes	No	Total	% Yes	% of studies ($N = 63$)
Age experience of females (A)	11	6	17	65	17
Body condition of female (FC)	2	3	5	40	3
Density of nests (D)	22	6	28	79	35
Nest limitation (NL)	10	8	18	56	16
Quality of nest sites (NQ)	3	1	4	75	5
Density/availability of hosts (H)	5	2	7	71	8
Mate limitation (ML)	3	0	3	100	5
Ecological conditions—drought (ECD)	3	0	3	100	5
Ecological conditions—flooding (ECF)	1	0	1	100	2
Synchrony of nesting (S)	2	2	4	50	3
Time in season (T)	19	3	22	86	30
Total	86	32	118		

Data are the frequency (number of studies) for which each variable was claimed to be correlated (yes) or not (no) with the occurrence of CBP ($N = 63$ studies with correlates reported). Multiple correlates may have been evaluated in a single study. Abbreviations in parentheses refer to the codes used in the online Appendix

barn swallow (*Hirundo rustica*), Møller (1989) demonstrated the effect of host availability by increasing the local occurrence of parasitism rate experimentally by adding empty nests.

It is much harder to experimentally evaluate the role of restraint, although in theory food supplementation experiments could assess some aspects of restraint. Inspecting the ecological correlates associated with the studies that invoke BOBJ provides indirect evidence that restraint may be relatively uncommon (Table 6.3). Only two studies report female body condition as a correlate of parasitism, although in general discussions it is often mentioned as a possible factor influencing CBP. In contrast, female age is a frequent correlate ($n = 11$), but this could either reflect young females that choose not to breed or that are competitively inferior in competition for limited nest sites or other resources. A couple of studies do provide fairly convincing evidence for restraint—in canvasbacks (*Aythya valisineria*) and redheads (*A. americana*)—rates of parasitism increased during drought conditions when the prospects for successful nesting were low (Sorenson 1991, 1993). These ducks build nests in emergent vegetation so nest sites should not be limiting.

Our survey reveals that nesting females that engage in parasitism infrequently do so because they are forced due to breeding interruption (Table 6.2); this explanation was reported for ten species, including two for which only experimental nest loss caused parasitism (Feare 1991; Shaw and Hauber 2012). One experimental study of nest loss suggests that these experiments should be interpreted with caution: experimental destruction of European starling nests caused parasitism, but few cases of naturally occurring parasitism are associated with nest loss because nest predation is rare (Stouffer and Power 1991).

For most nesting parasites, reproductive enhancement seems to be the mostly likely explanation for parasitism; it was reported in 15 studies of 12 species (Table 6.2). For seven species, nesting parasites lay more total eggs than nonparasitic nesting birds, indicating a fecundity enhancement. The critical question then becomes why the females do better by laying these eggs parasitically rather than in their own nests. This is essentially a clutch size question, and it is surprising how few studies of brood parasitism consider this issue. Because clutch size determination is thought to differ between birds that feed their offspring (altricial, semi-precocial) and those that do not (precocial), we consider this issue separately for these two groups of birds.

CBP as a clutch size decision has been explicitly assessed in only four species that feed their offspring. In three species, the observation that parents suffered brood reduction due to limited food suggests that brood parasitism allows nesting females to bypass parental care constraints on family size and increase total production of offspring (Jackson 1993; Lyon 1993; McRae 1996). In the cliff swallow, some nesting females transfer eggs in their beak from their own nests to host nests. Because egg transfer often occurs after the parasites have completed laying their own clutch, Brown and Brown (1988) suggested that the birds are unlikely to be increasing their fecundity (although they could have laid larger clutches in anticipation). Further investigation revealed that this form of parasitism targets host nests with above-average success rates (Brown and Brown 1991). It is also worth stressing that fitness comparisons of entire classes of eggs—parasite versus nonparasite—can be misleading, because it is the fitness gained from specific eggs that matters. Hatching success of parasitic eggs is typically lower than that of host eggs (19 of 24 studies). This does not rule out reproductive enhancement as an explanation for CBP, contrary to some suggestions. The key is how the parasitic eggs would have fared had they been laid in the parental nest; investigations from this perspective show that parasitism allows females to increase the success of marginal eggs that would have low success had they been laid at home (Jackson 1993; Lyon 1998).

One surprise from our survey is the preponderance of nesting parasites in the precocial waterfowl. Waterfowl do not feed their offspring, and it has long been assumed that their clutch size is strongly influenced by egg laying capacity rather than food for the chicks (Lack 1967; Ankney and MacInnes 1978; but see Arnold and Rohwer 1991 for a contrary perspective). The observation that female nesting parasites in several waterfowl species lay more total eggs than nonparasites (Eadie 1989; Sorenson 1991; Åhlund and Andersson 2001) suggests that some other factors must be at play in these species. Understanding CBP by nesting waterfowl is essentially a clutch size problem. However, clutch size determination in this group remains poorly understood although several factors have been identified that might play a role, including saving reserves for renesting (Milonoff 1989), decreased egg viability combined with length of nest exposure to nest predation (Arnold et al. 1987), nest exposure to predation alone (Andersson and Åhlund 2012), variability in nest predation risk (Pöysä 1999), incubation costs to the female (Hepp et al. 1990), length of the incubation period (Rohwer 1985), hatching success (Rohwer 1985), and fledging success (Eadie and Lyon 1998).

Most of the above factors have not yet been examined in the context of CBP. Two egg removal experiments with common goldeneye suggested that females can lay more eggs than they normally do (Andersson and Eriksson 1982; Milonoff and Paananen 1993), an observation that raises the question as to why nonparasitic females do not lay these extra eggs in their own nests. However, one complication is that these studies were done in populations with CBP, and the authors were not able to distinguish between host and parasite eggs (all eggs were assumed to be nonparasitic). Odell and Eadie (2010) showed that parasitic female wood ducks are more likely to target host nests with low numbers of eggs, so we need to be sure that increases in fecundity due to experiments actually reflect the nest owner's response.

In addition to the traditional ideas on clutch size limitation in precocial birds, two new hypotheses have been proposed for brood parasites. In a hypothesis aimed specifically at the puzzle of nesting parasites, Andersson and Åhlund (2012) suggested that parasitism might allow females to maximize their annual reproductive success by reducing the length of time that eggs are exposed to the risk of parasitism. The fitness advantage to this type of risk spreading is quite modest, and an unstated assumption is that parasites must have very similar hatching success compared to nonparasitic eggs. This would require almost perfect laying synchrony with the host; otherwise the fitness loss of eggs laid too late to hatch would swamp the modest anti-predation gains. In seven of nine studies of seven waterfowl species, parasitic eggs were less successful than host eggs, mainly because the parasite eggs were laid after the host began incubation (online Appendix). However, because the hypothesis applies specifically to nesting females, it will be critical to separately assess the success of the eggs laid by nesting females. In American coots (*Fulica americana*), parasitic eggs of nesting parasites were more than twice as successful as eggs of non-nesting parasites, entirely due to better synchrony with the host's laying schedule (Lyon 1993).

A second hypothesis could explain parasitism in precocial birds. Pöysä (1999) proposed that parasites specifically target safe nest sites with high survival rates. Non-nesting parasites may benefit by laying parasitically in a high-quality host nest, rather than establishing a nest of their own in an unsafe site. Nesting females forced to occupy relatively unsafe nest sites could benefit by laying as many as their eggs as possible in hosts with safe nests and, when hosts are no longer available, lay the remainder in their own nest. This idea has only been tested so far for one species (common goldeneyes); the results suggest that parasites do target safe sites, but it is unclear which females (nesting or non-nesting parasites) do so (Pöysä 1999, 2006).

6.5 Brood Parasitism Through Life

One question that has received little attention is how brood parasitism integrates into a female's entire lifetime life history strategy. It is clear from our review of studies that followed individual females across years (or breeding attempts) that females are flexible and change tactics across years, depending on factors such as age, population density, nest and territory availability, and ecological conditions like drought. Given this flexibility, one interesting question is what trajectories individual females follow through the course of their entire life. Do all females show similar trajectories (an age/experience-driven strategy) or do females differ in when, and how much, they invest in parasitism based on ecological or phenotypic contingencies?

In terms of the lifespan of a female, nesting and CBP are alternative reproductive tactics that, when combined together in all possible combinations, comprise a conditional life history strategy (Brockmann 2001; Gross 1996). Sorenson (1991) proposed one framework, based on reproductive effort, for thinking about how this socially and ecologically driven flexibility fits together in terms of a life history strategy. Alternatively, two studies illustrate how the effects of population dynamics on the resources needed for nesting, such as nest sites, might cause temporal variation in the conditions that favor nesting versus parasitism (Eadie and Fryxell 1992; Eadie et al. 1998). Finally, Jaatinen et al. (2011) use a strategy selection modeling approach to show how variation in female quality can influence the trajectory and combination of tactics adopted by females throughout life. To the extent that the drivers that favor CBP are unpredictable across time, these different models predict that we might expect considerable variation among females in their lifetime patterns of parasitism, nesting, and perhaps even sabbaticals from breeding entirely.

It is also possible that there are consistent differences among females in how they respond to the conditions that favor parasitism, either due to genetic differences or long-lasting effects from early life (ontogenetic effects). Intriguingly, two studies found that parasitism by nesting females was repeatable across breeding attempts (Møller 1987; Schielzeth and Bolund 2010). Why would females show consistent differences in when and how much they employ CBP? Two broad explanations are possible (these also apply to alternative male reproductive tactics, Brockmann 2001). First, conditions early in life could influence a female's phenotype (size, quality, etc.), which in turn affects her optimal pattern of investment in nesting versus parasitism throughout her life. Second, trade-offs between nesting and parasitism could favor the evolution of different traits that enable success at one of these tactics, but not both. In the extreme, these trade-offs might favor the evolution of lifetime specialists, as is commonly observed for male alternative tactics (Oliveira et al. 2008), but as noted above, there is no evidence for life specialist conspecific parasites. Alternatively, there may be genetic variation among females in the propensity to engage in CBP in particular contexts; a reaction norm approach could be used to examine this (Oliveira et al. 2008). This could be daunting for field studies but should be feasible for captive populations like the zebra finch (*Taeniopygia guttata*) where consistency in allocation to CBP has already been demonstrated (Schielzeth and Bolund 2010).

6.6 A Role for Kinship?

For some birds CBP might not be parasitism at all. Andersson and Eriksson (1982) were the first to note that CBP is disproportionately common in waterfowl (Anatidae), a group where natal philopatry is female-biased rather the more usual avian pattern of male-biased natal philopatry. Andersson and Eriksson (1982) suggested that female philopatry could result in hosts and parasites being related. Andersson (1984) followed with an elegantly simple model showing how female relatedness might facilitate the evolution of CBP, a finding that prodded researchers to reconsider CBP as a kin-selected, cooperative breeding system (see Chap. 12) rather than as a parasitic interaction. This stimulated a flurry of genetic studies, although a consensus has yet to be reached, and it may be relevant mostly to the waterfowl. Nonetheless, the possibility of cooperation, potentially facilitated by kinship, blurs the distinction between CBP and cooperative breeding such that they may simply represent points along a life history continuum (Zink 2000; Zink and Lyon 2016).

An important caveat, however, is that kin-selection can facilitate host acceptance of parasitic eggs (reducing selection on host resistance), but it does not explain why the parasites lay the eggs as brood parasites in the first place. For this reason, we did not focus on the kinship aspect in this paper but recognize it as an intriguing development in understanding CBP, especially for species such as waterfowl with female-biased philopatry (several reviews and theoretical papers address this topic, including Andersson 1984; Zink 2000; Lyon and Eadie 2000; Andersson 2001; Lopez-Sepulcre and Kokko 2002; Eadie and Lyon 2011; Anderson 2017).

Concluding Remarks and Future Directions

In slightly over three decades, we have moved from simply documenting the frequency of CBP among species to exploring, in a diversity of species, the question of why females pursue this behavior and what fitness benefits might accrue. With the advent of powerful molecular genetic methods and detailed field studies of marked individuals, we have learned that CBP is not simply a rare, aberrant, or accidental behavior as some early authors proposed. Rather, it is clear that CBP represents a flexible alternative female life history tactic that allows females to adjust reproductive effort to varying ecological and physiological conditions.

We offer four take-home messages and directions for future research. First, attention to the methods and analyses used to detect CBP is critical. Care in assigning maternity must be extended to how we exclude females as possible mothers because false exclusion could lead to "apparent" support for a variety of appealing (but possibly incorrect) hypotheses. The number of markers in molecular studies is sometimes low, and conclusions should be supported

(continued)

using multiple lines of evidence, including a focus on demographic aspects like egg laying rates at focal nests.

Second, our survey reveals that there are clearly two types of CBP—by non-nesting and by nesting females. For non-nesting parasites, there is no evidence for "pure" lifelong parasites, and most researchers conclude that this type of parasitism is a best of a bad job (BOBJ), although the underlying constraints or restraints are rarely known. Parasitism by nesting females was surprisingly common. This was unexpected, especially for waterfowl. Why don't these females lay all their eggs in their own nest rather than risk their eggs in a nest of another female? A number of hypotheses have recently been proposed, but much empirical and theoretical work remains, particularly in the context of adaptive clutch size. Most studies to date have been observational, and the clutch size aspect would clearly benefit from an experimental approach.

Third, the variation in the extent to which females pursue CBP within and among years points to a critical need to follow females over their entire lifespan. Virtually no studies have followed females throughout their life to explore the flexibility of parasitic behavior or the trade-offs that might be involved. Indeed, some females might be more specialized towards parasitism and others towards nesting, as might be expected if different skills, experience, physiological state, or personality influences a female's reproductive trajectory.

Fourth, we highlight the value of CBP as a model system to examine an array of ecological and evolutionary dynamics. CBP is unusual in being an alternative *female* reproductive tactic; most research on ARTs has focused on males, and CBP offers an exceptional opportunity to expand this framework. Moreover, given that females are typically the sex that drives population dynamics, there is rich potential to explore the links between the behavioral dynamics, population dynamics, and evolutionary dynamics of this intriguing behavior.

Acknowledgments We thank the following for funding: NSF IOS 1355208 (JME), D. G. Raveling Endowment (JME), and NSF IOS 1354894 (BEL). We thank M. Soler for the opportunity to contribute this chapter and M. Soler and Y. Yom-Tov for helpful comments.

References

Åhlund M (2005) Behavioural tactics at nest visits differ between parasites and hosts in a brood-parasitic duck. Anim Behav 70:433–440

Åhlund M, Andersson M (2001) Brood parasitism – female ducks can double their reproduction. Nature 414:600–601

Anderson M (2017) Helping relatives survive and reproduce: inclusive fitness and reproductive value in brood parasitism. Am Nat 189:138–152

Andersson M (1984) Brood parasitism within species. In: Barnard CJ (ed) Producers and scroungers: strategies of exploitation and parasitism. Croom Helm, London, pp 195–228

Andersson M (2001) Relatedness and the evolution of conspecific brood parasitism. Am Nat 158:599–614

Andersson M, Åhlund M (2001) Protein fingerprinting: a new technique reveals extensive conspecific brood parasitism. Ecology 82:1433–1442

Andersson M, Åhlund M (2012) Don't put all your eggs in one nest: spread them and cut time at risk. Am Nat 180:354–363. https://doi.org/10.1086/667191

Andersson M, Eriksson MOG (1982) Nest parasitism in goldeneyes *Bucephala clangula* – some evolutionary aspects. Am Nat 120:1–16

Ankney CD, MacInnes CD (1978) Nutrient reserves and reproductive performance of female Lesser Snow Geese. Auk 95:459–471

Arnold TW, Rohwer FC (1991) Do egg formation costs limit clutch size in waterfowl – a skeptical view. Condor 93:1032–1038

Arnold TW, Rohwer FC, Armstrong T (1987) Egg viability, nest predation, and the adaptive significance of clutch size in prairie ducks. Am Nat 130:643–653

Brockmann HJ (2001) The evolution of alternative strategies and tactics. In: Advances in the study of behavior, vol 30. Academic, San Diego, pp 1–51

Brown CR (1984) Laying eggs in a neighbors nest benefit and cost of colonial nesting in swallows. Science 224:518–519

Brown CR, Brown MB (1988) A new form of reproductive parasitism in cliff swallows. Nature 331:66–68

Brown CR, Brown MB (1991) Selection of high-quality host nests by parasitic cliff swallows. Anim Behav 41:457–465

Bulmer MG (1984) Risk avoidance and nesting strategies. J Theor Biol 106:529–535

Davies NB (2000) Cuckoos, cowbirds and other cheats. Poyser, London

Eadie JM (1989) Alternative female reproductive tactics in a precocial bird: the ecology and evolution of brood parasitism in Goldeneyes. PhD thesis, University of British Columbia, Vancouver, BC

Eadie JM (1991) Constraint and opportunity in the evolution of brood parasitism in waterfowl. Acta XX Congr Int Ornithol 20:1031–1040

Eadie JM, Fryxell JM (1992) Density-dependence, frequency-dependence, and alternative nesting strategies in goldeneyes. Am Nat 140:621–641

Eadie JM, Lyon BE (1998) Cooperation, conflict, and creching behavior in goldeneye ducks. Am Nat 151:397–408

Eadie JM, Lyon BE (2011) The relative role of relatives in conspecific brood parasitism. Mol Ecol 20:5114–5118. https://doi.org/10.1111/j.1365-294X.2011.05320.x

Eadie JM, Kehoe FP, Nudds TD (1988) Pre-hatch and post-hatch brood amalgamation in North American Anatidae: a review of hypotheses. Can J Zool 66:1709–1721

Eadie J, Semel B, Sherman PW (1998) Conspecific brood parasitism, population dynamics, and the conservation of cavity-nesting birds. In: Caro T (ed) Behavioral ecology and conservation biology. Oxford University Press, New York, pp 306–340

Eadie JM, Smith JNM, Zadworny D, Kuhnlein U, Cheng K (2010) Probing parentage in parasitic birds: an evaluation of methods to detect conspecific brood parasitism using goldeneyes *Bucephala islandica* and *Bl. clangula* as a test case. J Avian Biol 41:163–176. https://doi.org/10.1111/j.1600-048X.2009.04735.x

Feare CJ (1991) Intraspecific nest parasitism in starlings *Sturnus vulgaris* – effects of disturbance on laying females. Ibis 133:75–79. https://doi.org/10.1111/j.1474-919X.1991.tb04813.x

Field J (1992) Intraspecific parasitism as an alternative reproductive tactic in nest-building wasps and bees. Biol Rev Camb Philos Soc 67:79–126

Gibbons DW (1986) Brood parasitism and cooperative nesting in the moorhen, *Gallinula chloropus*. Behav Ecol Sociobiol 19:221–232

Gowaty PA, Bridges WC (1991) Nestbox availability affects extra-pair fertilizations and conspecific nest parasitism in eastern bluebirds, *Sialia sialis*. Anim Behav 41:661–675. https://doi.org/10.1016/s0003-3472(05)80904-2

Gowaty PA, Karlin AA (1984) Multiple maternity and paternity in single broods of apparently monogamous eastern bluebirds (*Sialia sialis*). Behav Ecol Sociobiol 15:91–95

Gross MR (1996) Alternative reproductive strategies and tactics: diversity within sexes. Trends Ecol Evol 11:263–263

Henson SA, Warner RR (1997) Male and female alternative reproductive behaviors in fishes: a new approach using intersexual dynamics. Annu Rev Ecol Syst 28:571–592

Hepp GR, Kennamer RA, Harvey IVWF (1990) Incubation as a reproductive cost in female Wood Ducks. Auk 107:756–764

Hopper KR, Rosenheim JA, Prout T, Oppenheim SJ (2003) Within-generation bet hedging: a seductive explanation? Oikos 101:219–222

Hötker H (2000) Conspecific nest parasitism in the Pied Avocet *Recurvirostra avosetta*. Ibis 142:280–288. https://doi.org/10.1111/j.1474-919X.2000.tb04867.x

Jaatinen K, Lehtonen J, Kokko H (2011) Strategy selection under conspecific brood parasitism: an integrative modeling approach. Behav Ecol 22:144–155

Jackson WM (1992) Estimating conspecific nest parasitism in the northern masked weaver based on within-female variability in egg appearance. Auk 109:435–443

Jackson WM (1993) Causes of conspecific nest parasitism in the northern masked weaver. Behav Ecol Sociobiol 32:119–126. https://doi.org/10.1007/bf00164044

Kimwele C, Graves J (2003) A molecular genetic analysis of the communal nesting of the ostrich (*Struthio camelus*). Mol Ecol 12:229–236

Lack D (1967) The significance of clutch-size in waterfowl. Wildfowl 18:125–128

Ligon JD, Burt DB (2004) Evolutionary origins. In: Koenig W, Dickinson JL (eds) Ecology and evolution of cooperative breeding in birds. Cambridge University Press, Cambridge

Lopez-Sepulcre A, Kokko H (2002) The role of kin recognition in the evolution of conspecific brood parasitism. Anim Behav 64:215–222

Lyon BE (1993) Conspecific brood parasitism as a flexible female reproductive tactic in American coots. Anim Behav 46:911–928

Lyon BE (1998) Optimal clutch size and conspecific brood parasitism. Nature 392:380–383

Lyon BE (2003) Egg recognition and counting reduce costs of avian conspecific brood parasitism. Nature 422:495–499. https://doi.org/10.1038/nature01505

Lyon BE, Eadie JM (2000) Family matters: kin selection and the evolution of conspecific brood parasitism. Proc Natl Acad Sci U S A 97:12942–12944

Lyon BE, Eadie JM (2008) Conspecific brood parasitism in birds: a life-history perspective. Annu Rev Ecol Evol Syst 39:343–363

Lyon, BE, Eadie JM (2017) A comprehensive survey of the life history aspects of conspecific brood parasitism in birds. Dataset posted at Figshare. https://doi.org/10.6084/m9.figshare.4787872

Lyon BE, Montgomerie R (2012) Sexual selection is a form of social selection. Philos Trans R Soc Lond B 367:2266–2273

McRae SB (1996) Brood parasitism in the Moorhen: brief encounters between parasites and hosts and the significance of an evening laying hour. J Avian Biol 27:311–320. https://doi.org/10.2307/3677262

McRae SB (1998) Relative reproductive success of female moorhens using conditional strategies of brood parasitism and parental care. Behav Ecol 9:93–100

McRae SB, Burke T (1996) Intraspecific brood parasitism in the moorhen: parentage and parasite-host relationships determined by DNA fingerprinting. Behav Ecol Sociobiol 38:115–129

Milonoff M (1989) Can nest predation limit clutch size in precocial birds? Oikos 55:424–427

Milonoff M, Paananen P (1993) Egg formation, brood survival, and cost of reproduction as clutch-size-determining factors in common goldeneyes. Auk 110:943–946

Møller AP (1987) Intraspecific nest parasitism and antiparasite behavior in swallows, *Hirundo rustica*. Anim Behav 35:247–254. https://doi.org/10.1016/s0003-3472(87)80230-0

Møller AP (1989) Intraspecific nest parasitism in the swallow *Hirundo rustica* – the importance of neighbors. Behav Ecol Sociobiol 25:33–38. https://doi.org/10.1007/bf00299708

Nielsen CR, Parker PG, Gates RJ (2006) Intraspecific nest parasitism of cavity-nesting wood ducks: costs and benefits to hosts and parasites. Anim Behav 72:917–926

Odell NS, Eadie JM (2010) Do wood ducks use the quantity of eggs in a nest as a cue to the nest's value? Behav Ecol 21:794–801. https://doi.org/10.1093/beheco/arq055

Oliveira RF, Taborsky M, Brockmann HJ (2008) Alternative reproductive tactics: an integrative approach. Cambridge University Press, Cambridge

Payne RB (1977) Ecology of brood parasitism in birds. Annu Rev Ecol Evol Syst 8:1–28

Pöysä H (1999) Conspecific nest parasitism is associated with inequality in nest predation risk in the common goldeneye (*Bucephala clangula*). Behav Ecol 10:533–540. https://doi.org/10.1093/beheco/10.5.533

Pöysä H (2003) Low host recognition tendency revealed by experimentally induced parasitic egg laying in the common goldeneye (*Bucephala clangula*). Can J Zool 81:1561–1565

Pöysä H (2006) Public information and conspecific nest parasitism in goldeneyes: targeting safe nests by parasites. Behav Ecol 17:459–465

Queller DC, Strassmann JE, Hughes CR (1993) Microsatellites and kinship. Trends Ecol Evol 8:285–288

Robertson GJ (1998) Egg adoption can explain joint egg-laying in common eiders. Behav Ecol Sociobiol 43:289–296. https://doi.org/10.1007/s002650050493

Rohwer FC (1985) The adaptive significance of clutch size in prairie ducks. Auk 102:354–361

Rohwer FC, Freeman S (1989) The distribution of conspecific nest parasitism in birds. Can J Zool 67:239–253

Rubenstein DI (1982) Risk, uncertainty and evolutionary strategies. In: Kings's College Sociobiology Group (ed) Current problems in sociobiolgy. Cambridge University Press, Cambridge, pp 91–111

Saitou T (2001) Floaters as intraspecific brood parasites in the grey starling *Sturnus cineraceus*. Ecol Res 16:221–231. https://doi.org/10.1046/j.1440-1703.2001.00388.x

Schielzeth H, Bolund E (2010) Patterns of conspecific brood parasitism in zebra finches. Anim Behav 79:1329–1337. https://doi.org/10.1016/j.anbehav.2010.03.006

Semel B, Sherman PW (2001) Intraspecific parasitism and nest-site competition in wood ducks. Anim Behav 61:787–803

Shaw RC, Hauber ME (2012) Linking nest predation with brood parasitism in captive zebra finches: a multi-pair study. J Ethol 30:255–262. https://doi.org/10.1007/s10164-011-0319-4

Sorenson MD (1991) The functional significance of parasitic egg laying and typical nesting in redhead ducks: an analysis of individual behaviour. Anim Behav 42:771–796

Sorenson MD (1993) Parasitic egg-laying in Canvasbacks – frequency, success, and individual behavior. Auk 110:57–69

Stouffer PC, Power HW (1991) Brood parasitism by starlings experimentally forced to desert their nests. Anim Behav 41:537–539

Wrege PH, Emlen ST (1987) Biochemical determination of parental uncertainty in white-fronted bee-eaters. Behav Ecol Sociobiol 20:153–160. https://doi.org/10.1007/bf00299728

Yom-Tov Y (1980) Intraspecific nest parasitism in birds. Biol Rev Camb Philos Soc 55:93–108

Yom-Tov Y (2001) An updated list and some comments on the occurrence of intraspecific nest parasitism in birds. Ibis 143:133–143

Zink AG (2000) The evolution of intraspecific brood parasitism in birds and insects. Am Nat 155:395–405

Zink AG, Lyon BE (2016) Evolution of conspecific brood parasitism versus cooperative breeding as alternative reproductive tactics. Am Nat 187:35–47

7 Evolution and Maintenance of Egg Rejection by Hosts as Adaptation Against Conspecific Brood Parasites: An Individual-Based Model

Fugo Takasu

Abstract

An individual-based simulation model is constructed to explore evolutionary dynamics of the three adaptive traits: (1) proportion of eggs allocated as conspecific parasitism, (2) rejection ability to reject unlike eggs in own nest, and (3) egg appearance. Simulation analysis suggests that egg rejection can logically evolve due to conspecific brood parasitism and that variability of egg appearance plays a key role to determine the evolutionary trajectory of the three adaptive traits. If variability of egg appearance is small enough, conspecific parasitism selects for lower inter-clutch variation, and all individuals have nearly identical eggs. The population converges to a state where ability to recognize and reject unlike eggs is neutral to selection, and a small cost ignored in this model will suffice to prevent the evolution of egg rejection. However, if egg variability is large enough, conspecific parasitism selects for rejection ability to increase but this in turn selects against conspecific parasitism. The population converges to a state where all individuals show rejection ability at a high level but invest fewer eggs to conspecific parasitism. I highlight the importance of variability of egg appearance for the evolution and the maintenance of egg rejection ability in conspecific brood parasites and call for quantitative measures of the variability of egg appearance in bird species that practice conspecific brood parasitism.

7.1 Introduction

It has been widely recognized that many bird species practice brood parasitism; parasitic individuals lay their eggs in the nest of other individuals as host and let the host incubate and take care of parasitic eggs (Payne 1977; Rothstein 1990; Davies

F. Takasu (✉)
Department of Information and Computer Science, Nara Women's University, Nara, Japan
e-mail: takasu@ics.nara-wu.ac.jp

M. Soler (ed.), *Avian Brood Parasitism*, Fascinating Life Sciences,
https://doi.org/10.1007/978-3-319-73138-4_7

2000). By definition, brood parasites reproduce at the expense of reproductive success of the host. Therefore, the parasite and the host are expected to evolve various adaptive traits and behaviors in a conflicting manner, such as an ability to recognize and reject parasitism (defense against parasitism by the host) (Davies and Brooke 1988; Moksnes et al. 1991) and egg mimicry to deceive the host (counter defense by the parasite) (Brooke and Davies 1988; Moksnes and Røskaft 1995; Honza et al. 2001), etc.

Conspecific brood parasitism (thereafter CBP) is one form of brood parasitism where parasitism is directed to conspecific individuals (Yom-Tov 1980). It has been demonstrated that more and more bird species are found to practice CBP (Yom-Tov 2001, Chap. 5). Lyon and Eadie have provided two thorough reviews on CBP in birds (Lyon and Eadie 2008, Chap. 6).

Different from interspecific parasitism, conspecific brood parasites enjoy fitness benefits from practicing parasitism, while they may have a risk of being parasitized; a parasitic individual who lays its eggs in the nests of conspecifics can be parasitized by them. This complicates theoretical explorations to understand the evolution of CBP because two fitness components have to be explicitly considered, i.e., the one for the reproductive success from its own nest by rejecting parasitic eggs and the other for the reproductive gain obtained from successful parasitism (Broom and Ruxton 2004).

There have been several theoretical studies aimed to explore how CBP can evolve under what conditions. Most of them focus on optimal allocation of eggs in own nest and in others' nests (Yamauchi 1993; Lyon 1998; Broom and Ruxton 2002; Ruxton and Broom 2002; Jaatinen et al. 2011). Valpine and Eadie (2008) give a comprehensive review on theoretical studies on CBP and provide a general framework to explore evolutionary dynamics of adaptive traits in focus and their impact on population dynamics. Although these models attempt to answer different questions, all ignore the variation of egg appearance in population, an important factor that should be critical for the evolution of egg rejection ability because recognition of parasitic egg is based on visual stimuli.

In this chapter, I build an individual-based model to explore how CBP can select for and maintain egg rejection ability. In this model, I explore evolutionary dynamics of three adaptive traits: (1) proportion of own eggs laid as parasitic (degree of CBP), (2) rejection ability against unlike eggs, and (3) egg appearance. Each individual is assigned a set of the three traits as quantitative trait, and it lays eggs parasitically and rears eggs in the clutch according to the assigned traits. Repeating hypothesized parasitic behaviors (laying eggs and rejecting unlike eggs) naturally result in a stochastic population dynamics and evolutionary dynamics of the adaptive traits; thereby evolutionary trajectories of the three adaptive traits can be explored.

Individual-based approach has an advantage that we can flexibly assume individual behaviors related to adaptive traits in focus. A drawback is the difficulty with which the model behaviors are comprehensively understood because the model is algorithmically described and we have to realize simulations for different parameters used. Despite the drawback, however, this approach will be valuable in the sense that it can clearly demonstrate possible consequences of the assumptions assumed in the model.

Based on simulation results of the model, I discuss how egg rejection can be maintained in CBP. Further research prospect is finally discussed to better understand CBP.

7.2 The Model

I assign each individual a set of the following three adaptive traits: (1) proportion p of eggs that are laid parasitically in the nests of other individuals ($0 \leq p \leq 1$), (2) rejection ability r to discriminate and reject unlike eggs laid in own nest ($0 \leq r \leq 1$), and (3) egg appearance e of own eggs measured in an arbitrary scale ($-\infty < e < \infty$). The proportion p describes the propensity to conspecific parasitism. The rejection ability r describes the ability to recognize and reject parasitic eggs. But this recognition ability critically depends on the similarity of parasitic eggs to its own, and hence I consider egg appearance e of the individual.

I assume that the number of eggs an individual can produce is fixed as an integer-valued constant n. An individual with (p, r, e) distributes C eggs with the appearance e one by one in other individual's nests that are chosen randomly and lays the rest of $n - C$ eggs in its own nest. Here, C is an integer-valued random number determined as follows:

$$C = [np] \quad \text{with the probability } 1 - np + [np],$$

$$C = [np] + 1 \quad \text{with the probability } np - [np],$$

where $[x]$ is the largest integer not exceeding x and the expected values of C is $E[C] = np$. I assume no intra-clutch variation in this model.

After all individuals finish laying their eggs, each individual's nest will have a certain number of eggs that were parasitically laid. I assume that an individual with (p, r, e) can recognize and reject a parasitic egg with the appearance e' laid in the clutch with the probability defined as

$$r[1 - A(e, e')] = r\left(1 - \exp\left[-\gamma(e - e')^2\right]\right),$$

where $A(e, e')$ describes the acceptance probability that the nest owner with the egg appearance e accepts the parasitic egg e' and the parameter γ controls the sensitivity against unlike eggs (Takasu 2003; Stokke et al. 2007; Yang et al. 2010). Note that $A(e, e')$ is in general a decreasing function of the difference in appearance $|e - e'|$ and $A(e, e) = 1$, i.e., the more dissimilar the parasite egg is, the more likely it is rejected, and even if an individual has high rejection ability $r \approx 1$, it cannot reject parasitic egg with identical appearance $e' = e$. This rejection rule is applied to all parasitic eggs in the clutch. For example, when the nest of an individual (p, r, e) has two parasitic eggs, e' and e'', each is rejected with the probability $r(1 - A(e, e'))$ and $r(1 - A(e, e''))$ but is

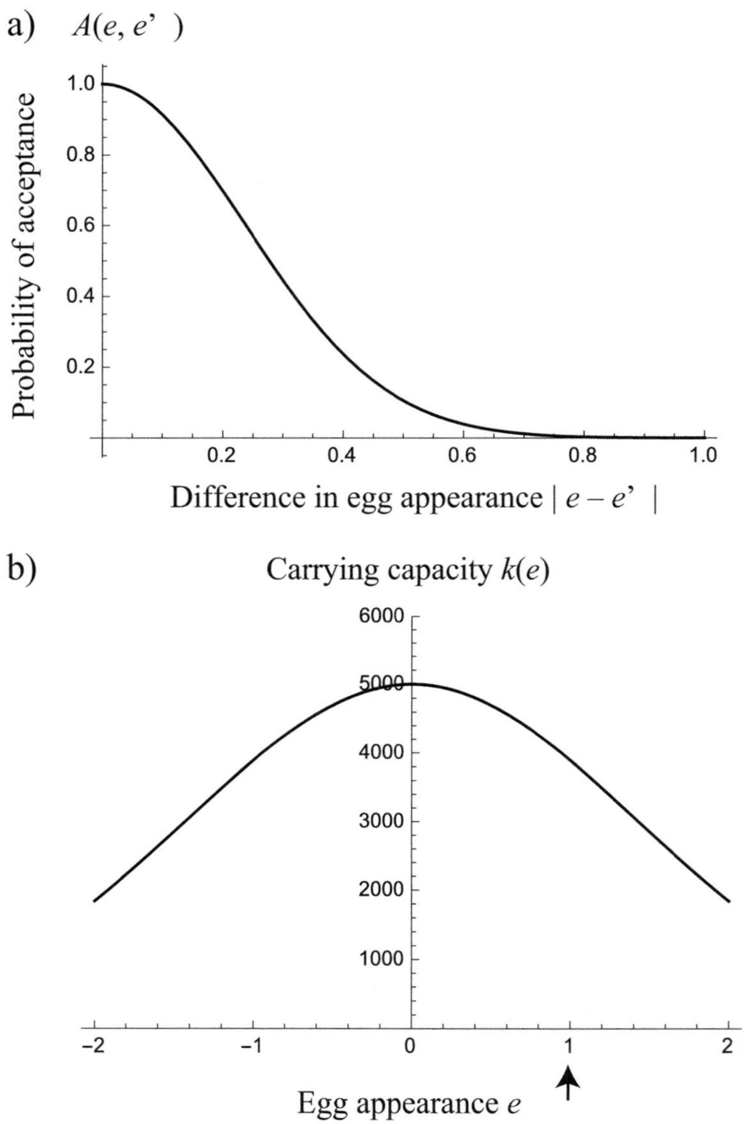

Fig. 7.1 The acceptance probability and the carrying capacity. (**a**) Acceptance probability $A(e, e')$ that the nest owner with egg appearance e accepts a parasitic egg with egg appearance e' as a function of the difference in the egg appearance $|e - e'|$. $A(e, e') = A(|e - e'|) = \exp[-\gamma(e - e')^2]$ with $\gamma = 9$. (**b**) Carrying capacity $k(e)$ as a function of the egg appearance e with $e = 0$ being most favored in the absence of conspecific and interspecific parasitism. $k = 5000$ and $\sigma_k = 2$. Egg appearance of interspecific brood parasites $e_i = 1$ is shown as arrow, very different from $e = 0$

accepted and remains in the clutch with the rest of the probability, respectively. The acceptance probability plotted against the difference in egg appearance is shown in Fig. 7.1.

Some conspecific brood parasites suffer from parasitism by interspecific brood parasites (Lahti 2005). Interspecific parasitism can drive the evolution of host rejection behavior if accepting parasitism incurs a larger cost for the host reproduction (Rothstein 1975; Kelly 1987; Brooker et al. 1990; Takasu et al. 1993; Takasu 1998). In order to compare which factor, interspecific or conspecific parasitism, drives the host rejection behavior more effectively, I assume that a fixed fraction i of individuals are parasitized interspecifically with parasitic egg with the appearance e_i. If interspecifically parasitized, the individual (p, r, e) rejects the parasitism with the probability $r(1 - A(e, e_i))$. I assume that a nest can be interspecifically parasitized only once.

After all individuals finish checking their nest contents according to their rejection ability, each individual's nest will have a certain number of own eggs, conspecific parasite egg(s), and an interspecific parasite egg, if any. I assume that if the nest contains an interspecific parasite egg, the nest owner has no reproductive output. Otherwise, the individual has some reproductive output. Here I assume that potential to rear a clutch is limited, and the probability that each of eggs successfully fledges depends on the total clutch size. Specifically, I assume the following probability of successful fledgling as a decreasing function of the clutch size x as

$$f(x) = s_{\text{egg}} \frac{1 + l}{1 + l \exp[mx]},$$

where s_{egg} is the baseline fledging probability of eggs, l controls the severity of the limited potential to rear the clutch, and m scales the maximum reproductive output as mentioned below ($l, m > 0$). The number of eggs that fledge and survive to the next breeding season from a clutch whose size is x is proportional to

$$R(x) = xf(x).$$

I assume that the clutch size in the absence of parasitism has been optimized so that it maximizes the reproductive output (Lack 1947), i.e., $R(x)$ is maximized at $x = n$ (Takasu 2004). The functional forms of $f(x)$ and $R(x)$ are shown in Fig. 7.2.

I have described the rules of reproduction within a breeding season (egg laying, rejection of unlike eggs, and fledging). I next describe the rules for survival of individuals to the next breeding season.

In order to prevent population explosion, I assume that a density-dependent effect works equally on fledglings and adults and that the survival of eggs to the next breeding season is given as

$$\frac{k(e)}{k(e) + H} f(x)$$

and that of adults as

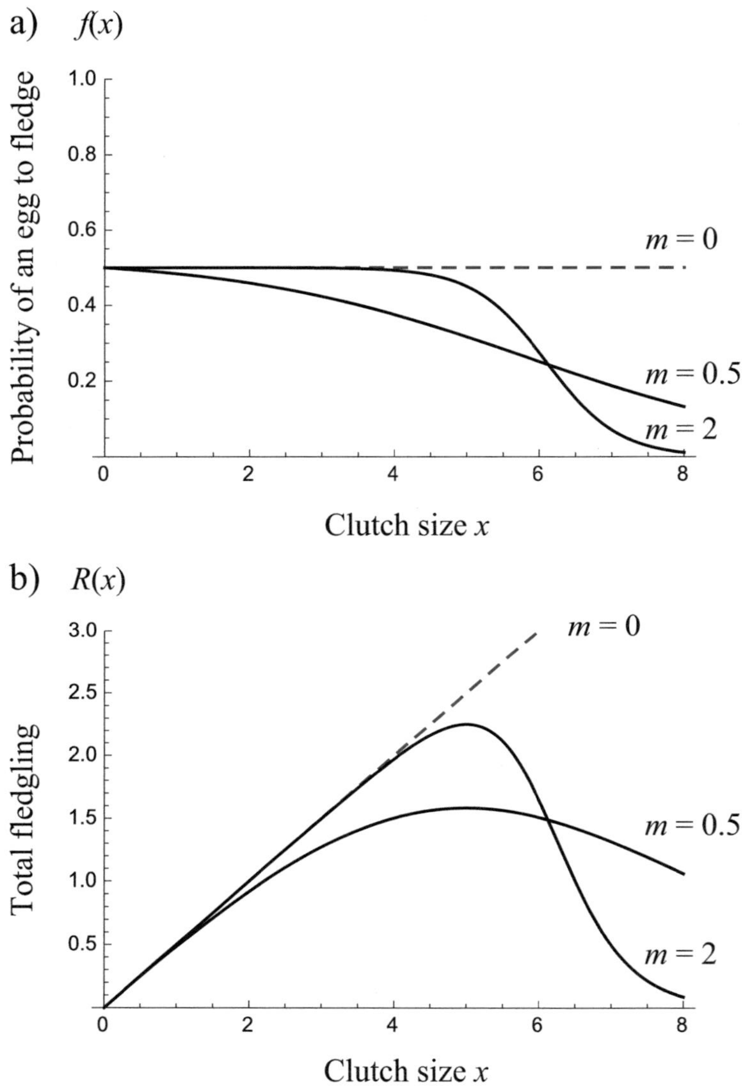

Fig. 7.2 Probability of successful fledging and the reproductive success. (**a**) Probability of successful fledging of an egg $f(x)$. (**b**) The total fledgling success $R(x) = x\,f(x)$ as a function of the clutch size x. $s_{\text{egg}} = 0.5$. Limited potential to rear a clutch $m = 0.5$ and $l = 5.472 \times 10^{-2}$ and $m = 1.5$ and $l = 8.509 \times 10^{-5}$. No limit $m = 0$

$$\frac{k(e)}{k(e) + H} S_H,$$

where $1/k(e)$ measures the density-dependent effect as a function of egg appearance e, or $k(e)$ as the carrying capacity for those with egg appearance e, H is the number of adult individuals censused at the beginning of the breeding season, and s_H is the intrinsic annual survival probability of adults. In order to avoid the case that the egg appearance e is neutral to selection and diverges indefinitely, I assume that $e = 0$ is most favored in the absence of interspecific and conspecific parasitism with

$$k(e) = k\exp\left[-\frac{e^2}{2\sigma_k^2}\right],$$

where the parameter σ_k controls the convergence of egg appearance to $e = 0$ (Fig. 7.1). In the limit $\sigma_k = \infty$, egg appearance is neutral in the absence of parasitism. The smaller σ_k becomes, the more favored the egg appearance $e = 0$ becomes by factors not related to parasitism, e.g., camouflage against predators.

I finally assume that the adaptive traits assigned to each individual (p, r, e) are inherited asexually to offspring with a certain mutation as

$$p_{\text{offspring}} = p_{\text{parent}} + \mu_p,$$

$$r_{\text{offspring}} = r_{\text{parent}} + \mu_r,$$

$$e_{\text{offspring}} = e_{\text{parent}} + \mu_e,$$

where μ is the random number that follows Gaussian $N(0, \sigma^2)$ and σ^2 controls the rate with which the trait in focus evolves. For example, setting $\sigma_p = 0$ means that the trait p is kept constant and its evolution is frozen (offspring has exactly the same p as its parent). The trait p and r are bounded within the range from 0 to 1. When offspring trait calculated is outside the range, the mirror image trait with the nearest boundary is taken.

These rules constitute the model. The model has been implemented by the programming language C with Xcode on OS X 10.8. Before I let the three adaptive traits evolve, I repeat the rules of egg laying and annual survival for 1000 years with the evolution of the proportion p and the rejection ability r being frozen (these are kept 0) in order to reach an equilibrium of population dynamics with a stable distribution of egg appearance e. And then I let p and r evolve to realize population dynamics (temporal change of the number of individuals) and evolutionary dynamics of the three traits (temporal change of the distributions of p, r, e in the population.

7.3 Results

7.3.1 Evolution of Conspecific Parasitism

I first analyze how conspecific parasitism can evolve. Figure 7.3 shows an example of the temporal change of the population size H, distributions of the proportion p, and egg appearance e in the population when the evolution of egg rejection is frozen, i.e., the rejection ability r is fixed as $r = 0$ for all individuals.

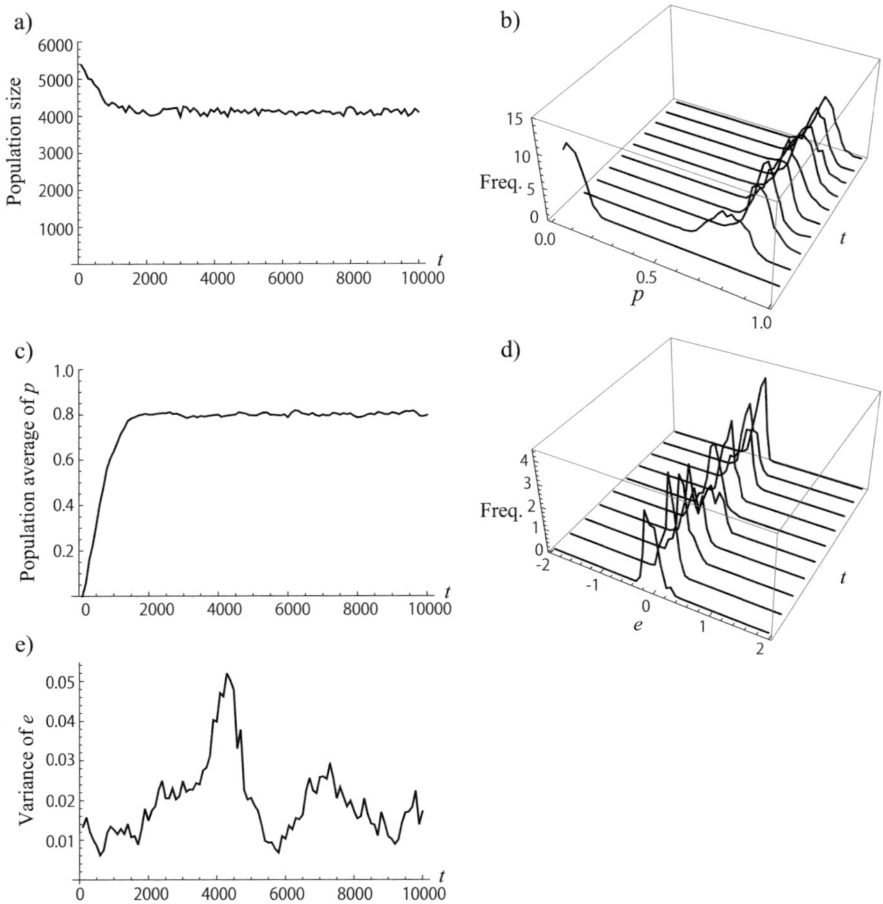

Fig. 7.3 Population and evolutionary dynamics when rejection ability is frozen. Temporal change of (**a**) the population size H, (**b**) the distribution of the proportion p, (**c**) the population average of the proportion p, (**d**) the distribution of the egg appearance e, and (**e**) the variance of the egg appearance e for 10,000 years. Parameters used are $\sigma_p = 0.005$, $\sigma_r = 0$, $\sigma_e = 0.005$, $n = 5$, $l = 5.472 \times 10^{-2}$, $m = 0.5$, $k = 5000$, $\sigma_k = 2$, $i = 0$, $s_H = 0.5$, $s_{\text{egg}} = 0.5$, and $\gamma = 9$

The population size declines as individuals allocate more eggs to conspecific parasitism and individuals rearing a large number of eggs have a reduced reproductive success (Fig. 7.3a, b) Eventually, the population average of the proportion p converges to around 0.8 (Fig. 7.3c). Distribution of the egg appearance e nearly remains the same (Fig. 7.3d) with its population average nearly kept at $e = 0$ (not shown). The variance of egg appearance fluctuates due to stochasticity (Fig. 7.3e). No individuals reject eggs, and egg appearance e is just influenced by the variability

σ_e and the carrying capacity $k(e)$. It has been previously demonstrated that conspecific parasitism can evolve under a certain condition in which potential to rear a clutch plays a key role (Takasu 2004).

7.3.2 Evolution of Conspecific Parasitism and Rejection Ability

I then allow the rejection ability r to evolve together with the proportion p and the egg appearance e. The population size declines as individuals allocate more eggs to conspecific parasitism (Fig. 7.4a, b). The population average of the proportion p converges to around 0.8 again (Fig. 7.4c), and the rejection ability tends to increase, but the distribution of r becomes broader with time (Fig. 7.4d). On the other hand, distribution of the egg appearance e becomes more sharply distributed around $e = 0$ (Fig. 7.4f) with its variance being considerably reduced (Fig. 7.4g). This means that all individuals practice conspecific parasitism at a high level (80% eggs as conspecific parasitism), but they come to have nearly identical eggs (inter-clutch variation is very low); thereby the rejection ability has become neutral to selection. Under such a situation, population average of the rejection ability just drifts randomly as shown in Fig. 7.4h.

When the egg appearance e is assumed to be more variable in the sense that it has larger variance in the offspring trait, evolutionary trajectory of (p, r, e) becomes very different from that shown in Fig. 7.4. The population size initially declines as individuals allocate more eggs to conspecific parasitism (Fig. 7.5a). However, the population average of the proportion p starts to decrease to a low level (Fig. 7.5b, c) as more individuals have higher rejection ability r being nearly 100% (Fig. 7.5d, e). In the meantime, distribution of the egg appearance e becomes narrowed initially, but eventually it becomes widely distributed around $e = 0$ with its variance being large again (Fig. 7.5f, g). Evolutionary trajectory of the population-averaged proportion p and rejection ability r converges to an equilibrium state where the degree of conspecific parasitism is low but the rejection ability is high (Fig. 7.5h). In this equilibrium, individuals have various egg appearances (inter-clutch variation being large), and this enables them to discriminate and reject conspecific parasitic eggs in the clutch.

7.3.3 Evolution of Conspecific Parasitism and Rejection Ability in the Presence of Interspecific Parasitism

I finally introduce interspecific parasitism, i.e., each individual can be interspecifically parasitized with a certain probability in addition to conspecific parasitism. Temporal change in the population size and the distributions of adaptive traits are similar to the previous cases without interspecific parasitism except for that the distribution of the rejection ability now keeps its unimodality and its population average converges to nearly 1. Figure 7.6 shows the evolutionary trajectories of the population average of the proportion p and the rejection ability r when all individuals

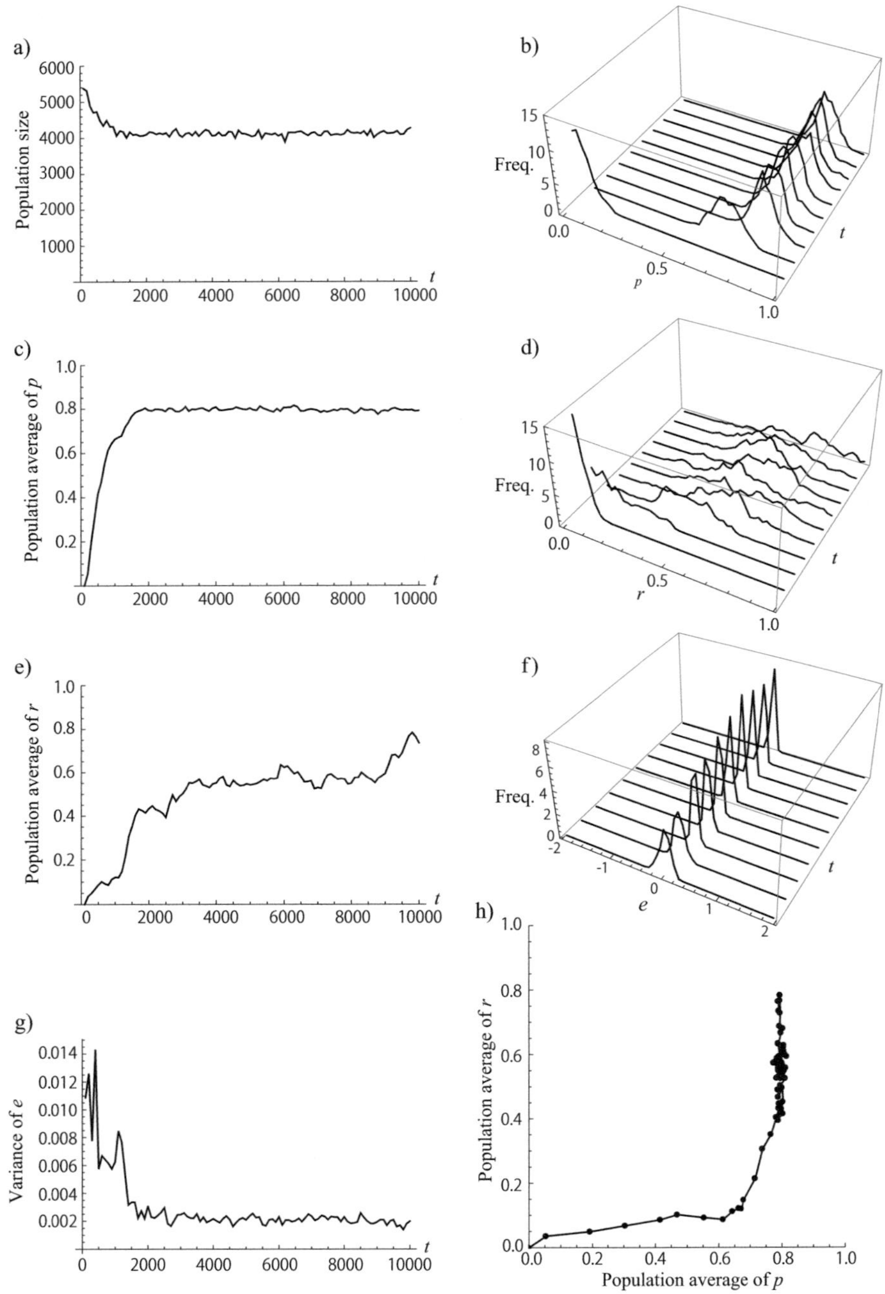

Fig. 7.4 Population and evolutionary dynamics of the three adaptive traits. Temporal change of (**a**) the population size H, (**b**) the distribution of the proportion p, (**c**) the population average of the proportion p, (**d**) the distribution of the rejection ability r, (**e**) the population average of the rejection ability r, (**f**) the distribution of the egg appearance e, and (**g**) the variance of the egg appearance e for 10,000 years. (**h**) Trajectory of the population average of the proportion p and the rejection ability

can be interspecifically parasitized with the probability $i = 0.05$ and 0.1. (On average 5% and 10% of individuals are interspecifically parasitized.) The higher the probability of interspecific parasitism is, the faster the rejection ability evolves. When variability of the egg appearance e is small enough, the variance of e becomes smaller, and all individuals become to have nearly identical egg appearance (inter-clutch variation being very low). But the rejection ability converges to nearly 1 because the egg appearance of interspecific parasites is assumed to be very different $e_i = 1$. However, when the egg appearance e is assumed to be variable, the variance of e is kept large, and the trajectory converges to an equilibrium state where individuals invest less eggs as conspecific parasitism but keep high rejection ability. In either case with interspecific parasitism, the rejection ability r evolves faster compared with the cases without interspecific parasitism, and it eventually converges towards 1, irrespective of the variability of egg appearance.

7.4 Discussion

An individual-based model has been constructed in which each individual is assigned a set of the three adaptive traits, the proportion of eggs laid as conspecific parasitism p, the rejection ability r, and the egg appearance e as quantitative trait, and individual's behaviors (egg laying and egg rejection) and survival have been algorithmically described. In this model, the number of individuals and the distributions of these traits in the population stochastically change with time. Simulation analysis has shown that rejection ability can evolve, and it can be maintained in the population.

Variability of egg appearance σ_e plays a key role for the evolution and the maintenance of the rejection behavior. If it is kept low and all individuals eventually have nearly identical eggs (inter-clutch variation being considerably reduced), rejection ability becomes neutral to selection; the population converges to an equilibrium state where individuals invest more eggs to conspecific parasitism, but egg rejection just drifts neutrally. However, if egg variability is greater, each individual has different egg appearance which enables rejection ability to have a foothold to evolve; the population converges to an equilibrium state where rejection ability is kept at a high level, but individuals invest less eggs as conspecific parasitism because parasitism is likely rejected. These results suggest that maintenance of egg rejection in CBP critically depends on the variability of egg appearance in the population.

In this model, variability of egg appearance in the population, or inter-clutch variation, is controlled mainly by two factors: the carrying capacity $k(e)$ as external

Fig. 7.4 (continued) r with closed circles representing a state at every 100 years. Parameters used are $\sigma_p = 0.005$, $\sigma_r = 0.005$, $\sigma_e = 0.005$, $n = 5$, $l = 5.472 \times 10^{-2}$, $m = 0.5$, $k = 5000$, $\sigma_k = 2$, $i = 0$, $s_H = 0.5$, $s_{egg} = 0.5$, and $\gamma = 9$

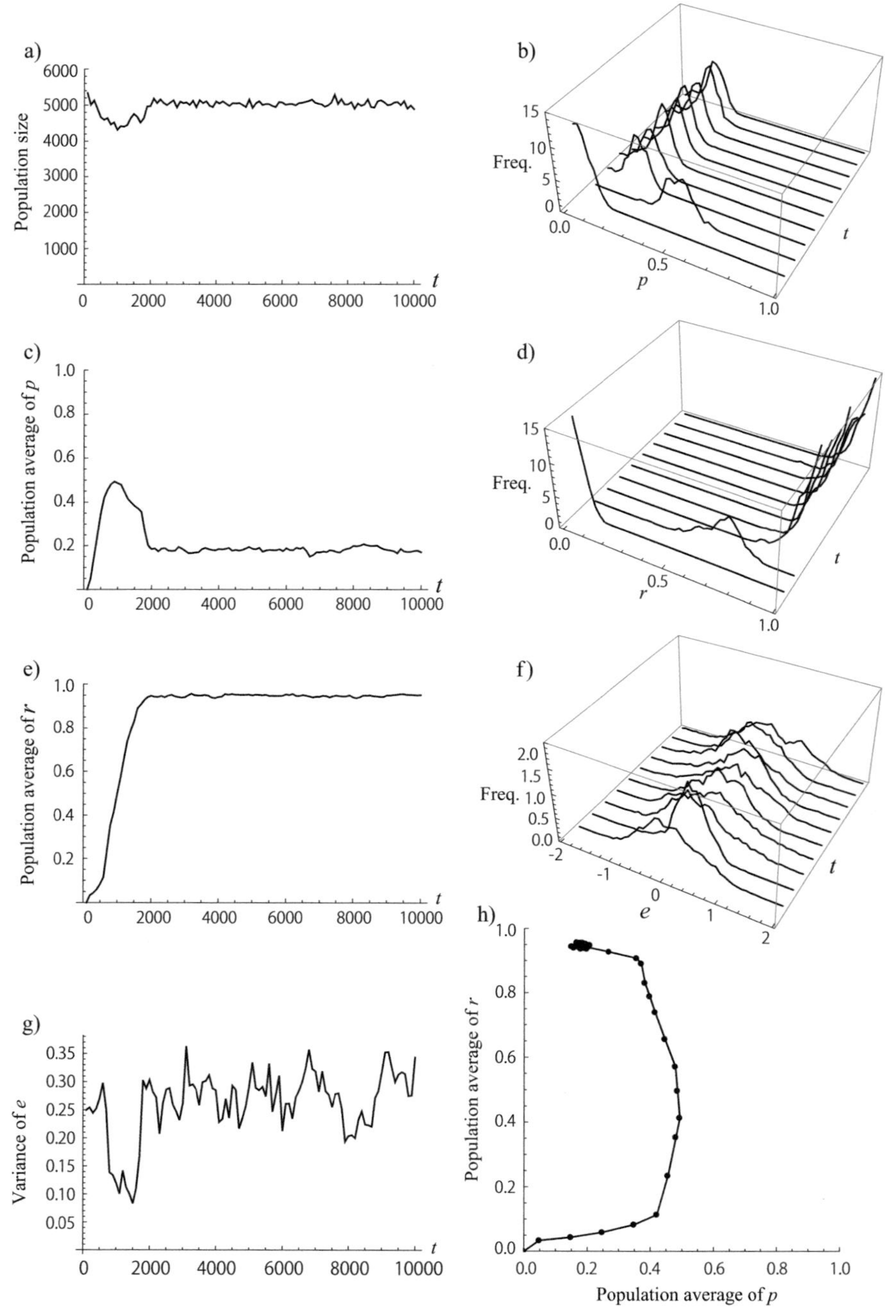

Fig. 7.5 Population and evolutionary dynamics of the three adaptive traits when egg appearance is more variable. Temporal change of (**a**) the population size H, (**b**) the distribution of the proportion p, (**c**) the population average of the proportion p, (**d**) the distribution of the rejection ability r, (**e**) the population average of the rejection ability r, (**f**) the distribution of the egg

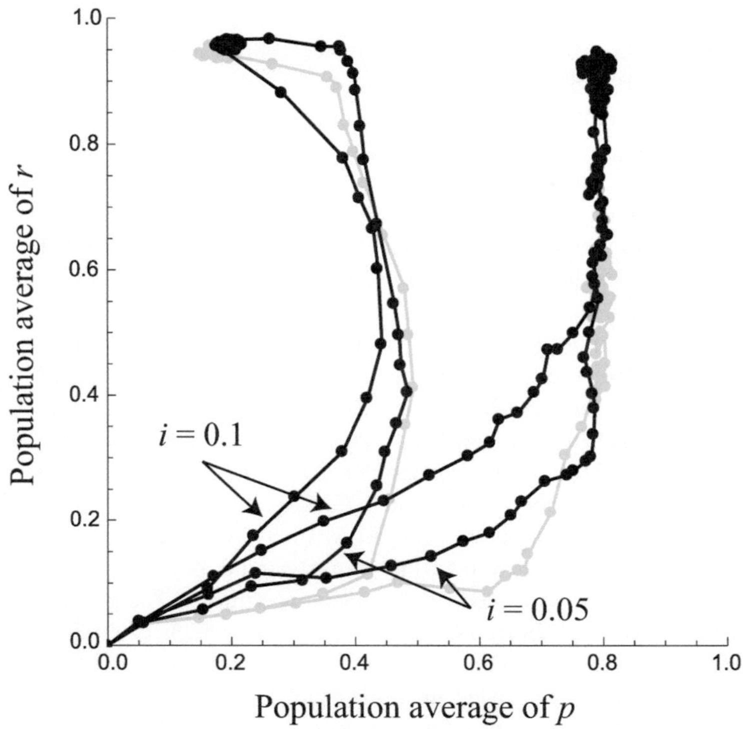

Fig. 7.6 Evolutionary trajectories in the presence of interspecific parasitism. Evolutionary trajectories of the population average of the proportion p and the rejection ability r in the presence of interspecific parasitism with $i = 0.05$ and 0.1. Closed circles represent a state at every 100 years. Trajectories in the absence of interspecific parasitism $i = 0$ are shown in gray (the same as Figs. 7.4h and 7.5h). Parameters used are $\sigma_p = 0.005$, $\sigma_r = 0.005$, $\sigma_e = 0.005$ and 0.05, $n = 5$, $l = 5.472 \times 10^{-2}$, $m = 0.5$, $k = 5000$, $\sigma_k = 2$, $s_H = 0.5$, $s_{\text{egg}} = 0.5$, and $\gamma = 9$

factor and the evolution of the three adaptive traits as internal factor. The former restricts unlimited divergence of egg appearance with selection pressure favoring the appearance $e = 0$ in the absence of parasitism. The latter is the evolutionary dynamics in focus that leads to the qualitatively different equilibria depending on the variability of egg appearance σ_e (Fig. 7.6). When σ_e is smaller, variability of egg appearance in the population decreases to a very low level (Fig. 7.4g), but when it is larger, the variability eventually recovers to the level similar to the initial level (Fig. 7.5g). This suggests that CBP does not select for a larger inter-clutch variation although it can select for higher rejection ability. In an extreme case that there is no

Fig. 7.5 (continued) appearance e, and (**g**) the variance of the egg appearance e for 10,000 years. (**h**) Trajectory of the population average of the proportion p and the rejection ability r with closed circles representing a state at every 100 years. Parameters used are $\sigma_p = 0.005$, $\sigma_r = 0.005$, $\sigma_e = 0.05$, $n = 5$, $l = 5.472 \times 10^{-2}$, $m = 0.5$, $k = 5000$, $\sigma_k = 2$, $i = 0$, $s_H = 0.5$, $s_{\text{egg}} = 0.5$, and $\gamma = 9$

external selection pressure favoring $e = 0$, i.e., $\sigma_k = \infty$, CBP may select for higher inter-clutch variation, but in this case, initial level of inter-clutch variation before the evolution of CBP and the rejection ability is allowed to start will be very high. Therefore, CBP seemingly may not select for high inter-clutch variation.

The model analysis has reconfirmed a natural consequence that interspecific parasitism can drive the evolution of egg rejection as has been already demonstrated by previous theoretical studies (Kelly 1987; Brooker et al. 1990; Takasu et al. 1993; Davies et al. 1996; Takasu 1998; Britton et al. 2007). The model analysis, however, shows that even in the absence of interspecific parasitism, rejection ability can evolve and be maintained as long as there exists variation in egg appearance and recognition errors and rejection costs do not exist. Considering both interspecific parasitism and CBP is certainly a big challenge to explore how egg rejection ability can evolve and be maintained in general (Yamauchi 1995). However, previous models did not consider variability of eggs in the population. I stress again the importance of variability of eggs (inter-clutch variation) for the evolution and the maintenance of egg rejection ability due to CBP.

Parameter values used in the simulation analysis are arbitrary without any biological background. Therefore, the results should be considered as logical consequences possible in the model as "thought experiment." There seemingly exists a threshold for the variability of egg appearance σ_e which determines to which equilibrium state the population converges. Exploring the threshold in relation to other parameter values is needed in further analysis. Egg appearance is inherited asexually to daughter by mother for some bird species (Gibbs et al. 2000; Gosler et al. 2000; Fossøy et al. 2016), but for other species, both males and females contribute to daughters' egg appearance (Collias 1993). Exploring other modes of inheritance of egg appearance, e.g., multiple loci with several alleles, is needed to test if the results obtained in this model apply to general cases including sexual inheritance of egg appearance.

In this model, the cost of recognizing and rejecting unlike eggs has been completely ignored. However, it has been found that birds may mistakenly consider own eggs as parasitic and reject some of them (recognition errors) and that they may mistakenly damage own eggs while rejecting a parasitic egg (cost of rejection) (Rothstein 1982; Davies and Brooke 1989; Lotem et al. 1992; Marchetti 1992; Davies et al. 1996; Stokke et al. 2007, 2016; Samas et al. 2014). These costs will prevent the rejection ability to evolve. Therefore, it should be noted that the evolution and maintenance of rejection ability are overestimated in this model. For example, if I introduce these costs in the model, rejection ability as nearly a neutral trait will not evolve anymore and could not be maintained in the case that variability of egg appearance is very low and all individuals have nearly identical eggs as shown in Fig. 7.4.

The model also ignores potential restrictions to perform CBP; individuals can freely lay a certain number of own eggs parasitically without paying any cost related to nest owner attack, appropriate timing in nesting stage, etc. This could have resulted in a higher investment of own eggs as parasitic (~80%), and hence the evolution of egg rejection might have been overestimated.

The model considered in this chapter is made of several rules given as a set of algorithm. As mentioned in the introduction, it has an advantage as well as disadvantage. Such an algorithmic model should be reconstructed as an "analytical" model with mathematical expressions to fully study under what conditions the different outcomes suggested in this model can be realized and to study parameter dependency. Further analytical study with explicit consideration of the distribution of egg appearance is worth challenging based on the general frameworks (Broom and Ruxton 2004; Valpine and Eadie 2008; Jaatinen et al. 2011).

Back to the real world, it has remained an open question if egg rejection has really evolved in response to CBP. It has been demonstrated that the house sparrow *Passer domesticus* that practices CBP has a fine ability to discriminate and reject conspecific eggs (Soler et al. 2011. But see Manna et al. 2017). If they can recognize and reject conspecific eggs, there should exist a measurable inter-clutch variation to a certain degree. Samas et al. (2014) suggested that egg rejection observed in *Turdus* thrushes has evolved due to conspecific parasitism. But Soler (2014) and Ruiz-Raya et al. (2016) suggest it unlikely that egg rejection has evolved as a response to conspecific parasitism in the thrushes. Because these thrushes could have been parasitized by the common cuckoo, egg rejection might have evolved due to interspecific parasitism.

Variability of eggs has been yet to be explored in depth both for the house sparrow and the thrushes. Recently, Hanley et al. (2017) demonstrate that egg recognition by a *Turdus* species is specifically tuned to the natural gradient of eggshell color. As the model analysis has shown, egg rejection can logically evolve and be maintained due to conspecific parasitism in the absence of interspecific parasitism depending on the variability of egg appearance in the population. Conspecific eggs are similar to each other by nature. Any visual difference either in eggshell color or pattern, or both, however, should exist if some individuals can recognize and reject conspecific parasitic eggs. Objectively measuring an egg's appearance is certainly needed (Spottiswoode and Stevens 2010, 2012; Stoddard and Stevens 2010, 2011). Estimation of the variability at a population level may reconcile the contrasted results and provide important information to understand how egg rejection has evolved in CBP. Egg appearance is a complex phenotypic trait, not easy to quantify. Development of measures with precision is certainly worth the challenge.

Concluding Remarks and Future Directions

More and more bird species have been found to practice CBP. However, it has been remained an open question if CBP per se can drive the evolution of egg rejection ability, as has been demonstrated in interspecific brood parasitism. In order to theoretically explore the evolution and maintenance of egg rejection ability driven by CBP, an individual-based model is constructed as an algorithmic model where individuals reproduce and survive according to a set of adaptive traits. The model analysis shows that egg rejection can logically

(continued)

evolve, and it can be maintained due to CBP if there exist a certain variability of egg appearance and an absence of both recognition errors and rejection costs in the population. Some important factors such as recognition errors and rejection costs when practicing egg rejection defense have been ignored in the model, and this could have resulted in overestimation of the evolution of egg rejection. Quantitative estimation of these costs that will prevent the evolution of egg rejection in CBP should be pursued. Further theoretical study to explicitly consider these factors in an analytical way and to quantify a parameter range that allows egg rejection ability to evolve is also needed. Objectively measuring variability of eggshell color and patterns may reconcile the contrast disputed for the origin of egg rejection ability in CBP. Further empirical study to quantify egg variability is needed.

Acknowledgments I thank Manolo Soler for giving me a chance to write a chapter for this book. I also thank N. Baran, T. Grim, and M. Soler for constructive comments and suggestions to improve the chapter content.

References

Britton NF, Planque R, Franks NR (2007) Evolution of defence portfolios in exploiter-victim systems. Bull Math Biol 69:957–988

Brooke MD, Davies NB (1988) Egg mimicry by cuckoos *Cuculus canorus* in relation to discrimination by hosts. Nature 335:630–632. https://doi.org/10.1038/335630a0

Brooker LC, Brooker MG, Brooker AMG (1990) An alternative population/genetic model for the evolution of egg mimesis and egg crypsis in cuckoos. J Theor Biol 146:123–143

Broom M, Ruxton GD (2002) A game theoretical approach to conspecific brood parasitism. Behav Ecol 13:321–327

Broom M, Ruxton GD (2004) A framework for modelling and analysing conspecific brood parasitism. J Math Biol 48:529–544

Collias EC (1993) Inheritance of egg-color polymorphism in the village weaver (*Ploceus cucullatus*). Auk 110:683–692

Davies NB (2000) Cuckoos, cowbirds and other cheats. Poyser, London

Davies NB, Brooke ML (1988) Cuckoos versus reed warblers. Adaptations and counter-adaptations. Anim Behav 36:262–284

Davies NB, Brooke ML (1989) An experimental study of co-evolution between the cuckoo, *Cuculus canorus*, and its hosts. II. Host egg markings, chick discrimination and general discussion. J Anim Ecol 58:225–236

Davies NB, Brooke ML, Kacelnik A (1996) Recognition errors and probability of parasitism determine whether reed warblers should accept or reject mimetic cuckoo eggs. Proc R Soc Lond B 263:925–931

Fossøy F, Sorenson MD, Liang W, Ekrem T, Moksnes A, Møller AP, Rutila J, Røskaft E, Takasu F, Yang C, Stokke BG (2016) Ancient origin and maternal inheritance of blue cuckoo eggs. Nat Commun 7:10272. https://doi.org/10.1038/ncomms10272

Gibbs HL, Sorenson MD, Marchetti K, Brooke ML, Davies NB, Nakamura H (2000) Genetic evidence for female host-specific races of the common cuckoo. Nature 407:183–186

Gosler AG, Barnett PR, Reynolds SJ (2000) Inheritance and variation in eggshell patterning in the great tit *Parus* major. Proc R Soc Lond B 267:2469–2473

Hanley D, Grim T, Igic B, Samas P, Lopez AV, Shawkey MD, Hauber ME (2017) Egg discrimination along a gradient of natural variation in eggshell coloration. Proc R Soc Lond B 284:20162592. https://doi.org/10.1098/rspb.2016.2592

Honza M, Moksnes A, Røskaft E, Stokke BG (2001) How are different common cuckoo *Cuculus canorus* egg morphs maintained? An evaluation of different hypotheses. Ardea 89:341–352

Jaatinen K, Lehtonen J, Kokko H (2011) Strategy selection under conspecific brood parasitism: an integrative modeling approach. Behav Ecol 22:144–155

Kelly C (1987) A model to explore the rate of spread of mimicry and rejection in hypothetical populations of cuckoos and their hosts. J Theor Biol 125:283–299

Lack D (1947) The significance of clutch size. Ibis 89:309–352

Lahti DV (2005) Evolution of bird eggs in the absence of cuckoo parasitism. PNAS 102:18057–18062

Lotem A, Nakamura H, Zahavi A (1992) Rejection of cuckoo eggs in relation to host age: a possible evolutionary equilibrium. Behav Ecol 3:128–132

Lyon BE (1998) Optimal clutch size and conspecific brood parasitism. Nature 392:380–383

Lyon BE, Eadie JM (2008) Conspecific brood parasitism in birds: a life-history perspective. Annu Rev Ecol Evol Syst 39:343–363

Manna T, Copper C, Baylis S, Shawkey MD, Waterhouse GIN, Grim T, Hauber ME (2017) Does the house sparrow *Passer domesticus* represent a global model species for egg rejection behavior? J Avian Biol 48:346–352. https://doi.org/10.1111/jav.01193

Marchetti K (1992) Costs to host defence and the persistence of parasitic cuckoos. Proc R Soc Lond B 248:41–45

Moksnes A, Røskaft E (1995) Egg-morphs and host preference in the common cuckoo (*Cuculus canorus*): an analysis of cuckoo and host eggs from European museum collections. J Zool 236:625–648. https://doi.org/10.1111/j.1469-7998.1995.tb02736.x

Moksnes A, Røskaft E, Braa AT, Korsnes L, Lampe HM, Pedersen HC (1991) Behavioural responses of potential hosts towards artificial cuckoo eggs and dummies. Behaviour 116:64–89

Payne RB (1977) The ecology of brood parasitism in birds. Annu Rev Ecol Syst 8:1–28

Rothstein SI (1975) Evolutionary rates and host defenses against avian brood parasitism. Am Nat 109:161–176

Rothstein SI (1982) Mechanisms of avian egg recognition: which egg parameters elicit responses by rejecter species? Behav Ecol Sosiobiol 11:229–239

Rothstein SI (1990) A model system for co-evolution: avian brood parasitism. Annu Rev Ecol Syst 21:481–508

Ruiz-Raya F, Soler M, Roncali G, Abaurrea T, Ibanes-Alamo JD (2016) Egg rejection in blackbirds *Turdus merula*: a by-product of conspecific parasitism or successful resistance against interspecific brood parasites? Front Zool 13:16. https://doi.org/10.1186/s12983-016-0148-y

Ruxton GD, Broom M (2002) Intraspecific brood parasitism can increase the number of eggs that an individual lays in its own nest. Proc R Soc Lond B 269:1989–1992

Samas P, Hauber ME, Cassey P, Grim T (2014) Host responses to interspecific brood parasitism: a by-product of adaptations to conspecific parasitism? Front Zool 11:34

Soler M (2014) No evidence of conspecific brood parasitism provoking egg rejection in thrushes. Front Zool 11:68. https://doi.org/10.1186/s12983-014-0068-7

Soler M, Ruiz-Castellano C, Fernandez-Pinos MC, Rösler A, Ontanilla J, Perez-Contreras T (2011) House sparrows selectively eject parasitism conspecific eggs and incur very low rejection costs. Behav Ecol Sociobiol 65:1998–2005

Spottiswoode CN, Stevens M (2010) Visual modeling shows that avian host parents use multiple visual cues in rejecting parasitic eggs. PNAS 107:8672–8676

Spottiswoode CN, Stevens M (2012) Host-parasite arms races and rapid changes in bird egg appearance. Am Nat 179:633–648

Stoddard MC, Stevens M (2010) Pattern mimicry of host eggs by the common cuckoo, as seen through a bird's eye. Proc R Soc Lond B 277:1387–1393

Stoddard MC, Stevens M (2011) Avian vision and the evolution of egg color mimicry in the common cuckoo. Evolution 65:2004–2013

Stokke BG, Takasu F, Moksnes A, Røskaft E (2007) The importance of clutch characteristics and learning for antiparasite adaptations in hosts of avian brood parasites. Evolution 61:2212–2228

Stokke BG, Røskaft E, Moksnes A, Møller AP, Antonov A, Fossøy F, Liang W, López-Iborra G, Moskát C, Shykoff J, Soler M, Vikan JR, Yang C, Takasu F (2016) Disappearance of eggs from nonparasitized nests of brood parasite hosts: the evolutionary equilibrium hypothesis revisited. Biol J Linn Soc 118:215–225

Takasu F (1998) Why do all host species not show defense against avian brood parasitism: evolutionary lag or equilibrium? Am Nat 151:193–205

Takasu F (2003) Co-evolutionary dynamics of egg appearance in avian brood parasitism. Evol Ecol Res 5:345–362

Takasu F (2004) How many eggs should be laid in one's own nest and others' in intraspecific brood parasitism? Popul Ecol 46:221–229

Takasu F, Kawasaki K, Nakamura H, Cohen JE, Shigesada N (1993) Modeling the population dynamics of a cuckoo-host association and the evolution of host defenses. Am Nat 142:819–839

Valpine P, Eadie JM (2008) Conspecific brood parasitism and population dynamics. Am Nat 172:547–562

Yamauchi A (1993) Theory of intraspecific nest parasitism in birds. Anim Behav 46:335–345

Yamauchi A (1995) Theory of evolution of nest parasitism in birds. Am Nat 145:434–456

Yang C, Liang W, Cai Y, Shi S, Takasu F, Møller AP, Antonov A, Fossøy F, Moksnes A, Røskaft E, Stokke BG (2010) Coevolution in action: disruptive selection on egg colour in an avian brood parasite and its host. PLoS One 5:e10816

Yom-Tov Y (1980) Intraspecific nest parasitism in birds. Biol Rev 55:93–108

Yom-Tov Y (2001) An updated list and some comments on the occurrence of intraspecific nest parasitism in birds. Ibis 143:133–143

Part III

Brood Parasites as Study Models

Interspecific Avian Brood Parasitism as a Model System for Exploring Ecological and Evolutionary Questions Related to Habitat Selection

8

Juan J. Soler

Abstract

Characteristics of avian brood parasitism make it an appropriate system to explore general ecological and evolutionary questions. This is mainly because the selection pressures due to environmental conditions that are working on hosts and on brood parasites are asymmetric, and the fact that hosts and brood parasites develop in similar environmental conditions, which allows for testing exclusive predictions associated with ecological and evolutionary hypotheses. I here concentrate on the appropriateness of brood parasitism for exploring questions directly or indirectly related to habitat selection and dispersal. I introduce the importance of social information and risk of parasitism on habitat selection by hosts and parasites, which, among others, would influence the geographic distribution and abundance of counterparts. I also introduce the ecological implications of evolutionary outcomes of the interactions between parasites and their hosts, which will partially determine the coevolutionary dynamics between parasite and hosts. Furthermore, to exemplify the advantages of using brood parasitism to explore these questions, I discuss some published papers that address each treated subject, mainly those from the research group I have been working with.

8.1 Introduction

Brood parasitism is a breeding strategy in which the parasite takes advantage of the parental behaviour of their hosts. The parasitic individuals lay their eggs in nests of others who will care for and raise the parasitic offspring. Therefore, brood parasites

J. J. Soler (✉)
Departamento de Ecología Funcional y Evolutiva, Estación Experimental de Zonas Áridas (EEZA-CSIC), Almería, Spain
e-mail: jsoler@eeza.csic.es

M. Soler (ed.), *Avian Brood Parasitism*, Fascinating Life Sciences,
https://doi.org/10.1007/978-3-319-73138-4_8

reduce their expenses in reproduction while inflicting costs on their hosts (Davies 2000). The brood parasitism reproductive tactic is taxonomically widespread as it has been described in a variety of arthropod, fish and bird taxa; it may occur intra- and interspecifically, and brood parasites may be facultative or obligate (Roldán and Soler 2011).

Costs of parasitism select for antiparasitic defences in hosts that depending on the strength of parasitism selection pressure will rapidly spread within the host population. On the other hand, most of the host defences would incur costs to brood parasites (i.e. defences of resistance) that select for parasite counter-adaptations that hosts should overcome again and again in the evolutionary arms races in which most parasites and hosts are involved (Rothstein 1990; Davies 2000; Feeney et al. 2014; Soler 2014). Obligate interspecific avian brood parasites (hereafter brood parasites) impose more severe costs on their hosts than facultative or intraspecific brood parasites. Thus, adaptations and counter-adaptations are more easily detected in obligate parasite/host systems, which have been proposed as one of the most appropriate models for studying coevolutionary dynamics (Rothstein 1990).

Brood parasite–host systems serve as models for studying evolutionary questions (Rothstein 1990; Feeney et al. 2014), particularly those related with coevolution. Avian brood parasitism is also appropriate to explore other evolutionary questions related to sexual selection and the evolution of signals and more traditional ecological questions including those related to climate change, optimal habitat selection and dispersal of hosts and parasites associated with the strength of selection in the parasite–host dynamic. Here I review some of these works highlighting characteristics of avian brood parasitism demonstrating that they are appropriate systems to explore general ecological and evolutionary questions.

The appropriateness of avian brood parasitism to explore general ecological questions is due to the asymmetric selection pressures that environmental conditions exert on hosts and on brood parasites, which allows exclusive predictions dealing with differential effects (i.e. interactions) depending on environmental conditions. Moreover, in contrast to other parasites, avian brood parasites share with their avian hosts most of the evolutionary history. Avian brood parasites and their hosts are within the subclass Neornithae, which allow the interpretation of some characteristics of brood parasites as consequences of the parasitic lifestyle more directly than when comparing drastic morphological and physiological differences between parasites and hosts belonging to different phyla. In addition, because avian brood parasites use nests and parental care of their hosts for reproduction, hosts and brood parasites experience similar environmental conditions during growing, both abiotic (i.e. temperature and humidity) and biotic (parental care, predators, ectoparasites, microbial environments, etc.). The key point here is that detected differential effects of environmental conditions in hosts and in their brood parasites have to be interpreted as consequence of particularities of the brood parasite–host system. In statistical terms, testing predictions of the effects of environmental conditions on brood parasites and on host system is similar to a pairwise comparison approach where host-parasite pairs at the individual, population or species level are the pairwise items with which predictions of the effects of environmental conditions can be tested.

To demonstrate these points, I first deeply explain reasons underlying the asymmetric selection pressures in brood parasites and hosts. Then, to show the advantage of using brood parasitism to explore such effects, I discuss some published results on the differential effects of climate change on both magpie (*Pica pica*) and their brood parasite great spotted cuckoo (*Clamator glandarius*) nestlings. Later, I introduce the importance of habitat selection in evolutionary ecology and discuss the advantage of exploring related hypotheses and predictions in brood parasite–host systems. I concentrate on the role of social information and risk of parasitism on habitat selection but also on the ecological implications of the evolutionary outcomes of the interactions between brood parasites and their hosts. Defences of hosts and counter-defences of parasites should influence habitat selection of each and, therefore, geographic distribution and abundance of both brood parasites and their hosts. This non-random habitat selection should affect coevolutionary dynamics, and I develop possible consequences of this interaction at the end of the chapter.

8.2 Asymmetric Selection Pressures on Host and Brood Parasites

Selection pressures on hosts and brood parasites for defences and counter-defences are typically asymmetric since parasites need a host for reproduction, but not all hosts will suffer from parasitism (e.g. Schmid-Hempel 2011). Thus, selection for parasite responses to hosts is usually stronger than for host responses to parasites, a situation that are classically described as the 'life-dinner principle' (Dawkins and Krebs 1979).

Several types of environmental conditions also exert asymmetric selection pressures on hosts and on brood parasites (Soler 1999). For instance, food for developing offspring is likely a more limited resource for host than for parasitic offspring. This is mainly because parasitic nestlings often evict, kill or outcompete host offspring and monopolize parental care at the nests (Davies 2000). Moreover, parasitic nestlings are not genetically related to foster parents or foster siblings and, thus, are in general more efficient than host nestlings at manipulating feeding efforts of their parents for their own benefit (Caro et al. 2016). Finally, parasitism from ectoparasites and microorganisms also contributes to asymmetric selection pressures due to resource availability for growth. Host nestlings suffer from generalist and species-specific parasites (i.e. vertically transmitted), whereas brood parasitic nestlings only suffer from generalist parasites (Soler et al. 1999c). Below I explain why this asymmetry in selection pressures related to resource availability is an advantage to explore the effects of environmental conditions on distribution, abundance and evolution of animal populations.

8.2.1 Opportunity for Exploring Exclusive Predictions of Environmental Effects

In theory, through their effects on resource availability, climatic conditions would affect animal phenotypes as well as the interaction between animals including parasites (Møller et al. 2013; Northfield and Ives 2013). These should be species-specific, a prediction that needs testing in different organisms experiencing similar environmental conditions as occur for brood parasite and host nestlings in the same nest or even in the same study area. This hypothesis was tested recently in the great spotted cuckoo-magpie system by comparing climatic conditions and phenotypes of host and parasitic nestlings within the same study population sampled during 11 years (Soler et al. 2014b). The main results were that (1) nestling phenotypes varied among years but differentially for great spotted cuckoo and magpies, (2) nestling phenotypes showed significant among-year covariation with breeding climatic conditions (temperature and precipitation), and (3) these associations differed for cuckoos and magpies for some phenotypic traits. In addition, the strength of the within-year relationships between the probability of ectoparasitism by *Carnus hemapterus* flies and laying date (used as an estimate of the within-year variation in climatic conditions) was negatively affected by the annual accumulated precipitation in April, but no significant differences were detected for both magpies and cuckoos. These results strongly support the hypothesis that variation in climatic conditions result in asymmetric effects on different species. Because the study was performed in a brood parasite–host system, the detected differential effect on probability of ectoparasitism, immunity and body size cannot be explained by different species using different habitat for growing (Soler et al. 2014b), which is a clear advantage when exploring predictions of climate change on interacting species.

8.3 Brood Parasitism as a Model System to Study Habitat Selection

The major aims in ecology are to understand the distribution and abundance of different types of organisms and the physical, the chemical and especially the biological features and the interactions that determine these distributions and abundances (Begon et al. 2006). Habitat selection of animals through dispersal, and associated factors, plays a central role explaining such questions (Clobert et al. 2001).

Increased risk of brood parasitism reduces habitat quality and should affect the habitat selection and/or dispersal behaviour of hosts (Boulinier and Lemel 1996). On the other hand, hosts are the 'habitat' for parasite reproduction, and thus, host selection by parasites should follow general theory of habitat selection (Price 1980). Importantly, because habitat selection determines the selective environment encountered by individuals, they should evolve local adaptations that may result in specialization and speciation of animals. This is the reason why habitat selection is considered one of the effects of natural selection (Endler 1986). Because each of the counterparts in parasite–host interactions is part of the environment of the other, detecting characteristics of hosts and parasites that determine habitat quality for each would help to understand animal decisions about dispersal and habitat selection (Boulinier and Lemel 1996). Moreover, since host and parasite populations usually vary geographically and temporally in characteristics that affects optimal habitat selection of their counterpart, parasite–host systems can be used to explore factors affecting temporal dynamics and geographical patterns of habitat selection (Boulinier and Lemel 1996; Danchin et al. 1998).

8.3.1 Impact of Habitat Selection Studies in Brood Parasite–Host Systems (WOS)

The search in WOS of the topic (i.e. 'habitat selection and parasit*') results in 815 articles published from 1970, when the first article on this matter appeared, to the end of 2015. The number of papers published has increased during this period in a sigmoidal shape (Fig. 8.1). These papers were cited 24,269 times, and the number of citations has increased exponentially during this period (Fig. 8.1). Published articles on habitat selection in relation to brood parasitism ('habitat selection and brood parasit*'), as well as the number of citations, increase at a rate similar to those dealing with parasitism in general (Fig. 8.1). Interestingly, although the number of published articles with brood parasites is decreasing since 2007, the number of citations is increasing exponentially, which suggests that these articles are being cited in studies not directly related to brood parasitism. Studies with brood parasites represent 16.9% of the total number of articles published on habitat selection, and their citations corresponded to the 17.5% of total citations. The average number of citations of papers published with brood parasites (31.3) is very similar to those with parasites in general (29.8). Therefore, studies of brood parasites have contributed to the understanding of habitat selection in the general context of parasitism, and the scientific utility of brood parasite studies, as measured in term of citations, is similar to that of studies performed with other parasites.

8.3.2 Breeding Habitat Selection by Brood Parasites: The Role of Social Information

Most of the work performed within brood parasite–host systems dealing with breeding habitat selection involves host selection by parasites. This literature highlights the importance of risk of predation (Lima 2009), nest site characteristics (Burhans and Thompson 1998), nest defence by hosts (Quinn and Ueta 2008) and antiparasitic defences in hosts (Soler 2014). Particularly interesting is the use of information on habitat quality that parasites may gather from the environment to select particular habitat (Parejo and Avilés 2007). Individuals may collect such information by direct interactions with the environment (i.e. personal information) or by observing the interactions of others (social information), which includes

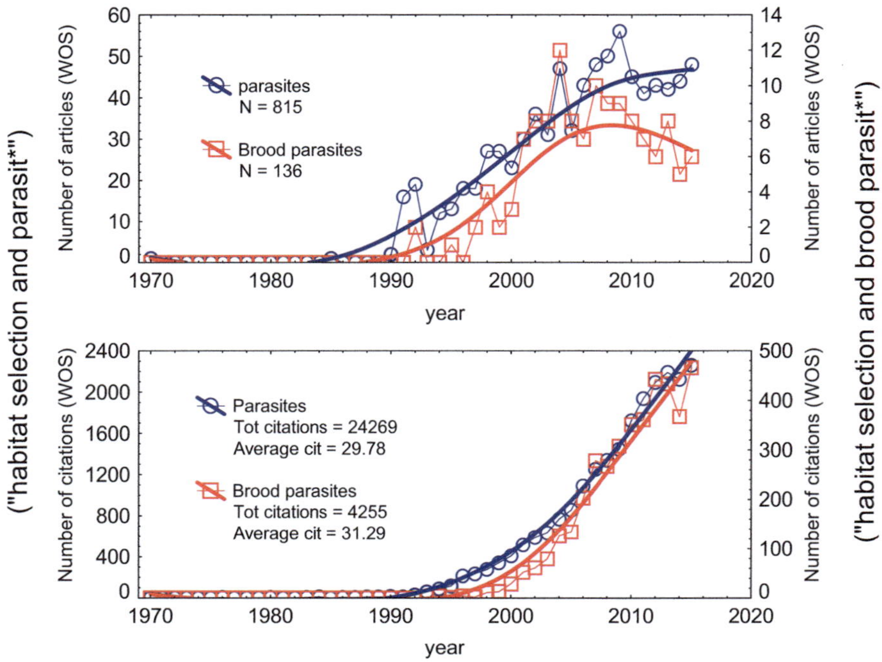

Fig. 8.1 Number of articles published, and number of citations, since 1970 to the end of 2015 on habitat selection and parasitism in general (i.e. 'habitat selection and parasit*') or brood parasitism ('habitat selection and brood parasit*') in particular (source Web of Knowledge)

information arising for cues inadvertently produced by the behaviour of other individuals with similar requirements (Danchin et al. 2004). Considering the role of social information is actually contributing to a better understanding of the process of breeding habitat selection and its evolutionary implications in terms of, for instance, specialization and speciation (Doligez et al. 2003). Brood parasite–host systems are good study models for exploring such implications.

An example of information use by a brood parasite is the use of inadvertent social information by the great spotted cuckoos associated with their magpie hosts for host selection. This system was one of the first where the hypothesis that nest characteristics cue parental abilities of builders was demonstrated (Soler et al. 1998b). Nest size of magpies decreases as the season progresses (Soler et al. 1995) and is positively related to clutch size and later commencement of incubation (i.e. more synchronous hatching) (Soler et al. 2001a). Moreover, immune responses of nestlings depended on nest size (De Neve et al. 2004) and probability of escaping from predation (Quesada 2007; Soler et al. 2014a). Thus, because breeding habitat selection by brood parasites should minimize probability of predation (Avilés et al. 2006) and maximize resource availability for their offspring (i.e. parental quality of their hosts) (Soler et al. 1995), great spotted cuckoos should select magpie nests of large size for parasitism. In accordance with this prediction, probability of brood parasitism was related to nest size in magpies (Soler et al. 1995, 2014a; Molina-Morales et al. 2012b, 2016). Similarly to great spotted cuckoos, Asian koels (*Eudynamys scolopacea*) also select bigger host nests for parasitism (Begum et al. 2011), and shiny cowbirds (*Molothrus bonariensis*) prefer to parasitize more concealed nests, where they could lay undetected by their mockingbird (*Mimus saturninus*) host (Fiorini et al. 2009). However, although a relationship between nest size and phenotypic quality of great reed warbler (*Acrocephalus arundinaceus*) females has been recently detected (Jelinek et al. 2016), preferences of common cuckoos (*Cuculus canorus*) parasitizing large nests of several other host species, including the great reed warbler, have not been detected (Antonov et al. 2007; Avilés et al. 2009). The use of social information by brood parasites to decide breeding habitat selection has also been demonstrated in brown-headed cowbirds parasitizing prothonotary warbler (*Protonotaria citrea*). Female cowbirds decide about which hosts to use based on information of past experimentally manipulated cowbird reproductive success (Louder et al. 2015). These results therefore highlight the importance of social information for breeding habitat selection by brood parasites.

Host traits that are sexually selected will reflect phenotypic conditions of parents directly or indirectly through differential investment in relation to the exaggeration of attractiveness of partner (Burley 1986, 1988; Soler et al. 1998a). Cuing on such traits of hosts by parasites would be of selective advantage for brood parasites (Garamszegi and Avilés 2005; Parejo and Avilés 2007), while for hosts, it would constrain the exaggeration of sexual traits or any other traits that brood parasites might observe via eavesdropping. This interesting possibility can be easily detected in natural conditions within brood parasite–host systems. We know, for instance, that brood parasites may constrain the evolution of life history strategies (Krüger 2007) and the exaggeration of sexual signals related to parental ability (Soler et al. 1999b, 2012) or may reduce risk of nest predation (Soler et al. 2014a). Therefore, brood parasitism is an ideal system for exploring the importance of social information and eavesdropping driving both habitat selection by parasites and the evolution of sexual signals in hosts (Danchin et al. 2004).

8.3.3 Breeding Habitat Selection by Hosts of Brood Parasites: The Role of Risk of Parasitism

Hosts should select breeding habitats of high quality in terms of resource availability and accessibility, but risk of parasitism could also influence host habitat use decisions (Boulinier and Lemel 1996; Boulinier et al. 2001; Johnson 2007). Evidence supporting the predicted relationship between parasitism and habitat selection (i.e. hosts avoid habitat where parasitism risk is high) is however limited and came from data on return rates to the breeding area (Hoover and Reetz 2006; Kolecek et al. 2015) or on breeding dispersal (Boulinier et al. 2001; Møller et al. 2004). Thus, brood parasitism may be a key factor influencing dispersal of animals (Van Baalen and Hochberg 2001; Møller et al. 2004). Hosts may decide whether to repeat breeding in a given location in subsequent reproductive attempts depending on experience with parasites during previous reproductive events or social information of risk of parasitism (Thorogood and Davies 2016). Evidences of a link between breeding dispersal and brood parasitism came from studies with hosts of the parasitic brown-headed cowbird (*Molothrus ater*) (Sedgwick 2004) and of the common cuckoo (Lindholm 1999; Kolecek et al. 2015). However, dispersal of prothonotary warblers (Hoover 2003; Hoover and Hauber 2007) and of magpies was not related to parasitism by the brown-headed cowbird (Hoover 2003; Hoover and Hauber 2007) and the great spotted cuckoo (Molina-Morales et al. 2012a), respectively (although long-distance movements were not monitored in magpies).

Changes in population density of hosts of brood parasites can be explained by increased breeding dispersal of individuals that breed in heavily parasitized areas (Soler et al. 1998c; Martín-Gálvez et al. 2007). If this were true, population dynamics of hosts could then be influenced by the prevalence of brood parasitism. This scenario may be the reason for the very rapid population dynamics reported for magpie subpopulations where magpies even completely disappeared from areas heavily parasitized by the great spotted cuckoos during the previous years (Soler et al. 1994, 1998c; Soler and Soler 2000). Meanwhile, magpie density in geographically closed populations with low prevalence of brood parasitism increased (Soler et al. 1994, 1998c; Soler and Soler 2000). Changes in population densities are so rapid that it seems likely that magpie breeding dispersal in response to parasitism risk is driving the changes in density rather than some other factor.

Changes in host population abundance in association with brood parasitism have also been detected for hosts of brown-headed cowbirds, for which parasitism is considered one of the main factors of host population extinctions (e.g. Wilsey et al. 2014 and references therein). However, in this case, decline of host populations in relation to parasitism is not as rapid as that described for magpies (Ladin et al. 2016) and is mainly explained as the consequence of a reduced fecundity, immigration and recruitment rates and survival of adult hosts (Ladin et al. 2016). An increase rate of nest predation by brood parasites associated with an increased rate of parasitism would also explain detected negative associations between parasitism and host population density (Arcese et al. 1996; Hoover and Robinson 2007). Although differential breeding dispersal in relation to brood parasitism by brown-headed cowbirds (Sedgwick 2004) could help to explain population declines experienced by hosts at the local scales, data supporting this possibility is scarce. This is mainly because host population declines are also observed at larger geographical scales including populations that greatly differ in parasitism selection pressure (Robinson et al. 1995; Ladin et al. 2016).

8.4 Ecological and Evolutionary Implications of Optimal Habitat Selection and Distribution of Hosts and Parasites

Host defences influence reproductive success of parasites, and parasite's counter-defences, including virulence, affect host fitness (see Sect. 8.1). Accordingly, host defences should affect habitat (i.e. host) selection by parasites, and the level of parasite counter-defences and virulence should influence habitat selection by hosts.

The costs of parasitism however would be lower for adaptive hosts with efficient antiparasitic defences, and thus, antiparasitic capabilities should affect habitat selection by hosts. Similarly, habitat selection by parasites with appropriate counter-defence would be independent of the level of host defence. Thus, the geographic and temporal variation in defences and counter-defences of hosts and parasites would not only depend on differential migration (i.e. gene flow) among populations due to variation in selection pressures of hosts and parasites but also of the antiparasitic capabilities of hosts and counter-defences of individuals parasites (Thompson 2005). This differential migration of hosts and parasites depending on traits related to antagonistic interaction would have important consequences for the coevolutionary dynamics of hosts and parasites (Hochberg and Van Baalen 1998; Van Baalen and Hochberg 2001). In this section, I further explain these associations and highlight the appropriateness of brood parasite–host systems to explore them.

8.4.1 Antiparasitic Capabilities and Habitat Selection by Hosts

Because parasites feed on host resources, habitat quality for hosts, as indicated by host productivity in terms of reproductive success in the absence of parasitism, would be an important factor determining the distribution and abundance of hosts and parasites within their geographic range. Moreover, if parasites select populations with high density of hosts (e.g. Brown and Lawes 2007; Stokke et al. 2007), they would be most influential in these areas rather than in areas scarcely used by hosts (Hochberg and Van Baalen 1998). In this scenario, host productivity is directly related to the level of antiparasitic defences that allow hosts to exploit good-quality habitats while diminishing the negative effects of parasitism (Hochberg and Van Baalen 1998). When effects of parasitism are costly enough, hosts are confronted with choosing between leaving the patch where risk of parasitism is high or staying in the patch and resisting. Because dispersal may entail costs (disperser cannot be sure of finding safer patch), the optimal decision of this dilemma would be related to antiparasitic resistance of individual hosts (Boulinier et al. 2001). Dispersing to poorer habitats, where parasitism risk is low, would be the optimal strategy for nondefensive hosts, while investing in local adaptations (i.e. augmenting antiparasitic defences in relation to counter-adaptation of parasites) would be the optimal strategy for hosts with such ability, at least temporally (Van Baalen and Hochberg 2001; Martín-Gálvez et al. 2007; Soler et al. 2013). This scenario would therefore predict a geographically, but not necessarily genetically, structured population of hosts that should differ in level of antiparasitic defences against parasites (Soler et al. 1999a).

Exploring the role that host capabilities of resistance to parasitism play in host movements and population structure is quite difficult. This is mainly because hosts may use different kinds of resistance against the same parasite (i.e. immune responses) or because host resistance to some parasite species can increase susceptibility to other species or even other strains of parasites of the same hosts (Sorci et al. 1997). Brood parasites elicit specific responses of resistance in their hosts (i.e. recognizing and ejecting parasitic eggs), which are unlikely related to resistance capabilities against other kinds of parasites. Thus, brood parasitism is an ideal system to test the predicted associations between parasitism, resistance and dispersal or distribution of animals (Boulinier et al. 2001). In fact, some of the evidences of such association came from avian brood parasitism literature.

We know, for instance, that the enormous among-populations variability in prevalence of defensive resistance in great reed warblers against their brood parasite, the European cuckoo, was positively associated with clutch size [i.e. reflecting availability of resource (Lack 1968; Stokke et al. 2008)]. Moreover, the meta-population of great reed warblers is geographically structured in terms of defensive abilities, but not genetically structured (Procházka et al. 2011). Therefore, it is likely that differential structuring of hosts associated with resistance capabilities explained the association between resistance and resource availability. Similar association between clutch size and defensive resistance of host populations in the absence of genetic population structure was also detected in a meta-population of magpies parasitized by the great spotted cuckoos (Martín-Gálvez et al. 2007).

Within the meta-population of magpies and great spotted cuckoos, further evidence suggests a link between parasitism selection pressure and magpies selecting habitat of high resource availability. Geographic variation of environmental conditions is non-randomly distributed, and individuals and environmental characteristics of geographically adjacent populations are, on average, more similar to each other than those of geographically distant populations (Lawton 1993; Legendre 1993; Hochberg and Van Baalen 1998). Resources are themselves spatially autocorrelated, resulting in spatial gradients of survival and reproduction, i.e. demography (Hochberg and Van Baalen 1998). By taking advantage of the spatial autocorrelations of resource availability in nature, Soler et al. (2013) found support for the prediction that great spotted cuckoo parasitism and magpie host defences were also spatially autocorrelated. In accordance with the assumed spatial autocorrelation of resources, they also showed that clutch sizes of magpies were spatially autocorrelated. Interestingly, an elevated risk of parasitism eliminated the spatial autocorrelation for clutch size, which then became randomly distributed (Soler et al. 2013). Thus, not only parasitism but also defensive abilities should influence geographic distribution of hosts, and brood parasite–host systems may be ideal for exploring predictions derived from this hypothesis.

8.4.2 Host Defence Capabilities and Habitat Selection by Parasites

Parasites should also prefer geographic areas where their hosts experience the highest reproductive success in the absence of parasitism, not only because hosts would be more abundant in these areas but also because resources for individual parasites would also be of better quality (i.e. brood parasitic nestlings will receive abundant good-quality food). This would particularly apply to parasites exploiting host subpopulations with relatively low frequency of phenotypes with effective antiparasitic resistance (Martín-Gálvez et al. 2007). Because in situations of high risk of parasitism hosts with no effective antiparasitic defences may prefer to disperse to geographic locations with lower risk (see above), parasites may also move to new geographical locations once the abundance of hosts became critically low in a target area (Van Baalen and Hochberg 2001). Furthermore, because dispersing hosts are expected to be those with relatively low level of antiparasitic defences, parasites with relatively low level of counter-defences could experience the highest fitness rewards of following host movements (Van Baalen and Hochberg 2001). Therefore, similar to hosts, this scenario could lead to geographically structured populations of parasites that differ in level of counter-adaptation (Soler et al. 1999d).

8.4.3 Habitat Selection by Hosts and Parasites and Coevolutionary Dynamics

Empirical studies of the influence of dispersal and/or habitat selection on antagonistic coevolutionary outcomes between parasites and hosts are relatively scarce in the literature, but in general, the concept parallels predictions of source-sink dynamics. Parasitism selection pressure is an important determinant of the source-sink dynamics of their hosts, and defensive and counter-defensive coevolutionary outcomes in hosts and parasites determine at least partially source-sink dynamics of parasites (Holt 1985). Geographic distribution of hosts and parasites relative to their defensive and counter-defensive skills has important evolutionary implications (Hochberg and Van Baalen 1998; Thompson 2005) (see below), and the brood parasite–host systems offer a poorly used but extraordinary opportunity to study these potential geographic dynamics of coevolutionary interactions (Møller and Soler 2012).

In agreement with the importance of host population density for the distribution and dispersal of brood parasites, European cuckoo parasitism on reed warblers in Europe depends on host density (Stokke et al. 2007). This result indicated that if host density is below a specific threshold, cuckoo parasitism is absent regardless of the state of other potentially confounding variables including host defences (Stokke et al. 2007). Parasites selecting hosts from populations of high density may impose important evolutionary consequences by influencing the probability of specialization and speciation of parasites on more abundant host populations and, thus, coevolutionary dynamics (Jaenike 1990; Norton and Carpenter 1998; Soler et al. 2009). In agreement with this hypothesis, European cuckoos mimic the eggs of the most abundant potential hosts in Europe (Soler et al. 2009). On the other hand, antiparasitic defences of magpie hosts depend on selection pressure exerted by their brood parasite, the great spotted cuckoo, even after correcting for gene flow among subpopulations (Soler et al. 2001b). Thus, the geographic variation of coevolutionary relationships (the geographic mosaic of coevolution) largely depends on factors determining habitat selection by hosts and by parasites (Thompson 2005). Recently published experimental results supported this scenario and showed that the strength and the mode of coevolutionary dynamics depend on resource availability; arms race dynamic between bacteria and virus mainly occurred in experimental scenarios of high resource availability, while fluctuating selection dynamics were detected in artificial arenas of intermediate availability of resources (Lopez Pascua et al. 2014). Arms race and fluctuating selection dynamics occur in different brood parasite–host systems (Soler 2014), and therefore the expected association with proxies of resources availability can be easily tested.

Concluding Remarks and Future Directions

- The asymmetric selection pressures due to environmental conditions on hosts and on brood parasites open a great opportunity for further exploring the effect of climate and global changes on the abundance and distribution of animals, as well as on their interactions. Scientific community would benefit from works testing exclusive predictions in brood parasite–host systems.
- Exploring the effect of previous experience of hosts with brood parasites, or of social cues, is needed for a further understanding of the role of risk of parasitism driving habitat selection of hosts and potentially explaining their population dynamics including risk of extinction. Brood parasitic cowbirds and their hosts are excellent systems to perform such studies.
- Theoretically, parasitism selection pressure is a prime factor determining habitat selection and dispersal of animals. Similarly, antiparasitic defence of hosts would affect habitat selection and dispersal of parasites. Antagonistic interaction between parasites and hosts are therefore main determinants of the geographic distribution and population abundance of animals. Avian brood parasites and their hosts have similar environmental requirements, and thus, the detection of differential habitat selection and/or dispersal for hosts or for their brood parasites would have great impact in the scientific community since it should be interpreted as a consequence of the dynamic of the interaction. Future studies should try to detect such effects in different brood parasite–host systems.
- Interestingly, defence and counter-defence capabilities of hosts and parasites would theoretically affect their habitat selection, geographic distribution and dispersal. Therefore, adaptive phenotypes of parasites and hosts would not be evenly distributed within their geographic range, which would explain the geographic mosaic of coevolution between parasites and hosts. Resource availability determining habitat selection would therefore play a central role explaining dynamics and different outcomes of host-parasite interaction that can be tested in natural condition in meta-populations of avian brood parasites and their hosts.

I hope this review of important subjects in ecology and evolution encourages further research in brood parasite–host systems that attract the attention of a broader audience interested in fundamental ecological and evolutionary questions.

Acknowledgements I thank the constructive comments by Jeff Hoover, Merche Molina-Morales and Manuel Soler on a previous version of the manuscript, which considerably improved and clarified the text. While writing this paper, I benefited from the funding by the Spanish Ministerio de Ciencia e Innovación, Ministerio de Economía y Competitividad and European funds (FEDER) (project CGL2013-48193-C3-1-P).

References

Antonov A, Stokke BG, Moksnes A, Røskaft E (2007) Factors influencing the risk of common cuckoo *Cuculus canorus* parasitism on marsh warblers *Acrocephalus palustris*. J Avian Biol 38:390–393

Arcese P, Smith JNM, Hatch MI (1996) Nest predation by cowbirds and its consequences for passerine demography. Proc Natl Acad Sci U S A 93:4608–4611

Avilés JM, Stokke BG, Parejo D (2006) Relationship between nest predation suffered by hosts and brown-headed cowbird parasitism: a comparative study. Evol Ecol 20:97–111

Avilés JM, Moskat C, Ban M, Hargitai R, Parejo D (2009) Common cuckoos (*Cuculus canorus*) do not rely on indicators of parental abilities when searching for host nests: the importance of host defenses. Auk 126:431–438. https://doi.org/10.1525/auk.2009.08162

Begon M, Townsend CR, Harper JL (2006) Ecology. From individuals to ecosystems, 4th edn. Blackwell, Oxford

Begum S, Moksnes A, Røskaft E, Stokke BG (2011) Factors influencing host nest use by the brood parasitic Asian Koel (*Eudynamys scolopacea*). J Ornithol 152:793–800

Boulinier T, Lemel JY (1996) Spatial and temporal variations of factors affecting breeding habitat quality in colonial birds: some consequences for dispersal and habitat selection. Acta Oecol 17:531–552

Boulinier T, McCoy KD, Sorci G (2001) Dispersal and parasitism. In: Clobert J, Nichols JA, Dhoudt A, Danchin E (eds) Dispersal. Oxford University Press, Oxford, pp 169–179

Brown M, Lawes MJ (2007) Colony size and nest density predict the likelihood of parasitism in the colonial Southern Red Bishop Euplectes orix – Diderick Cuckoo Chrysococcyx caprius system. Ibis 149:321–327

Burhans DE, Thompson FR (1998) Effects of time and nest-site characteristics on concealment of songbird nests. Condor 100:663–672. https://doi.org/10.2307/1369747

Burley N (1986) Sexual selection for aesthetic traits in species with biparental care. Am Nat 127:415–445

Burley N (1988) The differential-allocation hypothesis: an experimental test. Am Nat 132:611–628

Caro SM, West SA, Griffin AS (2016) Sibling conflict and dishonest signaling in birds. Proc Natl Acad Sci U S A 113:13803–13808. https://doi.org/10.1073/pnas.1606378113

Clobert J, Danchin E, Dhondt AA, Nichols JD (2001) Dispersal. Oxford University Press, Oxford

Danchin E, Boulinier T, Massot M (1998) Conspecific reproductive success and breeding habitat selection: implications for the study of coloniality. Ecology 79:2415–2428. https://doi.org/10.1890/0012-9658(1998)079[2415:crsabh]2.0.co;2

Danchin E, Giraldeau LA, Valone TJ, Wagner RH (2004) Public information: from nosy neighbors to cultural evolution. Science 305:487–491

Davies NB (2000) Cuckoos, cowbirds, and others cheats. T & A D Poyser, London

Dawkins D, Krebs JR (1979) Arms races between and within species. Proc R Soc Lond B Biol Sci 205:489–511

De Neve L, Soler JJ, Soler M, Pérez-Contreras T (2004) Nest size predicts the effect of food supplementation to magpie nestlings on their immunocompetence: an experimental test of nest size indicating parental ability. Behav Ecol 15:1031–1036

Doligez B, Cadet C, Danchin E, Boulinier T (2003) When to use public information for breeding habitat selection? The role of environmental predictability and density dependence. Anim Behav 66:973–988. https://doi.org/10.1006/anbe.2002.2270

Endler JA (1986) Natural selection in the wild. Monographs in population biology. Princeton University Press, Princeton, NJ
Feeney WE, Welbergen JA, Langmore NE (2014) Advances in the study of coevolution between avian brood parasites and their hosts. Annu Rev Ecol Evol Syst 45:227–246. https://doi.org/10.1146/annurev-ecolsys-120213-091603
Fiorini VD, Tuero DT, Reboreda JC (2009) Host behaviour and nest-site characteristics affect the likelihood of brood parasitism by shiny cowbirds on chalk-browed mockingbirds. Behaviour 146:1387–1403. https://doi.org/10.1163/156853909x433338
Garamszegi LZ, Avilés JM (2005) Brood parasitism by brown-headed cowbirds and the expression of sexual characters in their hosts. Oecologia 143:167–177
Hochberg ME, Van Baalen M (1998) Antagonistic coevolution over productivity gradients. Am Nat 152:620–634
Holt RD (1985) Population dynamics in two-patch environments: some anomalous consequences of an optimal habitat distribution. Theor Popul Biol 28:181–208. https://doi.org/10.1016/0040-5809(85)90027-9
Hoover JP (2003) Multiple effects of brood parasitism reduce the reproductive success of prothonotary warblers, *Protonotaria citrea*. Anim Behav 65:923–934. https://doi.org/10.1006/anbe.2003.2155
Hoover JP, Hauber ME (2007) Individual patterns of habitat and nest-site use by hosts promote transgenerational transmission of avian brood parasitism status. J Anim Ecol 76:1208–1214
Hoover JP, Reetz MJ (2006) Brood parasitism increases provisioning rate, and reduces offspring recruitment and adult return rates, in a cowbird host. Oecologia 149:165–173. https://doi.org/10.1007/s00442-006-0424-1
Hoover JP, Robinson SK (2007) Retaliatory mafia behavior by a parasitic cowbird favors host acceptance of parasitic eggs. Proc Natl Acad Sci U S A 104:4479–4483
Jaenike J (1990) Host specialization in phytophagous insects. Annu Rev Ecol Evol Syst 21:243–273
Jelinek V, Pozgayova M, Honza M, Prochazka P (2016) Nest as an extended phenotype signal of female quality in the great reed warbler. J Avian Biol 47:428–437. https://doi.org/10.1111/jav.00700
Johnson MD (2007) Measuring habitat quality: a review. Condor 109:489–504. https://doi.org/10.1650/8347.1
Kolecek J, Jelinek V, Pozgayova M, Trnka A, Baslerova P, Honza M, Prochazka P (2015) Breeding success and brood parasitism affect return rate and dispersal distances in the great reed warbler. Behav Ecol Sociobiol 69:1845–1853
Krüger O (2007) Cuckoos, cowbirds and hosts: adaptations, trade-offs and constraints. Philos Trans R Soc Lond Ser B Biol Sci 362:1873–1886
Lack D (1968) Ecological adaptations for breeding in birds. Methuen, London
Ladin ZS, D'Amico V, Baetens JM, Roth RR, Shriver WG (2016) Long-term dynamics in local host-parasite interactions linked to regional population trends. Ecosphere 7:e01420. https://doi.org/10.1002/ecs2.1420
Lawton JH (1993) Range, population abundance and conservation. Trends Ecol Evol 8:409–413
Legendre P (1993) Spatial autocorrelation: trouble or new paradigm. Ecology 74:1659–1673
Lima SL (2009) Predators and the breeding bird: behavioral and reproductive flexibility under the risk of predation. Biol Rev 84:485–513. https://doi.org/10.1111/j.1469-185X.2009.00085.x
Lindholm AK (1999) Brood parasitism by the cuckoo on patchy reed warbler populations in Britain. J Anim Ecol 68:293–309
Lopez Pascua L, Hall AR, Best A, Morgan AD, Boots M, Buckling A (2014) Higher resources decrease fluctuating selection during host-parasite coevolution. Ecol Lett 17:1380–1388
Louder MIM, Schelsky WM, Albores AN, Hoover JP (2015) A generalist brood parasite modifies use of a host in response to reproductive success. Proc R Soc Lond B Biol Sci 282. https://doi.org/10.1098/rspb.2015.1615
Martín-Gálvez D, Soler JJ, Martínez JG, Krupa AP, Soler M, Burke T (2007) Cuckoo parasitism and productivity in different magpie subpopulations predict frequencies of the 457bp allele: a

mosaic of coevolution at a small geographic scale. Evolution 61:2340–2348. https://doi.org/10.1111/j.1558-5646.2007.00194.x

Molina-Morales M, Martínez JG, Avilés JM (2012a) Factors affecting natal and breeding magpie dispersal in a population parasitized by the great spotted cuckoo. Anim Behav 83:671–680

Molina-Morales M, Martínez JG, Martín-Gálvez D, Dawson A, Rodríguez-Ruiz J, Burke T, Avilés JM (2012b) Evidence of long-term structured cuckoo parasitism on individual magpie hosts. J Anim Ecol 82:389–398. https://doi.org/10.1111/1365-2656.12022

Molina-Morales M, Martínez JG, Avilés JM (2016) Criteria for host selection in a brood parasite vary depending on parasitism rate. Behav Ecol 27:1441–1448. https://doi.org/10.1093/beheco/arw066

Møller AP, Soler JJ (2012) A coevolutionary framework based on temporal and spatial ecology of host parasite-interactions: a missing link in studies of brood parasitism. Chin Birds 3:259–273

Møller AP, Martin-Vivaldi M, Soler JJ (2004) Parasitism, host immune defence and dispersal. J Evol Biol 17:603–612

Møller AP et al (2013) Assessing the effects of climate on host-parasite interactions: a comparative study of European birds and their parasites. PLoS One 8:e82886

Northfield TD, Ives AR (2013) Coevolution and the effects of climate change on interacting species. PLoS Biol 11:e1001685. https://doi.org/10.1371/journal.pbio.1001685

Norton DA, Carpenter MA (1998) Mistletoes as parasites: host specificity and speciation. Trends Ecol Evol 13:101–105

Parejo D, Avilés JM (2007) Do avian brood parasites eavesdrop on heterospecific sexual signals revealing host quality? A review of the evidence. Anim Cogn 10:81–88

Price PW (1980) Evolutionary biology of parasites. Monographs in population biology. Princeton University Press, Princeton, NJ

Procházka P et al (2011) Low genetic differentiation among reed warbler Acrocephalus scirpaceus populations across Europe. J Avian Biol 42:103–113

Quesada J (2007) The different roles of the roof density and nest size in the Iberian magpie nest. Acta Ethol 10:41–45

Quinn JL, Ueta M (2008) Protective nesting associations in birds. Ibis 150:146–167. https://doi.org/10.1111/j.1474-919X.2008.00823.x

Robinson SK, Rothstein SI, Brittingham MC, Petit LJ, Grzybowski JA (1995) Ecology and behavior of cowbirds and their impact on host populations. In: Finch DM, Martin TE (eds) Ecology and management of neotropical migratory birds. Oxford University Press, New York, pp 428–460

Roldán M, Soler M (2011) Parental-care parasitism: how do unrelated offspring attain acceptance by foster parents? Behav Ecol 22:679–691

Rothstein SI (1990) A model system for coevolution: avian brood parasitism. Annu Rev Ecol Evol Syst 21:481–508

Schmid-Hempel P (2011) Evolutionary parasitology. The integrated study of infections, immunology, ecology, and genetics. Oxford University Press, Oxford

Sedgwick JA (2004) Site fidelity, territory fidelity, and natal philopatry in willow flycatchers (Empidonax traillii). Auk 121:1103–1121. https://doi.org/10.1642/0004-8038(2004)121[1103:sftfan]2.0.co;2

Soler JJ (1999) Brood parasites: the advantages of being different species. In: Adams NJ, Slotow R (eds) Proceeding of the 22nd international ornithological congress, Durban. BirdLife South Africa, Johannesburg, pp 3098–3106

Soler M (2014) Long-term coevolution between avian brood parasites and their hosts. Biol Rev 89:688–704

Soler JJ, Soler M (2000) Brood-parasite interactions between great spotted cuckoos and magpies: a model system for studying coevolutionary relationships. Oecologia 125:309–320

Soler M, Soler JJ, Martínez JG, Møller AP (1994) Micro-evolutionary change in host response to a brood parasite. Behav Ecol Sociobiol 35:295–301

Soler JJ, Soler M, Møller AP, Martínez JG (1995) Does the great spotted cuckoo choose magpie hosts according to their parenting ability? Behav Ecol Sociobiol 36:201–206

Soler JJ, Cuervo JJ, Møller AP, de Lope F (1998a) Nest building is a sexually selected behaviour in the barn swallow. Anim Behav 56:1435–1442

Soler JJ, Møller AP, Soler M (1998b) Nest building, sexual selection and parental investment. Evol Ecol 12:427–441

Soler M, Soler JJ, Martínez JG, Pérez-Contreras T, Møller AP (1998c) Micro-evolutionary change and population dynamics of a brood parasite and its primary host: the intermittent arms race hypothesis. Oecologia 117:381–390

Soler JJ, Martínez JG, Soler M, Møller AP (1999a) Genetic and geographic variation in rejection behavior of cuckoo eggs by European magpie populations: an experimental test of rejecter-gene flow. Evolution 53:947–956

Soler JJ, Martínez JG, Soler M, Møller AP (1999b) Host sexual selection and cuckoo parasitism: an analysis of nest size in sympatric and allopatric magpie *Pica pica* populations parasitized by the great spotted cuckoo *Clamator glandarius*. Proc R Soc Lond B Biol Sci 266:1765–1771

Soler JJ, Møller AP, Soler M, Martínez JG (1999c) Interactions between a brood parasite and its host in relation to parasitism and immune defence. Evol Ecol Res 1:189–210

Soler JJ, Sorci G, Soler M, Møller AP (1999d) Change in host rejection behavior mediated by the predatory behavior of its brood parasite. Behav Ecol 10:275–280. https://doi.org/10.1093/beheco/10.3.275

Soler JJ, De Neve L, Martínez JG, Soler M (2001a) Nest size affects clutch size and the start of incubation in magpies: an experimental study. Behav Ecol 12:301–307

Soler JJ, Martínez JG, Soler M, Møller AP (2001b) Coevolutionary interactions in a host-parasite system. Ecol Lett 4:470–476. https://doi.org/10.1046/j.1461-0248.2001.00247.x

Soler JJ, Martín-Vivaldi M, Møller AP (2009) Geographic distribution of suitable hosts explains the evolution of specialized gentes in the European cuckoo *Cuculus canorus*. BMC Evol Biol 9:88

Soler JJ, Avilés JM, Møller AP, Moreno J (2012) Attractive blue-green egg coloration and cuckoo-host coevolution. Biol J Linn Soc 106:154–168

Soler JJ, Martin-Galvez D, De Neve L, Soler M (2013) Brood parasitism correlates with the strength of spatial autocorrelation of life history and defensive traits in Magpies. Ecology 94:1338–1346

Soler JJ, Aviles JM, Martin-Galvez D, De Neve L, Soler M (2014a) Eavesdropping cuckoos: further insights on great spotted cuckoo preference by magpie nests and egg colour. Oecologia 175:105–115

Soler JJ, De Neve L, Martin-Galvez D, Molina-Morales M, Perez-Contreras T, Ruiz-Rodríguez M (2014b) Do climatic conditions affect host and parasite phenotypes differentially? A case study of magpies and great spotted cuckoos. Oecologia 174:327–338

Sorci G, Møller AP, Boulinier T (1997) Genetics of host-parasite interactions. Trends Ecol Evol 12:196–200

Stokke BG et al (2007) Host density predicts presence of cuckoo parasitism in reed warblers. Oikos 116:913–922

Stokke BG, Hafstad I, Rudolfsen G, Moksnes A, Møller AP, Røskaft E, Soler M (2008) Predictors of resistance to brood parasitism within and among reed warbler populations. Behav Ecol 19:612–620

Thompson JN (2005) The geographic mosaic of coevolution. The University of Chicago Press, Chicago

Thorogood R, Davies NB (2016) Combining personal with social information facilitates host defences and explains why cuckoos should be secretive. Sci Rep 6:19872

Van Baalen M, Hochberg ME (2001) Dispersal in antagonistic interactions. In: Clobert J, Danchin E, Dhondt AA, Nichols JD (eds) Dispersal. Oxford University Press, Oxford, pp 299–310

Wilsey CB, Lawler JJ, Cimprich D, Schumaker NH (2014) Dependence of the endangered black-capped vireo on sustained cowbird management. Conserv Biol 28:561–571. https://doi.org/10.1111/cobi.12176

The Brown-Headed Cowbird: A Model Species for Testing Novel Research Questions in Animal Ecology, Evolution, and Behavior

9

Brian D. Peer, James W. Rivers, Loren Merrill, Scott K. Robinson, and Stephen I. Rothstein

Abstract

Although the brown-headed cowbird (*Molothrus ater*) is the most intensively studied brood parasite in the world, much of the research on cowbirds has focused on the negative effects of parasitism. Here, we argue that negative attitudes toward the cowbird have overshadowed opportunities this species provides for advancing our understanding of social behavior, physiology, evolution, and ecology and conservation of birds. Cowbirds are widely distributed, amenable to captive study, and easy to study in areas where they are abundant. Cowbird nestlings must communicate to unrelated host parents, but unlike some parasitic nestlings, they have no specialized adaptations for doing so. In some areas they often share nests with relatives, which may influence the degree of virulence host experience. The generalist strategy of the cowbird can be used to answer questions about the impact of high reproductive output on female cowbirds, maternal allocation of resources into eggs, and the consequences of exposure to a range of pathogens while visiting host nests. Cowbirds and their hosts provide a contrast to cuckoo-host systems because they are at an earlier stage of

B. D. Peer (✉)
Department of Biological Sciences, Western Illinois University, Moline, IL, USA
e-mail: BD-Peer@wiu.edu

J. W. Rivers
Department of Forest Ecosystems and Society, Oregon State University, Corvallis, OR, USA

L. Merrill
Illinois Natural History Survey, Prairie Research Institute, University of Illinois, Champaign, IL, USA

S. K. Robinson
Florida Museum of Natural History, Gainesville, FL, USA

S. I. Rothstein
Department of Ecology, Evolution and Marine Biology, University of California, Santa Barbara, CA, USA

M. Soler (ed.), *Avian Brood Parasitism*, Fascinating Life Sciences,
https://doi.org/10.1007/978-3-319-73138-4_9

coevolution, and only a minority of hosts shows effective defenses against parasitism. Cowbirds serve as a model species for studying song learning as aspects of their complex vocalizations are dependent on experience to different degrees. Cowbirds also challenge assumptions of the link between mating systems and parental care because cowbirds are often socially monogamous. Finally, cowbirds are unique among brood parasites for their effects on endangered host species. In this chapter, we discuss the value of cowbirds as a model species in these areas and suggest avenues for future research.

9.1 The Cowbird as a Misconstrued Species

The brown-headed cowbird (*Molothrus ater*, hereafter "cowbird") is the most widespread obligate brood parasite in North America. Cowbirds are a coevolutionary contrast to parasitic cuckoos because they are of a more recent origin (Rothstein et al. 2002), and their interactions with hosts are at an earlier stage of coevolution. Because they are generalists and most of their hosts have evolved limited defenses against parasitism, cowbird parasitism has exacerbated the effects of habitat loss on populations of several endangered bird species (Rothstein and Peer 2005; Peer et al. 2013a). This, combined with a view that cowbirds are misconstrued as "lazy," makes the animosity directed toward cowbirds from laypeople and scientists alike unique among brood parasites (Peer et al. 2013b; Peer and Abernathy 2017).

The negative attitudes displayed toward cowbirds obscure the fact that the cowbird serves as a model species for the study of learning and social behavior, physiology, evolution, and ecology and conservation due to its brood parasitic lifestyle. Indeed, the cowbird is the most intensively studied brood parasite in the world in a range of biological disciplines in part because of its large geographic range, relatively high abundance, and the ease with which it can be studied in captivity. For example, in the past 15 years, there has been an average of 32 publications each year on the common cuckoo (*Cuculus canorus*), whereas there has been an average of 41 publications on the cowbird (Web of Science 2017). Prior to the 1980s–1990s, most research on cowbirds focused on their basic biology, but this changed when concern increased over the cowbird's role in declining neotropical migrant bird populations (Peer et al. 2013a). Rothstein and Peer (2005) clarified a series of myths concerning cowbirds and their relationship to endangered songbird hosts and found that in almost every case, habitat loss was a major factor in population declines. In this chapter, we emphasize this brood parasite as a model species in the hope of stimulating new research. We begin by examining begging behaviors by cowbirds, followed by cowbird physiology, cowbird–host coevolution, cowbird social behavior, and the ecological and conservation implications of cowbird parasitism; lastly, we suggest future research directions.

9.2 Cowbird as a Model System for Studying the Development and Evolution of Nestling Begging Displays

9.2.1 The Influence of Proximate Factors on Cowbird Begging Displays

Like all obligate brood parasites, cowbird nestlings face the challenge of communicating their needs to unrelated host "parents." Critical in this endeavor is their begging behavior, the combination of physical movements and vocalizations to solicit food from parents, as well as responding to appropriate stimuli, such as parental food calls given to elicit begging. Cowbirds typically exhibit a higher response rate to inappropriate stimuli and beg more intensely than host young (Hauber 2003a; Rivers 2009), suggesting that they have evolved an especially rapid begging response to initiate food solicitation more quickly than nestmates and to be first to receive food from parents (Rivers 2007). The cost of this rapid response is that it entails errors and begging at inappropriate times, which may be costly via wasted energy or increased rates of nest predation (Dearborn 1998; Haskell 2002).

Although nestling begging by nonparasitic species changes in response to short-term need (i.e., hunger; Clark 2002), only a handful of studies have evaluated the hunger-begging relationship in cowbirds (Hauber and Ramsey 2003; Lichtenstein and Dearborn 2004; Rivers 2007; Rivers et al. 2013, 2014) despite the prediction that parasites would exhibit a begging response that is invariant relative to short-term need (Lichtenstein and Dearborn 2004). Such studies have mostly found that begging intensity increases with short-term need in cowbirds (Hauber and Ramsey 2003; Lichtenstein and Dearborn 2004; Rivers et al. 2013, 2014). The lone exception came from Rivers (2007) who found that cowbird begging was not strongly influenced by food deprivation and supplementation; however, this could have resulted from a methodological approach whereby begging displays were averaged across a 90-min period during which hunger levels changed. Indeed, additional work with the same population found strong differences in begging intensity as short-term need varied (Rivers et al. 2013, 2014), suggesting that cowbird begging intensity may be an informative signal of need across populations.

Studies have also examined how begging displays change relative to the size of host offspring against which cowbirds compete for food. Cowbirds have fledged from the nests of >140 host species and have been raised by hosts ranging from 8 to 110 g, a >17-fold range that spans 12–210% of the mass of an adult cowbird (Peer and Bollinger 1997; Ortega 1998). Most studies have evaluated cowbirds competing against smaller hosts (Dearborn 1998; Lichtenstein and Sealy 1998) and confirmed that begging is more intense in cowbirds than in hosts. Host parents biased feeding to the largest nestling in the brood, which is typically a cowbird (Dearborn 1998; Lichtenstein and Sealy 1998; Rivers 2007). Only one study (Rivers 2007) quantified begging across the range of host sizes cowbirds encounter, finding that begging intensity was greatest when cowbirds competed against both small and large hosts, with a reduction in begging intensity in similar-sized host nests. This resulted from constraints on the amount of food cowbirds obtained in hosts of different sizes. Although cowbirds outcompeted nestmates of small hosts, food acquisition of small hosts was constrained by low parental feeding abilities. In contrast, parents of large hosts provided adequate food, but cowbirds were outcompeted by larger host young. Cowbirds that competed against nestlings of similar size obtained the greatest amount of food and ultimately reduced begging (Rivers 2007).

One area that remains virtually unexplored is the extent to which proximate factors influencing begging displays (e.g., nestmate size) have long-term consequence on cowbirds that carry into adulthood. Although the long-term consequences of postnatal environments have received widespread attention in birds (Monaghan 2008), only one study has examined how rearing environments influence begging displays downstream in brood parasites (Liu et al. 2016). That study found that cowbird fledgling vocalizations resulted in changes to vocal motor pathway of the forebrain used for vocal imitation in adulthood (Liu et al. 2016). Fledgling cowbirds produced begging vocalizations similar to host nestlings with which they were reared, suggesting they matched host offspring vocalizations (Liu et al. 2016). Therefore, host-specific variation in natal environments may lead to changes in adult vocalizations, which could affect song learning and reproductive success. More research in this area is warranted.

9.2.2 The Influence of Ultimate Factors on Cowbird Begging Displays

Theory posits that three types of costs, predation (Haskell 2002), physiological (Chappell and Bachman 2002), and inclusive fitness (Godfray 1995), collectively shape the expression of offspring begging. Theoretical models of begging in brood parasites assume that parasitic young do not incur inclusive fitness costs because they are raised by unrelated hosts and compete against unrelated host young (Motro 1989; Godfray 1995; Holen et al. 2001). In turn, this leads to the expectation that cowbirds are only constrained by predation and physiological costs and therefore should exhibit more intense begging than nonparasitic young. Few studies have evaluated cowbird begging relative to a nonparasitic close relative, and most have failed to assess links between parental provisioning and cowbird and host begging in nests of different host species (Hauber 2003b; Madden et al. 2005; Rivers 2007). In contrast, two recent studies tested begging displays by cowbirds and red-winged blackbird (*Agelaius phoeniceus*) when both species were raised in standardized heterospecific environments while also evaluating host feeding (Rivers et al. 2013, 2014). Those studies focused on two cowbird hosts that harbor the majority of cowbirds within a heavily parasitized host community in Kansas (Rivers et al. 2012, 2014): the Bell's vireo (*Vireo bellii*), a species that only raises a single cowbird host when parasitized (Kosciuch and Sandercock 2008), and the dickcissel (*Spiza americana*), a preferred cowbird host that typically raises several of its own young when they share the nest with parasitic young (Zimmerman 1983). Despite providing markedly different rearing environments, both studies found that blackbird begging displays were more intense than cowbirds in both host species, against theoretical predictions, raising questions regarding the upper constraints on cowbird begging displays (Rivers et al. 2012, 2014).

A related study conducted at the same location found that multiple parasitism was common within the host community and that >70% of cowbird offspring shared the nest with at least one other cowbird (Rivers et al. 2010, 2012). Genetic analysis revealed the likelihood that two cowbirds sharing a nest were full siblings was 40.4%, indicating that cowbird offspring often competed against close relatives (Rivers et al. 2012), and suggests that the begging of Kansas cowbirds may have been reduced via competition with close kin. Identical experiments were conducted in a separate population (Illinois) where cowbirds are typically reared without other cowbirds to test whether begging intensity was linked to population-level variation in within-brood cowbird relatedness as predicted by theory (Godfray 1995). In support of theoretical expectations, Illinois cowbirds begged more intensively in four distinct components of the begging display: latency to beg, begging score (a composite measure of begging posture and time spent begging), call rate, and call amplitude (Rivers and Peer 2016). Moreover, the two other factors that could explain population-level differences in begging intensity, predation and physiological costs, were similar in the two populations. Therefore, the results from these experiments support the hypothesis that begging intensity in Kansas cowbirds has been reduced by inclusive fitness costs arising from kin to kin competition in host nests. These results are especially noteworthy because they illustrate that the begging displays of an obligate brood parasite, a group that has been viewed as representing the pinnacle of selfishness in developing birds (Davies 2000), can be constrained in the same manner as nonparasitic offspring.

9.2.3 Cowbird Begging Displays and the Evolution of Virulence

Study of nestling virulence has been limited almost exclusively to investigations of directed killing of host young (nestmate killers; Rivers and Peer 2016), which occurs in honeyguides and some cuckoos (Davies 2000). Nestmate killers are not known to exhibit variation in host-killing behavior; in contrast, non-evictors do not cause harm to hosts through direct aggression but instead outcompete host young for food (non-evictors; Rivers and Peer 2016). Begging intensity is constrained, in part, by relatedness (Godfray 1995), so begging serves as the mechanism by which virulence is mediated in non-evicting parasitic young (Kilner 2005; Peer et al. 2013a). In turn, the expression of virulence is influenced by the same factors that shape the intensity of begging displays, including the degree of relatedness between parasitic offspring and their competitors (Kilner 2005; Buckling and Brockhurst 2008). Theoretical models have found that virulence is reduced when related parasites compete for limiting resources (Kilner 2005; Buckling and Brockhurst 2008), but nearly all empirical tests of this idea have been limited to microparasites (bacteria, viruses) that reproduce asexually and are behaviorally and structurally simple (Frank 1994; Rumbaugh et al. 2012). In addition, prior studies have failed to assess low- and high-relatedness parasite populations to evaluate putative mechanisms responsible for how virulence changes over ecological and evolutionary timescales (Buckling and Brockhurst 2008). One exception is the study by Rivers and Peer (2016) that assessed the consequences of population-level variation in begging intensity by evaluating cowbirds in the two populations described above. They found that the more intense begging of cowbirds in the low-relatedness population resulted in >10% reduction in body mass of dickcissel nestlings after only 5 days. This difference was detected despite there being no differences between the populations in dickcissel body mass at hatching when experimental broods were created. Body mass at fledging has been linked to post-fledging survival (Naef-Daenzer et al. 2001), so the population-level differences in virulence observed have the potential to affect host offspring survival and recruitment. Cowbird densities are high throughout much of the Great Plains of the USA, so the relationship between parasite relatedness, begging intensity, and cowbird virulence may hold in other high-density areas and should be the focus of future work. More broadly, it is common for a range of brood parasitic offspring to compete against conspecifics in host nests (*Clamator* cuckoos: Martínez et al. 1998; *Vidua* finches: Schuetz 2005; *Anomalospiza* cuckoo finch: Stevens et al. 2013), so parasitic begging displays that are constrained by relatedness may be present in other species as well.

9.3 Life History, Maternal Effects, and Parasites in Cowbirds

9.3.1 Reproduction and Life History

Female cowbirds are thought to be highly fecund and may lay 25–60 eggs per year (Scott and Ankney 1980), although genetic studies suggest this number is less (Alderson et al. 1999). Their reproductive strategy allows researchers to study the costs of egg production independent from parental care. These costs occur via resource limitation (energy/nutrition) or pleiotropic effects of elevated reproductive hormones over an extended period (Williams 2005; Nager 2006). For example, corticosterone (CORT) levels rise during egg production in birds (Wingfield and Farner 1978), and Dufty and Wingfield (1986) documented a prolonged period of elevated CORT during the breeding season in female cowbirds compared to females of other species. CORT is immunosuppressive when elevated for extended periods (Sapolsky 1992) and could lead to reduced immune function in female cowbirds (Merrill et al. 2013). There is increasing evidence that female cowbirds have adopted a fast pace-of-life strategy resulting in high annual reproductive effort, low levels of immune function (Merrill et al. 2013), reduced hematocrit (Keys et al. 1986), and low annual survivorship compared to male conspecifics and females of the related, nonparasitic red-winged blackbird (Ortega and Ortega 2009; Hahn and Smith 2011).

9.3.2 Maternal Effects

Maternal effects are any effect the maternal genotype or phenotype has on an offspring's phenotype (Wolf and Wade 2009). Females can influence offspring phenotypes in a non-genetic manner via differential allocation of resources such as hormones, antioxidants, antibodies, and other proteins and fats (Schwabl 1996; Grindstaff et al. 2003; Royle et al. 2011), as well as via differences in incubation behavior (DuRant et al. 2010) and resource provisioning to the chicks (Ghalambor et al. 2013). Because brood parasites such as cowbirds exhibit no parental behavior, they are limited in their capacity to influence the development of their offspring to pre-laying maternal effects that include (1) selecting among available hosts and depositing eggs in the nests of the highest-quality hosts, (2) adjusting resources to their eggs based on perceived differences in host nest conditions (*sensu* Merrill et al. 2017), or both. There is some evidence that female cowbirds discriminate among potential host nests and preferentially parasitize some species over others regardless of relative host abundance (Merrill et al. 2017). Moreover, there is some evidence that female cowbirds discriminate among nests within a species and preferentially parasitize nests with eggs that are small relative to the host species' average but only for host species with eggs larger than those of the cowbird (Merrill et al. 2017). Such selectivity could confer a competitive advantage to the cowbird in the nests of species which may challenge the cowbird chick for access to resources. There is also emerging evidence that female cowbirds adjust resources to their eggs based on environmental conditions and features of the host nest (Merrill personal observation). Maternal resource allocation likely influences the competitive capabilities of nestling birds, and this may be especially important for nestling cowbirds. Cowbird eggs have a short incubation period (Briskie and Sealy 1990); they beg more intensely compared to hosts (see above) and develop rapidly (Kilpatrick 2002). A nestling cowbird may compete for resources against many small nestmates or a few large nestmates depending on the host species. Maternal effects may thus be especially important for appropriately preparing the cowbird chick for its host nest. Schwabl (1996) proposed that females could differentially allocate testosterone levels among eggs, thereby manipulating the competitive abilities of the young in the nest, and while there is some support for this idea (Eising et al. 2001), there is no indication that cowbird chicks have consistently higher levels of testosterone compared to their hosts (Hauber and Pilz 2003; Hahn et al. 2005). Hahn et al. (2017), however, found that cowbird eggs had higher levels of yolk androgens in multiply parasitized nests, indicative of a potential response to elevated levels of nest competition from other cowbird nestmates. Moreover, Royle et al. (2011) found that levels of the antioxidant vitamin E in cowbird eggs positively covaried with levels of vitamin E in the eggs of the host red-winged blackbird. They proposed that these antioxidants are critical for combating elevated levels of reactive oxygen species resulting from high begging activities and accelerated growth rates of the nestling cowbirds and that female cowbirds should allocate antioxidants to their eggs at levels comparable to the host (Royle et al. 2011). These results suggest that female cowbirds may be capable of affecting their offspring via traditional (e.g., differential allocation of egg resources) and nontraditional forms (e.g., host nest selection) of maternal effects.

9.3.3 Cowbirds as Hosts

Cowbirds have been recorded parasitizing nearly 250 species (Lowther 2013). Most avian species are host to numerous ectoparasites (Clayton and Johnson 2001) and potentially a suite of other parasites. The host-generalist nature of cowbirds may expose them to a much broader range of parasites and pathogens than their nonparasitic relatives. Hahn et al. (2000) found that cowbirds acquired lice from some hosts, although it is unclear if cowbirds are suitable long-term hosts for the majority of ectoparasites they encounter in host nests because many are host-specific (Clayton and Johnson 2001).

In support of the hypothesis that cowbirds should have more robust immune systems due to increased exposure, Reisen and Hahn (2007) and Hahn et al. (2013) examined responses to West Nile virus and innate immune function among cowbirds and related icterid species and found that cowbirds were the most immunocompetent species examined. However, Merrill et al. (2013) found that levels of immune function in cowbirds varied by sex and season and that during the breeding season females exhibited lower levels of bacteria-killing ability (BKA) and a reduced response to an injection of phytohemagglutinin (PHA) compared to male cowbirds and female red-winged blackbirds. Discrepancies among studies could be a result of different assays assessing different immune system components.

9.4 Coevolution

9.4.1 Cowbird–Host Coevolution

The interactions between cowbirds and their hosts differ from other parasite–host systems because they are at an earlier stage of coevolution (Rothstein et al. 2002). Cowbirds have been parasites for approximately 3–4 mya, whereas lineages such as the common cuckoo have been for 6–8 mya and *Viduidae* and *Anomalospiza* for 13 mya (Rothstein et al. 2002; Sorenson et al. 2004). This may account for one of the most enigmatic feature of cowbird–host interactions which is the widespread acceptance of parasitism by most cowbird hosts. Models suggest that when selection from parasitism is particularly strong, egg rejection can evolve rapidly (Soler 2014). However, there are many cowbird hosts that show no adaptive response to parasitism and yet raise none of their young when parasitized (Peer et al. 2013a, b). Cowbird hosts also show relatively little intraspecific variation in rejection response compared to Old World cuckoo hosts (Rothstein 1990).

9.4.2 Why Do so Many Hosts Accept?

Numerous hypotheses have been proposed to explain acceptance of parasitism by cowbird hosts including (1) evolutionary lag, (2) evolutionary equilibrium, (3) mafia enforcement, and (4) source–sink population dynamics. Rothstein (1975) concluded that evolutionary lag was the best explanation for acceptance by cowbird hosts, and this continues to be the case (see Peer and Sealy 2004a). Hosts that reject cowbird eggs nest in open habitats or along forest edges and have had the longest time to coevolve with cowbirds (Peer et al. 2000; Peer and Sealy 2004a; Soler 2014). Rejecter species may also be more responsive to multiple parasitism that represents a greater threat (Lang et al. 2014), whereas accepters accept parasitism regardless of whether there is single or multiple cowbird eggs in the nest (Ward et al. 1996).

Although lag can account for some acceptance, there are hosts that have been parasitized for very long periods of time and yet accept parasitism (e.g., red-winged blackbird; Peer and Sealy 2004a). The evolutionary equilibrium hypothesis proposes that egg rejection entails costs which make acceptance of parasitism an option with greater net benefits. Potential costs include recognition errors, whereby hosts with eggs that resemble those of the cowbird accidentally reject their eggs, and rejection errors in which a host damages its egg when attempting to remove the thick-shelled cowbird egg. Recognition difficulties occur in grassland hosts and the northern cardinal (*Cardinalis cardinalis*), which have eggs resembling those of the cowbird (Peer and Sealy 2004a). Grassland birds reject non-mimetic eggs more frequently than cowbird eggs (Peer et al. 2000). Cardinal eggs can be similar to cowbird eggs, and cardinals are more likely to reject immaculate eggs that differ from their own, and they also eject some eggs with reduced ultraviolet reflectance (Abernathy and Peer 2016). Cardinal clutches have a high degree of intraclutch egg variation, and this, combined with the poor survival of cowbirds in cardinal nests, may indicate that rejection costs make acceptance of the optimal strategy (Abernathy and Peer 2016). Hosts with small bills risk rejection errors because they are incapable of grasp-ejecting cowbird eggs and instead must puncture-eject them. Because cowbirds lay thick-shelled eggs, host bills may bounce off the cowbird eggs into their own causing damage to them (Sealy 1996). However, these costs appear to be negligible in hosts such as the warbling vireo, which has a small bill and only loses 0.3 eggs for every cowbird egg ejected (Sealy 1996).

Another hypothesis for acceptance of parasitism is that hosts are forced to due to mafia tactics. A brood parasite revisits a nest it has parasitized, and if its egg has been rejected, it will destroy the nest contents (see Chap. 15). For this scenario to occur, the following must be true: (1) the parasite must revisit nests they have parasitized (Soler et al. 1995), (2) the parasite must be able to distinguish between its eggs and those of the host (Sealy and Underwood 2012), and (3) the host must be able to raise some of its young when parasitized because acceptance of a parasitic egg should be more beneficial than ejecting the egg and then having the nest subsequently destroyed (Soler et al. 1995). Hoover and Robinson (2007) simulated egg ejection by removing cowbird eggs from prothonotary warbler (*Protonotaria citrea*) nests and then used an experimental approach to allow cowbirds to reenter some nestboxes or prevented reentry into others. They found that failure was significantly greater at nests that cowbirds could access compared to those they could not enter and concluded that cowbirds forced warblers to accept their eggs (Hoover and Robinson 2007).

The one criterion listed above that was satisfied in the Hoover and Robinson (2007) study was that warblers raise some of their young when parasitized. It was unclear whether the same females that laid the eggs revisited and destroyed the nest contents following cowbird egg removal. It is also unclear if female cowbirds can recognize whether their eggs have been ejected because of the similar appearance of warbler and cowbird eggs (Peer and Sealy 2004a). Although there is evidence female cowbirds can distinguish between cowbird and host eggs (Dubina and Peer 2013), differentiating between two similar eggs in a dark cavity might be difficult. Likewise, mafia tactics should evolve after a host begins ejecting the parasite's eggs. However, it would also be challenging for a female warbler to recognize a cowbird egg in her dark nest, and there is no evidence that warblers eject foreign eggs (Hoover 2003). There have been no other reports of mafia effects in cowbirds (McLaren and Sealy 2000).

One final possibility is that large-scale source–sink population dynamics may be responsible for the lack of defenses against cowbirds (Robinson et al. 2013). Because there are still large areas of cowbird-free forests in North America (Fry et al. 2011), many hosts reside in areas where there is no selective pressure for evolving defenses (Robinson et al. 2013). The lack of host defenses may not be a case of evolutionary lag; rather, it may reflect the extent to which evolution acts slowly in landscapes in which habitats are population sinks (Holt 1996). Much of the Midwestern USA, where cowbird parasitism is extensive, is also characterized by high nest predation rates (Robinson et al. 1995b). There is so little reproduction in these woodlots that defenses would evolve slowly because the few young produced would be swamped by young produced in large forests in which parasitism is rare (Barabas et al. 2004; Robinson et al. 2013). Paradoxically, the lack of host defenses may be an indicator of overall population health. As long as there are large unfragmented habitat patches that can act as population sources, there may be no need for management practices such as cowbird trapping.

9.4.3 What Do We Know About Hosts That Have Evolved Egg Rejection Behavior?

Hosts that have evolved rejection nest in open habitats and along forest edges, tend to be large with bills (>17 mm) that facilitate cowbird egg ejection, and have eggs that differ in appearance from those of cowbirds (Rothstein 1975; Peer and Sealy 2004a). Hosts reject eggs based on differences in size, background color, spotting, and ultraviolet reflectance (Rothstein 1982; Abernathy and Peer 2015, Chap. 22), but reduction of intraclutch egg variation does not appear to facilitate rejection (Abernathy and Peer 2014). Furthermore, some host species such as the cedar waxwing and yellow warbler (Rothstein 1976; Guigueno and Sealy 2011) do not always eject when parasitized (see Chap. 25). These hosts decrease efforts to remove cowbird eggs when the costs of doing so are prohibitive relative to the risk and costs of being parasitized. Once a host evolves egg rejection, it may be retained up to several million years in the absence of parasitism (Rothstein 2001; Peer and Sealy 2004b; Peer et al. 2007, 2011a, b), although there are cases of phenotypic plasticity (Peer and Rothstein 2010; Kuehn et al. 2016) and partial but never complete loss of rejection (Peer and Sealy 2004b; Kuehn et al. 2014).

9.4.4 Other Responses by Hosts to Cowbird Parasitism

Smaller hosts may respond to parasitism by desertion or by burying the cowbird egg in the nest lining. Our understanding of nest desertion and how often it is a *bona fide* response to parasitism is lacking in cowbirds. Desertion of naturally parasitized nests is more common than desertion of experimentally parasitized nests, suggesting that observing a cowbird at the nest may make a host more responsive (Strausberger and Burhans 2001). Hosoi and Rothstein (2000) found that hosts nesting in non-forested habitat were more likely to desert parasitized nests than those in forests.

9.4.5 Cowbird Response to Host Defenses

There has been little selection on cowbirds to modify the appearance of their eggs in response to host discrimination. Numerous hosts lay eggs that resemble cowbird eggs, especially those nesting within the core of the cowbird's range, and many of these hosts are more likely to eject non-mimetic eggs. One study examined hosts that nest largely outside of grasslands but found no evidence of mimicry (Underwood and Sealy 2008). Cowbirds may cause rejecters to misimprint on and accept cowbird eggs as their own by parasitizing naïve individuals, but it is unclear if this is a by-product of high local cowbird abundance or an evolved strategy (Strausberger and Rothstein 2009). A response to host egg discrimination occurs in the brown-headed cowbird's sister species, the shiny cowbird, which has wider eggs where a favored host uses egg width to detect parasitic eggs (Mason and Rothstein 1986). Finally, cowbirds may multiply parasitize hosts, and this may be a strategy to increase host acceptance (e.g., Gloag et al. 2012).

In contrast to egg appearance, eggshell structure has been modified in at least two ways. First, while not a defense against host egg rejection, cowbird eggshells have greater porosity that appears to enhance embryonic respiration and allow cowbird nestlings to hatch sooner than most hosts (Jaeckle et al. 2012). Second, cowbirds lay eggs with thick shells which resist host egg puncture (Spaw and Rohwer 1987) and may reduce damage to cowbird eggs when contacting host eggs during laying (Rothstein 1990). Less than one-third of the rejecter hosts in North America are known to puncture-eject, which suggests that thick eggshells evolved for another reason, perhaps to protect the cowbird egg during laying, with resistance to puncture-ejection being a benefit for the few hosts that demonstrate it.

9.5 Song, Mating and Spatial Systems, and Social Behavior

9.5.1 Spatial Behavior

Most birds choose a breeding habitat that meets the needs of their young and themselves. But cowbirds can uncouple breeding and feeding activities by carrying them out in different places (Rothstein et al. 1984; Chace et al. 2003), which allows them to exploit a larger range of habitats than nonparasitic birds (Rothstein 1994). For example, in the Sierra Nevada of California cowbirds occupied breeding habitat where egg laying and mating take place in forests and meadows from sunrise until late morning/early afternoon and then congregated at feeding sites (Rothstein et al. 1980, 1984). Commuting distances between breeding and feeding sites ranged up to 7 km in the Sierra (Rothstein et al. 1984) and up to 16 km in New Mexico (Curson and Mathews 2003), so a single feeding area can be visited by cowbirds breeding over a large area (Anderson et al. 2012). Cowbirds are social at afternoon feeding sites, but both males and females are alone or in small groups in breeding areas where they respond aggressively to playback of sex-specific vocalizations (Dufty 1982; Rothstein et al. 1988; Yokel 1989). This alone does not mean that cowbirds are territorial because breeding ranges can overlap (Rothstein et al. 1984). However, the aggression suggests some attempt to defend breeding areas.

Cowbirds are highly interactive in the morning, and vocalizations typically occur when >1 individual is present. Unlike most passerines, male cowbirds sing close to others (<1 m) while facing them, so it is often possible to determine the individual to whom a song is directed. Males sing to and court females in the morning if no other males are present, and males sing to each other if no females are present. However, if a male is courting a female and other males arrive, nearly all subsequent songs will be between males (Rothstein et al. 1988). Copulations occur only if no other cowbirds are nearby (Gorney-Labinger and Rothstein 2002), but female choice is likely influenced by the numbers of songs males direct toward them at other times and in male counter-singing sessions. Experiencing male song stimulates female reproductive activity in captivity (West et al. 2002) but it is unknown whether this occurs in the wild.

9.5.2 Song Development and Function

Cowbirds have played a major role in understanding vocal development and function in birds, a subject that has provided insights into the development of human speech (Brainard and Doupe 2002). Similarities between human and songbird vocal development include a sensitive period early in life and the need for auditory feedback. Partly because of their easy adjustment to captivity, cowbirds have enabled researchers to study the extent to which different aspects of vocalizations are dependent on experience. A long-standing need in bird song research was a way to assess female choice. Although males in most passerines respond aggressively to field playbacks of male song, playbacks elicit little change in females. This problem was partially solved by King and West (1977) who found that captive female cowbirds deprived of male contact respond to playback of male song with the lordosis posture, i.e., copulation solicitation display (CSD). These results led to the discovery that administration of estradiol could cause females of many species to respond to male songs with CSDs (Searcy and Marler 1981).

The vocalization type with which female cowbirds were first tested is perched song (PS), which is primarily used when males are near conspecifics (Rothstein et al. 1988). It has introductory notes below the frequency of most bird song followed by components above the frequency of most species' songs. The basic structure is innate (King and West 1977), but males build repertoires by copying the PS types of older males (O'Loghlen and Rothstein 2010a, 2012a). PS is accompanied by a display involving wing spreads, bowing and tail cocking, and often ending in bill wiping (O'Loghlen and Rothstein 2010a, b). These movements are coordinated with the production of sound (Cooper and Goller 2004), are innate (Hoepfner and Goller 2013), and are more intense when directed at a male (O'Loghlen and Rothstein 2010b, 2012b). Video and sound of a male displaying and singing elicits stronger CSD responses than sound alone (O'Loghlen and Rothstein 2012b).

Male cowbirds also perform flight whistles (FW) composed of two to four mostly pure tone syllables (Rothstein et al. 1988) that are used in long-distance communication, in flight, or by a perched male that has landed or is about to fly away. FWs show extraordinarily well-defined spatial dialects, and this is another reason cowbirds have been key in birdsong studies (Rothstein and Fleischer 1987). FW dialects often differ in overt ways such as the three-syllable FW found in the Mammoth Lakes area of the Sierra Nevada, California, and the adjacent four-syllable FW found at Convict Lake. FW dialects occur throughout the cowbird's range (Rothstein and Fleischer 1987; O'Loghlen and Rothstein 2002a). See Fig. 1 for examples of male vocalizations.

Although oscine passerines show intraspecific song variation, it was thought that cowbirds would be an exception and have innate unvarying vocalizations due to their brood parasitism. But the high degree of variation in FWs suggests that learning is involved (O'Loghlen and Rothstein 1993, 1995, 2010a, b). Most cowbirds do not complete vocal development until they are 2 years old because adult males stop singing when female cowbirds stop laying, which limits the opportunity for young birds to memorize song during their hatch year (O'Loghlen and Rothstein 1995). Yearling males are sexually mature (Gros-Louis et al. 2006) and perform FWs but do not have complete local FWs until the second year (O'Loghlen 1995).

Adult males within a FW dialect have PS repertoires assembled from a pool of about ten song types (O'Loghlen and Rothstein 2002a, b). Some PS occurs over areas that encompass multiple FW dialects, but adults within a FW dialect have greater PS sharing than with males from other dialects (O'Loghlen 1995). As with FWs, PS repertoires of many yearling males are deficient and have a larger proportion of unique song types (O'Loghlen and Rothstein 1993, 2002a, b). Yearling males that hatched early in the previous breeding season may perform the appropriate FW and PS types having experienced local song types as juveniles (O'Loghlen and Rothstein 2002b).

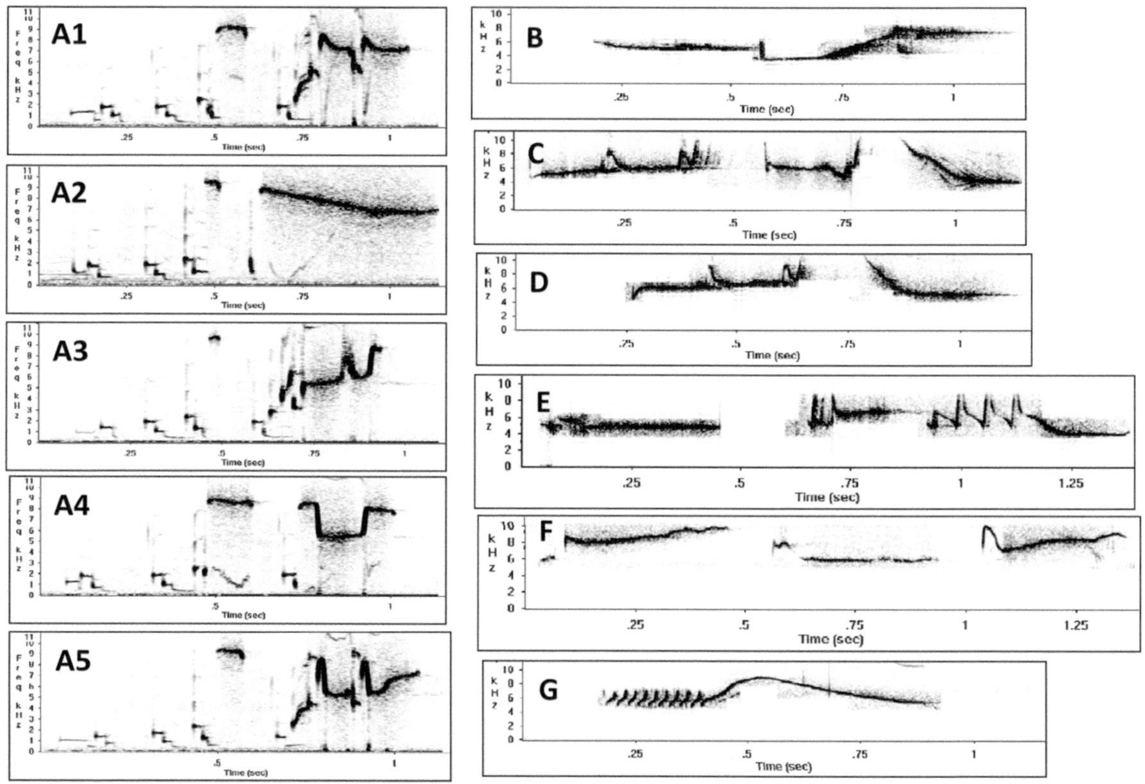

Fig. 1 Perched songs (PSs) and flight whistles (FWs). Panels A1–A5 are the PS types in the repertoire of a single male from Santa Barbara, CA. Panels B–G are FWs done in different populations. Note that PSs always follow distinctive structural rules, very low frequency elements followed by mush higher sounds whereas FWs have a much narrower frequency band but are variable in duration and sound type (usually modulated pure tones but with some FWs containing buzzes or trills). FW B is the type done in the Santa Barbara, CA area. FWs C, D, and E are from the Convict, Mammoth and Lee Vining dialects of the eastern Sierra of California (Rothstein and Fleischer 1987; O'Loghlen and Rothstein 2010a). FWs F and G show further diversity and are from dialects in Oklahoma and Washington State

9.5.3 Mating System

Although cowbirds do not provide parental care, observations and genetic studies show that most cowbirds are monogamous, but there are populations that exhibit polygyny (Teather and Robertson 1986; Alderson et al. 1999) and promiscuity (Elliott 1980). Female choice among potential mates clearly occurs because females choose the time and place of mating by attracting a male by broadcasting their chatter or rattle call, which explains why chatter playbacks attract males (Rothstein et al. 1988). Females also use this call to signal aggression when sung to by males and when competing with other females (Dufty 1982).

Indirect evidence indicates that female choice is based on male genetic quality. In an experiment in which a female's consort was removed, females tended to form a pair bond with an already-mated male suggesting that "widowed females" did not chose new mates randomly (Yokel and Rothstein 1991). Age is one indicator of genetic quality because all male consorts were at least 2 years old (Yokel 1989). Yearling males and some adults fail to mate, but all females breed (Fleischer et al. 1987). Females can identify adult males (>2 years old) by their shinier black plumage and absence of brownish underwing coverts (Ortega 1998). These coverts are revealed if males spread their wings during a PS (O'Loghlen and Rothstein 2010a, b), which makes the display an honest signal of age and may explain why it evolved. The FW type and PS repertoire furnish additional evidence of age since adult males are more likely to have vocalizations that match a local dialect (O'Loghlen and Rothstein 1993, 2002a, b).

Young males memorize the local FW and PS types done by older males. Yearling males are attracted to playback of adult vocalizations in the spring (Snyder-Mackler and White 2011) but do not perform the local FW and PS types until they are 2 years old (O'Loghlen and Rothstein 1993). Females hear more songs from dominant males (Rothstein et al. 1986), and the amount of singing correlates with male reproductive success in captivity (Gros-Louis et al. 2006). Song development in young males is also affected by older females that exert more influence on the vocal development than juvenile females (Miller et al. 2008). In nature, females experience FWs and PS repertoires of local males, and males respond to female chatter calls with FWs. CSD experiments in the lab show that females find the local FW and PS types more stimulating than those from other dialects (O'Loghlen and Rothstein 1995, 2003).

A further indication of the importance of male vocalizations occurs during copulation. Males produced FWs as they fly to a chattering female and rush through their vocal repertoire as they mount the female, which allows females a final assessment of a male's vocalizations. Mating males perform these loud vocalizations with virtually no intervening silence, something done only during copulations (Gorney-Labinger and Rothstein 2002) and thus advertise copulations, as in other animals (Hauser 1993).

9.5.4 Studies Conducted in Captivity

Because cowbirds readily adapt to captivity, complex questions about social behavior can be tested by mixing birds in various combinations. However, these results can be difficult to relate to natural situations where cowbirds move large distances and experience social dynamics that cannot be replicated in captivity. For example, White et al. (2012) found that captive male cowbirds introduced to different individuals developed different behaviors than males in static groups which never occur in nature. Captive groups have also been used to assess whether individuals show assortative mating (Freeberg et al. 2008) by mixing birds from distant locations, but such mixing is unlikely in nature, and distant populations that show assortative mating in captivity may be linked genetically by a series of populations found between the distant localities (Fleischer and Rothstein 1988). It is unclear how captive studies relate to events in nature as even large cages cannot replicate the social flux that cowbirds experience (Rothstein et al. 1984). Moreover, captivity alters behaviors and endocrine systems (Calisi and Bentley 2009). The most meaningful results are likely when captivity is used to test explicit questions prompted by field studies (O'Loghlen and Rothstein 2003).

9.6 Cowbirds as Tests of Hypotheses Concerning the Impact of Brood Parasites on Host Population Dynamics

9.6.1 Effects of Cowbirds on Host Fitness

Parasitism by cowbirds reduces host fitness in nearly all species parasitized (Ortega 1998). These fitness costs result from multiple behaviors including (1) host egg removal (Sealy 1992); (2) damage to host eggs during laying by cowbirds (see above); (3) early hatching and rapid development of cowbird nestlings, which enables them to monopolize the food delivered by their foster parents (see above); (4) the choice of mostly smaller host species, which means that the larger cowbird nestlings monopolize food delivered to the nest (see above); and (5) loud begging by cowbird nestlings, which may increase conspicuousness of nests to some predators and decrease food delivered to host nestlings (Dearborn 1999). These costs are multiplied in nests that receive more than one cowbird egg per nest, a common occurrence in parts of North America with high cowbird densities (Robinson 1992; Rivers et al. 2010). There is evidence that raising cowbirds may be more costly for adults than raising their own young, which suggests that cowbird nestlings may cause parents to work harder to feed them (Hoover and Reetz 2006) and to renest more often because females that raise a cowbird are more likely to initiate a second brood than those that raise a brood without a cowbird possibly to recoup the losses after raising a cowbird (Louder et al. 2014).

The extent to which cowbirds threaten host populations varies greatly among hosts, habitats, years, landscape cover, and regions (Robinson et al. 1995a). Small hosts such as flycatchers (family Tyrannidae) that have slow developmental rates and small clutches tend to fledge very few host young when parasitized (Peer et al. 2013a), whereas larger hosts such as wood thrushes (*Hylocichla mustelina*) can raise their young alongside cowbirds and only incur costs in regions where parasitism levels are extremely high such as the Midwest (Robinson et al. 1995b). Host species with small ranges and fragmented remaining habitat patches are likely to be particularly threatened by cowbird parasitism because they lack refugia from parasitism (Smith et al. 2000; Wilsey et al. 2013). Cowbirds have also been blamed for causing population declines in widespread species, even those in which exposure to cowbirds varies throughout their breeding ranges (Mayfield 1977).

One mechanism that has been hypothesized to cause these long-term declines is an increase in the proportion of "sink" habitat (*sensu* Pulliam 1988) in landscapes in which breeding habitat is fragmented and therefore exposed to cowbird parasitism (Donovan et al. 1995; Robinson and Hoover 2011). Small forest fragments surrounded by agriculture and pasture have abundant cowbird feeding habitat in close proximity (Thompson 1994). These areas tend to have very high cowbird populations (Cummings and Veech 2013) and parasitism levels (Robinson et al. 1995b), which can drive host productivity below levels necessary to compensate for mortality (Donovan et al. 1995). Because cowbirds search for nests in areas that can be up to 16 km from where they feed (see above), only a few foraging sites such as pastures, stockyards, or agricultural fields can support cowbird populations in large proportions of the landscape (Morse and Robinson 1999, but see Stoklosa et al. 2014). Decreases in nesting success resulting from cowbird parasitism have been implicated in creating sink habitat (Robinson and Hoover 2011; Robinson et al. 2013). These regions, however, also typically have very high levels of edge-associated nest predation (Donovan et al. 1997), which can make it difficult to disentangle the effects of losses to cowbird parasitism from losses to nest predators. Nevertheless, there is little doubt that the combined effects of predation and parasitism create steep source–sink habitat gradients and that these gradients are related to habitat fragmentation (Robinson and Hoover 2011).

Some host species have complex and seemingly adaptive "decision rules" that enable them to avoid areas with high nest predation rates and return to areas where most nests escape predation (Hoover 2003). Most hosts, however, do not have comparable decision rules with respect to cowbird parasitism. They are just as likely to return to an area where they raised cowbirds than to an area where they raised their young as long as those nests escaped predation (Hoover 2003). Therefore, cowbird parasitism may threaten populations because habitats with high cowbird parasitism, but low rates of nest predation, continue to attract and retain breeding populations of hosts, even if parasitism drops nesting success below the source–sink threshold. These habitats have been called "ecological traps" (*sensu* Gates and Gysel 1978), which can be operationally defined as attractive sink habitat (Robinson and Hoover 2011).

Management plans that reduce habitat fragmentation to lessen cowbird parasitism have become a core element in landscape-level conservation (Llewellyn et al. 1996; Beissinger et al. 2000). Preservation of very large, mostly forested tracts reduces both nest predation and cowbird parasitism in the Midwest (Robinson et al. 1995b).

Such large tracts may act as source habitat that produces surplus individuals that recolonize small patches in which local populations are not viable in the absence of immigration (Donovan et al. 1995; Robinson et al. 1995b). For critically endangered host species, cowbird control has also been recommended as a more direct method of reducing the negative consequences of cowbird parasitism (Smith et al. 2000), although this approach is not without controversy (Rothstein and Peer 2005).

9.6.2 Role of Cowbirds in the Conservation of Endangered Host Populations

Cowbird parasitism has been at least partly blamed for the endangerment of several songbird species or subspecies that have small remaining populations in geographical ranges in which cowbird habitat dominates large proportions of the remaining landscape (Smith et al. 2000). Cowbird trapping has been used and has been shown to increase nesting success for the Kirtland's warbler (*Setophaga kirtlandii*), least bell's vireo (*Vireo bellii pusillus*), southwestern willow flycatcher (*Empidonax traillii extimus*), and black-capped vireo (*Vireo atricapilla*) (Smith et al. 2000). At least some of these populations recovered or stabilized after cowbird control was initiated. Kirtland's warblers and black-capped vireos, for example, experienced parasitism levels of well over 50% of nests, levels that the populations were unlikely to sustain over long timescales. As such, they appear to be conservation-dependent species (Rockwell et al. 2012; Wilsey et al. 2013). All of these endangered species, however, also have small population sizes resulting from extreme habitat loss and disturbance, which reduces habitat quality and likely also increases nest predation, and may suffer from winter habitat loss as well (Rothstein and Peer 2005; Rockwell et al. 2012). Many population recoveries were associated with improved habitat management at the same time that cowbird control was being practiced. The recovery of the Kirtland's warbler, in particular, may have been driven more by an accidental habitat burn that greatly enlarged the amount of suitable breeding habitat rather than by cowbird control per se (Rothstein and Peer 2005; Rockwell et al. 2012). Therefore, cowbird control appears to work best in association with intensive habitat management and may not be necessary once populations have recovered and occur over wide enough area to have refugia from cowbird parasitism (Rothstein and Peer 2005). Some species, however, may still require more focused cowbird control efforts, especially in landscapes dominated by cowbird feeding habitat such as rangelands in which the entire landscape matrix provides feeding habitat, including much of the habitat occupied by the endangered hosts themselves (Smith et al. 2000; Brodhead et al. 2007; Wilsey et al. 2013).

9.6.3 Landscape Management

The most sustainable long-term strategy for reducing the negative effects of cowbird parasitism on their hosts is to reduce habitat fragmentation by enlarging habitat patches (Brunswig et al. 2016). This strategy would potentially benefit hosts by increasing their populations, decreasing cowbird populations via a reduction in afternoon feeding habitat, and increasing "interior" habitat that is far from cowbird feeding areas that might act as source habitat. Although evidence for this strategy is mixed (see above), this approach should also reduce excessive levels of nest predation associated with edge-adapted predators such as raccoons (*Procyon lotor*) and feral species. For landscapes such as Eastern North America, the most important element of any conservation strategy would be to maintain the large forested areas in regions such as those in the Appalachians that can act as source habitat where there are few cowbirds. Landscape-level management that reduces the negative effects of habitat loss and fragmentation may eliminate the need for large-scale cowbird control.

Concluding Remarks and Future Directions

The cowbird is an ideal species for examining questions regarding host–parasite coevolution, maternal effects and physiology, song learning and patterns of cultural transmission, and ecology and conservation due to its life history as a generalist brood parasite. Avenues for future research in host–parasite coevolution include continuing to investigate alternatives to evolutionary lag for the widespread acceptance of parasitism by cowbird hosts. Many hosts have been exposed to parasitism for relatively long periods and yet do not appear to respond adaptively. Further examination of cowbird counter-strategies including multiple parasitism, farming, and whether mafia behavior is a widely used strategy is needed. Studies of cowbird nestlings when competing against large hosts across a range of age differentials will elucidate how early hatching by cowbirds enhances survival when raised by a large host. We also need to determine how rearing environments influence adult phenotype, which components are robust against early competitive environments, and whether these relationships differ from nonparasitic relatives. Additional experiments on whether cowbird begging is linked to population-level factors, including within-brood relatedness, are also needed. Equally valuable are studies of whether cowbirds discriminate between conspecific and heterospecific young to modulate begging.

Data on hormonal and immunological profiles of cowbirds would increase our understanding of the trade-offs of individual life history strategies. Experimental manipulation of egg production would be an effective way to examine costs of egg laying, including CORT levels, immune function, oxidative stress, and telomere length. Another major question is the reason female cowbirds in diverse regions lay eggs at the same rate of ~0.7 per day.

The degree to which cowbirds are territorial, especially at low densities, is still unclear. We know that male song stimulates female reproductive activity in captivity (West et al. 2002), but we need to determine if this occurs in nature. Testing whether female sexual responses are influenced by differences

(continued)

in yearling versus adult male plumages would provide valuable information on female choice. Similarly, indirect evidence indicates that female choice is based on male genetic quality, and testing this hypothesis with direct genetic evidence is needed. Additional studies are also necessary to assess variation in population levels of social and genetic monogamy.

One promising approach to cowbird management is to use trapping programs as experiments to identify factors that limit breeding populations. We need better data on adult survival from marked host populations because relatively small increases in survival estimates for forest hosts can have a profound effect on the source–sink threshold and therefore the extent to which parasitism limits populations. Once understood, we will have a better understanding of the levels of landscape management needed to create sustainable regional populations. We also need to use large-scale habitat management projects as experimental tests of the effects of landscape fragmentation on cowbird parasitism. We need a better understanding of the role of different kinds of landscape matrices in providing cowbird feeding habitat, and we need to study whether cowbirds themselves may have source–sink habitats based on host community composition and the average nesting success of their hosts.

Acknowledgments We thank Alex Cruz, Gabriel Jamie, and Manolo Soler for their comments on the manuscript. LM thanks T.J. Benson and the Illinois Natural History Survey for support, and Tara Stewart Merrill for feedback. JWR's work has been supported by the Animal Behavior Society, the Chapman Fund of the American Museum of Natural History, Helmetcamera.com, the Los Angeles Audubon Society, the Graduate Division at UC-Santa Barbara, the Academic Senate at UC-Santa Barbara, the NSF-Doctoral Dissertation Improvement Grant program, and the NSF-LTER Program at Konza Prairie.

References

Abernathy VA, Peer BD (2014) Intraclutch variation in egg appearance of brown-headed cowbird hosts. Auk 131:467–475

Abernathy VA, Peer BD (2015) Mechanisms of egg recognition in brown-headed cowbird hosts: the role of ultraviolet reflectance. Anim Behav 109:73–79

Abernathy VA, Peer BD (2016) Reduced ultraviolet reflectance does not affect egg rejection by northern cardinals (*Cardinalis cardinalis*). Wilson J Ornithol 128:328–336

Alderson GW, Gibbs HL, Sealy SG (1999) Determining the reproductive behaviour of individual brown-headed cowbirds using microsatellite DNA markers. Anim Behav 58:895–905

Anderson KE, Fujiwara M, Rothstein SI (2012) Demography and dispersal of juvenile and adult brown-headed cowbirds (*Molothrus ater*) in the eastern Sierra Nevada, California, estimated using multistate models. Auk 129:307–318

Barabas L, Gilicze B, Takasu F, Moskat C (2004) Survival and anti-parasite defense in a host metapopulation under heavy brood parasitism: a source-sink dynamic model. J Ethol 22:143–151

Beissinger SR, Reed JM, Wunderle Jr JM, Robinson SK, Finch DM (2000) Report of the AOU conservation committee on the partners in flight species prioritization plan. Auk 117:549–561
Brainard MS, Doupe AJ (2002) What songbirds teach us about learning. Nature 417:351–358
Briskie JV, Sealy SG (1990) Evolution of short incubation periods in the parasitic cowbirds, *Molothrus* spp. Auk 107:789–794
Brodhead KM, Stoleson SH, Finch DM (2007) Southwestern willow flycatchers (*Empidonax traillii extimus*) in a grazed landscape: factors influencing brood parasitism. Auk 124:1213–1228
Brunswig N, Richardson S, Johnson M, Heitkamp B (2016) Bird-friendly recommendations for bottomland forests in the Carolinas: birds and people on common ground. In: Proceedings of 18th biennial southern silvicultural research conference. GTR SRS-212, Asheville, NC. USDA Forest Service, Southern Research Station
Buckling A, Brockhurst MA (2008) Kin selection and the evolution of virulence. Heredity 100:484–488
Calisi RM, Bentley GE (2009) Lab and field experiments: are they the same animal? Horm Behav 56:1–10
Chace JF, Walsh JJ, Cruz A, Prather JW, Swanson H (2003) Spatial and temporal activity patterns of the parasitic brown-headed cowbird at an urban/wildland interface. Landsc Urban Plan 64:179–185
Chappell MA, Bachman GC (2002) Energetic costs of begging behavior. In: Wright J, Leonard ML (eds) The evolution of begging. Kluwer, Dordrecht, pp 143–162
Clark AB (2002) Appetite and the subjectivity of nestling hunger. In: Wright J, Leonard ML (eds) The evolution of begging. Kluwer, Dordrecht, pp 173–198
Clayton DH, Johnson KP (2001) What's bugging brood parasites? Tree 16:9–10
Cooper BG, Goller F (2004) Multimodal signals: enhancement and constraint of song motor patterns by visual display. Science 303:544–546
Cummings K, Veech JA (2013) Assessing the influence of geography, land cover and host species on the local abundance of a generalist brood parasite, the brown-headed cowbird. Divers Distrib 20:396–404
Curson DR, Mathews NE (2003) Reproductive costs of commuting flights in brown-headed cowbirds. J Wildl Manage 67:520–529
Davies NB (2000) Cuckoos, cowbirds and other cheats. T & AD Poyser, London
Dearborn DC (1998) Begging behavior and food acquisition by brown-headed cowbird nestlings. Behav Ecol Sociobiol 43:259–270
Dearborn DC (1999) Brown-headed cowbird nestling vocalizations and the risk of nest predation. Auk 116:448–457
Donovan TM, Thompson III FR, Faaborg J, Probst JR (1995) Reproductive success of migratory birds in habitat sources and sinks. Conserv Biol 9:1380–1395
Donovan TM, Jones PW, Annand EM, Thompson III FR (1997) Variation in local scale edge effects: mechanisms and landscape context. Ecology 78:2064–2075
Dubina KM, Peer BD (2013) Egg pecking and discrimination by female and male brown-headed cowbirds. J Ornithol 154:553–557
Dufty AM (1982) Response of brown-headed cowbirds *Molothrus ater* to simulated conspecific intruders. Anim Behav 30:1043–1052
Dufty AM, Wingfield JC (1986) Temporal patterns of circulating LH and steroid hormones in a brood parasite, the brown-headed cowbird, *Molothrus ater*. J Zool 208:205–214
DuRant SE, Hepp GR, Moore IT, Hopkins BC, Hopkins WA (2010) Slight differences in incubation temperature affect early growth and stress endocrinology of wood duck (*Aix sponsa*) ducklings. J Exp Biol 213:45–51
Eising CM, Eikenaar C, Schwabl H, Groothuis TGG (2001) Maternal androgens in black-headed gull (*Larus ridibundus*) eggs: consequences for chick development. Proc R Soc B 268:839–846
Elliott PF (1980) Evolution of promiscuity in the brown-headed cowbird. Condor 82:138–141

Fleischer RC, Rothstein SI (1988) Known secondary contact and rapid gene flow among subspecies and dialects in the brown-headed cowbird. Evolution 42:1146–1158

Fleischer RC, Smyth SP, Rothstein SI (1987) Temporal and age-related variation in the laying rate of the parasitic brown-headed cowbird in the eastern Sierra Nevada, California. Can J Zool 65:2724–2730

Frank SA (1994) Kin selection and virulence in the evolution of protocells and parasites. Proc R Soc B 258:153–161

Freeberg TM, King AP, West M (2008) Behavioral compatibility and reproductive outcomes in two distant populations of captive brown-headed cowbirds (*Molothrus ater*). J Comp Psychol 122:437–440

Fry J, Xian G, Dewitz J, Homer C, Yang L, Barnes C, Herold N, Wickham J (2011) Completion of the 2006 national land cover database for the conterminous United States. Photogramm Eng Remote Sens 77:858–864

Gates JE, Gysel LE (1978) Avian nest dispersion and fledging success in field-forest ecotones. Ecology 59:871–883

Ghalambor CK, Peluc SI, Martin TE (2013) Plasticity of parental care under the risk of predation: how much should parents reduce care? Biol Lett 9:20130154

Gloag R, Fiorini VF, Reboreda JC, Kacelnik A (2012) Brood parasite eggs enhance egg survivorship in a multiply parasitized host. Proc R Soc B 279:1831–1839

Godfray HCJ (1995) Evolutionary theory of parent-offspring conflict. Nature 376:133–138

Gorney-Labinger E, Rothstein SI (2002) Brown-headed cowbirds in eastern and western North America show similar use of courtship songs in copulation contexts. Condor 104:866–872

Grindstaff JL, Brodie ED, Ketterson ED (2003) Immune function across generations: integrating mechanism and evolutionary process in maternal antibody transmission. Proc R Soc B 270:2309–2319

Gros-Louis J, White DJ, King AP, West MJ (2006) Do juvenile males affect adult males' reproductive success in brown-headed cowbirds (*Molothrus ater*)? Behaviour 143:219–237

Guigueno MF, Sealy SG (2011) Aggression towards egg-removing cowbird elicits clutch abandonment in parasitized yellow warblers, *Dendroica petechia*. Anim Behav 81:211–218

Hahn DC, Smith GW (2011) Life history trade-offs between longevity and immunity in the parasitic brown-headed cowbird. Open Evol J 5:8–13

Hahn DC, Price RD, Osenton PC (2000) Use of lice to identify cowbird hosts. Auk 117:943–951

Hahn DC, Hatfield JS, Abdelnabi MA, JM W, Igl LD, Ottinger MA (2005) Inter-species variation in yolk steroid levels and a cowbird-host comparison. J Avian Biol 36:40–46

Hahn DC, Summers SG, Genovese KJ, He H, Kogut MH (2013) Obligate brood parasites show more functionally effective innate immune responses: an eco-immunological hypothesis. Evol Biol 40:554–561

Hahn DC, Fox DM WJC, Walker BG, Thomley JE (2017) Maternal androgens in avian brood parasites and their hosts: Responses to parasitism and competition? Gen Comp Endrocrinol 240:143–152

Haskell DG (2002) Begging behavior and nest predation. In: Wright J, Leonard ML (eds) The evolution of begging. Kluwer, Dordrecht, pp 163–172

Hauber ME (2003a) Lower begging responsiveness of host versus parasitic brown-headed cowbird (*Molothrus ater*) nestlings is related to species identity but not to early social experience. J Comp Psychol 117:24–30

Hauber ME (2003b) Hatching asynchrony, nestling competition, and the cost of interspecific brood parasitism. Behav Ecol 14:227–235

Hauber ME, Pilz KM (2003) Yolk testosterone levels are not consistently higher in the eggs of obligate brood parasites than their hosts. Am Midl Nat 149:354–362

Hauber ME, Ramsey CK (2003) Honesty in host-parasite communication signals: the case for begging by fledgling brown-headed cowbirds *Molothrus ater*. J Avian Biol 34:339–344

Hauser MD (1993) Rhesus monkey copulation calls: honest signals for female choice? Proc R Soc B 254:93–96

Hoepfner AR, Goller F (2013) Atypical song reveals spontaneously developing coordination between multi-modal signals in brown-headed cowbirds (Molothrusater). PLoS ONE 8(6): xe65525. https://doi.org/10.1371/journal.pone.0065525

Holen OH, Saetre GPP, Slagsvold T, Stenseth NC (2001) Parasites and supernormal manipulation. Proc R Soc B 268:2551–2558

Holt RD (1996) Adaptive evolution in source-sink environments: direct and indirect effects of density-dependence on niche evolution. Oikos 75:182–192

Hoover JP (2003) Multiple effects of brood parasitism reduce the reproductive success of prothonotary warblers *Protonotaria citrea*. Anim Behav 65:923–937

Hoover JP, Reetz MJ (2006) Brood parasitism increases provisioning rate and reduces offspring recruitment and adult return rates in a cowbird host. Oecologia 149:165–173

Hoover JP, Robinson SK (2007) Retaliatory mafia behavior by a parasitic cowbird favors host acceptance of parasitic eggs. PNAS 104:4479–4483

Hosoi SA, Rothstein SI (2000) Nest desertion and cowbird parasitism: evidence for evolved responses and evolutionary lag. Anim Behav 59:823–840

Jaeckle WB, Kiefer M, Childs B, Harper RG, Rivers JW, Peer BD (2012) Comparison of eggshell porosity and estimated gas flux between the brown-headed cowbird and two common hosts. J Avian Biol 43:486–490

Keys GC, Fleischer RC, Rothstein SI (1986) Relationships between elevation, reproduction and the hematocrit level of brown-headed cowbirds. Comp Biochem Physiol A: Molec Integr Physiol 83:765–769

Kilner RM (2005) The evolution of virulence in brood parasites. Ornithol Sci 4:55–64

Kilpatrick AM (2002) Variation in growth of brown-headed cowbird (*Molothrus ater*) nestlings and energetic impacts on their host parents. Can J Zool 80:145–153

King AP, West MJ (1977) Species identification in the North American cowbird: appropriate responses to abnormal song. Science 195:1002–1004

Kosciuch KL, Sandercock BK (2008) Cowbird removals unexpectedly increase productivity of a brood parasites and the songbird host. Ecol Appl 18:537–548

Kuehn MJ, Peer BD, Rothstein SI (2014) Variation in host response to brood parasitism reflects evolutionary differences and not phenotypic plasticity. Anim Behav 88:21–28

Kuehn MJ, Peer BD, McCleery RA, Rothstein SI (2016) Yellow warbler defenses are retained in the absence of brood parasitism but enhanced by experience with cowbirds. Behav Ecol 27:279–286

Lang AK, Bollinger EK, Peer BD (2014) Effect of parasite-to-host egg ratio on egg rejection by a brown-headed cowbird host. Auk 131:694–701

Lichtenstein G, Dearborn DC (2004) Begging and short-term need in cowbird nestlings: how different are brood parasites? Behav Ecol Sociobiol 56:352–359

Lichtenstein G, Sealy SG (1998) Nestling competition, rather than supernormal stimulus, explains the success of brown-headed cowbird chicks in yellow warbler nests. Proc R Soc B 265:249–254

Liu WC, Rivers JW, White DJ (2016) Vocal matching and intensity of begging calls are associated with a forebrain song circuit in a generalist brood parasite. Dev Neurobiol 76:615–625

Llewellyn DW, Shaffer GP, Craig NJ, Creasman L, Pashley DN, Swan M, Brown C (1996) A decision support system for prioritizing restoration sites on the Mississippi River Alluvial Plain. Conserv Biol 10:1446–1455

Louder MIM, Schelsky WM, Benson TJ, Hoover JP (2014) Brown-headed cowbirds exploit a host's compensatory behavioral response to fecundity reduction. Behav Ecol 26:255–261

Lowther PE (2013) Lists of victims and hosts of the parasitic cowbirds (Molothrus). Field Museum version 26 Aug 2013. http://fm1.fieldmuseum.org/aa/Files/lowther/CBList.pdf. Accessed 2 Jan 2014

Madden JR, Kilner RM, Davies NB (2005) Nestling responses to adult food and alarm calls: 2. Cowbirds and red-winged blackbirds reared by eastern phoebe hosts. Anim Behav 70:629–637

Martínez JG, Soler JJ, Soler M, Burke T (1998) Spatial patterns of egg laying and multiple parasitism in a brood parasite: a non-territorial system in the great spotted cuckoo (*Clamator glandarius*). Oecologia 117:286–294

Mason P, Rothstein SI (1986) Coevolution and avian brood parasitism: cowbird eggs show evolutionary response to host discrimination. Evolution 40:1207–1214

Mayfield H (1977) Brown-headed cowbird: agent of extermination? Am Birds 31:107–113

McLaren CM, Sealy SG (2000) Are nest predation and brood parasitism correlated in yellow warblers? A test of the cowbird predation hypothesis. Auk 117:1056–1060

Merrill L, O'Loghlen AL, Wingfield JC, Rothstein SI (2013) Immune function in an avian brood parasite and its nonparasitic relative. Physiol Biochem Zool 86:61–72

Merrill L, Chiavacci SJ, Paitz RT, Benson TJ (2017) Rates of parasitism, but not egg resources, vary among and within hosts of a generalist avian brood parasite. Oecologia 184. https://doi.org/10.1007/s00442-017-3870-z

Miller JL, King AP, West MJ (2008) Female social networks influence male vocal development in brown-headed cowbirds, *Molothrus ater*. Anim Behav 76:931–941

Monaghan P (2008) Early growth conditions, phenotypic development and environmental change. Philos Trans R Soc Lond B 363:1635–1645

Morse SF, Robinson SK (1999) Nesting success of a Neotropical migrant in a multiple-use forested landscape. Conserv Biol 13:327–337

Motro U (1989) Should a parasite expose itself? (Some theoretical aspects of begging and vigilance behavior). J Theor Biol 140:279–287

Naef-Daenzer B, Widmer F, Nuber M (2001) Differential post-fledging survival of great and coal tits in relation to their condition and fledging date. J Anim Ecol 70:730–738

Nager RG (2006) The challenges of making eggs. Ardea 94:323–346

O'Loghlen AL (1995) Delayed access to local songs prolongs vocal development in dialect populations of brown-headed cowbirds. Condor 97:402–414

O'Loghlen AL, Rothstein SI (1993) An extreme example of delayed vocal development: song learning in a population of wild brown-headed cowbirds. Anim Behav 46:293–304

O'Loghlen AL, Rothstein SI (1995) Culturally correct song dialects are correlated with male age and female song preferences in wild populations of brown-headed cowbirds. Behav Ecol Sociobiol 36:251–259

O'Loghlen AL, Rothstein SI (2002a) Ecological effects on song learning: delayed development is widespread in wild populations of brown-headed cowbirds. Anim Behav 63:475–486

O'Loghlen AL, Rothstein SI (2002b) Vocal development is correlated with an indicator of hatching date in brown-headed cowbirds. Condor 104:761–771

O'Loghlen AL, Rothstein SI (2003) Female preference for the songs of older males and the maintenance of dialects in brown-headed cowbirds (*Molothrus ater*). Behav Ecol Sociobiol 53:102–109

O'Loghlen AL, Rothstein SI (2010a) Delayed sensory learning and development of dialect songs in brown-headed cowbirds (*Molothrus ater*). Anim Behav 79:299–311

O'Loghlen AL, Rothstein SI (2010b) Multimodal signalling in a songbird: male audiovisual displays vary significantly by social context in brown-headed cowbirds. Anim Behav 79:1285–1292

O'Loghlen AL, Rothstein SI (2012a) Delayed vocal ontogeny in songbirds: a laboratory study validates a model for delayed development derived from field studies. J Ethol 30:369–378

O'Loghlen AL, Rothstein SI (2012b) When less is best: female brown-headed cowbirds prefer less intense male displays. PLoS One 7:e36130. https://doi.org/10.1371/journal.pone.0036130

Ortega CP (1998) Cowbirds and other brood parasites. University of Arizona Press, Tucson

Ortega JC, Ortega CP (2009) Sex ratios and survival probabilities of brown-headed cowbirds (*Molothrus ater*) in southwest Colorado. Auk 126:268–277

Peer BD, Abernathy VE (2017) The brown-headed cowbird: ecology and management of an avian brood parasite. In: Linz GM, Avery ML, Dolbeer RA (eds) Ecology and management of blackbirds (Icteridae) in North America. CRC Press, Boca Raton, pp 77–99

Peer BD, Bollinger EK (1997) Explanations for the infrequent cowbird parasitism on common grackles. Condor 99:151–161

Peer BD, Rothstein SI (2010) Phenotypic plasticity in common grackles in response to repeated brood parasitism. Auk 127:293–299

Peer BD, Sealy SG (2004a) Correlates of egg rejection in hosts of the brown-headed cowbird. Condor 106:580–599

Peer BD, Sealy SG (2004b) Fate of grackle defenses in the absence of brood parasitism: implications for long-term brood parasite-host coevolution. Auk 121:1172–1186

Peer BD, Robinson SK, Herkert JR (2000) Egg rejection by cowbird hosts in grasslands. Auk 117:892–901

Peer BD, Rothstein SI, Delaney KS, Fleischer RC (2007) Defence behaviour against brood parasitism is deeply rooted in mainland and island scrub-jays. Anim Behav 73:55–63

Peer BD, Kuehn MJ, Rothstein SI, Fleischer RC (2011a) Persistence of host defence behaviour in the absence of brood parasitism. Biol Lett 7:670–673

Peer BD, McIntosh CE, Kuehn MJ, Rothstein SI, Fleischer RC (2011b) Complex biogeographic history of shrikes and its implications for the evolution of defenses against avian brood parasitism. Condor 113:385–394

Peer BD, Rivers JW, Rothstein SI (2013a) Cowbirds, conservation, and coevolution: potential misconceptions and directions for future research. Chin Birds 4:15–30

Peer BD, Rivers JW, Rothstein SI (2013b) The brown-headed cowbird: North America's avian brood parasite. Chin Birds 4:93–98

Pulliam HR (1988) Sources, sinks, and population regulation. Am Nat 132:652–661

Reisen WK, Hahn DC (2007) Comparison of immune responses of brown-headed cowbird and related blackbirds to West Nile and other mosquito-borne encephalitis viruses. J Wildl Dis 43:439–449

Rivers JW (2007) Nestmate size, but not short-term need, influences begging behavior of a generalist brood parasite. Behav Ecol 18:222–230

Rivers JW (2009) Parent-absent begging in the brown-headed cowbird (*Molothrus ater*): the role of short-term need and nestmate size. Behav Ecol Sociobiol 63:707–717

Rivers JW, Peer BD (2016) Relatedness constrains virulence in an obligate avian brood parasite. Ornithol Sci 15:191–201

Rivers JW, Jensen WE, Kosciuch KL, Rothstein SI (2010) Community-level patterns of host use by the brown-headed cowbird (*Molothrus ater*), a generalist brood parasite. Auk 127:263–273

Rivers JW, Young S, Gonzalez EG, Horton B, Lock J, Fleischer RC (2012) High levels of relatedness between brown-headed cowbird (*Molothrus ater*) nestmates in a heavily-parasitized host community. Auk 129:623–631

Rivers JW, Blundell MA, Loughin TM, Peer BD, Rothstein SI (2013) The exaggerated begging behavior of an obligate avian brood parasite is shared with a nonparasitic close relative. Anim Behav 86:529–536

Rivers JW, Blundell MA, Rothstein SI (2014) Mismatched begging displays between foreign and host offspring reduced brood parasite fitness. Behav Ecol 25:785–793

Robinson SK (1992) Population dynamics of breeding Neotropical migrants in a fragmented Illinois landscape. In: Hagan JM III, Johnston DW (eds) Ecology and conservation of Neotropical migrant land birds. Smithsonian Institution Press, Washington, DC, pp 408–418

Robinson SK, Hoover JP (2011) Does forest fragmentation and loss generate sources, sinks, and ecological traps in migratory songbirds? In: Liu J, Hull V, Morzillo A, Wiens JA (eds) Sources, sinks and sustainability across landscapes. Cambridge University Press, Cambridge, pp 423–449

Robinson SK, Rothstein SI, Brittingham MC, Petit LJ, Grzybowski Jr JA (1995a) Ecology and behavior of cowbirds and their impact on host populations. In: Martin TE, Finch DM (eds) Ecology and management of neotropical migratory landbirds. Oxford University Press, New York, pp 428–460

Robinson SK, Thompson FR III, Donovan TM, Whitehead DR, Faaborg J (1995b) Regional fragmentation and the nesting success of migratory birds. Science 267:1987–1990
Robinson SK, Schelsky WM, Hoover JP (2013) Does landscape structure contribute to the long-term maintenance of generalized parasitism in cowbirds? Chin Birds 4:31–38
Rockwell SM, Bocetti CI, Marra PP (2012) Carry-over effects of winter climate on spring arrival date and reproductive success in an endangered migratory bird, Kirtland's warbler (*Setophaga kirtlandii*). Auk 129:744–752
Rothstein SI (1975) An experimental and teleonomic investigation of avian brood parasitism. Condor 77:50–271
Rothstein SI (1976) Cowbird parasitism of the cedar waxwing and its evolutionary implications. Auk 93:498–509
Rothstein SI (1982) Mechanisms of avian egg recognition: which egg parameters elicit responses by rejecter species? Behav Ecol Sociobiol 11:229–239
Rothstein SI (1990) A model system for coevolution: avian brood parasitism. Annu Rev Ecol Evol Syst 21:481–508
Rothstein SI (1994) The brown-headed cowbird's invasion of the Far West: history, causes and consequences experienced by host species. Stud Avian Biol 15:301–315
Rothstein SI (2001) Relic behaviors, coevolution and the retention versus loss of host defenses after episodes of avian brood parasitism. Anim Behav 61:95–107
Rothstein SI, Fleischer RC (1987) Vocal dialects and their possible relation to honest status signalling in the brown-headed cowbird. Condor 89:1–23
Rothstein SI, Peer BD (2005) Conservation solutions for threatened and endangered cowbird (*Molothrus* spp.) hosts: separating fact from fiction. Ornithol Monogr 57:98–114
Rothstein SI, Verner J, Stevens E (1980) Range expansion and diurnal changes in dispersion of the brown-headed cowbird in the Sierra Nevada. Auk 97:253–267
Rothstein SI, Verner J, Stevens E (1984) Radio-tracking confirms a unique diurnal pattern of spatial occurrence in the parasitic brown-headed cowbird. Ecology 65:77–88
Rothstein SI, Yokel DA, Fleischer RC (1986) Social dominance, mating and spacing systems, female fecundity, and vocal dialects in captive and free-ranging brown-headed cowbirds. Curr Ornithol 3:127–185
Rothstein SI, Yokel DA, Fleischer RC (1988) The agonistic and sexual functions of vocalizations of male brown-headed cowbirds (*Molothrus ater*). Anim Behav 36:73–86
Rothstein SI, Patten MA, Fleischer RC (2002) Phylogeny, specialization and brood parasite-host coevolution: some possible pitfalls of parsimony. Behav Ecol 13:1–10
Royle NJ, Hall ME, Blount JD, Forbes S (2011) Patterns of egg yolk antioxidant co-variation in an avian brood parasite-host system. Behav Ecol Sociobiol 65:313–323
Rumbaugh KP, Trivedi U, Watters C, Burton-Chellew MN, Diggle SP, West SA (2012) Kin selection, quorum sensing, and virulence in pathogenic bacteria. Proc R Soc B 279:3584–3588
Sapolsky RM (1992) Neuroendocrinology of the stress response. In: Becker JB, Breedlove SM, Crew D (eds) Behavioral endocrinology. MIT Press, Cambridge, MA, pp 287–324
Schuetz JG (2005) Low survival of parasitic chicks may result from their imperfect adaptation to hosts rather than expression of defenses against parasitism. Evolution 59:2017–2024
Schwabl H (1996) Maternal testosterone in the avian egg enhances postnatal growth. Comp Biochem Physiol A Physiol 114:271–276
Scott DM, Ankney CD (1980) Fecundity of the brown-headed cowbird *Molothrus ater* in southern Ontario Canada. Auk 97:677–683
Sealy SG (1992) Removal of yellow warbler eggs in association with cowbird parasitism. Condor 94:40–54
Sealy SG (1996) Evolution of host defenses against brood parasitism: implications of puncture-ejection by a small passerine. Auk 113:346–355
Sealy SG, Underwood TJ (2012) Egg discrimination by hosts and obligate brood parasites: a historical perspective. Chin Birds 3:274–294

Searcy WA, Marler P (1981) A test for responsiveness to song structure and programming in female sparrows *Melospiza melodia*. Science 213:926–928

Smith JNM, Cook TL, Rothstein SI, Robinson SK, Sealy SG (2000) Ecology and management of cowbirds and their hosts. University of Texas Press, Austin

Snyder-Mackler N, White DJ (2011) The developmental ecology of acoustic reactions approaches to song playbacks by male cowbirds change across their first year of life. Behaviour 148:747–764

Soler M (2014) Long-term coevolution between avian brood parasites and their hosts. Biol Rev Camb Philos Soc 89:688–704

Soler M, Soler JJ, Martinez JG, Møller AP (1995) Magpie host manipulation by great spotted cuckoos: evidence for an avian mafia? Evolution 49:770–775

Sorenson MD, Balakrishnan CN, Payne RB (2004) Clade-limited colonization in brood parasitic finches (*Vidua* spp.) Syst Biol 53:140–153

Spaw CD, Rohwer S (1987) A comparative study of eggshell thickness in cowbirds and other passerines. Condor 89:307–318

Stevens M, Troscianko J, Spottiswoode CN (2013) Repeated targeting of the same host by a brood parasite compromises host egg rejection. Nat Commun 4:2475

Stoklosa SK, Kearns LJ, Rodewald AD (2014) Risky edges: temporal variation in brood parasitism of northern cardinals. Wilson J Ornithol 126:94–97

Strausberger BM, Burhans DE (2001) Nest desertion by field sparrows and its possible influence on the evolution of cowbird behavior. Auk 118:770–776

Strausberger BM, Rothstein SI (2009) Parasitic cowbirds may defeat host defense by causing rejecters to misimprint on cowbird eggs. Behav Ecol 20:691–699

Teather KL, Robertson RJ (1986) Pair bonds and factors influencing the diversity of mating systems in brown-headed cowbirds *Molothrus ater*. Condor 88:63–69

Thompson FR III (1994) Temporal and spatial patterns of breeding brown-headed cowbirds in the Midwestern US. Auk 111:979–990

Underwood TJ, Sealy SG (2008) UV reflectance of eggs of brown-headed cowbirds (*Molothrus ater*) and accepter and rejecter hosts. J Ornithol 149:313–321

Ward D, Lindholm AK, Smith JNM (1996) Multiple parasitism of the red-winged blackbird: further experimental evidence of evolutionary lag in a common host of the brown-headed cowbird. Auk 113:408–413

Web of Science (2017) Clarivate Analytics, Philadelphia, Pennsylvania

West MJ, White DJ, King AP (2002) Female brown-headed cowbirds' (*Molothrus ater*) organization and behaviour reflects male social dynamics. Anim Behav 64:377–385

White DJ, Gersick AS, Snyder-Mackler N (2012) Social networks and the development of social skills in cowbirds. Philos Trans R Soc B 367:1892–1900

Williams TD (2005) Mechanisms underlying the costs of egg production. BioScience 55:39–48

Wilsey CB, Lawler JJ, Cimprich D, Schumaker NH (2013) Dependence of the endangered black-capped vireo on sustained cowbird management. Conserv Biol 28:561–571

Wingfield JC, Farner DS (1978) The endocrinology of a natural breeding population of the white-crowned sparrow (*Zonotrichia leucophrys pugetensis*). Physiol Zool 51:188–205

Wolf JB, Wade MJ (2009) What are maternal effects (and what are they not)? Philos Trans R Soc B: Biol Sci 364:1107–1115

Yokel DA (1989) Intrasexual aggression and the mating behavior of brown-headed cowbirds: their relation to population densities and sex ratios. Condor 91:43–51

Yokel DA, Rothstein SI (1991) The basis for female choice in an avian brood parasite. Behav Ecol Sociobiol 29:39–45

Zimmerman JL (1983) Cowbird parasitism of dickcissels in different habitats and at different nest densities. Wilson Bull 95:7–22

Cuckoos as Indicators of Biodiversity

10

Anders Pape Møller, Federico Morelli, and Piotr Tryjanowski

Abstract

Brood parasites constitute a small fraction of all individuals and species of birds in any given community. Cuckoos may constitute unique bioindicators of species richness because the presence of the cuckoo is positively correlated with high values of taxonomic diversity, functional richness and functional evenness of bird communities. A number of studies of common cuckoos *Cuculus canorus* in Europe and Asia have shown that cuckoos are particularly successful surrogates of biodiversity. These relationships between cuckoo abundance and species richness of birds apply both to cuckoo hosts and non-hosts. Such an ability to indicate biodiversity may arise from unique aspects of the biology of cuckoos including their diet of caterpillar larvae, their parasitic behaviour and their adaptation to a large diversity of hosts. Cuckoo songs are widely known to the general public, and they are repeated ceaselessly during the breeding season. The common cuckoo has since ancient times been considered to reveal the quality of human habitats by reflecting the quality of local living conditions and hence human longevity. In fact, we have documented that this might be the case. The number of repeats of syllables in such songs reliably reflects intrinsic individual quality but also extrinsic habitat quality. Thus different aspects of indicator ability are revealed by different aspects of the ecology of cuckoos. We provide a number of ways in which this approach may be extended to brood parasites and parasites in general.

A. P. Møller (✉)
Ecologie Systématique Evolution, Université Paris-Sud, CNRS, AgroParisTech, Université Paris-Saclay, Orsay Cedex, France
e-mail: anders.moller@u-psud.fr

F. Morelli
Faculty of Environmental Sciences, Department of Applied Geoinformatics and Spatial Planning, Czech University of Life Sciences Prague, Prague 6, Czech Republic

P. Tryjanowski
Institute of Zoology, Poznan University of Life Sciences, Poznań, Poland

M. Soler (ed.), *Avian Brood Parasitism*, Fascinating Life Sciences,
https://doi.org/10.1007/978-3-319-73138-4_10

10.1 Introduction

Biodiversity is defined as 'the variety and variability of life on Earth' (Gaston and Spicer 2004). The United Nations Environment Programme, Conservation Monitoring Centre, defined it in terms of variability within species, between species and between ecosystems (UNEP 2015). Independent of the exact definition biodiversity at the local, continental or global scale can only be assessed from extensive sampling (Gaston and Spicer 2004). While numerous studies have investigated biodiversity, often with great difficulty because of the potential for arriving at biased estimates, there has since long been a trend towards the use of biodiversity surrogates that by definition reflect biodiversity by assessing abundance and distribution of a few key species rather than all co-occurring taxa (Gaston and Spicer 2004). The effectiveness of biodiversity surrogates continues to be debated (Grantham et al. 2010; Marfil-Daza et al. 2013), although it is widely acknowledged that more efficient and reliable surrogates of biodiversity than those currently used are needed (e.g. Caro 2010; Sattler et al. 2014). The reason why biodiversity surrogates are so important is because biodiversity data are generally difficult to collect in the field (Gaston and Spicer 2004). Thus, the use of a tool that indirectly allows estimation of biodiversity in an ecosystem may prove helpful (Magurran 2004; Noss 1990).

Raptors have repeatedly been used as bioindicators but with relatively limited success (e.g. Sergio et al. 2005, 2008; Cabeza et al. 2007). We have for the last years developed the presence and the abundance of cuckoos as an alternative surrogate of biodiversity. Although there are a total of 144 cuckoo species worldwide, cuckoos are generally rare in terms of number of species and individuals (Erritzøe et al. 2012). The common cuckoo *Cuculus canorus* (hereafter cuckoo) is abundant and occurs across most parts of Europe and Asia. Thus, cuckoos are widely distributed (Erritzøe et al. 2012), which allows for the use of the same surrogate across large spatial and temporal scales.

Our recent studies of avian biodiversity have indicated that common cuckoos are remarkably reliable bioindicators of species richness, biodiversity and different aspects of surrogacy (Morelli et al. 2015, 2017b; Møller et al. 2016a, 2017a). Why is such surrogacy by a single species effective? We have hypothesized that it may be the breeding habitat, diet and brood parasitic breeding behaviour that accounts for this effect. These effects also go beyond cuckoos and their hosts because cuckoo abundance reveals biodiversity of both hosts and non-hosts. However, these effects also go beyond cuckoos and their hosts because cuckoo abundance reveals biodiversity of both hosts and overall species richness (Morelli et al. 2015).

The common cuckoo is widely known by humans across Europe and Asia, and hence any study revealing links between distribution and abundance of cuckoos and environmental conditions can readily be transmitted to the general public. Indeed, there is a huge literature on the folklore of cuckoos dating back more than 1000 years (e.g. Brøndegaard 1985; Martin 1993; Loyd 2010; Yuzieva 2014). In China, Korea and Japan, but also in Europe, cuckoos are the subject of numerous superstitions and beliefs and equally many poems and other documents (e.g. Brøndegaard 1985; Martin 1993; Loyd 2010; Yuzieva 2014). Thus there are great opportunities to popularize science and develop citizen science projects as for assessment of spatial and temporal patterns of biodiversity and patterns of effects of climate change and global change in general.

This chapter reviews recent research addressing how cuckoo abundance, distribution and songs may reliably reflect biodiversity and condition of the environment. First, we review different aspects of biodiversity revealed by the presence of cuckoos. Second, we review how cuckoo songs may reveal different aspects of habitat quality, conspecific competition and spatial distribution of cuckoos. These effects must be taken into account in order to avoid problems with confounding factors causing these relationships. Third, we show that cuckoo songs may reveal habitat quality for humans and human viability. Fourth, we explain how the common cuckoo and other brood parasites reliably reflect biodiversity. Fifth, we discuss the potentials and the advantages offered by the use of the cuckoo and similar species for environmental monitoring, mainly through citizen science. Finally, we briefly address how cuckoo diversity may affect the temporal and spatial distribution of cuckoo songs.

10.2 Indicators of Biodiversity

Biodiversity of an ecosystem can be studied at different levels (Duffy et al. 2007; Rodrigues and Brooks 2007). There is a long history of diversity indices dating back to diversity measures such as Shannon-Weaver, Simpson and others (Magurran 2004). While these indices have been used extensively in ecology, they have mostly been used as indicators of how diverse is a given community, and they have only rarely been used as indicators of biodiversity. There has in recent years been a transformation towards measures of biodiversity that directly reflect biodiversity or components of biodiversity. Currently, the three most commonly studied components of biodiversity are taxonomic diversity, functional diversity and phylogenetic diversity (Meynard et al. 2011; Venail et al. 2015).

Species richness (the number of species inhabiting an area) constitutes the most popular diversity index to characterize biodiversity in ecology, several times preferred over species abundance (Venail et al. 2015). In fact, data on species richness were often used as support for a wide range of high-profile conservation efforts, particularly when information on endemism and changes in land use and land cover were also available (e. g. Fleishman et al. 2005; Kareiva and Marvier 2003).

Functional diversity is an essential aspect of biodiversity, because it links species composition in communities to ecosystem functioning and environmental constraints (Mouchet et al. 2010). Functional diversity can reveal species coexistence processes, variation in the roles that different organisms play in the ecosystem and assembly rules driven by functional traits (Petchey and Gaston 2006). Functional diversity metrics are often based on approaches derived from species-specific trait, and they constitute an additional tool to the traditional taxonomic approach (de Bello et al. 2010).

Finally, phylogenetic diversity is a measure of a species' evolutionary distances and relationships to other closely and less closely related taxa. Therefore, phylogenetic diversity can be considered an indicator of the extent to which each species is evolutionarily unique (Zupan et al. 2014).

Because each diversity metric can express a different component of overall biodiversity, it is necessary to focus on all these different aspects in order to achieve a more accurate representation of the ecosystem.

The presence of common cuckoos is a surrogate of avian biodiversity but potentially also biodiversity of other taxa such as caterpillar larvae that constitute the main diet of cuckoos and the insects that constitute the diet of hosts. Morelli et al. (2015) hypothesized that common cuckoos are reliable surrogates of bird species richness because they should mirror the abundance of potential avian hosts. Since different host races co-occur in sympatry, only separated by timing of breeding and habitat (Møller et al. 2011a, b), a larger number of sympatric host races should reveal stronger positive relationships between cuckoo and host abundance. High bird species richness was strongly positively correlated with cuckoo abundance in seven European countries (Fig. 10.1; Morelli et al. 2015). Interestingly, there were similar patterns for different habitats, implying that cuckoo surrogacy was not specific to any particular habitat. Furthermore, cuckoo abundance was positively correlated with bird species richness and the Shannon-Weaver diversity index (Morelli et al. 2015). This implies that not only species richness but also diversity of species is predicted by the presence of the common cuckoo.

Top predators have been hypothesized to constitute prime bioindicators of biodiversity (Sergio et al. 2005, 2008; Cabeza et al. 2007). However, there have been few attempts to compare the ability of different bioindicators to predict biodiversity.

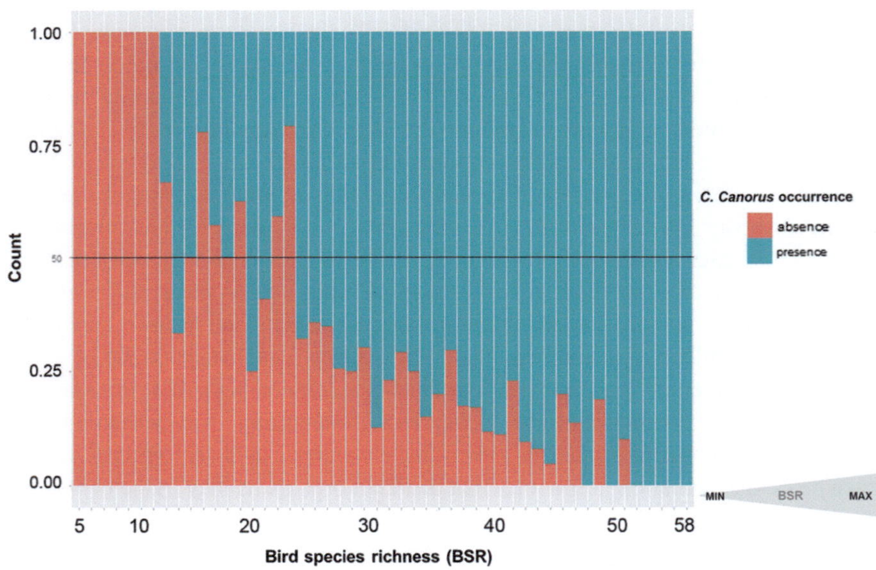

Fig. 10.1 Count of common cuckoo occurrence in sampled sites ordered by values of bird species richness (BSR) for the large-scale analysis in France ($N = 1153$). The heights of the bars represent the counts of cases in each group, where 1 is the maximum height of the bar corresponding to 100% of the cases. Cuckoos were absent from all sites with low bird species richness. Adapted from Morelli et al. (2015)

Species richness and abundance of birds were positively correlated with the abundance of cuckoos and top predator raptors (Tryjanowski and Morelli 2015; Fig. 10.1). Importantly, an analysis of the strength of these relationships for cuckoos and top predators revealed a stronger relationship for the former group of species, showing that the abundance of common cuckoos outperforms the abundance of top predators as bioindicators of biodiversity.

Because different components of biodiversity may be reflected by the presence of cuckoos, we investigated the ability of the presence of common cuckoos to reflect six measures of biodiversity. Morelli et al. (2017a, 2018) investigated these relationships between six measures of biodiversity (taxonomic diversity, host species richness, functional richness, functional evenness, functional divergence and evolutionary distinctiveness) in relation to the occurrence of common cuckoos. The presence of cuckoos was positively correlated with values of taxonomic diversity, functional richness and functional evenness (Fig. 10.2).

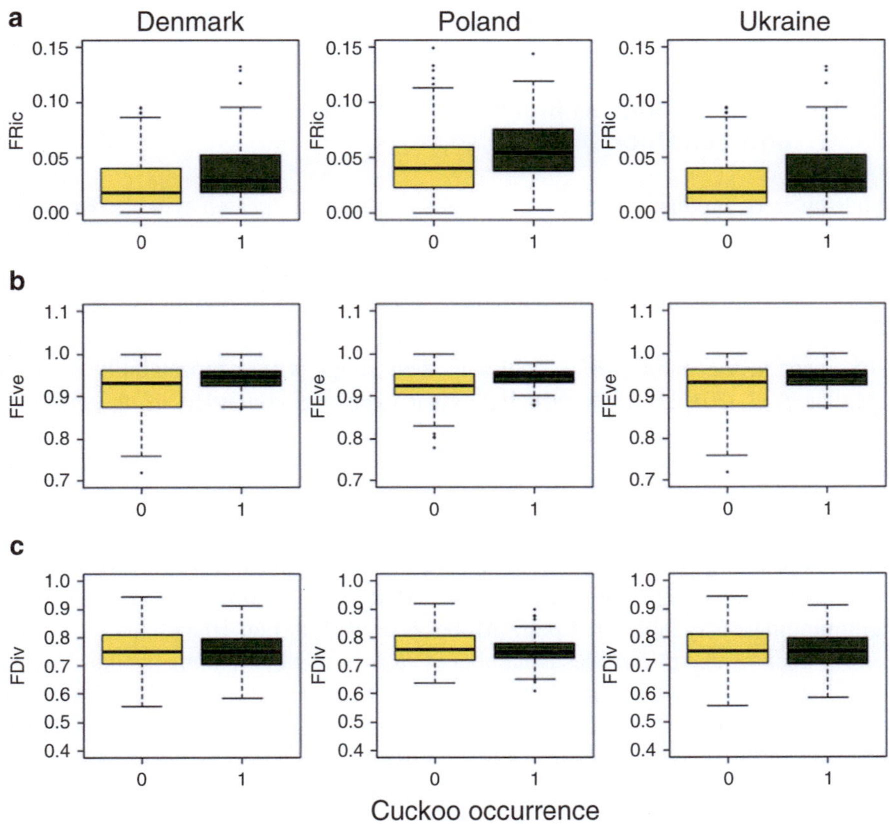

Fig. 10.2 Functional diversity indices [(**a**) functional richness (FRic), (**b**) functional evenness (FEve) and (**c**) functional diversity (FDiv)] for breeding bird communities in relation to presence or absence of common cuckoo. Adapted from Morelli et al. (2017a)

10.3 Bioindicators and Their Correlates

10.3.1 Timing of Migration and Surrogacy

Cuckoos are long-distance migrants that generally follow the timing of migration of passerines, and spring arrival of both cuckoos, but also other migrants, can thus be assumed to reflect spring phenology linked to climate change (Saino et al. 2009; Møller et al. 2011a, b). Hence we hypothesize that the timing of cuckoo migration matches that of their most common hosts resulting in a temporal link between phenology of cuckoos and phenology of hosts. Thus cuckoos should be bioindicators of migrating birds. Therefore, cuckoos and hosts should match their timing of arrival and hence the timing of reproduction, as shown by differences in timing of breeding by different host races of the common cuckoo (Møller et al. 2011a). Common cuckoos are long-distance migrants that breed in the northern hemisphere and migrate to the tropics of Africa and Asia for winter (Payne 2005; Erritzøe et al. 2012). Hence we hypothesize that common cuckoos but potentially also other cuckoo species time their migration with that of hosts.

10.3.2 Cuckoo Songs as Bioindicators of Habitat Quality and Condition

Male cuckoos produce songs with a variable number of repeated syllables followed by pauses without vocalizations (Møller et al. 2016a). Cuckoos also have individually recognizable songs (Zsebők et al. 2017). We followed male cuckoos when producing two or more songs allowing us the estimate repeatability and hence consistency in the number of syllables. Indeed, 80 male cuckoos were highly consistent in the number of syllables in their songs with a repeatability of 0.61 (Møller et al. 2016a). Thus each individual male was consistent in its ability to produce songs with repeated syllables.

The diurnal pattern of songs is important for methodological reasons for quantifying song traits when songs are the most common and the best bioindicators. Diurnal patterns of song activity may also be important because they mirror diurnal patterns of change in activity of the biodiversity of birds. There was a clear diurnal pattern of cuckoo songs with the number of syllables decreasing from 19 at 4 a.m. to a minimum of 6 syllables at 4 p.m. (Møller et al. 2016a; Fig. 10.3).

We documented differences in number of cuckoo syllables among habitats implying differences in biodiversity. Cuckoo songs contained on average 11.2 syllables in forest compared to 8.6 syllables in meadows (Møller et al. 2016a; Fig. 10.3). There were also differences in number of syllables in songs produced by cuckoos on black soil with a mean of 21.4, while only 12.9 syllables were produced on average on sandy soil (Møller et al. 2016a; Fig. 10.3). This might imply that cuckoos with different songs are restricted to specific habitats, or that cuckoos in better condition sing longer songs, but are also better at competing for such limiting habitats. Higher-quality soil increases primary productivity (Begon et al. 2005), which in turn may affect the abundance of herbivores such as butterflies (Begon et al. 2005), and their

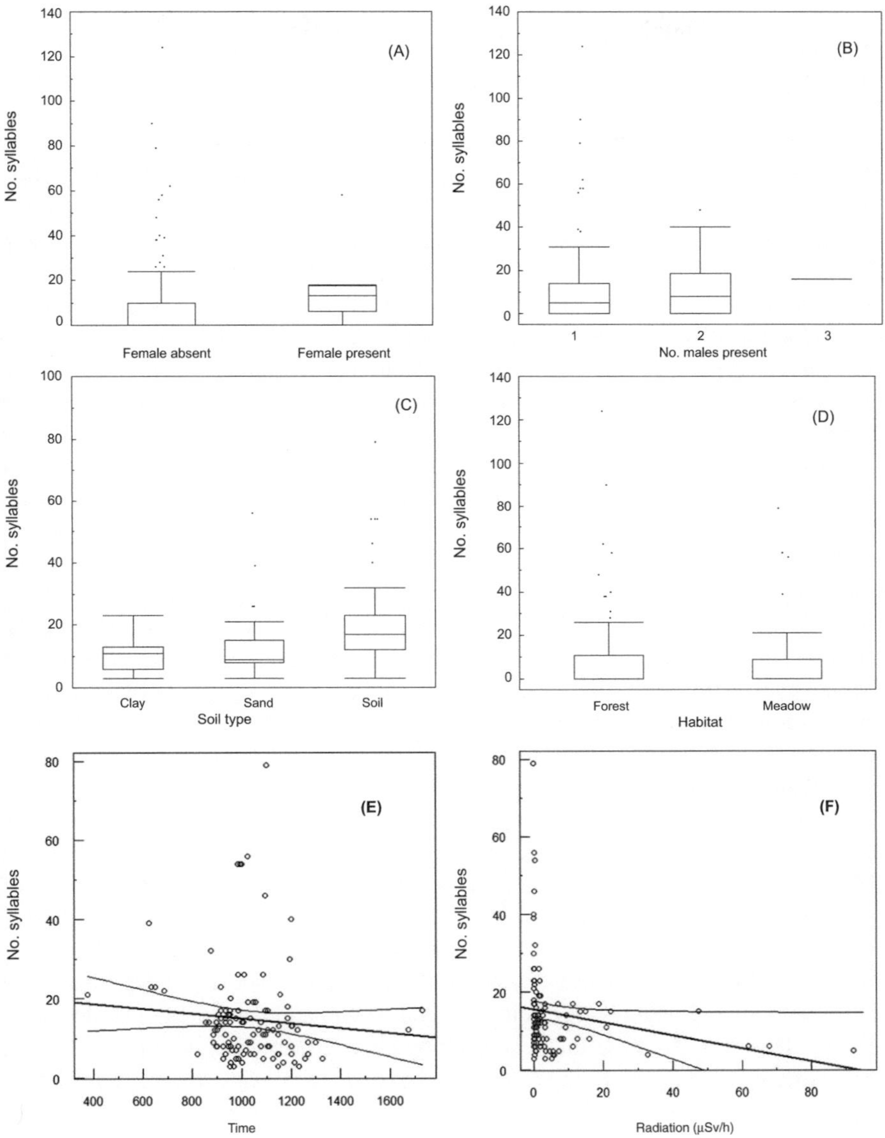

Fig. 10.3 Number of syllables in a common cuckoo song in relation to (**a**) presence of a female cuckoo, (**b**) number of male cuckoos present, (**c**) soil type, (**d**) habitat, (**e**) time of day and (**f**) background radiation level (μSv/h). Adapted from Møller et al. (2016a)

caterpillars in turn constitute a major component of the diet of cuckoos (Cramp and Perrins 1986). In other words, more cuckoos with songs containing more syllables should occur on more fertile soil, but such soil should also imply higher biodiversity as reflected by the greater abundance of cuckoos and the larger number of syllables in cuckoo songs.

Not only common cuckoos but also most other birds show a strong preference for proximity to human habitation with a distribution aggregated in such areas with more than 70% of all birds being restricted to a distance of 100 m or less from human habitation (Møller and Díaz 2016). Human habitation may be considered a refuge against raptors, because they keep long distances from humans. Human habitation is also a refuge against the common cuckoo that breeds at long distances away from human habitation (Møller et al. 2016b). This may be because cuckoos mimic the appearance of *Accipiter* hawks (Gluckman and Mundy 2013; Trnka and Prokop 2012; Welbergen and Davies 2011). If cuckoos compete particularly intensely for access to hosts near human habitation, where the density of hosts is the highest and where food availability may be the highest as well, we should expect that cuckoo songs near human habitation on average contain more syllables. Indeed, that turned out to be the case (A. P. Møller et al., in preparation). This pattern is not just correlational because playback of cuckoo songs varying in number of syllables revealed stronger responses to cuckoo songs with more syllables but also stronger responses to playback of cuckoo songs near human habitation (P. Tryjanowski et al., in preparation).

Most bird song is produced in the context of the two components of sexual selection, male-male competition and female choice (Catchpole and Slater 2008). Cuckoos sang songs with more syllables when there were a larger number of male cuckoos singing in the neighbourhood but also in the presence of female cuckoos (Møller et al. 2016a; Fig. 10.3). This suggests that cuckoo song indeed plays a role in competition among males and in attraction of females, the two components of sexual selection. It is against this background that the songs of cuckoos can be used as bioindicators of biodiversity.

We have used the unique opportunity to test whether mutations as reflected by the level of ionizing radiation cause a decrease in the quality of songs at Chernobyl (Møller et al. 2016a; Fig. 10.3). Indeed, we documented a decline in the number of syllables with increasing level of background radiation (Møller et al. 2016a). Furthermore, we observed six male cuckoos with aberrant songs at Chernobyl, Ukraine, following the nuclear disaster at the Chernobyl Nuclear Power Plants, all males with such deviant songs being located in contaminated areas (Møller et al. 2016a). This is many more than expected by chance (Møller et al. 2016a). In fact, males singing such deviant songs all produced songs with many fewer syllables than expected by chance, suggesting that the number of syllables reflects phenotypic quality (Møller et al. 2016a).

10.3.3 Cuckoos as Indicators of Environmental Conditions for Humans

Cuckoos play a strong role in human folklore across Europe and Asia, with cuckoos believed to turn into hawks during fall and the reverse in spring (e.g. Brøndegaard 1985; Yuzieva 2014). The amount of beliefs and superstitions surrounding cuckoos includes the suggestion that cuckoos reveal longevity of any human counting the number of syllables in a cuckoo song (Møller et al. 2017a). This widespread belief

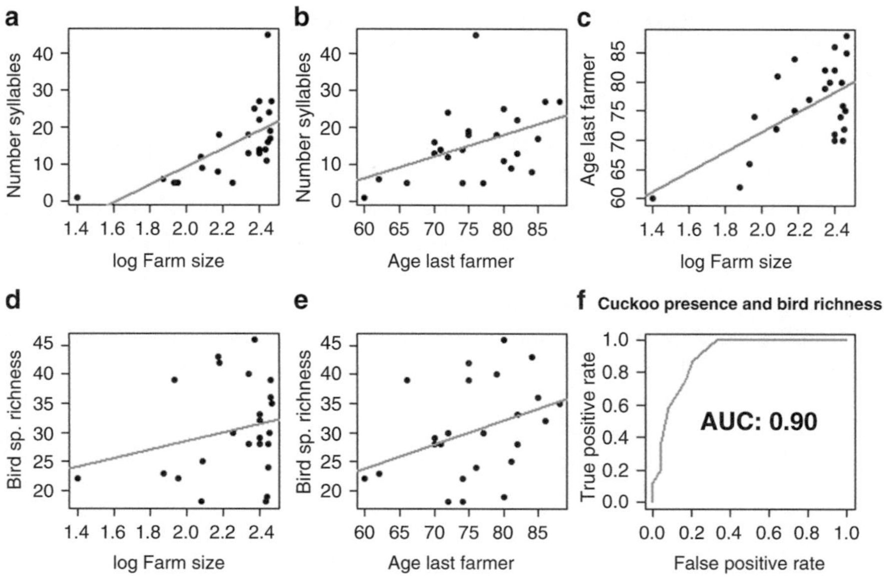

Fig. 10.4 (**a**) Number of syllables in relation to farm size, (**b**) number of syllables in relation to age of farmers (years), (**c**) age of farmers (years) in relation to farm size, (**d**) bird species richness in relation to farm size, (**e**) bird species richness in relation to age of farmer (years) and (**f**) true-positive rate in relation to false-positive rate. Adapted from Møller et al. (2017b)

occurs across most of Central and Eastern Europe (Møller et al. 2017b). Why that should be the case remained a mystery until recently when Møller et al. (2017b) showed that the number of syllables in cuckoo songs was positively related to the longevity of farmers due to strong relationships between large farm size and a large number of syllables in cuckoo songs, which in turn reflected biodiversity (Møller et al. 2017b; Fig. 10.4). Indeed, biodiversity as reflected by the species richness of breeding birds predicted the presence of common cuckoos and the number of syllables in cuckoo songs (Møller et al. 2017b; Fig. 10.4). Thus cuckoo calls may actually reveal human longevity as revealed by the longevity of farmers. This provides a unique link between folklore and nature conservation.

10.3.4 Multiple Species of Cuckoos and Their Effects on Cuckoo Songs

The temporal pattern of song should depend on the intensity of intraspecific and interspecific competition (Catchpole and Slater 2008). Therefore, we predict that diurnal versus nocturnal song in cuckoos should depend on the number of sympatric cuckoos. Indeed, song during night is rare in common cuckoos in Europe, while it is very common in the common cuckoo in Southern China where multiple species of sympatric cuckoos compete for access to hosts (W. Liang et al., in preparation). This has important implications for the use of cuckoos as bioindicators because the duration of songs in cuckoos should in this situation not only reflect biodiversity and environmental conditions but also the abundance and the diversity of other cuckoos and the vocal activity of these different cuckoo species.

We also hypothesize that the seasonal pattern of song should be extended in the presence of multiple species of cuckoos each competing for access to hosts, as suggested for isolation of host races of the common cuckoo by timing of breeding and by breeding habitat. Such an extended seasonal pattern of song should reflect the pattern of interspecific competition. Indeed, while the common cuckoo sings from late April to late June in Europe, it sings from early April to early July in Japan and from early April to late July in southern China (A. P. Møller et al., in preparation). Such differences in spatial and temporal distribution of songs should have consequences for the efficiency of cuckoos as bioindicators.

10.4 Future Prospects

May similar effects as those described for the common cuckoo here occur in other classes of animals than birds? We have shown that the presence of common cuckoos acts as an efficient bioindicator in Europe and Asia. Therefore, we hypothesize that this may also be the case for brood parasites in other continents such as Africa, Australia and the Americas. If that is the case, we may also hypothesize that these arguments apply to other brood parasites such as cowbirds, honeyguides and parasitic finches. Finally, we hypothesize that similar arguments can be applied to parasites in general because parasites constitute key elements in ecological communities (Poulin and Morand 2005).

Cuckoos may be prime bioindicators of temporal changes in bird communities but also of changes in the abundance of individual species. We suggest that the trend in population size of cuckoos can mirror the overall population trends of different species of birds in America and Europe (Morelli et al. 2017b), using data provided in Stephens et al. (2016) as a basis for our analyses. Finally, the climate suitability trend, defined as 'the slope of the regression between logit annual probability of occurrence for each species, under the bioclimate variables measured for each year' (Stephens et al. 2016), for the common cuckoo was positively correlated with the overall climate suitability trend of all other bird species (Morelli et al. 2017a). This suggests that the common cuckoo is also a bioindicator of response of common birds to climate change scenarios.

The timing of migration by cuckoo is also known to follow the timing of migration of passerines, and thus they are likely bioindicators of spring phenology. Hence we hypothesize that common cuckoos but potentially also other cuckoo species in their timing of migration follow the timing of migration by host species (Møller et al. 2011a, b). If common cuckoos are efficient surrogates, we should predict such an association between timing of migration by common cuckoos and migration by host species. We emphasize that such effects can most readily be determined from responses of spring phenology of cuckoos to climate change (Saino et al. 2009; Møller et al. 2011a, b).

Concluding Remarks

Cuckoos are prime candidates as bioindicators because they play a key role in interactions among many different categories of species such as other brood parasites, hosts, plant communities exploited by cuckoos and their hosts for food and the primary components of their diet such as large butterfly caterpillars that have become increasingly rare in agricultural habitats.

Cuckoos are prime surrogates of avian biodiversity as reflected by different components of biodiversity, bioindicators of environmental quality, easy detectability and a clear association with climate change scenarios. Cuckoos and their calls are known to everybody, and as such they are easy to identify, locate and map. These close associations among a diversity of taxa open the possibility that cuckoos may serve as tools in popularizing science, but also as prime targets of citizen science to assess biodiversity, to collect unprecedented amounts of information on bioindicators and to analyse spatial and temporal distributions of this biodiversity.

Acknowledgements To referees, J.D. Ibáñez-Álamo, J.J. Soler and M. Soler, who provided useful comments.

References

Begon M, Townsend CR, Harper JL (2005) Ecology, 4th edn. Wiley-Blackwell, London

Brøndegaard VJ (1985) Folk og fauna, vol 2. Rosenkilde og Bagger, Copenhagen

Cabeza M, Arponen A, Van Teeffelen A (2007) Top predators: hot or not? A call for systematic assessment of biodiversity surrogates. J Appl Ecol 45:976–980

Caro TM (2010) Conservation by proxy: indicator, umbrella, keystone, flagship, and other surrogate species. Island Press, Washington, DC

Catchpole CK, Slater PJ (2008) Bird song: biological themes and variations. Cambridge University Press, Cambridge

Cramp S, Perrins CM (eds) (1986) The birds of the Western Palearctic, vol 5. Oxford University Press, Oxford

de Bello F, Lavorel S, Gerhold P, Reier Ü, Pärtel M (2010) A biodiversity monitoring framework for practical conservation of grasslands and shrublands. Biol Conserv 143:9–17. https://doi.org/10.1016/j.biocon.2009.04.022

Duffy JE, Cardinale BJ, France KE, McIntyre PB, Thébault E, Loreau M (2007) The functional role of biodiversity in ecosystems: incorporating trophic complexity. Ecol Lett 10:522–538. https://doi.org/10.1111/j.1461-0248.2007.01037.x

Erritzøe J, Mann CF, Brammer FP, Fuller RA (2012) Cuckoos of the world. Christopher Helm, London

Fleishman E, Thomson JR, MacNally R, Murphy DD, Fay JP (2005) Using indicator species to predict species richness of multiple taxonomic groups. Conserv Biol 19:1125–1137. https://doi.org/10.1111/j.1523-1739.2005.00168.x

Gaston KJ, Spicer JI (2004) Biodiversity: an introduction, 2nd edn. Blackwell, Cornwall

Gluckman TL, Mundy NI (2013) Cuckoos in raptors' clothing: barred plumage illuminates a fundamental principle of Batesian mimicry. Anim Behav 86:1165–1181

Grantham HS, Pressey RL, Wells JA, Beattie AJ (2010) Effectiveness of biodiversity surrogates for conservation planning: different measures of effectiveness generate a kaleidoscope of variation. PLoS One 5:e11430. https://doi.org/10.1371/journal.pone.0011430

Kareiva P, Marvier M (2003) Conserving biodiversity coldspots. Am Sci 91:344–351

Loyd LR (2010) Bird facts and fallacies: bird legend and superstitions. Hutchinson, London

Magurran A (2004) Measuring biological diversity. Blackwell, Oxford

Marfil-Daza C, Pizarro M, Moreno-Rueda G (2013) Do hot spots of breeding birds serve as surrogate hot spots of wintering birds? An example from central Spain. Anim Conserv 16:60–68. https://doi.org/10.1111/j.1469-1795.2012.00569.x

Martin LC (1993) The folklore of birds. Globe Pequot Press, Old Saybrook

Meynard CN, Devictor V, Mouillot D, Thuiller W, Jiguet F, Mouquet N (2011) Beyond taxonomic diversity patterns: how do α, β and γ components of bird functional and phylogenetic diversity respond to environmental gradients across France? Glob Ecol Biogeogr 20:893–903. https://doi.org/10.1111/j.1466-8238.2010.00647.x

Møller AP, Díaz M (2016) Avian preference for close proximity to human habitation and its ecological consequences. Curr Zool. https://doi.org/10.1093/cz/zox073

Møller AP, Antonov A, Stokke BG, Fossøy F, Moksnes A, Røskaft E, Takasu F (2011a) Isolation by time and habitat and coexistence of distinct host races of the common cuckoo. J Evol Biol 24:676–684

Møller AP, Saino N, Adamík P, Ambrosini R, Antonov A, Campobello D, Stokke BG, Fossøy F, Lehikoinen E, Martin-Vivaldi M, Moksnes A, Moskat C, Røskaft E, Rubolini D, Schulze-Hagen K, Soler M, Shykoff JA (2011b) Rapid change in host use of the common cuckoo *Cuculus canorus* linked to climate change. Proc R Soc Lond B 278:733–738

Møller AP, Morelli F, Mousseau TA, Tryjanowski T (2016a) The number of syllables in Chernobyl cuckoo calls reliably indicate habitat, soil and radiation levels. Ecol Indic 66:592–597

Møller AP, Díaz M, Liang W (2016b) Brood parasitism and proximity to human habitation. Behav Ecol 27:1314–1319

Møller AP, Benedetti Y, Liang W, Morelli F, Su T, Tryjanowski P, Zhou B, Yang C (2017a) Multiple species of cuckoos are superior predictors of bird species richness. Ecosphere 8(11): e02003. https://doi.org/10.1002/ecs2.2003

Møller AP, Morelli F, Tryjanowski P (2017b) Cuckoo folklore and human well-being: cuckoo calls predict how long farmers live. Ecol Indic 72:766–768

Morelli F, Jiguet F, Reif J, Plexida S, Valli AS, Indykiewicz P, Simová P, Tichit M, Moretti M, Tryjanowski P (2015) Cuckoo and biodiversity: testing the correlation between species occurrence and bird species richness in Europe. Biol Conserv 190:123–132

Morelli F, Møller AP, Nelson E, Benedetti Y, Liang W, Simova P, Moretti M, Tryjanowski P (2017a) The common cuckoo is an effective indicator of high bird species richness in Asia and Europe. Sci Rep 7:4376

Morelli F, Møller AP, Nelson E, Benedetti Y, Tichit M, Símova P, Jerzak L, Moretti M, Tryjanowski P (2017b) Cuckoo as indicator of high functional diversity in bird communities: a new paradigm for biodiversity surrogacy. Ecol Indic 72:565–573

Morelli F, Benedetti Y, Møller AP, Liang W, Carrascal LM (2018) Cuckoos host range is associated positively with distribution range and negatively with evolutionary uniqueness. J Anim Ecol (in press)

Mouchet MA, Villéger S, Mason NWH, Mouillot D (2010) Functional diversity measures: an overview of their redundancy and their ability to discriminate community assembly rules. Funct Ecol 24:867–876. https://doi.org/10.1111/j.1365-2435.2010.01695.x

Noss RF (1990) Indicators for monitoring biodiversity: a hierarchical approach. Conserv Biol 4:355–364. https://doi.org/10.1111/j.1523-1739.1990.tb00309.x

Payne RB (2005) The cuckoos. Oxford University Press, Oxford

Petchey OL, Gaston KJ (2006) Functional diversity: back to basics and looking forward. Ecol Lett 9:741–758. https://doi.org/10.1111/j.1461-0248.2006.00924.x

Poulin R, Morand S (2005) Parasite biodiversity. Smithsonian Institution Press, Washington, DC

Rodrigues ASL, Brooks TM (2007) Shortcuts for biodiversity conservation planning: the effectiveness of surrogates. Annu Rev Ecol Evol Syst 38:713–737

Saino N, Rubolini D, Lehikoinen E, Sokolov LV, Bonisoli-Alquati A, Ambrosini R, Boncoraglio G, Møller AP (2009) Climate change effects on migration phenology may mismatch brood parasitic cuckoos and their hosts. Biol Lett 5:539–541

Sattler T, Pezzatti GB, Nobis MP, Obrist MK, Roth T, Moretti M (2014) Selection of multiple umbrella species for functional and taxonomic diversity to represent urban biodiversity. Conserv Biol 28:414–426. https://doi.org/10.1111/cobi.12213

Sergio F, Newton I, Marchesi L (2005) Conservation: top predators and biodiversity. Nature 436:192

Sergio F, Newton I, Marchesi L (2008) Top predators and biodiversity: much debate, few data. J Appl Ecol 45:992–999

Stephens PA, Mason LR, Green RE, Gregor RD, Sauer JR, Alison J, Aunins A, Brotons L, Butchart SHM, Campedelli T, Chodkiewicz T, Chylarecki P, Crowe O, Elts J, Escandell V, Foppen RPB, Heldbjerg H, Herrando S, Husby M, Jiguet F, Lehikoinen A, Lindström Å, Noble DG, Paquet J-Y, Reif J, Sattler T, Szép T, Teufelbauer N, Trautmann S, van Strien AJ, van Turnhout CAM, Vorisek P, Willis SG (2016) Consistent response of bird populations to climate change on two continents. Science 352:84–87

Trnka A, Prokop P (2012) The effectiveness of hawk mimicry in protecting cuckoos from aggressive hosts. Anim Behav 83:263–268

Tryjanowski P, Morelli F (2015) Presence of Cuckoo reliably indicates high bird diversity: a case study in a farmland area. Ecol Indic 55:52–58. https://doi.org/10.1016/j.ecolind.2015.03.012

United Nations Environment Programme, Conservation Monitoring Centre (2015) What is biodiversity? UNEP, New York, NY

Venail P, Gross K, Oakley TH, Narwani A, Allan E, Flombaum P, Isbell F, Joshi J, Reich PB, Tilman D, van Ruijven J, Cardinale BJ (2015) Species richness, but not phylogenetic diversity, influences community biomass production and temporal stability in a re-examination of 16 grassland biodiversity studies. Funct Ecol 29:615–626. https://doi.org/10.1111/1365-2435.12432

Welbergen JA, Davies NB (2011) A parasite in wolf's clothing: Hawk mimicry reduces mobbing of cuckoos by hosts. Behav Ecol 22:574–579

Yuzieva K (2014) Bird image in the traditional vision of the Mari people (ethnolinguistic aspect): the cuckoo. Esuka-Jeful 5–2:205–221

Zsebők S, Moskát C, Bán M (2017) Individually distinctive vocalization in Common Cuckoos (*Cuculus canorus*). J Ornithol 157:213–222

Zupan L, Cabeza M, Maiorano L, Roquet C, Devictor V, Lavergne S, Mouillot D, Mouquet N, Renaud J, Thuiller W (2014) Spatial mismatch of phylogenetic diversity across three vertebrate groups and protected areas in Europe. Divers Distrib 20:674–685. https://doi.org/10.1111/ddi.12186

Hippocampus and Spatial Memory in Brood Parasitic Cowbirds

11

Mélanie F. Guigueno and David F. Sherry

Abstract

Natural selection can modify cognition and its neural mechanisms if these modifications enhance fitness. Brood parasites are ideal subjects to study sex-specific adaptations in cognition and the brain because it is often females that search for potential host nests, and some species as a result show a reversal of sex-typical space use usually seen in mammals. Research from North and South America shows that female brown-headed (*Molothrus ater*) and shiny (*M. bonariensis*) cowbirds have a larger hippocampus than males, and female brown-headed cowbirds exhibit more hippocampal neurogenesis than males. Female cowbirds have better spatial memory than males in some tasks, especially tasks requiring long-term spatial memory in an ecologically relevant context. The hippocampus and spatial memory in cowbirds appear to be specialized for brood parasitism. Because of their diversity and unusual breeding biology, brood parasites offer many opportunities for investigating general questions about the adaptive modification of cognition and the brain.

11.1 The Evolutionary Importance of Spatial Memory

Natural selection produces adaptations that increase reproductive success. Spatial memory, for example, can promote individual survival and reproduction (Sherry 2006; Roth and Pravosudov 2009). Brood parasites like brown-headed cowbirds

M. F. Guigueno (✉)
Department of Natural Resource Sciences, McGill University, Sainte-Anne-de-Bellevue, QC, Canada
e-mail: melanie.guigueno@mail.mcgill.ca

D. F. Sherry
Advanced Facility for Avian Research, Department of Psychology, Western University, London, ON, Canada
e-mail: sherry@uwo.ca

M. Soler (ed.), *Avian Brood Parasitism*, Fascinating Life Sciences,
https://doi.org/10.1007/978-3-319-73138-4_11

(*Molothrus ater*) remember the locations of suitable host nests in which to lay their eggs (Sherry 2006). Presumably, female brood parasites that failed to remember the locations of target nests were less successful at reproducing than those that remembered. If such variation in spatial memory has a genetic basis, the outcome would be selection for enhanced memory for host nests. In other words, spatial memory may evolve because of the fitness-promoting effects of improved recall of spatial information (Nairne and Pandeirada 2008). The exact mechanisms by which selection affects spatial memory, however, are not always clear.

11.2 Cowbird Neuroecology

11.2.1 Ecology and Space Use in Cowbirds

Cowbirds (*Molothrus* spp.) and other brood parasites are ideal models for testing ideas about the adaptive specialization of memory and the brain because some species exhibit both sex and seasonal differences in how much they rely on spatial memory for host nests. The five parasitic species of cowbirds (*Molothrus ater*, *M. aeneus*, *M. bonariensis*, *M. oryzivora*, *M. rufoaxillaris*) form a monophyletic group that excludes all other non-parasitic icterids and the non-parasitic grayish baywing (*Agelaiodes badius*; previously bay-winged "cowbird" or *Molothrus badius*; Johnson and Lanyon 1999; Lanyon and Omland 1999). Icterids in the genera *Agelaius* and *Quiscalus* are more closely related to *Molothrus* than grayish baywings are (Johnson and Lanyon 1999; Lanyon and Omland 1999). Grayish baywings belong to a South American icterid clade that is a sister clade to both the *Molothrus* and *Agelaius* clades (Johnson and Lanyon 1999; Lanyon and Omland 1999). Comparative analyses incorporating grayish baywings, parasitic *Molothrus* species, and non-parasitic icterid species (*Agelaius* and *Quiscalus* spp.) make for ideal comparative research as described by Garland and Adolph (1994).

Brown-headed cowbirds and the ecologically similar South American shiny cowbird (*Molothrus bonariensis*) show a female-biased sex difference in space use, with females searching for host nests in the breeding season unassisted by males (Rothstein et al. 1986; Mason 1987). Both species of cowbird are extreme host generalists. Female brown-headed and shiny cowbirds parasitize nests during a brief 1-h window before sunrise, when it is still quite dark, so they are not searching for new host nests at this time (Gloag et al. 2013). Instead, they find suitable host nests at least 1 day before parasitizing them (Rothstein et al. 1986; Scardamaglia et al. 2017).

Brown-headed and shiny cowbirds are rare examples of species in which females—because they search for host nests—experience greater demand on spatial memory than males, a reversal of the sex difference often found in mammals, in which males experience greater demands on spatial memory as a result of greater male range size (Sherry et al. 1992). Female shiny cowbirds, in fact, have smaller home ranges than males (Scardamaglia and Reboreda 2014). Any sex difference in spatial ability favouring female shiny cowbirds, as described below (Astié et al. 2015), would thus not be associated with spatial mapping of a larger home range but instead with remembering the locations of multiple host nests (Scardamaglia and Reboreda 2014).

In addition to finding host nests, female brown-headed cowbirds monitor hosts' nesting and laying to ensure that parasitism occurs by early incubation, allowing enough time for the cowbird eggs to hatch (White et al. 2009). Although shiny cowbirds rarely return to nests they have already parasitized, they will visit target nests and puncture host eggs before and/or during the laying event (Gloag et al. 2014; Scardamaglia et al. 2017). Female brown-headed cowbirds likely also make repeated visits to host nests although previous research has not examined the behaviour of individually marked females (Sealy 1992). Female cowbirds may thus make multiple visits to host nests and are under strong selection pressure to remember the locations and nesting stages of potential hosts to ensure that their eggs hatch and that their young are successfully raised (Rothstein et al. 1986; Gates and Evans 1998; White et al. 2009).

11.2.2 Cowbird Mating Systems

A long history of field research on brown-headed cowbirds has shown that their mating system varies among populations and can be monogamous, promiscuous, polygynous, or polyandrous (Lowther 1993; Woolfenden et al. 2002; Strausberger and Ashley 2003). Available data indicate that shiny cowbirds are polygynous and promiscuous and only females search for nests of multiple host species, whereas screaming cowbirds are socially monogamous and both sexes search for the nests of their single host species, the grayish baywing (Fraga 1986; Mason 1987; Scardamaglia and Reboreda 2014).

11.2.3 Sex and Species Differences in the Cowbird Hippocampus

The hippocampus processes spatial information in both birds and mammals (O'Keefe and Burgess 1996; Smulders 2006; Moser et al. 2008; Pravosudov and Roth 2013). Sex differences in the volume of the hippocampus have been reported in some cowbirds (Table 11.1). The avian hippocampus is located on the dorsal surface of the brain, and its boundaries can be identified by changes in cell size, density, and distribution in Nissl-stained tissue or tissue immunolabelled for the neuronal marker NeuN (Fig. 11.1; Sherry et al. 1989, 1993; Guigueno et al. 2016). Consistent with sex differences in spatial behaviour observed in the wild and spatial memory for host nests, females have a larger hippocampus than males in brown-headed cowbirds (Sherry et al. 1993; Guigueno et al. 2016) and shiny cowbirds (Reboreda et al. 1996; Clayton et al. 1997). No sex difference in hippocampal volume was reported in two other South American species, the monogamous host specialist screaming cowbird and the non-brood parasitic grayish baywing, (Reboreda et al. 1996; Clayton et al. 1997). Mean hippocampal volume was larger in parasitic species (shiny and screaming cowbirds) than the non-parasitic species (Reboreda et al. 1996). Sherry et al. (1993)

Table 11.1 Summary of sex and seasonal differences in the hippocampus of brown-headed, shiny, and screaming cowbirds, and grayish baywings, common grackles, and red-winged blackbirds (F=female, M=male, B=breeding, NB=non-breeding)

Sex differences in the hippocampus							
Species	**Hippocampus volume**			**Hippocampal neurogenesis**		**Substance P terminal field volume**	
Brood Parasites							
Brown-headed cowbird	F > M	Sherry et al. (1993)	Guigueno et al. (2016)	F > M	Guigueno et al. (2016)		
Shiny cowbird	F > M	Reboreda et al. (1996)	Clayton et al. (1997)			F = M	Nair-Roberts et al. (2006)
Screaming cowbird	F = M	Reboreda et al. (1996)	Clayton et al. (1997)			F = M	Nair-Roberts et al. (2006)
Non-parasites							
Grayish baywing	F = M	Reboreda et al. (1996)	Clayton et al. (1997)			F = M	Nair-Roberts et al. (2006)
Red-winged blackbird	F = M	Sherry et al. (1993)		F = M	Guigueno et al. (2016)		
	F > M	Guigueno et al. (2016)					
Common grackle	F = M	Sherry et al. (1993)					
Seasonal differences in the hippocampus							
Species	**Hippocampus volume**			**Hippocampal neurogenesis**			
Brood Parasites							
Brown-headed cowbird	B = NB	Guigueno et al. (2016)		NB > B	Guigueno et al. (2016)		
Shiny cowbird	B > NB	Clayton et al. (1997)					
Screaming cowbird	B > NB	Clayton et al. (1997)					
Non-parasites							
Red-winged blackbird	B = NB	Guigueno et al. (2016)		B = NB	Guigueno et al. (2016)		

Species differences in the hippocampus	
Hippocampus volume	
Brown-headed cowbird > red-winged blackbird	Sherry et al. (1993), Guigueno et al. (2016)
Brown-headed cowbird > common grackle	Sherry et al. (1993)
Shiny cowbird > grayish baywing	Reboreda et al. (1996)
Screaming cowbird > grayish baywing	Reboreda et al. (1996)
Hippocampal neurogenesis	
Brown-headed cowbird > red-winged blackbird	Guigueno et al. (2016)
Substance P terminal field volume	
Shiny cowbird > grayish baywing	Nair-Roberts et al. (2006)
Screaming cowbird > grayish baywing	Nair-Roberts et al. (2006)
Shiny cowbird > screaming cowbird	Nair-Roberts et al. (2006)

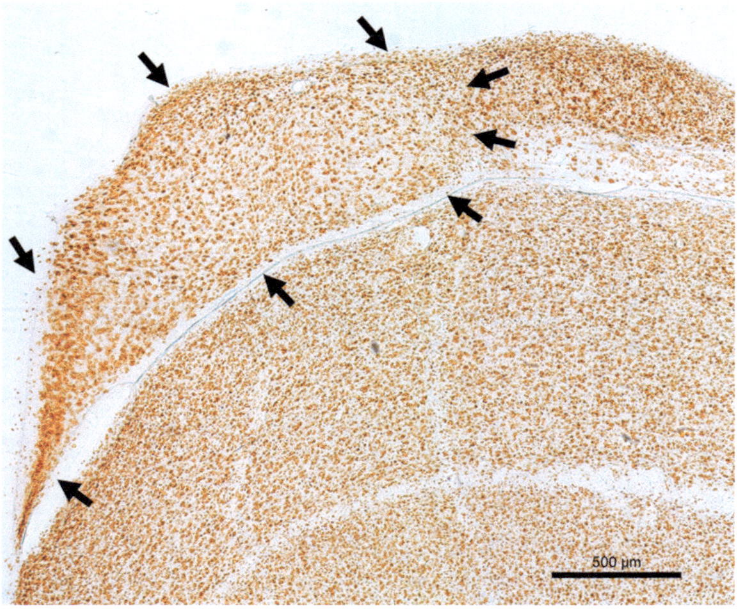

Fig. 11.1 A coronal section of the hippocampus in the right dorsomedial forebrain of a brown-headed cowbird (*Molothrus ater*), with hippocampal boundaries indicated by arrows. Mature neurons have been labelled for NeuN immunoreactivity. Figure modified from Guigueno et al. (2016)

previously reported no sex difference in the volume of the hippocampus of non-brood parasitic relatives of cowbirds, the red-winged blackbird, and the common grackle (*Quiscalus quiscula*). The results of studies of hippocampus volume, however, have not all been consistent. No sex differences in hippocampus volume were detected in a subsequent study of all three South American cowbird species (Nair-Roberts et al. 2006). In addition, Guigueno et al. (2016) found a sex difference favouring females in the volume of the hippocampus in red-winged blackbirds.

Some studies have examined the cowbird hippocampus in more detail (Table 11.1). Immunohistochemical labelling of neural proteins has been used to determine the size of regions within the hippocampus (Nair-Roberts et al. 2006) and to describe adult neurogenesis in the cowbird hippocampus (Guigueno et al. 2016). Neuropeptide substance P (SP), which has memory-promoting effects (Huston and Hasenöhrl 1995), was measured in the hippocampus of two South American cowbird species and the non-parasitic grayish baywing (Nair-Roberts et al. 2006). Nair-Roberts et al. (2006) described an area within the hippocampus rich in SP which they named the hippocampus SP terminal field (SPh). The two parasitic species (shiny and screaming cowbirds) had a larger SPh than the non-brood parasitic grayish baywing, consistent with the idea that SPh may promote memory in the parasitic species that rely on spatial memory to find and return to host nests (Nair-Roberts et al. 2006). In addition, the generalist shiny cowbird had a larger SPh than the specialist screaming cowbird, indicating that SPh may be increasingly involved in

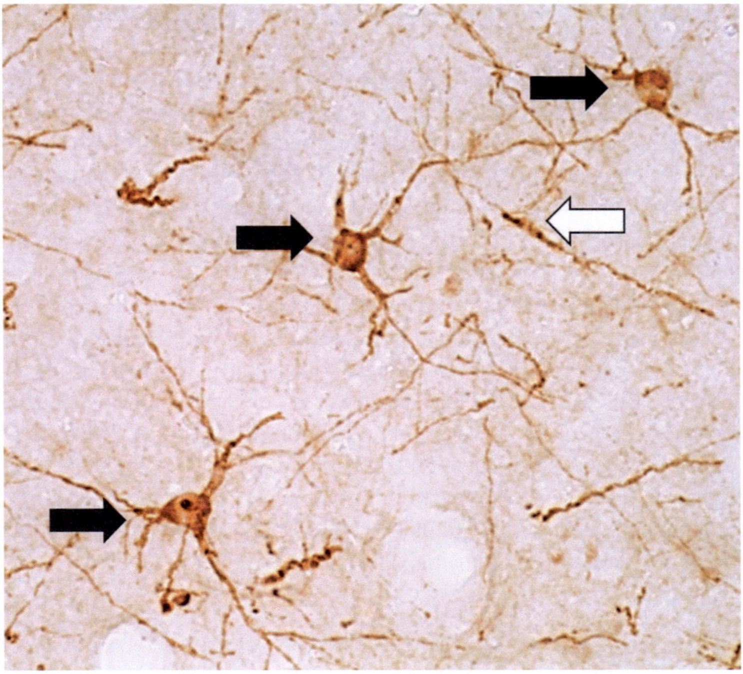

Fig. 11.2 Doublecortin (DCX) immunoreactivity in the brain of a brown-headed cowbird (*Molothrus ater*), with white arrow indicating an immature migrating cell and black arrows indicating immature differentiating cells

behavioural and cognitive processes as the number of different species of hosts that are parasitized increases (Nair-Roberts et al. 2006). Although Nair-Roberts et al. (2006) found robust species differences in SP labelling, they did not find sex differences in any of the cowbird species they examined.

Doublecortin (DCX) is an endogenous marker of neurogenesis. It is a microtubule-associated protein only expressed in immature neurons, specifically in migrating and differentiating neurons (Francis et al. 1999; Gleeson et al. 1999). DCX is a reliable marker of neurogenesis in birds (Balthazart and Ball 2014a, b), although some limitations have been discussed (Vellema et al. 2014). Guigueno et al. (2016) found greater DCX+ labelling in the hippocampus of cowbirds compared to blackbirds (Table 11.1; Fig. 11.2). There was a significant interaction between sex and species with neurogenesis being greater in female than in male cowbirds with no sex difference in blackbirds. This significant sex by species interaction in adult hippocampal neurogenesis provides strong support for the idea that the hippocampus is specialized in female brown-headed cowbirds.

11.2.4 Seasonal Differences in the Cowbird Hippocampus

The cowbird hippocampus has been shown to vary seasonally in volume and in neurogenic plasticity (Table 11.1). Shiny and screaming cowbirds had a larger hippocampus in the breeding season than in the non-breeding season (Clayton et al. 1997). Although breeding condition did not affect the volume of the hippocampus in brown-headed cowbirds, it did affect hippocampal neurogenesis (Guigueno et al. 2016). Neurogenesis peaked in non-breeding condition, when cowbirds are not searching for host nests and any associated memory demands are low. Seasonal changes in hippocampal neurogenesis in cowbirds described by Guigueno et al. (2016) resemble patterns exhibited by polygynous Richardson's ground squirrels (*Urocitellus richardsonii*; Burger et al. 2014). Although home ranges are larger in the breeding season, especially for males that mate with multiple females, hippocampal neurogenesis is at a peak in non-breeding condition (Burger et al. 2014). Heightened hippocampal neurogenesis in post-breeding cowbirds may prepare their brains to acquire, process, and retain new spatial information in the upcoming breeding season. Hippocampal neurogenesis may facilitate the forgetting of past memories, such as the locations of past host nests (Akers et al. 2014; Epp et al. 2016; Guigueno et al. 2016).

11.2.5 Cognition in Cowbirds

Spatial cognition, presumed to be hippocampus dependent, and nonspatial cognition have been studied in both shiny and brown-headed cowbirds. This research shows a complex pattern of results.

In a first study on shiny cowbirds, Astié et al. (1998) tested whether females or males would more quickly learn that one site in an 8 × 8 array, either in a specific location or identified by a specific colour, was baited with food. Birds had to learn to associate the baited site with a colour or with a location that remained constant between trials. No sex difference was observed for the spatial task. Females, however, learned to recover food faster than males when it was associated with a colour. A female-biased sex difference was thus observed for the task in which a sex difference was not expected (colour task), but no sex difference was found for the task in which a female-biased sex difference was expected (spatial task).

In a recent study, Astié et al. (2015) used an operant device with two illuminated response keys to test for sex differences in learning, reversal, and retention after extinction on colour and spatial tasks in shiny cowbirds. Astié et al. (2015) reported no sex differences in the learning and reversal phases of testing for both the spatial and colour tasks. However, during the extinction phase, up to 50 days after a correct choice was last rewarded for colour and spatial tasks, females performed significantly better than chance on both tasks (85–90% of trials were correct), whereas males did not (Astié et al. 2015). Female shiny cowbirds seem to have much better long-term memory than males for both colour and spatial cues.

A recent study suggests that female brown-headed cowbirds, unlike shiny cowbirds, do not have better colour memory than males (Guigueno et al. 2015) but

Table 11.2 Summary of sex and seasonal differences in cognition of brown-headed and shiny cowbirds (F=female, M=male, B=breeding, NB=non-breeding)

Spatial memory/learning			
Sex differences			
Task type	*Brown-headed cowbird*	*Shiny cowbird*	*Reference*
Food cups[a]	F > M		Guigueno et al. (2014)
Touchscreens[b,c]	M > F		Guigueno et al. (2015)
Testing board in cage		F = M	Astié et al. (1998)
Two response keys: learning and reversal		F = M	Astié et al. (2015)
Two response keys: extinction after 50 days		F > M	Astié et al. (2015)
Seasonal differences			
Task type	*Brown-headed cowbird*		*Reference*
Food cups	B = NB		Guigueno et al. (2014)
Touchscreens	F: B = NB M: B > NB		Guigueno et al. (2015)
Color memory/learning			
Sex differences			
Task type	*Brown-headed cowbird*	*Shiny cowbird*	*Reference*
Touchscreens[b,c]	F = M		Guigueno et al. (2015)
Testing board in cage		F > M	Astié et al. (1998)
Two response keys: learning and reversal		F = M	Astié et al. (2015)
Two response keys: extinction after 50 days		F > M	Astié et al. (2015)
Seasonal differences			
Task type	*Brown-headed cowbird*		*Reference*
Touchscreens	F: B > NB M: B = NB		Guigueno et al. (2015)

[a]Task required subjects to move through space and locate a previously baited food cup
[b]Task required subjects to remember a location or a colour in their immediate visual field
[c]Both females and males performed better on the spatial touchscreen task than on the colour touchscreen task

do have better long-term spatial memory (Guigueno et al. 2014). Sex differences in brown-headed cowbird spatial memory are dependent on task type (Table 11.2). Females performed better than males on a task in which birds had to move through space and remember a location for 24 h (Fig. 11.3; Guigueno et al. 2014), whereas males performed better than females on a stationary touchscreen task in which they had to remember a location on a screen for up to 60 s (Fig. 11.4; Guigueno et al. 2015). In the task that most closely resembles female search for host nests, brown-headed

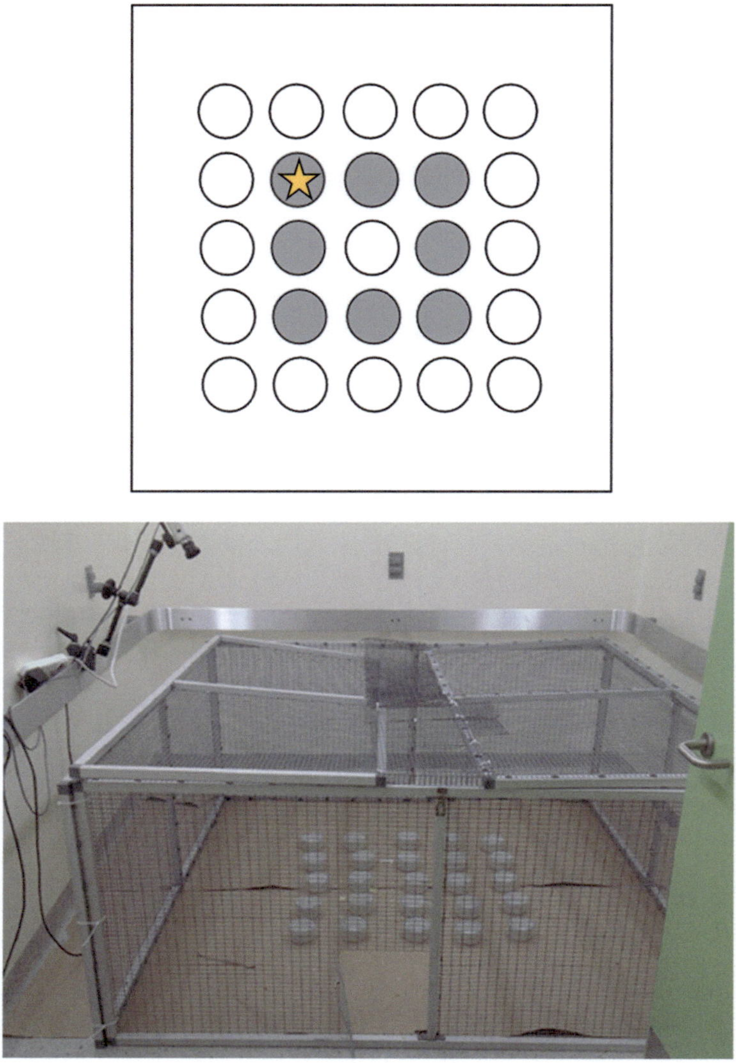

Fig. 11.3 Top: Schematic diagram of testing apparatus with potential rewarded cups shaded. Only one cup was rewarded per trial, with an example indicated by the star. Diagram modified from Guigueno et al. (2014). Bottom: The testing apparatus

cowbird females perform better than males, a reversal of the typical sex difference in spatial memory generally found in mammals (Gaulin and Fitzgerald 1986; Williams et al. 1990; Silverman et al. 2000; Postma et al. 2004).

The basis for the better performance by male brown-headed cowbirds on the touchscreen task is difficult to understand (Guigueno et al. 2015). There may be a trade-off associated with specialization in a particular form of memory, because enhanced cognitive function has energetic and life history costs, with the result that

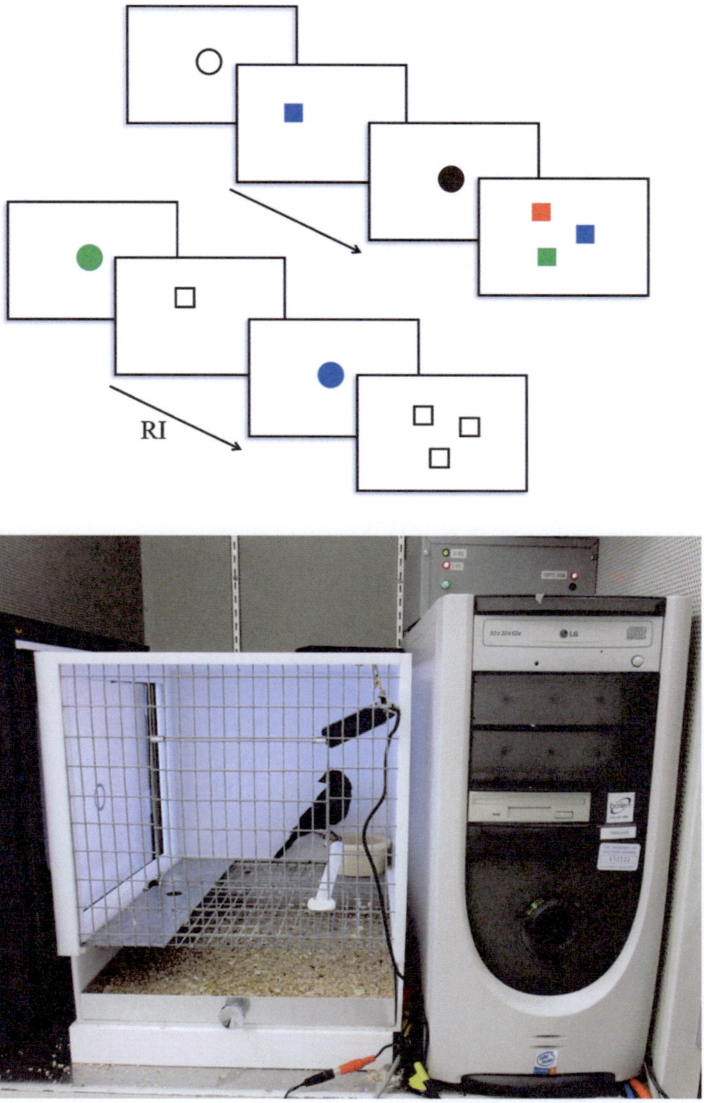

Fig. 11.4 Top: Colour (upper) and spatial (lower) delayed matching to sample touchscreen tasks used by Guigueno et al. (2015). The retention interval (RI) varied from 5 to 60 s. Bottom: Operant chamber and touchscreen

females that perform better than males on one kind of spatial task but do less well on another (Hasenstaub et al. 2010; Burns et al. 2011; Cole et al. 2012). Alternatively, there could be functional incompatibility between these two types of spatial memory (Sherry and Schacter 1987). In addition to these functional or evolutionary considerations, there may also be proximate reasons for better male performance on the touchscreen task: improved male spatial performance on the touchscreen task could be due to increased androgen concentrations. Male androgen concentrations increase more between non-breeding and breeding conditions than do female androgen concentrations, and it was only in breeding condition that males outperformed females (Guigueno et al. 2015). Elevated androgens have been shown to increase spatial memory performance in mammals and songbirds and may have caused male cowbirds to improve in spatial performance from non-breeding to breeding condition and outperform breeding females (Galea et al. 1996; Hodgson et al. 2008). In rodents, elevated levels of estradiol, which are produced from testosterone via the enzyme aromatase, increase the number of dendritic spines and the number of synapses on these spines in hippocampal neurons (Woolley and McEwen 1994; Murphy and Segal 1996; Yankova et al. 2001). Less is known about the effect of hormones on the avian hippocampus, but several studies have shown that aromatase is enriched in the hippocampus in songbirds (Shen et al. 1994; Saldanha et al. 1998; Metzdorf et al. 1999; Fusani et al. 2000). In sum, there are multiple possible explanations for the male-biased sex difference on the touchscreen task including trade-offs between different forms of spatial memory and changes in circulating androgen concentrations due to breeding (Guigueno et al. 2015).

Another breeding condition effect in brown-headed cowbirds was enhanced colour memory in female cowbirds during breeding when females search for host nests in the wild (Guigueno et al. 2015). Female shiny cowbirds outperformed males on visual memory tasks (Astié et al. 1998, 2015). Likewise, improvement in performance on another visual ask increased from non-breeding to breeding conditions in female but not in male brown-headed cowbirds (Guigueno et al. 2015). Research on both brown-headed and shiny cowbirds thus suggests that visual memory, a form of cognition not dependent on the hippocampus, may play an important role along with long-term spatial memory in cowbirds' brood parasitic mode of reproduction.

Concluding Remarks and Future Directions

Many unanswered questions remain regarding cognition and the brain of avian brood parasites. To test whether brood parasitism has specifically caused evolutionary change in the hippocampus and hippocampal neurogenesis, it is necessary to conduct larger scale phylogenetic analyses, similar to those that have examined cognition and the brain of food-storing birds. Because the hippocampus can change seasonally, and because search for host nests is clearly associated with the breeding season, data on cognition and the brain in both breeding and non-breeding conditions are most valuable.

There is little detailed information on space use by brood parasites in the wild. Determining exactly how females perform spatial search for potential host nests and how frequently they visit host nests before, during, and after parasitism would help define the spatial problems these birds face. Techniques

(continued)

such as MRI that would allow repeatable measures of hippocampal anatomy would also make it possible to examine changes in the hippocampus within the same individuals between breeding and non-breeding or even within the breeding season. Individual differences in the hippocampus may also occur among females that make more or fewer visits to host nests or that navigate larger or smaller home ranges.

Finally, more detailed descriptions of the cowbird hippocampus could provide valuable information on species differences, sex differences, and seasonal change. Variation in hippocampal cell number, for example, may help explain subtle changes in behaviour, and many techniques are available for investigating hippocampal connectivity and plasticity.

In conclusion, shiny and brown-headed cowbirds have a hippocampus and spatial memory specialized for brood parasitism. In brown-headed cowbirds, females outperformed males on a navigational memory task resembling host nest search and had a larger hippocampus with greater neurogenesis than males. In shiny cowbirds, females had better long-term memory for spatial and visual cues and a larger hippocampus than males. Finally, brown-headed cowbirds had more hippocampal neurogenesis in post-breeding condition, indicating that hippocampal neurogenesis, if it indeed contributes to female memory for host nests, takes place in advance of the coming breeding season, or is a process of hippocampal modification that follows breeding. There is good evidence for specialization of spatial memory and the hippocampus in cowbird brood parasites, providing a strong test of ideas in neuroecology, specifically the origin of sex-specific adaptive modifications of cognition and the brain.

Acknowledgements We thank the reviewers, Juan C. Reboreda, Romina C. Scardamaglia, and Manolo Soler, for constructive comments on earlier drafts of this chapter.

References

Akers KG, Martinez-Canabal A, Restivo L, Yiu AP, De Cristofaro A, Hsiang HLL, Wheeler AL, Guskjolen A, Niibori Y, Shoji H, Ohira K, Richards BA, Miyakawa T, Josselyn SA, Frankland PW (2014) Hippocampal neurogenesis regulates forgetting during adulthood and infancy. Science 344:598–602

Astié AA, Kacelnik A, Reboreda JC (1998) Sexual differences in memory in shiny cowbirds. Anim Cogn 1:77–82

Astié AA, Scardamaglia RC, Muzio RN, Reboreda JC (2015) Sex differences in retention after a visual or a spatial discrimination learning task in brood parasitic shiny cowbirds. Behav Process 119:99–104

Balthazart J, Ball GF (2014a) Doublecortin is a highly valuable endogenous marker of adult neurogenesis in canaries. Commentary on Vellema M et al (2014): Evaluating the predictive value of doublecortin as a marker for adult neurogenesis in canaries (*Serinus canaria*). Journal of Comparative Neurology 522, 1299–1315. Brain Behav Evol 84:1–4

Balthazart J, Ball GF (2014b) Endogenous versus exogenous markers of neurogenesis in canaries and other birds: advantages and disadvantages. J Comp Neurol 522:4100–4120

Burger DK, Gulbrandsen T, Saucier DM, Iwaniuk AN (2014) The effects of season and sex on dentate gyrus size and neurogenesis in a wild rodent, Richardson's ground squirrel (*Urocitellus richardsonii*). Neuroscience 272:240–251

Burns JG, Foucaud J, Mery F (2011) Costs of memory: lessons from 'mini' brains. Proc R Soc B Biol Sci 278:923–929

Clayton NS, Reboreda JC, Kacelnik A (1997) Seasonal changes of hippocampus volume in parasitic cowbirds. Behav Process 41:237–243

Cole EF, Morand-Ferron J, Hinks AE, Quinn JL (2012) Cognitive ability influences reproductive life history variation in the wild. Curr Biol 22:1808–1812

Epp JR, Mera RS, Köhler S, Josselyn SA, Frankland PW (2016) Neurogenesis-mediated forgetting minimizes proactive interference. Nat Commun 7:10838

Fraga RM (1986) The bay-winged cowbirds (*Molothrus badius*) and it brood parasites: interactions, coevolution and comparative efficiency. Ph. D. dissertation, University of California, Santa Barbara

Francis F, Koulakoff A, Boucher D, Chafey P, Schaar B, Vinet MC, Friocourt G, McDonnell N, Reiner O, Kahn A, McConnell SK, Berwald-Netter Y, Denoulet P, McConnell SK (1999) Doublecortin is a developmentally regulated, microtubule-associated protein expressed in migrating and differentiating neurons. Neuron 23(2):247–256

Fusani L, Van't Hof T, Hutchison JB, Gahr M (2000) Seasonal expression of androgen receptors, estrogen receptors, and aromatase in the canary brain in relation to circulating androgens and estrogens. J Neurobiol 43:254–268

Galea LA, Kavaliers M, Ossenkopp KP (1996) Sexually dimorphic spatial learning in meadow voles *Microtus pennsylvanicus* and deer mice *Peromyscus maniculatus*. J Exp Biol 199:195–200

Garland T, Adolph SC (1994) Why not to do two-species comparative studies: limitations on inferring adaptation. Physiol Zool 67:797–828

Gates JE, Evans DR (1998) Cowbirds breeding in the central Appalachians: spatial and temporal patterns and habitat selection. Ecol Appl 8:27–40

Gaulin SJ, FitzGerald RW (1986) Sex differences in spatial ability: an evolutionary hypothesis and test. Am Nat 127:74–88

Gleeson JG, Lin PT, Flanagan LA, Walsh CA (1999) Doublecortin is a microtubule-associated protein and is expressed widely by migrating neurons. Neuron 23:257–271

Gloag R, Fiorini VD, Reboreda JC, Kacelnik A (2013) The wages of violence: mobbing by mockingbirds as a frontline defence against brood-parasitic cowbirds. Anim Behav 86:1023–1029

Gloag R, Fiorini VD, Reboreda JC, Kacelnik A (2014) Shiny cowbirds share foster mothers but not true mothers in multiply parasitized mockingbird nests. Behav Ecol Sociobiol 68:681–689

Guigueno MF, Snow DA, MacDougall-Shackleton SA, Sherry DF (2014) Female cowbirds have more accurate spatial memory than males. Biol Lett 10:20140026

Guigueno MF, MacDougall-Shackleton SA, Sherry DF (2015) Sex differences in spatial memory in brown-headed cowbirds: males outperform females on a touchscreen task. PloS One 10(6): e0128302

Guigueno MF, MacDougall-Shackleton SA, Sherry DF (2016) Sex and seasonal differences in hippocampal volume and neurogenesis in brood-parasitic brown-headed cowbirds (*Molothrus ater*). Dev Neurobiol 76:1275–1290

Hasenstaub A, Otte S, Callaway E, Sejnowski TJ (2010) Metabolic cost as a unifying principle governing neuronal biophysics. Proc Natl Acad Sci U S A 107:12329–12334

Hodgson ZG, Meddle SL, Christians JK, Sperry TS, Healy SD (2008) Influence of sex steroid hormones on spatial memory in a songbird. J Comp Physiol A 194:963–969

Huston JP, Hasenöhrl RU (1995) The role of neuropeptides in learning: focus on the neurokinin substance P. Behav Brain Res 66:117–127

Johnson KP, Lanyon SM (1999) Molecular systematics of the grackles and allies, and the effect of additional sequence (cyt b and ND2). Auk 116:759–768

Lanyon SM, Omland KE (1999) A molecular phylogeny of the blackbirds (Icteridae): five lineages revealed by cytochrome-b sequence data. Auk 116:629–639

Lowther PE (1993) Brown-headed cowbird (*Molothrus ater*). In: Poole A (ed) The birds of North America Online. Cornell Lab of Ornithology, Ithaca. Retrieved from the Birds of North America Online http://bna.birds.cornell.edu.proxy2.lib.uwo.ca/bna/species/047

Mason P (1987) Pair formation in cowbirds: evidence found for screaming but not shiny cowbirds. Condor:349–356

Metzdorf R, Gahr M, Fusani L (1999) Distribution of aromatase, estrogen receptor, and androgen receptor mRNA in the forebrain of songbirds and nonsongbirds. J Comp Neurol 407:115–129

Moser EI, Kropff E, Moser M-B (2008) Pace cells, grid cells, and the brain's spatial representation system. Annu Rev Neurosci 31:69–89

Murphy DD, Segal M (1996) Regulation of dendritic spine density in cultured rat hippocampal neurons by steroid hormones. J Neurosci 16:4059–4068

Nairne JS, Pandeirada JN (2008) Adaptive memory: remembering with a stone-age brain. Curr Dir Psychol Sci 17:239–243

Nair-Roberts RG, Erichsen JT, Reboreda JC, Kacelnik A (2006) Distribution of substance P reveals a novel subdivision in the hippocampus of parasitic South American cowbirds. J Comp Neurol 496:610–626

O'Keefe J, Burgess N (1996) Geometric determinants of the place fields of hippocampal neurons. Nature 381:425–428

Postma A, Jager G, Kessels RP, Koppeschaar HP, van Honk J (2004) Sex differences for selective forms of spatial memory. Brain Cogn 54:24–34

Pravosudov VV, Roth TCI (2013) Cognitive ecology of food hoarding: the evolution of spatial memory and the hippocampus. Annu Rev Ecol Syst 44:173–193

Reboreda JC, Clayton NS, Kacelnik A (1996) Species and sex differences in hippocampus size in parasitic and non-parasitic cowbirds. Neuroreport 7:505–508

Roth TC, Pravosudov VV (2009) Hippocampal volumes and neuron numbers increase along a gradient of environmental harshness: a large-scale comparison. Proc R Soc B Biol Sci 276:401–405

Rothstein SI, Yokel DA, Fleischer RC (1986) Social dominance, mating and spacing systems, female fecundity, and vocal dialects in captive and free-ranging brown-headed cowbirds. In: Johnston RF (ed) Current ornithology. Plenum, New York, pp 127–185

Saldanha CJ, Popper P, Micevych PE, Schlinger BA (1998) The passerine hippocampus is a site of high aromatase: inter-and intraspecies comparisons. Horm Behav 34:85–97

Scardamaglia RC, Reboreda JC (2014) Ranging behavior of female and male shiny cowbirds and Screaming Cowbirds while searching for host nests. Auk 131:610–618

Scardamaglia RC, Fiorini VD, Kacelnik A, Reboreda JC (2017) Planning host exploitation through prospecting visits by parasitic cowbirds. Behav Ecol Sociobiol 71:23

Sealy SG (1992) Removal of yellow warbler eggs in association with cowbird parasitism. Condor 94:40–54

Shen P, Campagnoni CW, Kampf K, Schlinger BA, Arnold AP, Campagnoni AT (1994) Isolation and characterization of a zebra finch aromatase cDNA: in situ hybridization reveals high aromatase expression in brain. Mol Brain Res 24:227–237

Sherry DF (2006) Neuroecology. Annu Rev Psychol 57:167–197

Sherry DF, Schacter DL (1987) The evolution of multiple memory systems. Psychol Rev 94(4):439

Sherry DF, Vaccarino AL, Buckenham K, Herz RS (1989) The hippocampal complex of food-storing birds. Brain Behav Evol 34:308–317

Sherry DF, Jacobs LF, Gaulin SJ (1992) Spatial memory and adaptive specialization of the hippocampus. Trends Neurosci 15:298–303

Sherry DF, Forbes MR, Khurgel M, Ivy GO (1993) Females have a larger hippocampus than males in the brood-parasitic brown-headed cowbird. Proc Natl Acad Sci U S A 90:7839–7843

Silverman I, Choi J, Mackewn A, Fisher M, Moro J, Olshansky E (2000) Evolved mechanisms underlying wayfinding: further studies on the hunter-gatherer theory of spatial sex differences. Evol Hum Behav 21:201–213

Smulders TV (2006) A multi-disciplinary approach to understanding hippocampal function in food-hoarding birds. Rev Neurosci 17:53–69

Strausberger BM, Ashley MV (2003) Breeding biology of brood parasitic brown-headed cowbirds (Molothrus ater) characterized by parent-offspring and sibling-group reconstruction. Auk 120 (2):433–445

Vellema M, Hertel M, Urbanus SL, Linden A, Gahr M (2014) Evaluating the predictive value of doublecortin as a marker for adult neurogenesis in canaries (*Serinus canaria*). J Comp Neurol 522(6):1299–1315

White DJ, Ho L, Freed-Brown G (2009) Counting chicks before they hatch: female cowbirds can time readiness of a host nest for parasitism. Psychol Sci 20:1140–1145

Williams CL, Barnett AM, Meck WH (1990) Organizational effects of early gonadal secretions on sexual differentiation in spatial memory. Behav Neurosci 104:84–97

Woolfenden BE, Gibbs LH, Sealy SG (2002) High opportunity for sexual selection in both sexes of an obligate brood parasitic bird, the brown-headed cowbird (*Molothrus ater*). Behav Ecol Sociobiol 52(5):417–425

Woolley CS, McEwen BS (1994) Estradiol regulates hippocampal dendritic spine density via an N-methyl-D-aspartate receptor-dependent mechanism. J Neurosci 14:7680–7687

Yankova M, Hart SA, Woolley CS (2001) Estrogen increases synaptic connectivity between single presynaptic inputs and multiple postsynaptic CA1 pyramidal cells: a serial electron-microscopic study. Proc Natl Acad Sci U S A 98:3525–3530

Brood Parasitism and Cooperative Breeding: Seeking an Evolutionary Link

12

Vittorio Baglione and Daniela Canestrari

Abstract

From an evolutionary point of view, cooperative breeding and brood parasitism seem to have little in common. In avian cooperative species, individuals help feed young that are not their offspring, whereas brood parasites exploit the parental behaviour of a host, thus avoiding the costs of raising their progeny. Recent evidence, however, uncovered an evolutionary link between the two systems. On the one hand, it has been shown theoretically and empirically that conspecific brood parasitism (CBP) and cooperative breeding can be extremes of a continuum of parental care, where kinship among interacting females and ecological constraints on independent reproduction shape individual decisions. On the other hand, cooperative breeding and interspecific brood parasitism (IBP) might co-evolve, because the benefits of extra-care may select for a preference of nests with helpers in parasites, whereas a defensive function of helpers against the parasites would promote cooperative breeding in hosts. According to this hypothesis, the richness of cooperative species and brood parasites worldwide shows striking similarities, and within the two major host spots (Australasia and sub-Saharan Africa), passerine species that host obligate brood parasites are more likely to breed cooperatively. Although we have just begun to unravel the link between cooperative breeding and brood parasitism, the results have already widened our perspective on the evolution of avian breeding systems. Future studies that aim at building this bridge should therefore be encouraged.

V. Baglione (✉)
Department of Biodiversity and Environmental Management, University of León, León, Spain
e-mail: vbag@unileon.es

D. Canestrari
Department of Biology of Organisms and Systems, University of Oviedo, Oviedo, Spain

Unidad Mixta de Investigación en Biodiversidad (CSIC, UO, PA), Mieres, Spain

M. Soler (ed.), *Avian Brood Parasitism*, Fascinating Life Sciences,
https://doi.org/10.1007/978-3-319-73138-4_12

12.1 Introduction

Cooperative breeding, where three or more individuals raise young in a single brood, occurs in approximately 9% of bird species, which exhibit a large variety of social organizations and mating systems (Cockburn 2006). In many species, cooperation at the nest takes place when non-breeding members of a social group act as "helpers", raising young that are not their offspring. Besides provisioning the brood (Cockburn 1998) and, in some cases, the incubating breeder (Radford 2004), helpers may contribute to nest/territory defence (Golabek et al. 2012), incubating eggs (Woxvold et al. 2006) and cleaning nest/chicks (Bolopo et al. 2015a). Assisted breeders generally have higher reproductive success than pairs that lack aid (Canestrari et al. 2008; Ridley and Van Den Heuvel 2012) and/or decrease the level of costly nest provisioning when there are auxiliaries that compensate for this reduction. This load-lightening effect (Meade et al. 2010) increases survival so that breeders can eventually benefit from the presence of helpers by raising more offspring per reproductive attempt and/or by reproducing more times throughout their lives. On the other hand, in most species, helpers can gain indirect fitness benefits by assisting the breeders (Dickinson and Hatchwell 2004). Helpers are typically retained offspring of the dominant breeding pair that delay dispersal and remain in the natal territory, assisting their parents to raise new siblings (Cockburn 1998). The positive fitness consequences of kin-directed helping for both breeders and helpers strongly support a role of kin selection in shaping alloparental care (Dickinson and Hatchwell 2004). In some species, however, helpers are unrelated to the young they contribute to feed (e.g. McDonald et al. 2008), showing that other evolutionary mechanism can promote the seemingly altruistic behaviour of non-breeding group members.

Cooperative breeding can also arise among co-breeders that share a single brood and simultaneously care for their own offspring and those of other group members. Typically, "communal breeding" occurs among males that share in reproduction with a single female (cooperative polyandry; see e.g. Davies 1985). However, there are also species, particularly among ratites, passerines and members of the families Rallidae and Crotophaginae, where several monogamous or polygamous females can lay eggs in a single nest (joint laying systems; Vehrencamp and Quinn 2004). The variety of cooperative systems in birds is large and includes complex societies where individuals can derive different benefits from cooperation even within the same social group (Canestrari et al. 2005). This sometimes generates some confusion on terminology. Throughout this chapter, we will use "helpers" to indicate non-breeding auxiliaries and "co-breeders" to refer to cases where reproduction is shared among multiple breeders. This distinction between helping systems and communal nesting/joint laying systems is particularly relevant and shall be kept in mind throughout this chapter.

Both cooperation among co-breeders and helping by non-breeders appear diametrically opposed to the strategy of avian brood parasites, which avoid the cost of chick caring by laying their eggs in nests of others and have their offspring raised by foster parents (Davies 2000). The difference, however, may not be so striking. In this chapter, we will address the evolution of cooperative breeding and brood parasitism by reviewing theoretical and empirical insights that suggest a tight evolutionary link between the two reproductive systems. First, we will focus on how communal nesting and interspecific brood parasitism might be endpoints on a continuum of parental care, rather than evolutionary distinct reproductive strategies. Second, we will look at the matching geographic distribution of avian obligate brood parasites and cooperatively breeding passerines (where helping-at-the-nest is far more common than communal nesting) and will explore why, within a given geographic region, species that suffer brood parasitism are more likely to breed cooperatively than non-host species. Finally, we will focus on how helpers react to the presence of parasitic nestlings and how brood parasitism and cooperative breeding interact in determining the levels of nest provisioning.

12.2 The Evolution of Conspecific Brood Parasitism and Communal Nesting

Conspecific brood parasitism (CBP), where females lay eggs in nests of conspecifics but then do not provide any care to the clutch, either to nest on their own or to live as floater, is relatively common in birds. It has been described in 256 species (Chap. 5), mostly Anseriformes and Passeriformes. Among CBP birds, a few species also nest communally, with multiple females laying and helping at the same nest, each mating either with a single male or with multiple males. These species can be either semiprecocial or altricial, suggesting that the amount of care that chicks demand does not play a key role in shaping female decisions. Examples are the common moorhen *Gallinula chloropus* (McRae 1996), some members of the family Crotophaginae (groove-billed ani *Crotophaga sulcirostris*, smooth-billed ani Crotophaga ani and Guira cuckoo *Guira guira*; Vehrencamp and Quinn 2004), the white-fronted bee-eater *Merops bullockoides* (Emlen and Wrege 2010) and the magpie geese *Anseranas semipalmata* (Whitehead and Tschirner 1991). In these species, a secondary female can pursue different tactics: laying and deserting the clutch (parasitic), establishing a social bond with a breeding female (the primary female) and working cooperatively to raise a communal brood or leaving to nest on her own (see e.g. the white-fronted bee-eater; Emlen and Wrege 2010). These cases, reviewed and modelled by Zink (2000), exemplify the connection between CBP and communal breeding, showing that similar evolutionary and ecological processes can influence the expression of either reproductive option (Zink and Lyon 2016). On the one hand, kin selection (Hamilton 1964) is known to promote cooperation among members of a social group (see an example in Baglione et al. 2003) and modulates the contribution that each individual allocates in raising the young (e.g. Hatchwell et al. 2001). However, high levels of kinship have been also found between interspecific brood parasites and their hosts, especially in waterfowls (e.g. Andersson and Ahlund 2000; Jaatinen et al. 2009; Tiedemann et al. 2011), and increasing evidence indicates that indirect fitness benefits to the host can favour acceptance of alien eggs in species that possess mechanisms of kin recognition. On the other hand, a shortage of breeding options or potential mates may either force females to lay eggs in nests of conspecifics or to establish social bonds with another female and raise the brood communally, making the best of a bad job in circumstances where independent reproduction is unlikely (Sandell and Diemer 1999; Lyon and Eadie 2008). Based on these similarities between CBP and communal nesting and the fact that, in some species like anis and pukeko *Porphyrio porphyrio*, brood parasitism may be mechanism for joining established breeding groups, Vehrencamp and Quinn (2004) suggested that cooperative joint nesting evolved "*via an intermediate step of conspecific brood parasitism*".

Other theoretical models that recognize the common evolutionary and ecological bases of communal nesting and CBP consider the two strategies as extremes on a continuum of parental care by a secondary female. Zink (2000) analysed the life history context of species where females can parasitize conspecifics but retain the capacity of nesting on their own in the same year. The model predicts that relatedness between primary (nesting) and secondary females sets the stage for cooperation (communal nesting), whereas CBP is more likely when the parasite is unrelated to the host. A similar model developed specifically to waterfowl, however, makes opposite predictions (Andersson 2001), showing that kinship might in fact promote CBP, by providing indirect fitness benefits to a primary female that accepts the eggs of a related parasitic female. Lyon and Eadie (2008) explain this seemingly divergence by the different assumptions of the two models. Zink (2000) postulated a high cost of CBP that would reduce the indirect fitness of a parasitic female that targets a related host, whereas Andersson (2001) assumes a minimal cost, which is most likely in precocial species like the waterfowls. Taking together, these two models therefore encompass a larger variety of systems where costs and benefits of CBP can be very different (Zink and Lyon 2016). Empirical data indeed support both sets of predictions with species where hosts and parasites are related (Andersson and Ahlund 2000; Nielsen 2006; Waldeck and Andersson 2006) and others where they are not (Semel and Sherman 2001; Pöysä 2004).

Zink and Lyon (2016) followed up the previous models by analysing a context where the only alternative available to female conspecific brood parasites is communal breeding (i.e. parasitic females are unable to nest independently). The new model incorporates a third "actor", i.e. an alternative host female that the secondary female can target if she does not cooperate with the primary female. Interestingly, the model predicts the case where the secondary female becomes "helper", giving up all reproduction in order to assist the primary female (complete reproductive skew). The conditions for cooperation without reproduction (helping), however, are very restrictive (very high group productivity for cooperating females and very low value of the CBP option), which might explain the paucity of female helpers in birds (Cockburn 1998), as compared to males. Indeed, the model shows that a slight increase in the payoff of CBP would lead females to abandon the helping strategy. Conversely males, for which CBP is not an option, may be more likely to cooperate even in the absence of direct fitness incentives.

The model also applies to macroevolutionary processes and predicts that CBP and communal nesting may not always co-occur in a population, but, being evolutionary alternatives that influence each other, they should co-occur in some clades significantly more often than in others (Zink and Lyon 2016). This link may also extend beyond CBP. One hypothesis on the evolution of obligate brood parasitism in birds considers CBP as a "stepping stone" towards the use of an interspecific host (Payne 1977; Lyon and Eadie 1991). The scenario, depicted by Vehrencamp and Quinn (2004), may therefore widen by including communal nesting and interspecific brood parasitism (IBP) as alternative evolutionary endpoints of a process that originated with CBP. However, it must be noted that comparative analyses provided only partial support for a transition from CBP to IBP. Whereas this may have happened in the Anatidae (Lyon and Eadie 2000), it has been shown that, in altricial birds, IBP is more likely to have evolved directly from a non-parasitic nesting ancestor (Yom-Tov and Geffen 2006).

12.3 Overlapping Geographical Distribution of Avian Obligate Brood Parasites and Cooperatively Breeding Passerines

A close geographic association in the richness of brood parasite species and cooperatively breeding species has been recently uncovered (Feeney et al. 2013), being Australasia and sub-Saharan Africa two major host spots for both kinds of species. This pattern might reveal a direct association between IBP and cooperative breeding, but it may also be the result of a confounding third factor, which simultaneously promotes both reproductive systems. However, the fact that within a given geographic region host species of brood parasites breed cooperatively significantly more than non-host species (Table 12.1) strongly supports a direct causal link. Feeney et al. (2013) explored the process behind this pattern by examining three non-mutually exclusive hypotheses: (1) brood parasites preferentially target CB species because helpers increase chick survival (see also Poiani and Elgar 1994; Monadjem 1996), (2) nests attended by multiple carers are more easily detected and parasitized as compared to nests of unassisted pairs, and (3) helpers at the nest may contribute to defend the nest against brood parasitism, selecting for cooperative breeding in hosts. Australian superb fairy wrens (*Malurus cyaneus*), which are the primary host of the Horsfield's bronze cuckoo *Chalcites basalis,* strongly support hypotheses 1 and 3 (Feeney et al. 2013). Cuckoo chicks grow faster and survive better when reared by larger groups, because they are better fed and more protected against nest predators. However, large wren groups are also less likely to be parasitized, because helpers improve cuckoo detection and contribute to mob the parasite.

Table 12.1 The occurrence of cooperative breeding (number of species, percentage in brackets) is significantly higher among hosts than non-hosts of avian obligate brood parasites, in two host spots of richness for both kinds of species

	Host		Non-host	
	Cooperatively breeding (%)	Non-cooperatively breeding (%)	Cooperatively breeding (%)	Non-cooperatively breeding (%)
Australian passerine species	46 (53)	41 (47)	5 (12)	37 (88)
Southern African bird species	39 (28)	99 (72)	57 (8)	697 (92)

Data obtained from Feeney et al. (2013), where details on statistical analyses can be found

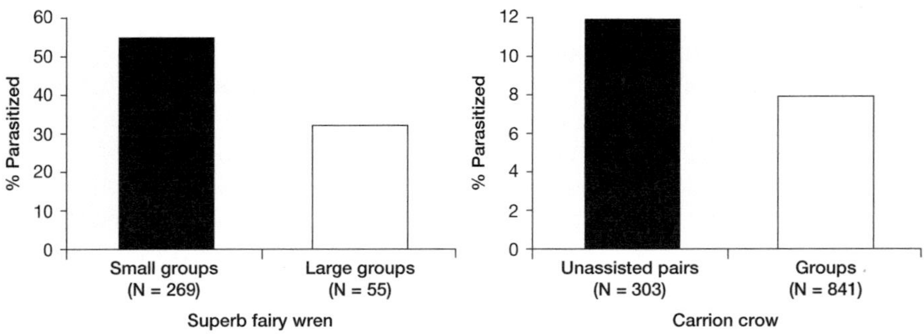

Fig. 12.1 Difference in parasitism rate between small and large groups of superb fairy wrens (left) and between unassisted pairs and groups with helpers in carrion crows (right). Data obtained from Feeney et al. (2013) and Canestrari et al. (2009), respectively. Statistical analyses, available in the original papers, proved significant at 0.05 level

Cooperatively breeding carrion crows (*Corvus corone*) provide similar evidence (Fig. 12.1). Depending on environmental factors (Baglione et al. 2017), this corvid species can be either the primary or the secondary host of a non-evicting brood parasite, the great spotted cuckoo *Clamator glandarius*. In Spain, carrion crows breed cooperatively in groups that comprise a reproductive pair, retained offspring and male immigrants, whereas unassisted pairs are found only on approximately 25% of territories (Baglione et al. 2002, 2003; Roldán et al. 2013). Cuckoo chick survival improves with increasing number of helpers, but this benefit is realized only if female cuckoos lay their eggs timely within the laying sequence of the host (Canestrari et al. 2009). When this occurs, the shorter incubation time of cuckoo eggs (Arias de Reyna 1998) grants the parasitic chicks an age advantage over the much larger crow nest mates that is crucial for competing efficiently for the food provided by the host parents (Canestrari et al. 2009). However, this is seldom achieved because cuckoos parasitizing larger groups often mismatch the laying period of the crow host, wasting the potential age advantage for their chicks. In addition, groups with helpers at the nest are significantly less parasitized than unassisted pairs (Canestrari et al. 2009). Crow helpers may therefore hinder cuckoo parasitism although, unlike the superb fairy wrens, they do not do it actively. This host species lacks defences against the cuckoo (Soler 1990; Soler et al. 2002; Canestrari et al. 2014), but helpers contribute to feed the incubating female, allowing her to spend more time at the nest and therefore decreasing the laying opportunities for the parasite (Canestrari et al. 2009). Taken together, the fairy wren and the carrion crow study cases indicate that helpers can protect against brood parasitism, providing a plausible mechanism for the observed geographic association between cooperative breeding and IBP.

Interestingly, Roldán et al. (2013) found that the protection offered by helpers against the great spotted cuckoo was lost in a crow population in southern Spain where the parasite reaches very high density. Here, unassisted pairs and groups with helpers were parasitized at a similar rate over the 4-year study period. The interaction between group size and parasitism rate seems also to vary in the superb fairy wren/ Horsfield's bronze cuckoo system. The defensive function of host helpers against the brood parasite described by Feeney et al. (2013) was not detectable in an earlier work (Langmore and Kilner 2007). The two studies only partly overlapped in time and field study places. Taken together, the cases of the carrion crows and the superb fairy wren suggest that current conditions might influence the relationship between cooperatively breeding hosts and their parasites.

12.4 Alien Chicks in Nests with Helpers: Unexplored Effects of Breeding Auxiliaries on Parasites and Host Parents

Although host species frequently breed cooperatively (Feeney et al. 2013), surprisingly little is known about the interaction between parasitic chicks and helpers at the nest. Individual effort in nest provisioning is typically very variable among members of a cooperative group (Cockburn 1998). In complex avian societies, different group members can pursue different benefits from cooperating at the nest and may have contrasting interests (Canestrari et al. 2005). Factors like reproductive status, sex and age therefore influence the effort that carers are willing to allocate in feeding the young and the way they respond to both intrinsic conditions (e.g. body condition) or environmental conditions (e.g. food availability). It is therefore interesting to ask how helpers, as compared to breeders, react to the presence of a brood parasite in the nest and how this affects the cost that brood parasitism imposes to host parents. The available information, however, is very scant and does not yet allow looking for general patterns. Our aim here is therefore that of deriving ideas and hypotheses from the few single species studies available, hoping to stimulate future research.

Bolopo et al. (2015b) found that carrion crow breeders and helpers follow the same rule to allocate food between crow chicks and great spotted cuckoo chicks. At equal intensity of begging, all group members preferentially feed crow nestlings. However, this seldom occurs because cuckoos beg much more intensively than crows and hence compensate the disadvantage, ultimately receiving on average the same number of feeds than host nestlings. Therefore, at least in this species, helpers and breeders represent for the cuckoo an equally valuable "source" of care (Fig. 12.2). Helpers, however, also have "hidden effects" that may affect brood parasites. It has been shown in several species, including the carrion crows (Canestrari et al. 2011), that assisted females in groups with helpers reduce the size of their eggs as compared to unaided females (Russell et al. 2007; Paquet et al. 2013, but see Koenig et al. 2009). Along the nestling period, helpers' contribution to nest provisioning compensates for the initial size handicap of the hatchlings, which eventually do not undergo long-lasting effects of the energy-saving strategy of their mothers. From the point of view of a parasite chick however, a group with helpers might mean lighter host eggs/hatchlings to eject from the nest or, in case of non-evicting species, smaller host nest mates that may be easier to outcompete, particularly at the beginning of the nesting period. The hidden effects of helpers may therefore translate to the parasite, making cooperatively breeding group more "appealing" for the latter.

A neglected aspect of coevolution between avian brood parasites and their host is the long-term effects of raising a parasite for host adult survival and future fecundity.

Fig. 12.2 Mean (±SE) proportion of feeds given to cuckoos by carrion crow breeders (black bars) and helpers (white bars), according to the ratio of cuckoo chicks/crow chicks in the brood. Differences are non-significant at 0.05 level (statistical details available in Bolopo et al. 2015b)

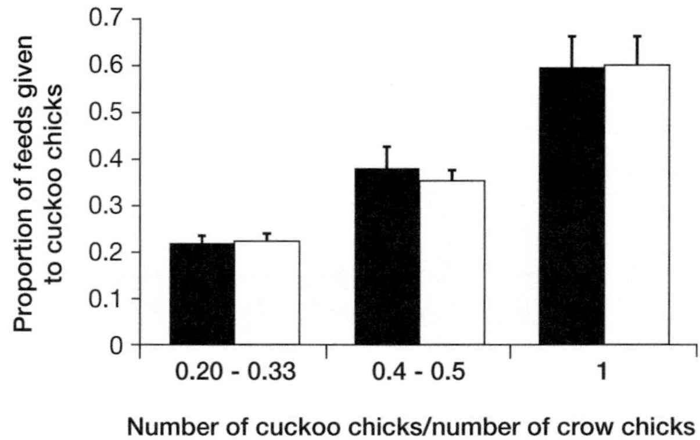

Focusing on cooperative breeding, one further question may be whether helpers at the nest buffer such hypothetical cost, reducing the effect on adult lifetime reproductive success. Interestingly, brood parasitism by screaming cowbird *Molothrus rufoaxillaris* and shiny cowbird *Molothrus bonariensis* increases helper recruitment in the grayish baywing *Agelaioides badius.* Ursino et al. (2011) experimentally parasitized nests of baywings with hatchlings of both cowbird species and found that the size of the cooperatively breeding group increased as compared to nests were only host chicks or screaming cowbird chicks were present. It is unclear whether helpers moved into action because they directly responded to the vehement begging of the parasites or the breeding pair recruited them as the brood demand increased beyond a certain threshold. In either case, the extra contribution by helpers might have mitigated the cost of parasitism to the breeding pair. This study suggests that a further evolutionary link between cooperative breeding and brood parasitism is theoretically possible. By reducing the load that foster parents bear when raising alien nestlings, recruited helpers might increase the survival of the breeders, allowing them future nesting attempts that may ameliorate the fitness consequences of the loss of the parasitized clutch (or part of it). This might reduce the selective pressure for the evolution of defences against the parasite, contributing to explain why cooperatively breeding species host brood parasites more often that non-cooperative ones. Intriguingly, in two cooperative host species that do not reject parasite's eggs or nestlings, the carrion crow and the splendid fairy wren *Malurus splendens*, breeders do not pay a survival cost for raising, respectively, a great spotted cuckoo or a Horsfield's bronze cuckoo (Brooker and Brooker 1996; Canestrari et al. 2014), whereas socially monogamous prothonotary warblers *Protonotaria citrea* that raised multiple broods showed reduced survival, if they had been parasitized by brown-headed cowbirds *Molothrus ater* (Hoover and Reetz 2006). These studies suggest at least that a load-lightening effect of helpers is plausible, although direct evidence is still lacking and the information available is far too scarce. It must also be noted that lack of survival consequences on parasitized adults has been reported also for a non-cooperatively breeding species, the indigo bunting *Passerina cyanea* raising brown-headed cowbirds (Payne and Payne 1998), which therefore would not fit the suggested scenario. However, simple comparison between parasitized and non-parasitized adults can be misleading because phenotypic correlations (i.e. adults of better quality being more likely to be parasitized; see e.g. Soler et al. 1995) can mask a true effect of parasitism on survival and lifetime reproductive success (Krüger 2007). More long-term individual-based studies are therefore needed, and, importantly, they should be backed up with experimental manipulations (Krüger 2007). Regardless of the implications for studying the interaction between cooperative breeding and brood parasitism, this kind of studies will fill a gap in our understanding of co-evolution between avian brood parasites and their hosts.

Concluding Remarks and Future Directions

Several non-mutually exclusive hypotheses have been proposed to explain why cooperative breeding passerines and interspecific brood parasites are so tightly associated (summarized in Table 12.2). However, the number of studies that addressed the issue is surprisingly little, and evidence is mixed so that no firm conclusions can be drawn at this stage. In spite of the enormous interest that both cooperative breeding and brood parasitism have attracted in the past decades, the study of the fascinating interaction between the two reproductive systems is still in its infancy. New work that helps in building this bridge has to be encouraged. On the one hand, we need to explore the phylogenetic roots of the association between communal nesting and CBP, with comparative analyses that test whether they co-occur in some avian clades more often than expected by chance. Future research should also address how individuals move through the parental care continuum that connects CBP and communal nesting and how these endpoints can become fixed, giving rise to macroevolutionary transitions among breeding systems. Regardless of whether CBP is a precursor or an evolutionary alternative to communal or cooperative breeding, the link between the two strategies needs to be taken into account to understand avian life histories.

On the other hand, we must assess the generality of current evidence on the benefits that helpers can accrue to the parasite (increased chick survival) and/or their defensive function against brood parasitism, with empirical studies on a large number of species. Whether brood parasites benefit from reduced investment in egg size by assisted host females also needs experimental testing, ideally through brood manipulations that would allow comparing parasite survival depending on the weight of host eggs or hatchlings.

Finally, attention should be paid to the effect of helpers in reducing the cost of parasitism, by looking at the survival and fecundity of breeders that have been sometimes parasitized during their life. Ideally, studies should look at multiple populations within each species, to investigate how environmental conditions shape the interaction between parasites and hosts with helpers. Indeed, spatio-temporal variability of the ecological relationship between species is being increasingly reported, including in brood parasite/host systems, and needs to be taken into account for a most comprehensive understanding of avian breeding systems.

Table 12.2 Summary of current hypotheses on the association between cooperative breeding species and interspecific brood parasites and supporting evidence

Hypothesis	Species	Type of evidence	References
Brood parasites choose cooperative breeding species	Superb fairy wren Carrion crow	In both cases, cuckoo would do better in nest with helpers, although the benefit is seldom achieved	Feeney et al. (2013); Canestrari et al. (2009)
Nests with helpers are more conspicuous to brood parasites	None	Absent	Poiani and Elgar (1994); Monadjem (1996)
Selection for cooperative breeding in hosts because of better defence against the parasite	Superb fairy wren Carrion crow	A defensive function of helpers has been shown, either active or passive, but might be influenced by current environmental conditions	Feeney et al. (2013); Canestrari et al. (2009)
Reduced selection for evolving defences in cooperatively breeding species, because helpers lighten the cost of parasitism to breeders	Splendid fairy wren Carrion crow Prothonotary warbler	No survival effect of being parasitized on cooperatively breeding splendid fairy wrens and carrion crows. Negative effects on monogamous prothonotary warbler. But see indigo buntings for a seemingly mismatching case	Brooker and Brooker (1996); Canestrari et al. (2014); Hoover and Reetz (2006); Payne and Payne (1998)

Acknowledgements We are very grateful to Bruce Lyon, Eli Geffen and Manuel Soler for the useful comments on an earlier draft of this chapter. Financial support was provided by the Spanish Ministry of Economy and Competitiveness (CGL2016-77636-P) to VB.

References

Andersson M (2001) Relatedness and the evolution of conspecific brood parasitism. Am Nat 158:599–614

Andersson M, Ahlund M (2000) Host-parasite relatedness shown by protein fingerprinting in a brood parasitic bird. Proc Natl Acad Sci U S A 97:13188–13193

Arias de Reyna L (1998) Coevolution of the great spotted cuckoo and its hosts. In: Rothstein SI (ed) Parasitic birds and their hosts, studies in coevolution. Oxford University Press, Oxford, pp 129–142

Baglione V, Marcos JM, Canestrari D (2002) Cooperatively breeding groups of carrion crow (*Corvus corone corone*) in northern Spain. Auk 119:790–799

Baglione V, Canestrari D, Marcos JM, Ekman J (2003) Kin selection in cooperative alliances of carrion crows. Science 300:1947–1949

Baglione V, Bolopo D, Canestrari D et al (2017) Spatiotemporal variation of host use in a brood parasite: the role of the environment. Behav Ecol 28:49–58

Bolopo D, Canestrari D, Marcos JM, Baglione V (2015a) Nest sanitation in cooperatively breeding carrion crows. Auk 132:604–612

Bolopo D, Canestrari D, Roldán M et al (2015b) High begging intensity of great spotted cuckoo nestlings favours larger-size crow nest mates. Behav Ecol Sociobiol 69:873–882

Brooker M, Brooker L (1996) Acceptance by the splendid fairy-wren of parasitism by Horsfield's bronze-cuckoo: further evidence for evolutionary equilibrium in brood parasitism. Behav Ecol 7:395–407

Canestrari D, Marcos JM, Baglione V (2005) Effect of parentage and relatedness on the individual contribution to cooperative chick care in carrion crows *Corvus corone corone*. Behav Ecol Sociobiol 57:422–428

Canestrari D, Marcos JM, Baglione V (2008) Reproductive success increases with group size in cooperative carrion crows, *Corvus corone corone*. Anim Behav 75:403–416

Canestrari D, Marcos JM, Baglione V (2009) Cooperative breeding in carrion crows reduces the rate of brood parasitism by great spotted cuckoos. Anim Behav 77:1337–1344

Canestrari D, Marcos JM, Baglione V (2011) Helpers at the nest compensate for reduced maternal investment in egg size in carrion crows. J Evol Biol 24:1870–1878

Canestrari D, Bolopo D, Turlings TCJ, Röder G, Marcos JM, Baglione V (2014) From parasitism to mutualism: unexpected interactions between a cuckoo and its host. Science 343:1350–1352

Cockburn A (1998) Evolution of helping behavior in cooperatively breeding birds. Annu Rev Ecol Syst 29:141–177

Cockburn A (2006) Prevalence of different modes of parental care in birds. Proc R Soc Lond B Biol Sci 273:1375–1383

Davies NB (1985) Cooperation and conflict among dunnocks, *Prunella modularis*, in a variable mating system. Anim Behav 33:628–648

Davies NB (2000) Cuckoos, cowbirds and other cheats. T & AD Poyser, London

Dickinson JL, Hatchwell BJ (2004) Fitness consequences of helping. In: Koenig W, Dickinson JL (eds) Ecology and evolution of cooperative breeding in birds. Cambridge University Press, Cambridge, pp 48–66

Emlen ST, Wrege PH (2010) Forced copulations and intra-specific parasitism: two costs of social living in the white-fronted bee-eater. Ethology 71:2–29

Feeney WE, Medina I, Somveille M, Heinsohn R, Hall ML, Mulder RA, Stein JA, Kilner RM, Langmore NE (2013) Brood parasitism and the evolution of cooperative breeding in birds. Science 342:1506–1508

Golabek KA, Ridley AR, Radford AN (2012) Food availability affects strength of seasonal territorial behaviour in a cooperatively breeding bird. Anim Behav 83:613–619

Hamilton WD (1964) The genetical evolution of social behaviour, I & II. J Theor Biol 7:1–52

Hatchwell BJ, Ross DJ, Fowlie MK, McGowan A (2001) Kin discrimination in cooperatively breeding long–tailed tits. Proc Biol Sci 268(1470):885–890

Hoover JP, Reetz MJ (2006) Brood parasitism increases provisioning rate, and reduces offspring recruitment and adult return rates, in a cowbird host. Oecologia 149:165–173

Jaatinen K, Jaari S, O'Hara RB, Öst M, Merilä J (2009) Relatedness and spatial proximity as determinants of host-parasite interactions in the brood parasitic Barrow's goldeneye (*Bucephala islandica*). Mol Ecol 18:2713–2721

Koenig WD, Walters EL, Haydock J (2009) Helpers and egg investment in the cooperatively breeding acorn woodpecker: testing the concealed helper effects hypothesis. Behav Ecol Sociobiol 63:1659–1665

Krüger O (2007) Cuckoos, cowbirds and hosts: adaptations, trade-offs and constraints. Philos Trans R Soc Lond Ser B Biol Sci 362:1873–1886

Langmore NE, Kilner RM (2007) Breeding site and host selection by Horsfield's bronze-cuckoos, *Chalcites basalis*. Anim Behav 74:995–1004

Lyon BE, Eadie JM (1991) Mode of development and interspecific avian brood parasitism. Behav Ecol 2:309–318

Lyon BE, Eadie JM (2000) Family matters: kin selection and the evolution of conspecific brood parasitism. Proc Natl Acad Sci 97:12942–12944

Lyon BE, Eadie JM (2008) Conspecific brood parasitism in birds: a life-history perspective. Annu Rev Ecol Evol Syst 39:343–363

McDonald PG, Kazem AJN, Clarke MF, Wright J (2008) Helping as a signal: does removal of potential audiences alter helper behavior in the bell miner? Behav Ecol 19:1047–1055

McRae SB (1996) Brood parasitism in the moorhen: brief encounters between parasites and hosts and the significance of an evening laying hour. J Avian Biol 27:311–320

Meade J, Nam K-B, Beckerman AP, Hatchwell BJ (2010) Consequences of 'load-lightening' for future indirect fitness gains by helpers in a cooperatively breeding bird. J Anim Ecol 79:529–537

Monadjem A (1996) Cooperative breeding in southern African passerines and brood parasitism: Is there a relationship? S Afr J Sci 92:295–297

Nielsen CR (2006) Host-parasite relatedness in wood ducks: patterns of kinship and parasite success. Behav Ecol 17:491–496

Paquet M, Covas R, Chastel O et al (2013) Maternal effects in relation to helper presence in the cooperatively breeding sociable weaver. PLoS One 8:e59336

Payne RB (1977) The ecology of brood parasitism in birds. Annu Rev Ecol Syst 8:1–28

Payne RB, Payne LL (1998) Brood parasitism by cowbirds: risks and effects on reproductive success and survival in indigo buntings. Behav Ecol 9:64–73

Poiani A, Elgar MA (1994) Cooperative breeding in the Australian avifauna and brood parasitism by cuckoos (Cuculidae). Anim Behav 47:697–706

Pöysä H (2004) Relatedness and the evolution of conspecific brood parasitism: parameterizing a model with data for a precocial species. Anim Behav 67:673–679

Radford AN (2004) Incubation feeding by helpers influences female nest attendance in the green woodhoopoe, *Phoeniculus purpureus*. Behav Ecol Sociobiol 55:583–588

Ridley AR, Van Den Heuvel IM (2012) Is there a difference in reproductive performance between cooperative and non-cooperative species? A southern African comparison. Behaviour 149:821–848

Roldán M, Martín-Gálvez D, Rodríguez J, Soler M (2013) Breeding biology and fledgling survival in a carrion crow *Corvus corone* population of southern Spain: a comparison of group and pair breeder. Acta Ornithol 48:221–235

Russell AF, Langmore NE, Cockburn A et al (2007) Reduced egg investment can conceal helper effects in cooperatively breeding birds. Science 317:941–944

Sandell M, Diemer M (1999) Intraspecific brood parasitism: a strategy for floating females in the European starling. Anim Behav 57:197–202

Semel B, Sherman PW (2001) Intraspecific parasitism and nest-site competition in wood ducks. Anim Behav 61:787–803

Soler M (1990) Relationships between the great spotted cuckoo *Clamator glandarius* and its corvid hosts in a recently colonized area. Ornis Scand 21:212–223

Soler JJ, Soler M, Møller AP, Martínez JG (1995) Does the great spotted cuckoo choose magpie hosts according to their parenting ability? Behav Ecol Sociobiol 36:201–206

Soler M, Soler J, Pérez-Contreras T, Martinez J (2002) Differential reproductive success of great spotted cuckoos *Clamator glandarius* parasitizing magpies *Pica pica* and carrion crows *Corvus corone*: the importance of parasitism costs and host defences. Avian Sci 1:1–9

Tiedemann R, Paulus KB, Havenstein K et al (2011) Alien eggs in duck nests: brood parasitism or a help from grandma? Mol Ecol 20:3237–3250

Ursino CA, De Mársico MC, Sued M et al (2011) Brood parasitism disproportionately increases nest provisioning and helper recruitment in a cooperatively breeding bird. Behav Ecol Sociobiol 65:2279–2286

Vehrencamp SL, Quinn JS (2004) Joint laying systems. In: Koenig W, Dickinson JL (eds) Ecology and evolution of cooperative breeding in birds. Cambridge University Press, Cambridge, pp 177–196

Waldeck P, Andersson M (2006) Brood parasitism and nest takeover in common eiders. Ethology 112:616–624

Whitehead P, Tschirner K (1991) Patterns of egg-laying and variation in egg size in the magpie goose *Anseranas semipalmata*: evidence for intra-specific nest parasitism. Emu 91:26–31

Woxvold IA, Mulder RA, Magrath MJL (2006) Contributions to care vary with age, sex, breeding status and group size in the cooperatively breeding apostlebird. Anim Behav 72:63–73

Yom-Tov Y, Geffen E (2006) On the origin of brood parasitism in altricial birds. Behav Ecol 17:196–205

Zink AG (2000) The evolution of intraspecific brood parasitism in birds and insects. Am Nat 155:395–405

Zink AG, Lyon BE (2016) Evolution of conspecific brood parasitism versus cooperative breeding as alternative reproductive tactics. Am Nat 187:35–47

Part IV

Host Use by Brood Parasites

Host Resource Partitioning Among Sympatric Molothrine Generalist Brood Parasites

13

Jameson F. Chace and Alexander Cruz

Abstract

We examine host partitioning among sympatric brood parasites in southeastern Arizona and South Florida. In Florida, range expansion of three cowbirds (*Molothrus bonariensis* shiny cowbird, *M. ater* brown-headed cowbird, and *M. aeneus* bronzed cowbird) has brought them together in a region where host species have never been parasitized before. Whereas in southeastern Arizona, bronzed cowbird and brown-headed cowbird have been in sympatry longer. That sympatry and the larger size separation between the smallest race of the brown-headed cowbird (*M. a. obscurus*) and bronzed cowbird are correlated with the larger cowbird primarily parastizing larger host species. Furthermore, morphometric analysis shows that brown-headed cowbirds are significantly smaller in sympatry than allopatry, suggesting that the size variation across their range is not only a function of clinal variation but driven, perhaps, by long-term interspecific competitive interactions for nests to parasitize (i.e., character displacement model). Interspecific competition in southeastern Arizona for nests to parasitize has led to some partitioning of the host resources. In South Florida, recent contact between three parasitic species suggests a different outcome. In South Florida, the shiny cowbird was recorded mainly from coastal areas, whereas brown-headed cowbirds were found in coastal and inland areas leading to a degree of habitat partitioning. In Florida, brown-headed cowbirds parasitized species ranging from 6 to 100 g, with the majority being smaller species (20 g or less). There are far fewer breeding records for shiny cowbirds, but the similarities in size and species use outside of Florida suggest host overlap with the brown-headed. The longer breeding season of the shiny cowbird suggests that, while there will be host

J. F. Chace (✉)
Department of Biology and Biomedical Sciences, Salve Regina University, Newport, RI, USA
e-mail: Jameson.Chace@salve.edu

A. Cruz
Department of Ecology and Evolutionary Biology, University of Colorado, Boulder, CO, USA

M. Soler (ed.), *Avian Brood Parasitism*, Fascinating Life Sciences,
https://doi.org/10.1007/978-3-319-73138-4_13

overlap with the brown-headed, a temporal separation may occur. The larger bronzed cowbird parasitized the larger hosts in the community, e.g., red-winged blackbird and spot-breasted oriole.

13.1 Introduction

Fitness of obligate brood parasites is directly related to host attributes, and the costs of being parasitized have selected for counter-adaptations in many hosts (Rothstein 1990; Chap. 1). To ensure their own reproductive success, brood parasites must gain access to the nests of suitable hosts that lack antiparasitic behaviors during the appropriate nest stage (host laying or early incubation). Interference by other brood parasites, either in the form of egg puncturing, egg removal, or multiple parasitism, reduces a parasite's success (Peer and Sealy 1999; Nakamura and Cruz 2000; Trine 2000).

Bronzed, brown-headed, and shiny cowbirds have been recorded parasitizing at least 102, 248, and 267 different host species, respectively (Sealy et al. 1997; Ortega 1998; Lowther 2013). Many of these recorded hosts are potentially unsuitable or of low quality as typically less than 50% of the hosts listed are known to fledge cowbirds. Female reproductive success is directly related to choosing appropriate hosts and synchronizing laying during host clutch initiation (Ortega 1998; Guigueno and Sealy 2010). Optimal hosts accept cowbird eggs, are insectivorous, have open cup nests, and have an incubation period longer than the cowbird (Mason 1986; Johnsgard 1997; Ortega 1998). Cowbird size may reflect host choice because larger hosts can grasp and eject smaller eggs (Spaw and Rohwer 1987; Rohwer and Spaw 1988; Rasmussen and Underwood 2010) and outcompete parasitic nestlings (Dearborn and Lichtenstein 2002; Rivers et al. 2010).

Among the molothrine cowbirds, nonoverlapping territories occur in areas where host density is high and cowbird density is low, effectively reducing scramble competition for nests to parasitize; intraspecific competition increases for nests to parasitize when hosts become rare, or more clumped, among a relatively high density of parasites (Elliott 1980; Dufty 1982; Rothstein et al. 1984). Partitioning of the host community, or "alloxenia," has been documented to reduce interspecific competition for nests to parasitize among sympatric brood parasitic cuckoos in Africa, Australia, and Japan (Friedmann 1967; Payne and Payne 1967; Brooker and Brooker 1990, 1992; Higuchi 1998).

Resource partitioning is taken as evidence of interspecific competition among species with similar niche requirements (Lack 1971; Schluter 1986; Wiens et al. 2014). Sympatric brood parasites could partition host resources through niche divergence in host selection or habitat selection. It is not entirely clear from Friedmann's (1967) work whether host partitioning drives the observed divergent habitat selection or specific cuckoo habitat requirements have led to host partitioning, as the primary hosts may breed in distinct but adjacent habitats. Nevertheless, either mechanism would reduce competitive interactions for nests to parasitize.

Host size appears to be an important consideration of host selection when brood parasitic cowbirds are sympatric. Character displacement is a known evolutionary response to interspecific competition (Brown and Wilson 1956; Schulter et al. 1985; Goldberg and Lande 2006), and in sympatric cowbirds, divergent body size may facilitate host partitioning by host body size.

Here we examine the partitioning in host and habitat use among sympatric cowbird brood parasites in two regions in the United States: southern Florida and southeastern Arizona.

13.1.1 Molothrine Cowbirds in South Florida: A Three-Species System

Recent range expansion of three molothrine cowbirds has brought them into contact with avian communities in South Florida that have never experienced brood parasitism. The shiny cowbird (*Molothrus bonariensis*), originally confined to South America and Trinidad and Tobago, has spread dramatically into the West Indies during the past century (Post and Wiley 1977; Cruz et al. 1985), and since 1985 it has been found in Florida (Smith and Sprunt 1987; Post et al. 1993; Cruz et al. 1998, 2000). From the opposite direction, the North American brown-headed cowbird (*M. ater*) and bronzed cowbird (*M. aeneus)* has spread into Florida (Cruz et al. 1998, 2000; Pranty and Nelson 2010). Here we examine the status, distribution, host use, and potential interactions among these generalist brood parasites in an area where formerly no cowbirds were present. This topic is particularly timely as these brood parasites are using new species and host populations (Robinson et al. 1995; Cruz et al. 1998).

13.1.2 Molothrine Cowbirds in Southeastern Arizona: A Two-Species System

Cowbird body size varies across the range of both the brown-headed and bronzed cowbird such that the smallest subspecies of the brown-headed cowbird (*M. ater obscurus)* is sympatric with the largest subspecies of bronzed cowbird (*M. aeneus*) in southern Texas and New Mexico, and *loyei* in Arizona (Johnsgard 1997). Divergent parasite body size may reflect resource partitioning. Avian body mass is strongly correlated with egg volume (Gill 1994), therefore, if concordant with host size partitioning, divergent body sizes among sympatric parasites would be evidence of importance of competition in the evolution of the *Molothrus* lineage and the host-parasite coevolutionary relationship.

In the Huachuca Mountains of southeastern Arizona, 107 nests of seven host species were monitored, 45% of which were parasitized (Chace 2005). Overall, bronzed cowbird parasitism (17%) was less than the parasitism by brown-headed cowbirds (31%). Bronzed cowbirds parasitized larger hosts, which were not parasitized by brown-headed cowbirds, while brown-headed cowbirds parasitized smaller hosts which were not parasitized by the bronzed cowbird. Thus, in southeastern Arizona there appears to be a partitioning of host size by bronzed and brown-headed cowbirds based primarily on the size of the hosts (Chace 2005). There was overlap in some moderate-sized hosts with three plumbeous vireo nests having been parasitized by both cowbirds (Chace 2005). In Chace's (2005) study, there was no parasitism of tanagers by the smaller brown-headed cowbird, a host group that is commonly parasitized in allopatric portions of the brown-headed cowbird range (Goguen and Mathews 1998; Rosenberg et al. 1999; Fischer et al. 2002).

Host size appears to be an important consideration of host selection when brood parasitic cowbirds are sympatric. Character displacement is a known evolutionary response to interspecific competition (Brown and Wilson 1956; Schluter et al. 1985; Goldberg and Lande 2006), and in sympatric cowbirds, divergent body size may facilitate host partitioning. Here we report on an investigation of cowbird body size in allopatry and sympatry to determine if body size divergence might be related to host size partitioning.

13.2 Methods

13.2.1 Cowbirds in South Florida

South Florida as defined here includes the following counties: Broward, Collier, Hendry, Lee, Miami-Dade, Monroe, and Palm Beach (Fig. 13.1a). We gathered information on distribution, breeding biology of cowbirds, and their potential and actual host species. Documentation of the distribution of cowbirds is based on literature review, including the first and second Florida Breeding Bird Atlas (BBA) project (U.S. Geological Survey 2017), eBird, sightings posted to the Miami Bird Board (Tropical Audubon Society), the Florida Ornithological Society Field Observations Committee (FOC) seasonal reports, communications with individuals knowledgeable of the avifauna, and field work.

Fieldwork in South Florida was conducted between 1989 and 1997. Major study areas were in Everglades National Park, Sanibel Island, and the Florida Keys and included mangroves, freshwater marshes, upland hammocks, and human-modified habitats, such as agricultural and disturbed areas. Our work included extensive searches for cowbirds in the different areas, including point counts of cowbirds and potential host species. We gathered information on the breeding biology of cowbirds by observations of courtship behavior and examination of nests of potential host species for the presence of cowbird eggs or nestlings (see Prather and Cruz 1995, 2002; and Prather et al. 1999 for habitat description and nest search information).

13.2.2 Cowbirds in Southeastern Arizona

To test whether sympatric cowbird populations in southeastern Arizona exhibited a shift in body size, potentially in response to divergent selection pressure, 168 brown-headed cowbirds (77 female, 91 male) and 204 bronzed cowbirds (106 female,

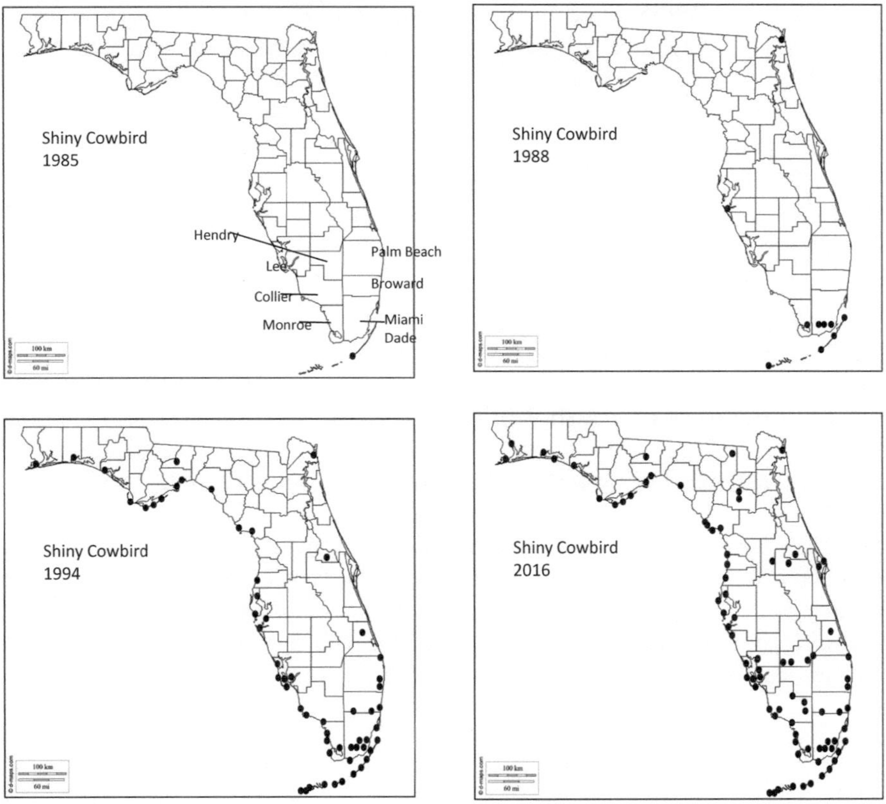

Fig. 13.1 Shiny cowbird breeding season distribution in Florida in 1985 (**a**), 1988 (**b**), 1994 (**c**), and 2016 (**d**), with emphasis on South Florida counties

98 male) were measured from collections of the University of Colorado Museum ($n = 14$), Denver Museum of Natural History ($n = 16$), Louisiana State University Museum ($n = 36$), Western Foundation of Vertebrate Zoology ($n = 39$), University of Arizona Museum ($n = 43$), National Museum of Natural History ($n = 65$), and American Museum of Natural History ($n = 169$). We measured length of wing, tail, tarsus, culmen, and distance from the tip of the maxilla to the anterior end of the nares (hereafter nares) following Baldwin et al. (1931). Wing (flat) and tail measurements were taken with a wing rule (±0.5 mm), and bill and tarsus measurements were taken with Vernier dial calipers (±0.01 mm). Specimens measured that met the following criteria: (1) Birds were collected in reproductive condition or during the breeding season, (2) males in at least their third year and females in at least their second year, and (3) no appreciable feather wear on the primaries or rectrices. Whether each specimen bred in sympatric or allopatric site was based on locality information overlaid on range limits.

Partial correlation analysis revealed significantly high correlations between culmen and nares. Culmen was removed from analysis because of the two measures the percent variation in culmen was greater than in nares. Remaining variables were tested for normality, and skewed variables were log-transformed and retested for normality. The prediction that body size differed significantly between allopatric and sympatric sites was tested with Student's *t*-test for normally distributed variables or Wilcoxon rank sum test for skewed variables. Because combined measures may be a better indicator of body size (Freeman and Jackson 1990), all measures were combined in a Bonferroni-adjusted logistic regression to predict the sympatric-allopatric classification.

13.3 Results and Discussion

13.3.1 Cowbirds in South Florida

13.3.1.1 Molothrine Cowbird Distribution in South Florida

Shiny Cowbird Shiny cowbirds colonized Florida from the West Indies and are now resident in South Florida, where they have been observed during the breeding season (March to August) in primarily coastal localities (Cruz et al. 2000; Fig. 13.1), occurring in habitats, ranging from residential and agricultural areas, modified habitats, and mangroves in the mainland and the Keys (pers. obs.).

Brown-headed Cowbird Brown-headed cowbirds were first observed breeding in North Florida in 1956 (Newman 1957) and now breed throughout the state (Stevenson and Anderson 1994; Cruz et al. 1998, 2000; Fig. 13.2). Similar to shiny cowbirds, brown-headed cowbirds were seen in the Keys and the mainland in residential areas, agricultural areas, and other modified habitats. We observed brown-headed cowbirds in areas frequented by shiny cowbirds and bronzed cowbirds and were often seen foraging together.

Bronzed Cowbird Bronzed cowbirds were first observed in Florida in 1962 (Truchot 1962) and are now permanent residents in southeastern Florida (Pranty and Nelson 2010; Fig. 13.3). Bronzed cowbirds can be found in urban and agricultural areas between Key Largo and West Palm Beach and Flamingo in Everglades National Park and mixed with other blackbirds throughout southern Florida.

13.3.1.2 Cowbird Host Use in South Florida

Molothrine cowbirds are generalist obligate brood parasites with a wide range of hosts (Sealy et al. 1997; Ortega 1998; Ellison and Lowther 2009; Lowther 2013). However, despite conservation implications surrounding recent cowbird expansion into South Florida, there are few breeding records of cowbirds.

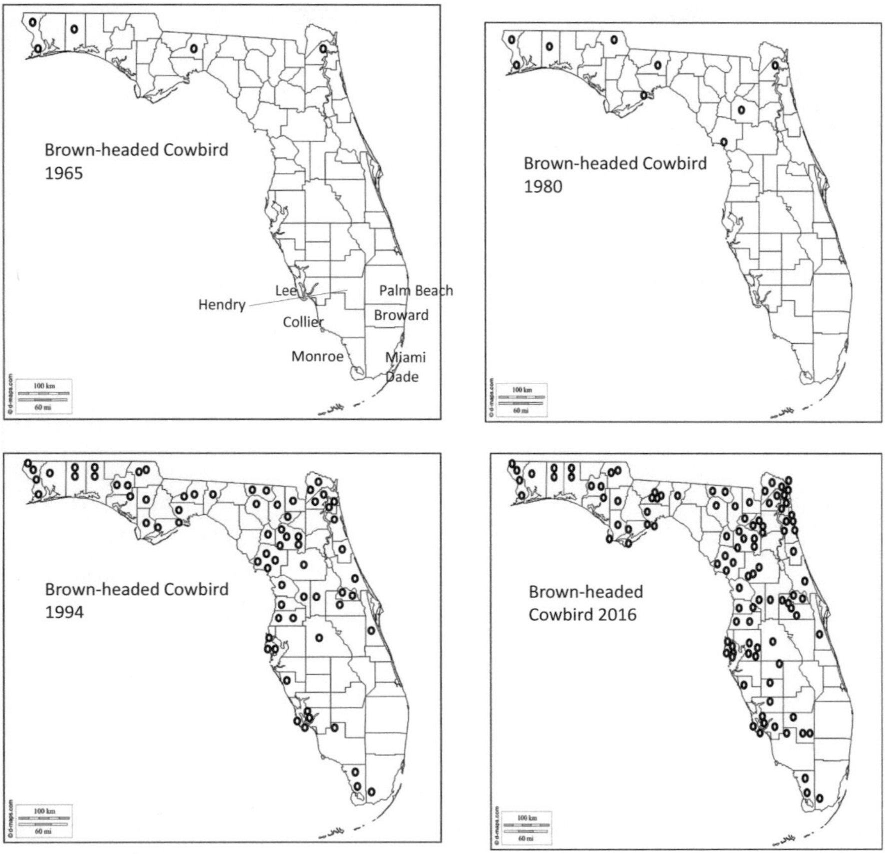

Fig. 13.2 Brown-headed cowbird breeding season confirmed nesting activity in Florida in 1965 (**a**), 1980 (**b**), 1994 (**c**), and 2016 (**d**), with emphasis on South Florida counties

Brown-headed Cowbird Although South Florida records are still widely scattered, data compiled by the Florida Breeding Bird Atlas Program, Reetz (2008), and our study found that 15 species that breed in South Florida have been confirmed as parasitized in Florida other than South Florida by cowbird eggs in nest, young in nest, or adults feeding a fledgling cowbird and that an additional five species have as yet not been confirmed as parasitized in South Florida but have been recorded as parasitized outside of Florida (Table 13.1) (Cruz et al. 1998, 2000; Ortega 1998; Lowther 1993, 2013, https://www.fieldmuseum.org/sites/default/files/Molothrus_hosts-26aug2013.pdf).

Shiny Cowbird Although there has been an increase in shiny cowbird abundance in Florida, there is little information about host species use (Lowther and Post 1999; Cruz et al. 2000). During May to July, we observed singing accompanied by display

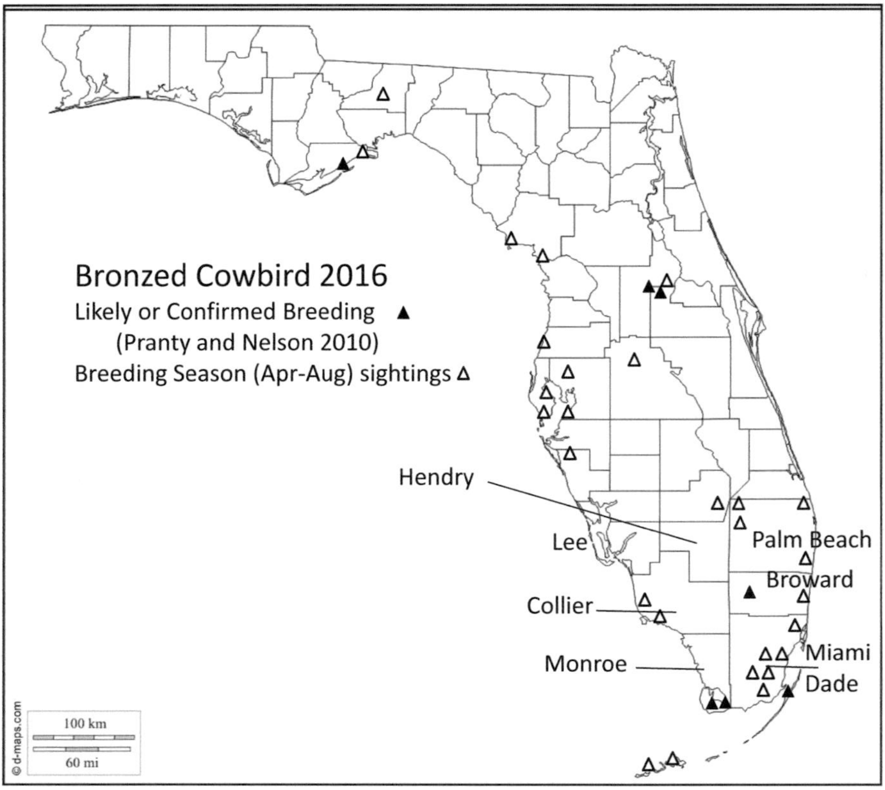

Fig. 13.3 Bronzed cowbird distribution in Florida based on breeding season observations and confirmed or likely breeding activity

(song spread) on several occasions, indicative of breeding activity; however copulation was not observed. There are four observations of fledgling shiny cowbirds from South Florida, including three of fledglings being fed by adults of other species (Smith and Sprunt 1987; Kale 1989; Stevenson and Anderson 1994).

Reetz et al. (2010) provided further evidence of breeding by shiny cowbirds through females collected in Central Florida. In neighboring Georgia, Sykes and Post (2001) captured a reproductive female shiny cowbird. Post and Sykes (2011) collected 17 (13 females, 4 males) shiny cowbirds in southwestern South Carolina. Five females were laying eggs. Prior to Post and Sykes (2011), only males had been collected, but most of them had enlarged testes.

Bronzed Cowbird Bronzed cowbirds have been reported in Florida during summer since 2003 (Fig. 13.4). In each instance of hosts feeding bronzed cowbird fledglings, the hosts were Icterids: three by spot-breasted orioles and one by a red-winged blackbird (Table 13.1). However, Sealy and Lorenzana (1997) have shown that

Table 13.1 Shiny, brown-headed, and bronzed cowbird hosts present in South Florida and known to be parasitized elsewhere, * denotes confirmed parasitism in Florida

	Cowbird		
Host species	Shiny (33 g)	Brown-headed (41 g)	Bronzed (57 g)
Blue-gray gnatcatcher *Polioptila caerulea* (6 g)		X*	
Prairie warbler *Setophaga discolor* (8 g)	X*[a]	X*	
Northern parula *Setophaga Americana* (9 g)		X*	
Cuban yellow warbler *Setophaga petechia* (10 g)	X	X*	
Hooded warbler *Setophaga citrine* (10 g)		X*	
Common yellowthroat *Geothlypis trichas* (10 g)		X*	
White-eyed vireo *Vireo griseus* (12 g)		X*	
Prothonotary warbler *Protonotaria citrea* (14 g)		X	
Red-eyed vireo *Vireo olivaceus* (17 g)		X*	
Black-whiskered vireo *Vireo altiloquus* (20 g)	X*[a]	X*	
Bachman's sparrow *Peucaea aestivalis* (20 g)		X*	
Carolina wren *Thryothorus ludovicianus* (20 g)		X	
Blue grosbeak *Passerina caerulea* (28 g)		X*	
Summer tanager *Piranga rubra* (29 g)		X*	
Great-crested flycatcher *Myiarchus crinutis* (34 g)		X	
Eastern towhee *Pipilo erythrophthalmus* (40 g)		X*	
Red-winged blackbird *Agelaius phoeniceus* (42 g)	X*[a]	X*	X
Northern cardinal *Cardinalis cardinalis* (45 g)	X*[a]	X*	X
Gray kingbird *Tyrannus dominicensis* (47 g)	X		
Northern mockingbird *Mimus polyglottos* (47 g)	X	X*	X
Spot-breasted oriole *Icterus pectoralis* (50 g)			X*
Common grackle *Quiscalus quiscula* (94 g)		X	X
Eastern meadowlark *Sturnella magna* (100 g)		X	

[a]No definite record of parasitism by this species is known yet for Florida, but red-winged blackbird has been claimed to have been a shiny cowbird host, and three other species (prairie warbler, black-whiskered vireo, and northern cardinal) are known to have served as cowbird hosts within the Florida range of brown-headed and shiny cowbirds (Paul 1989; Stevenson and Anderson 1994; Cruz et al. 1998, 2000)

some cowbird fledglings are fed by more than the putative host species, making identification of hosts based on fledglings unreliable.

13.3.1.3 Host Partitioning in South Florida

The published data suggest that there may be a greater overlap in host nest use by brown-headed and shiny cowbirds, whereas not as great with bronzed cowbirds (Table 13.1). In Florida, brown-headed cowbirds parasitized species ranging from 6 to 100 g, with the majority being smaller species (20 g or less, Table 13.1). Bronzed cowbirds are larger (56.9 ± 5.1 g; Carter 1986) and generally parasitize larger hosts than brown-headed (41.3 ± 2.7 g; Ortega and Cruz 1992) and shiny

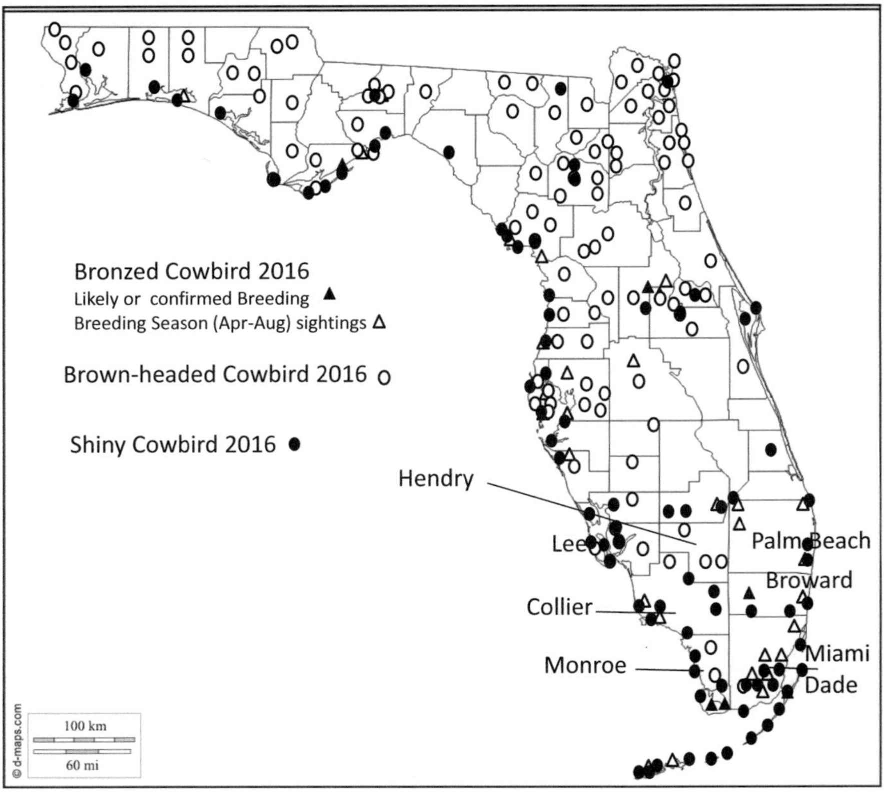

Fig. 13.4 Current (2016) combined breeding distributions of bronzed, brown-headed, and shiny cowbirds in Florida, with emphasis on South Florida counties

cowbirds (33.3 ± 3.1 g; Wiley 1988), but some overlap was noted with brown-headed cowbirds (Chace 2004, 2005; Ellison 2004; Ellison and Lowther 2009; pers. obs.).

In its broad range outside of Florida (northern Colombia to Louisiana), bronzed cowbirds use different host species with larger hosts such as icterines figuring prominently (Friedmann 1963; Ellison and Lowther 2009; Ortega 1998; Chace 2004). However, in South Florida, spot-breasted oriole and red-winged blackbirds are the only species recorded as host.

Differences in habitat preferences and breeding cycles among the cowbird species may minimize host use overlap. The shiny cowbird, for example, has been recorded primarily from coastal areas in Florida, whereas brown-headed cowbirds have been found in both coastal and inland areas. The shiny cowbird breeding season in the West Indian region extends from mid-March to September, whereas the brown-headed and bronzed cowbird breeds from May to early July (Cruz et al. 1989, 1998; Chace et al. 2000; Ellison and Lowther 2009; Walsh et al. 2015). In Florida, brown-headed cowbird eggs have been found from May through July (Kale et al. 1992).

These data suggest broad breeding season overlap. The more extended breeding season of the shiny cowbird suggests that, while there is overlap, temporal separation can also exist with populations that breed in July and August being parasitized by the shiny cowbird.

Implications for Florida, based on studies in Texas and Arizona, suggest that initially, when cowbird densities are low, host partitioning is likely. However, as cowbird densities in Florida increase, competition for nests to parasitize will result in greater overlap of host use and more frequent multiple parasitism events at single nests.

An additional challenge in Florida is that the shiny cowbird is relatively rare and that distinguishing juvenile cowbirds by species is difficult. The limited documentation of breeding by shiny cowbirds in Florida suggests that difficulties in documenting breeding and/or identifying females or juveniles resulted in underreporting of breeding by this species (Reetz et al. 2010). Focused collection studies of cowbirds may be helpful in identifying the extent of breeding by shiny cowbirds in these initial decades of colonization of mainland North America (Reetz et al. 2010).

13.3.2 Cowbirds in Southeastern Arizona

In the Huachuca Mountains of southeastern Arizona, 107 nests of seven host species were monitored, 45% of which were parasitized (Chace 2005). Overall, bronzed cowbird parasitism (17%) was less than parasitism by brown-headed cowbirds (31%). Bronzed cowbirds parasitized larger hosts, which were not parasitized by brown-headed cowbirds, while brown-headed cowbirds parasitized smaller hosts, which were not parasitized by the bronzed cowbird. Thus, in southeastern Arizona there appears to be a partitioning of host by bronzed and brown-headed cowbirds based primarily on the size of the hosts (Chace 2005). There was overlap in some moderate-sized hosts with three plumbeous vireo nests having been parasitized by both cowbirds (Chace 2005).

13.3.2.1 Host Partitioning in Southeastern Arizona

In southeastern Arizona bronzed and brown-headed cowbirds partition use of some host species, overlapping in a few species. This pattern might be related to the relatively low breeding densities of cowbirds and the scattered distribution of foraging sites in urban areas, irrigated agricultural land and golf courses among a much larger matrix of potential nesting habitat (Chace 2004).

Alloxenia may be present when the relative abundance of cowbirds is low compared to the diversity of hosts. The cowbird–host ratio in Arizona was 0.07 (Chace 2001) compared to 0.28 in Texas (Carter 1986) where there is greater overlap in host use by sympatric cowbirds (Carter 1986; Peer and Sealy 1999). Sympatric brood parasites may also partition host use by phenology of egg laying. In Arizona, bronzed cowbirds' peak of egg laying is slightly later than brown-headed cowbirds and corresponds more closely to the peak of egg laying by tanagers, the larger sized hosts parasitized by bronzed cowbirds (Chace 2005).

Work in Texas (Carter 1986; Peer and Sealy 1999; Ellison et al. 2006; Ellison and Sealy 2007), where bronzed and brown-headed cowbirds are sympatric, shows a greater degree of host overlap and multiple parasitism than we observed in Arizona. Arizona has lower cowbird densities and higher host richness than at the specific study sites in Texas. Differences in overlap in host use likely reflect aspects of cowbird densities and songbird community composition (Ellison and Lowther 2009).

13.3.2.2 Is There Evidence of Character Displacement?

Morphological measurements of cowbird specimens provided an opportunity to examine whether the size difference between brown-headed cowbirds and bronzed cowbirds in Arizona is greater in sympatric than in allopatric populations (Table 13.2). Male and female brown-headed cowbirds are significantly smaller in sympatry than in allopatry based on the four measures of body size (Table 13.2). However, no significant differences were detected for bronzed cowbirds despite the fact that, for most measures, males and females are larger in sympatric sites than in allopatric sites (Table 13.2).

Where bronzed and brown-headed cowbirds are sympatric in southeastern Arizona, the larger bronzed cowbird parasitizes larger host species than the smaller brown-headed cowbird (Chace 2005). As shown here, sympatric brown-headed cowbirds are significantly smaller than in allopatry (Table 13.2) corresponding with brown-headed cowbird host-size selection. The smaller size in sympatry presumably gives the brown-headed cowbird a slight advantage in smaller host nests. This character shift in brown-headed cowbirds in Arizona reduces competition for large host nests in sympatry where bronzed cowbirds parasitize preferentially these

Table 13.2 Mean (SE) body size (mm) of bronzed cowbirds and brown-headed cowbirds in sympatry and allopatry

	n	Wing	Tail	Tarsus	Nares
Brown-headed cowbird, female					
Allopatry	43	98 (0.63)*	65 (0.42)*	24.27 (0.22)*	10.19 (0.09)*
Sympatry	34	93 (0.70)*	62 (0.47)*	22.43 (0.25)*	9.72 (0.10)*
Brown-headed cowbird, male					
Allopatry	47	110 (0.67)*	75 (0.53)*	26.54 (0.023)*	11.84 (0.11)*
Sympatry	44	103 (0.69)*	70 (0.55)*	24.27 (0.24)*	11.03 (0.11)*
Bronzed cowbird, female					
Allopatry	36	104 (0.86)	70 (0.69)	27.90 (0.28)	13.51 (0.17)
Sympatry	72	104 (0.62)	71 (0.49)	27.29 (0.20)	13.85 (0.12)
Bronzed cowbird, male					
Allopatry	37	116 (0.98)	76 (0.87)	29.67 (0.34)	15.25 (0.20)
Sympatry	61	117 (0.75)	79 (0.68)	29.90 (0.27)	15.68 (0.16)

*Significant difference between allopatric and sympatric populations, Wilcoxon rank sums test with Bonferroni-adjusted alpha $P < 0.0125$

larger host species such as tanagers. Thus, larger brown-headed cowbirds would have lower breeding success than smaller brown-headed cowbirds, which would tend to parasitize smaller host species. This would drive the evolutionary change provoking a reduction of size in brown-headed cowbirds (cf. Schluter 1994).

Concluding Remarks and Future Directions

Friedmann's work in Africa was the first to describe host partitioning by sympatric brood parasites, a pattern he defined as alloxenia. Not surprising for highly host-specific brood parasites, further work with sympatric cuckoos has shown similar patterns in a number of studies throughout the Old World. Interestingly, molothrine cowbirds, despite being broad generalist brood parasites, have shown in Arizona to partition hosts by body size with a slight separation in the peak of egg laying. This has not been the case in all sympatric zones between brown-headed and bronzed cowbirds, such as several studies in Texas have shown. Florida presents a special case with recent range overlap of brown-headed, bronzed, and shiny cowbirds. Already some habitat and host specificity has been identified; however this is likely to change as populations of shiny, brown-headed, and bronzed cowbirds continue to grow.

Clearly, continued data collection on potential host species in Florida is needed as populations of all three cowbird species continue to grow. Future work involving careful observation and determining patterns of host use of sympatric cowbirds in South Florida is an important area for study. Continued monitoring of the duration of the breeding season for these three brood parasites is important to determine if there is any temporal shift in egg laying as observed in Arizona. Nests of potential host species require monitoring to determine incidence of parasitism and whether certain species are preferentially parasitized by the cowbird species. A potential problem for studies of shiny cowbird and brown-headed cowbirds is that of differentiating their eggs, fledglings, and females from one another. Determining patterns of host use of sympatric cowbirds by microsatellite DNA markers are needed (Ellison et al. 2006). Focused trapping and gonadal studies of cowbirds are needed in order to identify the extent of breeding by three cowbird species.

Acknowledgments Access to museum specimens was enabled by the American Museum of Natural History's Frank M. Chapman Memorial Fund, Sigma Xi Grants-in-Aid of Research, and the Walker Van Riper Fund of the University of Colorado Museum. The Florida work was supported by a grant from the U.S. Fish and Wildlife Service. For assistance in Florida, we thank the staff of the J. N. Ding Darling National Wildlife Refuge and the Sanibel-Captiva Conservation Foundation. We also thank the Everglades Science Center (National Audubon) for logistical help and H. W. Kale II and W. Pranty for providing cowbird records. A special thanks to J. W. Prather for assistance in the field. M. Soler, S.G. Sealy, and V. Fiorini provided valuable comments on earlier versions of this manuscript.

References

Baldwin SP, Oberholser HC, Worley LG (1931) Measurement of birds. Cleveland Museum of Natural History, Scientific Publication no 17

Breeding Bird Atlas Explorer (online resource) (2017) U.S. geological survey patuxent wildlife research center. http://www.pwrc.usgs.gov/bba

Brooker LC, Brooker MG (1990) Why are cuckoos host specific? Oikos 57:301–309

Brooker MG, Brooker LC (1992) Evidence for individual female host specificity in two Australian bronze-cuckoos (*Chrysococcyx spp.)*. Aust J Zool 40:485–493

Brown WL Jr, Wilson EO (1956) Character displacement. Syst Zool 5:49–64

Carter MD (1986) The parasitic behavior of the bronzed cowbird in south Texas. Condor 88:11–25

Chace JF (2001) Host and habitat partitioning by sympatric brood parasites in southeastern Arizona. Dissertation, University of Colorado-Boulder

Chace JF (2004) Habitat selection by sympatric brood parasites in southeastern Arizona: the influence of landscape, vegetation, and species richness. Southwest Nat 49:24–32

Chace JF (2005) Host use by sympatric cowbirds in southeastern Arizona. Wilson Bull 117:375–381

Chace JF, Cruz A, Marvil RE (2000) Reproductive interactions between the brown-headed cowbird and plumbeous vireo in Colorado. In: Smith JNM, Cooke TL, Rothstein SI, Robinson SK, Sealy SG (eds) Ecology and management of cowbirds and their hosts. University of Texas Press, Austin, pp 128–134

Cruz A, Manolis T, Wiley JW (1985) The shiny cowbird: a brood parasite expanding its range in the Caribbean region. In: Buckley PA, Foster MS, Morton ES, Ridgely RS, Buckley FG (eds) Neotropical ornithology, American Ornithologists Union Ornithological Monographs, no 36, pp 607–620

Cruz A, Wiley JW, Nakamura TK, Post W (1989) The shiny cowbird *Molothrus bonariensis* in the west Indian region – biogeography and ecological implications. In: Wood CA (ed) Biogeography of the West Indies. Sandhill Crane Press, Gainesville, FL, pp 519–540

Cruz A, Post W, Wiley JW, Ortega CP, Nakamura TK, Prather JW (1998) Potential impacts of cowbird range expansion in Florida. In: Rothstein SI, Robinson SK (eds) Parasitic birds and their hosts. Studies in coevolution. Oxford University Press, Oxford, pp 313–336

Cruz A, Prather JW, Post W, Wiley JW (2000) The spread of shiny and brown-headed cowbirds into the Florida region. In: Smith JNM, Cooke TL, Rothstein SI, Robinson SK, Sealy SG (eds) Ecology and management of cowbirds and their hosts. University of Texas Press, Austin, pp 47–57

Dearborn DC, Lichtenstein G (2002) Begging behaviour and host exploitation in parasitic cowbirds. In: Wright J, Leonard ML (eds) The evolution of begging. Kluwer Academic, Dordrecht, pp 361–388

Dufty AM Jr (1982) Movements and activities of radio-tracked brown-headed cowbirds. Auk 99:316–327

Elliott PF (1980) Evolution of promiscuity in the brown-headed cowbird. Condor 82:138–141

Ellison K (2004) Host use by sympatric cowbirds (*Molothrus aeneus* and *M. ater*). Dissertation, University of Manitoba

Ellison K, Lowther PE (2009) Bronzed cowbird (Molothrus aeneus), version 2.0. In: Rodewald PG (ed) The birds of North America. Cornell Lab of Ornithology, Ithaca, NY. https://doi-org.ez-salve.idm.oclc.org/10.2173/bna.144

Ellison K, Sealy SG (2007) Small hosts infrequently disrupt laying by brown-headed cowbirds and bronzed cowbirds. J Field Ornithol 78:378–398

Ellison K, Sealy SG, Gibbs HL (2006) Genetic elucidation of host use by individual sympatric bronzed cowbirds (*Molothrus aeneus*) and brown-headed cowbirds *(M. ater)*. Can J Zool 84:1269–1280

Fischer KN, Prather JW, Cruz A (2002) Nest site characteristics and reproductive success of the western tanager (*Piranga ludoviciana*) on the Colorado front range. West North Am Nat 62:479–483

Freeman S, Jackson WM (1990) Univariate metrics are not adequate to measure avian body size. Auk 107:69–74

Friedmann H (1963) Host relations of the parasitic cowbirds. US Natl Mus Bull 233:1–276

Friedmann H (1967) Alloxenia in three sympatric African species of *Cuculus*. Proc US Natl Mus 124:1–14

Gill FB (1994) Ornithology, 2nd edn. WH Freeman, New York

Goguen CB, Mathews NE (1998) Songbird community composition and nesting success in grazed and ungrazed pinyon-juniper woodlands. J Wildl Manag 62:474–484

Goldberg EE, Lande R (2006) Ecological and reproductive character displacement on an environmental gradient. Evolution 60:1344–1357

Guigueno MF, Sealy SG (2010) Clutch abandonment by parasitized yellow warblers: egg burial or nest desertion? Condor 112:399–406

Higuchi H (1998) Host use and egg color of Japanese cuckoos. In: Rothstein SI, Robinson SK (eds) Parasitic birds and their hosts. Oxford University Press, New York, pp 80–112

Johnsgard PA (1997) The avian brood parasites: deception at the nest. Oxford University Press, New York

Kale HW (1989) Florida birds. Fla Nat 62:14

Kale HW II, Pranty B, Stith BM, Biggs CW (1992) The atlas of the breeding birds of Florida. Final report. Florida Game and Fresh Water Fish Commission, Tallahassee

Lack D (1971) Ecological isolation in birds. Harvard University Press, Cambridge

Lowther PE (1993) Brown-headed cowbird (Molothrus ater), version 2.0. In: Rodewald PG (ed) The birds of North America. Cornell Lab of Ornithology, Ithaca, NY. https://doi-org.ezsalve.idm.oclc.org/10.2173/bna.47

Lowther PE (2013) Lists of victims and hosts of parasitic cowbirds (*Molothrus*). Version 26 Aug 2013. https://www.fieldmuseum.org/sites/default/files/Molothrus_hosts-26aug2013.pdf

Lowther PE, Post W (1999) Shiny cowbird (*Molothrus bonariensis*). In: Poole A, Gill F (eds) Birds of North America, no. 399. The Birds of North America, Philadelphia, PA

Mason P (1986) Brood parasitism in a host generalist, the shiny cowbird: I. The quality of different species as hosts. Auk 103:52–60

Nakamura TK, Cruz A (2000) The ecology of egg-puncture behavior by the shiny cowbird in southwestern Puerto Rico. In: Smith JNM, Cook TL, Rothstein SI, Robinson SK, Sealy SG (eds) Ecology and management of cowbirds and their hosts. University of Texas Press, Austin, TX, pp 178–186

Newman G (1957) The Florida region. Audubon Field Notes 11:412–413

Ortega CP (1998) Cowbirds and other brood parasites. University of Arizona Press, Tucson, AZ

Ortega CP, Cruz A (1992) Differential growth patterns of nestling brown-headed cowbirds and yellow-shouldered blackbirds. Auk 109:368–376

Paul RT (1989) Florida region. Am Birds 43:1307–1310

Payne RB, Payne K (1967) Cuckoo hosts in southern Africa. Ostrich 38:135–143

Peer BD, Sealy SG (1999) Parasitism and egg puncture behavior by bronzed and brown-headed cowbirds in sympatry. Stud Avian Biol 18:235–240

Post W, Sykes PW Jr (2011) Reproductive status of the shiny cowbird in North America. Wilson J Ornithol 123:151–154

Post W, Wiley JW (1977) Reproductive interactions of the shiny cowbird and the yellow-shouldered blackbird. Condor 79:176–184

Post W, Cruz A, McNair DB (1993) The North American invasion pattern of the shiny cowbird. J Field Ornithol 64:32–41

Pranty B, Nelson A (2010) First breeding record of the bronzed cowbird (*Molothrus aeneus*) in Florida. Fla Field Nat 38:1–7

Prather JW, Cruz A (1995) Breeding biology of Florida prairie warblers and Cuban yellow warblers. Wilson Bull 107:475–484

Prather JW, Cruz A (2002) Distribution, abundance, and breeding biology of potential cowbird hosts on Sanibel Island, Florida. Fla Field Nat 30:21–76

Prather JW, Ortega CP, Cruz A (1999) Aggressive responses of red-winged blackbirds (*Agelaius phoeniceus*) toward brown-headed cowbirds (*Molothrus ater*) in areas of recent and long-term sympatry. Bird Behav 13:1–7

Rasmussen JL, Underwood TJ, Sealy SG (2010) Functional morphology as a barrier to the evolution of grasp-ejection in hosts of the brown-headed cowbird (*Molothrus ater*). Can J Zool 88:1210–1217

Reetz MJ (2008) Patterns of brown-headed cowbird parasitism in a recently invaded are and potential mechanisms limiting cowbird reproduction. Dissertation, University of Florida-Gainesville

Reetz MJ, Musser JM, Kratter AW (2010) Further evidence of breeding by shiny cowbirds in North America. Wilson J Ornithol 122:365–369

Rivers JW, Loughin TM, Rothstein SI (2010) Brown-headed cowbird nestlings influence nestmate begging, but not parental feeding, in hosts of three distinct sizes. Anim Behav 79:107–116

Robinson SK, Rothstein SI, Brittingham MC, Petit LJ, Grzybowski JA (1995) Ecology and behavior of cowbirds and their impact on host populations. In: Ecology and management of neotropical migratory birds. Oxford University Press, New York, pp 428–460

Rohwer S, Spaw CD (1988) Evolutionary lag versus bill-size constraints: a comparative study of the acceptance of cowbird eggs by old hosts. Evol Ecol 2:27–36

Rosenberg KV, Lowe JD, Dhondt AA (1999) Effects of forest fragmentation on breeding tanagers: a continental perspective. Conserv Biol 13:568–583

Rothstein SI (1990) A model system for coevolution: avian brood parasitism. Ann Rev Ecol Syst 21:481–508

Rothstein SI, Verner J, Stevens E (1984) Radio-tracking confirms a unique diurnal pattern of spatial occurrence in the parasitic Brown-headed cowbird. Ecology 65:77–88

Schluter D (1986) Character displacement between distantly related taxa? Finches and bees in the Galapagos. Am Nat 127:95–102

Schluter D (1994) Experimental evidence that competition promotes divergence in adaptive radiation. Science 266:798–800

Schulter D, Price TD, Grant PR (1985) Ecological character displacement in Darwin's finches. Science 227:1056–1059

Sealy SG, Lorenzana JC (1997) Feeding of nestling and fledgling brood parasites by individuals other than the foster parents: a review. Can J Zool 75:1739–1752

Sealy SG, Sánchez JE, Campos RC, Marin M (1997) Bronzed cowbird hosts: new records, trends in host use, and cost of parasitism. Ornitol Neotrop 8:175–184

Smith PW, Sprunt A IV (1987) The shiny cowbird reaches the U.S. Am Birds 41:370–371

Spaw CD, Rohwer S (1987) A comparative study of eggshell thickness in cowbirds and other passerines. Condor 89:307–318

Stevenson HM, Anderson BH (1994) The birdlife of Florida. University of Florida Press, Gainesville, FL

Sykes PW Jr, Post W (2001) First specimen and evidence of breeding by the shiny cowbird in Georgia. Oriole 66:45–51

Trine CL (2000) Effects of multiple parasitism on cowbird and wood thrush nesting success. In: Smith JNM, Cook TL, Rothstein SI, Robinson SK, Sealy SG (eds) Ecology and management of cowbirds and their hosts. University of Texas Press, Austin, TX, pp 135–144

Truchot E (1962) Bronzed or red-eyed cowbird. Fla Nat 35:135

Walsh JJ, Tuff TA, Cruz A, Chace JF (2015) Differential parasitism between two suitable cowbird hosts. Open Ornithol J 8:32–38

Wiens JD, Anthony GR, Forsman ED (2014) Competitive interactions and resource partitioning between northern spotted owls and barred owls in western Oregon. Wildl Monogr 185:1–50

Wiley JW (1988) Host selection by the shiny cowbird. Condor 90:289–303

High Diversity of Brood Parasites in China and Coevolution Between Cuckoos and Their Hosts

14

Wei Liang, Canchao Yang, and Anders Pape Møller

Abstract

Ecosystems consisting of multiple cuckoo species and their hosts constitute interesting examples of complex coevolutionary interactions. Competition among cuckoo species may increase the intensity of selection for adaptation to specialization but also for specific host species to evolve particular defenses against a specific species of cuckoo. Such specialization is further exacerbated by the evolution of different color morphs of the plumage of cuckoos but also by the evolution of different color morphs of cuckoo eggs. Here we review host use by sympatric brood parasitic cuckoos and coevolution in such communities of cuckoos and hosts in China. There was intense competition and overlap in host use among cuckoo species in China, and competitive exclusion may play an important role in the evolution of these patterns. In addition, parasitic cuckoos of median body size showed a particularly strong flexibility in exploiting hosts differing in body sizes because cuckoos of intermediate body size are particularly widespread. We hypothesize that such sympatric assemblages of multiple species of cuckoo may serve as model systems for the study of frequency-dependent selection and its evolutionary consequences.

W. Liang (✉) · C. Yang
Ministry of Education Key Laboratory for Ecology of Tropical Islands, College of Life Sciences, Hainan Normal University, Haikou, P. R. China
e-mail: liangwei@hainnu.edu.cn

A. P. Møller
Ecologie Systématique Evolution, Université Paris-Sud, CNRS, AgroParisTech, Université Paris-Saclay, Orsay Cedex, France

M. Soler (ed.), *Avian Brood Parasitism*, Fascinating Life Sciences,
https://doi.org/10.1007/978-3-319-73138-4_14

14.1 Introduction

Interactions between a parasite and its hosts can lead to a coevolutionary arms race. Cuckoos and their hosts are among the best studied of all brood parasite systems (Rothstein 1990), and they constitute textbook examples of coevolutionary arms races. However, most of this research has focused on the common cuckoo (*Cuculus canorus*) and its hosts in Europe, where it is the only brood parasite in most of the continent with the great spotted cuckoo (*Clamator glandarius*) being the only additional brood parasite. In contrast, a number of sympatric brood parasitic cuckoo species breed in Asia and Africa, and there are particularly many in China, India, and parts of East Africa. Many of these species have attracted little attention, and they are much less well known than the common cuckoo (Rothstein and Robinson 1998; Davies 2000; Payne 2005; Soler 2014).

Ecosystems consisting of multiple cuckoo species and their hosts constitute interesting examples of complex coevolutionary interactions. Not only do different cuckoo species potentially compete for access to hosts but hosts may likewise have evolved specific defenses against different species of cuckoos. Competition among sympatric cuckoo species may result in increased intensity of selection for adaptation to exploitation of particular hosts but also for such host species to evolve particular defenses against a given species of cuckoo. Such specialization is further exacerbated by the evolution of different color morphs of cuckoo species but also by the evolution of different color morphs of cuckoo eggs. Here we provide an updated review of host use by brood parasitic cuckoos in an ecosystem with multiple cuckoo species in China, summarizing information on different species of cuckoos and their hosts while discussing the coevolutionary interactions among these cuckoo and host species.

14.2 Distributions and Species Richness of Parasitic Cuckoos in China

The Chinese name of the cuckoo, “杜鹃” or “布谷鸟,” which means the farming season in spring, is mentioned as early as in the Kingdom of Han (202–220 BC). Since then, cuckoos have often been described in many Chinese poems and folklore during the last 1800 years (Røskaft et al. 2012). Unlike the English language, the name cuckoo has never been adopted into the Chinese language as “cuckold” or “cuckolded” or been compared with birds of prey (Røskaft et al. 2012).

So far, at least 17 parasitic cuckoo species, belonging to six genera, have been recorded in China (MacKinnon and Phillipps 1999; Zheng 2011; Yang et al. 2012a; Zheng 2018; see Fig. 14.1). By using the Get Trees function from BirdTree.org (http://birdtree.org) that is based on data and methodology from Jetz et al. (2012), a phylogenetic tree of cuckoos in China, including parasitic and nesting cuckoos, showed that *Clamator* cuckoos that are non-evictor cuckoos cluster in an

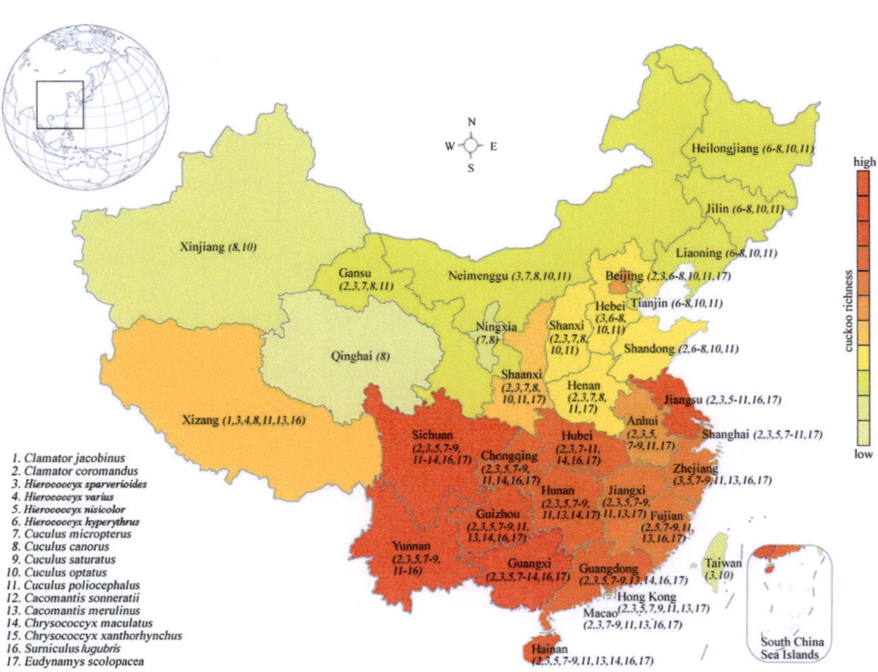

Fig. 14.1 The diversity of cuckoo species in China (modified from Yang et al. 2012a)

independent group apart from other species of evicting cuckoos (Fig. 14.2). However, there is little information on the breeding biology and host use for many of these cuckoo species (Payne 2005; Erritzøe et al. 2012; Yang et al. 2012a).

As in Europe, the common cuckoo is the most widespread brood parasite species in China that breeds across China except for some highland areas in Northwest China (Fig. 14.1). The Oriental cuckoo (*Cuculus optatus*) and Himalayan cuckoo (*C. saturatus*) were previously considered to be the same species (Erritzøe et al. 2012), but they have recently been considered separate species that breed in North and South China, respectively (Xia et al. 2016). Four parasitic cuckoo species, the banded bay cuckoo (*Cacomantis sonneratii*), violet cuckoo (*Chrysococcyx xanthorhynchus*), Jacobin cuckoo (*Clamator jacobinus*), and common hawk-cuckoo (*Hierococcyx varius*), are very rare in China and can only be found in a small area at the western edge. The diversity of cuckoo species is clearly greater in South than in North China, and they are most abundant in Southwest China (Yang et al. 2012a; Fig. 14.1).

Most cuckoo species can be easily identified by their calls. Moreover, it is particularly easy to identify cuckoo species in single-cuckoo systems such as in Europe because few cuckoo species are breeding there. However, in an area with multiple cuckoo species and their hosts such as China, it is difficult to identify

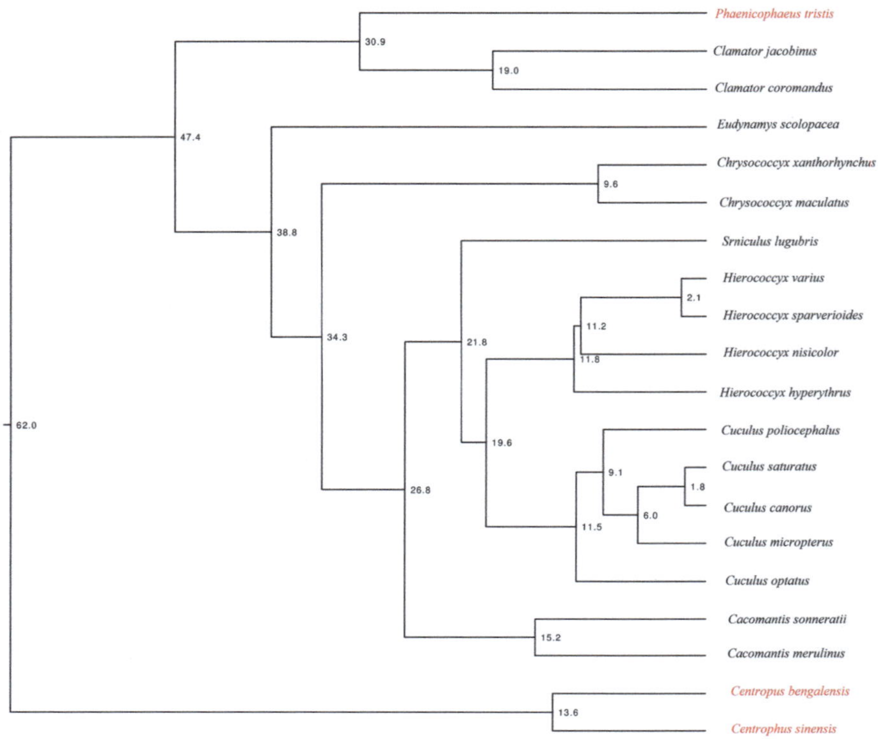

Fig. 14.2 Phylogenetic tree of parasitic and nesting cuckoos in China with nesting cuckoo species highlighted in red

cuckoo species, especially the *Cuculus* species. Given differences in calls among species of cuckoos, we can hypothesize that not only conspecific but perhaps also heterospecific cuckoos and hosts may impose strong selection on the calls of focal cuckoo species. Thus, knowledge about cuckoo morphs and species identification is important and helpful.

With respect to cuckoo nestlings, generally, the most reliable method in the field for identification of cuckoo species in host nests is the morphology of cuckoo chicks, but in some cases blood samples can be obtained to ensure species identity in the laboratory. For example, in an area of sympatry where the common cuckoo, Himalayan cuckoo, Indian cuckoo (*Cuculus micropterus*), and lesser cuckoo (*C. poliocephalus*) coexist, Himalayan cuckoo chicks can be easily spotted by their two triangular patches on the palate adjoining the cutting edges of the upper mandible, which are absent from other species of cuckoos (Fig. 14.3). Furthermore, plumage traits can be used to distinguish among nestlings of different cuckoo species (Fig. 14.4).

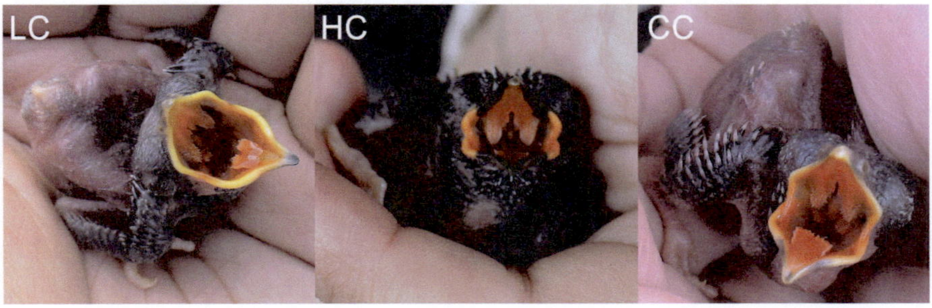

Fig. 14.3 Photographs of 6-day-old chicks of lesser cuckoo *Cuculus poliocephalus* (LC), Himalayan cuckoo *Cuculus saturatus* (HC), and common cuckoo *Cuculus canorus* (CC) (modified from Yang et al. 2012b)

Fig. 14.4 Photographs of nestlings of lesser cuckoo *Cuculus poliocephalus* (LC), Himalayan cuckoo *Cuculus saturatus* (HC), common cuckoo *Cuculus canorus* (CC), and Indian cuckoo *Cuculus micropterus* (IC) (modified from Yang et al. 2012b)

14.3 Egg Color and Egg Morphs of Cuckoos and Their Hosts in China

The common cuckoo has evolved numerous host-specific races (termed "gentes"), each laying eggs with color and pattern mimicking the eggs of their respective host species (Moksnes and Røskaft 1995; Davies 2000). In Europe, more than 20 host-specific egg morphs are recognized, although cuckoo eggs have been found in the nests of more than 100 different host species (Moksnes and Røskaft 1995). Some of these host races also occur in China, while many others appear to be unique to China. In addition, the genetic architecture of egg coloration and the evolutionary maintenance of host-specific races of the common cuckoo remain largely unknown (Gibbs et al. 2000; Fossøy et al. 2011; but see Spottiswoode et al. 2011; Fossøy et al. 2016). Therefore, egg color and egg morphs of cuckoos in China would help us understand the evolution and the maintenance of host-specific races (Fossøy et al. 2016).

In Europe, the common cuckoo generally lays spotted eggs with two egg morphs being immaculate: pale blue eggs of the redstart (*Phoenicurus phoenicurus*) cuckoo gens and plain immaculate white cuckoo eggs found occasionally in different hosts but mimicking black redstart (*P. ochruros*) eggs (Makatsch 1937). However, both immaculate and spotted eggs are common in China (Table 14.1).

Immaculate cuckoo egg morphs are common in China, especially immaculate blue eggs. A recent study (Fossøy et al. 2016) suggested that the blue egg morph of the common cuckoo originated in Asia and subsequently expanded to Europe as female cuckoos that lay blue eggs interbred with the existing European population and thus introduced a new host race that favors the expansion of host use. Furthermore, the blue egg morph of the common cuckoo was found to be maternally inherited (Fossøy et al. 2016).

As in Europe, the common cuckoo in China is also a generalist that uses a variety of hosts and lays a number of egg morphs, including immaculate and spotted traits. The egg morphs of its hosts are generally matched by its mimetic eggs, indicating that different gentes of the common cuckoo lay specific egg morphs that have been adapted to their hosts (Yang et al. 2012a). According to the egg morphs of cuckoos and their hosts in China (Table 14.1), although common cuckoos utilize most host species, their eggs are highly mimetic or with intermediate mimicry to hosts. In other words, their gentes are more specialized than many other cuckoo species, which have not developed gentes. At least five hosts of the common cuckoo have evolved strong egg rejection capacities, including ashy-throated parrotbill (*Paradoxornis alphonsianus*) (Yang et al. 2010, 2014d), vinous-throated parrotbill (*Paradoxornis webbianus*) (Yang et al. 2015a), Oriental reed warbler (*Acrocephalus orientalis*) (Yang et al. 2014a, 2016c), reed parrotbill (*Paradoxornis heudei*) (Li et al. 2016), and Daurian redstart (*Phoenicurus auroreus*) (Yang et al. 2016a). Correspondingly, the cuckoo eggs parasitizing these host species are all mimetic to the host eggs. Therefore, there are different gentes related to these host species, respectively. The egg recognition abilities of other host species are unknown. Therefore, as in Europe, the common cuckoo is a generalized obligate brood parasite as a species, but for different host races, they

Table 14.1 Egg phenotype of cuckoos and their hosts in China (modified from Yang et al. 2012a)

Parasite	Egg morph of parasite	Host	Egg morph of host	Egg mimicry
Plaintive cuckoo *Cacomantis merulinus*	White or pale green with light rufous spots	Common tailorbird *Orthotomus sutorius*	White or pale green with light rufous spots	High
Asian emerald cuckoo *Chrysococcyx maculatus*	White with brown spots	Chestnut-crowned warbler *Seicercus castaniceps*	White	Low
Asian emerald cuckoo *Chrysococcyx maculatus*	White with brown spots	Bianchi's warbler *Seicercus valentini*	White	Low
Chestnut-winged cuckoo *Clamator coromandus*	Blue	Oriental magpie robin *Copsychus saularis*	Pale green with brown spots	No
Chestnut-winged cuckoo *Clamator coromandus*	Blue	Greater necklaced laughingthrush *Garrulax pectoralis*	Blue	High
Chestnut-winged cuckoo *Clamator coromandus*	Blue	Masked laughingthrush *Garrulax perspicillatus*	Pale blue with faint reddish-brown spots	Low
Common cuckoo *Cuculus canorus*	Light cyan with brown spots	Black-browed reed warbler *Acrocephalus bistrigiceps*	Off-white with olive spots	Intermediate
Common cuckoo *Cuculus canorus*	Off-white with olive spots	Oriental reed warbler *Acrocephalus orientalis*	Off-white with olive spots	High
Common cuckoo *Cuculus canorus*	Light cyan with brown spots	Azure-winged magpie *Cyanopica cyanus*	Light cyan with brown and purple spots	Intermediate
Common cuckoo *Cuculus canorus*	Dark blue	White-bellied redstart *Hodgsonius phoenicuroides*	Dark blue	High
Common cuckoo *Cuculus canorus*	Blue, pale blue, or white	Ashy-throated parrotbill *Paradoxornis alphonsianus*	Blue, pale blue, or white	High
Common cuckoo *Cuculus canorus*	Off-white with olive spots	Reed parrotbill *Paradoxornis heudei*	Off-white with olive spots	Intermediate
Common cuckoo *Cuculus canorus*	Blue, pale blue, or white	Vinous-throated parrotbill *Paradoxornis webbianus*	Blue, pale blue, or white	High

(continued)

Table 14.1 (continued)

Parasite	Egg morph of parasite	Host	Egg morph of host	Egg mimicry
Common cuckoo *Cuculus canorus*	White or pale blue with firebrick spots	Daurian redstart *Phoenicurus auroreus*	White or pale blue with firebrick spots	High
Common cuckoo *Cuculus canorus*	Pinkish white with sandy brown spots	Pale-legged willow warbler *Phylloscopus tenellipes*	Pinkish white	Intermediate
Common cuckoo *Cuculus canorus*	Dark blue	Grey bushchat *Saxicola ferrea*	Blue	Intermediate
Indian cuckoo *Cuculus micropterus*	Pinkish white with firebrick spots	Black drongo *Dicrurus macrocercus*	Pale cream with few brown spots	Low
Indian cuckoo *Cuculus micropterus*	Off-white with firebrick spots	Common blackbird *Turdus merula*	Blueish green with reddish-brown spots	No
Oriental cuckoo *Cuculus optatus*	Blue with red spots	Lesser shortwing *Brachypteryx leucophrys*	Olive green to sea green with light reddish-brown spots	Intermediate
Oriental cuckoo *Cuculus optatus*	White	Hume's warbler *Phylloscopus humei*	White	High
Oriental cuckoo *Cuculus optatus*	Pink with dense red spots or white with brown spots	Yellow-bellied prinia *Prinia flaviventris*	Pink with dense red spots	High
Oriental cuckoo *Cuculus optatus*	Brown	Red-headed tree babbler *Stachyridopsis ruficeps*	White	No
Lesser cuckoo *Cuculus poliocephalus*	Brown	Strong-footed bush warbler *Cettia fortipes*	Brown	High
Lesser cuckoo *Cuculus poliocephalus*	Light cyan	Blue-and-white flycatcher *Cyanoptila cyanomelana*	White	Low
Himalayan cuckoo *Cuculus saturatus*	White with few brown spots	Strong-footed bush warbler *Cettia fortipes*	Brown	No
Himalayan cuckoo *Cuculus saturatus*	White with few brown spots	Yellow-throated bunting *Emberiza elegans*	Off-white with black spots	Intermediate

Himalayan cuckoo *Cuculus saturatus*	White with few brown spots	Blyth's leaf warbler *Phylloscopus reguloides*	White	Intermediate
Himalayan cuckoo *Cuculus saturatus*	White with few brown spots	Collared finchbill *Spizixos semitorques*	Pink with dense violet spots	No
Common koel *Eudynamys scolopaceus*	Sage green	Black-collared starling *Sturnus nigricollis*	Blue to blue green	Intermediate
Large hawk-cuckoo *Hierococcyx sparverioides*	White	Chinese babax *Babax lanceolatus*	Blue	No
Large hawk-cuckoo *Hierococcyx sparverioides*	White	Hwamei *Garrulax canorus*	Blue	No
Large hawk-cuckoo *Hierococcyx sparverioides*	Off-white with brown spots	Elliot's laughingthrush *Garrulax elliotii*	Light sky blue with brown spots and steaks	Low
Large hawk-cuckoo *Hierococcyx sparverioides*	Off-white with brown spots	Masked laughingthrush *Garrulax perspicillatus*	Light cyan	No
Large hawk-cuckoo *Hierococcyx sparverioides*	White	White-browed laughingthrush *Garrulax sannio*	White or pale blue	High
Large hawk-cuckoo *Hierococcyx sparverioides*	Olive gray with brown spots	Common magpie *Pica pica*	Greenish gray with brown spots	Intermediate
Large hawk-cuckoo *Hierococcyx sparverioides*	Off-white with brown spots	Spot-breasted scimitar babbler *Pomatorhinus mcclellandi*	White	Low

are specialists. Similarly, the plaintive cuckoo (*Cacomantis merulinus*) and lesser cuckoo are both specialists. They parasitize common tailorbirds (*Orthotomus sutorius*) and brownish-flanked bush warblers (*Cettia fortipes*), respectively, with mimetic eggs, while these hosts are rejecters correspondingly (Yang et al. 2016b; Table 14.1). In contrast, the Himalayan cuckoo is a generalist that only lays one egg morph but parasitizes different host species that lay different egg morphs (Yang et al. 2011, 2012a; Table 14.1). Compared to common cuckoos, large hawk-cuckoos (*Hierococcyx sparverioides*) generally lay nonmimetic or low-mimetic eggs in their hosts' nests (Yang et al. 2015c). They lay white eggs in nests of Chinese babaxes (*Babax lanceolatus*) that lay blue eggs. Therefore, the cuckoo eggs are nonmimetic. Correspondingly, the babaxes are intermediate rejecters. However, they also lay mimetic blue eggs in nests of Elliot's laughingthrush (*Garrulax elliotii*) (Hu et al. 2013). In summary, it looks like cuckoo egg mimicry in China is related to the egg recognition capacities of hosts to a large extent.

14.4 Competition Among Cuckoo Host Races and Cuckoo Species in China

Ecological niche segregation must occur among sympatric cuckoo species, and host segregation probably has evolved as a result of interspecific competition among cuckoo species (Friedmann 1928; Southern 1954; Lack 1971; Wyllie 1981). Baker's (1942) work in India suggested that cuckoo species in sympatry avoid competing with each other by utilizing different hosts or by breeding at different times of the year. This could result in two closely related species coming together and staying in sympatry without competing for the same hosts. Møller et al. (2011) showed that different gentes of the common cuckoo in Europe are isolated by timing of breeding and by habitat. Brooker and Brooker (1989) and Higuchi (1989) investigated interactions between sympatrically occurring cuckoo species in Australia and Japan, respectively, and also obtained some support for the niche segregation hypothesis.

The mechanisms behind and the generality of such niche segregation, however, are at present poorly known because there have been few studies aiming to disclose the importance of interspecific competition among cuckoo species and no studies investigating the importance of both intraspecific and interspecific competition in combination. Asia, and China in particular, provides an ideal opportunity for studying competition among cuckoo host races and cuckoo species. Interspecific competition for host utilization should exist among sympatric cuckoo species in areas with several cuckoo species. Parasitic cuckoos lay eggs smaller than those of nonparasitic cuckoos with a similar body mass, suggesting that during coevolution with their hosts they generally parasitize passerine birds that are much smaller (Davies 2000). In China, for most cuckoo species with a specific body mass, they cannot exploit very small or very large hosts because small hosts may not provide enough food for proper growth of cuckoo chicks. As for large hosts, however, host nestlings may be too large for allowing successful eviction or outcompeting hosts by

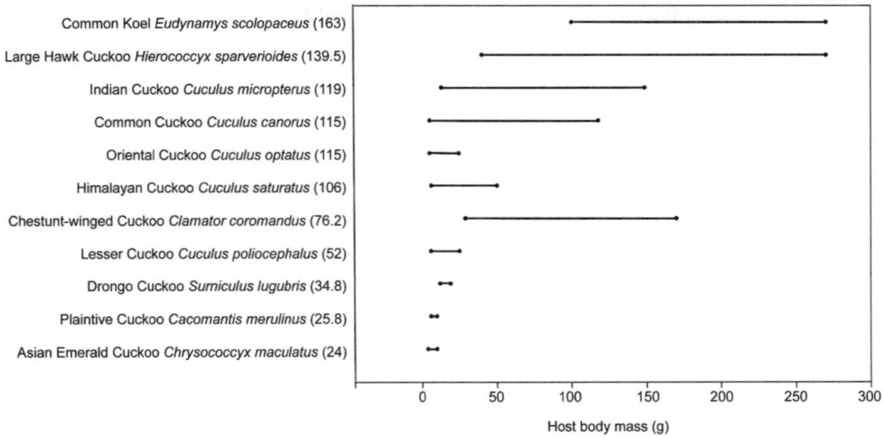

Fig. 14.5 Overlap in body mass range among hosts (range) that are utilized by cuckoos in China. Numbers in brackets refer to the median body mass of cuckoos

the cuckoo chick. According to the species list by Yang et al. (2012a), we extracted data on body mass of both cuckoos and hosts and showed that there was much competition and overlap in host use among cuckoo species in China (Fig. 14.5). Therefore, evictor parasitic cuckoos may have to trade the benefits of food supply against eviction or competition. In other words, generally cuckoos with a specific range in body mass may exploit hosts with a specific range of body mass, and thus different cuckoo species with the same or similar body mass may compete more over host use. However, *Clamator* cuckoos may be an exception. For example, the chestnut-winged cuckoo, one of the only two non-evictor cuckoo species in China, is of medium size but utilizes hosts that have a larger range than the Indian cuckoo. The body mass ranges of its hosts almost overlap with that of the large hawk-cuckoo. Finally, evicting medium-sized cuckoos (from Indian cuckoo to Himalayan cuckoo in Fig. 14.5) can also utilize small-sized hosts that are similar in size with hosts of small-sized cuckoos. This implies that parasitic cuckoos of medium size are most flexible in use of differently sized hosts, and this explains why medium-sized cuckoos are more widespread. In summary, the relation between cuckoo and host body mass and overlap in host use among sympatric cuckoo species is well documented in China. Competitive exclusion may have played an important role in the evolution of these patterns.

14.5 Coevolutionary Interactions Between Chinese Cuckoos and Their Hosts in an Area with Multiple Cuckoo Species

The arms race between brood parasites and their hosts seems confined to the egg stage of the breeding cycle (Davies 2000). However, it is widely accepted that hosts have evolved many different defensive strategies that can be classified into different lines of defense operating at each stage of the breeding cycle of the host and that brood parasites have evolved counter-adaptations against all of these lines of host defense (Soler 2014).

Four lines of defense could be used by hosts: (1) nest defense, (2) egg rejection, (3) chick discrimination, and (4) fledgling abandonment (Soler 2014). Therefore, information on the four lines of defense used by hosts in the ecosystem of multiple cuckoos in China could contribute to our knowledge of reciprocal adaptations and counter-adaptations in cuckoos and their hosts. There exists preliminary information about the first three lines of defense in the cuckoo–host systems in China.

14.5.1 Nest Defense

This line of defense is regarded as a frontline defense to avoid cuckoo parasitism. The great reed warbler (*Acrocephalus arundinaceus*) is well-known for defending its nests by aggressively attacking cuckoos (Davies 2000). In China, Oriental reed warblers also showed the highest level of aggression to the co-occurring common cuckoos, with female warblers showing better ability to distinguish between parasites and nest predators than males (Li et al. 2015). However, some cuckoo host species in China did not aggressively attack cuckoos. For example, barn swallows (*Hirundo rustica*) strongly attacked cuckoos and could distinguish between common cuckoo and sparrowhawk (*Accipiter nisus*) in Denmark but did not differ in behavior when facing sparrowhawk and cuckoo dummies in China (Liang and Møller 2015), although actually they were able to distinguish between common cuckoo and sparrowhawk and respond accordingly in their vocal behavior (Yu et al. 2016). A similar case was also found in the great tit (*Parus major*), which showed strong egg rejection in China but not in Europe (Liang et al. 2016). The Chinese great tits did not attack cuckoos but were able to distinguish sparrowhawks from common cuckoos and convey relevant information in alarm calls by adjusting the number and combinations of notes of a single call type (Yu et al. 2017).The yellow-bellied prinia in mainland China, where many parasitic cuckoos coexist, does not attack common cuckoos or Oriental cuckoos, while the island population in Taiwan where only the Oriental cuckoo breeds sympatrically showed attack behavior toward cuckoos and distinguished Oriental cuckoo from common cuckoos despite the fact that they are very similar as judged by humans (Yang et al. 2014b). Furthermore, the island population that is more aggressive toward the cuckoo also possesses a higher capacity of egg rejection (Fig. 14.6; see also Moksnes et al. 1991). One possible explanation is that the yellow-bellied prinia in the island of Taiwan has been parasitized for a longer time than the population of mainland China. A similar pattern of defense was found in a single population of brown-breasted bulbul (*Pycnonotus xanthorrhous*) in which rejecters behave more aggressively toward cuckoos than acceptors (Su et al. 2016). These findings are consistent with the hypothesis that one line of defense may facilitate the evolution of another stage of defense (Kilner and Langmore 2011; Feeney et al. 2012). However, this result is opposite to the strategy-blocking hypothesis positing that selection at the former stage of defense may block the defense in later stages (Soler et al. 1999).

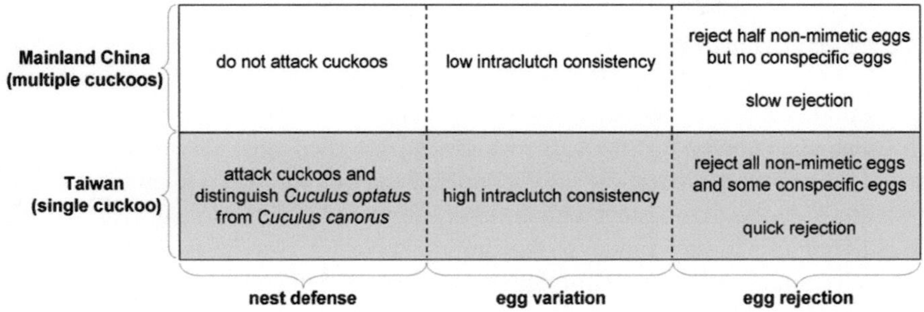

Fig. 14.6 Comparison of defenses against cuckoo parasitism between mainland and island populations of yellow-bellied prinia *Prinia flaviventris* (modified from Yang et al. 2014b)

14.5.2 Egg Rejection

Compared to nest defense, egg rejection behavior by cuckoo hosts in China is more consistent with that of cuckoo hosts in Europe. That is, egg rejection is generally related to the extent of current or past cuckoo parasitism (Davies 2000; Yang et al. 2014c, 2015a, b; Soler 2014). However, this general principle cannot explain why some host species accept cuckoo eggs, and then the evolutionary equilibrium and the evolutionary lag hypothesis were proposed (Takasu 1998). Furthermore, the egg recognition mechanism is an important trait that has received less attention. Previous studies found that cuckoo hosts show true egg recognition based on knowledge of the hosts' own egg appearance for rejecting alien eggs, although another mechanism termed discordancy suggested that cuckoo hosts simply recognize minority eggs as alien eggs (Rothstein 1975, 1982; Hauber and Sherman 2001; Bán et al. 2013; Stevens et al. 2013). A recent study has demonstrated that both mechanisms may exist in a single population of ashy-throated parrotbill, which is one of the major hosts of the common cuckoo in China (Yang et al. 2014d). Such mixed mechanisms may be explained by different strategies adopted by male and female hosts, respectively, for countering cuckoo parasitism and avoiding erroneous rejection (Liang et al. 2012). Because the ashy-throated parrotbill is a strong rejecter of nonmimetic eggs, but females lay polymorphic eggs, males that mate with females laying different egg morphs should use discordancy recognition to avoid rejection error caused by mis-imprinting (Liang et al. 2012). However, the mis-imprinting hypothesis has not been supported in recent experimental studies (Soler et al. 2013; Moskát et al. 2014).

14.5.3 Chick Discrimination

Compared to egg rejection, chick discrimination is considerably rarer. Theoretical models have suggested that chick recognition by learning might be maladaptive because of the high costs of recognition errors (Lotem 1993). However, this hypothesis cannot generally explain the scarcity of chick recognition because hosts typically do not learn chick appearance (e.g., Langmore et al. 2003). The "rarer enemies" hypothesis seems to generally explain the scarcity of chick recognition (Langmore et al. 2003; Grim 2006, 2011). It suggested that egg recognition weakens selection for evolution of chick recognition. Yang et al. (2015b) compared egg and chick recognition capacities between barn swallow and red-rumped swallow (*H. daurica*) in China and showed that the barn swallow rejects alien eggs but not alien chicks, while the red-rumped swallow fully accepts alien eggs but rejects alien chicks, thus providing evidence consistent with the rarer enemies hypothesis (Yang et al. 2015b). This hypothesis is very similar to the strategy-blocking hypothesis, which suggests that selection at earlier stages of defense may block defenses at later stages (Soler et al. 1999).

Concluding Remarks and Future Directions

The coevolutionary interactions between cuckoos and their hosts in China have made some progress in recent years but still need more investigation. We hypothesize that such a multiple-cuckoo system may serve as a model system for the study of frequency-dependent selection and its evolutionary consequences.

We showed that parasitic cuckoos of medium size in China are most flexible in parasitizing hosts of different body sizes and that they are particularly widespread. This may arise from the fact that cuckoos of intermediate body size experience stronger interspecific competition than the smallest or the largest cuckoos that are mainly competing with cuckoos of the smallest or the largest body size, respectively. Attempts to study overlap in host use and competitive exclusion through specialization by breeding at particular times of the year or by specialization on specific breeding habitats may help us understand why cuckoo diversity is so high in particular parts of Asia.

Acknowledgments We would like to thank M. Soler, A. Moksnes, and C. Moskát for providing constructive comments and helpful suggestions. This work was supported by the National Natural Science Foundation of China (Nos. 31272328, 31472013 and 31772453 to WL and 31672303 and 31260514 to CY,) and the Program for New Century Excellent Talents in University (NCET-13-0761).

References

Baker ECS (1942) Cuckoo problems. Witherby, London

Bán M, Moskát C, Barta Z, Hauber ME (2013) Simultaneous viewing of own and parasitic eggs is not required for egg rejection by a cuckoo host. Behav Ecol 24:1014–1021

Brooker MG, Brooker LC (1989) Cuckoo hosts in Australia. Aust Zool Rev 2:1–67

Davies NB (2000) Cuckoos, cowbirds and other cheats. T & AD Poyser, London

Erritzøe J, Mann CF, Brammer FP, Fuller RA (2012) Cuckoos of the world. Christopher Helm, London

Feeney WE, Welbergen JA, Langmore NE (2012) The frontline of avian brood parasite–host coevolution. Anim Behav 84:3–12

Fossøy F, Antonov A, Moksnes A, Røskaft E, Vikan JR, Møller AP, Shykoff JA, Stokke BG (2011) Genetic differentiation among sympatric cuckoo host races: males matter. Proc R Soc Lond B 278:1639–1645

Fossøy F, Sorenson MD, Liang W, Ekrem T, Moksnes A, Møller AP, Rutila J, Røskaft E, Takasu F, Yang C, Stokke BG (2016) Ancient origin and maternal inheritance of blue cuckoo eggs. Nat Commun 7:10272

Friedmann H (1928) Social parasitism in birds. Q Rev Biol 3:554–569

Gibbs HL, Sorenson MD, Marchetti K, Brooke M de L, Davies NB, Nakamura H (2000) Genetic evidence for female host-specific races of the common cuckoo. Nature 407:183–186

Grim T (2006) The evolution of nestling discrimination by hosts of parasitic birds: why is rejection so rare? Evol Ecol Res 8:785–802

Grim T (2011) Ejecting chick cheats: a changing paradigm? Front Zool 8:14

Hauber ME, Sherman PW (2001) Self-referent phenotype matching: theoretical considerations and empirical evidence. Trends Neurosci 24:609–616

Higuchi H (1989) Responses of the bush warbler *Cettia diphone* to artificial eggs of *Cuculus* cuckoos in Japan. Ibis 131:94–98

Hu Y, Wang X, Chang H, Sun Y (2013) Brood parasitism on Elliot's laughingthrush by large hawk cuckoo. Chin J Zool 48:292–293

Jetz W, Thomas GH, Joy JB, Hartmann K, Mooers AO (2012) The global diversity of birds in space and time. Nature 491:444–448

Kilner RM, Langmore NE (2011) Cuckoos versus hosts in insects and birds: adaptations, counter-adaptations and outcomes. Biol Rev 86:836–852

Lack DL (1971) Ecological isolation in birds. Blackwell, London

Langmore NE, Hunt S, Kilner RM (2003) Escalation of a coevolutionary arms race through host ejection of brood parasitic young. Nature 422:157–160

Li D, Wei H, Zhang Z, Liang W, Stokke BG (2015) Oriental reed warbler (*Acrocephalus orientalis*) nest defence behaviour towards brood parasites and nest predators. Behaviour 152:1601–1621

Li D, Zhang Z, Grim T, Liang W, Stokke BG (2016) Explaining variation in brood parasitism rates between potential host species with similar habitat requirements. Evol Ecol 30:905–923

Liang W, Møller AP (2015) Hawk mimicry in cuckoos and anti-parasitic aggressive behavior of barn swallows in Denmark and China. J Avian Biol 46:216–223

Liang W, Yang C, Antonov A, Fossøy F, Stokke BG, Moksnes A, Røskaft E, Shykoff JA, Møller AP, Takasu F (2012) Sex roles in egg recognition and egg polymorphism in avian brood parasitism. Behav Ecol 23:397–402

Liang W, Møller AP, Stokke BG, Yang C, Kovařík P, Wang H, Yao C-T, Ding P, Lu X, Moksnes A, Røskaft E, Grim T (2016) Geographic variation in egg ejection rate by great tits across 2 continents. Behav Ecol 27:1405–1412

Lotem A (1993) Learning to recognize nestling is maladaptive for cuckoo *Cuculus canorus* hosts. Nature 362:743–744

MacKinnon J, Phillipps K (1999) A field guide to the birds of China. Oxford University Press, Oxford

Makatsch W (1937) Der Brutparasitismus der Kuckuckvögel. Quelle & Meyer, Leipzig

Moksnes A, Røskaft E (1995) Egg-morphs and host preference in the common cuckoo *Cuculus canorus*: an analysis of cuckoo and host eggs from European museum collections. J Zool 236:625–648

Moksnes A, Røskaft E, Braa AT, Korsnes L, Lampe HM, Pedersen HC (1991) Behavioural responses of potential hosts towards artificial cuckoo eggs and dummies. Behaviour 116:64–89

Møller AP, Antonov A, Stokke BG, Fossøy F, Moksnes A, Røskaft E, Takasu F (2011) Isolation by time and habitat and coexistence of distinct host races of the common cuckoo. J Evol Biol 24:676–684

Moskát C, Bán M, Hauber ME (2014) Naïve hosts of avian brood parasites accept foreign eggs, whereas older hosts fine-tune foreign egg discrimination during laying. Front Zool 11:45

Payne RB (2005) The cuckoos. Oxford University Press, Oxford

Røskaft E, Liang W, Stokke BG (2012) Avian brood parasitism–a growing research area in behavioral ecology. Chin Birds 3:243–244

Rothstein SI (1975) Mechanisms of avian egg-recognition: do birds know their own eggs? Anim Behav 23:269–278

Rothstein SI (1982) Mechanisms of avian egg recognition: which egg parameters elicit responses by rejecter species? Behav Ecol Sociobiol 11:229–239

Rothstein SI (1990) A model system for coevolution: avian brood parasitism. Annu Rev Ecol Syst 21:481–508

Rothstein SI, Robinson SK (1998) Parasitic birds and their hosts: studies in coevolution. Oxford University Press, Oxford

Soler M (2014) Long-term coevolution between avian brood parasites and their hosts. Biol Rev 89:688–704

Soler JJ, Soler M, Pérez-Contreras T, Aragón S, Møller AP (1999) Antagonistic antiparasite defenses: nest defense and egg rejection in the magpie host of the great spotted cuckoo. Behav Ecol 10:707–713

Soler M, Ruiz-Castellano C, Carra LG, Ontanilla J, Martín-Galvez D (2013) Do first-time breeding females imprint on their own eggs? Proc R Soc B 280:20122518

Southern HN (1954) Mimicry in the cuckoo eggs. In: Huxley JS, Hardy AC, Ford EB (eds) Evolution as a process. George Allen & Unwin, London, pp 219–232

Spottiswoode CN, Stryjewski KF, Quader S, Colebrook-Robjent JFR, Sorenson MD (2011) Ancient host specificity within a single species of brood parasitic bird. Proc Natl Acad Sci U S A 108:17738

Stevens M, Troscianko J, Spottiswoode CN (2013) Repeated targeting of the same hosts by a brood parasite compromises host egg rejection. Nat Commun 4:2475

Su T, Yang C, Liang G, Ding C, Liang W (2016) Association between nest defense and egg rejection behavior of brown-breasted bulbuls (*Pycnonotus xanthorrhous*). Auk 133:231–235

Takasu F (1998) Why do all host species not show defense against avian brood parasitism: evolutionary lag or equilibrium? Am Nat 151:193–205

Wyllie I (1981) The cuckoo. Batsford, London

Xia C, Liang W, Carey GJ, Zhang Y (2016) Song characteristics of oriental cuckoo *Cuculus optatus* and Himalayan cuckoo *Cuculus saturatus* and implications for distribution and taxonomy. Zool Stud 55:38

Yang C, Liang W, Cai Y, Shi S, Takasu F, Møller AP, Antonov A, Fossøy F, Moksnes A, Røskaft E, Stokke BG (2010) Coevolution in action: disruptive selection on egg colour in an avian brood parasite and its host. PLoS One 5:e10816

Yang C, Cai Y, Liang W (2011) Visual modeling reveals cryptic aspect in egg mimicry of Himalayan cuckoo (*Cuculus saturatus*) on its host Blyth's leaf warbler (*Phylloscopus reguloides*). Zool Res 32:451–455

Yang C, Liang W, Antonov A, Cai Y, Stokke BG, Fossøy F, Moksnes A, Røskaft E (2012a) Diversity of parasitic cuckoos and their hosts in china. Chin Birds 3:9–32

Yang C, Cai Y, Liang W (2012b) Species identification of sympatric cuckoo nestlings in a multiple-cuckoo system, China. Chin Birds 3:108–112

Yang C, Li D, Wang L, Liang G, Zhang Z, Liang W (2014a) Geographic variation in parasitism rates of two sympatric cuckoo hosts in China. Zool Res 35:67–71

Yang C, Wang L, Cheng SJ, Hsu YC, Liang W, Møller AP (2014b) Nest defenses and egg recognition of yellow-bellied prinia against cuckoo parasitism. Naturwissenschaften 101:727–734

Yang C, Liu Y, Zeng L, Liang W (2014c) Egg color variation, but not egg rejection behavior, changes in a cuckoo host breeding in the absence of brood parasitism. Ecol Evol 4:2239–2246

Yang C, Møller AP, Røskaft E, Moksnes A, Liang W, Stokke BG (2014d) Reject the odd egg: egg recognition mechanisms in parrotbills. Behav Ecol 25:1320–1324

Yang C, Wang L, Cheng SJ, Hsu YC, Stokke BG, Røskaft E, Moksnes A, Liang W, Møller AP (2015a) Deficiency in egg rejection in a host species as a response to the absence of brood parasitism. Behav Ecol 26:406–415

Yang C, Wang L, Chen M, Liang W, Møller AP (2015b) Nestling recognition in red-rumped and barn swallows. Behav Ecol Sociobiol 69:1–6

Yang C, Su T, Liang W, Møller AP (2015c) Coevolution between the large hawk-cuckoo (*Cuculus sparverioides*) and its two sympatric Leiothrichidae hosts: evidence for recent expansion and switch in host use? Biol J Linn Soc 115:919–926

Yang C, Li Z, Zhang Y, Wang H, Liang W, Møller AP (2016a) Egg polymorphism and egg discrimination in the daurian redstart *Phoenicurus auroreus*, a host of the common cuckoo *Cuculus canorus*. Ornithol Sci 15:127–132

Yang C, Huang Q, Wang L, Jiang A, Stokke BG, Fossøy F, Tunheim OH, Røskaft E, Liang W, Møller AP (2016b) Plaintive cuckoos do not select tailorbird hosts that match the phenotypes of their own eggs. Behav Ecol 27:835–841

Yang C, Wang L, Liang W, Møller AP (2016c) Egg recognition as antiparasitism defence in hosts does not select for laying of matching eggs in parasitic cuckoos. Anim Behav 122:177–181

Yu J, Wang L, Xing X, Yang C, Ma J, Møller AP, Wang H, Liang W (2016) Barn swallows (*Hirundo rustica*) differentiate between common cuckoo and sparrowhawk in China: alarm calls convey information on threat. Behav Ecol Sociobiol 70:171–178

Yu J, Xing X, Jiang Y, Liang W, Wang H, Møller AP (2017) Alarm call-based discrimination between common cuckoo and Eurasian sparrowhawk in a Chinese population of great tits. Ethology 123:542–550

Zheng G (2011) A checklist on the classification and distribution of the birds of China, 2nd edn. Science Press, Beijing

Zheng G (2018) A checklist on the classification and distribution of the birds of China, 3rd edn. Science Press, Beijing

Part V

Coevolutionary Interactions at Any Stage

Brood Parasites as Predators: Farming and Mafia Strategies

15

M. Soler, T. Pérez-Contreras, and J. J. Soler

Dedicated to Amoz Zahavi, the brilliant evolutionary biologist who envisaged the mafia hypothesis, deceased a few months ago (May 12, 2017), with our maximum admiration and respect.

Abstract

Avian brood parasites may depredate unparasitized host nests in advanced stage (farming strategy) or those of hosts that have rejected the parasitic offspring (mafia strategy). Both predatory strategies induce host renesting, but the benefit of the former is to increase availability of host nests for future parasitism, while the latter imply extra fitness costs to rejecters and, thus, induce hosts to accept parasitism in replacement clutches. Despite clear expected benefits of these strategies, they have been reported only rarely. Mafia has been demonstrated in only two parasitic species, but not all available results confirm it. Results from recent studies indicate that magpies (*Pica pica*) reject real great spotted cuckoo (*Clamator glandarius*) eggs less frequently and that cuckoos punished rejecter magpies less frequently now than about 25 years ago. We suggest that costs of retaliatory behaviours together with phenotypic plasticity of magpie and great spotted cuckoo's defences and counter-defences are the crucial points explaining these differences. Reported geographical and temporal variation in parasitism

M. Soler (✉) · T. Pérez-Contreras
Departamento de Zoología, Facultad de Ciencias, Universidad de Granada, Granada, Spain

Grupo Coevolución, Unidad Asociada al CSIC, Universidad de Granada, Granada, Spain
e-mail: msoler@ugr.es

J. J. Soler
Grupo Coevolución, Unidad Asociada al CSIC, Universidad de Granada, Granada, Spain

Departamento de Ecología Funcional y Evolutiva, Estación Experimental de Zonas Áridas (CSIC), Almería, Spain

M. Soler (ed.), *Avian Brood Parasitism*, Fascinating Life Sciences,
https://doi.org/10.1007/978-3-319-73138-4_15

rates suggests that parasitism by great spotted cuckoos would depend on environmental variables, intensity of host defences or propensity to compliance of magpies. This highly variable risk of parasitism, together with the facultative virulence of cuckoos against rejecters, makes plasticity of antiparasitic defences adaptive in magpies. Understanding the mechanisms underlying the mafia remains elusive. Future studies should concentrate on examining how often and in which context brood parasites prey upon host nests in different host populations and in different brood parasite–host systems.

15.1 Introduction

Parasites and hosts are involved in antagonistic interactions in which parasitic manipulation frequently reduces host fitness (e.g. Schmid-Hempel 2011). Given that parasites pose a serious threat to host fitness, arms race coevolutionary theory predicts that hosts should resist parasite manipulation and evolve counter-defences that minimize the negative effects of parasites (Moore 2002; Rolff and Siva-Jothy 2003; Thompson 2005). When host defences reduce parasite fitness, the parasite evolves new ways to escape from host defences, which gives rise to a coevolutionary arms race that involves subsequent adaptations and counter-adaptations (Dawkins and Krebs 1979; Thompson 2005). The main defensive strategy used by hosts against their parasites is resistance, which reduces parasite burden by preventing its establishment or inhibiting its growth; therefore, this implies negative effects for parasites. There is another important component of host defence that has been largely overlooked in animal studies: tolerance, which consists of responses that do not imply negative effects for parasites but minimize the negative impact of parasitism on host fitness (Svensson and Råberg 2010; Medzhitov et al. 2012). Only the former kind of host defences would induce coevolutionary arm races among counterparts (Soler and Soler 2017).

Tolerance defences are favoured by natural selection and nowadays are considered as adaptive for a host species as resistance (Svensson and Råberg 2010; Medzhitov et al. 2012; Råberg 2014). Hosts of brood parasites (i.e. those that evade all parental care by laying their eggs in the host's nests), for instance, as strategies of tolerance, can adjust their life history traits and reduce clutch size while increasing reproductive attempts, which increase the probability of escaping parasitism (Medina and Langmore 2016). Tolerance has been scarcely studied in brood parasite-host systems (Medina and Langmore 2016), but a positive association between levels of tolerance by magpies (*Pica pica*) and of parasitism by great spotted cuckoos (*Clamator glandarius*) has been demonstrated at the population level (Soler et al. 2011). This association was predicted because of the particular predatory behaviour of cuckoos, and we here discuss the general association between tolerance of hosts and predation of brood parasites.

There is extensive evidence that brood parasites frequently cause partial or complete predation on both parasitized and unparasitized host nests (see below), and this predatory behaviour can be used to drive parasite-induced alteration of host behaviour. Avian brood parasites have been shown to induce behavioural changes in their hosts by retaliatory behaviour (Soler et al. 1999a), and thus, we discuss the importance of a link between parasite predation and its influence on host behaviour driving the evolution of farming and mafia behaviour of parasites. Parasite-induced alterations of host behaviour include fascinating examples in which host behaves in ways absolutely different from what is the normal behaviour of the species, but that enhance parasite transmission or survival (see reviews by Moore 2002; Lefèvre et al. 2008, 2009; Poulin 2010).

In this chapter, we first review the role of brood parasites as predators and later the two cases where predatory behaviour by brood parasites provokes modifications of host strategies: farming and mafia behaviour. Finally, we discuss possible reasons why these strategies are scarce in nature in relation to factors such as parasite virulence, manipulation of host behaviour, tolerance and phenotypic plasticity in host responses, which may be related to each other and influence the evolution of the relationships between hosts and brood parasites.

15.2 Brood Parasites and Depredation of Host Nests

Brood parasites have been reported to cause partial or complete predation on both parasitized and unparasitized host nests. Brood parasites provoke partial nest failure by either puncture or removal of one or more host eggs, mainly before or during laying one of their own (e.g. Wyllie 1981; Soler et al. 1997; Peer 2006; Fiorini et al. 2009). Partial or complete host nest failures by brood parasites have been reported in some species, such as the common cuckoo (*Cuculus canorus*; Wyllie 1981), the great spotted cuckoo (Soler et al. 1995) and the brown-headed cowbird (*Molothrus ater*). By far, predatory behaviour is more frequent in the later species (Arcese et al. 1996; Peer and Sealy 1999; Hoover and Robinson 2007; Swan et al. 2015).

Brood parasites may cause a biased predation pressure over either parasitized or unparasitized host nests, which would depend on particular characteristics of the brood parasite–host system and of the host species. On the one hand, brood parasites would enhance the probability of nest predation because (1) the typical increased intensity and loudness of the begging calls emanating from parasitized nests directly attract predators (Dearborn 1999; Kosciuch and Sandercock 2008; Hannon et al. 2009; Ibáñez-Álamo et al. 2011). It could also be that parasitism affects nest detectability by depredators because (2) host parents increase provisioning rate of nestlings in parasitized nests (McLaren and Sealy 2000; Payne and Payne 1998) or because (3) brood parasites near host nests facilitate nest detectability by depredators (Arcese and Smith 1999). Biased depredation of parasitized nests would also occur if (4) nests that are easier to parasitize are also easier to depredate (i.e. nest vulnerability, Arcese and Smith 1999) or because (5) of a coincidental preference by brood parasites and nest predators for particular habitat features (McLaren and Sealy 2000). On the other hand, biased depredation over unparasitized nests would be expected when (1) brood parasites select for nests with a lower probability of becoming predated (Hauber 2000), a possibility that has recently been experimentally demonstrated (Soler et al. 2014a). (2) Adult brood parasites could also defend parasitized nests against predators (Soler et al. 1999b), and/or (3) characteristic of brood parasitic nestlings may deter depredators (Canestrari et al. 2014, but see Soler et al. 2017), which are examples of brood parasites directly contributing to host nest defence. The most clear direct effect of biased predation pressure over unparasitized nests is (4) the predatory behaviour of brood parasites on unparasitized nests (Wyllie 1981; Peer and Sealy 1999; Granfors et al. 2001; Stake and Cimprich 2003; Swan et al. 2015). In this chapter, we concentrate on this last possibility.

15.3 Predation by Brood Parasites and the Farming Strategy

The behaviour of removing one or more eggs from host nests is very common in many avian brood parasites including cuckoos of several genera, cowbirds (*Molothrus* sp.), widowbirds (*Vidua* sp.) and possibly honeyguides (Indicatoridae) (Friedmann 1948, 1968; Wyllie 1981; Davies 2000). The possibility that a brood parasite could destroy unparasitized host nests was first suggested by Chance as early as 1940 when he proposed that common cuckoos could do this to enhance their opportunities of successful parasitism later in the breeding season using replacement clutches (Chance 1940; in Arcese et al. 1996). Later, trying to explain that in a population of song sparrows (*Melospiza melodia*) parasitized by brown-headed cowbirds, the frequency of nest failure was positively related to the frequency of cowbird parasitism, Arcese et al. (1992) proposed the cowbird predation hypothesis, which is basically the same idea as suggested by Chance. The main prediction of this hypothesis is that unparasitized nests should suffer a higher predation rate than parasitized nests, a prediction that was experimentally tested a few years later (Arcese et al. 1996). However, although most subsequent studies confirmed this prediction (Johnson and Temple 1990; Donovan et al. 1997; Arcese and Smith 1999; Hauber 2000; Clotfelter and Yasukawa 1999; Hoover and Robinson 2007; Swan et al. 2015), some others did not (Wolf 1987; Rogers et al. 1997; Payne and Payne 1998; McLaren and Sealy 2000).

The crucial idea behind the brood parasite predation hypothesis is that predation of unparasitized nests by brood parasites has the function of forcing hosts to renest, increasing future opportunities for parasitism. This selective predation of unparasitized nests can be considered a brood parasite manipulation of host behaviour that provokes an increase in the number of host nests available for parasites. For this reason, this predatory behaviour of brood parasites is known as the “farming” strategy. Thus, depredated unparasitized nests should be those found too late in the nesting cycle to be suitable for parasitism. This prediction of the farming hypothesis has been experimentally demonstrated by Swan et al. (2015) in a brown-headed cowbird population breeding in captivity. Briefly, researchers presented cowbirds with nests simulating early and late stages of embryo development and found that cowbirds were able to discriminate nests at the late incubation period because these nests were more often predated. A key prediction of the farming hypothesis is that direct benefits for the individuals that scan host nests and depredate them in the late incubation stage have to be detected. Hoover and Robinson (2007) indirectly demonstrate this prediction in a population of brown-headed cowbirds parasitizing their prothonotary warbler (*Protonotaria citrea*) host. The authors found that most (85%) of the subsequent renesting attempts of prothonotary warblers that lost their nests due to cowbird predation were subsequently parasitized, while only 36% of renesting attempts not linked to cowbird predation were parasitized. This result demonstrated that predation of unparasitized nests in this brood parasite–host system is an adaptive farming strategy that increases future laying opportunities of the brown-headed cowbird.

Clear evidence of farming behaviour has only been found in the brown-headed cowbird (Arcese et al. 1996; Clotfelter and Yasukawa 1999; Hoover and Robinson 2007; Swan et al. 2015), but it has also been suggested to occur in the bronzed cowbird (*Molothrus aeneus*; Peer and Sealy 1999). For some other brood parasites as the great spotted cuckoo (Soler et al. 1995), the screaming cowbird (*M. rufoaxillaris*; De Mársico and Reboreda 2014; Fiorini et al. 2014) and the shiny cowbird (*M. bonariensis*; Massoni and Reboreda 1998; Tuero et al. 2012; Fiorini et al. 2009, 2014), no associations between predation of host nests and embryo stage have been detected. Furthermore, this species punctures host eggs during the laying stage, which is contrary to the farming hypothesis (see references above), and thus, it is unlikely that the predatory behaviour of this species increases availability of host nests for future parasitism.

15.4 Predation by Brood Parasites and the Mafia-like Strategy

In 1979 Amotz Zahavi proposed an exciting idea trying to explain the fact that many host species do not discriminate against cuckoos. His verbal model suggested that cuckoos would revisit the parasitized nests and, if the cuckoo nestling had been rejected, adult cuckoos would prey upon the host nestlings (Zahavi 1979). This hypothesis was later criticized as unworkable (Guildford and Read 1990), but it was experimentally demonstrated. Soler and coworkers removed great spotted cuckoo eggs from magpie nests, and these experimental nests suffered significantly higher predation rates (73%) than control nests (11%) in which the parasitic eggs were not removed but were visited at the same frequency than experimental nests (Soler et al. 1995). These authors also collected prints of cuckoo bill on plasticine model eggs in magpie nests after removing cuckoo eggs, which proved the predatory behaviour of cuckoos in experimental nests (Soler et al. 1995). In addition, they also reported one case of direct observation of predation: just after one magpie nest was checked, one great spotted cuckoo went into the nest; the authors climbed again to the nests and observed that magpie eggs had been destroyed (Soler et al. 1995). Other piece of evidence showing that cuckoos are the predators of magpie nests is that Soler and coauthors also found on the ground a recently dead (i.e. still warm) magpie nestling that was supposedly killed by the great spotted cuckoo. The multiple hematomas and injuries that can be observed in this nestling (Fig. 15.1) indicate that this predation

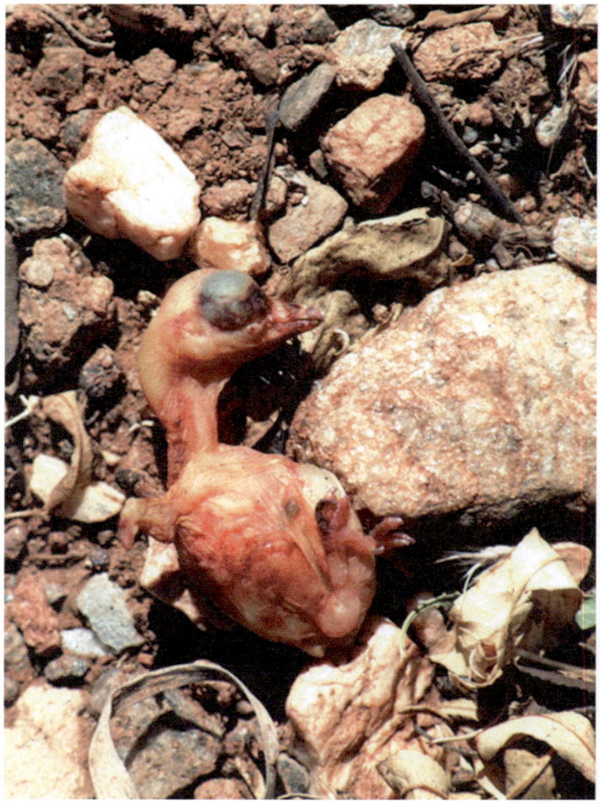

Fig. 15.1 Magpie nestling recently (i.e. still warm) killed by great spotted cuckoos. The nestling was not eaten, and the multiple hematomas and injuries that can be observed indicate that this predation event had not been perpetrated by a well-armed predator

event had not been perpetrated by a well-armed predator. Retaliatory cuckoos benefited themselves because they induced magpies [which only have one clutch per year (Birkhead 1991)] to renest, which would be the last opportunity of yearly reproduction, and then accepting the parasitic egg is the adaptive strategy for magpies (Soler et al. 1995, 1999a). Thus, the retaliatory behaviour of parasites selects for collaborative behaviour (i.e. acceptance of parasitism) in their hosts by imposing extra fitness costs on noncollaborative magpies. That is the reason why this particular predatory behaviour in cuckoos was called “avian mafia” by Soler et al. (1995). The mafia hypothesis later received considerable theoretical support (Pagel et al. 1998; Soler et al. 1998a; Robert et al. 1999; Abou-Chakra et al. 2014, 2016, but see Hauber 2014) and was experimentally demonstrated in another brood parasite–host system (brown-headed cowbirds–prothonotary warblers). Hoover and Robinson (2007) simulated ejection of cowbird eggs, and they found that 56% of nests in which the cowbird egg was removed were depredated. Instead, only 6% of the nests in which the parasitic egg was not removed were depredated. By comparing depredation events occurring in normal and in experimental nest boxes that prevent cowbirds to go in, they convincingly demonstrated that cowbird predation provoked that ejector nests produced 60% fewer host fledglings than accepter nests. Currently, mafia behaviour is considered one of the extreme examples of parasite-induced alteration of host behaviour (Lefèvre et al. 2008, 2009).

At least two conditions have to be fulfilled in parasite–host systems for the mafia-type tactic of host exploitation to evolve (Soler et al. 1998a). First, behavioural defences of the host and the level of virulence of the parasite have both to be plastic and negatively related to each other. Second, that host compliance must have fitness advantages once the mafia system is established. This second condition limits the possibility that the mafia tactic evolved in systems where parasitized and completely predated nests of hosts have the same reproductive success, which is invariably the case of those involving evictor brood parasites. In other words, for mafia tactic evolution in parasites, hosts should have the opportunity to successfully rear some of their nestlings in parasitized nests. Thus, the mafia-like behaviour would not evolve in common cuckoos. Accordingly, when cuckoo eggs were experimentally removed from the nests of their great reed warbler (*Acrocephalus arundinaceus*) host, these nests did not suffer higher predation rates (sample size not provided; Nakamura, pers. comm., in Davies 2000).

The experimental removal of parasitic eggs or chicks provoked significantly higher rates of predation in experimental than in control nests (the main prediction of the mafia hypothesis) in another study carried out with the brown-headed cowbird parasitizing song sparrows (*Melospiza melodía*; Hauber 2009), but not when parasitizing other host species (Arcese et al. 1996; McLaren and Sealy 2000). The same negative results have been found in several recent studies involving the great spotted cuckoo and its two main host species (Roldán et al. 2013; Soler and de Neve 2013; Canestrari et al. 2014; Soler et al. 2014b, c, 2017) (see a detailed discussion below). This is also the case in the only system not involving brood parasites in which an experimental test of mafia hypothesis has been made (the cricket *Nemobius sylvestris* and its parasitic hairworm *Paragordius tricuspidatus*; Biron et al. 2005). From the point of view of the brood parasites, retaliatory behaviour against hosts that eject parasitic eggs from their nests selects against the evolution of discriminating mechanisms of parasitic offspring in the hosts. Discriminating hosts will have no advantage over non-discriminating hosts because, at a shorter term, the retaliatory brood parasites would force the host to accept the parasitic egg and to learn that it is better to accept the parasitic egg and try to rear one of their own chicks than to reject it (Soler et al. 1998a; Hauber et al. 2004). Experimental evidence suggested that this is the case in magpies given that magpie pairs, which were experimentally parasitized, and had their nests depredated after rejection of the parasitic egg, changed their behaviour from rejection to acceptance (Soler et al. 1999a). Magpie responses to experimental predation in relation to rejection of experimental parasitic eggs varied among study plots differing in parasitism selection pressure. Changes from rejection in first breeding attempts to acceptance in replacement clutches occurred only in study plots with high levels of parasitism in which two or more cuckoo females parasitized the same nest (Soler et al. 1999a). This means that magpies can plastically modulate their ejection defence.

From the host's point of view, acceptance of the brood parasitic eggs is less costly than resistance in terms of reproductive success because the latter would likely result in complete nest predation. Thus, as predicted by the mafia hypothesis, by imposing extra fitness costs, retaliatory brood parasites have the potential to select for host compliance (Soler et al. 1998a; Ponton et al. 2006). Acceptance of the parasitic egg by the host can benefit the host regardless of whether it also benefits the parasite. In this scenario, taking into account that virulence is costly (Soler et al. 1998a), it could be expected that the parasite will adjust its facultative virulence to the degree of compliance shown by the host (Soler et al. 1998a; Ponton et al. 2006).

15.5 Differences and Similarities Between Farming and Mafia Strategies of Brood Parasites

Both farming and mafia predatory strategies may be profitable for brood parasites because both create future opportunities for parasitism by forcing the host to renest. These behaviours are similar because they imply predation of host nests. Farming behaviour, however, implies the predation of host nests that cannot be successfully parasitized, while mafia behaviour implies the predation of nests in which their offspring were ejected (Fig. 15.2). Therefore, farming would not induce hosts to accept parasitism, while the main objective of the mafia strategy is to punish rejecter hosts. However, the mafia behaviour implies subsequent visits to host nests, and thus, should be considered costlier than farming behaviour. In addition, mafia also has a cognitive cost: must recognize/remember their own eggs/chicks and location of the nest site.

The advantage of the farming strategy is to increase future reproductive possibilities for the parasite, and thus, there are few potential systems other than brood parasite–hosts in which it could work. The advantage of a mafia-like strategy is to make the acceptance of parasitism a response of adaptive value. This kind of advantage would be of interest not only for brood parasites but for any parasites exploiting resistant hosts. Apart from some cognitive abilities, the only necessary character in parasites for the evolution of mafia behaviour is facultative virulence, which should be expressed in relation to the level of resistance shown by hosts (Soler et al. 1998a). Thus, the mafia strategy could potentially evolve in a large range of parasite–host systems and may be a widespread mechanism in biological processes at a number of different levels (Soler et al. 1998a; Thomas et al. 2005; Wellnitz 2005; Ponton et al. 2006), but there is little evidence for it.

15.6 Discussion: Why Mafia Strategy Is Less Frequent than Expected?

Accepting parasitism while minimizing its negative effects may be more adaptive than increasing resistance for a host species. Contrary to resistance, tolerance does not select for counter-defences or virulence in their parasites (Miller et al. 2006; Boots 2008). In this case, to be tolerant of the manipulative parasite by accepting its egg would have the advantage of not suffering retaliation by cuckoos (Soler et al. 1995, 1998a; Ponton et al. 2006). On the contrary, if the host opposes resistance that

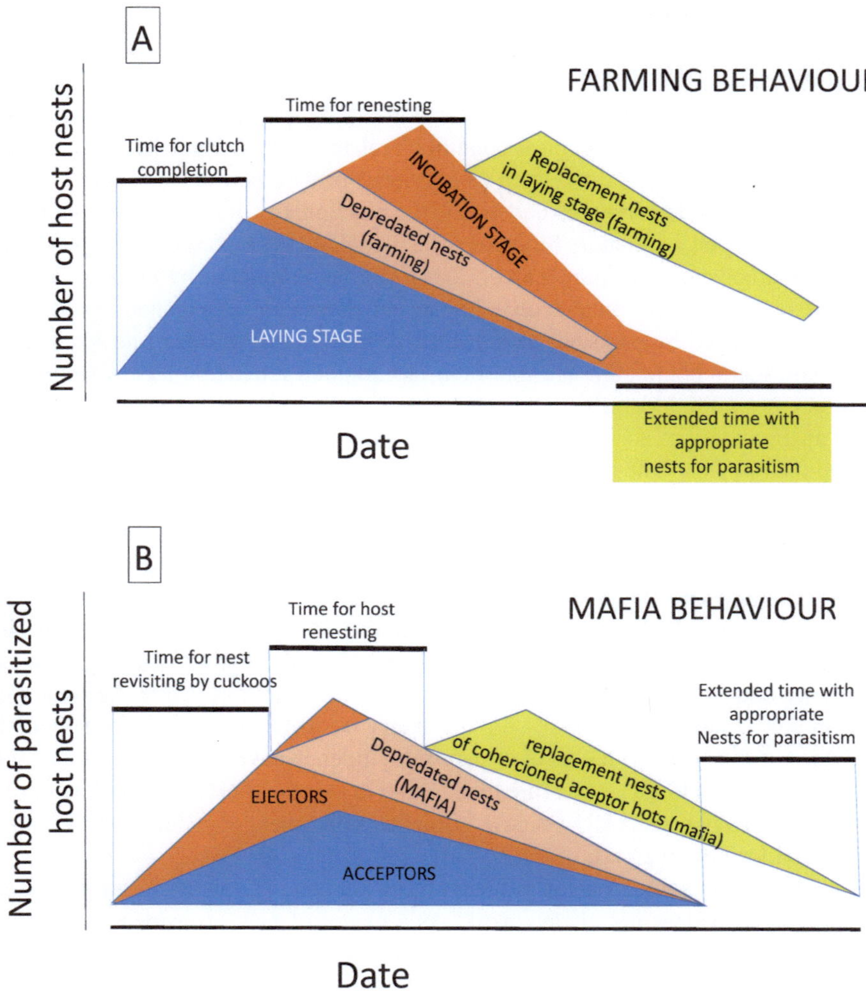

Fig. 15.2 Diagrams showing the number of first and replacement clutches available for the brood parasite along the laying and incubation periods in both farming (**a**) and mafia (**b**) strategies. (**a**) *Farming strategy*. Host nests are classified as being in the laying (blue) or in the incubation stage (brown). Depredated nests by farming brood parasites only occur in those in incubation stage (i.e. not appropriate for parasitism). The same number of depredated nests will be available for parasitism after the time of renesting. Number of nests available for parasitism at a particular date would then be the length of a perpendicular line to the X-axis that crosses blue and/or yellow area. Thus, the extra number of nests and time for laying that brood parasites obtain due to the farming behaviour are represented by the yellow polygon. (**b**) *Mafia strategy*. Diagram only considered parasitized nests during the egg-laying stage and differentiates between nests that accept (blue) or reject (brown) parasitic offspring. Nests that were later depredated by retaliatory cuckoos only occur in a proportion of nests that ejected parasitic offspring. The same number of depredated nests will be available for parasitism (acceptors during the laying stage) after the time of cuckoos revisiting previously parasitized host nests and of host renesting. These ejected hosts should accept parasitism in replacement clutches as a consequence of retaliatory behaviour. Number of acceptor nests and extended dates that brood parasites obtain due to the mafia behaviour are showed, respectively, by the height and length of the yellow polygon at each breeding date. Both figures assume that all depredated nests renested with identical elapsed time (thus area and shape of polygons are identical)

minimizes the fitness of the parasite, the facultatively virulent parasite could increase its negative effects on the host, making any defensive response less profitable for the host than cooperation (Soler et al. 1995, 1998a; Ponton et al. 2006). This means that hosts should also be favoured by selection if, by plastically modifying their behaviour according to ecological conditions related to risk, intensity and/or virulence of parasitism, they can reduce the virulence of their parasites.

Theoretical models trying to understand the functionality of mafia behaviour have concluded (1) that a predatory brood parasite would always be favoured by selection over a nonpredatory parasite (Hauber 2014), (2) that retaliation by brood parasites against rejecter hosts could evolve just as an extension of a general predatory strategy displayed by parasites (Pagel et al. 1998), (3) that farming and retaliation may represent a continuum and that both farming and retaliation tactics would be seen in cases of high parasitism rates (Hauber 2014) and (4) that the transition from pure manipulation of host behaviour to mafia strategies (i.e. interactive strategies of manipulation) should be favoured by selection (Ponton et al. 2006). All these points imply important fitness advantages of mafia strategies and, thus, should be frequent in nature. However, as we have seen above, the mafia hypothesis has received empirical or experimental support in very few brood parasite–host systems and studies, being more frequent those in which predictions were not supported (see references above). The high rates of studies failing to support the mafia hypothesis are likely the consequence of three possibilities: First, facultative virulent parasites (the only ones that can use mafia-like strategies) are scarce in nature. Second, the demanding conditions necessary for the mafia tactic to evolve (i.e. behavioural defences of the host and the level of parasitic virulence have both to be plastic and negatively related to each other) are hard to find in parasite–host systems. And third, the mafia behaviour is costly because it implies subsequent visits to host nests and considerable cognitive costs.

In addition, in some systems in which those demanding conditions are fulfilled, mafia behaviour does not always work. As a matter of fact, no evidence of mafia behaviour has been detected in recent studies of the great spotted cuckoo–magpie system. The great spotted cuckoo is one of two brood parasite species in which the mafia hypothesis was experimentally demonstrated by removing parasitic eggs from magpie nests (Soler et al. 1995). Recent experimental work has implied that removing parasitic nestlings from magpie nests (for scientific purposes other than testing the mafia hypothesis) did not increase predation rates of such nests (Roldán et al. 2013; Soler and de Neve 2013; Canestrari et al. 2014; Soler et al. 2014b, c, 2017). Similarly, recently performed experiments also implied that the removal of parasitic eggs from magpie nests did not either increase predation rates. During 2013–2014, cuckoo eggs were removed from 24 magpie nests; none of them suffered total predation and only 10% partial predation, which does not differ from predation rates occurring in non-parasitized nests or those parasitized with cuckoo eggs (5 and 10%), respectively (T. Pérez-Contreras and M. Soler, unpublished data).

Why is retaliation much scarcer in the great spotted cuckoo–magpie system about 25 years later? Several parameters related to both spatial and temporal variation affect both egg rejection by magpies and probability of retaliation by great spotted cuckoos (Soler et al. 1999a). Some of these factors changed over the last 20 years. First, these studies have not been carried out in exactly the same study plots as the original one. Our Guadix magpie population is composed of several nearby plots, which are separated by uncultivated areas with few or no potential nest sites for the magpie (Soler et al. 1994, 1998b; Martín-Galvez et al. 2007; Soler et al. 2013). These plots have been demonstrated to differ significantly in ecological conditions including food availability, magpie density and brood parasitism rates (Soler et al. 1994, 1998b, 2013; Martín-Galvez et al. 2007). Dramatic variation in parameters such as abundance of cuckoos, parasitism rate and even rejection rate has also been detected among study plots (Soler et al. 1998c). The recent studies in which we removed cuckoo nestlings and cuckoo eggs were done in different plots than the original experiment, in which the mafia hypothesis was tested, and that may partially explain the detected differences in frequency of cuckoo retaliation.

Second, the parasitism rate is enormously variable, both spatially and temporarily. When our mafia experiment was carried out (1990–1992), 176 out of 277 magpie nests were parasitized (parasitism rate = 63.5%), but later we reported both lower (45%, $n = 342$ nests, breeding seasons 2006–2009; Soler and de Neve 2012) and higher (83.2%, $n = 131$, breeding season 2011; Soler et al. 2015) parasitism rates. More remarkably, parasitism rates greatly differ between plots very close to each other during the same year (Soler et al. 2013) and between consecutive years in the same plots (Martín-Galvez et al. 2007; Soler et al. 2011, 2015).

Third, we also observed some apparent changes in great spotted cuckoo behaviour, particularly in those related to the defence of parasitized magpie nests. This behaviour is particularly interesting here because it is likely related to the probability of nest revisit by cuckoos to check whether parasitic offspring disappeared from host nests. During the fieldwork for our experimental test of the mafia hypothesis (1990–1992), great spotted cuckoos scolded us on 25% of our visits to parasitized magpie nests, whereas we were scolded by magpies in only 5% of our visits to these same nests (Soler et al. 1995). In 1992 we started systematically to collect information on nest defence by magpies and great spotted cuckoos on both parasitized and unparasitized nests, and we were witness to a very rapid decrease in this behaviour: 6.1 and 12.5% in 1992, 3.2 and 4.8% in 1993 and 0 and 0.9% in 1994, for laying and incubation periods, respectively (Soler et al. 1999b). This change in defence behaviour was not accompanied by a change in parasitism rates (66.4%, 63.5% and 52.0% for 1992, 1993 and 1994, respectively). Nest defence behaviour by great spotted cuckoos after 1994 has remained very scarcely detected.

Fourth, with respect to magpie ejection of real cuckoo eggs and retaliation by great spotted cuckoos between 1990 and 1992, seven (5.2%) cuckoo eggs were ejected, and most (86%) of these magpie nests were depredated, while only 12% of nests with accepted cuckoo eggs and 22.5% of unparasitized magpie nests were depredated (Soler et al. 1995). However, between 2011 and 2014, only four (2.06%; $n = 194$) cuckoo eggs were ejected by magpies, and 50% of them were depredated, but 50% ($n = 190$) of the nests in which the cuckoo egg was not ejected and 28.7% ($n = 150$) of unparasitized nests were also depredated (T. Pérez-Contreras and M. Soler, unpublished data). These natural history observations suggest that magpies reject real great spotted cuckoo eggs less frequently now than about 25 years ago and that cuckoos "punished" hosts that removed cuckoo eggs more often about 25 years ago than today.

Fifth, another important difference is that in 1990–1992 acceptors raised as many of their own offspring as did abandoners or ejectors (0.43, 0.0, 0.29, respectively; Table 2 in Soler et al. 1995). During 2011–2014, the number of fledglings produced per season was 0.32 ($n = 182$) for acceptors, 0.0 ($n = 11$) for abandoners and 2.91 ($n = 11$) for ejectors. Thus, currently, naturally parasitized magpie nests in which the parasitic egg was ejected produced tenfold more magpie fledglings than 25 years ago (because of mafia behaviour being currently less common), but in those nests in which the great spotted cuckoo egg was accepted produced fewer of their own nestlings than about 25 years ago.

What can we conclude from all these data about great spotted cuckoos and magpies? Understanding the mechanisms underlying the mafia strategy remains elusive. However, considering, first, that it was unequivocally demonstrated in a concrete population and time (Soler et al. 1995) and, second, that changes in magpie defences are too rapid to be considered the result of selection processes, we suggest that phenotypic plasticity by both great spotted cuckoos and magpies is the crucial point here. Great spotted cuckoos are not territorial, and they move among breeding areas during the same breeding season (Martínez et al. 1998), exploring different magpie populations. Thus, they may be able to decide where to lay their eggs according to environmental variables, intensity of host defences (Martín-Galvez et al. 2007) or propensity to compliance. In our opinion, this is the most likely explanation to the enormous variation in parasitism rate reported in different magpie plots and in different years (see above). At the same time, this highly variable risk of parasitism, and probably the fact that magpies are interacting with a cuckoo with facultative virulence, makes phenotypic plasticity adaptive in magpies (Soler 2014). We have strong evidence supporting the assertion that magpies plastically reject parasitic eggs from their nests: (1) they are able to modulate their propensity to eject parasitic eggs as a function of previous experience (Soler et al. 1999a) and (2) cuckoo density (Soler et al. 1999a). Furthermore, in relation to plasticity, it is worth mentioning that the magpie is a long-lived species with high cognitive abilities (Birkhead 1991) that might learn how to behave in order to limit parasitic costs to a minimum.

Concluding Remarks and Future Directions

The scarce evidence of the existence of a mafia-like strategy in manipulative parasites in general and in brood parasites in particular could be the consequence of a lack of appropriate studies or because it is very difficult for such a strategy to evolve and be maintained, at least in avian brood parasites. Regarding brood parasites, future studies should concentrate on examining how often and in which context brood parasites prey upon host nests in different host

(continued)

populations and in different brood parasite–host systems. The future studies should also focus on detecting evidences supporting that both great spotted cuckoos and brown-headed cowbirds recognize their own eggs and chicks (before these were removed by hosts). Information on geographical and temporal variability in this predatory behaviour, as well as on the factors affecting such variation, is of prime importance. Regarding farming, it should also be documented whether renesting attempts are parasitized more frequently than from random expectation or not. Again, exploring possible geographical and temporal variation is essential. Finally, regarding mafia strategies, it is needed to determine the frequency with which nest predation is related to previous ejection of the parasitic egg by the hosts, mainly in evictor species. This last objective would preferentially be addressed experimentally by removing parasitic eggs from parasitized nests and following the fate of those nests in comparison to appropriate control nests.

Acknowledgements We thank Mark Hauber and Anders P. Møller for their constructive and helpful comments, which have considerably improved our manuscript. While writing this paper, the authors benefit from founding by the Spanish Ministerio de Ciencia e Innovación, Ministerio de Economía y Competitividad, European funds (FEDER) (CGL2013-48193-C3-1-P to JJS and the Junta de Andalucía (project RNM610) to MS).

References

Abou Chakra M, Hilbe C, Traulsen A (2014) Plastic behaviors in hosts promote the emergence of retaliatory parasites. Sci Rep 4:4251

Abou Chakra M, Hilbe C, Traulsen A (2016) Coevolutionary interactions between farmers and mafia induce host acceptance of avian brood parasites. R Soc Open Sci 3:160036. https://doi.org/10.1098/rsos.160036

Arcese P, Smith JNM (1999) Impacts of nest predation and brood parasitism on the productivity of North American passerines. In: Adams NJ, Slotow RH (eds) Proceedings of the 22nd international ornithological congress. Durban, 1998. BirdLife South Africa, Johannesburg, pp 2953–2966

Arcese P, Smith JMN, Hochachka WM, Rogers CM, Ludwig D (1992) Stability, regulation, and the determination of abundance in an insular song sparrow population. Ecology 73:805–822

Arcese P, Smith JNM, Hatch MI (1996) Nest predation by cowbirds, and its consequences for passerine demography. Proc Natl Acad Sci USA 93:4608–4611

Birkhead T (1991) The magpies. The ecology and behaviour of black-billed and yellow-billed magpies. Poyser, London

Biron DG, Ponton F, Joly C, Menigoz A, Hanelt B, Thomasa F (2005) Water seeking behavior in insects harboring hairworms: should the host collaborate? Behav Ecol 16:656–660

Boots M (2008) Fight or learn to live with the consequences? Trends Ecol Evol 23:248–250

Canestrari D, Bolopo D, Turlings TCJ, Röoder G, Marcos JM, Baglione V (2014) From parasitism to mutualism: unexpected interactions between a cuckoo and its host. Science 343:1350–1352

Clotfelter ED, Yasukawa K (1999) Impact of brood parasitism by brown-headed cowbirds on red-winged blackbird reproductive success. Condor 101:105–114

Davies NB (2000) Cuckoos, cowbirds and other cheats. T. & A.D. Poyser, London

Dawkins R, Krebs JR (1979) Arms races between and within species. Proc R Soc Lond B 205:489–511

De Mársico MC, Reboreda JC (2014) High frequency but low impact of brood parasitism by the specialist Screaming Cowbird on its primary host, the Baywing. Emu 114:309–316

Dearborn DC (1999) Brown-headed cowbird nestling vocalizations and risk of nest predation. Auk 116:448–457

Donovan TM, Jones PW, Annand EM, Thompson IIIFR (1997) Variation in local-scale edge effects: mechanisms and landscape context. Ecology 78:2064–2075

Fiorini VD, Tuero DT, Reboreda JC (2009) Shiny Cowbirds synchronize parasitism with host laying and puncture host eggs according to host characteristics. Anim Behav 77:561–568

Fiorini VD, Gloag R, Reboreda JC, Kacelnik A (2014) Strategic egg destruction by brood parasitic cowbirds? Anim Behav 93:229–235

Friedmann H (1948) The parasitic cuckoos of Africa. Washington Academy of Sciences, Washington

Friedmann H (1968) The evolutionary history of the Avian genus *Chrysococcyx*. Smithsonian Institution Press, Washington

Granfors DA, Pietz PJ, Joyal LA (2001) Frequency of egg and nestling destruction by female brown-headed cowbirds at grassland nests. Auk 118:765–769

Guilford T, Read AF (1990) Zahavian cuckoos and the evolution of nestling discrimination by hosts. Anim Behav 39:600–601

Hannon SJ, Wilson S, McCallum CA (2009) Does cowbird parasitism increase predation risk to American redstart nests? Oikos 118:1035–1043

Hauber ME (2000) Nest predation and cowbird parasitism in Song Sparrows. J Field Ornithol 71:389–398

Hauber ME (2009) Does the removal of avian brood parasite eggs increase host productivity? A case study with brown-headed cowbirds *Molothrus ater* and song sparrows *Melospiza melodia* near Ithaca, New York, USA. Conserv Evid 6:83–88

Hauber ME (2014) Mafia or farmer? Coevolutionary consequences of retaliation and farming as predatory strategies upon host nests by avian brood parasites. Coevolution 2:1–8

Hauber ME, Yeh PJ, Roberts JOL (2004) Patterns and coevolutionary consequences of repeated brood parasitism. Proc R Soc Lond B 271:S317–S320

Hoover JP, Robinson SK (2007) Retaliatory mafia behavior by a parasitic cowbird favors host acceptance of parasitic eggs. Proc Natl Acad Sci USA 104(180):4479–4483

Ibáñez-Álamo JD, Arco L, Soler M (2011) Experimental evidence for a predation cost of begging using active nests and real chicks. J Ornithol 153:801–807

Johnson RG, Temple SA (1990) Nest predation and brood parasitism of tallgrass prairie birds. J Wild Manag 54:106–111

Kosciuch KL, Sandercock BK (2008) Cowbird removals unexpectedly increase productivity of a brood parasite and the songbird host. Ecol Appl 18:537–548

Lefèvre T, Roche B, Poulin R, Hurd H, Renaud F, Thomas F (2008) Exploiting host compensatory responses: the ⍰must⍰ of manipulation? Trends Parasitol 24:435–439

Lefèvre T, Adamo SA, Biron DG, Misse D, Hughes D, Thomas F (2009) Invasion of the body snatchers: the diversity and evolution of manipulative strategies in host-parasite interactions. Adv Parasitol 68:45

Martínez JG, Burke T, Dawson D, Soler JJ, Soler M, Møller AP (1998) Microsatellite typing reveals mating patterns in the brood parasitic great spotted cuckoo (*Clamator glandarius*). Mol Ecol 7:289–297

Martín-Galvez D, Soler JJ, Martínez JG, Krupa AP, Soler M, Burke T (2007) Cuckoo parasitism and productivity in different magpie subpopulations predict frequencies of the 457bp allele: a mosaic of coevolution at a small geographic scale. Evolution 61:2340–2348

Massoni V, Reboreda JC (1998) Costs of brood parasitism and the lack of defenses on the yellow-winged blackbird-shine cowbird system. Behav Ecol Sociobiol 42:273–280

McLaren CM, Sealy SG (2000) Are nest predation and brood parasitism correlated in yellow warblers? A test of the cowbird predation hypothesis. Auk 117:1056–1060

Medina I, Langmore NE (2016) The evolution of acceptance and tolerance in hosts of avian brood parasites. Biol Rev 91:569–577

Medzhitov R, Schneider DS, Soares MP (2012) Disease tolerance as a defense strategy. Science 335:936–941

Miller MR, White A, Boots M (2006) The evolution of parasites in response to tolerance in their hosts: the good, the bad, and apparent commensalism. Evolution 60:945–956

Moore J (2002) *Parasites and the behavior of animals*. Oxford University Press, Oxford

Pagel M, Møller AP, Pomiankowski A (1998) Reduced parasitism by retaliatory cuckoos selects for hosts that rear cuckoo nestlings. Behav Ecol 9:566–572

Payne RB, Payne LL (1998) Brood parasitism by cowbirds: risks and effects on reproductive success and survival in indigo buntings. Behav Ecol 9:64–73

Peer BD (2006) Egg destruction and egg removal by avian brood parasites: adaptiveness and consequences. Auk 123:16–22

Peer BD, Sealy SG (1999) Parasitism and egg puncture behaviour by bronzed and brown-headed cowbirds in sympatry. In: Morrison ML, Hall LS, Robinson SK, Rothstein SI, Hahn DC, Rich TD (eds) Research and management of the brown-headed cowbird in western landscapes, studies in avian biology, vol 18. Cooper Ornithological Society, Camarillo, pp 235–240

Ponton F, Biron DG, Moore J, Møller AP, Thomas F (2006) Facultative virulence: a strategy to manipulate host behaviour? Behav Processes 72:1–5

Poulin R (2010) Parasite manipulation of host behavior: an update and frequently asked questions. Adv Study Behav 41:151–186

Råberg L (2014) How to live with the enemy: understanding tolerance to parasites. PLoS Biol 12 (11):e1001989

Robert M, Sorci G, Møller AP, Hochberg M, Pomiankowski A, Pagel M (1999) Retaliatory cuckoos and the evolution of host resistance to brood parasites. Anim Behav 58:817–824

Rogers CM, Taitt MJ, Smith JNM, Jongejan G (1997) Nest predation and cowbird parasitism create a demographic sink in wetland-breeding Song Sparrows. Condor 99:622–633

Roldán M, Soler M, Márquez R, Soler JJ (2013) The vocal begging display of Great Spotted Cuckoo *Clamator glandarius* nestlings in nests of its two main host species: genetic differences or developmental plasticity? Ibis 155:867–876

Rolff J, Siva-Jothy MT (2003) Invertebrate ecological immunology. Science 301:472–475

Schmid-Hempel P (2011) *Evolutionary parasitology: the integrated study of infections, immunology, ecology, and genetics*. Oxford University Press, Oxford

Soler M (2014) Long-term coevolution between avian brood parasites and their hosts. Biol Rev 89:688–704

Soler M, de Neve L (2012) Great spotted cuckoo nestlings but not magpie nestlings starve in experimental age-matched broods. Ethology 118:1036–1044

Soler M, de Neve L (2013) Brood mate eviction or brood mate acceptance by brood parasitic nestlings? An experimental study with the non-evictor great spotted cuckoo and its magpie host. Behav Ecol Sociobiol 67:601–607

Soler M, Soler JJ, Martínez JG, Møller AP (1994) Micro-evolutionary change in host response to a brood parasite. Behav Ecol Sociobiol 35:295–301

Soler M, Soler JJ, Martínez JG, Møller AP (1995) Magpie host manipulation by great spotted cuckoos: evidence for an avian mafia? Evolution 49:770–775

Soler M, Soler JJ, Martínez JG (1997) Great spotted cuckoos improve their reproductive success by damaging magpie host eggs. Anim Behav 54:1227–1233

Soler JJ, Møller AP, Soler M (1998a) Mafia behaviour and the evolution of facultative virulence. J Theor Biol 191:267–277

Soler M, Soler JJ, Martínez JG (1998b) Duration of sympatry and coevolution between the great spotted cuckoo (*Clamator glandarius*) and its primary host, the magpie (*Pica pica*). In: Rothstein SI, Robinson SK (eds) Parasitic birds and their hosts: studies in coevolution. Oxford University Press, New York, pp 113–128

Soler M, Soler JJ, Martínez JG, Pérez-Contreras T, Møller AP (1998c) Micro-evolutionary change and population dynamics of a brood parasite and its primary host: the intermittent arms race hypothesis. Oecologia 117:381–390

Soler JJ, Sorci G, Soler M, Møller AP (1999a) Change in host rejection behavior mediated by the predatory behavior of its brood parasite. Behav Ecol 10:275–280
Soler M, Martínez JG, Soler JJ (1999b) Nest defence by Magpies *Pica pica* and the brood parasitic great spotted cuckoos *Clamator glandarius* in parasitized and unparasitized nests. J Ornithol 140:199–205
Soler JJ, Soler M (2017) Evolutionary change: facultative virulence by brood parasites and tolerance and plastic resistance by hosts. Anim Behav 125:101–107
Soler JJ, Martín-Gálvez D, Martínez JG, Soler M, Canestrari D, Abad-Gómez M, Møller AP (2011) Evolution of tolerance by magpies to brood parasitism by great spotted cuckoos. Proc R Soc B 278:2047–2052
Soler JJ, Martín-Gálvez D, de Neve L, Soler M (2013) Brood parasitism correlates with the strength of spatial autocorrelation of life history and defensive traits in magpies. Ecology 94:1338–1346
Soler JJ, Avilés JM, Martín-Gálvez D, de Neve L, Soler M (2014a) Eavesdropping cuckoos: further insights on great spotted cuckoo preference by magpie nests and egg colour. Oecologia 175:105–115
Soler M, de Neve L, Roncalli G, Macías-Sánchez E, Ibáñez-Álamo JD, Pérez-Contreras T (2014b) Great spotted cuckoo fledglings are disadvantaged by magpie host parents when reared together with magpie nestlings. Behav Ecol Sociobiol 68:333–342
Soler M, Pérez-Contreras T, Ibáñez-Álamo JD, Roncalli G, Macías-Sánchez E, de Neve L (2014c) Great spotted cuckoo fledglings often receive feedings from other magpie adults than their foster parents: which magpies accept to feed foreign cuckoo fledglings? PLoS One 9(10):e107412
Soler M, Pérez-Contreras T, Soler JJ (2015) Synchronization of laying by great spotted cuckoos and recognition ability of magpies. J Avian Biol 46:608–615
Soler M, de Neve L, Roldán M, Pérez-Contreras T, Soler JJ (2017) Great spotted cuckoo nestlings have no antipredatory effect on magpie or carrion crow host nests in southern Spain. PLoS One 12(4):e0173080. https://doi.org/10.1371/journal.pone.0173080
Stake MM, Cimprich DA (2003) Using video to monitor predation at black-capped vireo nests. Condor 105:348–357
Svensson EI, Råberg L (2010) Resistance and tolerance in animal enemy-victim coevolution. Trends Ecol Evol 25:267–274
Swan DC, Zanette LC, Clinchy M (2015) Brood parasites manipulate their hosts: experimental evidence for the farming hypothesis. Anim Behav 105:29–35
Thomas F, Adamo S, Moore J (2005) Parasitic manipulation: where are we and where should we go? Behav Process 68:185–199
Thompson JN (2005) *The geographic mosaic of coevolution*. University of Chicago Press, Chicago
Tuero DT, Fiorini VD, Reboreda JC (2012) Do shiny cowbird females adjust egg pecking behaviour according to the level of competition their chicks face in host nests? Behav Processes 89:137–142
Wellnitz T (2005) Parasite-host conflicts: winners and losers or negotiated settlements? Behav Processes 68:245–246
Wolf L (1987) Host-parasite interactions of Brown-headed Cowbirds and Dark-eyed Juncos in Virginia. Wilson Bull 99:338–350
Wyllie I (1981) The Cuckoo. Universe, New York
Zahavi A (1979) Parasitism and nest predation in parasitic cuckoos. Am Nat 113:157–159

Cuckoo–Host Coevolutionary Interactions Across All Breeding Stages: Unusual Ecological Setting of a Cavity-Nesting Host

16

Tomáš Grim and Jarkko Rutila

Abstract

The great majority of brood parasitism studies focused on a single ontogenetic stage, typically egg stage, and on open-nesting hosts, especially those of the common cuckoo. Using extensive data from the cuckoo's only known regular cavity-nesting host, the common redstart, we highlight the importance of a comprehensive approach when all ontogenetic stages are studied. In contrast to open-nesting hosts, only minority of the cuckoo eggs are a threat to redstart hosts: most are laid outside the host nest cup and perish. Contrary to previous claims, we found that the impact of parasitism per host nest was virtually the same between this only regular cuckoo cavity-nesting host and a typical open-nesting host (the reed warbler): in both species, fitness of an average non-parasitized host nest was by an order of magnitude higher than fitness of an average parasitized host nest. This was partly because of uniquely low eviction success of cuckoo chicks and resulting cohabitation of parasite and host progeny in mixed broods. Data from post-fledging period, which remains the least known stage of parasite–host coevolution in any study system globally, were crucial because they showed that data from nestling period greatly overestimated cuckoo fitness. We suggest that metareplication of these approaches (i.e. integrative study of laying, incubation, nestling, fledgling and migration stages) across various parasite–host systems is the most important task for future coevolutionary studies in the context of brood parasite–host coevolution.

T. Grim (✉)
Department of Zoology and Laboratory of Ornithology, Palacký University, Olomouc, Czech Republic
e-mail: tg.cuckoo@gmail.com

J. Rutila
Näätäkarankatu 14 B 9, Lappeenranta, Finland

M. Soler (ed.), *Avian Brood Parasitism*, Fascinating Life Sciences,
https://doi.org/10.1007/978-3-319-73138-4_16

16.1 Introduction

Just like in any field of science, studies of brood parasite–host coevolution focus on a few traditional model systems. Therefore they necessarily neglect other, consequently underused, models. Naturally, researchers also pay most attention to life stages that are methodologically easier to deal with. Therefore they inevitably create knowledge accumulation bias against other life stages.

Our aim in this contribution is twofold. First, we summarize literature data and our detailed field studies of parasite–host interactions in common redstarts (*Phoenicurus phoenicurus*, hereafter redstarts), the single documented regular cavity-nesting hosts of common cuckoos (*Cuculus canorus*, hereafter cuckoos) in Europe. We highlight the importance of extending research to less known brood parasite–host systems and specifically to those differing ecologically from typical (open nesting) hosts. Second, by addressing parasite–host interactions across all life stages, we illustrate how integrating data across life stages can improve our understanding of parasite–host arms races (Dawkins and Krebs 1979).

Redstart males are among the most colourful and attractive European birds, while females are drab. Such clear sexual dimorphism is unusual among cuckoo hosts and provides a clear logistic advantage over typical cuckoo hosts which are mostly monomorphic, given that it allows studying any parasite–host interactions at the level of individual sexes even without marking the birds (Grim et al. 2009a).

The redstart is the main cuckoo host in Finland (Haikola and Rutila 2008; Mikulica et al. 2017), and its congeners serve as cuckoo hosts in Asia (Yang et al. 2013, 2016). The redstart is a suitable cuckoo host since it feeds its chicks predominantly with invertebrate diet and it is fairly common. However, unlike other frequently used cuckoo hosts, the redstart is a cavity nester. Based on nest card records from the Helsinki Museum of Natural History, 50% of all the cuckoo eggs found in Finland (1931–2000) were in the redstart nests ($n = 127$; Rutila 2004).

Scientists classified cavity nesters as unsuitable cuckoo hosts (e.g. Davies and Brooke 1989), since many cavities are too small for the cuckoo female to lay its eggs effectively or the young cuckoo to fledge successfully. Löhrl (1979) showed that cavity entrances with diameters <50 mm are too small for young cuckoos to fledge. Redstarts however use a variety of different types of natural cavities and crevices, including those with large entrances (von Haartman 1969). This makes redstarts more vulnerable to cuckoo parasitism compared to strict cavity nesters (von Haartman 1981; Grim et al. 2014; Grim 2016; Liang et al. 2016).

Our study site is located in Ruokolahti, South Karelia in Southeast Finland (Samaš et al. 2016; hereafter, data provided without a bibliographic reference are unpublished results). All the results we present here are from several partly isolated study plots spread over the area of 25 × 7 km (Fig. 1, Samaš et al. 2016). This spatial metareplication decreases risks of pseudoreplicated sampling from the same females which is a common problem in typical studies which are based on spatial sampling 10–20 times smaller (details in Samaš et al. 2016). The population has been monitored over three decades (since 1983, J. Haikola, pers. comm.). All redstarts we studied bred in nest boxes with large entrances (60–80 mm, mostly 70 mm), which were created to reflect the situation in natural cavities (for details see Grim and Samaš 2016; Samaš et al. 2016).

Although country-wise parasitism rate is 3% (Rutila 2004), the parasitism rate in our study area is 33% (yearly variation: 17–50%), similarly to two other intensely studied populations in Joensuu (21%, yearly variation: 0–58%; Rutila et al. 2002) and Oulu (31%, yearly variation 13–47%; Thomson et al. 2016). This variation among sites is remarkably low compared to other hosts (for examples see Samaš et al. 2016). This may reflect very homogeneous breeding habitat of redstarts, i.e. rather open cultivated coniferous forests.

We have collected detailed data throughout the redstart and cuckoo breeding cycle from egg laying (Samaš et al. 2016), through incubation (Rutila et al. 2002), nestling (Grim and Samaš 2016) and post-fledgling periods (this chapter) to migration to wintering grounds (Vega et al. 2016). Further, we addressed questions at several levels of biological complexity, from chemical (Igic et al. 2012) and structural levels (including methods of material sciences; Igic et al. 2011), through molecular (Fossøy et al. 2016), to behavioural levels (Grim et al. 2009a, b). We also quantified fitness for parasites and hosts in respect to the constraints related to cavity nesting (this chapter). This has been done rarely with similar extent for any brood parasite–host system. We believe that such complex and comprehensive approach is beneficial because some conclusions (see below) could not be reached without integrating results from several life stages (as advocated by Grim 2007a and Grim et al. 2011) or levels of biological organization (Igic et al. 2012).

16.2 Laying Stage: Nest Design Constraints

Strikingly, from the 213 cuckoo eggs, that we have found over 11 intensely studied breeding seasons (2006–2016), only 36% were found inside the host nest cup (Samaš et al. 2016). The rest was found inside the nest box but at the nest rim (54%) or even on the ground under the nest box (5%) (Samaš et al. 2016). Some eggs were even found in incomplete nests (5%) (Samaš et al. 2016). Were these eggs ejected by hosts or are they the result of imperfect laying by female cuckoos?

We tested this hypothesis directly. For the first time, we video recorded a cavity-nesting hosts during the egg-laying period. Previously, the egg-laying behaviour of cuckoos has been reliably documented only rarely and only in open-nesting hosts (Wyllie 1981; Moksnes et al. 2000; Andou et al. 2005). We found that all eggs found outside the nest cup on the nest rim ($n = 12$) were mislaid by cuckoo females and did not result from redstart egg ejection. Redstarts ignored all cuckoo eggs, both those laid outside the nest cup (above) and those laid inside it ($n = 14$) (Samaš et al. 2016). The video recordings lasted for similar periods (mean = 5 days) after the cuckoo laid her egg as is the standard period used in experimental studies to score cuckoo host individuals as acceptors or rejecters (6 days; Moksnes et al. 1991; Grim et al. 2011).

16.3 Laying Stage: Host Front-Line Defences

Aggression against adult cuckoo represents the front line of host defences against brood parasitism. However, *Lanius* shrikes and the great reed warblers (*Acrocephalus arundinaceus*) are the only regular cuckoo hosts that were documented to harm adult parasites (Trnka and Grim 2013). Similarly to some other hosts (Moksnes et al. 2000), redstarts also showed low and ineffective nest defence, i.e. they were unable to chase laying cuckoo females away. Only at 15% nests ($n = 26$) redstart pairs showed any response against a cuckoo dummy (the dummy was placed at the top of the nest box during the egg-laying stage for 60 min). Responses included alarm calls and dives above the dummy. As a control, we exposed redstarts to a mistle thrush (*Turdus viscivorus*) dummy; none out of four tested pairs showed any response. Thus, redstarts seemed to recognize the cuckoo as an enemy but their responses were weak.

By video recording the cuckoo laying attempts, we directly estimated redstart aggression levels under natural conditions. Redstarts showed some responses towards female cuckoos, mainly giving alarm calls (at 17% of the video-recorded nests, $n = 46$). This proportion is virtually the same as that observed during dummy experiments, which suggests that dummy experiments well reflect the biological reality in this host species.

Overall, across host species, egg rejection rates and dummy aggression rates correlate positively (Moksnes et al. 1991). Low aggression combined with low egg rejection rates of natural parasite eggs (11% across 5 populations, all rejections by desertion; Samaš et al. 2016) in redstarts fits well these interspecific patterns.

16.4 Incubation Stage: Perfect Mimicry

Redstart-cuckoos strikingly differ from majority of other gentes of the common cuckoo in having immaculate eggs (but see Fuchs 1957; Yang et al. 2010). The redstart-cuckoo eggs are plain blue (Fig. 16.1a), only rarely (6%, $n = 54$ eggs) contain sparse and tiny rusty spots (Fig. 16.1b; see also Čapek 1896). Host redstart clutches (18%, $n = 110$) also contain rusty spots at variable proportion of eggs within a clutch (subsample of clutches with at least some spotted eggs: 17–100%, mean = 73%).

Blue cuckoo eggs represent a perfect match of host redstart eggs (Fig. 16.1a). This is irrespective of the assessment method, be it a naked human eye (Moksnes et al. 1995), spectrometry (Igic et al. 2012) or visual modelling (Avilés 2008). Surprisingly, Stoddard and Stevens (2011) reported no overlap in background egg colour of blue cuckoo and blue redstart eggs. What was the cause of this outlying result is unclear. Based on our personal field experience with natural freshly laid eggs (which were also used by Igic et al. 2012), we note that while the inset photo of

Fig. 16.1 (**a**) Immaculate blue cuckoo eggs represent one of the highest levels of mimicry achieved by any cuckoo host race (top right-hand side egg is the cuckoo's). (**b**) Rare eggs with rusty spots on plain blue background (a redstart egg is pictured; cuckoo eggs may show similar spots). (**c**) Cuckoo chick evicted two redstart chicks and one egg but currently still shares the nest with four redstart chicks and another egg. Photo credits: T. Grim

the redstart egg in Fig. 1 in Stoddard and Stevens (2011) looks typical, the redstart-cuckoo egg is aberrant and not representative as for both background colour and the presence of faint spots. Therefore we suspect that Stoddard and Stevens' (2011) conclusion on redstart-cuckoo mimicry is a by-product of using old museum eggs.

Although host redstarts possess cognitive abilities to eject nonmimetic experimental eggs, they never ejected any naturally laid cuckoo eggs at any of several study sites across Fennoscandia ($n = 330$ eggs from 4 populations; Samaš et al. 2016). This is not because redstarts would be puncture-ejecters, and increased thickness of cuckoo eggshells would prevent them to break the parasite eggs (Igic et al. 2011)—redstarts are grasp-ejecters as inferred indirectly from ejection of hard artificial models (Hauber et al. 2014; Dinets et al. 2015) and directly from video recordings (Samaš et al. 2016). This implies that it is indeed the parasite–host egg similarity which is responsible for complete absence of ejection of natural cuckoo eggs (see also Hanley et al. 2017). These lines of evidence and quantitatively very similar results (spectrometry, visual modelling) of Avilés (2008) and Igic et al. (2012) lead us to conclude that redstart-cuckoo eggs indeed do represent an example of perfect mimicry.

The uniqueness of the plain blue egg phenotype of redstart-cuckoos has a genetic basis (Fossøy et al. 2016). Cuckoo females that produce blue eggs form a ~2.6 Myr old monophyletic lineage as for both their mitochondrial DNA and W-chromosome DNA. In contrast, they do not differ from other gentes as for nuclear DNA which confirms that males mate irrespective of their female partner's genetic origin. These data also suggest that genes controlling blue egg colour are at female W chromosome.

Cuckoo eggs show perfect colour mimicry (Igic et al. 2012) thus pre-empting egg ejection as a viable host defence strategy (Thomson et al. 2016). This explains why the only redstart response to cuckoo parasitism is nest desertion (also in other populations; Rutila et al. 2006; Samaš et al. 2016). Specifically, in our study population, nest desertion rates in naturally parasitized nests were statistically significantly higher (19%, $n = 43$) than desertion rates in both naturally non-parasitized nests (6%, $n = 89$) and nests experimentally parasitized by us (1%, $n = 73$) (Samaš et al. 2016). This suggests that desertion is a specific defence against natural parasitism (but not a specific response to artificial parasitism, see above) and that a sight of a laying cuckoo female might be an important additional cue that alerts hosts and triggers their egg discrimination behaviour (see also Moksnes et al. 2000). However, experimental data are needed to test these suggestions directly.

How perfect colour mimicry could have evolved if the ejection rate of even nonmimetic eggs is only intermediate? Mimicry is not a result of current egg rejection rates but of past host egg discrimination (Davies 2000). If a parasite evolved perfect mimicry, then hosts that attempt to reject parasite eggs might suffer egg rejection costs and errors (Samas et al. 2014; Stokke et al. 2016) which leads to lower fitness of rejecter alleles and may translate into a decline of host anti-parasite adaptations (Davies 2000). Thus, at the egg stage, redstarts might have reached stage 4 in a coevolutionary sequence depicted by Davies (2000, p. 119), i.e. cuckoo egg mimicry that evolved due to *past* redstart egg rejection is now so precise that they force *current* redstarts to accept today's cuckoo eggs (Avilés et al. 2005).

However, this does not mean that redstart-cuckoo coevolution reached a dead end. The presence of rusty spots on some redstart eggs (Fig. 16.1b) may represent a new line of defence whereby hosts diverge from ancestral egg phenotypes (both their own and mimetic parasite) to facilitate egg recognition and rejection (as generally hypothesized before; Davies 2000). Preliminary analysis ($n = 47$ parasitized nests) did not find covariation between nest desertion (yes/no) and the presence (yes/no) of rusty spots on cuckoo eggs (Fisher's exact test: $p = 1.00$) or redstart eggs (Fisher's exact test: $p = 0.39$); however, only three nests in this dataset were deserted leading to low power of the analysis.

16.5 Nestling Stage: Unsuccessful Evictions and Mixed Broods

Before hatching, cuckoo and redstart eggs were incubated for very similar time periods (mean ± SD: 13.2 ± 0.2 and 13.7 ± 0.2, respectively, $n = 51$ clutches) (Samaš et al. 2016). This contrasts with typical cuckoo hosts where the cuckoo egg is incubated for 1–2 days shorter period than a host clutch (Davies 2000). Also in absolute terms, the incubation period of redstart-cuckoo eggs is unusually long (cf. 11.6 days in reed warbler nests; Wyllie 1981). This has fundamental consequences: every second cuckoo egg hatched on the same day or even later than redstart eggs. This unusual pattern, perhaps augmented by low concentrations of energy reserve lipids in cuckoo eggs (Igic et al. 2015) and nest cup design (see further), then translates into uniquely low success in killing host offspring by the cuckoo (Fig. 16.1c).

The hallmark of the common cuckoo biology is the instinct shown by the young parasite to evict its nestmates (Wyllie 1981). Cuckoo chicks virtually always succeed in pushing all host's eggs and nestlings over the nest rim, managing to quickly "clean" the nest of host offspring in the nests of open-cup breeding hosts (Honza et al. 2007 and references therein). The redstart-cuckoos represent a striking exception to this rule: redstart-cuckoo chicks often fail to evict some host progeny (20% under natural conditions; Samaš et al. 2016; Thomson et al. 2016), and the process of eviction is prolonged (even a week after hatching) and arduous (decreasing cuckoo chick growth by cc 25%; Grim et al. 2009b; see also Anderson et al. 2009). Such a prolonged eviction instinct in redstart-cuckoos might even be an adaptation against the cavity-nesting habits of its host: nest cup design (steep inner walls of redstart nests) decreases the chances that the cuckoo will succeed in its eviction attempts (Fig. 5 in Grim et al. 2009b; see also Honza et al. 2007; Grim et al. 2011). Therefore, we hypothesize that cuckoo individuals with prolonged eviction instinct might reach higher fitness than those whose eviction instinct dissipates earlier. However, the timing of eviction instinct dissipation remains unknown in all hosts other than the redstarts (because the data from other hosts were always observational; Honza et al. 2007) and should be tested experimentally in the future (as in Grim et al. 2009b).

Although in all cases the cuckoo has an open space where to push host offspring (typically the area between the nest cup and the front wall of the box), in some cases the nest cups are built just next to the rear wall of the box with little space to hold evicted eggs and chicks which may, after having been evicted, even fall back into the nest cup. Cuckoo chicks are blind for the most of the eviction period (Wyllie 1981) and seem to evict in random directions (but this was not tested so far). Thus, we hypothesize that the location of the nest cup relative to the box walls may affect cuckoo eviction success, growth and survival.

Although cavity nesting may bring some eviction costs that do not materialize in open nests (see above), it may prevent other costs associated with open nest design. In open-cup nests, cuckoo chicks sometimes evict themselves and die; although this happens rarely, the phenomenon was reported from several host species (Wyllie 1981). We video recorded a case when a cuckoo chick climbed out of the nest cup, crawled ca. 15 cm route across the nest box interior and finally climbed back into the nest cup (Grim et al. 2009b). Additionally, three cuckoo chicks (two in Finland, one in Czech Republic) repeatedly evicted themselves while evicting redstarts even though the observer returned them into the nest cup each time; all these self-evictors finally died, probably due to exhaustion and hypothermia (P. Samaš, M. Kysučan, pers. comm.). Thus, cavity nesting prevents self-eviction and may increase cuckoo survival prospects.

Surprisingly frequently (25%, $n = 44$ parasitized broods) we also recorded cases when an evicted redstart chick (or several chicks) fell back into the nest cup, even after spending hours in a distant part of the nest box, several cm from the rim of the nest cup (and therefore ignored by its parents). Such cases increased the workload for the cuckoo chick which was forced to evict the same individual host chicks repeatedly. In some cases, the cuckoo was unable to evict the returning host chick again and consequently shared the nest with it (experimentally manipulated nest sharing has a strong negative effects on cuckoo—but not redstart—growth and survival; Grim et al. 2009a).

Such complex temporal dynamics of brood composition in the nest cup are impossible in open-cup nests. One could speculate that cavity nest environment creates a potential for natural selection to favour more active host chicks (more crawling, even undirected one, would increase chances of accidentally climbing back into the nest cup) or even an ability in host chicks to eavesdrop on begging calls from the nest cup to find the way directly back into the only place where chicks are recognized as such by redstarts or any other passerines (Grim 2006c; see also p. 81 in Mikulica et al. 2017).

Although evicting cuckoo chicks suffer reduced growth during the eviction phase, they are able to compensate for this detrimental effect of their eviction efforts later (Anderson et al. 2009; Grim et al. 2009b). They increase their growth rate, perhaps via exaggerated begging (compared to solitary chicks, but this needs to be tested), and manage to reach similar fledging masses and success as solitary chicks (i.e. chicks that were relieved from the eviction costs experimentally; Grim et al. 2009a). Still, they fledge significantly later but the effect size is small (1 day later than solitary cuckoos). These plastic growth patterns do not seem to be a specific adaptation to cavity hosts as they were found in open-nesting hosts (Anderson et al. 2009; see also Geltsch et al. 2012).

Both observational (Rutila et al. 2002) and experimental data (Grim et al. 2009a) from redstart-cuckoos revealed that cuckoos are extremely poor competitors for parental care as it has also been showed in other host species (Martín-Gálvez et al. 2005; Grim 2006b; Hauber and Moskát 2008; Grim et al. 2011). Even a single cohabiting host chick inflicts massive negative effects on cuckoo chick growth (Geltsch et al. 2012). Cuckoos in mixed broods (i.e. sharing care with host chicks) show dramatically lower fledging masses (reaching only 74% mass of solitary cuckoos), delayed fledging (by 15% later than solitary cuckoos) and extremely poor survival till fledging (only 44% success of solitary cuckoos; estimates based on data from Grim et al. 2009a).

16.6 Nestling Stage: Behaviour, Host Provisioning and Unusual Diet

When begging, older cuckoo chicks perform asymmetrical wing shaking (Grim 2008): they raise single wing at a time and shiver it when fosterer arrives at the nest. Similarly to cuckoo chicks in reed warbler nests, preliminary observations show that also redstart-cuckoos wing shake only when fosterers are feeding them and mostly shake the wing towards the provisioning "parent" (see also Tyller et al. 2018). This suggests that asymmetrical wing shaking might be a species-level trait (i.e. not specific to a reed warbler-cuckoo gens where it was documented: Grim 2008) that is expressed in both open and cavity nests.

Avian chicks typically respond to parental alarm calls by reducing begging. Davies et al. (2006) showed that reed warbler-cuckoo chicks respond specifically to reed warbler alarms, and additionally to ceasing calling, the cuckoos show a specific response of gaping which might be a defensive action. In contrast, redstart-cuckoos (and even redstart chicks) do not respond to host alarm calls (Davies et al. 2006). The reason for this lack of responsiveness to parental alarm calls might be safe nesting site in cavities where predation rates are small compared to open-cup nests.

Redstart females generally invested more in the care for cuckoo chicks: compared to males, they provided ca. 50% more of their feedings to the cuckoo than to their own chicks in mixed broods (29% in females vs. 21% in males; Grim et al. 2009a; for a similar conclusion in a different cuckoo host see Požgayová et al. 2015). Both females and males increased proportions of their feeding to the parasite throughout the nestling period (Grim et al. 2009a). This suggests that both parents increased their provisioning in response to increasing body size ratio between the cuckoo vs. host chicks.

Cuckoo diet was rarely analysed in detail among cuckoo gentes with majority of the data confined to *Acrocephalus* warblers (Grim and Honza 2001 and references therein). Similarly to other hosts, we found that redstart-cuckoo chick diet is numerically dominated by insects and spiders (data from video recordings of hosts feeding redstart broods, $n = 87$ nests, and cuckoo nestlings, $n = 80$ nests; Grim et al. 2017). However, both redstart-cuckoos (at 11% of nests) and host's own broods (at 5% of nests), were also fed with berries. Previously, fruits were reported in the diet of cuckoo chicks only once (Martín-Gálvez et al. 2005). Both cuckoos (at 5% of nests) and host's own broods (at 7% of nests), were also fed with lizards, which were never before reported in the cuckoo nestling diet (Grim et al. 2017). Lizards were numerically rare (<1% of food items in both cuckoo vs. redstart chick diet), but their body size was ca. three orders of magnitude higher than any other items. Therefore, the mass dominance and, by implication, energy content of lizards were in fact much larger than mere numerical dominance would suggest.

This unique diet composition had fitness consequences for cuckoos: parasite, but not host chicks, fed with fruits and lizards had lower fledging masses (by 20% in both fruit- and lizard-fed cuckoos) and prolonged nestling periods (again by 20%). Although fledging success was not affected by diet composition (only invertebrate fed: 92%; also plant and vertebrate fed: 82%), the fledgling survival could still be decreased as a negative effect of lowered fledgling mass on post-fledging survival is the norm in birds (see also below). Thus, cuckoo chicks can digest plant and vertebrate material, but such unusual dietary components might cause ontogenetic stress (i.e. slower growth and delayed fledging). These results for the first time show that food composition may affect fitness of parasites not only at the host species level (Yang et al. 2013) but also at the level of individual host pairs. Although host diet selection most likely does not evolve as a specific anti-parasite defence in general (see discussion in Yang et al. 2013), it could be an important general life-history trait that affects brood parasite fitness (Grim et al. 2011).

Cuckoo chicks raised alone, i.e. after successfully evicting redstart progeny (thus not competing with host chicks), reached higher fledging masses (sometimes over 120 g; Grim et al. 2009a; Grim and Samaš 2016) than in any other host studied so far (reviewed in Grim 2006a). For example, at fledging a redstart-cuckoo can weigh twice as much as reed warbler-cuckoo (e.g. 60 g; Grim 2006a). This may be a consequence of cavity nesting: cavities protect chicks from inclement weather and partly release parents from the stressful influences of perceived predation risk (Zanette et al. 2011) (but not perceived parasitism risk, see Tolvanen et al. 2017). The length of the nestling period (20.6 ± 0.4 days, $n = 18$; Grim and Samaš 2016) was within the range (18–22 days) of previously reported values across cuckoos raised by 14 host species (reviewed in Grim 2006a; Grim and Samaš 2016). Additionally, the new redstart data support previous conclusions (Grim 2006a) that host species body mass does not influence either cuckoo chick fledging mass or age (Grim and Samaš 2016).

Longer nestling period of cuckoos (21 days) than redstarts (14 days) may lead to negative fitness consequences for the cuckoos. We detected two cases of mixed broods ($n = 16$) where both parents (or female, because males often stop to care for their brood before it fledges) focused on their own fledged brood and stopped to visit the nest where the cuckoo was still too young to fledge. Such cuckoo chicks perished due to starvation (Fig. 29.1c). It remains to be determined experimentally if such chick desertions are due to preprogrammed length of parental care in at least some redstart individuals (cf. Grim et al. 2003; Grim 2007b).

16.7 Fledgling Stage: Enigmatic in General

Although the post-fledging period and transition to independence are critical periods in the life of birds, these important aspects of parasite biology remain virtually unknown in any brood parasites (but see Soler et al. 2014 and references therein). In the common cuckoo, they have been studied only once. Wyllie (1981) reported that cuckoos fledged from reed warbler nests become able to fly clumsily 4–5 days after leaving the host nest and became independent on average 16 days after fledging. Host reed warbler fledglings reach independence earlier, after 10–14 days on average (Wyllie 1981). Data from a cuckoo chick attended by the chaffinch (*Fringilla coelebs*) showed quantitatively the same patterns (Tyller et al. 2018).

In contrast, our data (Kysučan et al., in prep.) show that at least some cuckoos fledged from redstart nests are able to fly immediately after departing from the nest (30%, $n = 10$) and can fly remarkably well—even directly up to forest canopy! This may be because some redstart-cuckoo chicks have a longer nestling period that allows cuckoos to mature more than in nests of other hosts (Grim and Samaš 2016). However, the majority (70%, $n = 10$) of the redstart-cuckoo fledglings observed by us mostly crawled in the herbaceous layer for the first 2–3 days after leaving the host nest and started to fly clumsily only afterwards. Surprisingly, radio tracking of young showed very similar lengths of post-fledging periods in both redstarts (18.5 ± 1.0 days, $n = 10$) and cuckoos (17.3 ± 0.8 days, $n = 10$).

Radio tracking of cuckoo fledglings revealed that post-fledging survival of cuckoo chicks that shared a nest with redstart chicks was 0% ($n = 5$; post-fledging survival to independence of cuckoos that were reared alone: 45%, $n = 11$; Kysučan et al. in prep.). This contrasts with the fledging success rates of cuckoos from experimental mixed broods (44%, $n = 9$; Grim et al. 2009a). Because all cuckoos that fledged from mixed broods seem to die within the first week after fledging (but see the low sample size), the nest sharing provides the strongest possible selection pressure on the maintenance of eviction instinct in cuckoo chicks in any host nests.

This finding also highlights the crucial importance of studying parasite–host interactions across all breeding stages (see Grim et al. 2011): without the data from post-fledging period, we (Grim et al. 2009a) severely underestimated the negative effects of cohabitation with host chicks for the parasite.

16.8 Post-independent Stage: Migrating Parasites

Satellite tracking (Vega et al. 2016) showed that first-year redstart-cuckoos initiate their migration the whole month after adults have already departed (means: August 14 vs. July 11). Young differed from adults in slower and less consistently directed movements after leaving the breeding grounds. Despite different directions and extensive crossings of the Baltic Sea (3 out of 5 fledglings) at the initiation of migration, the young independently arrived to similar wintering area in Angola ($n = 1$ fledgling surviving till arrival at wintering grounds).

The redstart-cuckoos thus provide the first direct evidence on migration route of any young common cuckoos and confirm for the first time that naïve parasites are able to reach correct wintering areas independently, solely via an innate migration programme (Vega et al. 2016). This may not hold in other brood parasites where there is a contact between adult parasites and their fledglings, creating a potential for joint migration of young and adult birds (Soler and Soler 1999).

16.9 Parasite Fitness: Little Effects of Cavity Nesting

Cavities have many disadvantages from the cuckoo's point of view. Two-thirds of cuckoo eggs are laid outside the nest cup and cannot produce a cuckoo chick. Eviction failures occurred at 20% of nests producing mixed broods. Cuckoo chicks that manage to fledge from mixed broods have no future, since they will die before they reach independence.

Surprisingly, cuckoos' reproductive success (no. of fledglings produced per egg laid) may not differ from that in common open-nesting hosts. Specifically, in our study population, the average reproductive success was 0.16. Similar reproductive success was reported in perhaps the most frequent (or at least most frequently studied) among cuckoo hosts, the reed warbler: 0.16 (Kleven et al. 2004). In other warblers, the cuckoo reproductive success varied from 0.04 to 0.30 (Kleven et al. 2004).

How is it possible that all the disadvantages of cavities do not translate into decreased fitness of the parasite? Cavities have also advantages: they protect the parasite chicks from inclement weather and predation. Redstarts, in contrast to typical cuckoo hosts, reject only minority of cuckoo eggs. These positive effects (from cuckoo's point of view) seem to balance the negative ones, resulting in virtually the same fitness of redstart-cuckoos compared to reed warbler-cuckoos.

16.10 Host Fitness: Reduced Costs of Parasitism?

Cavities have many advantages from the redstart's point of view, especially compared to open-nesting hosts. Only one third of cuckoo eggs end up inside the nest cup and produce a cuckoo chick; in contrast, in open-nesting hosts, all cuckoo eggs end up in the nest cup (Wyllie 1981; Moksnes et al. 2000; Andou et al. 2005; Mikulica et al. 2017). Only small proportion of cuckoo females (23%, $n = 26$) remove any redstart eggs prior to cuckoo's own laying, while egg removal is a rule in open-nesting hosts. Across populations, 15–45% of cuckoo chicks fail to kill all redstart progeny; in contrast, in open-nesting hosts, all host progeny dies due to cuckoo eviction (Davies 2000). In mixed redstart-cuckoo broods, at least some host's own chicks typically survive and fledge, while cuckoo chicks often die.

The cohabitation with a cuckoo chick decreased the average host fitness (no. of fledged redstart chicks) approximately twice: non-parasitized redstart nests fledged on average 5.7 redstart chicks, whereas parasitized nests with mixed broods fledged on average 2.4 redstart chicks (Samaš et al. 2016). Adjusting these values to original clutch size leads to quantitatively almost identical conclusion (i.e. a ratio between the fitness of the two types of broods): non-parasitized redstart nests fledged on average (mean ± SE) 0.84 ± 0.03 chick per egg laid ($n = 87$ broods), whereas parasitized nests with mixed broods fledged on average 0.38 ± 0.05 chick per egg laid ($n = 14$ broods). Overall host fitness across all parasitized broods where the cuckoo has hatched (i.e. mixed and solitary cuckoos) was 0.10 ± 0.03 chick per egg laid ($n = 75$ broods). Thus, an average non-parasitized brood has 8.4 times higher fitness than an average parasitized brood.

Comparing these values to available data from reed warblers leads to an unexpected result. Non-parasitized nests fledged on average 0.48 chick per egg laid, whereas parasitized nests fledged on average only 0.05 chick per egg laid (Øien et al. 1998). Thus, an average non-parasitized nest has 9.6 times higher fitness that an average parasitized brood. Therefore, the impact of parasitism per host nest is very similar between the only regular cuckoo cavity-nesting host known and a typical open-nesting host. However, at population level the selection exerted by cuckoos should be higher in redstarts than reed warblers: an average parasitism rate across regularly parasitized host populations is much higher in redstarts (29%, $n = 8$ populations; Samaš et al. 2016) than in reed warblers (12%, $n = 7$ populations; Stokke et al. 2008; non-parasitized populations excluded; see also Grim 2017).

Concluding Remarks and Future Directions

Peculiar biology and ecology of the redstart-cuckoo system provides specific suggestions for future research in this unique brood parasite–host pair. The mechanistic simplicity of redstart-cuckoo mimicry (plain eggs, typically no spots) makes it a more tractable system to study mimicry than most other gentes which show a highly complex egg markings. Future studies should also

(continued)

test if redstarts use spots (Fig. 16.1b), together with witnessing the laying cuckoo female (as already documented), as a cue to desert parasitized nests. Cohabitation of unsuccessful "evictors" with their surviving "victims" creates a stage for fosterer-offspring communication that is not paralleled in any other common cuckoo gens and which might be comparable to what happens in non-evictor cuckoo species. Fledglings from such mixed broods create a unique (in the common cuckoo) opportunity to study mechanisms of brood division which results in host male and female communicating with and feeding conspecific vs. heterospecific chick(s).

Different type of questions relates to redstart nest type. If nest design (nest cup size, shape and position within the nest box) is an adaptation against parasitism, does nest design prevent the cuckoo female from successful laying (egg stage), or does it constrain the success of cuckoo chick's eviction attempts (chick stage), or both? Future studies should quantify whether some aspects of nest design could play a role in eliminating parasite success at the two breeding stages, or whether the different stages could inflict opposite selection pressures on nest design.

Still other questions relate to studying all developmental stages of parasite–host relationships, namely, to the so far least studied post-fledging period. The focus should be on chick (both cuckoo and host) development of flying abilities, parental/fosterer provisioning, defence of young and even growth outside the nest cavity—it is possible to catch and measure majority of chicks during the first week after they leave host nests because they firstly do not fly at all and later only poorly.

Apart from these mostly redstart-cuckoos-specific topics, there are also various questions that reflect general interests of evolutionary and behavioural biologists in communication, coevolution, competition and parental investment. Cavity nesting, in both redstarts and other hosts breeding in cavities or dome nests, provides a different setting from typical passerine open nests and allows to realize experimental tests that are unfeasible in open nests. For example, both eviction success and growth performance of redstart-cuckoos show large variation. Is this variation caused by seasonally varying weather, host pair quality, nest cup design and composition, eviction of eggs vs. chicks or the proportion of time when the cuckoo chick pushed the host progeny against a nest-box wall? The nest sharing raises another question: are cuckoo chicks from the redstart gens more competitive than cuckoo chicks from other gentes where cuckoos almost never share a nest with host nestlings? This might be tested by cross-fostering, for example, reed warbler-cuckoo chicks into redstart nests, and examining their performance in mixed broods. Similarly, are host redstart chicks more competitive than typical passerine chicks because of their evolutionary experience with a large alien "sibling"?

Acknowledgements We dedicate this contribution to J. Haikola who initiated the long-term cuckoo research in our study area. We are grateful to M. Kysučan, P. Samaš and Z. Tyller for their work in the field. P. Samaš, R. L. Thomson and J. Tolvanen commented on drafts. The manuscript benefitted from comments by L. de Neve, W. Liang and M. Soler. Our research was supported by the Human Frontier Science Program (RGY69/2007 and RGY83/2012), the Czech Science Foundation (P506/12/2404) and Internal Grant Agency of Palacký University (PrF_2014_018, PrF_2015_018, IGA_PrF_2016_017).

References

Anderson MG, Moskát C, Bán M, Grim T, Cassey P, Hauber ME (2009) Egg eviction imposes a recoverable cost of virulence in chicks of a brood parasite. PLoS One 4:e7725

Andou D, Nakamura H, Oomori S, Higuchi H (2005) Characteristics of brood parasitism by common cuckoos on azure-winged magpies, as illustrated by video recordings. Ornithol Sci 4:43–48

Avilés JM (2008) Egg colour mimicry in the common cuckoo *Cuculus canorus* as revealed by modelling host retinal function. Proc R Soc Lond B 275:2345–2352

Avilés JM, Rutila J, Møller AP (2005) Should the redstart *Phoenicurus phoenicurus* accept or reject cuckoo *Cuculus canorus* eggs? Behav Ecol Sociobiol 58:608–617

Čapek V (1896) Beiträge zur fortpflanzungsgeschichte des Kuckucks. Ornithol Jahrb 7:41–72, 102–117, 146–157, 165–183

Davies NB (2000) Cuckoo, cowbirds and other cheats. T&AD Poyser, London

Davies NB, Brooke ML (1989) An experimental study of co-evolution between the cuckoo, *Cuculus canorus*, and its hosts. I. Host egg discrimination. J Anim Ecol 58:207–224

Davies NB, Madden JR, Butchart SHM, Rutila J (2006) A host-race of the cuckoo *Cuculus canorus* with nestlings attuned to the parental alarm calls of the host species. Proc R Soc Lond B 273:693–699

Dawkins R, Krebs JR (1979) Arms races between and within species. Proc R Soc Lond B 205:489–511

Dinets V, Samaš P, Croston R, Grim T, Hauber ME (2015) Predicting the responses of native birds to transoceanic invasions by avian brood parasites. J Field Ornithol 86:244–251

Fossøy F, Sorenson MD, Liang W, Ekrem T, Moksnes A, Møller AP, Rutila J, Røskaft E, Takasu F, Yang C, Stokke BG (2016) Ancient origin and maternal inheritance of blue cuckoo eggs. Nat Commun 7:10272

Fuchs W (1957) Funde von weissen Kuckuckseiern bei Sempach und Einsiedeln. Ornithol Beob 54:136–137

Geltsch N, Hauber ME, Anderson MG, Bán M, Moskát C (2012) Competition with a host nestling for parental provisioning imposes recoverable costs on parasitic cuckoo chick's growth. Behav Proc 90:378–383

Grim T (2006a) Cuckoo growth performance in parasitized and unused hosts: not only host size matters. Behav Ecol Sociobiol 60:716–723

Grim T (2006b) Low virulence of brood parasitic chicks: adaptation or constraint? Ornithol Sci 5:237–242

Grim T (2006c) The evolution of nestling discrimination by hosts of parasitic birds: why is rejection so rare? Evol Ecol Res 8:785–802

Grim T (2007a) Equal rights for chick brood parasites. Ann Zool Fenn 44:1–7

Grim T (2007b) Experimental evidence for chick discrimination without recognition in a brood parasite host. Proc R Soc Lond B 274:373–381

Grim T (2008) Wing-shaking and wing-patch as nestling begging strategies: their importance and evolutionary origins. J Ethol 26:9–15

Grim T (2016) Are cavity nesters really unsuitable hosts for the common cuckoo (*Cuculus canorus*)? An experiment with the blue tit (*Cyanistes caeruleus*) and collared flycatcher (*Ficedula albicollis*). Sylvia 52:53–66

Grim T (2017) Chick discrimination versus adaptive parasitic egg acceptance: the egg dilution effect hypothesis revisited. Ornithol Sci 16:163–170

Grim T, Honza M (2001) Does supernormal stimulus influence parental behaviour of the cuckoo's host? Behav Ecol Sociobiol 49:322–329

Grim T, Samaš P (2016) Growth performance of nestling cuckoos *Cuculus canorus* in cavity nesting hosts. Acta Ornithol 51:175–188

Grim T, Kleven O, Mikulica O (2003) Nestling discrimination without recognition: a possible defence mechanism for hosts towards cuckoo parasitism? Proc R Soc Lond B 270:S73–S75

Grim T, Rutila J, Cassey P, Hauber ME (2009a) Experimentally constrained virulence is costly for common cuckoo chicks. Ethology 115:14–22

Grim T, Rutila J, Cassey P, Hauber ME (2009b) The cost of virulence: an experimental study of egg eviction by brood parasitic chicks. Behav Ecol 20:1138–1146

Grim T, Samaš P, Moskát C, Kleven O, Honza M, Moksnes A, Røskaft E, Stokke BG (2011) Constraints on host choice: why do parasitic birds rarely exploit some common potential hosts? J Anim Ecol 80:508–518

Grim T, Samaš P, Procházka P, Rutila J (2014) Are tits really unsuitable hosts for the common cuckoo? Ornis Fenn 91:166–177

Grim T, Tyller Z, Samaš P (2017) Unusual diet of brood parasite nestlings and its fitness consequences. Auk 134(3):732–750

Haikola J, Rutila J (2008) Käki. Otava, Helsinki

Hanley D, Grim T, Igic B, Samaš P, López AV, Shawkey MD, Hauber ME (2017) Egg discrimination along a gradient of natural variation in eggshell coloration. Proc R Soc Lond B 284:20162592

Hauber ME, Moskát C (2008) Shared parental care is costly for nestlings of common cuckoos and their great reed warbler hosts. Behav Ecol 19:79–86

Hauber ME, Samaš P, Anderson MG, Rutila J, Low J, Cassey P, Grim T (2014) Life-history theory predicts host behavioural responses to experimental brood parasitism. Ethol Ecol Evol 26:349–364

Honza M, Vošlajerová K, Moskát C (2007) Eviction behaviour of the common cuckoo *Cuculus canorus* chicks. J Avian Biol 38:385–389

Igic B, Braganza K, Hyland MM, Silyn-Roberts H, Cassey P, Grim T, Rutila J, Moskát C, Hauber ME (2011) Alternative mechanisms of increased eggshell hardness of avian brood parasites relative to host species. J R Soc Interface 8:1654–1664

Igic B, Cassey P, Grim T, Greenwood DR, Moskát C, Rutila J, Hauber ME (2012) A shared chemical basis of avian host–parasite egg colour mimicry. Proc R Soc Lond B 279:1068–1076

Igic B, Zarate E, Sewell MA, Moskát C, Cassey P, Rutila J, Grim T, Shawkey MD, Hauber ME (2015) A comparison of egg yolk lipid constituents between parasitic common cuckoos and their hosts. Auk 132:817–825

Kleven O, Moksnes A, Røskaft E, Rudolfsen G, Stokke BG, Honza M (2004) Breeding success of common cuckoos *Cuculus canorus* parasitising four sympatric species of *Acrocephalus* warblers. J Avian Biol 35:394–398

Liang W, Møller AP, Stokke BG, Yang C, Kovařík P, Wang H, Yao C-T, Ding P, Lu X, Moksnes A, Røskaft E, Grim T (2016) Geographic variation in egg ejection rate by great tits across 2 continents. Behav Ecol 27:1405–1412

Löhrl H (1979) Untersuchungen am Kuckuck, *Cuculus canorus* (Biologie, Ethologie und Morphologie). J Ornithol 120:139–173

Martín-Gálvez D, Soler M, Soler JJ, Martín-Vivaldi M, Palomino JJ (2005) Food acquisition by common cuckoo chicks in rufous bush robin nests and the advantage of eviction behaviour. Anim Behav 70:1313–1321

Mikulica O, Grim T, Schulze-Hagen K, Stokke BG (2017) The cuckoo: the uninvited guest. Wild Nature Press, Plymouth

Moksnes A, Røskaft E, Braa AT, Korsnes L, Lampe HM, Pedersen HC (1991) Behavioural responses of potential hosts towards artificial cuckoo eggs and dummies. Behaviour 116:64–89

Moksnes A, Røskaft E, Tysse T (1995) On the evolution of blue cuckoo eggs in Europe. J Avian Biol 26:13–19

Moksnes A, Røskaft E, Hagen LG, Honza M, Mørk C, Olsen PH (2000) Common cuckoo *Cuculus canorus* and host behaviour at reed warbler *Acrocephalus scirpaceus* nests. Ibis 142:247–258

Øien IJ, Moksnes A, Røskaft E, Honza M (1998) Costs of cuckoo *Cuculus canorus* parasitism to reed warblers *Acrocephalus scirpaceus*. J Avian Biol 29:209–215

Požgayová M, Beňo R, Procházka P, Jelínek V, Abraham MM, Honza M (2015) Lazy males and hardworking females? Sexual conflict over parental care in a brood parasite host and its consequences for chick growth. Behav Ecol Sociobiol 69:1053–1061

Rutila J (2004) Brood parasitism in birds: coevolutionary adaptations in two cuckoo host systems. Dissertation, University of Joensuu

Rutila J, Latja R, Koskela K (2002) The common cuckoo *Cuculus canorus* and its cavity nesting host, the redstart *Phoenicurus phoenicurus*: a peculiar cuckoo-host system? J Avian Biol 33:414–419

Rutila J, Jokimäki J, Avilés JM, Kaisanlahti-Jokimäki ML (2006) Responses of parasitized and unparasitized common redstart (*Phoenicurus phoenicurus*) populations against artificial cuckoo parasitism. Auk 123:259–265

Samas P, Hauber ME, Cassey P, Grim T (2014) Host responses to interspecific brood parasitism: a by-product of adaptations to conspecific parasitism? Front Zool 11:34

Samaš P, Rutila J, Grim T (2016) The common redstart as a suitable model to study cuckoo-host coevolution in a unique ecological context. BMC Evol Biol 16:255

Soler M, Soler JJ (1999) Innate versus learned recognition of conspecifics in great spotted cuckoos *Clamator glandarius*. Anim Cogn 2:97–102

Soler M, de Neve L, Roncalli G, Macías-Sánchez E, Ibáñez-Álamo JD, Pérez-Contreras T (2014) Great spotted cuckoo fledglings are disadvantaged by magpie host parents when reared together with magpie nestlings. Behav Ecol Sociobiol 68:333–342

Stoddard MC, Stevens M (2011) Avian vision and the evolution of egg color mimicry in the common cuckoo. Evolution 65:2004–2013

Stokke BG, Hafstad I, Rudolfsen G, Moksnes A, Møller AP, Røskaft E, Soler M (2008) Predictors of resistance to brood parasitism within and among reed warbler populations. Behav Ecol 19:612–620

Stokke BG, Røskaft E, Moksnes A, Møller AP, Antonov A, Fossøy F, Liang W, López-Iborra G, Moskát C, Shykoff JA, Soler M, Vikan JR, Yang C, Takasu F (2016) Disappearance of eggs from nonparasitized nests of brood parasite hosts: the evolutionary equilibrium hypothesis revisited. Biol J Linn Soc 118:215–225

Thomson RL, Tolvanen J, Forsman JT (2016) Cuckoo parasitism in a cavity nesting host: near absent egg-rejection in a northern redstart population under heavy apparent (but low effective) brood parasitism. J Avian Biol 47:363–370

Tolvanen J, Forsman JT, Thomson RL (2017) Reducing cuckoo parasitism risk via informed habitat choices. Auk 134:553–563

Trnka A, Grim T (2013) Color plumage polymorphism and predator mimicry in brood parasites. Front Zool 10:25

Tyller Z, Kysučan M, Grim T (2018) Post-fledging behavior of the common cuckoo (*Cuculus canorus*) attended by the chaffinch (*Fringilla coelebs*): a comprehensive approach to study the least known stage of brood parasite-host coevolution. Wilson J Ornithol 130

Vega ML, Willemoes M, Thomson RL, Tolvanen J, Rutila J, Samaš P, Strandberg R, Grim T, Fossøy F, Stokke BG, Thorup K (2016) First-time migration in juvenile common cuckoos documented by satellite tracking. PLoS One 11:e0168940

von Haartman L (1969) The nesting habits of Finnish birds. I. Passeriformes. Soc Scient Fenn Comm Biologicae 32:1–187
von Haartman L (1981) Coevolution of the cuckoo *Cuculus canorus* and a regular cuckoo host. Ornis Fenn 58:1–10
Wyllie I (1981) The cuckoo. Batsford, London
Yang C, Liang W, Cai Y, Shi S, Takasu F, Møller AP, Antonov A, Fossøy F, Moksnes A, Røskaft E, Stokke BG (2010) Coevolution in action: disruptive selection on egg colour in an avian brood parasite and its host. PLoS One 5:e10816
Yang C, Stokke BG, Antonov A, Cai Y, Shi S, Moksnes A, Røskaft E, Møller AP, Liang W, Grim T (2013) Host selection in parasitic birds: are open-cup nesting insectivorous passerines always suitable cuckoo hosts? J Avian Biol 44:216–220
Yang C, Li Z, Zhang Y, Wang H, Liang W, Møller AP (2016) Egg polymorphism and egg discrimination in the Daurian redstart *Phoenicurus auroreus*, a host of the common cuckoo *Cuculus canorus*. Ornithol Sci 15:127–132
Zanette LY, White AF, Allen MC, Clinchy M (2011) Perceived predation risk reduces the number of offspring songbirds produce per year. Science 334:1398–1401

Part VI

Coevolutionary Interactions at the Prelaying Stage

Evidence of Adaptations and Counter-Adaptations Before the Parasite Lays Its Egg: The Frontline of the Arms Race

17

William E. Feeney

Abstract

The interactions between avian brood parasites and their hosts provide tractable model systems to study coevolutionary processes under natural conditions. Here, I review evidence of reciprocal adaptations and counter-adaptations in brood parasites and hosts that are deployed prior to the parasite depositing its egg in the host nest: the 'frontline' of the arms race. Unlike interactions at latter stages of the nesting cycle, frontline interactions primarily concern adult brood parasites and adult hosts, offering opportunities to study how exchanges between these species influence adult phenotypes. Placing emphasis on recent advances, I discuss how frontline interactions have shaped the life histories, behaviours, morphologies and physiologies of adult brood parasites and adult hosts. Similar to latter stages of the nesting cycle, frontline interactions comprise diverse adaptations and counter-adaptations that appear to be a product of coevolution and are important for determining the outcome of the exchanges between these species. Further investigation of these interactions is essential for categorizing the diversity of adaptations and counter-adaptations at this stage of the nesting cycle and expanding our understanding of how adaptations and counter-adaptations at all stages of the nesting cycle evolve in relation to one another.

W. E. Feeney (✉)
School of Biological Sciences, University of Queensland, Brisbane, QLD, Australia
e-mail: w.feeney@uq.edu.au

M. Soler (ed.), *Avian Brood Parasitism*, Fascinating Life Sciences,
https://doi.org/10.1007/978-3-319-73138-4_17

17.1 Introduction

Humans have demonstrated a deep and lasting interest in the natural habits of the avian brood parasites. Their unmistakeable calls are recognized as synonymous with seasonal change throughout Europe, Asia and Australia (Lai 1998; Davies 2000; Gray and Fraser 2013); one acts as foraging partner with humans in sub-Saharan Africa (Spottiswoode et al. 2016); and references to their parasitic breeding habits are evident throughout the world's folklores, languages and mythologies (Lai 1998; Davies 2000; Gray and Fraser 2013; Møller et al. 2017). Throughout history myriad explanations have been proposed to rationalize why these species never tend to their offspring (Davies 2000). Of these, natural selection prevailed (Darwin 1859), and these systems have since become models for studying the ecological and evolutionary ramifications of species interactions under natural conditions (Rothstein 1990; Davies 2000; Feeney et al. 2014).

Research is currently being conducted on parasite–host systems that include representatives from each of the seven brood parasite lineages (Sorenson and Payne 2005) and on species pairs that inhabit every continent except for Antarctica (where none occur, Davies 2000). We know that the cost of hosting a parasite varies across species pairs (Kilner 2005) and that the benefits of hosting a parasite can outweigh the costs under some circumstances (Canestrari et al. 2014). The ecological consequences of these interactions can be globally conspicuous, predicting broad patterns of trait diversity such as small clutch sizes (Hauber 2003) and cooperative breeding (Feeney et al. 2013) in hosts, as well as hawk mimicry (Gluckman and Mundy 2013) and plumage polymorphisms (Thorogood and Davies 2013) in parasitic cuckoos. The depth of coevolution between these species is also becoming clear, with data now demonstrating that coevolved adaptations and counter-adaptations are evident at all stages of the host's nesting cycle (reviewed in Feeney et al. 2014; Soler 2014)

In this chapter, I will discuss the interactions between avian brood parasites and hosts that occur prior to the parasite laying its egg in the host's nest: the 'frontline' of the arms race (Welbergen and Davies 2009). These exchanges are of utmost importance, as a parasite's reproductive output hinges on its ability to successfully deposit eggs in host nests, and successful deterrence of adult parasites offers the first opportunity for hosts to defend their nests. Unlike reciprocal adaptations and counter-adaptations at latter stages of the nesting cycle, frontline interactions primarily concern adult brood parasites and hosts, offering opportunities for selection to shape adult phenotypes of the involved species as well as their natural histories. Following a recent review of this topic (Feeney et al. 2012), I will discuss evidence of how coevolution has affected the life histories, behaviours, morphologies and physiologies of brood parasites and hosts. However, to minimize any unnecessary overlap, this chapter will place emphasis on new research as well as research that was not heavily discussed in Feeney et al. (2012). Discussion of traits will be conducted in a roughly chronological order.

17.2 Life Histories

17.2.1 Breeding Phenology

Parasitism can only occur if there is breeding synchrony between parasite and host. Despite being inherently difficult to demonstrate, evidence suggests that some host species, or populations of hosts within species, may have shifted their breeding phenologies to avoid breeding in synchrony with parasites. Brooker and Brooker (1989) were the first to compare breeding records of Australian cuckoos and their hosts and found numerous examples of hosts starting breeding before their cuckoos arrived or ending after they left, consistent with the hypothesis that hosts shift their breeding phenology to minimize overlap with cuckoos. Medina and Langmore (2016) built on this and found that early-laying yellow-rumped thornbills (*Acanthiza chrysorrhoa*) were parasitized significantly less than late breeders. Similar trends have been found in other host species (see Middleton 1977; Campomizzi et al. 2013 for similar evidence in hosts of the brown-headed cowbird (*Molothrus ater*); Table 17.1).

Research into the effects that climate change has had on bird breeding behaviours may offer insights into effectiveness and potential evolutionary consequences of this putative defence. Notably, migratory parasites are likely more vulnerable to shifts in host breeding phenology compared to their resident counterparts, as they are less able to track their host's breeding schedule (Saino et al. 2009; Møller et al. 2011b; Péron et al. 2016). Those that are unable to effectively track host laying may be forced to switch hosts (Møller et al. 2011b; Péron et al. 2016) or face a reduction in their abundance (Saino et al. 2009). Host switching, in turn, may promote hybridization of closely related parasite species (Péron et al. 2016) or divergence into distinct 'races' (gentes) (Møller et al. 2011a) or species (Sorenson et al. 2003; Péron et al. 2016). Together, these studies suggest that shifts in host breeding phenology could provide an extremely effective defence against brood parasitism, especially in parasite–host systems that comprise migratory parasites and resident hosts, and could have dire evolutionary repercussions for the parasites.

If parasites and hosts do breed in synchrony, hosts may tactically adjust their laying dates to minimize their likelihood of parasitism or exhibit compensatory breeding behaviours to mitigate the cost of parasitism if it does occur. As parasites are constrained by their laying schedules, synchronous breeding in host communities reduces the likelihood of any individual being parasitized through a 'swamping' or 'dilution' effect (Clark and Robertson 1979; Martínez et al. 1996; Massoni and Reboreda 2001). Parasites can counter these traits with behaviours, such as the tactical destruction (i.e. 'farming') of host nests (e.g. Hoover and Robinson 2007; Swan et al. 2015), which forces renesting and offers an opportunity for subsequent parasitism (see Chap. 15; Table 17.2). If parasitism does occur, hosts may double-brood in an attempt to compensate for their lost reproductive success. Recently, Louder et al. (2015b) found that parasitism by brown-headed cowbirds prompts

Table 17.1 Life-history phenotypes that affect the interactions between brood parasites and hosts

	Trait	Purpose	Key reference(s)
Host	Shifted breeding phenology	Mismatch breeding synchrony between parasite and host	Álvarez (1996)* Middleton (1977) Brooker and Brooker (1989) Soler et al. (2012)* Campomizzi et al. (2013) Boves et al. (2014) Medina and Langmore (2016)*
	Compensatory breeding	Dilute the cost of parasitism	Louder et al. (2015b)*
	Synchronous breeding by multiple hosts	Dilute the likelihood of parasitism	Clark and Robertson (1979) Martínez et al. (1996) Massoni and Reboreda (2001)
	Deceptive nest architecture	Decrease the perceived quality of the nest	Soler et al. (1999)
	Defensive nest architecture	Obstruct parasite access/ minimize their likelihood of successful parasitism	Rutila et al. (2002)* Grim et al. (2011)*
	Cryptic nest architecture/ placement	Minimize the likelihood of the parasite discovering the nest	Burhans (1997) Clarke et al. (2001) Jelínek et al. (2014)
	Unappealing nest location	Minimize the likelihood of parasitism through placement near aggressive or noxious heterospecifics	Clark and Robertson (1979)
	Breeding mode	Facilitates nest defence	Brown and Lawes (2007) Wiley (2012) Trnka and Prokop (2011) Canestrari et al. (2009)

Asterisk (*) indicates a manipulative experimental component. Parasite counter-adaptations to these adaptations tend to be behavioural (e.g. strategic targeting of nests and nest farming) and are therefore in Table 17.2

Table 17.2 Behaviours that affect the interactions between brood parasites and hosts

	Trait	Purpose	Key reference(s)
Parasite	Increased spatial memory	Facilitates effective parasitism	White et al. (2009)* Guigueno et al. (2014*, 2015*)
	Strategic targeting of host nests	Increases the likelihood of successful parasitism	Soler et al. (1995a*, 1999) Moskát and Honza (2000) Honza et al. (2002) Grant and Sealy (2002) McLaren and Sealy (2003) Skjelseth et al. (2004) Langmore and Kilner (2007) Polačiková et al. (2009)
	Flexible adjustment of nest searching strategies	Optimizes nest location	Woolfenden et al. (2003) Strausberger and Ashley (2005) Jelínek et al. (2014)
	Discreet nest monitoring behaviour	Lowers the likelihood of the host detecting the parasite	Álvarez (1993) Øien et al. (1996) Honza et al. (2002)
	Fast egg laying	Lowers the likelihood of the host detecting the parasite at the nest	Davies (2000)
	Nest farming	Destroy host nests to force renesting	Hoover and Robinson (2007)* Swan et al. (2015)*
	Tactical destruction of host eggs during laying	Provide a less competitive environment for the brood parasite's offspring	Spottiswoode and Colebrook-Robjent (2007) Fiorini et al. (2014)* Gloag et al. (2013)
	Mafia behaviours	Force acceptance of parasite egg (s) through retaliatory nest destruction following egg rejection	Soler et al. (1995b)* Hoover and Robinson (2007)*
Host	Specific recognition and denotation of brood parasites	Facilitates parasite-specific defences	Hobson and Sealy (1989)* Uyehara and Narins (1995)*

(continued)

Table 17.2 (continued)

	Trait	Purpose	Key reference(s)
			Gill and Sealy (2004)* Gill et al. (1997*, 2008*), Welbergen and Davies (2008)* Feeney et al. (2013)*
	Tactical use of social and personal information	Optimizes defences against brood parasites	Hauber and Russo (2000) Davies et al. (2003)* Davies and Welbergen (2009)* Campobello and Sealy (2011)* Welbergen and Davies (2012)* Feeney and Langmore (2015)* Thorogood and Davies (2016)*
	Decreasing time spent near the nest	Provides fewer opportunities for parasites to locate their nest	Banks and Martin (2001)
	Mobbing behaviours	Physically deter brood parasites from accessing host nests	Welbergen and Davies (2009)* Feeney et al. (2013)* Gloag et al. (2013)
	Nest protection behaviours	Physically blocking access to the nest	Hobson and Sealy (1989)*

Asterisk (*) indicates a manipulative experimental component compensatory double-brooding in prothonotary warblers (*Protonotaria citrea*). They also found that double-brooding females are often parasitized during their subsequent breeding attempt, suggesting that cowbirds may be promoting and then exploiting compensatory breeding behaviours to maximize their own reproductive success (Table 17.2).

17.2.2 Nest Architecture and Placement

When hosts and parasites breed sympatrically, hosts can also lower their likelihood of parasitism with defensive nest architectures or by building their nests in areas that are less likely to be located by parasites. In an elegant series of studies, Soler et al. (1995a, 1998, 1999, 2001) demonstrated that deceptive nest architectures can be employed by hosts in response to brood parasitism. They showed that in magpies (*Pica pica*), nest size is a sexually selected trait that correlates with reproductive success (Soler et al. 2001), that great spotted cuckoos (*Clamator glandarius*) preferentially parasitize larger magpie nests (Soler et al. 1995a), and that magpies that breed in sympatry with cuckoos have smaller nests than those in allopatry (Soler et al. 1999), together suggesting that brood parasitism is influencing sexually selected traits in these populations.

Hosts can also choose nesting sites that lower their likelihood of being parasitized. Individual parasites vary in their nest-searching behaviours and patterns of host use and can vary these behaviours according to the abundance of suitable host nests (Table 17.2). Hosts can respond by building nests that counter typical parasitic searching strategies, such as building nests that are removed from vantage points (e.g. 'perch proximity' hypothesis, Øien et al. 1996; Hauber and Russo 2000) or by increasing their defences when nesting in higher-risk areas (Welbergen and Davies 2012). Alternatively, hosts may place nests in locations that are less desirable to potential parasites, such as near aggressive interspecifics (Clark and Robertson 1979). Whether these kinds of interspecific associations lower the likelihood of brood parasitism awaits experimental investigation.

Alternatively, hosts could build secretive nests (e.g. Burhans 1997; Clarke et al. 2001; Jelínek et al. 2014) or nests that obstruct parasite access to minimize their likelihood of parasitism (Rutila et al. 2002) (Table 17.1). Notably, cavity nests also appear to obstruct parasites from depositing their eggs, and their design may also inhibit the nestling parasite's competitiveness after hatching (e.g. Rutila et al. 2002, also see Grim et al. 2011). This may explain why cavity nesters are so rarely hosts of the common cuckoo (*Cuculus canorus*) (Moksnes and Røskaft 1995).

17.2.3 Breeding Modes

A recent series of studies has demonstrated that host breeding mode can facilitate defences against brood parasitism (Table 17.1). While the behavioural mechanism that underpins parasite deterrence varies (e.g. decreasing the opportunity for parasitism, Canestrari et al. 2009; or facilitating host aggression, Feeney et al. 2013), they cumulatively suggest that more defence is better defence (however see Ursino et al. 2011). Recently, Feeney et al. (2013) found that brood parasitism and cooperative breeding evolve together; however, they were not able to determine the direction of the relationship.

17.3 Behaviour

17.3.1 Cognitive Adaptations

Considerable research has been and continues to be conducted into the cognition of brood parasites and hosts. In parasites, investigation of brain morphologies tends to show that they have relatively large hippocampal regions areas that are generally associated with spatial memory (see Chap. 11). A recent series of experimental and field-based studies on the brown-headed cowbird built on these findings and has demonstrated that females have a more accurate spatial memory than males (Guigueno et al. 2014), are able to assess host nest readiness to maximize laying synchrony (White et al. 2009) and that they can remember and target particular host individuals, within and between seasons, that have a history of successfully raising cowbird offspring (Louder et al. 2015a; also see Guigueno et al. 2015; Astié et al. 2015; de la Colina et al. 2016). Similarly, a variety of studies have demonstrated that parasites can strategically choose nests according to the perceived quality of the host parents or nest (Table 17.2) in addition to those that are perceived to be less likely to elicit defences (Avilés et al. 2006) or have a lower likelihood of being depredated (Soler et al. 2014a, b). Some also appear able to change their nest-searching strategy according to the availability of nests (e.g. Jelínek et al. 2014). Together, these studies suggest that cognitive adaptations have played an important role in facilitating parasitic life histories and that these traits may have been further shaped through interactions with hosts.

Similar evidence of cognitive adaptations is evident in hosts. For example, it is becoming clear that at least some hosts recognize brood parasites as unique threats and produce alarm calls that denote parasites in order to elicit specific defences (Table 17.2). At least some can also learn to respond to the sight of a parasite (Davies and Welbergen 2009; Langmore et al. 2012; Feeney and Langmore 2013) and use personal and social information to adjust their behaviours in order to minimize their likelihood of being parasitized (Table 17.2). To date, these studies have tended to focus on the interactions between one host and its primary parasite. However, most brood parasite–host pairs coexist with other parasites and hosts (Davies 2000; Feeney et al. 2013), and while they may have preferred hosts, a variety of parasite species rarely use one host species exclusively (Davies 2000). Therefore, hosts species that exist in these environments may eavesdrop on one another to acquire information about the local risk of parasitism and/or cooperate with one another when defending their nests.

17.3.2 Stealth and Detection

By and large, parasites and hosts appear to increase their respective likelihoods of successfully parasitizing host nests, or dodging parasitism, by minimizing the likelihood of interacting with one another. Evidence suggests that hosts can lower the likelihood of parasites discovering their nests by decreasing the amount of time they spend around them (Banks and Martin 2001) and by producing complex versus simple vocalizations, which may be difficult for parasites to use as location cues (Garamszegi and Avilés 2005). Parasites also tend to be secretive around host nests (however, not always Gloag et al. 2013; Soler et al. 2014a, b). For example, a combination of monitoring, experimental, video and radio telemetry studies suggests the common cuckoo monitors nests from distant perches (Álvarez 1993; Øien et al. 1996), increases the amount of time spent monitoring a nest the day of host laying (Honza et al. 2002), lays eggs extremely quickly (Davies 2000) and adjusts nest-searching strategies according to the availability of host nests (Table 17.2). In turn, hosts use various sources of direct and indirect information to gauge the risk of being parasitized (Table 17.2). Notably, Thorogood and Davies (2016) recently demonstrated that reed warblers (*Acrocephalus scirpaceus*) use a combination of personal and social information to inform their deployment of defences at latter stages of the nesting cycle. Parasites can further minimize the likelihood of host recognition through a variety of morphological adaptations, which will be discussed in more detail below (see Sects. 17.4.1 and 17.4.2).

17.3.3 Aggression

Hosts and parasites also use physical aggression to defend or gain access to the nest. While not universal, most host species aggressively mob brood parasites that approach their nests (Table 17.2). This behaviour can deter parasitism from occurring (Welbergen and Davies 2009; Feeney et al. 2013) and can even result in the death of the parasite (Gloag et al. 2013). In general, physical aggression appears to be an effective defence against brood parasitism (Welbergen and Davies 2009; Feeney et al. 2013; Fig. 17.1); however, at least some parasites appear able to withstand vicious mobbing while they lay their eggs (e.g. Neudorf and Sealy 1992; Gloag et al. 2013; Soler et al. 2014a, b) or may even incite mobbing in order to gain information about the location of a nest (e.g. Robertson and Norman 1976; Strausberger 1998, but see Gill et al. 1997). Species that do not exhibit evidence of mobbing tend to be small-bodied species that rush to and sit on their nest when the risk of parasitism is high in an attempt to block parasite access (e.g. Hobson and Sealy 1989; Medina and Langmore 2016; but see Gloag et al. 2013; Soler et al. 2014a, b for notable exceptions).

Adult parasites can also use aggressive and violent behaviours to increase the likelihood of their offspring surviving. These adaptations have been best studied in the cowbirds, which have been shown to puncture host eggs to assess nest suitability for parasitism (Massoni and Reboreda 1999), destroy host clutches in order to force

Fig. 17.1 Australian superb fairywrens (*Malurus cyaneus*) mobbing a taxidermied shining bronze-cuckoo (*Chalcites lucidus*) in Canberra, Australia. Photo: William Feeney

renesting (Peer and Sealy 1999; Hoover and Robinson 2007; Dubina and Peer 2012; Swan et al. 2015) and destroy eggs during laying to minimize the competition faced by their chicks after hatching (e.g. Gloag et al. 2013; Fiorini et al. 2014; also see Spottiswoode and Colebrook-Robjent 2007 for a reference to similar behaviours in the greater honeyguide (*Indicator indicator*) and Soler and Martínez 2000 for references to similar behaviours in the great spotted cuckoo). Several parasites have also been suggested to monitor host nests following parasitism and destroy clutches that reject their eggs (e.g. 'mafia' behaviours, Soler et al. 1995b; Hoover and Robinson 2007). These kinds of manipulative behaviours appear to be more common in less virulent parasite species, where hosts do not always surrender their entire reproductive effort through hosting a parasite (but see Peer et al. 2013) and the parasite may therefore not be required to prioritize discretion near the nest (see Chap. 15).

17.4 Morphology

17.4.1 Camouflage

Parasites that are detected near host nests risk being vigorously attacked (Table 17.2) and forfeiting information about the host's risk of parasitism, which may result in an increased likelihood of defences being deployed at latter stages of the nesting cycle (Davies and Brooke 1988; Peer and Sealy 2004; Langmore et al. 2009). Similarly, conspicuous hosts risk becoming easy targets (Banks and Martin 2001; Garamszegi and Avilés 2005). Selection should therefore favour cryptic phenotypes in adult brood parasites and hosts.

Brood parasites exhibit a suite of behaviours that minimize their likelihood of being detected while monitoring host nests (Øien et al. 1996; Honza et al. 2002), and the effectiveness of these behaviours could be increased through cryptic morphological phenotypes. Recent phylogenetic analyses have provided some support to the observation that brood parasites tend to have dull plumages, suggesting that these attributes may be a product of coevolution with hosts (Payne 1967; Krüger et al. 2007; Medina and Langmore 2015). Notably, Krüger et al. (2007) found that, in contrast to other cuckoos, the *Clamator* cuckoos had evolved showier plumages, which may reflect their more conspicuous behaviours around host nests (Macías-Sánchez et al. 2013; Soler et al. 2014a, b; but see Medina and Langmore 2015). Breast barring, which is associated with hawk mimicry in a variety of parasitic cuckoos (Gluckman and Mundy 2013), may also serve to facilitate camouflage while monitoring host nests from tree branches.

17.4.2 Deceptive Plumages

The common cuckoo has become a model for understanding the ecology and evolution of deceptive plumage phenotypes in avian brood parasites. This species exhibits two distinct plumage polymorphisms: a grey and rufous morph (Davies 2000). The grey morph bears a striking resemblance to Eurasian sparrowhawks (*Accipiter* spp.), while the rufous morph has been noted to resemble Eurasian kestrels (*Falco tinnunculus*) (Davies and Welbergen 2008; Thorogood and Davies 2012; Trnka et al. 2015). Numerous studies have demonstrated that both host and non-host species are cautious when presented with models of grey morph cuckoos (Table 17.3), confirming that this cuckoo has evolved mimicry to facilitate parasitism. By contrast, mixed and negative support for mimicry has been found regarding mimicry in the rufous morph (Honza et al. 2006; Trnka et al. 2015). These studies instead support findings that plumage polymorphisms can evolve as a further counter-adaptation against hosts transmitting knowledge of cuckoos throughout a

Table 17.3 Morphological phenotypes that affect the interactions between brood parasites and hosts

	Trait	Purpose	Key reference(s)
Parasite	Dull plumages	Suspected to decrease parasite conspicuousness	Krüger et al. (2007) Medina and Langmore (2015)
	Showy plumages in males	Suspected to assist in male distraction behaviours	Krüger et al. (2007)
	Plumage mimicry	Promotes mistaken identity between the cuckoo and the harmful (Batesian) or harmless (Aggressive) model	Davies and Welbergen (2008)* Trnka and Prokop (2012)* Gluckman and Mundy (2013) Feeney et al. (2015)* Møller et al. (2015)
	Plumage polymorphisms	Decreases the likelihood of information transmission about the presence of the rare morph	Honza et al. (2006)* Thorogood and Davies (2012*, 2013) Trnka et al. (2015)*

Asterisk (*) indicates a manipulative experimental component population (Tables 17.2 and 17.3). This correlated evolution of mimicry and polymorphisms across a wide array of parasite species suggests that this series of coevolved adaptations and counter-adaptations is relatively common among parasite–host systems (Krüger et al. 2007; Thorogood and Davies 2013; Gluckman and Mundy 2013; Tanaka 2016).

Deceptive plumages also appear to exist in other brood parasite species (Table 17.3); however, they remain largely unstudied. Putative examples include drongo-cuckoos (*Surniculus* spp.) mimicking drongos (*Dicrurus* spp.) (Duckworth 1997), Wahlberg's honeyguide (*Prodotiscus regulus*) mimicking small grey flycatchers (Muscicapidae) (Payne 1967) and cuckoo-finches (*Anomalospiza imberbis*) mimicking female *Euplectes* weavers (Feeney et al. 2015). These putative examples of plumage mimicry largely await formal examination (for an exception, see Feeney et al. 2015).

17.4.3 Weapons and Armoury

Despite numerous studies suggesting that violent interactions between brood parasites and hosts may be common at host nests (Table 17.2), almost no research has investigated whether these interactions have selected morphological weaponry and armoury. Adaptations such as thicker skin or denser bones may be present in a variety of brood parasites (Moksnes et al. 2000; Gloag et al. 2013; Soler et al. 2014a, b); however, they await investigation.

17.5 Physiology

Evidence of physiological adaptations resulting from brood parasite–host coevolution remains scarce. Investigation of these kinds of adaptations at this stage of the nesting is largely limited to fast parasite laying, internal egg incubation, and investigation of immune function and hormone profiles in parasites (Dufty et al. 1987; Davies and de Brooke 1988; Kattan 1997; Mermoz and Reboreda 2003; Birkhead et al. 2011; Hahn et al. 2013; Merrill et al. 2013). Building on this, a recent study by Jung et al. (2016) investigated the testosterone profiles of adult common and lesser cuckoos (*C. poliocephalus*) throughout the breeding season and found that they largely resembled those that would be expected in non-brood parasites (also see Dufty et al. 1987).

Concluding Remarks and Future Directions

Frontline interactions, similar to those at latter stages of the nesting cycle, provide tractable opportunities to study coevolutionary processes under natural conditions. Though, unlike interactions at the egg, chick and fledgling

(continued)

stages, frontline interactions primarily concern adult brood parasites and adult hosts, offering opportunities to investigate how selection imposed on hosts by brood parasites, and vice versa, shapes adult phenotypes of the involved species. Recent research has and continues to demonstrate the ecological and evolutionary repercussions of frontline coevolution between these species, such as their ability to explain broad patterns of trait diversity in hosts (Feeney et al. 2013) and parasites (Thorogood and Davies 2013; Gluckman and Mundy 2013). However, they also offer unique avenues for future research, such as further investigation of the causal links between brood parasitism and host breeding systems (Table 17.1), how interactions between adult brood parasites and hosts affect behaviours at latter stages of the nesting cycle (Table 17.2) and whether the cost of brood parasitism has selected cryptic vocal or plumage phenotypes in hosts (Table 17.3). Ultimately, like interactions at other stages of the nesting cycle, a deeper understanding of frontline interactions is key to realizing the ecological and evolutionary repercussions of these iconic relationships.

Acknowledgements WEF would like to thank M. Soler for an invitation to contribute a chapter to this book, as well as B.D. Peer, M. Soler, S.G. Sealy and T. Ryan for helpful comments on the draft. WEF is funded by the Hermon Slade Foundation and the University of Queensland.

References

Álvarez F (1993) Proximity of trees facilitates parasitism by cuckoos *Cuculus canorus* on rufous warblers *Cercotrichas galactotes*. Ibis 135:331–331

Álvarez F (1996) Model cuckoo *Cuculus canorus* eggs accepted by rufous bush chats *Cercotrichas galactotes* during the parasite's absence from the breeding area. Ibis 138:340–342

Astié AA, Scardamaglia RC, Muzio RN, Reboreda JC (2015) Sex differences in retention after a visual or a spatial discrimination learning task in brood parasitic shiny cowbirds. Behav Process 119:99–104

Avilés JM, Stokke BG, Moksnes A et al (2006) Rapid increase in cuckoo egg matching in a recently parasitized reed warbler population. J Evol Biol 19:1901–1910

Banks AJ, Martin TE (2001) Host activity and the risk of nest parasitism by brown-headed cowbirds. Behav Ecol 12:31–40

Birkhead TR, Hemmings N, Spottiswoode CN et al (2011) Internal incubation and early hatching in brood parasitic birds. Proc R Soc Lond [Biol] 278:1019–1024

Boves TJ, Sperry JH, Comolli K, Weatherhead PJ (2014) Brood parasitism of Black-capped Vireos: frontline and post-laying behavioral responses and effects on productivity. J Field Ornithol 85:364–378

Brooker MG, Brooker LC (1989) Cuckoo hosts in Australia. Aust Zool Rev 2:1–67

Brown M, Lawes MJ (2007) Colony size and nest density predict the likelihood of parasitism in the colonial southern red bishop *Euplectes orix* – diederik cuckoo *Chrysococcyx caprius* system. Ibis 149:321–327

Burhans DE (1997) Habitat and microhabitat features associated with cowbird parasitism in two forest edge cowbird hosts. Condor 99:866–872

Campobello D, Sealy SG (2011) Use of social over personal information enhances nest defense against avian brood parasitism. Behav Ecol 22:422–428

Campomizzi AJ, Mathewson HA, Morrison ML et al (2013) Understanding nest success and brood parasitism in the endangered black-capped vireo: comparisons with two sympatric songbirds. Wilson J Ornithol 125:709–719

Canestrari D, Marcos JM, Baglione V (2009) Cooperative breeding in carrion crows reduces the rate of brood parasitism by great spotted cuckoos. Anim Behav 77:1337–1344

Canestrari D, Bolopo D, Turlings TCJ et al (2014) From parasitism to mutualism: unexpected interactions between a cuckoo and its host. Science 343:1350–1352

Clark KL, Robertson RJ (1979) Spatial and temporal multi-species nesting aggregations in birds as anti-parasite and anti-predator defenses. Behav Ecol Sociobiol 5:359–371

Clarke AL, Oien IJ, Honza M et al (2001) Factors affecting reed warbler risk of brood parasitism by the common cuckoo. Auk 118:534

Darwin C (1859) On the origin of species. Murray, London

Davies NB (2000) Cuckoos, cowbirds and other cheats. T. & A.D. Poyer, London

Davies NB, Brooke M de L (1988) Cuckoos versus reed warblers: adaptations and counteradaptations. Anim Behav 36:262–284

Davies NB, Welbergen JA (2008) Cuckoo-hawk mimicry? An experimental test. Proc R Soc Lond [Biol] 275:1817–1822

Davies NB, Welbergen JA (2009) Social transmission of a host defense against cuckoo parasitism. Science 324:1318–1320

Davies NB, Butchart SHM, Burke TA et al (2003) Reed warblers guard against cuckoos and cuckoldry. Anim Behav 65:285–295

de la Colina MA, Hauber ME, Strausberger BM et al (2016) Molecular tracking of individual host use in the shiny cowbird – a generalist brood parasite. Ecol Evol 6:4684–2696

Dubina KM, Peer BD (2012) Egg pecking and discrimination by female and male brown-headed cowbirds. J Ornithol 154:553–557

Duckworth JW (1997) Mobbing of a drongo cuckoo, *Surniculus lugubris*. Ibis 139:190–192

Dufty AM, Goldsmith AR, Wingfield JC (1987) Prolactin secretion in a brood parasite, the brown-headed cowbird, *Molothrus ater*. J Zool 212:669–675

Feeney WE, Langmore NE (2013) Social learning of a brood parasite by its host. Biol Lett 9:20130443

Feeney WE, Langmore NE (2015) Superb Fairy-wrens (*Malurus cyaneus*) increase vigilance near their nest with the perceived risk of brood parasitism. Auk 132:359–364

Feeney WE, Welbergen JA, Langmore NE (2012) The frontline of avian brood parasite–host coevolution. Anim Behav 84:3–12

Feeney WE, Medina I, Somveille M et al (2013) Brood parasitism and the evolution of cooperative breeding in birds. Science 342:1506–1508

Feeney WE, Welbergen JA, Langmore NE (2014) Advances in the study of coevolution between avian brood parasites and their hosts. Annu Rev Ecol Evol Syst 45:227–246

Feeney WE, Troscianko J, Langmore NE, Spottiswoode CN (2015) Evidence for aggressive mimicry in an adult brood parasitic bird, and generalized defences in its host. Proc R Soc Lond [Biol] 282:20150795

Fiorini VD, Gloag R, Kacelnik A, Reboreda JC (2014) Strategic egg destruction by brood-parasitic cowbirds? Anim Behav 93:229–235

Garamszegi L, Avilés J (2005) Brood parasitism by brown-headed cowbirds and the expression of sexual characters in their hosts. Oecologia 143:167–177

Gill SA, Sealy SG (2004) Functional reference in an alarm signal given during nest defence: seet calls of yellow warblers denote brood-parasitic brown-headed cowbirds. Behav Ecol Sociobiol 56:71–80

Gill SA, Neudorf DL, Sealy SG (1997) Host responses to cowbirds near the nest: cues for recognition. Anim Behav 53:1287–1293

Gill SA, Neudorf DLH, Sealy SG (2008) Do hosts discriminate between sexually dichromatic male and female brown-headed cowbirds? Ethology 114:548–556

Gloag R, Fiorini VD, Reboreda JC, Kacelnik A (2013) The wages of violence: mobbing by mockingbirds as a frontline defence against brood-parasitic cowbirds. Anim Behav 86:1023–1029

Gluckman T-L, Mundy NI (2013) Cuckoos in raptors' clothing: barred plumage illuminates a fundamental principle of Batesian mimicry. Anim Behav 86:1165–1181

Grant ND, Sealy SG (2002) Selection of red-winged blackbird (*Agelaius phoeniceus*) hosts by the brown-headed cowbird (*Molothrus ater*). Bird Behav 15:21–30

Gray J, Fraser I (2013) Australian bird names: a complete guide. Csiro Publishing, Collingwood

Grim T, Samaš P, Moskát C et al (2011) Constraints on host choice: why do parasitic birds rarely exploit some common potential hosts? J Anim Ecol 80:508–518

Guigueno MF, Snow DA, MacDougall-Shackleton SA, Sherry DF (2014) Female cowbirds have more accurate spatial memory than males. Biol Lett 10:20140026

Guigueno MF, MacDougall-Shackleton SA, Sherry DF (2015) Sex differences in spatial memory in brown-headed cowbirds: males outperform females on a touchscreen task. PLoS One 10: e0128302

Hahn DC, Summers SG, Genovese KJ et al (2013) Obligate brood parasites show more functionally effective innate immune responses: an eco-immunological hypothesis. Evol Biol 40:544–561

Hauber ME (2003) Interspecific brood parasitism and the evolution of host clutch sizes. Evol Ecol Res 5:559–570

Hauber ME, Russo SA (2000) Perch proximity correlates with higher rates of cowbird parasitism of ground nesting song sparrows. Wilson Bull 112:150–153

Hobson KA, Sealy SG (1989) Responses of yellow warblers to the threat of cowbird parasitism. Anim Behav 38:510–519

Honza M, Taborsky B, Taborsky M et al (2002) Behaviour of female common cuckoos, *Cuculus canorus*, in the vicinity of host nests before and during egg laying: a radiotelemetry study. Anim Behav 64:861–868

Honza M, Šicha V, Procházka P, Ležalová R (2006) Host nest defense against a color-dimorphic brood parasite: great reed warblers (*Acrocephalus arundinaceus*) versus common cuckoos (*Cuculus canorus*). J Ornithol 147:629–637

Hoover JP, Robinson SK (2007) Retaliatory mafia behavior by a parasitic cowbird favors host acceptance of parasitic eggs. Proc Natl Acad Sci USA 104:4479–4483

Jelínek V, Procházka P, Požgayová M, Honza M (2014) Common cuckoos *Cuculus canorus* change their nest-searching strategy according to the number of available host nests. Ibis 156:189–197

Jung W-J, Kim M-S, Noh H-J et al (2016) Hormone profiles of obligate avian brood parasites during the breeding season. Ibis 158:371–379

Kattan GH (1997) Shiny cowbirds follow the 'shotgun' strategy of brood parasitism. Anim Behav 53:647–654

Kilner RM (2005) The evolution of virulence in brood parasites. Ornithol Sci 4:55–64

Krüger O, Davies NB, Sorenson MD (2007) The evolution of sexual dimorphism in parasitic cuckoos: sexual selection or coevolution? Proc R Soc Lond [Biol] 274:1553–1560

Lai CM (1998) Messenger of spring and morality: cuckoo lore in Chinese sources. J Am Orient Soc 118:530–542

Langmore NE, Kilner RM (2007) Breeding site and host selection by Horsfield's bronze-cuckoos, *Chalcites basalis*. Anim Behav 74:995–1004

Langmore NE, Cockburn A, Russell AF, Kilner RM (2009) Flexible cuckoo chick-rejection rules in the superb fairy-wren. Behav Ecol 20:978–984

Langmore NE, Feeney WE, Crowe-Riddell J et al (2012) Learned recognition of brood parasitic cuckoos in the superb fairy-wren *Malurus cyaneus*. Behav Ecol 23:798–805

Louder MIM, Schelsky WM, Albores AN, Hoover JP (2015a) A generalist brood parasite modifies use of a host in response to reproductive success. Proc R Soc Lond B Biol Sci 282:20151615

Louder MIM, Schelsky WM, Benson TJ, Hoover JP (2015b) Brown-headed cowbirds exploit a host's compensatory behavioral response to fecundity reduction. Behav Ecol 26:255–261

Macías-Sánchez E, Martínez JG, Avilés JM, Soler M (2013) Sexual differences in colour and size in the great spotted cuckoo *Clamator glandarius*. Ibis 155:605–610

Martínez JG, Soler M, Soler JJ (1996) The effect of magpie breeding density and synchrony on brood parasitism by great spotted cuckoos. Condor 98:272–278

Massoni V, Reboreda JC (1999) Egg puncture allows shiny cowbirds to assess host egg development and suitability for parasitism. Proc R Soc Lond [Biol] 266:1871–1874

Massoni V, Reboreda JC (2001) Number of close spatial and temporal neighbors decreases the probability of nest failure and shiny cowbird parasitism in colonial yellow-winged blackbirds. Condor 103:521–529

McLaren CM, Sealy SG (2003) Factors influencing susceptibility of host nests to brood parasitism. Ethol Ecol Evol 15:343–353

Medina I, Langmore NE (2015) Coevolution is linked with phenotypic diversification but not speciation in avian brood parasites. Proc R Soc Lond [Biol] 282:20152056

Medina I, Langmore NE (2016) Batten down the thatches: front-line defences in an apparently defenceless cuckoo host. Anim Behav 112:195–201

Mermoz ME, Reboreda JC (2003) Reproductive success of shiny cowbird (*Molothrus bonariensis*) parasitizing the larger brown-and-yellow marshbird (*Pseudoleistes virescens*) in Argentina. Auk 120:1128–1139

Merrill L, O'Loghlen AL, Wingfield JC, Rothstein SI (2013) Immune function in an avian brood parasite and its nonparasitic relative. Physiol Biochem Zool 86:61–72

Middleton ALA (1977) Effect of cowbird parasitism on American Goldfinch nesting. Auk 94:304–307

Moksnes A, Røskaft E (1995) Egg-morphs and host preference in the common cuckoo (*Cuculus canorus*): an analysis of cuckoo and host eggs from European museum collections. J Zool 236:625–648

Moksnes A, Røskaft E, Hagen LG et al (2000) Common cuckoo *Cuculus canorus* and host behaviour at reed warbler *Acrocephalus scirpaceus* nests. Ibis 142:247–258

Møller AP, Antonov A, Stokke BG et al (2011a) Isolation by time and habitat and coexistence of distinct host races of the common cuckoo. J Evol Biol 24:676–684

Møller AP, Saino N, Adamík P et al (2011b) Rapid change in host use of the common cuckoo *Cuculus canorus* linked to climate change. Proc R Soc Lond [Biol] 278:733–738

Møller AP, Stokke BG, Samia DSM (2015) Hawk models, hawk mimics, and antipredator behavior of prey. Behav Ecol 26:1039–1044

Møller AP, Morelli F, Tryjanowski P (2017) Cuckoo folklore and human well-being: cuckoo calls predict how long farmers live. Ecol Indic 72:766–768

Moskát C, Honza M (2000) Effect of nest and nest site characteristics on the risk of cuckoo *Cuculus canorus* parasitism in the great reed warbler *Acrocephalus arundinaceus*. Ecography 23:335–341

Neudorf DL, Sealy SG (1992) Reactions of four passerine species to threats of predation and cowbird parasitism: enemy recognition or generalized responses. Behaviour 123:84–105

Øien IJ, Honza M, Moksnes A, Røskaft E (1996) The risk of parasitism in relation to the distance from reed warbler nests to cuckoo perches. J Anim Ecol 65:147–153

Payne RB (1967) Interspecific communication signals in parasitic birds. Am Nat 101:363–375

Peer BD, Sealy SG (1999) Parasitism and egg puncture behaviour by bronzed and brown-headed cowbirds in sympatry. Stud Avian Biol 18:235–240

Peer BD, Sealy SG (2004) Correlates of egg rejection in hosts of the brown-headed cowbird. Condor 106:580–599

Peer BD, Rivers JW, Rothstein SI (2013) Cowbirds, conservation, and coevolution: potential misconceptions and directions for future research. Chin Birds 4:15–30

Péron G, Altwegg R, Jamie GA, Spottiswoode CN (2016) Coupled range dynamics of brood parasites and their hosts responding to climate and vegetation changes. J Anim Ecol 85:1191–1199

Polačiková L, Procházka P, Cherry M, Honza M (2009) Choosing suitable hosts: common cuckoos *Cuculus canorus* parasitize great reed warblers *Acrocephalus arundinaceus* of high quality. Evol Ecol 23:879–891

Robertson RJ, Norman RF (1976) Behavioral defenses to brood parasitism by potential hosts of the brown-headed cowbird. Condor 78:166–173

Rothstein SI (1990) A model system for coevolution: avian brood parasitism. Annu Rev Ecol Evol Syst 21:481–508

Rutila J, Latja R, Koskela K (2002) The common cuckoo *Cuculus canorus* and its cavity nesting host, the redstart *Phoenicurus phoenicurus*: a peculiar cuckoo-host system? J Avian Biol 33:414–419

Saino N, Rubolini D, Lehikoinen E et al (2009) Climate change effects on migration phenology may mismatch brood parasitic cuckoos and their hosts. Biol Lett 5:539–541

Skjelseth S, Moksnes A, Røskaft E et al (2004) Parentage and host preference in the common cuckoo *Cuculus canorus*. J Avian Biol 35:21–24

Soler M (2014) Long-term coevolution between avian brood parasites and their hosts. Biol Rev Camb Philos Soc 89:688–704

Soler M, Martínez JG (2000) Is egg-damaging behaviour by great spotted cuckoos an accident or an adaptation? Behav Ecol 11:495–501

Soler JJ, Soler M, Moller AP, Martínez JG (1995a) Does the great spotted cuckoo choose magpie hosts according to their parenting ability? Behav Ecol Sociobiol 36:201–206

Soler M, Soler JJ, Martínez JG, Moller AP (1995b) Magpie host manipulation by great spotted cuckoos: evidence for an avian mafia? Evolution 49:770–775

Soler J, Møller A, Soler M (1998) Nest building, sexual selection and parental investment. Evol Ecol 12:427–441

Soler JJ, Martínez JG, Soler M, Møller AP (1999) Host sexual selection and cuckoo parasitism: an analysis of nest size in sympatric and allopatric magpie *Pica pica* populations parasitized by the great spotted cuckoo *Clamator glandarius*. Proc R Soc Lond [Biol] 266:1765–1771

Soler JJ, de Neve L, Martínez JG, Soler M (2001) Nest size affects clutch size and the start of incubation in magpies: an experimental study. Behav Ecol 12:301–307

Soler M, Martín-Vivaldi M, Fernández-Morante J (2012) Conditional response by hosts to parasitic eggs: the extreme case of the rufous-tailed scrub robin. Anim Behav 84:421–426

Soler JJ, Avilés JM, Martin-Gálvez D et al (2014a) Eavesdropping cuckoos: further insights on great spotted cuckoo preference by magpie nests and egg colour. Oecologia 175:105–115

Soler M, Pérez-Contreras T, de Neve L (2014b) Great spotted cuckoos frequently lay their eggs while their magpie host is incubating. Ethology 120:965–972

Sorenson MD, Payne RB (2005) A molecular genetic analysis of cuckoo phylogeny. In: Payne RB (ed) The cuckoos. Oxford University Press, Oxford, pp 68–94

Sorenson MD, Sefc KM, Payne RB (2003) Speciation by host switch in brood parasitic indigobirds. Nature 424:928–931

Spottiswoode CN, Colebrook-Robjent JFR (2007) Egg puncturing by the brood parasitic Greater Honeyguide and potential host counteradaptations. Behav Ecol 18:792–799

Spottiswoode CN, Begg KS, Begg CM (2016) Reciprocal signaling in honeyguide-human mutualism. Science 353:387–389

Strausberger BM (1998) Evident nest-searching behavior of female brown-headed cowbirds while attended by males. Wilson Bull 110:133–136

Strausberger BM, Ashley MV (2005) Host use strategies of individual female brown-headed cowbirds *Molothrus ater* in a diverse avian community. J Avian Biol 36:313–321

Swan DC, Zanette LY, Clinchy M (2015) Brood parasites manipulate their hosts: experimental evidence for the farming hypothesis. Anim Behav 105:29–35

Tanaka KD (2016) Polymorphism in avian brood parasitism: a coevolutionary perspective. Ornithol Sci 15:133–140

Thorogood R, Davies NB (2012) Cuckoos combat socially transmitted defenses of reed warbler hosts with a plumage polymorphism. Science 337:578–580

Thorogood R, Davies NB (2013) Hawk mimicry and the evolution of polymorphic cuckoos. Chin Birds 4:39–50

Thorogood R, Davies NB (2016) Combining personal with social information facilitates host defences and explains why cuckoos should be secretive. Sci Rep 6:19872

Trnka A, Prokop P (2011) Polygynous great reed warblers *Acrocephalus arundinaceus* suffer more cuckoo *Cuculus canorus* parasitism than monogamous pairs. J Avian Biol 42:192–195

Trnka A, Prokop P (2012) The effectiveness of hawk mimicry in protecting cuckoos from aggressive hosts. Anim Behav 83:263–268

Trnka A, Trnka M, Grim T (2015) Do rufous common cuckoo females indeed mimic a predator? An experimental test. Biol J Linn Soc 116:134–143

Ursino CA, Mársico MCD, Sued M et al (2011) Brood parasitism disproportionately increases nest provisioning and helper recruitment in a cooperatively breeding bird. Behav Ecol Sociobiol 65:2279–2286

Uyehara JC, Narins PM (1995) Nest defense by willow flycatchers to brood-parasitic intruders. Condor 97:361–368

Welbergen JA, Davies NB (2008) Reed warblers discriminate cuckoos from sparrowhawks with graded alarm signals that attract mates and neighbours. Anim Behav 76:811–822

Welbergen JA, Davies NB (2009) Strategic variation in mobbing as a front line of defense against brood parasitism. Curr Biol 19:235–240

Welbergen JA, Davies NB (2012) Direct and indirect assessment of parasitism risk by a cuckoo host. Behav Ecol 23:783–789

White DJ, Ho L, Freed-Brown G (2009) Counting chicks before they hatch female cowbirds can time readiness of a host nest for parasitism. Psychol Sci 20:1140–1145

Wiley JW (2012) Anti-brood parasite strategies of naïve populations of nesting birds in Puerto Rico. J Caribbean Ornithol 25:41–63

Woolfenden BE, Gibbs HL, Sealy SG, McMaster DG (2003) Host use and fecundity of individual female brown-headed cowbirds. Anim Behav 66:95–106

Parasitic Behaviour of Interspecific Brood Parasitic Females 18

Juan C. Reboreda, Vanina D. Fiorini, María C. De Mársico, Ros Gloag, and Romina C. Scardamaglia

Abstract

Interspecific avian brood parasites have to solve unique problems associated with their reproductive habit: they need to recognize potential hosts, search for and locate their nests, monitor nests progress and return to them at the appropriate time for egg laying. In addition, parasitic females may improve the survival of their own eggs and chicks by removing or destroying part of the clutch content. Lastly, they should remember the nests in which they have laid eggs to avoid laying two or more eggs in the same nest to prevent harming their own previously laid eggs and generating competition between their own offspring. In this chapter, we summarize information on the behaviour of parasitic females from the moment they start searching for host nests until they parasitize them. We review the different hypotheses for explaining the recognition of hosts and the cues used to search for and locate their nests. We also review the different adaptive explanations for the removal or destruction of eggs as well as the information on competition among females for host nests and repeat parasitism.

J. C. Reboreda (✉) · V. D. Fiorini · M. C. De Mársico
Facultad de Ciencias Exactas y Naturales, Departamento de Ecología, Genética y Evolución & IEGEBA-CONICET, Universidad de Buenos Aires, Buenos Aires, Argentina
e-mail: reboreda@ege.fcen.uba.ar

R. Gloag
Faculty of Science, Behaviour and Genetics of Social Insects Lab, School of Life and Environmental Sciences, University of Sydney, Sydney, NSW, Australia

R. C. Scardamaglia
Facultad de Ciencias Exactas y Naturales, Departamento de Ecología, Genética y Evolución & IEGEBA-CONICET, Universidad de Buenos Aires. Ciudad Universitaria, Buenos Aires, Argentina

M. Soler (ed.), *Avian Brood Parasitism*, Fascinating Life Sciences,
https://doi.org/10.1007/978-3-319-73138-4_18

18.1 Introduction

Unlike other birds, avian brood parasites must locate host nests where to lay their eggs and then decide whether to parasitize them. They should also synchronize parasitism with host laying, as this reduces the probability of rejection and increases the hatching success of parasite eggs and the survival of parasite chicks. Besides, although brood parasites do not provide parental care, they could increase the survival of their eggs and chicks by removing or destroying part of the host clutch. Thus, during the breeding season, parasitic females must take several hierarchical decisions since they start searching for host nests until they parasitize them (Fig. 18.1). To do that they should be capable of recognizing potential hosts (Brooke and Davies 1991; Slagsvold and Hansen 2001), locate and prospect their nests (Wyllie 1981; Clotfelter 1998; Moksnes et al. 2000; Soler and Pérez-Contreras 2012), return to them at the time nests are suitable for parasitism (Scardamaglia et al. 2017) and, at that time, circumvent host frontline defences to gain access to the nest (Feeney et al. 2012). During nest visits they should decide whether to remove or destroy part of the clutch (Davies and Brooke 1988; Sealy 1992; Soler et al. 1997; Peer 2006). They also should avoid parasitizing a nest repeatedly to prevent harm to their own previously laid eggs and competition among their offspring (Hahn et al. 1999; Ellison et al. 2006; Gloag et al. 2014a). All these behaviours affect directly the fitness of parasitic females, and therefore we expect they have been shaped by natural selection.

In this chapter, we review different hypotheses to explain how parasite females recognize their hosts, their spatial behaviour during the breeding season, the cues they use to find host nests, whether they prospect nests before parasitism and the timing of parasitism. We also review the different adaptive explanations for the removal or destruction of eggs and analyse whether parasitic females avoid competition with other females (i.e. multiparasitism) or competition between their own offspring (i.e. repeat parasitism). Finally we discuss future directions of research on parasitic behaviour of brood parasitic females.

18.2 Recognizing Hosts

The most accepted hypothesis to explain host recognition by brood parasites is that females imprint on their foster parents and, once mature, search for nests of the same species in which they were reared ("host preference hypothesis"; Brooke and Davies 1991; Slagsvold and Hansen 2001). This hypothesis predicts that even in generalist parasites (i.e. species that parasitize many hosts), individual females should parasitize preferentially one host (i.e. they should be specialists at individual level). This hypothesis has been directly supported by experiments with village indigobirds, *Vidua chalybeata*, bred in captivity and foster-reared by their normal host or by an experimental foster species and tested as adults for host choice (Payne et al. 2000). Indirect evidence in support of this hypothesis includes the association between host species and parasite's mitochondrial but not nuclear DNA in common cuckoos, *Cuculus canorus* (Gibbs et al. 2000), and greater honeyguides, *Indicator indicator*

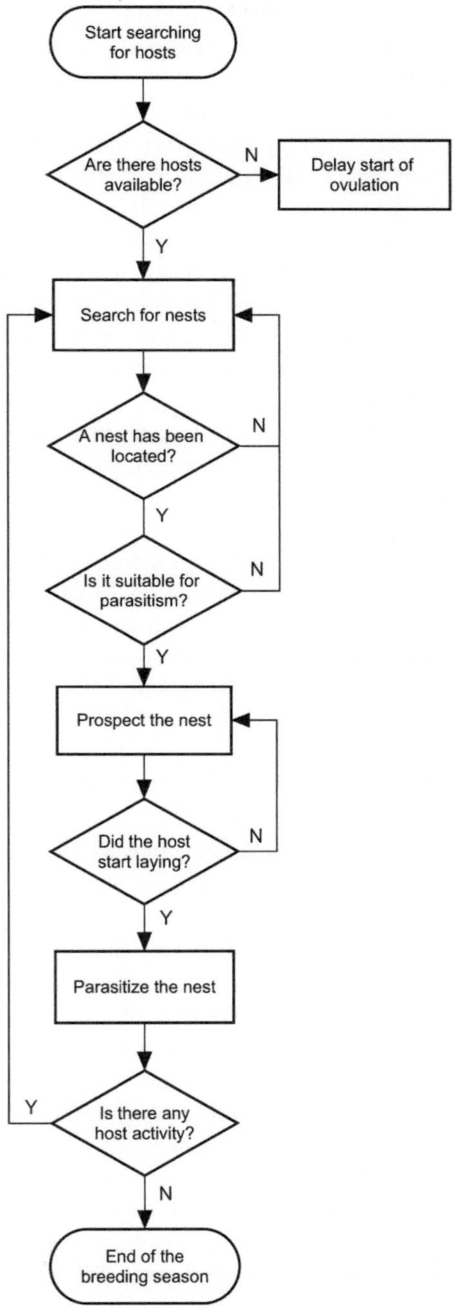

Fig. 18.1 Flow chart showing some of the decisions parasitic females take during the breeding season from the moment they start searching for host nests until they parasitize them. The chart does not include decisions related to the removal or destruction of eggs when they visit host nests

(Spottiswoode et al. 2011). Similarly, Mahler et al. (2007, 2009) have shown an association between host species and parasite's mitochondrial DNA in the shiny, *Molothrus bonariensis*, and the screaming, *M. rufoaxillaris*, cowbirds, which indicates non-random laying behaviour of these parasites.

Host preferences by parasitic females have been directly assessed by parentage analysis using microsatellite DNA markers. Marchetti et al. (1998) and Skjelseth et al. (2004) found that most common cuckoo females were specialized in parasitizing one single host. However, Martínez et al. (1998) found that great spotted cuckoo, *Clamator glandarius*, females use nests of two hosts in the same season and genetic data ruled out the presence of different host-specific races in this species (Baglione et al. 2017). Similarly, studies in the brown-headed cowbird, *M. ater*, showed contrasting results, with females of the same population using both specialist and generalist laying strategies (Alderson et al. 1999; Woolfenden et al. 2003; Strausberger and Ashley 2005), and studies in the shiny cowbird (de la Colina et al. 2016) found that females were mostly faithful to one particular host species throughout a reproductive season but that a few females parasitized more than one host. Thus, parentage analysis indicates that host preferences are not absolute and parasitic females may use more than one host in the same season.

Other hypotheses proposed to explain how parasitic females (mainly common cuckoos) find host nests are "natal philopatry" (females return to where they were born and choose hosts randomly; Brooke and Davies 1991), "nest site choice" (females choose host species with similar eggs and nest sites; Moksnes and Røskaft 1995) and "habitat imprinting" (females learn the characteristics of the habitat in which they grow up by an imprinting-like process and establish their reproductive home ranges in areas that resemble the habitat they had experienced as nestlings; Teuschl et al. 1998; Vogl et al. 2002).

18.3 Searching for Host Nests

The study of spatial behaviour by brood parasitic females during the reproductive season may help to understand how they search for hosts' nests. Unfortunately, data on space use and home ranges of radio-tagged females during the breeding season are only available for cuckoos and cowbirds.

Female common cuckoos spent significantly more time at areas of high host density on laying days than on non-laying days, and reproductive activities were observed predominantly in these areas, with individuals commuting to other areas for foraging (Vogl et al. 2002, 2004). Home ranges of female common cuckoos overlapped, and territories boundaries were not well defined when parasitism frequency was high (Nakamura and Miyazawa 1997). Besides, females seemed to know the location of every nest in their range but laid selectively in some of them (Nakamura et al. 2005). Similarly, in the great spotted cuckoo, territoriality is absent (Martínez et al. 1998), and there is a substantial overlap among home ranges of breeding females (Bolopo et al. 2017).

Female brown-headed cowbirds usually spend the morning in host-rich breeding areas and commute several kilometres to feeding areas for the rest of the day (Rothstein et al. 1984; Thompson 1994; Gates and Evans 1998). Home ranges for this species have generally been estimated by recording one location per animal per day during several days (Dufty 1982; Rothstein et al. 1984; Teather and Robertson 1985; Thompson 1994; Gates and Evans 1998; Hahn et al. 1999), and therefore it has not been possible to evaluate if females maintain stable daily ranges throughout the breeding season. However, females were more closely related to young cowbirds in nests inside than outside their home ranges, which would indicate that they lay eggs within stable home ranges (Hahn et al. 1999).

Shiny and screaming cowbirds showed considerable overlap in the female morning ranges during consecutive days, and the addition of new area to their ranges decreased over time indicating that the size of home ranges tends to an asymptotic value (Scardamaglia and Reboreda 2014). Shiny cowbirds showed sex differences in home ranges, with greater daily and cumulative ranges for males than for females, while these sex differences were not present in screaming cowbirds (Scardamaglia and Reboreda 2014). These data are consistent with the nest-searching behaviour of these species, as female shiny cowbirds search for nests alone while female screaming cowbirds search for host nests together with the male (Mason 1987). In addition to home range fidelity within the breeding season, studies in the brown-headed (Hahn et al. 1999) and the shiny (Scardamaglia and Reboreda 2014) cowbirds have shown range fidelity over successive breeding seasons.

18.4 Locating Host Nests

Various nonmutually exclusive mechanisms of nest location have been proposed (Table 18.1). These mechanisms have received varying support from observational and experimental studies conducted primarily in cuckoos and cowbirds. For instance, whereas the proximity of vantage points was positively related to parasitism risk in some host–parasite systems (i.e. Álvarez 1993; Øien et al. 1996; Clotfelter 1998; Hauber and Russo 2000; Begum et al. 2011), other studies failed to find such effect (i.e. Avilés et al. 2009; Fiorini et al. 2009a; Jelínek et al. 2014). Similarly, the level of host aggression towards parasitic females at the nest was

Table 18.1 Different hypotheses proposed to explain how brood parasitic females locate host nests

Hypothesis	Proposed mechanism	Parasite species
Perch proximity	Parasitic females scan their habitats from vantage points to spot suitable host nests	Common cuckoo (*Cuculus canorus*)[a,b,c] Brown-headed cowbird (*Molothrus ater*)[d,e] Asian koel (*Eudynamys scolopacea*)[f]
Host activity	Parasitic females cue on the level of host activity near the nest	Brown-headed cowbird[g] Great spotted cuckoo (*Clamator glandarius*)[h]
Nest exposure	Nest location is driven by the extent of visual conspicuousness of host nests to parasites	Common cuckoo[b,c,g,i,j,k,l]

References: [a]Álvarez (1993); [b]Øien et al. (1996); [c]Antonov et al. (2007); [d]Hauber and Russo (2000); [e]Clotfelter (1998); [f]Begum et al. (2011); [g]Banks and Martin (2001); [h]Soler and Pérez-Contreras (2012); [i]Moskát and Honza (2000); [j]Clarke et al. (2001); [k]Avilés et al. (2009); [l]Jelínek et al. (2014)

positively related (Robertson and Norman 1977; Smith et al. 1984), negatively related (Briskie et al. 1990) or unrelated (Gill et al. 1997; Clotfelter 1998) to the likelihood of parasitism. These conflicting results suggest that locating appropriate host nests in which to lay may be a complex and hierarchical process and that the mechanisms and cues employed by parasitic females can be influenced by many ecological factors such as host species, habitat structure, nest density and/or time of the breeding season (Martínez et al. 1996; Burhans 1997; Langmore and Kilner 2007; Patten et al. 2011; Jelínek et al. 2014).

In parasite species other than cuckoos and cowbirds, nest-searching strategies are poorly known. Anecdotal observations of captive and free-living individuals of the only precocial obligate brood parasite, the black-headed duck, *Heteronetta atricapilla*, indicate that this species searches for host nests in pairs and monitors host's nesting activities either by passing close to the nests or by sitting quietly in the surrounding vegetation (Rees and Hillgarth 1984; Lyon and Eadie 2013).

Females of the common cuckoo have been observed monitoring host's nesting activity from nearby trees (i.e. Chance 1922; Wyllie 1981). The effect of perch proximity on parasitism risk by cuckoos has received considerable attention (Álvarez 1993; Øien et al. 1996; Moskát and Honza 2000; Antonov et al. 2007), but fewer studies have examined the effect of host activity on cuckoo parasitism. Soler and Pérez-Contreras (2012) tested the effect of host activity on the risk of parasitism of magpie, *Pica pica*, nests by the great spotted cuckoo. They found that cuckoo parasitism was significantly more frequent in natural, active magpie nests during the pre-laying stage than in nests with no eggs and parental activity, whereas nest exposure had no clear effect on the likelihood of parasitism (Soler and Pérez-Contreras 2012). By contrast, the probability of host nests being parasitized by common cuckoos increases with nest exposure (Øien et al. 1996; Moskát and Honza 2000; Clarke et al. 2001; Antonov et al. 2007; Avilés et al. 2009; Jelínek et al. 2014). It is possible that common cuckoo females are first attracted by host activity and then find suitable nests by closer visual inspection or, conversely, they detect more exposed nests first and then cue on host activity to decide where to lay their eggs (Moksnes et al. 2000). Interestingly, the effect of nest exposure on the risk of parasitism may depend on the availability of host nests (Jelínek et al. 2014). More concealed nests of the great reed warbler were less likely to escape parasitism when host pairs breed at lower densities, which indicates that cuckoo females would search for host nests more intensively when suitable nests are in short supply (Jelínek et al. 2014).

Brood parasitic cowbirds exhibit similar nest-searching behaviours with both nest site characteristics and host activity influencing parasitism risk. The perch proximity hypothesis is relatively well supported in the brown-headed cowbird (Clotfelter 1998; Hauber and Russo 2000; Patten et al. 2011). The host activity hypothesis also received support in the brown-headed cowbird (Clotfelter 1998; Banks and Martin 2001; Robinson and Robinson 2001) and the shiny cowbird (Fiorini and Reboreda 2006; Svagelj et al. 2009). However, the effect of host activity arose more clearly in experimental studies that compared parasitism rates between active and inactive host nests (Robinson and Robinson 2001; Fiorini and Reboreda 2006; Svagelj et al. 2009) than in observational studies analysing the relationship between the level of host activity at the nest and the likelihood of parasitism (e.g. McLaren and Sealy 2003; Fiorini et al. 2009a). This would indicate that whereas host activity is key to elicit parasitism, parasitic females might rely primarily on habitat or nest characteristics for host nest detection (i.e. Fiorini and Reboreda 2006). The role of host's nest defensive behaviours towards visiting cowbirds as a cue for nest location is even more controversial (i.e. Gill et al. 1997).

Intriguingly, the social behaviour of some parasite species suggests other potential mechanism of nest location, namely, that individual females eavesdrop on each other's nest-searching activities or follow conspecific females to host nests from communal roosting sites (De Mársico et al. 2010; Gloag et al. 2013). This possibility has some indirect support from studies on the shiny cowbird showing that females overlap extensively in territory use (Scardamaglia and Reboreda 2014), and multiparasitism on a single day often involves a second laying female that tails closely a preceding conspecific on her arrival to the nest (Gloag et al. 2013). Nevertheless, whether eavesdropping provides an alternative nest-searching strategy to brood parasitic females awaits further research.

18.5 Prospecting Host Nests

After searching for and locating a host nest, parasitic females should monitor its progress to ensure they synchronize parasitism with host laying, as this increases the success of the parasite's egg and chick. Some studies have used the observation of the removal or destruction of host eggs before the event of parasitism as indirect evidence of prospecting behaviour by the parasitic female (e.g. Massoni and Reboreda 1999). However, to assume that the same female was responsible for both nest visits could be an important source of error in species at which there are no territories and multiparasitism is common. On the other hand, the use of video records in nest-centred studies (e.g. Moksnes et al. 2000; Gloag et al. 2013) cannot assess the activity of parasitic females around the nest, which increases the likelihood of false negatives (i.e. prospecting visits not recorded). Lastly, traditional tracking of radio-tagged individuals (e.g. Honza et al. 2002) has the drawback that it is quite difficult to continuously monitor the activity of the female, also giving an underestimate of the amount of prospecting behaviour.

Even with these limitations, there is some evidence that common cuckoo females prospect host nests before parasitism. In a study based on continuous video recordings made at nests of reed warblers, *Acrocephalus scirpaceus*, Moksnes et al. (2000) found that there were cases at which the female visited the nest prior to laying one or several times. Honza et al. (2002) examined the behaviour of radio-tagged cuckoo females in the area surrounding host nests during the pre-laying and laying periods and showed that only half of cuckoo nest visits resulted in laying, suggesting that cuckoo females visited host nests before parasitizing them.

Scardamaglia et al. (2017) studied visits to potential host nests by shiny and screaming cowbirds in the periods preceding and overlapping the laying period of their hosts. They recorded the presence of radio-tagged females within a 20 m area around nests of chalked-browed mockingbirds, *Mimus saturninus* (a common host of shiny cowbirds), and baywings, *Agelaioides badius* (the main host of screaming cowbirds), using proximity data loggers placed at host nests. They found that in all cases, females of both species visited potential host nests prior to laying. The number of prospecting visits was higher in screaming than in shiny cowbirds, likely because the host of this parasite begins laying at a less predictable interval after completion of the nest (De Mársico and Reboreda 2008), creating a need for more prospecting visits.

18.6 Timing of Parasitism

In addition to coinciding with the laying period of their hosts, two other aspects of the timing of parasitism can influence a parasite's success: the time of the day that parasitism occurs and the speed of egg laying itself.

Many hosts will actively attack female parasites at their nests, and by doing so may thwart a parasite's attempt to lay (Welbergen and Davies 2009; Gloag et al. 2013). The likelihood that parasites encounter host parents at the nest depends in part on the time of day that parasitism occurs. Cuckoos lay eggs anytime within a broad window of daylight hours (common cuckoo, afternoon-evening, Chance 1940; bronze-cuckoos, morning to early afternoon, *Chrysococcyx sp.*, Brooker et al. 1988; *Chalcites minutillus*, Gloag et al. 2014b; great spotted cuckoo, morning to afternoon, Soler et al. 2014). As hosts must periodically leave the nest during the day for feeding and other activities, female cuckoos thus have opportunities to approach an unoccupied nest. While such large laying windows do not guarantee that egg laying occurs by stealth (e.g. great spotted cuckoos regularly encounter hosts at the nest; Soler et al. 2014, and have behavioural strategies to evade host attention when laying; Álvarez and Arias de Reyna 1974), the unpredictable timing of cuckoo egg laying presumably makes the task of nest guarding against parasitism more difficult for hosts.

Cowbirds, in contrast, lay during a narrow interval of the 30 min or so before sunrise (Peer and Sealy 1999; Ellison and Sealy 2007; Gloag et al. 2013; Scardamaglia et al. 2017). Such a short, predictable time frame for parasitism favours a corresponding short bout of nest vigilance by hosts. For example, chalk-browed mockingbirds assume sentinel positions near their nests prior to sunrise and strive to intercept shiny cowbirds approaching their nests, such that almost all cowbirds at or near nests are aggressively mobbed (Gloag et al. 2013). Yellow warblers, *Dendroica petechia* (Tewksbury et al. 2002), and some orioles (Ellison and Sealy 2007) instead sit tight on their nests during the "parasitism hour" to deter brown-headed cowbird laying or egg removal. Similarly, baywings sit tight on the nest before sunrise, the time at which screaming cowbirds parasitize (De Mársico et al. 2013). At the same time, the low light levels prior to sunrise may provide some advantage to cowbirds, by making them hard for hosts to detect until they are already in the nest. The time of day of egg laying typical in other brood parasitic lineages is not well reported, but it is likely that in all cases this aspect of timing influences the probability that parasites evade host frontline defences and in turn the defensive strategies adopted by hosts.

For all parasites where the act of parasitism has been observed, egg laying occurs exceptionally rapid. Most bird species require 20 min or more to lay an egg (Sealy et al. 1995), but typical times among parasites vary from 5 to 41 s in common cuckoos (Moksnes et al. 2000; Honza et al. 2002; Andou et al. 2005), 4–40 s in great spotted cuckoos (Soler et al. 2014), 41s in brown-headed cowbirds (Sealy et al. 1995), 5–10 s in bronzed cowbirds, *M. aeneus* (Peer and Sealy 1999), and 2–16 s in shiny cowbirds (Gloag et al. 2013). While the actual time the parasite spends at the host nest may be prolonged by the clutch-reduction behaviours that precede laying or by interactions with hosts, the total visit time of these parasites to host nests rarely exceeds 1 min. Such rapid egg laying both reduces the time that a female parasite must endure host attack if detected and increases the chances that she evades host detection entirely.

18.7 Removing and Puncturing Eggs

Many brood parasites reduce host clutch by removing or by pecking and puncturing their eggs (Davies and Brooke 1988; Sealy 1992; Soler et al. 1997; Peer 2006). The host, as part of its nest sanitation behaviour, then removes the punctured egg (Kemal and Rothstein 1988; Soler et al. 1999). These egg removal or egg damaging behaviours can occur at the time the nest is parasitized, in a visit not associated with parasitism or after the nest has been parasitized (Sealy 1992; Gloag et al. 2014a). Different adaptive hypotheses have been proposed for explaining these behaviours including direct benefits for the parasitic female or benefits for the parasitic egg or chick (i.e. higher survival) (Table 18.2).

The first group (direct benefits) include the hypothesis that there are nutritional benefits to parasitic females from eating the removed egg (Davies and Brooke 1988; Scott et al. 1992; Sealy 1992). This hypothesis predicts that egg removal will occur in visits both associated and not associated with parasitism. Scott et al. (1992) found that less than half of the eggs removed by the brown-headed cowbird were eaten, and according to Sealy (1992) evidence supporting this hypothesis is scarce and equivocal.

An alternative direct benefit of egg destruction is the “farming hypothesis” (Arcese et al. 1996; Swan et al. 2015), which states that when a parasitic female finds a nest late in the nesting cycle, by destroying or removing most of the nest content, she may induce hosts to renest, creating new opportunities for parasitism. In this case, egg destruction should occur only in nests that have not been parasitized by the “farming” female (see Chap. 15).

Egg destruction might also confer indirect benefits to female parasites (i.e. to increase the survival of the parasite egg or chick). The “test incubation hypothesis” (Livesey 1936; Massoni and Reboreda 1999) states that by puncturing host eggs, the parasitic female may assess the degree of development of the embryo and decide whether to parasitize the nest or not (i.e. avoid nests at which incubation is advanced).

Table 18.2 Different adaptive hypotheses to explain the removal or puncture of eggs when brood parasitic females visit host nests

Hypotheses	Proposed benefits	Potential costs	Timing of removal or punctures	Occurrence of parasitism	# of eggs removed or punctured
Nutritional benefits[1,2,3]	Direct. Nutritional resources for the female	None	Before and after[a]	Yes or no	One or few
Farming[4]	Direct. New opportunities for parasitism	None	Before	No	Most
Test incubation[5,6]	Indirect. Higher egg and chick survival	None, or nest abandonment	Before	Yes or no	One or few
Incubation efficiency[7–12]	Indirect. Higher hatching success	Nest abandonment	Before	Yes	One or few
Deception[13]	Indirect. Lower probability of egg rejection	None	Before	Yes	One
Reduction of competition[14]	Indirect. Higher chick growth rate and survival	Nest abandonment	Before	Yes	One or few
Mafia[15,16]	Indirect. To enforce the acceptance of parasite eggs	None	After	Yes	Most

Direct benefits are those received by the female, while indirect benefits are those that increase the survival of the parasitic eggs and chicks. Each hypothesis makes different predictions regarding the timing of egg removal or egg punctures (before or after parasitism), whether the nests at which egg removal or egg punctures occur should be parasitized at any time and if clutch destruction should affect one egg, few eggs or most eggs

[a]This would require that parasites recognize and avoid eating their own eggs

References: [1]Davies and Brooke (1988); [2]Scott et al. (1992); [3]Sealy (1992); [4]Arcese et al. (1996); [5]Livesey (1936); [6]Massoni and Reboreda (1999); [7]Davies and Brooke (1988); [8]Lerkelund et al. (1993); [9]McMaster and Sealy (1997); [10]Peer and Bollinger (1997); [11]Peer and Bollinger (2000); [12]Soler et al. (1997); [13]Sealy (1992); [14]Sealy (1992); [15]Zahavi (1979); [16]Soler et al. (1995)

This hypothesis predicts that egg punctures should occur always before the event of parasitism, either in the same visit or in previous visits. It also predicts the occurrence of punctures in unparasitized nests, if such nests contained eggs at an advanced stage of development. Correlational support for this hypothesis has been found in nests of yellow-winged blackbirds, *Agelasticus thilius*, where punctures by shiny cowbirds occurred in advance or simultaneously with parasitism and nests were more frequently parasitized when host eggs were punctured during egg laying or early incubation than during mid or late incubation, when embryos were more developed (Massoni and Reboreda 1999). A similar correlational result was found in nests of creamy-bellied thrushes parasitized by shiny cowbirds (Astié and Reboreda 2009).

The "incubation limit" or "incubation efficiency hypothesis" (Davies and Brooke 1988; Lerkelund et al. 1993; McMaster and Sealy 1997; Peer and Bollinger 1997; Soler et al. 1997) states that by removing or destroying host eggs, the parasitic female increases the efficiency of incubation of its egg. This hypothesis predicts the removal or destruction of host eggs in parasitized nests and applies mainly to hosts larger than the parasite, as the removal of host eggs in these cases improves the contact of smaller parasitic eggs with the brood patch of the host and increases its hatching success. Peer and Bollinger (1997) found that the hatching success of brown-headed cowbird eggs was higher in nests of a large host, the common grackle (*Quiscalus quiscula*), when host eggs were removed. This benefit was reduced or absent when the cowbird egg was larger than host eggs as in the case of the yellow warbler (McMaster and Sealy 1997). Correlational studies conducted in hosts of the shiny cowbird larger than the parasite did not find an effect of the puncture of host eggs on the incubation efficiency of the parasite egg (Astié and Reboreda 2009; Fiorini et al. 2009b).

The "host deception hypothesis" (Sealy 1992) states that egg removal may dupe a host, reducing the likelihood that it detects parasitism and abandons the nest or ejects the parasitic egg. This hypothesis predicts that egg removal should occur in association with parasitism and assumes that some hosts may have the cognitive ability to assess the number of eggs (Hamilton and Orians 1965; Lyon 2003). Although some correlational evidence seems consistent with this hypothesis (Hamilton and Orians 1965; Moksnes and Røskaft 1987), other studies have failed to find supporting evidence (Sealy 1992; Moksnes and Røskaft 1989).

Another hypothesis for explaining the destruction of host eggs in parasitized nests is the "reduction of competition hypothesis" which states that by removing or puncturing host eggs, parasite females can improve the survival of their offspring as a result of the reduction of competition for food with nestmates (Sealy 1992). This hypothesis applies to parasites that do not eject nestmates, and it is most relevant to parasites whose chicks are smaller than host chicks and therefore have to compete for food with larger nestmates. In agreement with this hypothesis, Soler et al. (1997) showed that by damaging magpie eggs, great spotted cuckoos increase the likelihood that late-laid cuckoo eggs hatch. In greater honeyguides, although parasite chicks kill host chicks and therefore they do not have to compete for food with their nestmates, females puncture host eggs. Because chick killing has high energetic costs for the parasitic chick, egg puncture behaviour may benefit the parasitic chick through decreasing the number of nestlings it has to kill (Spottiswoode and Koorevaar 2012). The reduction of competition hypothesis was also proposed as an explanation of egg puncture behaviour in the bronzed cowbird (Carter 1986) and egg removal behaviour in the brown-headed cowbird (Sealy 1992).

Experimental work carried out in hosts of shiny cowbirds found that the reduction of host clutch size benefits the parasitic chick in nests of chalk-browed mockingbird (a host larger than the parasite) by increasing growth rate and chick survival (Fiorini et al. 2009b; Gloag et al. 2012). On the contrary, brown-headed cowbirds chicks reared in nests of eastern phoebes, *Sayornis phoebe*, and shiny cowbirds reared in nests of house wrens, *Troglodytes aedon*, both cases where the host chicks are smaller than the parasite chicks, benefit from the presence of host chicks by increasing their food intake and growth rate (Kilner et al. 2004; Gloag et al. 2012). Thus, nestling parasites may face a "provisions trade-off", whereby the presence of host nestlings can increase or decrease a parasite's food intake depending on whether host young cause parents to supply more extra food than they consume and the ability of the parasitic chick to monopolize those extra feedings (Kilner et al. 2004; Gloag et al. 2012). This hypothesis predicts that generalist brood parasites should adjust the removal or destruction of host eggs according to host characteristics, for example, destroying host eggs when parasitizing large hosts but not when parasitizing small hosts. Tuero et al. (2012) found correlational evidence in support of this hypothesis. Also, experimental work showed that pecking behaviour of shiny and screaming cowbird females is flexible as they adjust the number (and likely also the intensity) of pecks according to the number and eggshell strength of eggs present in the nest (Fiorini et al. 2014; Cossa et al. 2017).

18.8 Competing with Other Females for Host Nests and Avoiding Repeat Parasitism

Multiparasitism, whereby two or more females lay eggs in the same host nest, is common however in those brood parasitic systems where females' territories or home ranges overlap (Moskát and Honza 2002; Spottiswoode 2013; Gloag et al. 2014a; Scardamaglia and Reboreda 2014; Zölei et al. 2015; Bolopo et al. 2017). In such cases, competitive behaviours between females may take place indirectly at host nests themselves. For example, the evolution of thicker eggshells in shiny and screaming cowbirds (Hudson 1874), egg mimicry in honeyguides (Spottiswoode 2013) and egg crypsis in little bronze-cuckoos (Gloag et al. 2014b) have likely been driven by female-female competition in multiply parasitized nests, where second-to-arrive parasites preferentially destroy eggs of previous parasites, in turn selecting for parasite egg traits that evade their competitor's detection. How female parasites adjust other behaviours associated with parasitism to account for female-female competition remains poorly studied.

In addition to avoiding competition with other females for host nests, parasitic females should avoid laying more than one egg in the same host nest (i.e. repeat parasitism). This serves to prevent removing or destroying their own previously laid eggs and to reduce competition among a female's own offspring. The latter is particularly important in evictor parasitic species. To avoid repeat parasitism females should remember the location and status of host nests within their home ranges, which implies special cognitive capabilities (see Chap. 11).

Hahn et al. (1999) and Ellison et al. (2006) found that brown-headed and bronzed cowbirds avoided laying more than one egg in a particular host nest. On the contrary, McLaren et al. (2003) found that repeat parasitism by brown-headed cowbirds was frequent, and Rivers et al. (2012) found, in a heavily parasitized host community, that the likelihood that an individual cowbird in a multiparasitized nest shared the nest with a full sibling was 40%. Likewise, studies in shiny cowbirds produced contrasting results. Gloag et al. (2014a) found that repeat parasitism could have occurred in less than 5% of all recorded events of parasitism, and Scardamaglia et al. (2017) found that shiny cowbird females rarely revisit host nests after parasitism, yet de la Colina et al. (2016) showed that in sites with low host nest density, some shiny cowbird females parasitize the same nest repeatedly. Screaming cowbirds also occasionally show repeat parasitism and revisit a nest repeatedly after laying (Scardamaglia et al. 2017). These results suggest that the incidence of repeat parasitism in non-evictor parasitic species may depend on the density of hosts and that parasitic females may face a trade-off between avoiding laying eggs in nests they have already parasitized and finding new nests.

Concluding Remarks and Future Directions

In this review we summarized information on the behaviour of parasitic females from the moment they start searching for nests until they parasitize them. Most available data comes from two species of cuckoos (common and great spotted) and three species of cowbirds (brown-headed, shiny and screaming). For the remaining parasitic species (most Old World and New World cuckoos, Vidua finches, honeyguides, giant and bronzed cowbirds and the black-headed duck), there is almost no information about their behaviour. The study of female parasitic behaviour in these less studied interspecific brood parasites will help us to identify different evolutionary solutions to similar biological problems. For example, how do most brood parasites recognize potential hosts? Is the proposed mechanism of host imprinting a "universal" solution that evolved independently in most species, or are there numerous species-specific solutions? It would be particularly interesting to study how a precocial obligate brood parasite, the black-headed duck, has solved the problem of host recognition.

The development of new technologies (e.g. miniaturized global positioning system loggers) will allow us to study the spatial behaviour of brood parasites with high spatial and temporal resolution. This will help us to better understand how females gather information on nest sites and nest status before laying their eggs and to understand social interactions between females. For example, to what extent do females use the information of conspecifics to find nests? How does competition between females modify parasitic strategies?

In summary, the behaviour of parasitic females from the moment they start searching for host nests until they parasitize them includes a sequence of hierarchical decisions that affects the fitness of the female: Which host species to parasitize? How to find their nests? When to parasitize them? How many eggs to remove or destroy at the time of parasitism? How to avoid competition with other females or competition between their own offspring? Further observational and experimental studies on the behaviour of brood parasitic females may help us to understand how they have solved these unique problems associated with this reproductive habit.

Acknowledgements We thank William Feeney, Tomás Pérez Contreras and Manuel Soler for very helpful comments on a previous version of this chapter.

References

Alderson GW, Gibbs HL, Sealy SG (1999) Determining the reproductive behaviour of individual brown-headed cowbirds using microsatellite DNA markers. Anim Behav 58:895–905

Álvarez F (1993) Proximity of trees facilitates parasitism by cuckoos *Cuculus canorus* on rufous warblers *Cercotrichas galactotes*. Ibis 135:331–331

Álvarez F, Arias de Reyna L (1974) Mecanismos de parasitación por *Clamator glandarius* y defensa por *Pica pica*. L. Doñana Acta Vertebrata 1:43–65

Andou D, Nakamura H, Oomori S, Higuchi H (2005) Characteristics of brood parasitism by common cuckoos on azure-winged magpies, as illustrated by video recordings. Ornithol Sci 4:43–48

Antonov A, Stokke BG, Moksnes A, Røskaft E (2007) Factors influencing the risk of common cuckoo *Cuculus canorus* parasitism on marsh warblers *Acrocephalus palustris*. J Avian Biol 38:390–393

Arcese P, Smith JN, James NM, Hatch MI (1996) Nest predation by cowbirds and its consequences for passerine demography. Proc Natl Acad Sci U S A 93:4608–4611

Astié AA, Reboreda JC (2009) Function of egg punctures by shiny cowbird in parasitized and unparasitized creamy-bellied thrush nests. J Field Ornithol 80:336–343

Avilés JM, Moskát C, Bán M, Hargitai R, Parejo D (2009) Common cuckoos (*Cuculus canorus*) do not rely on indicators of parental abilities when searching for host nests: the importance of host defenses. Auk 126:431–438

Baglione V, Bolopo D, Canestrari D, Martínez JG, Roldán M, Vila V, Soler M (2017) Spatiotemporal variation of host use in a brood parasite: the role of the environment. Behav Ecol 28:49–58

Banks AJ, Martin TE (2001) Host activity and the risk of nest parasitism by brown-headed cowbirds. Behav Ecol 12:31–40

Begum S, Moksnes A, Røskaft E, Stokke BG (2011) Factors influencing host nest use by the brood parasitic asian koel (*Eudynamys scolopacea*). J Ornithol 152:793–800

Bolopo D, Canestrari D, Martínez JG, Roldan M, Macías-Sanchez E, Vila M, Soler M, Baglione V (2017) Flexible mating patterns in an obligate brood parasite. Ibis 159:103–112

Briskie JV, Sealy SG, Hobson KA (1990) Differential parasitism of least flycatchers and yellow warblers by the brown-headed cowbird. Behav Ecol Sociobiol 27:403–410

Brooke M de L, Davies NB (1991) A failure to demonstrate host imprinting in the cuckoo (*Cuculus canorus*) and alternative hypotheses for the maintenance of egg mimicry. Ethology 89:154–166

Brooker MG, Brooker LC, Rowley I (1988) Egg deposition by the bronze-cuckoos *Chrysococcyx basalis* and *Ch. lucidus*. Emu 88:107–109

Burhans DE (1997) Habitat and microhabitat features associated with cowbird parasitism in two forest edge cowbird hosts. Condor 99:866–872

Carter MD (1986) The parasitic behavior of the bronzed cowbird *Molothrus aeneus* in South Texas USA. Condor 88:11–25

Chance EP (1922) The cuckoo's secret. Sidgwick & Jackson, London

Chance EP (1940) The truth about the cuckoo. Country Life, London

Clarke AL, Øien IJ, Honza M, Moksnes A, Røskaft E (2001) Factors affecting reed warbler risk of brood parasitism by the common cuckoo. Auk 118:534–538

Clotfelter ED (1998) What cues do brown-headed cowbirds use to locate red-winged blackbird host nests? Anim Behav 55:1181–1189

Cossa N, Tuero DT, Reboreda JC, Fiorini VD (2017) Egg pecking and puncturing behaviors in shiny and screaming cowbirds: effects of eggshell strength and degree of clutch completion. Behav Ecol Sociobiol 71:60. https://doi.org/10.1007/s00265-017-2289-1

Davies NB, Brooke MDL (1988) Cuckoos versus reed warblers: adaptations and counteradaptations. Anim Behav 36:262–284

de la Colina MA, Hauber ME, Strausberger BM, Reboreda JC, Mahler B (2016) Molecular tracking of individual host use in the shiny cowbird, a generalist brood parasite. Ecol Evol 6:4684–4696

De Mársico MC, Reboreda JC (2008) Egg-laying behaviour in screaming cowbirds. Why does a specialist brood parasite waste so many eggs? Condor 110:143–153

De Mársico MC, Mahler B, Chomnalez M, Di Giácomo AG, Reboreda JC (2010) Host use by generalist and specialist brood-parasitic cowbirds at population and individual levels. Adv Study Behav 42:83–121

De Mársico MC, Gloag R, Ursino CA, Reboreda JC (2013) A novel method of rejection of brood parasitic eggs reduces parasitism intensity in a cowbird host. Biol Lett 9:007

Dufty AM (1982) Movements and activities of radio-tracked brown-headed cowbirds. Auk 99:316–327

Ellison K, Sealy SG (2007) Small hosts infrequently disrupt laying by brown-headed cowbirds and bronzed cowbirds. J Field Ornithol 78:379–389

Ellison K, Sealy SG, Gibbs HL (2006) Genetic elucidation of host use by individual sympatric bronzed cowbirds (*Molothrus aeneus*) and brown-headed cowbirds (*M. ater*). Can J Zool 84:1269–1280

Feeney WE, Welbergen JA, Langmore NE (2012) The frontline of avian brood parasite-host coevolution. Anim Behav 84:3–12

Fiorini VD, Reboreda JC (2006) Cues used by shiny cowbirds (*Molothrus bonariensis*) to locate and parasitise chalk-browed mockingbird (*Mimus saturninus*) nests. Behav Ecol Sociobiol 60:379–385

Fiorini VD, Tuero DT, Reboreda JC (2009a) Host behaviour and nest-site characteristics affect the likelihood of brood parasitism by shiny cowbirds on chalk-browed mockingbirds. Behaviour 146:1387–1403

Fiorini VD, Tuero DT, Reboreda JC (2009b) Shiny cowbirds synchronize parasitism with host laying and puncture host eggs according to host characteristics. Anim Behav 77:561–568

Fiorini VD, Gloag R, Reboreda JC, Kacelnik A (2014) Strategic egg destruction by brood parasitic cowbirds? Anim Behav 93:229–235

Gates JE, Evans DR (1998) Cowbirds breeding in the central Appalachians: spatial and temporal patterns and habitat selection. Ecol Appl 8:27–40

Gibbs HL, Marchetti K, Sorenson MD, Brooke M d L, Davies NB, Nakamura H (2000) Genetic evidence for female host-specific races of the common cuckoo. Nature 407:183–186

Gill SA, Grieef PM, Staib LM, Sealy SG (1997) Does nest defence deter or facilitate cowbird parasitism? A test of the nesting-cue hypothesis. Ethology 103:56–71

Gloag R, Tuero DT, Fiorini VD, Reboreda JC, Kacelnik A (2012) The economics of nestmate-killing in avian brood parasites: a provisions trade-off. Behav Ecol 23:132–140

Gloag R, Fiorini VD, Reboreda JC, Kacelnik A (2013) The wages of violence: mobbing by mockingbirds as a frontline defence against brood-parasitic cowbirds. Anim Behav 86:1023–1029

Gloag R, Fiorini VD, Reboreda JC, Kacelnik A (2014a) Shiny cowbirds share nests, but not mothers, in multiply parasitized mockingbird broods. Behav Ecol Sociobiol 68:681–689

Gloag R, Keller LA, Langmore NE (2014b) Cryptic cuckoo eggs hide from competing cuckoos. Proc R Soc B 281:20141014–20141014

Hahn D, Sedgwick J, Painter I, Casna N (1999) A spatial and genetic analysis of cowbird host selection. Stud Avian Biol 18:204–217

Hamilton WJ, Orians GH (1965) Evolution of brood parasitism in altricial birds. Condor 67:361–382

Hauber ME, Russo SA (2000) Perch proximity correlates with higher rates of cowbird parasitism of ground nesting song sparrows. Wilson Bull 112:150–153

Honza M, Taborsky B, Taborsky M, Teuschl Y, Vogl W, Moksnes A, Roskaft E (2002) Behaviour of female common cuckoos, *Cuculus canorus*, in the vicinity of host nests before and during egg laying: a radiotelemetry study. Anim Behav 64:861–868

Hudson WH (1874) Notes on the procreant instincts of the three species of *Molothrus* found in Buenos Ayres. Proc Zool Soc London 42(1):153–174

Jelínek V, Procházka P, Požgayová M, Honza M (2014) Common cuckoos *Cuculus canorus* change their nest-searching strategy according to the number of available host nests. Ibis 156:189–197

Kemal RE, Rothstein SI (1988) Mechanism of avian egg recognition: adaptive responses to eggs with broken shells. Anim Behav 36:175–183

Kilner RM, Madden JR, Hauber ME (2004) Brood parasitic cowbird nestlings use host young to procure resources. Science 305:877–879

Langmore NE, Kilner RM (2007) Breeding site and host selection by Horsfield's bronze-cuckoos, *Chalcites basalis*. Anim Behav 74:995–1004

Lerkelund HE, Moksnes A, Roskaft E, Ringsby TH (1993) An experimental test of optimal clutch size of the fieldfare; with a discussion on why brood parasites remove eggs when they parasitize a host species. Ornis Scand 24:95–102

Livesey TR (1936) Cuckoo problems. J Bombay Nat Hist Soc 38:734–758

Lyon BE (2003) Egg recognition and counting reduce costs of avian conspecific brood parasitism. Nature 422:495–499

Lyon BE, Eadie JM (2013) Patterns of host use by a precocial obligate brood parasite, the black-headed duck: ecological and evolutionary considerations. Chin Birds 4:71–85

Mahler B, Confalonieri VA, Lovette IJ, Reboreda JC (2007) Partial host fidelity in nest selection by the shiny cowbird (Molothrus bonariensis), a highly generalist avian brood parasite. J Evol Biol 20:1918–1923

Mahler B, Sarquis Adamson Y, Di Giacomo AG, Confalonieri VA, Reboreda JC (2009) Utilization of a new host in the host-specialist brood parasite *Molothrus rufoaxillaris*: host switch or host acquisition? Behav Ecol Sociobiol 63:1603–1608

Marchetti K, Gibbs HL, Nakamura H (1998) Host-race formation in the common cuckoo. Science 282:471–472

Martínez JG, Soler M, Soler JJ (1996) The effect of magpie breeding density and synchrony on brood parasitism by great spotted cuckoos. Condor 98:272–278

Martínez JG, Burke T, Dawson D, Soler JJ, Soler M, Moller AP (1998) Microsatellite typing reveals mating patterns in the brood parasitic great spotted cuckoo (*Clamator glandarius*). Mol Ecol 7:289–297

Mason P (1987) Pair formation in cowbirds: evidence found for screaming but not shiny cowbirds. Condor 89:349–356

Massoni V, Reboreda JC (1999) Egg puncture allows shiny cowbirds to assess host egg development and suitability for parasitism. Proc R Soc B 266:1871–1874

McLaren CM, Sealy SG (2003) Factors influencing susceptibility of host nests to brood parasitism. Ethol Ecol Evol 15:343–353

McLaren CM, Woolfenden BE, Gibbs HL, Sealy SG (2003) Genetic and temporal patterns of multiple parasitism by brown-headed cowbirds (*Molothrus ater*) on song sparrows (*Melospiza melodia*). Can J Zool 81:281–286

McMaster DG, Sealy GS (1997) Host-egg removal by brown-headed cowbirds: a test of the host incubation limit hypothesis. Auk 114:212–220

Moksnes A, Røskaft E (1987) Cuckoo host interactions in Norwegian mountain areas. Ornis Scand 18:168–172

Moksnes A, Røskaft E (1989) Adaptations of meadow pipits to parasitism by the common cuckoo. Behav Ecol Sociobiol 24:25–30

Moksnes A, Røskaft E (1995) Egg-morphs and host preference in the common cuckoo (*Cuculus canorus*): an analysis of cuckoo and host eggs from European museum collections. J Zool 236:625–648

Moksnes A, Røskaft E, Hagen LG, Honza M, Mørk C, Olsen PH (2000) Common cuckoo *Cuculus canorus* and host behaviour at reed warbler *Acrocephalus scirpaceus* nests. Ibis 142:247–258

Moskát C, Honza M (2000) Effect of nest and nest site characteristics on the risk of cuckoo *Cuculus canorus* parasitism in the great reed warbler *Acrocephalus arundinaceus*. Ecography 23:335–341

Moskát C, Honza M (2002) European cuckoo (*Cuculus canorus*) parasitism and host's rejection behaviour in a heavily parasitized great reed warbler (*Acrocephalus arundinaceus*) population. Ibis 144:614–622

Nakamura H, Miyazawa Y (1997) Movements, space use and social organisation of radio tracked common cuckoos during the breeding season in Japan. Jpn J Ornithol 46:23–54

Nakamura H, Miyazawa Y, Kashiwagi K (2005) Behavior of radio-tracked common cuckoo females during the breeding season in Japan. Ornithol Sci 4:31–41

Øien IJ, Honza M, Moksnes A, Røskaft E (1996) The risk of parasitism in relation to the distance from reed warbler nests to cuckoo perches. J Anim Ecol 65:147–153

Patten MA, Reinking DL, Wolfe DH (2011) Hierarchical cues in brood parasite nest selection. J Ornithol 152:521–532

Payne RB, Payne LL, Woods JL, Sorenson MD (2000) Imprinting and the origin of parasite–host species associations in brood-parasitic indigobirds, *Vidua chalybeata*. Anim Behav 59:69–81

Peer BD (2006) Egg destruction and egg removal by avian brood parasites: adaptiveness and consequences. Auk 123:16–22

Peer BD, Bollinger EK (1997) Explanations for the infrequent cowbird parasitism on common grackles. Condor 99:151–161

Peer BD, Sealy SG (1999) Laying time of the bronzed cowbird. Wilson Bull 111:137–139

Peer BD, Bollinger EK (2000) Why do female brown-headed cowbirds remove host eggs? A test of the incubation efficiency hypothesis. In: Smith JNM, Cook TL, Rothstein SI, Robinson SK, Sealy SG (eds) Ecology and management of cowbirds and their hosts: studies in the conservation of North American Passerine birds. Smith University of Texas Press, Texas, pp 187–192

Rees EC, Hillgarth N (1984) The breeding biology of captive black-headed ducks and the behavior of their young. Condor 86:242–250

Rivers JW, Young S, Gonzalez EG, Horton B, Lock J, Fleischer RC (2012) High levels of relatedness between brown-headed cowbird (*Molothrus ater*) nestmates in a heavily parasitized host community. Auk 129:623–631

Robertson RJ, Norman RF (1977) The function and evolution of aggressive host behavior towards the brown-headed cowbird (*Molothrus ater*). Can J Zool 55:508–518

Robinson WD, Robinson TR (2001) Is host activity necessary to elicit brood parasitism by cowbirds? Ethol Ecol Evol 13:161–171

Rothstein S, Verner J, Stevens E (1984) Radio-tracking confirms a unique diurnal pattern of spatial occurrence in the parasitic brown-headed cowbird. Ecology 65:77–88

Scardamaglia RC, Reboreda JC (2014) Ranging behavior of female and male shiny cowbirds and screaming cowbirds while searching for host nests. Auk 131:610–618

Scardamaglia RC, Fiorini VD, Kacelnik A, Reboreda JC (2017) Planning host exploitation through prospecting visits by parasitic cowbirds. Behav Ecol Sociobiol 71:23. https://doi.org/10.1007/s00265-016-2250-8

Scott DM, Weatherhead PJ, Ankney CD (1992) Egg-eating by female brown-headed cowbirds. Condor 94:579–584

Sealy SG (1992) Removal of yellow warbler eggs in association with cowbird parasitism. Condor 94:40–54

Sealy SG, Neudorf DLH, Hill DP (1995) Rapid laying by brown-headed cowbirds *Molothrus ater* and other parasitic birds. Ibis 137:76–84

Skjelseth S, Moksnes A, Røskaft E, Kleven O, Gibbs HL, Taborsky M, Taborsky B, Honza M (2004) Parentage and host preference in the common cuckoo *Cuculus canorus*. J Avian Biol 35:21–24

Slagsvold T, Hansen BT (2001) Sexual imprinting and the origin of obligate brood parasitism in birds. Am Nat 158:354–367
Smith JN, Arcese P, McLean IG (1984) Age, experience, and enemy recognition by wild song sparrows. Behav Ecol Sociobiol 14:101–106
Soler M, Pérez-Contreras T (2012) Location of suitable nests by great spotted cuckoos: an empirical and experimental study. Behav Ecol Sociobiol 66:1305–1310
Soler M, Soler JJ, Martinez JG, Møller AP (1995) Magpie host manipulation by great spotted cuckoos: evidence for an avian mafia? Evolution 49:770–775
Soler M, Soler JJ, Martínez JG (1997) Great spotted cuckoos improve their reproductive success by damaging magpie host eggs. Anim Behav 54:1227–1233
Soler M, Soler JJ, Perez-Contreras T (1999) The cost of host egg damage caused by a brood parasite: experiments on great spotted cuckoos (*Clamator glandarius*) and magpies (*Pica pica*). Behav Ecol Sociobiol 46:381–386
Soler M, Pérez-Contreras T, de Neve L (2014) Great spotted cuckoos frequently lay their eggs while their magpie host is incubating. Ethology 120:965–972
Spottiswoode CN (2013) A brood parasite selects for its own egg traits. Biol Lett 9:20130573
Spottiswoode CN, Koorevaar J (2012) A stab in the dark: chick killing by brood parasitic honeyguides. Biol Lett 8:241–244
Spottiswoode CN, Stryjewskic KF, Quadera S, Colebrook-Robjentd JFR, Sorenson MD (2011) Ancient host specificity within a single species of brood parasitic bird. Proc Natl Acad Sci U S A 108:17738–17742
Strausberger BM, Ashley MV (2005) Host use strategies of individual female brown-headed cowbirds *Molothrus ater* in a diverse avian community. J Avian Biol 36:313–321
Svagelj WS, Fernández GJ, Mermoz ME (2009) Effects of nest-site characteristics and parental activity on cowbird parasitism and nest predation in brown-and-yellow marshbirds. J Field Ornithol 80:9–18
Swan DC, Zanette LY, Clinchy M (2015) Brood parasites manipulate their hosts: experimental evidence for the farming hypothesis. Anim Behav 105:29–35
Teather KL, Robertson RJ (1985) Female spacing patterns in brown-headed cowbirds. Can J Zool 63:218–222
Teuschl Y, Taborsky B, Taborsky M (1998) How do cuckoos find their hosts? The role of habitat imprinting. Anim Behav 56:1425–1433
Tewksbury JJ, Martin TE, Hejl SJ, Kuehn MJ, Jenkins JW (2002) Parental care of a cowbird host: caught between the costs of egg-removal and nest predation. Proc R Soc B 269:423–429
Thompson F (1994) Temporal and spatial patterns of breeding brown-headed cowbirds in the midwestern United States. Auk 111:979–990
Tuero DT, Fiorini VD, Reboreda JC (2012) Do shiny cowbird females adjust egg pecking behavior according to the competition their chicks face in host nests? Behav Process 89:137–142
Vogl W, Taborsky B, Teuschl Y, Taborsky M, Honza M (2002) Cuckoo females preferentially use specific habitats when searching for host nests. Anim Behav 64:843–850
Vogl W, Taborsky M, Taborsky B, Teuschl Y, Honza M (2004) Habitat and space use of European cuckoo females during the egg laying period. Behaviour 141:881–898
Welbergen JA, Davies NB (2009) Strategic variation in mobbing as a front line of defense against brood parasitism. Curr Biol 19:235–240
Woolfenden BE, Gibbs HL, Sealy SG, McMaster DG (2003) Host use and fecundity of individual female brown-headed cowbirds. Anim Behav 66:95–106
Wyllie I (1981) The cuckoo. Batsford, London
Zahavi A (1979) Parasitism and nest predation in parasitic cuckoos. Am Nat 113:157–159
Zölei A, Bán M, Moskát C (2015) No change in common cuckoo *Cuculus canorus* parasitism and Great Reed Warblers' *Acrocephalus arundinaceus* egg rejection after seven decades. J Avian Biol 46:570–576

Part VII

Coevolutionary Interactions at the Laying Stage

Egg Color Polymorphism in Brood Parasites and Their Hosts: Adaptation and Evolution 19

Canchao Yang, Wei Liang, and Anders Pape Møller

Abstract

Polymorphism refers to the diversity of clearly distinct phenotypes within biological populations, and it is one of the extraordinary phenomena in biodiversity that is well known in predator–prey interactions. In avian brood parasitism, coevolution between parasites and hosts can also cause evolution of polymorphism because the egg traits of both parties are subject to selection causing mimetic parasite eggs to evolve as a response to host egg rejection, which in turn selects for the evolution of egg polymorphism in hosts, which further promotes egg polymorphism in parasites. Here we review previous studies and demonstrate that (1) egg polymorphism is an adaptation and counteradaptation in both hosts and parasitic cuckoos, (2) egg polymorphism has fitness consequences for both hosts and cuckoos, and (3) egg polymorphism in the cuckoo-host system changes temporally and spatially. Egg polymorphism can evolve as an anti-parasite strategy in cuckoo hosts, and it provides effective defenses to dramatically reduce the success of cuckoo parasitism. However, compared to hosts, egg polymorphism in cuckoos is simply a result of frequency-dependent selection, but not an effective counteradaptation to increase the success of cuckoos. If we consider each cuckoo species as a global population, egg polymorphism is common in most cuckoos. However, egg polymorphism in hosts is much more complex since it changes temporally and varies spatially among populations. It

C. Yang · W. Liang (✉)
Ministry of Education Key Laboratory for Ecology of Tropical Islands, College of Life Sciences, Hainan Normal University, Haikou, P. R. China
e-mail: liangwei@hainnu.edu.cn

A. P. Møller
Ecologie Systématique Evolution, Université Paris-Sud, CNRS, AgroParisTech, Université Paris-Saclay, Orsay Cedex, France

M. Soler (ed.), *Avian Brood Parasitism*, Fascinating Life Sciences,
https://doi.org/10.1007/978-3-319-73138-4_19

may be explained by the history of the interaction between cuckoos and their hosts and the intensity of selection linked to cuckoo diversity.

19.1 Introduction

There is an extraordinary variety and variability in life on earth termed biodiversity, which can be expressed in numerous ways (Magurran 2010). Polymorphism is one kind of biodiversity among individuals within populations, which is defined as the occurrence of two or more distinct and different morphs or forms, also referred to as alternative phenotypes, in the population of a species (Ford 1965). Polymorphism in natural populations generally evolves as a consequence of frequency-dependent selection imposed by predators. For example, bird eggs vary considerably among species and among individuals within a population in color and patterns of their eggshells (Kilner 2006), which are driven by differential sensitivity of color morphs to the environment (Roulin 2004). In brood parasites and their hosts, egg appearances are important traits in the battlefield of arms races between parasites and hosts because parasites evolve mimetic eggs to cheat their hosts, while hosts evolve polymorphic eggs to increase the contrasts between parasite and host eggs, and this may in turn select for the evolution of egg polymorphism in cuckoos just as in predator-prey interactions.

Therefore, avian brood parasitism provides a unique model system for investigating the evolution of egg polymorphism, as egg color is the main trait under selection in both parasites and hosts (Davies 2000; Soler 2014). Theoretical models suggest that discrete egg morphs can coexist evolutionarily both in host and parasite populations (Takasu 1998; Liang et al. 2012). Thus, there may be marked temporal and spatial variation in coevolved adaptations among hosts and parasites in general and among brood parasites and their hosts in particular (Yang et al. 2010). Here, we provide a general review of these questions.

19.2 Egg Color Diversity and Its Functions

Since egg appearances have evolved into numerous colors and patterns during evolution, selective agents should favor divergence of eggs. Furthermore, different kinds of color and pattern that have evolved in birds would possess adaptations for a variety of reasons.

19.2.1 The Function of Egg Color

Female birds generate pigments and deposit them on the surface of their eggs before laying (Rao et al. 1993). Therefore, pigments on eggshells are responsible for egg color, including ground color and markings. Egg color varies considerably among species, and the ancestral color of bird eggs is believed to be immaculate white, while different colors and patterns of egg color have subsequently evolved as different specific adaptations (Wallace 1889). Logically, the formation and persistence of egg color are the result of natural and sexual selection and thus possess adaptive functions, as reptiles and most cavity-nesting birds produce white eggs, while most birds that lay eggs in exposed nests produce pigmented eggs with variable colors and patterns. According to these hypotheses, egg coloration may be selected for crypsis or aposematism to reduce predation risk (Wallace 1889; Swynnerton 1916) or selected by male mates (Moreno and Osorno 2003). Another hypothesis suggests that eggs can be selected in the context of thermoregulation to resist high or low temperatures that would influence embryo development (Montevecchi 1976; Bertram and Burger 1981).

Brood parasitism was proposed as a hypothesis to explain the evolution of egg coloration (Kilner 2006). However, there is no indication that brood parasitism causes specific directional change in host egg coloration (Davies and de Brooke 1989). Moreover, although brood parasitism can be regarded as a specific type of predation, the search for host nests is based on host activities rather than host nests or eggs (Banks and Martin 2001; Patten et al. 2011; Soler and Pérez-Contreras 2012; Yang et al. 2016b). We suggest that although the evolution of egg coloration in hosts may not be selected by brood parasites, it may impact on and cause changes in egg coloration in brood parasites because egg mimicry is undoubtedly an important adaptation for parasitism that increases the acceptance success by hosts. Predation risk is reasonably a much more general and robust explanation for evolution of egg color because the threat for the majority of bird nests is logically predation. However, the aposematism hypothesis, which suggests that pigmented eggs are unpalatable to reduce predation risk, has received only weak evidence (Swynnerton 1916; Cott 1948, 1951, 1953). The crypsis hypothesis seems to be a general explanation for the evolution of egg color due to predators, although experimental evidence is rather mixed (Tinbergen et al. 1962; Montevecchi 1976; Götmark 1992; Weidinger 2001). This is probably due to the flawed methodology that attempts to test the effect of egg color on predation rates by painting eggs because such artificial egg color looks different according to human and avian vision (Underwood and Sealy 2002; Kilner 2006). Nevertheless, some studies still provide persuasive evidence that birds were able to arrange the background in an attempt to make the phenotype of the eggs match that of the background (Solis and de Lope 1995; Lovell et al. 2013), and the degree of matching between eggs and the background is negatively correlated with predation rate (Lee et al. 2010; Stoddard et al. 2016).

In summary, there is little evidence to support the aposematism and the male investment hypotheses, while for the thermoregulation hypothesis, it is generally restricted to precocial birds that build exposed nests on the ground. The brood parasitism hypothesis may be partly supported by changes in host egg coloration not being driven by brood parasites, although they determine changes in eggs of brood parasites. Finally, the crypsis hypothesis seems to be a better alternative to explain the evolution of egg coloration.

19.2.2 Egg Color Diversity Selected by Predators

If crypsis is the only function of egg coloration, we should expect that all eggs were cryptically brown or green with spots that rendered eggs inconspicuous. That is clearly not the case because egg phenotypes vary enormously among species, but also among subspecies and even among individuals within a population (Yang et al. 2010, 2016a). Although the main function of egg color may be selected by nest predators (Wallace 1889), predators may also explain the diversity of egg coloration. To answer this question, we need to separate it into two parts. The diversity of egg color should be divided into (1) diversity among species and (2) diversity within species. The latter case also called egg polymorphism refers to the occurrence of two or more clearly distinct egg phenotypes in the populations of a species. Diversity among species may be reasonably explained by selection by predators because different species of birds build different types of nests in different environments and possess different kinds of behavior to care for their offspring. Thus, it is logical to argue that diversity in egg color among species is an adaptation to their specific microenvironments although they are all selected by predators. However, this could not explain the diversity of egg color within species (i.e., egg polymorphism). Predation risk is found to be strongly related to nest site rather than egg appearance, probably because egg appearance has released eggs from their original function of being cryptic due to the evolution of bird nests and site selection (Martin and Roper 1988; Forstmeier and Weiss 2004; Davis 2005; Kilner 2006). Therefore, this hypothesis cannot explain egg polymorphism within a species or a single population of a species.

19.2.3 Egg Color Diversity Selected by Brood Parasites

Host egg polymorphism is selected by brood parasites (Lahti 2005; Yang et al. 2010, 2016a). For example, plaintive cuckoos (*Cacomantis merulinus*) suffer a low success rate of parasitism because their common tailorbird (*Orthotomus sutorius*) hosts lay polymorphic eggs and reject all unmatched cuckoo eggs, while cuckoos cannot choose to lay eggs in host nests that match the appearance of their own eggs (Yang et al. 2016a). Logically, it is possible that brood parasites drive the evolution of egg color diversity among host species if egg polymorphism selected by brood parasites causes host individuals with different egg phenotypes to diverge and finally become different species. In other words, if egg polymorphism induces reproductive isolation among individuals of different egg phenotypes, this will cause divergence of novel species and thus contribute to the evolution of interspecific diversity of egg color.

19.2.4 Egg Color Diversity Selected by Hosts

Nest predators should select for homogeneous eggs because any deviant egg should be more easily detected by a predator when the eggs are not covered by the incubating nest owner. Thus, color monomorphism or lack of heterogeneity should be the ancestral state. Eventually, such monomorphism may result in polymorphism in host eggs because any host with a deviant egg phenotype will benefit in terms of avoidance of brood parasitism. In turn, monomorphism in cuckoo eggs should result in polymorphism because any novel egg phenotype in the host should favor brood parasites that could match such host eggs. Such polymorphism in host eggs should favor cuckoo parasitism with an increasing number of host morphs being matched by an increasing number of parasite morphs. Eventually, the number of host and cuckoo egg color morphs becomes so large that the small populations of each brood parasite morph may go extinct for stochastic reasons.

19.3 Egg Polymorphism as Anti-Parasite Defense in Cuckoo Hosts

During the arms race between parasitic cuckoos and their hosts, a variety of behaviors and traits have evolved in both parties as parasite adaptations and anti-parasite counteradaptations. Because one of the most intense interactions between cuckoos and hosts occurs at the egg stage, egg coloration and its changes are important weapons for hosts to counter cuckoo parasitism.

19.3.1 Monomorphism and Polymorphism in Host Eggs

At the stage of coevolution between cuckoos and hosts, selection from parasitism had not caused hosts to evolve polymorphic eggs because host egg polymorphism can only work under the precondition that hosts can recognize and reject nonmimetic and unmatched eggs. Therefore, in the early stage of egg recognition, this is an important and necessary adaptation under the strong selection pressure due to cuckoo parasitism. When hosts have evolved such a cognitive capacity, they in turn may exert selection on the cuckoos to evolve mimetic eggs that simulate the appearance of host eggs to increase the probability of being accepted (Soler et al. 2003). The arms race between cuckoos and hosts continues until egg recognition by hosts is ever stronger, while cuckoo eggs become more and more mimetic. Subsequently, the escalation of such an interaction may lead to the occurrence of egg polymorphism or other adaptations in hosts. Egg polymorphism can work as an effective strategy to counter cuckoo parasitism because essentially it further increases the contrast between appearance of cuckoo and host eggs, and thus it favors hosts that recognize and reject cuckoo eggs accurately (Kilner 2006; Yang et al. 2017).

19.3.2 Monomorphism and Polymorphism in Cuckoo Eggs

Cuckoo parasitism selects for egg polymorphism in hosts. In contrast, host egg polymorphism acts as stabilizing selection on cuckoos to evolve polymorphic eggs. Such polymorphism in both parties that evolves as a consequence of frequency-dependent selection has been described in predator-prey interactions (Cook et al. 1999; Galeotti et al. 2003; Galeotti and Rubolini 2004). However, for the brood parasite-host system, the evidence is rare although several cases were reported recently. For example, ashy-throated parrotbills (*Paradoxornis alphonsianus*) were found to lay polymorphic eggs, while common cuckoos (*Cuculus canorus*) that utilize this host species also evolved exactly corresponding egg phenotypes (Yang et al. 2010). Similarly, both plaintive cuckoos and their common tailorbird hosts have evolved polymorphic eggs with corresponding egg phenotypes (Yang et al. 2016a). Vikan et al. (2011) described a similar situation in common cuckoos and their *Fringilla* finch hosts. Furthermore, large hawk cuckoos (*Hierococcyx sparverioides*) that parasitize white-browed laughingthrushes (*Garrulax sannio*) have evolved polymorphic white or pale blue eggs corresponding to those of their laughingthrush hosts (Yang et al. 2015b). Although the phenomenon of egg polymorphism occurring in both brood parasites and hosts is uncommon, cuckoo eggs can also become polymorphic as a response to the frequency-dependent scenario.

19.3.3 Effect of Host Egg Polymorphism on Cuckoo Parasitism

Is host egg polymorphism really an effective strategy for countering cuckoo parasitism? Yang et al. (2017) answered this question by developing a simple, but informative, theoretical analysis to simulate the effect of host egg polymorphism on cuckoo parasitism.

Egg polymorphism in hosts can be regarded as a strategy that increases anew the contrast in egg phenotype with cuckoos after the latter have evolved eggs that clearly mimic the eggs of the host (Yang et al. 2017). Initially, if we assume that cuckoos can find and parasitize all host clutches (i.e., the rate or proportion of cuckoo parasitism (hereafter RCP) equals 1), the successful parasitism rate (hereafter SPR) for cuckoos is 100% when cuckoo eggs perfectly mimic host eggs while hosts have only one egg phenotype (let us say a white egg). However, in a scenario when hosts have evolved a second egg phenotype (let us say blue egg), SPR gradually decreases with the increasing proportion of blue egg phenotype in hosts and finally becomes zero when all host eggs are blue. In other words, if hosts lay dimorphic eggs with half of the original egg phenotype and half of the new egg phenotype, cuckoos have half of SPR. The SPR is negatively correlated with the proportion of host egg phenotypes in a specific RCP. If we consider all possible values of RCP (i.e., from 1 to 0), the relationship between SPR and the proportion of host egg phenotype constitutes an area in Fig. 19.1a.

Subsequently, if we simulate the scenario that cuckoos have also evolved dimorphic eggs corresponding to those of the hosts, SPR will gradually decrease to half

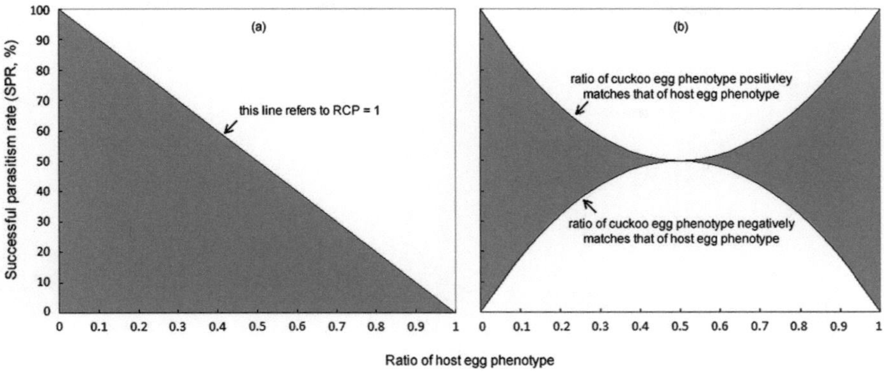

Fig. 19.1 Relationship between successful parasitism rate (SPR) and the proportion of host egg phenotypes based on the assumption that hosts lay dimorphic eggs while cuckoos lay monomorphic (**a**) or dimorphic (**b**) eggs (modified from Yang et al. 2017). RCP refers to the proportion of cuckoo parasitism to the proportion of host nests that were found and parasitized by the cuckoo in a host population. (**a**) simulates a situation that includes all values of RCP from 1 to 0, while (**b**) simulates a situation in a specific RCP of 1

and then increase to 100%, with the change in the host egg phenotype ratio ranging from 0 to 1 in a specific RCP set to 1 (Fig. 19.1b). In contrast, if the proportion of cuckoo egg phenotypes is assumed to be negatively matching that of host egg phenotypes (e.g., cuckoos lay white and blue eggs with an opposite ratio to that of host eggs), then SPR increases from zero to half and then decreases to zero, with the change in host egg phenotype ratio from 0 to 1 (Fig. 19.1b). Therefore, the area surrounded by these two curves indicates the extent of matching of the proportion of egg phenotypes between cuckoos and hosts. If we consider all possible values of RCP, the former situation (i.e., positively matching) and the latter situation (i.e., negatively matching) constitute areas presented in Fig. 19.2a, b, respectively.

According to the predictions above, (1) if cuckoos only lay monomorphic eggs, they suffer a continuous decline in SPR with an increasing proportion of a new egg phenotype in hosts; and (2) if both cuckoos and hosts have polymorphic eggs, hosts can minimize SPR of cuckoos by laying a symmetrical egg phenotype (i.e., the proportions of all phenotype are the same), (3) while cuckoos can maximize their SPR by laying a similar proportion of egg phenotypes corresponding to that of their hosts.

Yang et al. (2017) expanded the simulation to a situation in which hosts lay polymorphic eggs with more than two phenotypes. Firstly, if the cuckoos lay only one egg phenotype, and the RCP is set to 1, SPR declines rapidly in the beginning, but the decline becomes slower and slower later on (Yang et al. 2017). Thus, the efficiency of the negative effect of host egg polymorphism on SPR of cuckoos decreases with the number of host egg morphs, although the increase reduces the SPR of cuckoos (Yang et al. 2017). Interestingly, even if in the situation that cuckoos also lay polymorphic eggs with the number of cuckoo egg phenotypes being equal to that of host egg phenotypes, the pattern of relationship between SPR and number of

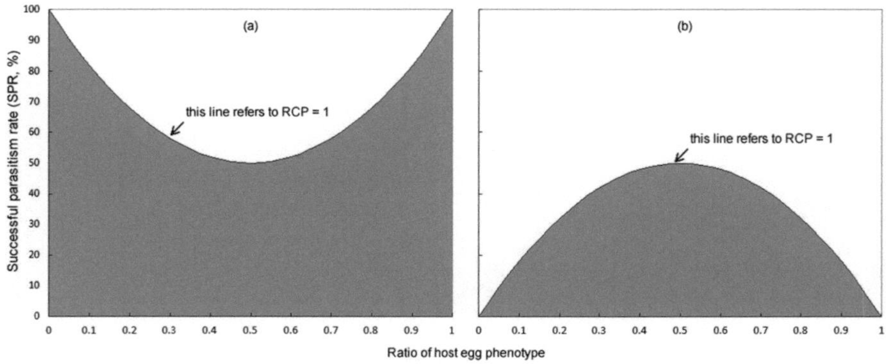

Fig. 19.2 Variation in successful parasitism rate (SPR) in cuckoos with the proportion of host egg phenotypes based on the assumption that both cuckoos and hosts lay dimorphic eggs (modified from Yang et al. 2017). All values of RCP, which refer to the proportion of cuckoo parasitism to the proportion of host nests that were found and parasitized by the cuckoo in a host population, are assumed to exist here. (**a**) and (**b**) refer to the situation that cuckoo eggs positively or negatively match those of host eggs in this dimorphic system, respectively

host egg phenotypes does not change because mathematically the change in the number of cuckoo egg phenotypes does not increase the total SPR in cuckoos (Yang et al. 2017).

19.4 Adaptation and Evolution of Egg Polymorphism in Time and Space

Egg polymorphism in cuckoo-host systems varies among geographically distinct populations. Polymorphic eggs are not ubiquitous, raising the question about the selective forces that contribute to such heterogeneity in polymorphism. One way of investigating heterogeneity in egg polymorphism is through analysis of the phylogenetic distribution of egg polymorphism in cuckoos and their hosts.

19.4.1 Phylogeny of Egg Polymorphism in Cuckoos

There are 142 species of birds in the order Cuculiformes, of which 59 (41.5%) are brood parasites. Egg polymorphism is common in host species, but what about parasitic cuckoos? In a broad sense, the common cuckoo is a famously polymorphic cuckoo, not only its gray or brown plumage but also its diversity in egg appearance that is adaptive for the many host species that it utilizes. If we consider each cuckoo species as a global population, egg polymorphism is common in most cuckoos except for the great spotted cuckoo *Clamator glandarius* (Soler et al. 2003). Therefore, as a global population that includes all host races, common cuckoos lay polymorphic eggs. However, what about other cuckoo species? Does parasitic breeding favor cuckoos that evolve polymorphic eggs? We established a phylogenetic tree of Cuculiformes species by using the Get Trees function from BirdTree.org (http://birdtree.org) that is based on the data and methodology from Jetz et al. (2012). Then we expressed the information of egg morphs on the tree by denoting monomorphism and polymorphism with black and red scientific names, respectively.

The classification of egg morphs (i.e., monomorphic or polymorphic) is based on the principle that only two or more clearly different egg morphs are classified as polymorphic eggs (Leimar 2005); otherwise, they are monomorphic eggs. Therefore, ambiguous descriptions of different egg morphs were not regarded as polymorphic. For example, blue and greenish blue or white and grayish white were not regarded as two egg phenotypes. Furthermore, the same type of egg maculation with continuous variation was not classified as polymorphism. However, different types of egg maculation (e.g., dot vs. line) were considered different egg phenotypes.

In total we extracted information on eggs from 119 species of Cuculiformes (accounting for 83.8%), while for the other species, such information still remains unknown. According to the phylogenetic tree of Cuculiformes, most of the parasitic cuckoos (84.3%) are aggregated in one clade, while the other species come from three other clades (Fig. 19.3). Therefore, the parasitic habit of cuckoos has evolved more than once from their common ancestor that is nesting. Most parasitic cuckoos lay polymorphic eggs, which account for 88.4% (n = 43, not including species without information on eggs), while only a small proportion of nesting cuckoos lay polymorphic eggs (7.6%, n = 53, not including species without information on eggs). The frequency of egg polymorphism in parasitic cuckoos was significantly higher than that of nesting cuckoos ($\chi^2 = 66.19$, df = 1, $P < 0.001$, likelihood ratio test). GLMM also showed that the breeding type of cuckoos significantly predicted egg polymorphism in Cuculiformes ($F_{2,\ 117} = 20.28$, $P < 0.001$, GLMM).

19.4.2 Phylogeny of Egg Polymorphism in Cuckoo Hosts

Does cuckoo parasitism lead to the evolution of egg polymorphism in hosts? Chen et al. (2016) established a parrotbill phylogenetic tree, which reveals two clades of species. One clade consists of small parrotbill species that lay immaculate eggs, while the other consists of large parrotbill species that lay maculate eggs (Fig. 19.4). This result indicates that egg polymorphism in parrotbills is linked to cuckoo parasitism, although parasitism does not always lead to host egg polymorphism (Fig. 19.4).

We obtained information on host species of common cuckoos from Lowther (2013), who has summarized the cuckoo hosts around the world. However, some obviously unsuitable host species were excluded from the list (e.g., precocial birds). We reviewed the information on host eggs from the electronic version of the *Handbook of Birds of the World* (http://www.hbw.com).

In total 181 species of common cuckoo hosts belonging to 34 families and 2 genera were included in the analyses. Host species that lay monomorphic eggs accounted for 61.9% of 181 species. The results of GLMM indicated that neither the

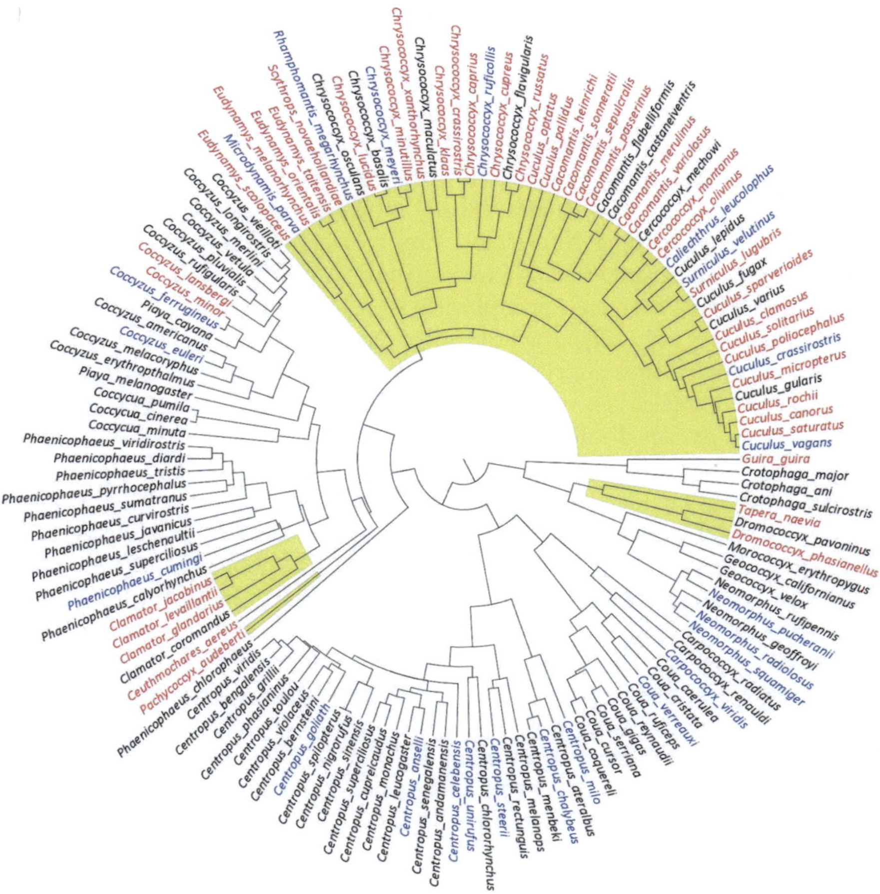

Fig. 19.3 The phylogenetic tree of Cuculiformes with parasitic species highlighted in yellow. Black and red colors of scientific names refer to species that lay monomorphic and polymorphic eggs, respectively, while blue-colored scientific names indicate that information on egg color of such species remains unknown

number of species nor the number of genera in host families predicted the frequency of egg polymorphism (number of host species: $F_{1,\ 32} = 1.14$, $P = 0.293$; number of genera: $F_{1,\ 32} = 0.28$, $P = 0.602$, GLMM). Most host species belonged to the family Muscicapidae, which accounts for 14.9% of the total number of host species, while all other families had less than 10% each (Fig. 19.5). Furthermore, half of the host families (17 out of 34) produced polymorphic eggs, varying from 18.2 to 100%, while the remaining families (50%) only had monomorphic eggs (Fig. 19.5). Four families of hosts possessed 100% polymorphic eggs, but they only contained one or two host species (Fig. 19.5). Families with many species of hosts (i.e., ≥ 10 species) also had a high percentage of egg polymorphism (45.85%, SD = 23.33, range: 18.18–90.91%, $n = 9$; Fig. 19.5). If we classified host families in two groups with or without egg polymorphism (i.e., the presence of egg polymorphism; Fig. 19.6), the

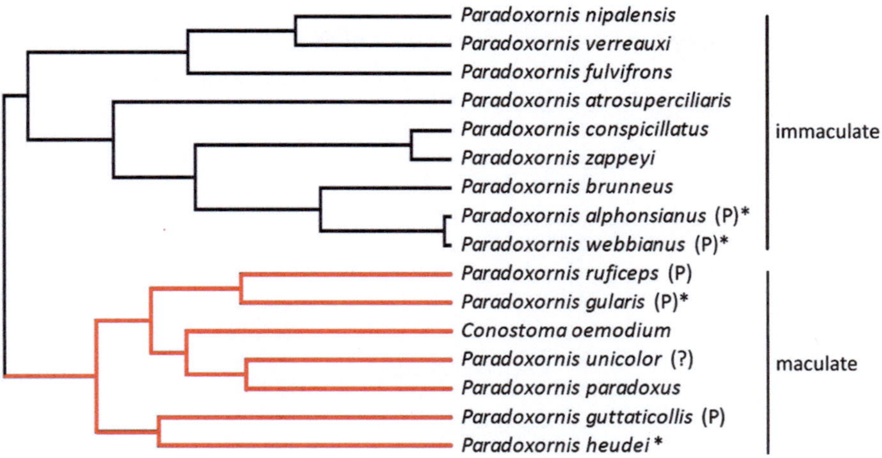

Fig. 19.4 The multilocus species tree of parrotbills (modified from Chen et al. 2016). Host species that are marked with P in brackets indicate that they lay polymorphic eggs, while all others lay monomorphic eggs. Host species that were recorded as cuckoo hosts are marked with asterisks. *P. gularis* was found to be parasitized by Hodgson's hawk cuckoo (*Hierococcyx nisicolor*), while other cases were all utilized by common cuckoo (*Cuculus canorus*). No information on eggs was available for *P. unicolor*

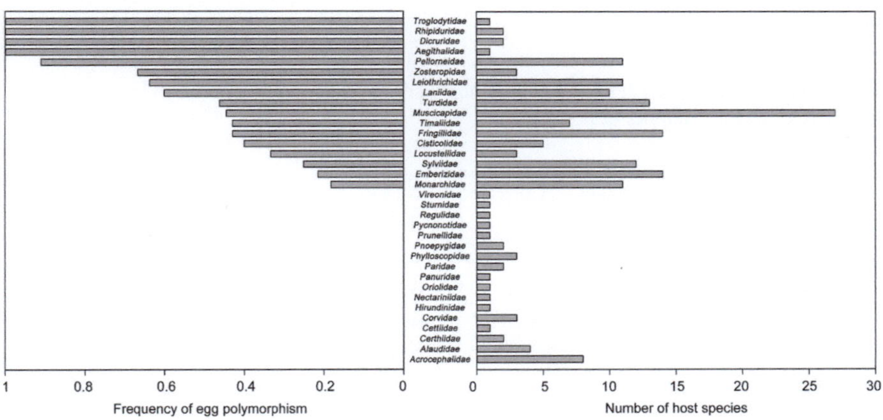

Fig. 19.5 Frequency of egg polymorphism and number of host species in different families of common cuckoo hosts

presence of egg polymorphism can be predicted by the number of species in different families ($F_{1,\ 32} = 5.48$, $P = 0.026$, GLMM; Table 19.1).

This result indicates that the number of species and genera in host families does not predict the frequency of egg polymorphism in families. However, if egg polymorphism of host families was classified as either absent or present, they can be predicted by the number of host species (Fig. 19.6). These findings imply that host egg polymorphism tends to evolve in host families that contain more species,

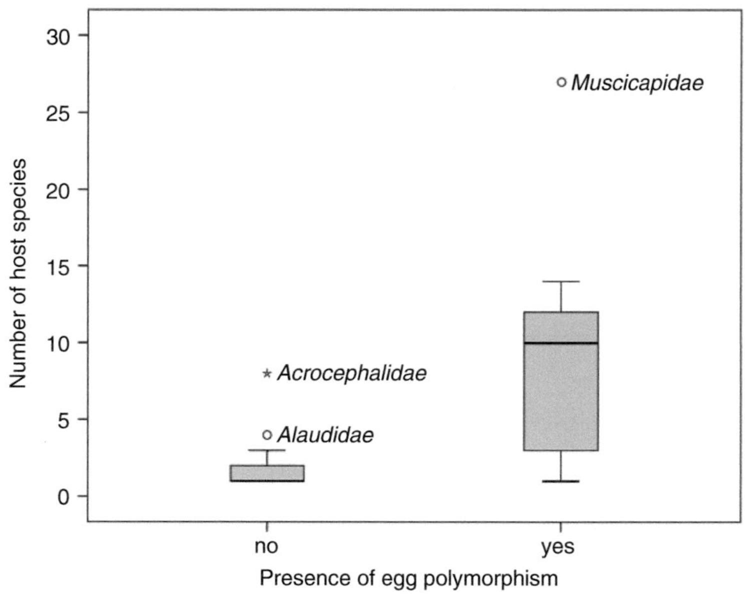

Fig. 19.6 Comparison of number of host species among families in which egg polymorphism is absent or present. Box plots show medians, quartiles, and 5th and 95th percentiles, while the circles and stars refer to outliers and an extreme outlier, respectively. The family names of the outliers were marked near the outliers

Table 19.1 Results of GLMM of egg polymorphism in cuckoo hosts in relation to the number of host species and the number of host genera

Target of the model	Source	F	df_1	df_2	P
Frequency of egg polymorphism	Corrected model	12.69	2	32	<0.001**
	Number of host species	1.14	1	32	0.293
	Number of host genera	0.28	1	32	0.602
Presence of egg polymorphism	Corrected model	6.04	2	32	0.006**
	Number of host species	5.48	1	32	0.026*
	Number of host genera	0.01	1	32	0.929

$^*P < 0.05$, $^{**}P < 0.01$

although the frequency of egg polymorphism in these families does not correlate with their species diversities themselves. Host families with more species tend to contain hosts laying polymorphic eggs because egg polymorphism is simply easy to detect in these families due to their many species. Nevertheless, according to our findings, egg polymorphism in common cuckoo hosts has evolved numerous times relatively recently, suggesting that egg polymorphism is an evolutionarily labile character that readily evolves given the intense selection pressures by cuckoos on their hosts.

Although brood parasites exert intense selection on their hosts resulting in the evolution of polymorphic eggs (Kilner 2006), the reason for why egg polymorphism occurs in some host species, but not in others, remains unclear. One possible explanation may be differences of coevolutionary histories among different cuckoo-host systems. However, although egg polymorphism may need long time to evolve, such long coevolution may not always result in egg polymorphism. Therefore, the absence of egg polymorphism among different host species may be explained by different mechanisms of anti-parasite behavior during coevolution different from egg rejection (i.e., not relying on egg recognition). For example, the Cape bulbul (*Pycnonotus capensis*) has strong mobbing behavior that effectively prevents the Jacobin cuckoo (*Clamator jacobinus*) from accessing its nest, and it has not even evolved egg recognition as an anti-parasitism defense (Liversidge 1970; Krüger 2011). Furthermore, the superb fairy wren (*Malurus cyaneus*) and red-rumped swallow (*Hirundo daurica*) can recognize cuckoo chicks and reject them (Langmore et al. 2003; Yang et al. 2015c), while the Asian verditer flycatcher (*Eumyias thalassinus*) feeds nestlings with poor quality and hard to digest food items that cannot ensure the survival of cuckoo nestlings (Yang et al. 2013). Therefore, different host species have evolved different mechanism to counter brood parasites.

19.4.3 Egg Polymorphism in Cuckoo-Host Systems Among Different Geographic Populations

Although long time of coevolutionary history may lead to the evolution of egg polymorphism in hosts and cuckoos, it is very difficult to witness such change in the process of interaction between cuckoos and their hosts. Therefore, studying different geographic populations of hosts that have interacted with cuckoos during different periods of time can serve as an alternative. For example, the mainland populations of vinous-throated parrotbill (*Paradoxornis webbianus*) are utilized by the common cuckoo and have evolved polymorphic eggs, while the island populations of hosts, which have escaped from cuckoo parasitism for more than two million years, only lay monomorphic eggs (Yang et al. 2015a). Similarly, the island population of plain prinia (*Prinia inornata*), which is not parasitized by cuckoos, lays monomorphic blue eggs with markings, while the mainland population that is utilized by cuckoos lays polymorphic blue or white eggs with markings (Wang et al. 2016). Lahti (2005) investigated variation in eggs among individuals between original and introduced populations of African village weaverbirds (*Ploceus cucullatus*), a host of diederik cuckoo (*Chrysococcyx caprius*), and found that introduced populations have significantly lower variation in eggs among individuals compared to original populations because of the absence of cuckoo parasitism for more than two centuries (Lahti 2005). A similar result was obtained by Yang et al. (2014), who surveyed variation in egg phenotypes of a common cuckoo host, the red-billed leiothrix (*Leiothrix lutea*), and compared them between original populations in China and introduced populations in Hawaii (Yang et al. 2014). Moreover, large hawk cuckoos parasitize white-browed laughingthrushes and

Chinese babaxes (*Babax lanceolatus*) in areas of sympatry, but their eggs are polymorphic and mimetic corresponding to that of laughingthrushes due to a longer coevolutionary history than babaxes (Yang et al. 2015c). Additionally, the frequencies of polymorphic eggs may vary among different geographic populations within a single host species, while the cuckoos may use an intermediate egg morph for parasitism. For example, Daurian redstarts (*Phoenicurus auroreus*) lay polymorphic white or blue eggs with reddish markings that vary in frequency among populations, while common cuckoos lay pale blue eggs with reddish markings that is an intermediate phenotype for parasitism of redstart nests because redstarts present intermediate recognition ability for nonmimetic eggs. Therefore, intermediate egg morphs may maximize the reproductive success for cuckoos in this case (Yang et al. 2016c).

Concluding Remarks and Future Directions

We have theoretically illustrated that host egg polymorphism considerably restricts successful parasitism by parasitic cuckoos. Comparative studies are encouraged and needed for testing these ideas. The classification of egg polymorphism is also important because avian vision differs from that of humans. Furthermore, host egg polymorphism with distinct phenotype contrasts may have effects on success rate of parasitism for brood parasites, but variation in eggs without such distinct contrasts is also necessary for such a test.

Host egg polymorphism is positively related to the diversity of cuckoos and is most common in continents with high diversity of parasitic cuckoos. More comparative studies should be performed to further test whether such relationships between cuckoos and hosts are a universal rule across different brood parasite-host systems. Moreover, the duration and intensity of coevolution between brood parasites and their hosts may not always lead to the evolution of egg polymorphism, as we have suggested. Further studies are needed to reveal why and how this happens.

In summary, host egg polymorphism is an effective strategy to counter cuckoo parasitism, and hosts can minimize the successful parasitism by cuckoos by laying polymorphic eggs in equal proportion if cuckoos also have evolved egg polymorphism as a counteradaptation and vice versa. Although cuckoos can also evolve polymorphic eggs that select for host egg polymorphism, the escalation in cuckoo egg morphs does not increase their total successful parasitism rate because one individual cuckoo can only lay eggs of one morph. Parasitic cuckoos are much more frequently laying polymorphic eggs than nesting cuckoos, suggesting that it is the parasitic habit that drives the evolution of egg polymorphism in cuckoos. Host egg polymorphism not only varies temporally but may also vary spatially. Coevolutionary histories may explain the evolution of egg polymorphism in hosts because egg polymorphism tends to evolve in those host species that have interacted with cuckoos for a longer time than others.

Acknowledgments M. Soler and D. Sherry provided constructive criticism and helpful suggestions. This work was supported by the National Natural Science Foundation of China (Nos. 31672303 and 31260514 to CY, 31272328, 31472013 and 31772453 to WL) and the Program for New Century Excellent Talents in University (NCET-13-0761).

References

Banks AJ, Martin TE (2001) Host activity and the risk of nest parasitism by brown-headed cowbirds. Behav Ecol 12:31–40

Bertram BCR, Burger AE (1981) Are ostrich *Struthio camelus* eggs the wrong colour? Ibis 123:207–210

Chen L, Zhu L, Yang X, Lyu N, Liu Y, Liang W, Sun Y (2016) Are egg colors and patterns signs of phylogenetic relatedness in parrotbills? Ornithol Sci 15:119–125

Cook LM, Dennis RLH, Mani GS (1999) Melanic morph frequency in the peppered moth in the Manchester area. Proc R Soc Lond B 266:293–297

Cott HB (1948) Edibility of the eggs of birds. Nature 161:8–11

Cott HB (1951) The palatability of the eggs of birds: illustrated by experiments on the food preferences of the hedgehog (*Erinaceus europaeus*). J Zool 121:1–41

Cott HB (1953) The palatability of the eggs of birds: illustrated by experiments on the food preferences of the ferret (*Putorius furo*) and cat (*Felis catus*); with notes on other egg-eating carnivora. J Zool 123:123–141

Davies NB (2000) Cuckoos, cowbirds and other cheats. T & A. D. Poyser, London

Davies NB, de Brooke ML (1989) An experimental study of co-evolution between the cuckoo, *Cuculus canorus*, and its hosts. II. Host egg markings, chick discrimination and general discussion. J Anim Ecol 58:225–236

Davis SK (2005) Nest-site selection patterns and the influence of vegetation on nest survival of mixed-grass prairie passerines. Condor 107:605–616

Ford EB (1965) Genetic polymorphism. Faber & Faber, London

Forstmeier W, Weiss I (2004) Adaptive plasticity in nest-site selection in response to changing predation risk. Oikos 104:487–499

Galeotti P, Rubolini D (2004) The niche variation hypothesis and the evolution of colour polymorphism in birds: a comparative study of owls, nightjars and raptors. Biol J Linn Soc 82:237–248

Galeotti P, Rubolini D, Dunn PO, Fasola D (2003) Colour polymorphism in birds: causes and functions. J Evol Biol 16:635–646

Götmark F (1992) Blue eggs do not reduce nest predation in the song thrush, *Turdus philomelos*. Behav Ecol Sociobiol 30:245–252

Jetz W, Thomas GH, Joy JB, Hartmann K, Mooers AO (2012) The global diversity of birds in space and time. Nature 491:444–448

Kilner RM (2006) The evolution of egg colour and patterning in birds. Biol Rev 81:383–406

Krüger O (2011) Brood parasitism selects for no defence in a cuckoo host. Proc R Soc Lond B 278:2777–2783

Lahti DC (2005) Evolution of bird eggs in the absence of cuckoo parasitism. Proc Natl Acad Sci U S A 102:18057–18062

Langmore NE, Hunt S, Kilner RM (2003) Escalation of a co-evolutionary arms race through host rejection of brood parasitic young. Nature 422:157–160

Lee WS, Kwon YS, Yoo JC (2010) Egg survival is related to the colour matching of eggs to nest background in black-tailed gulls. J Ornithol 151:765–770

Leimar O (2005) The evolution of phenotypic polymorphism: randomized strategies versus evolutionary branching. Am Nat 165:669–681

Liang W, Yang C, Stokke BG, Antonov A, Fossøy F, Vikan JR, Moksnes A, Røskaft E, Shykoff JA, Møller AP, Takasu F (2012) Modelling the maintenance of egg polymorphism in avian brood parasites and their hosts. J Evol Biol 25:916–929

Liversidge R (1970) The biology of the Jacobin cuckoo *Clamator jacobinus*. Ostrich 8 (Suppl):117–137

Lovell PG, Ruxton GD, Langridge KV, Spencer KA (2013) Egg-laying substrate selection for optimal camouflage by quail. Curr Biol 23:260–264

Lowther PE (2013) Host list of avian brood parasites – 2 – Cuculiformes – old world cuckoos. Field Museum, Chicago, IL

Magurran AE (2010) Q & A: what is biodiversity? BMC Biol 8:145

Martin TE, Roper JJ (1988) Nest predation and nest-site selection of a western population of the hermit thrush. Condor 90:51–57

Montevecchi WA (1976) Field experiments on the adaptive significance of avian eggshell pigmentation. Behaviour 58:26–39

Moreno J, Osorno JL (2003) Avian egg colour and sexual selection: does eggshell pigmentation reflect female condition and genetic quality? Ecol Lett 6:803–806

Patten MA, Dan LR, Wolfe DH (2011) Hierarchical cues in brood parasite nest selection. J Ornithol 152:521–532

Rao MS, Thorgeirsson S, Reddy MK, Lalwani ND, Evarts RE, Usman MI, Singh B, Reddy JK (1993) Observations of pigment accumulation in the epithelium of the shell gland and superficial pigmentation on the egg shell in Japanese quail. J Fac Agric Kyushu Univ 38:73–80

Roulin A (2004) The evolution, maintenance and adaptive function of genetic colour polymorphism in birds. Biol Rev 79:815–848

Soler M (2014) Long-term coevolution between avian brood parasites and their hosts. Biol Rev 89:688–704

Soler JJ, Aviles JM, Soler M, Møller AP (2003) Evolution of host egg mimicry in a brood parasite, the great spotted cuckoo. Biol J Linn Soc 79:551–563

Soler M, Pérez-Contreras T (2012) Location of suitable nests by great spotted cuckoos: an empirical and experimental study. Behav Ecol Sociobiol 66:1305–1310

Solis JC, de Lope F (1995) Nest and egg crypsis in the ground-nesting stone curlew *Burhinus oedicnemus*. J Avian Biol 26:135–138

Stoddard MC, Kupán K, Eyster HN, Rojas-Abreu W, Cruz-López M, Serrano-Meneses MA, Küpper C (2016) Camouflage and clutch survival in plovers and terns. Sci Rep 6:32059

Swynnerton CFM (1916) On the coloration of the mouths and eggs of birds: on the coloration of eggs. Ibis 58:529–606

Takasu F (1998) Why do all host species not show defense against avian brood parasitism: evolutionary lag or equilibrium? Am Nat 15:193–205

Tinbergen N, Broekhuysen GJ, Feekes F, Houghton JCW, Kruuk H, Szulc E (1962) Egg shell removal by the black-headed gull, *Larus ridibundus* L: a behavior component of camouflage. Behaviour 19:74–118

Underwood TJ, Sealy SG (2002) Adaptive significance of egg colouration. In: Deeming DC (ed) Avian incubation: behaviour, environment and evolution. Oxford University Press, Oxford, pp 280–289

Vikan JR, Fossøy F, Huhta E, Moksnes A, Røskaft E, Stokke BG (2011) Outcomes of brood parasite–host interactions mediated by egg matching: common cuckoos, *Cuculus canorus* versus *Fringilla* finches. PLoS One 6:604–607

Wallace AR (1889) Darwinism: an exposition of the theory of natural selection with some of its applications. Macmillan, London

Wang L, Liang W, Yang C, Cheng SJ, Hsu YC, Lu X (2016) Egg rejection and clutch phenotype variation in the plain prinia *Prinia inornata*. J Avian Biol 47:788–794

Weidinger K (2001) Does egg colour affect predation rate on open passerine nests? Behav Ecol Sociobiol 49:456–464

Yang C, Liang W, Cai Y, Shi S, Takasu F, Møller AP, Antonov A, Fossøy F, Moksnes A, Røskaft E, Stokke BG (2010) Coevolution in action: disruptive selection on egg colour in an avian brood parasite and its host. PLoS One 5:e10816

Yang C, Stokke BG, Antonov A, Cai Y, Shi S, Moksnes A, Røskaft E, Møller AP, Liang W, Grim T (2013) Host selection in parasitic birds: are open-cup nesting insectivorous passerines always suitable cuckoo hosts? J Avian Biol 44:216–220

Yang C, Liu Y, Zeng L, Liang W (2014) Egg color variation, but not egg rejection behavior, changes in a cuckoo host breeding in the absence of brood parasitism. Ecol Evol 4:2239–2246

Yang C, Wang L, Cheng SJ, Hsu YC, Stokke BG, Røskaft E, Moksnes A, Liang W, Møller AP (2015a) Deficiency in egg rejection in a host species as a response to the absence of brood parasitism. Behav Ecol 26:406–415

Yang C, Su T, Liang W, Møller AP (2015b) Coevolution between the large hawk-cuckoo (*Cuculus sparverioides*) and its two sympatric Leiothrichidae hosts: evidence for recent expansion and switch in host use? Biol J Linn Soc 115:919–926

Yang C, Wang L, Chen M, Liang W, Møller AP (2015c) Nestling recognition in red-rumped and barn swallows. Behav Ecol Sociobiol 69:1–6

Yang C, Huang Q, Wang L, Jiang A, Stokke BG, Fossøy F, Tunheim OH, Røskaft E, Liang W, Møller AP (2016a) Plaintive cuckoos do not select tailorbird hosts that match the phenotypes of their own eggs. Behav Ecol 27:835–841

Yang C, Wang L, Liang W, Møller AP (2016b) Egg recognition as anti-parasitism defense in hosts does not select for laying of matching eggs in parasitic cuckoos. Anim Behav 122:177–181

Yang C, Li Z, Zhang Y, Wang H, Liang W, Møller AP (2016c) Egg polymorphism and egg discrimination in the Daurian redstart *Phoenicurus auroreus*, a host of the common cuckoo *Cuculus canorus*. Ornithol Sci 15:127–132

Yang C, Liang W, Møller AP (2017) Why do hosts with obvious egg polymorphism suffer low parasitism rates under avian brood parasitism? A theoretical consideration. Behav Ecol Sociobiol 71:30

Adaptations of Brood Parasitic Eggs 20

Bård G. Stokke, Frode Fossøy, Eivin Røskaft, and Arne Moksnes

Abstract

This chapter deals with adaptations of brood parasitic eggs, which has received a lot of attention by researchers and naturalists for centuries. Both scientists and laymen are intrigued by the sophisticated traits that are enabling parasites to fool or force their hosts into accepting their eggs. A range of brood parasites have evolved mimetic or cryptic eggs to counter host egg rejection or to counter rejection from competing brood parasites. Mimetic eggs resemble host eggs in appearance. Cryptic eggs may appear non-mimetic to host eggs but tend to blend in with the nest lining or "disappear" in the darkness of the (cavity) nest. Rather than being mimetic or cryptic, parasites may also produce eggs that appear attractive to their hosts. In addition, a thick eggshell and a short incubation time may also be advantageous in relation to their brood parasitic lifestyle. The

B. G. Stokke (✉)
Department of Biology, Norwegian University of Science and Technology (NTNU), Trondheim, Norway

Department of Terrestrial Ecology, Norwegian Institute for Nature Research (NINA), Trondheim, Norway
e-mail: bard.g.stokke@ntnu.no

F. Fossøy
Department of Biology, Norwegian University of Science and Technology (NTNU), Trondheim, Norway

Department of Aquatic Ecology, Norwegian Institute for Nature Research (NINA), Trondheim, Norway
e-mail: frode.fossoy@ntnu.no

E. Røskaft · A. Moksnes
Department of Biology, Norwegian University of Science and Technology (NTNU), Trondheim, Norway
e-mail: eivin.roskaft@ntnu.no; arne.moksnes@ntnu.no

M. Soler (ed.), *Avian Brood Parasitism*, Fascinating Life Sciences,
https://doi.org/10.1007/978-3-319-73138-4_20

occurrence and importance of these traits among various brood parasites are discussed.

20.1 Introduction

In the co-evolutionary arms race between brood parasites and their hosts, adaptations have evolved on both sides, enabling smarter trickery in parasites and better defences in hosts. We discuss how egg mimicry, the resemblance of parasite eggs to host eggs in colour, spotting pattern, shape and size, varies within and between different parasite species (Sect. 20.2) and how it evolves and is maintained over time (Sect. 20.3). In other circumstances, it may be advantageous for parasites to produce eggs that blend in with the nest lining (i.e. cryptic eggs) instead of being mimetic (Sect. 20.4). Several brood parasites also lay eggs with remarkably thick eggshells, and the benefits of this trait are discussed (Sect. 20.5). Other topics covered relate to parasites tricking hosts to accept their eggs by laying irresistible (i.e. attractive) eggs (Sect. 20.6) and parasite egg physiology (Sect. 20.7).

20.2 Egg Mimicry

Egg mimicry, the resemblance of parasite eggs to host eggs in spotting pattern, ground and spot colour, brightness, size and shape, is often used as the textbook example of natural selection shaping adaptations in co-evolutionary arms races. The selective force behind the evolution of egg mimicry is rejection of unlike eggs by hosts (Brooke and Davies 1988; Davies and Brooke 1989; Moksnes et al. 1991b; Stoddard and Stevens 2010, 2011) or selective egg removal by competing parasites (Davies and Brooke 1988; Brooker and Brooker 1989).

After early researchers had noticed the similarity between the eggs of the common cuckoo, *Cuculus canorus*, and its hosts (Baldamus 1853; Rey 1892; Newton 1902), Pycraft (1910) was the first to consider this phenomenon in an evolutionary context. Since then the existence of host-specific female races, or gentes, each specializing on particular host species and often mimicking their eggs (Fig. 20.1), has received overwhelming support (Alvarez 1994; Moksnes and Røskaft 1995; Antonov et al. 2007).

The textbook example of good mimicry, however, is not always the reality. In many descriptions of common cuckoo eggs, the mimicry is described as poor or moderate (e.g. Moksnes and Røskaft 1995). Females belonging to the cuckoo gens parasitizing the dunnock, *Prunella modularis*, in Britain, for example, lay non-mimetic eggs (Brooke and Davies 1988). Among sympatric gentes parasitizing four species of *Acrocephalus* warblers in the Czech Republic, it was not with human vision alone possible to distinguish between cuckoo eggs belonging to different gentes (Edvardsen et al. 2001; see also Drobniak et al. 2014). These results could indicate that cuckoos in this area laid an intermediate egg type, some sort of

Fig. 20.1 Common cuckoo females are host specific and often lay eggs that mimic their particular hosts. This illustration shows eggs from 14 cuckoo females each specializing on different host species. Photo by Eivin Røskaft

"compromise" between the different host species, even though females mostly were host specific in their egg-laying (Honza et al. 2002; Skjelseth et al. 2004). Such "general" egg type has also been found in the Horsfield's bronze cuckoo, *Chalcites basalis*, a generalist parasite in Australia (Feeney et al. 2014a). A similar study on sympatric common cuckoo gentes was also carried out in Bulgaria (Antonov et al. 2010). Cuckoo and host egg phenotypes were in close accordance with two species of warblers with respect to both background colour and spots, but this was not the case for a third species where the cuckoo eggs showed poor mimicry with host eggs. On the other hand, there was a clear significant correlation between cuckoo and host egg size for the three species, but not in shape (see also Bán et al. (2011), Table 20.1). Finally, genetic analyses of both mitochondrial and nuclear DNA showed clear differentiation between the cuckoos from the three hosts, confirming that they belonged to different host races or gentes (Fossøy et al. 2011).

In some circumstances, one obvious reason for the lack of mimicry is that it has not evolved because selected host species do not reject non-mimetic eggs. Hence, this can explain why dunnock cuckoos lay brownish or greyish eggs that are easily distinguished from the blue dunnock eggs (Brooke and Davies 1988). The reason for poor mimicry in other circumstances may be that cuckoos are "forced" to lay in a "wrong" host species because of lack of nests in suitable stage of their favourite species. Alternatively, parasite eggs may appear non-mimetic according to human vision, even if they are mimetic from the birds' point of view. This can be disclosed by using methods that take bird vision into account (Cherry and Bennett 2001;

Table 20.1 Studies focusing on egg size and shape mimicry in avian obligate brood parasites and their hosts

Parasite	Host	Comment	Source
Cuculidae	–	Parasitic cuckoos lay smaller eggs than non-parasitic cuckoos of similar size	Payne (1974), Krüger and Davies (2002), Krüger and Davies (2004)
Cuculidae	–	Incubation time is positively correlated with egg size in parasitic cuckoos	Davies and Brooke (1988)
Cuculidae	–	Evolution of egg size in parasitic cuckoos is faster than in non-parasitic cuckoos	Medina and Langmore (2015)
Chrysococcyx basalis	*Malurus cyaneus*	Rejection decisions are based on differences in size between parasitic and host eggs, not colour Host is breeding in domed nest	Langmore et al. (2003)
Chrysococcyx caprius	*Euplectes orix*	Hosts do not reject parasite eggs based on size (only colour and spotting pattern are important). Host is breeding in domed nest	Lawes and Kirkman (1996)
Clamator glandarius	*Pica pica*	Spherical (cuckoo-like) eggs are rejected at higher frequency than oval (host-like) eggs. Oversized conspecific eggs are more readily accepted than normal-sized or undersized (cuckoo-like) eggs. Host is breeding in open or domed nest	Alvarez et al. (1976)
Cuculus canorus	2 host species	No difference in parasitic egg size, spotting pattern or colour in two sympatric host species	Drobniak et al. (2014)
Cuculus canorus	3 host species	Host egg phenotype matching in parasitic egg size, colour and spotting pattern. No significant difference in shape between eggs from parasites utilizing different hosts. Parasitic eggs differ from host in shape (rounder) in great reed warblers, but not in two other hosts	Antonov et al. (2010)
Cuculus canorus	6 host species	No difference in shape of parasitic eggs between those laid in various host species, but there is difference in shape between eggs found at two different continents (Japan and Hungary)	Bán et al. (2011)
Cuculus canorus	15 host species	Positive correlation in size between various cuckoo egg morphs and their corresponding host eggs	Moksnes and Røskaft (1995)

(continued)

Table 20.1 (continued)

Parasite	Host	Comment	Source
Cuculus canorus	*Acrocephalus arundinaceus*	Non-mimetic eggs (colour and spotting pattern) differing in shape are rejected at different frequencies	Zölei et al. (2012)
Cuculus canorus	*Acrocephalus palustris*	Differences in egg size between host and parasitic eggs are not important in egg rejection decisions. Colour and spotting pattern are important predictors of egg rejection	Antonov et al. (2006b)
Cuculus canorus	*Acrocephalus scirpaceus*	Greatly oversized mimetic eggs (colour and spotting pattern) are rejected more often than similarly coloured control- and "cuckoo-sized" eggs	Davies and Brooke (1988)
Cuculus canorus	*Acrocephalus scirpaceus*	Non-mimetic eggs (colour and spotting pattern) of different sizes are rejected at similar rates, but mode and speed of rejection vary	Stokke et al. (2010)
Cuculus canorus	*Acrocephalus scirpaceus*	Parasite egg size decreased over a 90-year period, while host egg size remained similar. Egg size could be influenced by both co-evolutionary interactions and climate	Berkowic et al. (2015)
Cuculus canorus	*Cercotrichas galactotes*	Greatly oversized "conspecific" eggs are accepted more often than host- or cuckoo-sized eggs	Alvarez (2000)
Cuculus canorus	*Cercotrichas galactotes*	Positive relationship found between weight and elongation of natural cuckoo eggs. Smaller eggs were more spherical than larger eggs	Alvarez (2003)
Cuculus canorus	*Phylloscopus bonelli*	Non-mimetic eggs (colour and spotting pattern) of different sizes are rejected at similar rates, but mode of rejection varies. Host is breeding in domed nest	Roncalli et al. (2017)
Cuculus canorus	*Phylloscopus trochilus*	Great tit eggs (larger than host eggs but smaller than cuckoo eggs) are rejected. Colour of warbler and tit eggs is approximately similar. Host is breeding in domed nest	Moksnes and Røskaft (1992)
Cuculus canorus	*Turdus merula*	Egg size is an important predictor of egg rejection of mimetic (colour and spotting pattern) eggs. Smaller (cuckoo-sized) eggs are more likely to be	Ruiz-Raya et al. (2015), Soler et al. (2015, 2017)

(continued)

Table 20.1 (continued)

Parasite	Host	Comment	Source
		rejected and also to be rejected faster than larger eggs. Regarding non-mimetic eggs, all eggs were recognized by the host, but heavier eggs were more likely to be accepted	
Cuculus poliocephalus	*Phylloscopus humei*	Egg size (relative size differences between own and foreign eggs) is more important than egg colour in egg rejection decisions. Host is breeding in domed nest	Marchetti (1992, 2000)
Cuculus saturatus	*Phylloscopus inornatus*	Parasitic eggs below a specific size threshold compared to host eggs are accepted, while the ones above are rejected. Most natural parasitic eggs are below the size threshold. Host is breeding in domed nest	Meshcheryagina et al. (2016)
Icteridae	–	Eggs of parasitic cowbirds and their non-parasitic relatives are not significantly different in size	Mermoz and Ornelas (2004)
Molothrus ater	2 host species	Egg-shaped objects are rejected less frequently than other objects. Various egg shapes are rejected infrequently and to same extent regardless of shape	Underwood and Sealy (2006a)
Molothrus ater	*Icterus galbula*	Colour and spotting pattern are important in rejection decisions, while egg size is not	Rothstein (1978)
Molothrus ater	*Setophaga petechia*	Both egg size and colour are important for egg rejection decisions. Larger eggs are rejected more frequently than smaller eggs. Tactile cues are known to be important for egg rejection in this host	Guigueno et al. (2014)
Molothrus ater	*Turdus migratorius*	Colour and spotting pattern are more important than egg size and shape in egg rejection decisions, but Rothstein found that size may elicit more rapid response than colour	Rothstein (1982); Igic et al. (2015a)
Molothrus ater	*Vireo gilvus*	Egg size is not an important predictor of egg rejection (spotting pattern is very important)	Underwood and Sealy (2006b)
Molothrus bonariensis	2 host species	Cowbird eggs are larger in the host with largest eggs, resulting in larger parasitic chick when	Tuero et al. (2012)

(continued)

Table 20.1 (continued)

Parasite	Host	Comment	Source
		hatching (better competitive abilities). Larger host is breeding in open nest, while the smallest one is breeding in cavities	
Molothrus bonariensis	*Furnarius rufus*	Cowbird eggs smaller than a specific ratio relative to host eggs are rejected, and those above (more similar in size to host eggs) are accepted. Width of eggs is also important for acceptance. Colour is not important. Host is breeding in domed nest	Mason and Rothstein (1986)
Molothrus bonariensis	*Paroaria coronata*	Colour is pivotal for egg rejection decisions, but shape is important for more subtle egg rejection decisions when colour mimicry is good. Wider eggs (more different from host eggs) are more likely to be rejected	Segura et al. (2016)
Indicatoridae	–	Evolution of egg size in parasitic honeyguides not faster than in non-parasitic relatives	Medina and Langmore (2015)
Indicator indicator	5 host species	Host-specific differentiation in egg size and shape. All hosts are breeding in cavities	Spottiswoode et al. (2011)
Indicator indicator	*Merops pusillus*	Competing parasites selectively rejected other parasite eggs instead of host eggs (based on size and shape, not colour). Hosts did not show any egg rejection. Host is breeding in cavities	Spottiswoode (2013)
Viduidae	–	Evolution of egg size in parasitic Vidua finches not faster than in non-parasitic relatives	Medina and Langmore (2015)

Honza et al. 2007; Stoddard and Stevens 2010). The pallid cuckoo, *Cuculus pallidus*, in Australia was earlier considered to lay only one egg type, based on human vision. However, by using reflectance spectrophotometry on cuckoo and host eggs, Starling et al. (2006) were able to demonstrate the existence of four gentes, each specializing on and closely mimicking the eggs of their specific host species.

The co-evolutionary arms race between avian brood parasites and their hosts is to a great extent a struggle linked to egg mimicry. When a brood parasite has evolved "perfect" mimicry with the host eggs, the next defence measures to take for the host could be the evolution of a more complicated pattern of spots or egg "signatures" (Stoddard et al. 2014; Caves et al. 2015). Another countermeasure would be to evolve greater diversification in egg appearance between different females, i.e. a higher interclutch variation. This would make it more difficult for the parasite to mimic host eggs and as a consequence easier for the host to recognize and reject the parasitic egg (Davies and Brooke 1989; Takasu 2003; Kilner 2006; Schulze-Hagen et al. 2009; Vikan et al. 2011; Spottiswoode and Stevens 2012; Caves et al. 2015; Medina et al. 2016). The latter ability could also be achieved by reducing the differences between eggs in the single clutch, i.e. a lower intraclutch variation. The support for such reduced intraclutch egg variation in hosts is, however, quite mixed (Kilner 2006; Medina et al. 2016).

Increased interclutch variation among host females may lead to increased egg polymorphism in both hosts and parasites. A striking example of how egg polymorphism may evolve in avian host—parasite systems is the interaction between the cuckoo finch, *Anomalospiza imberbis*, and its host, the tawny-flanked prinia, *Prinia subflava*, in Africa. By examining host and parasite eggs in museum collections and in the field, using spectrophotometry and models of avian vision, Spottiswoode and Stevens (2012) found that during a period of 40 years, there had been relatively "rapid" changes in egg appearance both in host and parasite. Host and parasite seemed to have "followed" each other quite closely because parasitic eggs showed best colour mimicry with those of contemporaneous hosts. These remarkably rapid changes in egg phenotypes show that strong selective forces can work in co-evolutionary arms races between avian brood parasites and their hosts. It seems that the cuckoo finch to some extent is able to counteract the host's egg polymorphism by the same female laying several eggs in the same host nest, sometimes outnumbering the host's eggs. Such repeated parasitism in combination with egg mimicry is disturbing the host's decision rules with the result that some hosts accept parasitic eggs (Stevens et al. 2013). Similar evolution of polymorphism was documented also in other host–parasite systems, like that between bramblings, *Fringilla montifringilla*, and common cuckoos in Northern Finland (Vikan et al. 2011).

As mentioned above, many hosts and parasites lay polymorphic eggs, but in most cases, the variation is continuous. There are, however, a few cases where the co-evolutionary arms race has resulted in discrete polymorphism. In East Asian ashy-throated parrotbills, *Paradoxornis alphonsianus*, parasitized by the common cuckoo, each female lays distinct immaculate blue, pale blue or white eggs, blue and white being the most common (Yang et al. 2010). The hosts show strong rejection of unlike parasite eggs and the cuckoo has evolved egg phenotypes that closely match those of the host (Yang et al. 2013). Here, co-evolution has resulted in a clear disruptive selection in both parties where the selection forces are cost of parasitism in the host and host defences in the cuckoo. Each egg type is predicted to fluctuate in frequency over time (Takasu 2003; Liang et al. 2012). Another example of such discrete polymorphism is the common tailorbird, *Orthotomus sutorius*, in China, which lays two distinct egg types, white and blue eggs with brownish spots, and is parasitized by the plaintive cuckoo, *Cacomantis merulinus*, which also lays similar dimorphic white and blue eggs (Yang et al. 2016).

Even if rarely documented, there are other selection pressures besides host rejection that may promote egg mimicry among avian brood parasites. Many brood parasites remove one or several eggs from host nests before they lay their own egg (Moksnes et al. 2000). Specific targeting of previously laid parasite eggs would be highly beneficial, especially in species where the parasite chick evicts all competitors. The selective pressure for evolution of egg mimicry due to competing parasites would only be prominent in cases where parasite density is high, resulting in high likelihood of multiple cuckoo females visiting the same host nests (see Davies and Brooke 1988). So far, studies have found no evidence of selective removal of parasite eggs among common cuckoos and Horsfield's bronze cuckoos (Davies 1999; Moskát and Honza 2002; Langmore and Kilner 2009; Šulc et al. 2016), but evolution of mimetic eggs in greater honeyguides, *Indicator indicator* (Spottiswoode 2013), may be due to competing parasites rather than host rejection. The greater honeyguide, which is parasitizing several host species in Africa, lays its eggs in hosts breeding in dark holes where colours are not important cues. The eggs are therefore immaculate both in parasite and hosts. The greater honeyguide lays white eggs that show a maternally inherited host-specific differentiation both in egg size and shape (Spottiswoode et al. 2011). Hosts do not discriminate parasitic eggs differing in size and shape, but the nests may be visited by more honeyguide females for whom it would pay to destroy eggs already laid by other conspecific females. Since honeyguide chicks kill all their nest mates, this would enhance survival of own offspring. Destruction by other females may, however, be counteracted by mimicking host eggs. Therefore, interference competition between the parasites themselves seems to be the driving force for evolving egg size and shape mimicry (Spottiswoode 2013).

Many studies have focused on egg mimicry in terms of ground colour and spotting pattern, which in many cases seems to be the most important traits used for egg recognition among hosts [for reviews, see, e.g. Kilner (2006); Feeney et al. (2014b)]. There are, however, far fewer studies dealing with co-evolved adaptations in terms of egg size and shape. We have searched the literature and compiled information on these traits (Table 20.1).

Parasitic cuckoos lay smaller eggs than their non-parasitic relatives (Payne 1974; Krüger and Davies 2002, 2004; Medina and Langmore 2015), but this pattern is not apparent in cowbirds (Mermoz and Ornelas 2004), honeyguides or parasitic finches (Medina and Langmore 2015). Furthermore, a positive correlation between host and parasite egg sizes has been found among different cuckoo egg morphs (Moksnes and Røskaft 1995), host-specific differentiation in egg size and shape has been found in honeyguides (Spottiswoode et al. 2011) and egg size in cuckoos has also been shown to decrease with time in specific populations so that they in the end became more similar in size to host eggs (Berkowic et al. 2015). Evolution of egg size in parasites may be driven by host egg rejection, but a smaller egg also reduces the amount of energy required to produce eggs allowing production of larger "clutches" (von Haartman 1981), increases incubation efficiency (for parasites targeting hosts smaller than themselves) and decreases incubation time (see Davies and Brooke 1988). Evolution of smaller eggs may be problematic in parasites that are non-evictors (e.g. cowbirds) since smaller eggs produce smaller nestlings, which may be competitively inferior to hosts (see, e.g. Tuero et al. 2012). It is also important to remember that egg size is difficult to assess for many birds in their nests relative to differences in ground colour and spot colour/pattern, since mostly only parts of the egg is visible in the nest. Hence, rejection only based on size or shape may be less reliable than rejection based on colour or spotting pattern. Egg shape too may be important in light of egg rejection but is also important for incubation efficiency and strength of eggshell (see, e.g. Zölei et al. 2012). Egg size and shape may be important as tactile cues in egg rejection, especially in dark nest environments where rejection decisions may be based on size and shape rather than colour (Alvarez et al. 1976; Mason and Rothstein 1986; Marchetti 1992; but see Lawes and Kirkman 1996; Marchetti 2000; Langmore et al. 2003; Spottiswoode et al. 2011; Meshcheryagina et al. 2016). Still, egg size is an important predictor of egg rejection also in some species with open nests (Guigueno et al. 2014; Ruiz-Raya et al. 2015; Soler et al. 2015, 2017). Subtle differences in egg shape between parasite and host eggs could serve as a recognition cue when overall mimicry in colour and spotting pattern is good (Segura et al. 2016). Finally, egg size may not influence rejection probabilities, but rather the mode and speed of rejection of eggs (Rothstein 1982; Stokke et al. 2010; Roncalli et al. 2017).

20.3 Evolution of Host-Specific Gentes and Maintenance of Egg Mimicry

The amazing diversity of egg colours and spotting patterns within avian brood parasites is simply breathtaking. But how have all of these egg colours evolved and what is the underlying genetic mechanism maintaining the mimetic cuckoo egg? This question has a long history and originates in the generalist parasitic nature of the common cuckoo which was noticed already by Edward Jenner, the inventor of vaccination, in 1788 (Jenner 1788): "The Cuckoo makes choice of the nests of a great variety of small birds. I have known its egg intrusted to the care of the Hedge-sparrow, the Water-wagtail, the Titlark, the Yellowhammer, the green Linnet, and the Whinchat". The selective advantage of mimetic eggs and the name "gens" was suggested by Alfred Newton in 1893, and the evolutionary conundrum of host specificity was identified and suggested to be sex-linked by Pycraft (1910). Later, this theory was updated by Wynne-Edwards (1933) and Punnett (1933) and thereafter by Jensen (1966). On this background, female-specific inheritance of egg mimicry and host specificity has been known as the "W-hypothesis".

Put simply, the W-hypothesis suggests that the gene(s) responsible for host-specific traits in a generalist parasite like the common cuckoo is located on the female-specific W sex chromosome. As males do not have W-chromosomes, they are effectually excluded from any gens affiliation, and host-specific egg mimicry will be inherited directly from mother to daughter. Female-specific inheritance has more or less been the accepted norm in cuckoo research for many decades with some rare exceptions of scepticism (Southern 1958; Fossøy et al. 2011). However, only recently has this hypothesis received direct support from genetic studies (Spottiswoode et al. 2011; Fossøy et al. 2016).

In the greater honeyguide, two highly distinct ancient maternal lineages parasitize ground or tree nesting host species, with clear differentiation in egg morphology (Spottiswoode et al. 2011). The first maternal lineage parasitizes hosts breeding in terrestrial burrows with both hosts and parasites having round eggs. The second maternal lineage parasitizes host species breeding in tree cavities and where both hosts and parasites have much more elongated eggs. In the common cuckoo, two highly distinct ancient maternal lineages show clear separation in egg colour with blue eggs only being produced in one of the two lineages (Fossøy et al. 2016). In both honeyguides and cuckoos, the maternal lineages have been separated for millions of years, with no females being able to switch lineage successfully. Strikingly, there is no evidence of any corresponding nuclear (i.e. genes that are shared among males and females) genetic divergence in either of the two species, and the data rather suggest high levels of male gene flow on a continental-wide scale. Thus, both studies provide strong evidence for strict maternal inheritance of egg traits and female-specific maintenance of host-specific traits in correspondence with the W-hypothesis.

Egg mimicry is found in a range of cuckoos species (Langmore and Spottiswoode 2012), but no other brood parasite can match the diversity of host species and the number of different egg morphs presented by the common cuckoo (Moksnes and Røskaft 1995; Kilner 2006; Yang et al. 2012). That being said, differences in egg morphology between females utilizing different hosts have also been found in brood parasites besides cuckoos (De La Colina et al. 2011; Spottiswoode et al. 2011; Langmore and Spottiswoode 2012).

20.4 Egg Crypsis

Egg crypsis may result from exactly the same selective pressures as egg mimicry and may be defined as eggs blending in with the background, making them literally invisible. Undoubtedly, many birds that lay eggs directly on the ground have evolved cryptic eggs to avoid nest predation (e.g. Kilner 2006; Stoddard et al. 2011; Hanley et al. 2013). In line with this, Wallace (1889) and later Harrison (1968) suggested that common cuckoos do not always necessarily mimic the eggs of their hosts, but rather produce eggs that do not stick out from the host eggs, making the whole clutch more difficult for predators to detect. However, in species that lay eggs in conspicuous nests, egg crypsis is often found to be of minor importance for predator avoidance (Mason and Rothstein 1987; Davies and Brooke 1988; Underwood and Sealy 2002) compared to nest crypsis (Kilner 2006; Cherry and Gosler 2010). Nonetheless, some studies have found support for colouration of eggs having an effect on risk of predation also in open, conspicuous nests (Westmoreland and Kiltie 2007; Westmoreland 2008). Cryptic eggs may evolve, just like mimetic eggs, to make the parasite egg harder to recognize for the host or for other parasites. One particular advantage of cryptic over mimetic eggs is that it requires less host specialization (one egg type fits all hosts with more or less similar nest structures), making it easier to start using naïve hosts and also prevent evolution of egg rejection in such hosts (Langmore et al. 2009). Marchant (1972) raised the question of crypsis being important in evolution of egg appearance in Australian bronze cuckoos, *Chalcites* spp., utilizing hosts breeding in domed nests where light conditions are generally poor. Bronze cuckoos are particularly interesting in that some species, like Horsfield's bronze cuckoo, lay eggs that are highly mimetic to host eggs (Langmore and Kilner 2009), while others apparently lay cryptic rather than mimetic eggs (Langmore et al. 2009). Modern visual modelling techniques have disclosed that these immaculate, matt dark olive or brown eggs are difficult to recognize in the dark, domed nests in which they are normally laid (Langmore et al. 2009; Kilner and Langmore 2011). Evolution of cryptic eggs in little bronze cuckoos, *Chalcites minutillus* (Gloag et al. 2014) may be due to competing cuckoos rather than host rejection.

Experimental studies addressing the potential advantage of cryptic eggs in other brood parasite-host systems are few and far between. The egg-nest contrast did not play an important role in egg discrimination in American robins, *Turdus migratorius*, or great reed warblers, *Acrocephalus arundinaceus*, hosts of brown-headed cowbirds, *Molothrus ater*, and common cuckoos, respectively (Aidala et al. 2015; Hauber et al. 2015). In nests placed in dim light, however, the egg-nest contrast may be of importance for egg rejection probabilities in potential hosts of common cuckoos (Antonov et al. 2011), and in general, egg rejection is often more prominent in well-lit than in dark nests (Langmore et al. 2005; Avilés et al. 2008, 2015; Honza et al. 2011, 2014).

20.5 Eggshell Strength (Thickness)

Phylogenetically, distant brood parasites like cuckoos, cowbirds and honeyguides have all evolved eggs with thicker shells than would be expected from their body size (Baker 1942; Lack 1968; Spaw and Rohwer 1987; Spottiswoode and Colebrook-Robjent 2007). The greater microhardness of cuckoo eggs relative to host eggs has been documented using scanning electron microscopy (Igic et al. 2011). Furthermore, a particularly round shape is also a prominent trait in several brood parasites, increasing the strength of the eggshell even further (Picman 1989; Brooker and Brooker 1991). Comparative studies have disclosed that these egg characteristics are more prominent in parasites than their non-parasitic relatives (Rahn et al. 1988; Mermoz and Ornelas 2004), pointing to that a strong eggshell has evolved due to the parasitic lifestyle. Furthermore, there must be a selective advantage for evolving thick eggshells, because there are obvious costs related to such traits for parasites. Eggs of cowbirds and cuckoos are also strong from the inside, making hatching from such eggs more difficult for the chick (Picman and Pribil 1997; Honza et al. 2001). Studies have shown that hatching from parasite eggs requires more time than from the thin-shelled host eggs, even though this is somewhat compensated by relatively stronger hatching muscle and a modified egg tooth (Honza et al. 2001, 2015).

There exist two major and not mutually exclusive hypotheses for the evolution of a strong eggshell in brood parasites (Antonov et al. 2012). The first hypothesis is often named "The laying damage hypothesis" (Lack 1968) and advocates the opinion that the strong eggshell has evolved to prevent that the parasite egg is broken during laying. Due to that many brood parasites are larger than their hosts, they are unable to sit properly on the nest and therefore have to "inject" their eggs from a distance (Antonov et al. 2012). Brood parasites that are utilizing hosts breeding in places that are hard to reach, e.g. in nest boxes or other cavities, have also been shown to lay their eggs in this fashion (Samaš et al. 2016). In addition, many brood parasites lay their eggs very rapidly to prevent them being observed (e.g. Davies 2000) or being attacked (Soler et al. 2014) by hosts at their nest. All these behaviours increase the risk that the parasite egg will be harmed at impact with the host eggs. A thicker-shelled parasite egg would therefore be advantageous. If some of the relatively thinner-shelled eggs of hosts are cracked in the process (e.g. Gaston 1976; Soler et al. 1997), this would also be advantageous both for evicting (fewer host eggs to evict requiring less energy used) and non-evicting (less competition with host offspring) parasites. Relevant to this last point, Soler et al. (1997, 1999) found that great spotted cuckoos, *Clamator glandarius*, a non-evicting parasite, often destroy some host eggs at laying without removing them and that this has a positive effect on parasite breeding success. Soler and Martinez (2000) found convincing support for the laying damage hypothesis, while others did not (Spaw and Rohwer 1987; Antonov et al. 2006a).

The hypothesis that has received most attention and support is "the puncture resistance hypothesis" (Swynnerton 1918; Spaw and Rohwer 1987; Rohwer and Spaw 1988). One of the most prominent host defences against brood parasitism is rejection of the parasitic egg, which in many species is undertaken by puncturing and ejecting the egg (Brooke and Davies 1989; Moksnes et al. 1991a). According to the puncture resistance hypothesis, the thick eggshell of parasite eggs may make this task more difficult for hosts, especially those with small bills that are not able to reject eggs by grasp ejection (Rohwer and Spaw 1988; Moksnes et al. 1991a).

Small-billed hosts may therefore suffer costs in terms of destroying own eggs in the process of puncture ejecting the parasite egg (Rohwer et al. 1989; Antonov et al. 2006a). Marsh warblers, *Acrocephalus palustris*, for instance, have experimentally been shown to suffer higher costs when ejecting thick-shelled compared to thinner-shelled eggs, and it also took longer to eject such eggs, but in the end they were not more likely to accept the thick-shelled eggs (Antonov et al. 2008a). Many small-billed cowbird hosts accept parasitic eggs, and it has been hypothesized that this is due to the high costs of rejecting such eggs (Rohwer and Spaw 1988). Since cowbirds are non-evictors, this scenario may be plausible if desertion and renesting is costly (see Krüger 2011). However, in hosts of evicting brood parasites, this is more unlikely since they will face high costs if the parasite egg is accepted (Moksnes et al. 1991a; Stokke et al. 2005). The most convincing support for the hypothesis in cuckoo hosts stems from studies of olivaceous warbler, *Hippolais pallida*, in Bulgaria. Video recordings revealed that this small-billed host clearly recognized cuckoo eggs and pecked them, but in many cases eventually accepted them due to eventual lack of motivation (Antonov et al. 2009). Noteworthy, hosts may also decide to accept eggs that they recognize even though they are not thick-shelled. Hence, there could be additional causes behind lack of motivation to reject eggs (Soler et al. 2012b; Ruiz-Raya et al.

2015). The puncture resistance hypothesis can be extended to include thick eggshells having evolved due to intraspecific egg destruction in non-parasites (Picman and Honza 2002) and in parasites (Brooker and Brooker 1991). A study has shown that the most common host species of the brood parasitic greater honeyguide do not reject parasitic eggs, but laying honeyguides preferentially puncture experimentally added "honeyguide" eggs. Thick eggshells in honeyguides have therefore most likely evolved due to competition between parasite females rather than host defences (Spottiswoode 2013).

It may be important not to look at evolution of eggshell thickness as a separate event but rather in the context of evolution of egg mimicry. Hence, only a thick eggshell may in many cases not be sufficient to avoid rejection, but may be so in combination with mimicry in egg colour and spotting pattern (Antonov et al. 2008a, b; Krüger 2011). Interestingly, common cuckoo gentes utilizing hosts with the best rejection abilities also have the thickest eggshells, and a similar correlation between cuckoo gentes utilizing different host species has also been found in the African diederik cuckoo, *Chrysococcyx caprius* (Spottiswoode 2010).

20.6 Attractive Eggs

It is well-known that apparently non-mimetic parasite eggs are sometimes accepted by the host. Alvarez (1999) suggested that brood parasites in some cases may produce "supernormal" eggs that appear particularly attractive to hosts and thereby tricking them to accept these eggs even though they are easily recognizable. Slightly oversized eggs may be attractive to hosts and may to some extent compensate for deficiency in colour and pattern mimicry (Davies and Brooke 1988; see also Alvarez 2000). Biliverdin is a pigment that is responsible for blue-green colouration on bird eggshells. Since the pigment is costly to produce, more intensely coloured eggs may reflect good quality individuals (Moreno et al. 2006). In line with this, Soler et al. (2012a) proposed that brood parasites may produce eggs with high biliverdin content in hosts that show a preference for blue-green egg pigmentation, because this would trick the hosts to accept their eggs. The "supernormal" egg hypothesis has remained largely untested since.

20.7 Egg Physiology

Early hatching compared to host eggs is advantageous in brood parasites, since they by doing so would have a competitive advantage over host chicks (non-evictors) or being able to get rid of the host progeny faster, eliminating competitors as quickly as possible (evictors). In line with this, several brood parasites lay eggs that require shorter incubation periods than expected for their size. Hence, the parasitic eggs usually hatch before host eggs (Payne 1977; Soler 1990; Kattan 1995; Davies 2000; Schulze-Hagen et al. 2009; Birkhead et al. 2011).

Several species of cuckoos and the greater honeyguide lay eggs with 48-h intervals, but at least in cuckoos, this does not seem to be a specific adaptation evolved in a co-evolutionary context, since several other non-parasitic cuckoos also lay eggs with such intervals (Birkhead et al. 2011). These adaptations, however, may facilitate brood parasitic habits, since the parasitic egg can be incubated internally for 24 h before laying, giving the cuckoo embryo a head start in the development compared to the host embryo. Support for internal incubation has been found in both cuckoos and honeyguides, but not in the cuckoo finch which lays one egg every day (Birkhead et al. 2011). Eggs of cowbirds, *Molothrus* spp., also require relatively short incubation time, but similarly to most other passerines, they lay an egg every 24 h, rendering internal incubation unlikely. Instead, they may lower the energy content in eggs leading to short incubation period since the yolk is depleted relatively quickly (Kattan 1995). The shorter incubation time of cowbird eggs relative to host eggs may be explained by higher eggshell porosity than host eggs, increasing the fluxes of respiratory gases and leading to more rapid development of the cowbird embryo (Jaeckle et al. 2012). In addition, the often larger cowbird egg may disrupt the incubation of the smaller host eggs, leading to postponed hatching of the latter (Mcmaster and Sealy 1998).

Besides internal incubation, several additional mechanisms have been proposed to explain the rapid development in common cuckoo embryos. Hence, cuckoo eggs contained greater yolk mass and nutrient/antioxidant contents (Török et al. 2004; Hargitai et al. 2010) and larger number of eggshell pores allowing gas exchange (Hargitai et al. 2010) than similarly sized great reed warbler host eggs. Portugal et al. (2014), however, found that gas exchange through the cuckoo eggshell was lower rather than higher compared to eggs of other species, including several cuckoo hosts. Instead, it was suggested that the lower gas exchange could result in slower depletion of yolk so that sufficient amounts of nutrients were available to the growing embryo throughout the development. This makes sense, since Igic et al. (2015b) found that cuckoo eggs in general contained less yolk lipids (energy deposits) than host eggs when controlling for differences in size.

Concluding Remarks and Future Directions

Brood parasites have evolved a range of adaptations at the egg stage to overcome host defences. The most general, and undoubtedly also the most well-known, of these is the laying of eggs mimicking the host eggs in shape, size, colour and spotting pattern. Many studies have convincingly shown that these traits have evolved to counter egg rejection by the hosts. Although a few studies have found support for maternal inheritance of egg traits and the "W-hypothesis", the genetic basis of egg traits in general is not known. Future genomic studies will be needed to finally locate the gene(s) for colour, size and shape of eggs and determine whether maternal inheritance is a general trait in the common cuckoo and other avian brood parasites. Further investigations are

(continued)

also needed to fully understand the occurrence of variation in egg morphology and host-specific gentes among brood parasites. On the other hand, relying on the fascinating studies on Australian cuckoos and their hosts, cryptic eggs blending in with the nest lining is an alternative way in which brood parasites can counteract host egg rejection. Egg crypsis could be more widespread than previously considered and should be investigated in other brood parasite-host systems. Importantly, both egg mimicry and crypsis may also evolve due to rejection from competing parasites. Several brood parasites lay eggs with relatively thicker and stronger shells than non-parasites. Studies have shown that this makes egg puncture more costly to hosts and more difficult for competing parasites. By observing nests using video recordings, egg recognition capabilities and the role of motivation in egg rejection can be studied in more detail. Thicker-shelled eggs may also prevent breakage during laying, but this possibility needs to be addressed more thoroughly in future studies. The "supernormal" egg hypothesis, claiming that non-mimetic parasite eggs may be attractive to hosts, has remained largely untested and hence represents a scope for future studies. Some brood parasites incubate their eggs internally before the actual egg laying, providing the embryo with a head start and leading to earlier hatching of the brood parasite chick. Such a scenario has been convincingly shown for a few parasite species, but how general this adaption is among parasites is still unknown.

Acknowledgements We want to thank Manuel Soler, Marcel Honza and Francisco Ruiz-Raya for their good and constructive comments that significantly improved the quality of the text.

References

Aidala Z, Croston R, Schwartz J, Tong L, Hauber ME (2015) The role of egg–nest contrast in the rejection of brood parasitic eggs. J Exp Biol 218:1126–1136

Alicia De La Colina M, Mahler B, Carlos Reboreda J (2011) Differences in morphology and colour pattern of shiny cowbird (*Molothrus bonariensis*) eggs found in nests of two hosts. Biol J Linn Soc 102:838–845

Alvarez F (1994) A gens of cuckoo *Cuculus canorus* parasitizing rufous bush chat *Cercotrichas galactotes*. J Avian Biol 25:239–243

Alvarez F (1999) Attractive non-mimetic stimuli in cuckoo *Cuculus canorus* eggs. Ibis 141:142–144

Alvarez F (2000) Response to common cuckoo *Cuculus canorus* model egg size by a parasitized population of rufous bush chat *Cercotrichas galactotes*. Ibis 142:683–686

Alvarez F (2003) Relationship between egg size and shape of an Iberian population of common cuckoo *Cuculus canorus* and its host. Ardea 91:127–129

Alvarez F, Arias de Reyna L, Segura M (1976) Experimental brood parasitism of the magpie (*Pica pica*). Anim Behav 24:907–916

Antonov A, Stokke BG, Moksnes A, Kleven O, Honza M, Røskaft E (2006a) Eggshell strength of an obligate brood parasite: a test of the puncture resistance hypothesis. Behav Ecol Sociobiol 60:11–18

Antonov A, Stokke BG, Moksnes A, Røskaft E (2006b) Egg rejection in marsh warblers (*Acrocephalus palustris*) heavily parasitized by common cuckoos (*Cuculus canorus*). Auk 123:419–430

Antonov A, Stokke BG, Moksnes A, Røskaft E (2007) First evidence of regular common cuckoo, *Cuculus canorus*, parasitism on eastern olivaceous warblers, *Hippolais pallida eldeica*. Naturwissenschaften 94:307–312

Antonov A, Stokke BG, Moksnes A, Røskaft E (2008a) Does the cuckoo benefit from laying unusually strong eggs? Anim Behav 76:1893–1900

Antonov A, Stokke BG, Moksnes A, Røskaft E (2008b) Getting rid of the cuckoo *Cuculus canorus* egg: why do hosts delay rejection? Behav Ecol 19:100–107

Antonov A, Stokke BG, Moksnes A, Røskaft E (2009) Evidence for egg discrimination preceding failed rejection attempts in a small cuckoo host. Biol Lett 5:169–171

Antonov A, Vikan JR, Stokke BG, Fossøy F, Ranke PS, Røskaft E, Moksnes A, Møller AP, Shykoff JA (2010) Egg phenotype differentiation in sympatric cuckoo *Cuculus canorus* gentes. J Evol Biol 23:1170–1182

Antonov A, Avilés JM, Stokke BG, Spasova V, Vikan JR, Moksnes A, Yang C, Liang W, Røskaft E (2011) Egg discrimination in an open nesting passerine under dim light conditions. Ethology 117:1128–1137

Antonov A, Stokke BG, Fossøy F, Liang W, Moksnes A, Røskaft E, Yang C, Møller AP (2012) Why do brood parasitic birds lay strong-shelled eggs? Chin Birds 3:245–258

Avilés JM, Pérez-Contreras T, Navarro C, Soler JJ (2008) Dark nests and conspicuousness in color patterns of nestlings of altricial birds. Am Nat 171:327–338

Avilés JM, Martín-Gálvez D, De Neve L, Soler M, Soler JJ (2015) Ambient light in domed nests and discrimination of foreign egg colors. Behav Ecol Sociobiol 69:425–435

Baker E (1942) Cuckoo problems. Witherby, London

Baldamus E (1853) Neue Beiträge zur Fortpflanzungsgeschichte des Kuckuchs, *Cuculus canorus*. Naumannia 3:307–325

Bán M, Barta Z, Muñoz AR, Takasu F, Nakamura H, Moskát C (2011) The analysis of common cuckoo's egg shape in relation to its hosts' in two geographically distant areas. J Zool 284:77–83

Berkowic D, Stokke BG, Meiri S, Markman S (2015) Climate change and coevolution in the cuckoo–reed warbler system. Evol Ecol 29:581–597

Birkhead TR, Hemmings N, Spottiswoode CN, Mikulica O, Moskát C, Bán M, Schulze-Hagen K (2011) Internal incubation and early hatching in brood parasitic birds. Proc R Soc B 278:1019–1024

Brooke MD, Davies NB (1988) Egg mimicry by cuckoos *Cuculus canorus* in relation to discrimination by hosts. Nature 335:630–632

Brooke MD, Davies NB (1989) Provisioning of nestling cuckoos *Cuculus canorus* by reed warbler *Acrocephalus scirpaceus* hosts. Ibis 131:250–256

Brooker MG, Brooker LC (1989) The comparative breeding behaviour of two sympatric cuckoos, Horsfield's bronze-cuckoo *Chrysococcyx basalts* and the shining bronze-cuckoo *C. lucidus*, in Western Australia: a new model for the evolution of egg morphology and host specificity in avian brood parasites. Ibis 131:528–547

Brooker MG, Brooker LC (1991) Eggshell strength in cuckoos and cowbirds. Ibis 133:406–413

Caves EM, Stevens M, Iversen ES, Spottiswoode CN (2015) Hosts of avian brood parasites have evolved egg signatures with elevated information content. Proc R Soc B 282

Cherry MI, Bennett ATD (2001) Egg colour matching in an African cuckoo, as revealed by ultraviolet-visible reflectance spectrophotometry. Proc R Soc B 268:565–571

Cherry MI, Gosler AG (2010) Avian eggshell coloration: new perspectives on adaptive explanations. Biol J Linn Soc 100:753–762

Davies NB (1999) Cuckoos and cowbirds versus hosts: co-evolutionary lag and equilibrium. Ostrich 70:71–79

Davies NB (2000) Cuckoos, cowbirds and other cheats. T & AD Poyser, London

Davies NB, Brooke MD (1988) Cuckoos versus reed warblers – adaptations and counteradaptations. Anim Behav 36:262–284

Davies NB, Brooke MD (1989) An experimental study of co-evolution between the cuckoo, *Cuculus canorus*, and its hosts. 1. Host egg discrimination. J Anim Ecol 58:207–224

Drobniak SM, Dyrcz A, Sudyka J, Cichoń M (2014) Continuous variation rather than specialization in the egg phenotypes of cuckoos (*Cuculus canorus*) parasitizing two sympatric reed warbler species. PLoS One 9:e106650

Edvardsen E, Moksnes A, Røskaft E, Øien IJ, Honza M (2001) Egg mimicry in cuckoos parasitizing four sympatric species of *Acrocephalus* warblers. Condor 103:829–837

Feeney WE, Stoddard MC, Kilner RM, Langmore NE (2014a) "Jack-of-all-trades" egg mimicry in the brood parasitic Horsfield's bronze-cuckoo? Behav Ecol 25:1365–1373

Feeney WE, Welbergen JA, Langmore NE (2014b) Advances in the study of coevolution between avian brood parasites and their hosts. Annu Rev Ecol Evol Syst 45:227–246

Fossøy F, Antonov A, Moksnes A, Røskaft E, Vikan JR, Møller AP, Shykoff JA, Stokke BG (2011) Genetic differentiation among sympatric cuckoo host races: males matter. Proc R Soc B 278:1639–1645

Fossøy F, Sorenson MD, Liang W, Ekrem T, Moksnes A, Møller AP, Rutila J, Røskaft E, Takasu F, Yang C, Stokke BG (2016) Ancient origin and maternal inheritance of blue cuckoo eggs. Nat Commun 7:10272

Gaston AJ (1976) Brood parasitism by the pied crested cuckoo *Clamator jacobinus*. J Anim Ecol 45:331–348

Gloag R, Keller L-A, Langmore NE (2014) Cryptic cuckoo eggs hide from competing cuckoos. Proc R Soc B 281:20141014

Guigueno MF, Sealy SG, Westphal AM (2014) Rejection of parasitic eggs in passerine hosts: size matters more for a non-ejecter. Auk 131:583–594

Hanley D, Cassey P, Doucet SM (2013) Parents, predators, parasites, and the evolution of eggshell colour in open nesting birds. Evol Ecol 27:593–617

Hargitai R, Moskát C, Bán M, Gil D, López-Rull I, Solymos E (2010) Eggshell characteristics and yolk composition in the common cuckoo *Cuculus canorus*: are they adapted to brood parasitism? J Avian Biol 41:177–185

Harrison CJO (1968) Egg mimicry in British cuckoos. Bird Study 15:22–28

Hauber ME, Tong L, Bán M, Croston R, Grim T, Waterhouse GIN, Shawkey MD, Barron AB, Moskát C (2015) The value of artificial stimuli in behavioral research: making the case for egg rejection studies in avian brood parasitism. Ethology 120:1–8

Honza M, Picman J, Grim T, Novák V, Čapek M, Mrlík V (2001) How to hatch from an egg of great structural strength. A study of the common cuckoo. J Avian Biol 32:249–255

Honza M, Taborsky B, Taborsky M, Teuschl Y, Vogl W, Moksnes A, Røskaft E (2002) Behaviour of female common cuckoos, *Cuculus canorus*, in the vicinity of host nests before and during egg laying: a radiotelemetry study. Anim Behav 64:861–868

Honza M, Polacikova L, Procházka P (2007) Ultraviolet and green parts of the colour spectrum affect egg rejection in the song thrush (*Turdus philomelos*). Biol J Linn Soc 92:269–276

Honza M, Procházka P, Morongová K, Čapek M, Jelínek V (2011) Do nest light conditions affect rejection of parasitic eggs? A test of the light environment hypothesis. Ethology 117:539–546

Honza M, Šulc M, Jelínek V, Požgayová M, Procházka P (2014) Brood parasites lay eggs matching the appearance of host clutches. Proc R Soc B 281:20132665

Honza M, Feikusová K, Procházka P, Picman J (2015) How to hatch from the common cuckoo (*Cuculus canorus*) egg: implications of strong eggshells for the hatching muscle (*Musculus complexus*). J Ornithol 156:679–685

Igic B, Braganza K, Hyland MM, Silyn-Roberts H, Cassey P, Grim T, Rutila J, Moskát C, Hauber ME (2011) Alternative mechanisms of increased eggshell hardness of avian brood parasites relative to host species. J R Soc Interface 8:1654–1664

Igic B, Nunez V, Voss HU, Croston R, Aidala Z, López AV, Van Tatenhove A, Holford ME, Shawkey MD, Hauber ME (2015a) Using 3D printed eggs to examine the egg-rejection behaviour of wild birds. PeerJ 3:e965

Igic B, Zarate E, Sewell MA, Moskát C, Cassey P, Rutila J, Grim T, Shawkey MD, Hauber ME (2015b) A comparison of egg yolk lipid constituents between parasitic common cuckoos and their hosts. Auk 132:817–825

Jaeckle WB, Kiefer M, Childs B, Harper RG, Rivers JW, Peer BD (2012) Comparison of eggshell porosity and estimated gas flux between the brown-headed cowbird and two common hosts. J Avian Biol 43:486–490

Jenner E (1788) Observations on the natural history of the cuckoo. By Mr. Edward Jenner. In a letter to John Hunter, Esq. F. R. S. Philos Trans R Soc Lond 78:219–237

Jensen RAC (1966) Genetics of cuckoo egg polymorphism. Nature 209:827

Kattan GH (1995) Mechanisms of short incubation period in brood-parasitic cowbirds. Auk 112:335–342

Kilner RM (2006) The evolution of egg colour and patterning in birds. Biol Rev 81:383–406

Kilner RM, Langmore NE (2011) Cuckoos versus hosts in insects and birds: adaptations, counter-adaptations and outcomes. Biol Rev 86:836–852

Krüger O (2011) Brood parasitism selects for no defence in a cuckoo host. Proc R Soc B 278:2777–2783

Krüger O, Davies NB (2002) The evolution of cuckoo parasitism: a comparative analysis. Proc R Soc B 269:375–381

Krüger O, Davies NB (2004) The evolution of egg size in the brood parasitic cuckoos. Behav Ecol 15:210–218

Lack D (1968) Ecological adaptations for breeding in birds. Methuen, London

Langmore N, Kilner R (2009) Why do Horsfield's bronze-cuckoo *Chalcites basalis* eggs mimic those of their hosts? Behav Ecol Sociobiol 63:1127–1131

Langmore N, Spottiswoode CN (2012) Visual trickery in avian brood parasites. In: Hughes D, Brodeur J, Thomas F (eds) Host manipulation by parasites. Oxford University Press, Oxford, pp 95–116

Langmore NE, Hunt S, Kilner RM (2003) Escalation of a coevolutionary arms race through host rejection of brood parasitic young. Nature 422:157–160

Langmore NE, Kilner RM, Butchart SHM, Maurer G, Davies NB, Cockburn A, Macgregor NA, Peters A, Magrath MJL, Dowling DK (2005) The evolution of egg rejection by cuckoo hosts in Australia and Europe. Behav Ecol 16:686–692

Langmore NE, Stevens M, Maurer G, Kilner RM (2009) Are dark cuckoo eggs cryptic in host nests? Anim Behav 78:461–468

Lawes MJ, Kirkman S (1996) Egg recognition and interspecific brood parasitism rates in red bishops (Aves: Ploceidae). Anim Behav 52:553–563

Liang W, Yang C, Stokke BG, Antonov A, Fossøy F, Vikan JR, Moksnes A, Røskaft E, Shykoff JA, Møller AP, Takasu F (2012) Modelling the maintenance of egg polymorphism in avian brood parasites and their hosts. J Evol Biol 25:916–929

Marchant S (1972) Evolution of the genus *Chrysococcyx*. Ibis 114:219–233

Marchetti K (1992) Costs to host defence and the persistence of parasitic cuckoos. Proc R Soc B 248:41–45

Marchetti K (2000) Egg rejection in a passerine bird: size does matter. Anim Behav 59:877–883

Mason P, Rothstein SI (1986) Coevolution and avian brood parasitism: cowbird eggs show evolutionary response to host discrimination. Evolution 40:1207–1214

Mason P, Rothstein SI (1987) Crypsis versus mimicry and the color of shiny cowbird eggs. Am Nat 130:161–167

Mcmaster DG, Sealy SG (1998) Short incubation periods of brown-headed cowbirds; how do cowbird eggs hatch before yellow warbler eggs? Condor 100:102–111

Medina I, Langmore NE (2015) Coevolution is linked with phenotypic diversification but not speciation in avian brood parasites. Proc R Soc B 282

Medina I, Troscianko J, Stevens M, Langmore NE (2016) Brood parasitism is linked to egg pattern diversity within and among species of Australian passerines. Am Nat 187:351–362

Mermoz ME, Ornelas JF (2004) Phylogenetic analysis of life-history adaptations in parasitic cowbirds. Behav Ecol 15:109–119

Meshcheryagina SG, Golovatin MG, Bachurin GN (2016) Experimental study of discrimination behavior in the yellow-browed warbler (*Phylloscopus inornatus*) brood-parasitized by the oriental cuckoo (*Cuculus* (*saturatus*) *optatus*). Russ J Ecol 47:104–106

Moksnes A, Røskaft E (1992) Responses of some rare cuckoo hosts to mimetic model cuckoo eggs and to foreign conspecific eggs. Ornis Scand 23:17–23

Moksnes A, Røskaft E (1995) Egg morphs and host preference in the common cuckoo (*Cuculus canorus*) – an analysis of cuckoo and host eggs from European museum collections. J Zool 236:625–648

Moksnes A, Røskaft E, Braa AT (1991a) Rejection behavior by common cuckoo hosts towards artificial brood parasite eggs. Auk 108:348–354

Moksnes A, Røskaft E, Braa AT, Korsnes L, Lampe HM, Pedersen HC (1991b) Behavioral responses of potential hosts towards artificial cuckoo eggs and dummies. Behaviour 116:64–89

Moksnes A, Røskaft E, Hagen LG, Honza M, Mork C, Olsen PH (2000) Common cuckoo *Cuculus canorus* and host behaviour at reed warbler *Acrocephalus scirpaceus* nests. Ibis 142:247–258

Moreno J, Lobato E, Morales J, Merino S, Tomas G, Martinez-de la Puente J, Sanz JJ, Mateo R, Soler JJ (2006) Experimental evidence that egg color indicates female condition at laying in a songbird. Behav Ecol 17:651–655

Moskát C, Honza M (2002) European cuckoo *Cuculus canorus* parasitism and host's rejection behaviour in a heavily parasitized great reed warbler *Acrocephalus arundinaceus* population. Ibis 144:614–622

Newton A (1902) Cuckoo. In: Encyclopedia Britannica, 10th edn. Times, London

Payne RB (1974) The evolution of clutch size and reproductive rates in parasitic cuckoos. Evolution 28:169–181

Payne RB (1977) Ecology of brood parasitism in birds. Annu Rev Ecol Syst 8:1–28

Picman J (1989) Mechanism of increased puncture resistance of eggs of brown-headed cowbirds. Auk 106:577–583

Picman J, Honza M (2002) Are house wren *Troglodytes aedon* eggs unusually strong? Test of the predicted effect of intraspecific egg destruction. Ibis 144:E57–E66

Picman J, Pribil S (1997) Is greater eggshell density an alternative mechanism by which parasitic cuckoos increase the strength of their eggs? J Ornithol 138:531–541

Portugal SJ, Hauber ME, Maurer G, Stokke BG, Grim T, Cassey P (2014) Rapid development of brood-parasitic cuckoo embryos cannot be explained by increased gas exchange through the eggshell. J Zool 293:219–226

Punnett RC (1933) Inheritance of egg-colour in the parasitic cuckoos. Nature 132:892–893

Pycraft W (1910) A history of birds. Methuen, London

Rahn H, Curran-Everett L, Booth DT (1988) Eggshell differences between parasitic and nonparasitic *Icteridae*. Condor 90:962–964

Rey E (1892) Altes und neues aus dem haushalte des kuckucks. Freese, Leipzig

Rohwer S, Spaw CD (1988) Evolutionary lag versus bill-size constraints: a comparative study of the acceptance of cowbird eggs by old hosts. Evol Ecol 2:27–36

Rohwer S, Spaw CD, Røskaft E (1989) Costs to northern orioles of puncture-ejecting parasitic cowbird eggs from their nests. Auk 106:734–738

Roncalli G, Ibáñez-Álamo JD, Soler M (2017) Size and material of model parasitic eggs affect the rejection response of western Bonelli's warbler *Phylloscopus bonelli*. Ibis 159:113–123

Rothstein SI (1978) Mechanisms of avian egg-recognition – additional evidence for learned components. Anim Behav 26:671–676

Rothstein SI (1982) Mechanisms of avian egg recognition – which egg parameters elicit responses by rejecter species. Behav Ecol Sociobiol 11:229–239

Ruiz-Raya F, Soler M, Sánchez-Pérez LL, Ibáñez-Álamo JD (2015) Could a factor that does not affect egg recognition influence the decision of rejection? PLoS One 10:e0135624

Samaš P, Rutila J, Grim T (2016) The common redstart as a suitable model to study cuckoo-host coevolution in a unique ecological context. BMC Evol Biol 16:255

Schulze-Hagen K, Stokke BG, Birkhead TR (2009) Reproductive biology of the European cuckoo *Cuculus canorus*: early insights, persistent errors and the acquisition of knowledge. J Ornithol 150:1–16

Segura LN, Di Sallo FG, Mahler B, Reboreda JC (2016) Red-crested cardinals use color and width as cues to reject shiny cowbird eggs. Auk 133:308–315

Skjelseth S, Moksnes A, Røskaft E, Gibbs HL, Taborsky M, Taborsky B, Honza M, Kleven O (2004) Parentage and host preference in the common cuckoo *Cuculus canorus*. J Avian Biol 35:21–24

Soler M (1990) Relationships between the great spotted cuckoo *Clamator glandarius* and its corvid hosts in a recently colonized area. Ornis Scand 21:212–223

Soler M, Martínez JG (2000) Is egg-damaging behavior by great spotted cuckoos an accident or an adaptation? Behav Ecol 11:495–501

Soler M, Soler JJ, Martinez JG (1997) Great spotted cuckoos improve their reproductive success by damaging magpie host eggs. Anim Behav 54:1227–1233

Soler M, Soler JJ, Pérez-Contreras T (1999) The cost of host egg damage caused by a brood parasite: experiments on great spotted cuckoos (*Clamator glandarius*) and magpies (*Pica pica*). Behav Ecol Sociobiol 46:381–386

Soler JJ, Avilés JM, Møller AP, Moreno J (2012a) Attractive blue-green egg coloration and cuckoo−host coevolution. Biol J Linn Soc 106:154–168

Soler M, Fernández-Morante J, Espinosa F, Martín-Vivaldi M (2012b) Pecking but accepting the parasitic eggs may not reflect ejection failure: the role of motivation. Ethology 118:662–672

Soler M, Pérez-Contreras T, de Neve L (2014) Great spotted cuckoos frequently lay their eggs while their magpie host is incubating. Ethology 120:1–8

Soler M, Ruiz-Raya F, Roncalli G, Ibáñez-Álamo JD (2015) Nest desertion cannot be considered an egg-rejection mechanism in a medium-sized host: an experimental study with the common blackbird *Turdus merula*. J Avian Biol 46:369–377

Soler M, Ruiz-Raya F, Roncalli G, Ibáñez-Álamo JD (2017) Relationships between egg-recognition and egg-ejection in a grasp-ejector species. PLoS One 12:e0166283

Southern H (1958) Mimicry in cuckoos' eggs. In: Huxley J, Hardy A, Ford E (eds) Evolution as a process. G. Allen and Unwin, London

Spaw CD, Rohwer S (1987) A comparative study of eggshell thickness in cowbirds and other passerines. Condor 89:307–318

Spottiswoode CN (2010) The evolution of host-specific variation in cuckoo eggshell strength. J Evol Biol 23:1792–1799

Spottiswoode CN (2013) A brood parasite selects for its own egg traits. Biol Lett 9:20130573

Spottiswoode CN, Colebrook-Robjent JFR (2007) Egg puncturing by the brood parasitic Greater Honeyguide and potential host counteradaptations. Behav Ecol 18:792–799

Spottiswoode CN, Stevens M (2012) Host-parasite arms races and rapid changes in bird egg appearance. Am Nat 179:633–648

Spottiswoode CN, Stryjewski KF, Quader S, Colebrook-Robjent JFR, Sorenson MD (2011) Ancient host specificity within a single species of brood parasitic bird. Proc Natl Acad Sci U S A 108:17738–17742

Starling M, Heinsohn R, Cockburn A, Langmore NE (2006) Cryptic gentes revealed in pallid cuckoos *Cuculus pallidus* using reflectance spectrophotometry. Proc R Soc B 273:1929–1934

Stevens M, Troscianko J, Spottiswoode CN (2013) Repeated targeting of the same hosts by a brood parasite compromises host egg rejection. Nat Commun 4:3475

Stoddard MC, Stevens M (2010) Pattern mimicry of host eggs by the common cuckoo, as seen through a bird's eye. Proc R Soc B 277:1387–1393

Stoddard MC, Stevens M (2011) Avian vision and the evolution of egg color mimicry in the common cuckoo. Evolution 65:2004–2013

Stoddard MC, Marshall KLA, Kilner RM (2011) Imperfectly camouflaged avian eggs: artefact or adaptation? Avian Biol Res 4:196–213

Stoddard MC, Kilner RM, Town C (2014) Pattern recognition algorithm reveals how birds evolve individual egg pattern signatures. Nat Commun 5:5117

Stokke BG, Moksnes A, Røskaft E (2005) The enigma of imperfect adaptations in hosts of avian brood parasites. Ornithol Sci 4:17–29

Stokke BG, Polačiková L, Dyrcz A, Hafstad I, Moksnes A, Røskaft E (2010) Responses of reed warblers *Acrocephalus scirpaceus* to non-mimetic eggs of different sizes in a nest parasitism experiment. Acta Ornithol 45:98–104

Šulc M, Procházka P, Capek M, Honza M (2016) Common cuckoo females are not choosy when removing an egg during parasitism. Behav Ecol 27:1642–1649

Swynnerton CFM (1918) VII. Rejections by birds of eggs unlike their own: with remarks on some of the cuckoo problems. Ibis:127–154

Takasu F (2003) Co-evolutionary dynamics of egg appearance in avian brood parasitism. Evol Ecol Res 5:345–362

Tuero DT, Fiorini VD, Mahler B, Reboreda JC (2012) Shiny cowbird *Molothrus bonariensis* egg size and chick growth vary between two hosts that differ markedly in body size. J Avian Biol 43:227–233

Török J, Moskat C, Michl G, Peczely P (2004) Common cuckoos (*Cuculus canorus*) lay eggs with larger yolk but not more testosterone than their great reed warbler (*Acrocephalus arundinaceus*) hosts. Ethol Ecol Evol 16:271–277

Underwood TJ, Sealy SG (2002) Adaptive significance of egg coloration. In: Deeming DC (ed) Avian incubation, behaviour, environment and evolution. Oxford University Press, Oxford, pp 280–298

Underwood TJ, Sealy SG (2006a) Influence of shape on egg discrimination in American robins and gray catbirds. Ethology 112:164–173

Underwood TJ, Sealy SG (2006b) Parameters of brown-headed cowbird *Molothrus ater* egg discrimination in warbling vireos *Vireo gilvus*. J Avian Biol 37:457–466

Vikan JR, Fossøy F, Huhta E, Moksnes A, Roskaft E, Stokke BG (2011) Outcomes of brood parasite-host interactions mediated by egg matching: common cuckoos *Cuculus canorus* versus *Fringilla* finches. PLoS One 6:e19288

von Haartman L (1981) Co-evolution of the cuckoo *Cuculus canorus* and a regular cuckoo host. Ornis Fennica 58:1–10

Wallace AR (1889) Darwinism: an exposition of the theory of natural selection, with some of its applications. Macmillan, London

Westmoreland D (2008) Evidence of selection for egg crypsis in conspicuous nests. J Field Ornithol 79:263–268

Westmoreland D, Kiltie RA (2007) Egg coloration and selection for crypsis in open-nesting blackbirds. J Avian Biol 38:682–689

Wynne-Edwards VC (1933) Inheritance of egg-colour in the parasitic cuckoo. Nature 132:822–823

Yang C, Cai Y, Liang W (2013) Eggs mimicry of common cuckoo (*Cuculus canorus*) utilizing ashy-throated parrotbill (*Paradoxornis alphonsianus*) host. Chin Birds 4:51–56

Yang C, Huang Q, Wang L, Jiang A, Stokke BG, Fossøy F, Tunheim OH, Røskaft E, Liang W, Møller AP (2016) Plaintive cuckoos do not select tailorbird hosts that match the phenotypes of their own eggs. Behav Ecol 27:835–841

Yang C, Liang W, Antonov A, Cai Y, Stokke BG, Fossøy F, Moksnes A, Røskaft E (2012) Diversity of parasitic cuckoos and their hosts in China. Chin Birds 3:9–32

Yang C, Liang W, Cai Y, Shi S, Takasu F, Møller A, Antonov A, Fossøy F, Moksnes A, Røskaft E, Stokke BG (2010) Coevolution in action: disruptive selection on egg colour in an avian brood parasite and its host. PLoS One 5:e10816

Zölei A, Hauber ME, Geltsch N, Moskát C (2012) Asymmetrical signal content of egg shape as predictor of egg rejection by great reed warblers, hosts of the common cuckoo. Behaviour 149:391–406

Implications of Nest Sanitation in the Evolution of Egg Rejection

21

Mélanie F. Guigueno and Spencer G. Sealy

Abstract

Nest sanitation or nest cleaning behaviour in Passeriformes has been widely reported in the literature. It was proposed that this behaviour is a preadaptation for the evolution of egg rejection behaviour in hosts of avian brood parasites, which are primarily passerines. Indeed, the same motor skills underlie nest sanitation and egg rejection behaviours. This "nest sanitation hypothesis" predicts that non-egg-shaped objects (i.e. simulated debris) are rejected more frequently than egg-shaped objects (i.e. simulated parasitic eggs), and the frequency of rejection mirrors the expected frequency of debris falling into the nest, rather than the risk of parasitism. We update information on nest sanitation processes (e.g. timing of defecation relative to faecal sac removal) and synthesize lists of objects removed from nests by passerines in non-experimental and experimental contexts. In the former, passerines remove various objects from their nests, including faecal sacs, parasitic invertebrates, vegetation and dead nestlings. Results of experiments reveal that non-egg-shaped objects, especially those most resembling debris, were more frequently rejected from nests, and that rejection of non-egg-shaped objects was generally constant among nesting stages prior to hatching.

M. F. Guigueno (✉)
Department of Natural Resource Sciences, McGill University, Sainte-Anne-de-Bellevue, QC, Canada
e-mail: melanie.guigueno@mail.mcgill.ca

S. G. Sealy
Department of Biological Sciences, University of Manitoba, Winnipeg, MB, Canada
e-mail: spencer.sealy@umanitoba.ca

M. Soler (ed.), *Avian Brood Parasitism*, Fascinating Life Sciences,
https://doi.org/10.1007/978-3-319-73138-4_21

21.1 Introduction

Passerine birds have long been reported to remove objects from their nests that are neither eggs nor live nestlings (hereafter "debris"), such as faecal sacs and eggshells (i.e. Plutarch *in* Skutch 1976). Nest sanitation behaviour, or nest cleaning behaviour, likely plays a crucial role in keeping nests dry and free of parasites, reduces attraction of predators and facilitates egg turning during incubation (Guigueno and Seay 2012). Birds may ingest debris or grasp it between their mandibles and drop it at a distant disposal site. Birds may push debris aside (Hoover 2003), which provides eggs or nestlings access to the main part of the cavity, or they may desert their nest and nest site altogether. These responses resemble egg rejection, a behaviour that is reported in many hosts of brood parasites. Nest sanitation behaviour may be a preadaptation to egg rejection behaviour, as hypothesized by Guigueno and Sealy (2012). Here, we review new information on nest sanitation behaviour and extend implications for its importance in the evolution of egg rejection behaviour within the context of avian brood parasitism.

How nests are ridded of debris depends on the type to be removed. Faecal sacs produced by younger nestlings may be consumed at the nest by the parents, who gain water (Morton 1979) and enhanced nutrition (McGowan 1995), in addition to conserving energy that would be required to carry faecal sacs far enough away to avoid attracting predators (McGowan 1995; Guigueno and Sealy 2012). As chicks age and become larger, with more effective digestive tracts, parents may carry the sacs away (Guigueno and Sealy 2012). Not all debris can be grasped and carried away, however, especially when it is large relative to the size of the birds. For example, yellow warblers (*Setophaga petechia*) buried experimentally added, large non-egg-shaped objects, but ejected smaller objects of the same shape (Guigueno and Sealy 2009). Burial is a form of clutch abandonment in which the adult bird buries the foreign object (debris or egg) along with its own eggs and superimposes a new nest above the previous nest and eggs (Mico 1998; Guigueno and Sealy 2010). Thus, the bird abandons its nest but not the nest site. Nest desertion is another form of clutch abandonment in which the bird deserts its nest and nest site (Guigueno and Sealy 2010). When cleaning their nests, therefore, passerines may ingest debris, carry it away, bury it in their nest (hereafter burial) or desert their nest and nest site (nest desertion).

Anti-parasite defences resemble responses to debris. Yellow warblers, hosts of brown-headed cowbirds (*Molothrus ater*), and great reed warblers (*Acrocephalus arundinaceus*), hosts of common cuckoos (*Cuculus canorus*), have been reported to bury and desert parasitic eggs and experimentally added non-egg-shaped objects in the context of studies on nest sanitation (Bártol et al. 2003; Moskát et al. 2003; Guigueno and Sealy 2009). In both yellow warblers and great reed warblers, burial and desertion are also considered anti-parasite strategies (Sealy 1995; Lotem et al. 1995; Moskát and Honza 2002). In a study of 60 populations from 35 host species of brown-headed cowbirds, Hosoi and Rothstein (2000) concluded that desertion is an evolved response to brood parasitism—hosts that incur large costs when parasitized have higher desertion rates among non-forest species, which more frequently encounter cowbirds, than among forest species. Few or no cowbird hosts desert solely in response to the cowbird egg (Sealy 1995; Hosoi and Rothstein 2000). Interaction with the parasite may be a necessary cue for nest desertion as has been reported in yellow warblers (Guigueno and Sealy 2011) and meadow pipits (*Anthus pratensis*), the latter, hosts of the common cuckoo (Moksnes and Røskaft 1989). Unlike hosts that eject parasitic eggs, visual egg discrimination is not necessary for hosts to bury or desert parasitized clutches. Recent research has demonstrated that these hosts rely more on tactile cues than ejectors to assess clutch volume before determining whether to bury or desert parasitized clutches (Guigueno and Sealy 2012; Guigueno et al. 2014). Tactile input from debris in the nest may also elicit nest sanitation, as yellow warblers that ejected star-shaped objects first settled on their clutch before removing them (Guigueno and Sealy 2009). As such, tactile input, that is, assessment of the volume and shape of debris or parasitic egg relative to the size of the nest cavity, may be implicated in determining whether debris or a parasitic egg is removed by ejection, burial or desertion, all methods used in the context of both nest sanitation and brood parasitism.

21.2 Non-experimental Objects Removed From the Nests

21.2.1 Faecal Sacs

Passerines constantly encounter debris in their nests that may or may not be removed. Faecal sacs are the most frequently reported debris removed from nests (Table 21.1). In fact, many researchers measure only faecal sac removal when quantifying nest sanitation behaviour. The removal of faecal sacs is widespread among passerines, with 227 *Birds of North America Online* species accounts out of 229 reporting faecal sac removal (Guigueno and Sealy 2012). In Table 21.1, we listed 37 passerine families in which faecal sac removal has been reported, more than any other type of non-experimental object removed from nests.

Parents remove faecal sacs by ingesting them or carrying them to a distant drop site. Faecal sacs of passerines are encapsulated in a mucous covering, which provides a physical barrier from the pathogenic enteric bacteria inside the sac (Ibáñez-Álamo et al. 2014a). Recently, Quan et al. (2015) demonstrated that nestling defecation is stimulated by feeding in red-whiskered bulbuls (*Pycnonotus jocosus*), rather than occurring at a set time after feeding. Such plasticity minimizes soiling of the nest and its contents and allows for cooperation between the adults and nestlings, described earlier by Blair and Tucker (1941). Nestlings wait for a feeding to defecate, at which time the parent is available to either ingest or carry away the faecal sac (Quan et al. 2015). Ingesting faecal sacs eliminates transportation and disposal costs, but because faecal sacs contain waste products, they may tax the adults' digestive tract (Hurd et al. 1991), even though the mucous covering of faecal sacs has been shown to have an antimicrobial function (Ibáñez-Álamo et al. 2014a). As such, adults normally shift from ingesting smaller faecal sacs early in the nestling period to carrying larger faecal sacs away when nestlings are older (Guigueno and Sealy 2012). Adults must carry

Table 21.1 Objects removed by passerines in a non-experimental context

Object	Passerine families		References
Faecal sacs	Tyrannidae	Sittidae	Guigueno and Sealy (2012)
	Furnariidae	Certhiidae	Bolopo et al. (2015)
	Menuridae	Nectariniidae	Markman et al. (1996)
	Acanthizidae	Mimidae	Quan et al. (2015)
	Neosittidae	Sturnidae	
	Laniidae	Muscicapidae	
	Vireonidae	Turdidae	
	Rhipiduridae	Muscicapidae	
	Corvidae	Cinclidae	
	Paridae	Passeridae	
	Remizidae	Ploceidae	
	Hirundinidae	Motacillidae	
	Alaudidae	Fringillidae	
	Pycnonotidae	Parulidae	
	Phylloscopidae	Icteridae	
	Leiothrichidae	Calcariidae	
	Regulidae	Emberizidae	
	Troglodytidae	Cardinalidae	
	Polioptilidae		
Invertebrates	Tyrannidae	Turdidae	Guigueno and Sealy (2012)
	Vireonidae	Prunellidae	Bolopo et al. (2015)
	Corvidae	Parulidae	Cantarero et al. (2013)
	Paridae	Icteridae	
	Sittidae	Cardinalidae	
	Mimidae		
Eggshells	Tyrannidae	Sturnidae	Guigueno and Sealy (2012)
	Corvidae	Prunellidae	
	Picathartidae	Fringillidae	
	Phylloscopidae	Parulidae	
	Leiothrichidae	Icteridae	
	Mimidae	Emberizidae	
	Turdidae		
Feathers	Acanthizidae		Guigueno and Sealy (2012)
Leaves, twigs and other vegetation	Corvidae		Guigueno and Sealy (2012)
	Acrocephalidae		Bolopo et al. (2015)
	Emberizidae		
Uneaten food	Tyrannidae	Turdidae	Guigueno and Sealy (2012)
	Corvidae		Bolopo et al. (2015)

(continued)

Table 21.1 (continued)

Object	Passerine families		References
Dead nestlings	Tyrannidae	Fringillidae	Guigueno and Sealy (2012)
	Vireonidae	Parulidae	Kirkpatrick et al. (2009)
	Troglodytidae	Icteridae	
	Mimidae	Calcariidae	
	Sturnidae	Emberizidae	
	Turdidae		

faecal sacs far enough from the nest to avoid attracting predators (but see Ibáñez-Álamo et al. 2014b), as predation has been shown experimentally to increase when the distance between eggs and faeces decreased (Petit et al. 1989). In addition, water pipits (*Anthus spinoletta*; Hendricks 1987) and northern wheatears (*Oenanthe oenanthe*; Brooke 1981) flew significantly farther from the nest when they carried a faecal sac than when they did not. Apparently, faecal sacs do not attract invertebrate parasites (Ibáñez-Álamo et al. 2016), but a potential antimicrobial function of the sacs suggests that their removal protects the nestlings and parents from harmful microorganisms (Ibáñez-Álamo et al. 2014a). In sum, removal of faecal sacs involves cooperation between nestlings and parents to keep the nest clean, potentially avoiding the attraction of predators and reduces exposure to harmful microorganisms.

21.2.2 Invertebrates

Passerines may also remove invertebrates from their nests and nestlings, which was reported in 11 families (Table 21.1). Invertebrates removed from the nest include blowflies (*Protocalliphora* spp.) (Hurtrez-Boussès et al. 2000) and hen fleas (*Ceratophyllus gallinae*) (Christie et al. 1996). Removal of invertebrates was highest in a species with the highest infestations levels (Cantarero et al. 2013). Invertebrate load may be so great that adults forage and sleep less to increase nest sanitation, which may reduce their survival after the breeding period (Christie et al. 1996; Tripet et al. 2002). As such, this high cost suggests that ridding nests of invertebrates is of paramount importance.

21.2.3 Eggshells

We reported eggshell removal from 12 passerine families (Table 21.1). It is beneficial for parents to remove eggshells from their nests as it prevents egg capping (i.e. eggshell adhering to intact, unhatched eggs) and injury to the hatched chicks, in addition to potentially reducing attraction of the nests to predators and invertebrate parasites. As with faecal sacs, eggshells may be eaten piecemeal at the nest or carried away by the parents before being dropped (Nethersole-Thompson and

Nethersole-Thompson 1942; Smith 1950; Winkler 2004; Guigueno and Sealy 2012). Because embryos extract about 80% of their skeletal calcium from the eggshell, the remaining shell probably provides limited nutrition (Simkiss 1961; Jones 1976), although eggshells may be used as grit (Nethersole-Thompson and Nethersole-Thompson 1942). Although other types of nest sanitation, namely, faecal sac removal, is "practically confined" to passerines and their allies, eggshell removal is frequently observed in other groups, such as gulls (Nethersole-Thompson and Nethersole-Thompson 1942; Tinbergen et al. 1962). Related to eggshell removal, passerines also remove broken eggs, but this declines abruptly during hatching, possibly so that parents do not remove their own pipped egg (Kemal and Rothstein 1988; Mallory et al. 2000; McMaster and Sealy 1997). As such, the removal of broken eggs is a type of nest sanitation behaviour that differs from other types because normally the removal of objects does not decline around pipping. In addition to laying their egg in hosts nests, some brood parasites, such as great spotted cuckoos (*Clamator glandarius*) and shiny cowbirds (*Molothrus bonariensis*), may puncture eggs already present in the nest (Soler et al. 1997, 1999; Gloag et al. 2012). The hosts, in turn, remove these punctured eggs (Soler et al. 1997, 1999; Gloag et al. 2012). In sum, although the removal of eggshells may provide only limited nutritional benefit when ingested, it seems to be a form of nest sanitation that is more widespread, occurring in non-passerines.

21.2.4 Dead Nestlings

The removal of dead nestlings was recorded in 11 families (Table 21.1). Removing dead nestlings eliminates the subsequent production of scents and bacteria associated with decomposition that could attract predators and invertebrate parasites. Dead nestlings are normally carried and dropped away from the nest. Brewer's blackbirds (*Euphagus cyanocephalus*) disposed of dead nestlings in areas normally used to drop faecal sacs (Welty 1982). Kirkpatrick et al. (2009) observed removal of entire broods of dead nestlings by yellow-eyed juncos (*Junco phaeonotus*) and red-faced warblers (*Cardellina rubrifrons*), which apparently had starved due to inclement weather. In these cases, empty nests do not necessarily indicate a predation event (Kirkpatrick et al. 2009). A partial predation event in which some dead chicks are left behind could also elicit their subsequent removal from the nest by the parents, leaving the nest empty and appearing as though the brood was completely depredated. In sum, some or all of the chicks may be removed from a nest if they are dead and carried away before being dropped.

21.2.5 Feathers, Vegetation and Uneaten Food

The removal of feathers, vegetation such as twigs and leaves, and uneaten food has been reported less frequently, but this type of behaviour was recorded in seven families (Table 21.1).

21.3 Nest Sanitation Hypothesis

As stated above, nest sanitation may reduce predation (and presumably disease), thereby increasing fitness. Removing foreign eggs also increases the host's fitness in the context of brood parasitism. There are at least three hypotheses related to the evolution of nest sanitation and egg rejection that could be expected: (1) There is no interaction between nest sanitation and egg rejection (i.e. they develop independently from each other); (2) nest sanitation is a preliminary stage for egg rejection, and well-developed nest sanitation behaviour results in the ability of hosts to reject parasitic eggs; and (3) egg rejection behaviour may affect nest sanitation as a side effect. The most plausible version may be (2) (but see Luro and Hauber 2017), although (2) and (3) are not mutually exclusive and little research has been conducted on (3). Many passerines demonstrate nest sanitation outside the context of brood parasitism, and, clearly, nest sanitation predates anti-parasitic defences in passerine birds. Consequently, nest sanitation behaviour may be a preadaptation for the evolution of egg rejection behaviour in hosts of avian brood parasites (hypothesis (2); Moskát et al. 2003; Guigueno and Sealy 2009). As proposed by Moskát et al. (2003), there are testable predictions associated with the nest sanitation hypothesis. First, the more an object departs from the shape of an egg, the more likely it will be rejected. Second, the frequency of rejection will be constant across pre-nestling stages (e.g. egg laying and incubation) because debris would be expected to occur at a similar frequency across them (Moskát et al. 2003; Guigueno and Sealy 2009, 2012). In sum, egg rejection by hosts of brood parasites is likely to evolve first by hosts rejecting all non-egg-shaped objects and second via the evolution of egg recognition (Moskát et al. 2003).

21.4 Non-egg-Shaped Objects Added to Nests in Experimental Contexts

Various types of non-egg-shaped objects have been experimentally added to nests of passerines (Table 21.2). Eight families have been included in such studies: Corvidae, Hirundinidae, Mimidae, Turdidae, Passeridae, Acrocephalidae, Parulidae and Icteridae (Table 21.2). Examples of object shapes include cubes, rectangles, stars, discs/other flat objects, cylinders and dumbbells (two partial spheres glued together at their flat ends) (Fig. 21.1; Table 21.2). The non-egg-shaped objects included in the experiments were all artificial except for peanut shells (Yang et al. 2015a, b), reed stems (Bártol et al. 2003), pine cone bract scale (Peer 2017) and Douglas fir leaf clipping (Luro and Hauber 2017) (Table 21.2).

Across studies, non-egg-shaped objects were rejected more frequently than round or egg-shaped objects (Table 21.2). Non-egg-shaped objects were rejected by ejection, burial or nest desertion at 6–100% of nests, whereas egg-shaped or round objects added to nests in the same studies were rejected at 0–69% of nests (Álvarez et al. 1976; Ortega and Cruz 1988; Ortega et al. 1993; Bártol et al. 2003; Hoover 2003; Moskát et al. 2003; Underwood and Sealy 2006; Guigueno and

Table 21.2 Non-egg-shaped objects experimentally added to nests of brood-parasitic hosts during the pre-nestling stages

Species	Non-egg-shaped object	% rejected (total *n*)	% egg-shaped or round object rejected (total *n*)	Reference
Eurasian magpie (*Pica pica*)	Cubes	Cubes ($n = 16$) rejected significantly faster than oval models ($n = 18$) and real eggs ($n = 20$)		Álvarez et al. (1976)
Barn swallow (*Hirundo rustica*)	Half peanut shell	100 (25)	34 (62)	Yang et al. (2015a)
Barn swallow (*Hirundo rustica*)	Stick model	80 (60)	48 (62)	Yang et al. (2015b)
	Coin model	74 (50)		
	Half peanut shell	93 (28)		
Red-rumped swallow (*Cecropis daurica*)	Stick model	20 (15)	0 (14)	Yang et al. (2015b)
	Half peanut shell	15 (13)		
Grey catbird (*Dumetella carolinensis*)	Cylinders, rectangles, cubes	58 (42)	21 (42)	Underwood and Sealy (2006)
American robin (*Turdus migratorius*)	Cylinders, rectangles, cubes	60 (42)	37 (43)	Underwood and Sealy (2006)
	Douglas fir leaf clipping	97 (32)	65 (35)	Luro and Hauber (2017)
Eurasian Tree sparrow (*Passer montanus*)	Flat cut-out models (egg and square shaped)	81.7 (60[a])	36.8 (19)	Poláček et al. (2013)
Great reed warbler (*Acrocephalus arundinaceus*)	Reed stems	88 (16)	Not measured	Bártol et al. (2003)
	Coins	91 (33)		
	Sticks	96 (50)	69 (51)	Moskát et al. (2003)
	Discs	91 (44)		
Yellow warbler (*Setophaga petechia*)	Stars	27 (139)	11 (200)	Guigueno and Sealy (2009)
	Dumbbells	6 (148)		
Prothonotary warbler (*Protonotaria citrea*)	Dice	45 (20)	40 (40)	Hoover (2003)

Red-winged blackbird (*Agelaius phoeniceus*)	Flagging tape	100 (23)	0 (53)	Peer (2017)
	Pine cone bract scale	100 (10)		
	Bead with hole	77 (26)	22[b] (89)	Ortega et al. (1993)
	Dowel (cylinder)	9 (23)		
	Star	95 (22)		
	Various non-egg-shaped objects (beads, pom-poms, oblong, stars)	98 (44)	21 (195)	Ortega and Cruz (1988)
Yellow-headed blackbird (*Xanthocephalus xanthocephalus*)	Various non-egg-shaped objects (beads, pom-poms, oblong, stars)	100 (6)	7 (83)	Ortega and Cruz (1988)

[a]60 nests each tested with two objects
[b]Round beads with hole filled in

Fig. 21.1 (**a**) Examples of non-egg-shaped objects added to the nests of a host of a brood parasite. These objects were added to the nests of yellow warblers, with larger objects on the top row and smaller objects on the bottom row (Guigueno and Sealy 2009). From left to right: Stars, dumbbells and real eggs (brown-headed cowbird egg on top, yellow warbler egg at bottom). (**b**) A large dumbbell buried with two warbler eggs, with a new clutch of four eggs laid on top. (**c**) A small star selectively buried, with the warbler eggs untouched

Sealy 2009; Poláček et al. 2013; Yang et al. 2015a; Luro and Hauber 2017; Peer 2017; Table 21.2). We counted each species in each study as a data point (i.e. two species in one study plus one species in another study as three data points). Out of the 15 data points in which the responses to non-egg-shaped objects and round or egg-shaped objects were measured simultaneously within the same study, all 15 data points indicated that all or some of the non-egg-shaped objects for each data point were rejected more frequently and/or more quickly than round or egg-shaped objects (Table 21.2). Only two data points recorded that other types of non-egg-shaped objects were not rejected more frequently than round or egg-shaped objects (Ortega et al. 1993; Guigueno and Sealy 2009). Objects diverge from the shape of an egg as the number of edges increases (Guigueno and Sealy 2009). For example, star-shaped objects are less similar to the shape of an egg than a cylinder or a dumbbell (Fig. 21.1a). Guigueno and Sealy (2009) reported that dumbbells, which may have appeared and felt like two host eggs, were not rejected more frequently than a cowbird egg, unlike stars (Fig. 21.1a). Similarly, Ortega et al. (1993) reported that cylinders were not rejected more frequently than round beads with holes filled in, unlike stars and beads with holes. The more an object resembles debris, therefore, the more likely, it seems, that it is to be rejected.

Rejection of non-egg-shaped objects remained constant between pre-incubation, when the risk of parasitism was high, and incubation, when the risk of parasitism was lower, if the cost of rejection also remained the same. Eurasian magpies (*Pica pica*), grey catbirds (*Dumetella carolinensis*), American robins (*Turdus migratorius*) and Eurasian tree sparrows (*Passer montanus*), whose bills were large enough relative to the size of non-egg-shaped objects, removed them by ejection but rejected the objects at similar frequencies between nesting stages (Álvarez et al. 1976; Underwood and Sealy 2006; Poláček et al. 2013). Great reed warblers and yellow warblers, however, resorted to alternate forms of rejection, depending on the object (Moskát et al. 2003; Guigueno and Sealy 2009). Both small (warbler egg volume) and large (double the volume of a warbler egg) non-egg-shaped objects were added to the nests of yellow warblers (Guigueno and Sealy 2009). Small objects were ejected or buried on their own (i.e. no warbler eggs were buried in the process; Fig. 21.1c), whereas large objects were buried with warbler eggs, which normally occurs when cowbird eggs are buried (Fig. 21.1b; Guigueno and Sealy 2009). As such, rejection of large objects was costlier than rejection of small objects, both in time and energy required to lay a new clutch. It follows that small objects were rejected at a similar frequency in pre-incubation and incubation stages, whereas large objects were rejected more frequently in pre-incubation (Guigueno and Sealy 2009). Similarly, there was no difference in the rejection frequency between pre-incubation and incubation in great reed warblers, but ejection replaced nest desertion in incubation when the clutch was complete (Moskát et al. 2003). Abandoning the entire clutch in incubation is costlier than abandoning one or two eggs in pre-incubation (Guigueno and Sealy 2010). In sum, the removal of non-egg-shaped objects is stable across pre-hatching stages, unless precluded by rejection costs, which is consistent with the nest sanitation hypothesis.

Egg ejection may be elicited by the process of keeping nests clean. Barn swallows (*Hirundo rustica*), which are intermediate egg rejecters, were more likely to reject a non-mimetic model egg if a half peanut shell was added to the nest at the same time (Yang et al. 2015a). Similarly, the frequency of egg rejection tended to increase with the increase in rejection frequency of non-egg-shaped objects across different barn swallow populations (Yang et al. 2015b). Red-rumped swallows (*Cecropis daurica*), which accept non-mimetic eggs, rejected non-egg-shaped objects less frequently than barn swallows, which suggests brood parasitism may in turn increase the rejection frequency of non-egg-shaped objects (Yang et al. 2015b). In contrast to Yang et al. (2015a), red-winged blackbirds (*Agelaius phoeniceus*) were *not* more likely reject a model cowbird egg, with its maculation but different background colour, if the model egg was simultaneously added to the nest with flagging tape or a pine cone bract scale (Peer 2017). Similarly, having American robins remove a Douglas fir leaf clipping from their nests did not influence whether they removed a model cowbird egg from their (Luro and Hauber 2017). Possibly, the half peanut shell used by Yang et al. (2015a) elicited a stronger sanitation response because it was more voluminous and roughly shaped than the tape or bract scale used by Peer (2017) or the fir leaf clipping in Luro and Hauber (2017), which then influenced subsequent egg rejection. In addition, the model egg added to barn swallow nests was immaculate blue, whereas barn swallow eggs are spotted against a white background. Barn swallows may have received a more salient visual cue than red-winged blackbirds or American robins in removing the model egg. All things considered, nest sanitation may elicit egg rejection.

Concluding Remarks and Future Directions

Costs and benefits of keeping a nest clean merit further study. For example, we know little of the costs of ingestion of faecal sacs versus carrying them away and whether these costs change over the nestling stage as the faecal sacs get larger, in the context of adult survival and reproduction. Identifying species with marked differences in nest sanitation behaviour, such as house finches (*Haemorhous mexicanus*), which do not clean their nests, and understanding the nature of these differences will allow us to more clearly test the importance of nests sanitation in the context of brood parasitism.

Nest sanitation seems to be complex, containing several different behavioural elements. Some of these elements are similar to each other (i.e. removing a twig versus a leaf from a nest) or completely different (i.e. removing faecal sacs versus invertebrate parasites). Future studies should take into consideration these differences, in respect to nesting stages (laying, incubation and nestlings).

Results from experiments with non-egg-shaped objects support the hypothesis that nest sanitation is a preadaptation for the evolution of egg rejection in hosts of brood parasites. Debris-like objects were rejected more frequently

(continued)

than egg-like objects, and rejection remained constant across egg laying and incubation. Nevertheless, answers to some questions are unclear, such as what is the degree to which nest sanitation influences egg rejection versus brood parasitism? Studies examining the responses of non-egg-shaped objects in parasitized versus unparasitized populations of the same host species may enhance our understanding of the potential side effect of brood parasitism on nest sanitation behaviour. Likewise, more studies addressing whether nest sanitation elicits egg rejection at the same nests, which use robust phylogenetic comparisons, are needed as they simultaneously test whether nest sanitation influences egg rejection in the same individual. Lastly, studies examining the responses of hosts to non-egg-shaped objects throughout the nesting stages (egg laying, incubation and nestling) will determine whether rejection of these objects is more pronounced when nest sanitation in a non-experimental context is increased at the nestling stage via the removal of faecal sacs. Although some evidence exists for the hypothesis that nest sanitation is a preadaptation for egg rejection in hosts of brood parasites, such future studies are essential to more clearly understand its potential role.

Acknowledgements We thank the reviewers, Csaba Moskát, Manolo Soler and Canchao Yang for constructive comments on the earlier drafts of this chapter.

References

Álvarez F, de Reyna LA, Segura M (1976) Experimental brood parasitism of the magpie (*Pica pica*). Anim Behav 24:907–916

Bártol I, Moskát C, Karcza Z, Kisbenedek T (2003) Great reed warblers bury artificial objects, not only cuckoo eggs. Acta Zool Acad Sci H 49:111–114

Blair RH, Tucker BW (1941) Nest sanitation. Br Birds 34:206–215, 226–235, 250–255

Bolopo D, Canestrari D, Marcos JM, Baglione V (2015) Nest sanitation in cooperatively breeding carrion crows. Auk 132:604–612

Brooke M d L (1981) How an adult wheatear (*Oenanthe oenanthe*) uses its territory when feeding nestlings. J Anim Ecol 50:683–696

Cantarero A, López-Arrabé J, Rodríguez-García V, González-Braojos S, Redondo AJ, Moreno J (2013) Factors affecting the presence and abundance of generalist ectoparasites in nests of three sympatric hole-nesting bird species. Acta Ornithol 48:39–54

Christie P, Richner H, Oppliger A (1996) Of great tits and fleas: sleep baby sleep. Anim Behav 52:1087–1092

Gloag R, Fiorini VD, Reboreda JC, Kacelnik A (2012) Brood parasite eggs enhance egg survivorship in a multiply parasitized host. Proc R Soc B Biol Sci 279:1831–1839

Guigueno MF, Sealy SG (2009) Nest sanitation plays a role in egg burial by yellow warblers. Ethology 115:247–256

Guigueno MF, Sealy SG (2010) Clutch abandonment by parasitized yellow warblers: egg burial or nest desertion? Condor 112:399–406

Guigueno MF, Sealy SG (2011) Aggression towards egg-removing cowbird elicits clutch abandonment in parasitized yellow warblers, *Dendroica petechia*. Anim Behav 81:211–218

Guigueno MF, Sealy SG (2012) Nest sanitation in passerine birds: implications for egg rejection in hosts of brood parasites. J Ornithol 153:35–52

Guigueno MF, Sealy SG, Westphal AM (2014) Rejection of parasitic eggs in passerine hosts: size matters more for a non-ejecter. Auk 131:583–594

Hendricks P (1987) Foraging patterns of water pipits (*Anthus spinoletta*) with nestlings. Can J Zool 65:1522–1529

Hoover JP (2003) Experiments and observations of prothonotary warblers indicate a lack of adaptive responses to brood parasitism. Anim Behav 65:935–944

Hosoi SA, Rothstein SI (2000) Nest desertion and cowbird parasitism: evidence for evolved responses and evolutionary lag. Anim Behav 59:823–840

Hurd PL, Weatherfaead PJ, McRac SB (1991) Parental consumption of nestling feces: good food or sound economics? Behav Ecol 2:69–76

Hurtrez-Boussès S, Renaud F, Blondel J, Perret P, Galan MJ (2000) Effects of ectoparasites of young on parents' behaviour in a Mediterranean population of blue tits. J Avian Biol 31:266–269

Ibáñez-Álamo JD, Ruiz-Rodríguez M, Soler JJ (2014a) The mucous covering of fecal sacs prevents birds from infection with enteric bacteria. J Avian Biol 45:354–358

Ibáñez-Álamo JD, Ruiz-Raya F, Roncalli G, Soler M (2014b) Is nest predation an important selective pressure determining fecal sac removal? The effect of olfactory cues. J Ornithol 155:491–496

Ibáñez-Álamo JD, Ruiz-Raya F, Rodríguez L, Soler M (2016) Fecal sacs attract insects to the nest and provoke an activation of the immune system of nestlings. Front Zool 13:3

Jones PJ (1976) The utilization of calcareous grit by laying *Quelea quelea*. Ibis 118:575–576

Kemal RE, Rothstein SI (1988) Mechanisms of avian egg recognition: adaptive responses to eggs with broken shells. Anim Behav 36:175–183

Kirkpatrick C, Conway CJ, Ali MH (2009) Sanitation of entire broods of dead nestlings may bias cause-specific nest failure rates. Ibis 151:207–211

Lotem A, Nakamura H, Zahavi A (1995) Constraints on egg discrimination and cuckoo-host co-evolution. Anim Behav 49:1185–1209

Luro AB, Hauber ME (2017) A test of the nest sanitation hypothesis for the evolution of foreign egg rejection in an avian brood parasite rejecter host species. Sci Nat 104:14

Mallory ML, Rendell WB, Robertson RJ (2000) Responses of birds to broken eggs in their nests. Condor 102:673–675

Markman S, Yom-Tov Y, Wright J (1996) The effect of male removal on female parental care in the orange-tufted sunbird. Anim Behav 52:437–444

McGowan KJ (1995) A test of whether economy or nutrition determines fecal sac ingestion in nesting corvids. Condor 97:50–56

McMaster DG, Sealy SG (1997) Red-winged blackbirds (*Agelaius phoeniceus*) accept prematurely hatching brown-headed cowbirds (*Molothrus ater*). Bird Behav 12:67–70

Mico MA (1998) Yellow warbler nests: structure, building materials and cowbird parasitism. MSc thesis, University of Manitoba, Winnipeg, 89 pp

Moksnes A, Røskaft E (1989) Adaptations of meadow pipits to parasitism by the common cuckoo. Behav Ecol Sociobiol 24:25–30

Morton ML (1979) Fecal sac ingestion in the mountain white-crowned sparrow. Condor 81:72–77

Moskát C, Honza M (2002) European cuckoo *Cuculus canorus* parasitism and host's rejection behaviour in a heavily parasitized great reed warbler *Acrocephalus arundinaceus* population. Ibis 144:614–622

Moskát C, Székely T, Kisbenedek T et al (2003) The importance of nest cleaning in egg rejection behaviour of great reed warblers *Acrocephalus arandinaceas*. J Avian Biol 34:16–19

Nethersole-Thompson C, Nethersole-Thompson D (1942) Egg-shell disposal by birds. Br Birds 35:162–169

Ortega CP, Cruz A (1988) Mechanisms of egg acceptance by marsh-dwelling blackbirds. Condor 90:349–358

Ortega JC, Ortega CP, Cruz A (1993) Does brown-headed cowbird egg coloration influence red-winged blackbird responses towards nest contents? Condor 95:217–219

Peer BD (2017) Nest sanitation does not elicit rejection in a brown-headed cowbird host. Anim Cogn 20:371–374

Petit KE, Petit LJ, Petit DR (1989) Fecal sac removal: do the pattern and distance of dispersal affect the chance of nest predation? Condor 91:479–482

Poláček M, Griggio M, Bartíková M, Hoi H (2013) Nest sanitation as the evolutionary background for egg ejection behaviour and the role of motivation for object removal. PLoS One 8:e78771

Quan RC, Li H, Wang B, Goodale E (2015) The relationship between defecation and feeding in nestling birds: observational and experimental evidence. Front Zool 12:21

Sealy SG (1995) Burial of cowbird eggs by parasitized yellow warblers: an empirical and experimental study. Anim Behav 49:877–889

Simkiss K (1961) Calcium metabolism and avian reproduction. Biol Rev 36:321–359

Skutch AF (1976) Parent birds and their young. University of Texas Press, Austin

Smith S (1950) The yellow wagtail. Collins, London

Soler M, Soler JJ, Martínez JG (1997) Great spotted cuckoos improve their reproductive success by damaging magpie host eggs. Anim Behav 54:1227–1233

Soler M, Soler JJ, Pérez-Contreras T (1999) The cost of host egg damage caused by a brood parasite: experiments on great spotted cuckoos (*Clamator glandarius*) and magpies (*Pica pica*). Behav Ecol Sociobiol 46:381–386

Tinbergen N, Broekhuysen GJ, Feekes F, Houghton JCW, Kruuk H, Szulc E (1962) Egg shell removal by the black-headed gull, *Larus ridibundus* L.; a behaviour component of camouflage. Behaviour 19:74–116

Tripet F, Glaser M, Richner H (2002) Behavioural responses to ectoparasites: time-budget adjustments and what matters to blue tits *Parus caeruleus* infested by fleas. Ibis 144:461–469

Underwood TJ, Sealy SG (2006) Influence of shape on egg discrimination in American robins and gray catbirds. Ethology 112:164–173

Welty JC (1982) The life of birds. WB Saunders, Philadelphia

Winkler DW (2004) Nests, eggs, and young: breeding biology of birds. In: Podulka S, Rohrbaugh RW Jr, Bonney R (eds) Handbook of bird biology, 2nd edn. Princeton University Press, Ithaca, pp 1–152

Yang C, Chen M, Wang L, Liang W, Møller AP (2015a) Nest sanitation elicits egg discrimination in cuckoo hosts. Anim Cogn 18:1373–1377

Yang C, Wang L, Liang W, Møller AP (2015b) Nest sanitation behavior in hirundines as a pre-adaptation to egg rejection to counter brood parasitism. Anim Cogn 18:355–360

Egg Characteristics Affecting Egg Rejection 22

Marcel Honza and Michael I. Cherry

Abstract

We have reviewed the most important results relating to particular egg characteristics responsible for recognition and subsequent rejection by hosts of brood parasites. Hosts remove a foreign egg after determining that it differs in one or more parameters. In turn, brood parasites have often evolved various mechanisms to confuse host defences and prevent egg recognition. The most conspicuous one is egg mimicry—imitation of the appearance of host eggs. We evaluate and discuss egg rejection experiments, particularly from a historical perspective, and the use of cameras in experiments. Further, we describe assessments of egg mimicry, and in particular we focus on the role played by particular characteristics in discrimination including egg colour, spottiness, chromatic versus achromatic cues, the role of UV spectra, the blunt egg pole, and the shape and volume of the parasitic egg. In addition, we discuss how research methodology and the application of experimental approaches to studying avian vision have affected studies on egg discrimination.

22.1 Introduction

A successful act of brood parasitism has a disastrous effect on host fitness because it decreases host reproduction considerably (Payne and Payne 1998; Davies 2000). However, in the face of this selection pressure, many hosts have evolved egg

M. Honza (✉)
Institute of Vertebrate Biology AS CR, Brno, Czech Republic
e-mail: honza@brno.cas.cz

M. I. Cherry
Department of Botany and Zoology, University of Stellenbosch, Matieland, South Africa

M. Soler (ed.), *Avian Brood Parasitism*, Fascinating Life Sciences,
https://doi.org/10.1007/978-3-319-73138-4_22

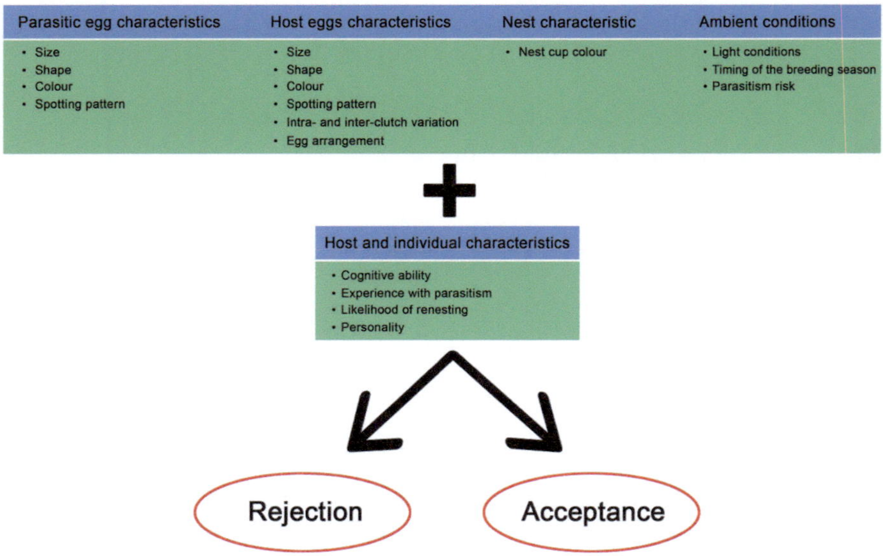

Fig. 22.1 Cues facilitating recognition of a foreign egg by avian brood parasite hosts

discrimination, as an antiparasite defence mechanism (Davies and Brooke 1989). Strong evidence that egg recognition can function as a defence against brood parasitism is provided by studies showing that egg rejection rates correlate spatially (Soler et al. 1999a; Lindholm and Thomas 2000) or temporally (Brooke et al. 1998; Nakamura et al. 1998) with the likelihood of cuckoo parasitism.

Egg discrimination is a two-stage process comprising perceptual (egg recognition) and operational (egg rejection) components (Hauber and Sherman 2001). With regard to perception, hosts can discern a foreign egg by at least two cognitive mechanisms (Moskát and Hauber 2007; Moskát et al. 2010), namely, (1) direct comparison (Rothstein 1974; Lahti and Lahti 2002) or (2) memory (Hauber et al. 2006; Moskát and Hauber 2007; see Chap. 24 for a discussion on how different mechanisms influence egg recognition and rejection). Many cues facilitating recognition of a foreign egg play an important role in the host recognition processes (see Fig. 22.1). After assessing the level of difference between foreign and its own eggs, the host takes into account other factors such as the costs associated with egg rejection and the risk of parasitism (Davies et al. 1996; Soler et al. 2012) and may then elect to remove the foreign egg. In turn, brood parasites have often evolved various tricks to confuse hosts and prevent egg recognition. The most conspicuous is egg mimicry—imitation of the appearance of host eggs—which evolves in response to selection pressure from host ejection of brood parasite eggs (see Chap. 20).

22.2 Egg Rejection Experiments

22.2.1 A Historical Perspective

After establishing that brood parasites lay eggs in host nests and that the egg sometimes disappears from the parasitized nests, naturalists turned their attention to answering questions about how the parasitic egg disappears and which factors caused its removal. Sealy and Underwood (2012) provide a detailed summary of observations by early naturalists who experimentally tested host responses towards real eggs originating from different species. Perhaps two names from a large number of experimenters should be highlighted: Swynnerton (1918)—who confirmed egg ejection—and Rensch (1925), who examined egg recognition by the hosts. A milestone in progress of the experimental approach is the series of papers published by Rothstein in the early 1970s (Rothstein 1974). These well and carefully designed studies based on experiments (using model eggs) provided both a baseline and inspiration for future studies. Early ones were Álvarez et al. (1976), Davies and Brooke (1988, 1989), Cruz and Wiley (1989), Higuchi (1989), Moksnes and Røskaft (1989), Soler (1990), and Sealy (1992), carried out in different areas studying different host–parasite systems. Since then, a numerous studies involving egg recognition experiments have been performed in 182 different host species (see Appendix in Soler 2014) based on different approaches using painted real, model or conspecific, or other natural parasitic eggs.

The next step involved the use of carefully designed colour manipulation of the both parasitic and host eggs: this approach has yielded new insights into underlying cognitive processes regarding discrimination of the parasitic egg (de la Colina et al. 2012). This has allowed the documentation of adaptive modulation of antiparasitic strategies through shifts in the acceptance threshold of hosts (Hauber et al. 2006) or the conditions under which hosts reduce discrimination of foreign eggs (Moskát et al. 2008a) and has demonstrated that rejection using a recognition template might be advantageous in populations with high rates of multiple parasitism (Moskát et al. 2010). A variety of other aspects of egg rejection behaviour have been tested experimentally, such as its repeatability (Honza et al. 2007a; Croston and Hauber 2014a), the costs of rejection (Martín-Vivaldi et al. 2002; Underwood and Sealy 2006a; Segura et al. 2016), responses to avian brood parasitism in sympatric and allopatric host populations (Soler and Møller 1990; Briskie et al. 1992; Soler et al. 1999b; Stokke et al. 2008; Vikan et al. 2010; Yang et al. 2014), methods of rejection of parasitic eggs (Sealy 1995; Soler et al. 2015), and timing of ejection (Požgayová et al. 2011).

Egg rejection experiments do not necessarily imply manipulation of egg colour alone. For example, tests of the "egg arrangement hypothesis" by Polačiková et al. (2013) and Hanley et al. (2015) and the evaluation of egg nest contrasts by Aidala et al. (2015) have brought new insights into additional cues driving parasitic egg rejection.

In addition, there have been many studies testing egg shape discrimination of the hosts of brood parasites using various non-egg-shaped objects, e.g. Ortega and Cruz (1988) and Moskát et al. (2003), as well as over a range of different-sized objects (Guigueno and Sealy 2009, 2012; Álvarez et al. 1976). Underwood and Sealy (2006b) concluded that rejection of odd-shaped objects most likely represents an expression of nest sanitation behaviour, where debris is removed from the nest. Egg shape recognition is predicted to be most advanced in birds which can differentiate between non-egg items in the nest and parasite or own eggs (Peer et al. 2007; Guigueno and Sealy 2009).

Igic et al. (2015) recently suggested a new method—3D printing technology—of producing experimental eggs which enables more precise manipulation of egg size and shape. Soler et al. (2015) also suggested a new method to reproduce the exact colour of both background and spot colours, in which a specialized company (Copingra Pinturas) used a laser scanner to produce paints for experimental eggs. In conclusion, there has been dramatic progress in experimental approaches over the last two decades in the use of egg rejection experiments, providing new insights into our understanding of the behaviour of hosts towards parasitic eggs.

22.2.2 Assessment of Egg Mimicry: From Human Vision to Model Avian Vision

Although in retrospect the method of the human scoring mimicry mostly from photographs using a three- to five-point scale (Moksnes and Røskaft 1993, 1995) seems crude, this method was successfully used in many studies on brood parasitism and certainly pushed the boundaries of our knowledge in this area. It must be noted that these studies did not take into consideration the ability of birds to detect UV wavelengths in the 300–400 nm range which is invisible to humans. In addition, eggs used in experimental studies were painted using acrylic paints, which have different spectral reflectances to natural eggshells—in particular not reflecting in the UV. Cherry and Bennett's (2001) paper, by placing emphasis on the potential role of the UV spectrum in avian egg discrimination, inspired a new wave of studies incorporating visual perception in brood parasitism research. These findings led to increased interest in studying the role of UV signals using portable spectrophotometers enabling measurements in the field; and the importance for human invisible spectra was also tested experimentally using UV blocking chemicals (see Sect. 22.3.3). Subsequent studies incorporated measurement of the UV visible range in the cuckoo (Avilés and Møller 2004; Cherry et al. 2007a) and pallid cuckoo (*Cacomantis pallidus*) hosts (Starling et al. 2006), but these studies measured the difference in reflectance between cuckoo and host eggs, ignoring how avian sensory systems process this information (Vorobyev et al. 1998; Cuthill et al. 2000). One should expect that the efficacy of a cuckoo egg in terms of matching would be influenced by the colour of the cuckoo egg itself, in contrast to the colour of the host eggs, the environment in which matching is perceived by the host, and the perceptual abilities of the host (Endler 1990; Vorobyev et al. 1998). This approach integrates reflectance spectra of cuckoo and host eggs and light conditions in the nest vicinity with published information on photoreceptor sensitivities, photoreceptor noise, and the transmission properties of avian colour vision media (Hart et al. 2000) to calculate differences in matching in host colour space (Vorobyev et al. 1998). The model, developed for the tetrachromatic visual system of birds in its long form (Vorobyev et al. 1998), provides a way of calculating the ability of a bird to distinguish between different colours while accounting for visual pigments absorbance and oil droplet transmittance (Hart et al. 2000; Hart 2001). Avilés (2008) developed a discrimination model approach that simulates host retinal functioning; Cassey et al. (2008) used a photoreceptor noise-limited colour opponent model of host perceptual physiology to predict behavioural rates of experimental egg discrimination of cuckoo or artificial eggs. Spectrometric measures have obvious advantages over human visual estimates of egg colouration, but they have the disadvantage of not assessing the spatial attributes of egg colouration. Stoddard and Stevens (2010) therefore extended this approach and developed a new technique to evaluate spotting patterns in several hosts of the common cuckoo, using digital image analysis to measure a range of attributes including marking size, diversity in size, contrast, coverage, and dispersion. This technique uses Fourier analysis of granularity based on early-stage, low-level visual processes. A refinement of this technique which explicitly tries to quantify recognisability in a way that mimics how a bird's brain works, rather than just extracting objective measures of pattern, was published recently (Stoddard et al. 2014).

22.2.3 Artificial Versus Real Eggs

Generally, researchers have used either real brood parasite eggs, conspecific host eggs, or similarly sized eggs of different host species. The usefulness of real eggs can be limited because the relative influences of egg shape, size, and colour on rejection decisions cannot be easily differentiated (Antonov et al. 2009). One great advantage of real eggs is that results more accurately reflect patterns of egg rejection particularly in species that are puncture ejectors. But apart from the fact that real brood parasite eggs are often not easily available, another obvious disadvantage is the fact that eggshell colouration may change quickly with time: Moreno et al. (2011), Navarro and Lahti (2014), and Hanley et al. (2016) have provided evidence that such changes impact brood parasite host eggshell colour mimicry during the incubation stage. So far published work has mostly neglected this fact, although it could be of significance. The use of natural eggs in experiments can also result in the destruction of viable eggs, which is ethically questionable, although if abandoned natural eggs are used (e.g. Cherry et al. 2007a), this is not problematic. A suitable alternative is the use of eggs from commercial breeds like Chinese quail (*Coturnix chinensis*) (Stokke et al. 2010) or non-fertilized eggs from exotic breeds like Bourke's parrot, (*Neophema bourkii*) (Honza et al. 2007a).

Artificial eggs are specifically useful in separating the relative influences of different phenotypic characteristics on rejection decisions (Álvarez et al. 1976; Rothstein 1982; Hauber et al. 2015; Roncalli et al. 2017 although see Lahti 2015). However, there are limitations to using artificial eggs, traditionally made of plasticine, plastic, wood, or plaster of Paris (Prather et al. 2007). For example, some researchers concede that rejection of hard-shelled eggs may underestimate rejection rate (Moksnes et al. 1991; Martín-Vivaldi et al. 2002; Prather et al. 2007), but see Honza and Moskát (2008) for opposing results. On the other hand, the softness of the material may overestimate ejection (Roncalli et al. 2017). The size of the host, which is manifest in its ability to grasp foreign eggs, plays a major role in this respect. Artificial material may also affect the mode of the responses towards such eggs, as unsuccessful attempts to puncture hard model eggs could increase the costs of rejection and/or provoke clutch desertion (Martín-Vivaldi et al. 2002).

In addition, shell thickness and strength of the parasitic egg (real or artificial) may also influence host reaction times (Antonov et al. 2008; Honza and Moskát 2008). Painting experimental or real eggs using acrylic paints, which have different spectral reflectances to natural eggshells (in particular not reflecting in the UV), is problematic as these eggs may be rejected based on features that were not experimentally manipulated or controlled (Prather et al. 2007). The advent of new acrylic paints reflecting in the UV opens new research possibilities (Šulc et al. 2016).

22.2.4 The Use of Cameras

The use of small cameras in well-planned experiments to minimize disturbance to breeding birds has recently become increasingly possible and should enable more detailed study of host responses towards parasitic eggs. Analyses of video recordings of experimentally parasitized nests in particular could contribute to a better understanding of cognitive processes in the context of egg discrimination.

Using cameras has enabled the discovery of birds aggressively pecking the experimental egg (Soler et al. 2002), indicating that egg recognition does not necessarily imply egg rejection. Further evidence of this was the study by Antonov et al. (2009), revealing that eastern olivaceous warblers (*Hippolais pallida*) frequently pecked real cuckoo eggs or experimental egg models but accepted almost half of them; the same behaviour exhibited by warbling vireos (*Vireo gilvus*) was recorded by Underwood and Sealy (2006a). Soler et al. (2012) suggested that pecking not followed by rejection should be considered as part of a stepwise discrimination process, in which accumulating motivation plays a key role in determining behavioural pathways shaping host response to parasitic eggs.

Furthermore, video recordings may be also useful in the detection of variation of the responses towards parasitic egg at the level of host populations, host species, or host individuals. These techniques have confirmed that in species where both sexes incubate and spend time in the nest, both males and females eject eggs (Soler et al. 2002; Požgayová et al. 2009), whereas unsurprisingly, in species when only females incubate, only females reject parasitic eggs (Požgayová et al. 2011). Furthermore, it has been demonstrated that increased egg mass provokes the acceptance of an experimental egg that has been previously recognized, because ejection of a heavy egg may imply higher rejection costs for hosts (Ruíz-Raya et al. 2015). So video recording has in particular allowed investigation of the relationship between egg recognition and egg ejection (Antonov et al. 2009; Soler et al. 2012, 2015, 2017; Ruíz-Raya et al. 2015), its timing (Antonov et al. 2008; Požgayová et al. 2011), and the consistency in egg rejection behaviour of hosts when parasitized repeatedly within one breeding attempt (Honza et al. 2007a).

In all these studies, hosts were typically confronted with an artificial, non-mimetic foreign egg placed into their nest. Because it is time-consuming, the majority of studies were conducted for only a short time after the host was "parasitized". Therefore, despite the difficulties involved, it would be very interesting to study the behaviour and discriminative processes towards real parasitic eggs, to try and resolve the puzzle of why some hosts delay their egg rejection decisions (Požgayová et al. 2011) or, even more important, they decide to accept a previously recognized foreign egg (Soler et al. 2017).

22.3 The Role of Different Factors Affecting Discrimination

22.3.1 Egg Colour and Spottiness

Inspired by the early experiments of Rothstein (1982), researchers have attempted to clarify the importance of these cues in recognition. Egg colouration is clearly vitally important as even relatively small perceivable differences in eggshell colouration can result in substantial increase in host rejection rates (Honza et al. 2011; Hauber et al. 2015). Generally, experiments have shown that hosts are able to recognize parasitic eggs on the basis of colour and spotting (Table 22.1): the larger the difference in colour between parasitic egg and host eggs and the greater the difference in spottiness, the greater the probability of rejection. There are some exceptions: for example, Honza et al. (2007b) in a study of song thrushes (*Turdus philomelos*) revealed that some colours of the parasitic eggs classified by humans as non-mimetic were accepted by the hosts. Some hosts may have strong rejection biases towards specific colours. Hanley et al. (2017), working on two Turdid species, the blackbird (*Turdus merula*) and the American robin (*Turdus migratorius*), found that across a natural colour gradient, both species were more likely to accept blue-green eggs and reject brown eggs, regardless of the perceived difference between foreign eggs and their own. By contrast, their responses did not vary across an artificial (green to purple) gradient, suggesting that in Turdids, at least, egg recognition is specifically tuned to the natural gradient of eggshell colouration.

Table 22.1 Experiments exploring colour and spottiness as cue for rejection (R) of avian brood parasite eggs by hosts

	Findings	Source
Common cuckoo (*Cuculus canorus*) hosts		
Bush warbler (*Cettia diphone*)	Adding spots to artificial egg reduced R by half compared to plain eggs	Higuchi (1989)
Rufous bush chat (*Cercotrichas galactotes*)	Test with pale and contrasted models, support for colours as cue for R, two types non-mimetic eggs R: 17–20%, two types mimetic eggs R: 10–62%	Álvarez (1999)
Song trush (*Turdus philomelos*)	Green part of the spectrum significant for rejection (R), group of mimetic eggs R: 20–77%; group of non-mimetic eggs R: 20–93%, colour is the cue	Honza et al. (2007b)
Great reed warbler (*Acrocephalus arundinaceus*)	Effect of experimentally decreased uniformity of eggs by adding extra spots on freshly laid egg affect R; egg with extra spots R: 60%, with no spots R: 95%	Moskát et al. (2008a)
Great reed warbler	Support for differential use of egg markers by manipulation of spottedness and colour, spot density 15–75% has no effect on R, the spots fully covered egg by dark brown R: 100%	Moskát et al. (2008b)
Great reed warbler	Several treatments with differently colouring eggs (green, orange), R is affected by overall clutch appearance	Moskát et al. (2014)
Oriental reed warbler (*Acrocephalus orientalis*)	Own eggs were painted with extra spots, effect on higher R	Li et al. (2016)
Diederik cuckoo (*Chrysococcyx caprius*) hosts		
Red bishop (*Euplectes oryx*)	Heavy maculation or darker ground colour significantly affected R	Lawes and Kirkman (1996)
Northern masked weaver (*Ploceus taeniopterus*)	Difference in egg colour was a strong predictor for R	Jackson (1993)
Village weaver (*Ploceus cucullatus*)	Use variable conspecific egg documented that colour and speckling contain the signature for R	Lahti and Lahti (2002)
Great spotted cuckoo (*Clamator glandarius*) hosts		
Black-billed magpie (*Pica pica*)	Difference in mean colour value between model and parasite egg affected R	Soler et al. (2000)
Shiny cowbird (*Molothrus bonariensis*) hosts		
Brown and yellow marshbird (*Pseudoleistes virescens*)	R is elicited by difference in colour between parasitic and host eggs	Mermoz and Reboreda (1994)
Mockingbird (*Mimus saturninus*)	Strong R of white cowbird morph, evidence for colour discrimination	Sackmann and Reboreda (2003)

(continued)

Table 22.1 (continued)

	Findings	Source
Mockingbird	R sharply decreased with spots and increased with brightness, R can be explained by spotting and brightness	de la Colina et al. (2012)
Rufous-bellied thrush (*Turdus rufiventris*)	Strong R of white cowbird morph, evidence for colour discrimination	Sackmann and Reboreda (2003)
Creamy-bellied thrush (*Turdus amaurochalinus*)	White eggs R more frequently than spotted eggs when parasitism was associated with the presentation of the cowbird model, but there were no differences when the model was absent	Astie and Reboreda (2005)
Red-crested cardinal (*Paroaria coronata*)	Groups of mimetic and non-mimetic egg types, R: 6–100%, colour is the cue for R, but also width of egg	Segura et al. (2016)
Brown-headed cowbirds (*Molothrus ater*) hosts		
American robins (*Turdus migratorius*)	R only those egg that differed at least two of the three tested trait	Rothstein (1982)
Gray catbirds (*Dumetella carolinensis*)	R solely on their white ground colour	Rothstein (1982)
Warbling vireo (*Vireo gilvus*)	Spot pattern significantly influenced the probability of R	Underwood and Sealy (2006a)
Cuckoo finch (*Anomalospiza imberbis*) hosts		
Tawny-flanked prinia (*Prinia subflava*)	Important predictions for R: colour and aspects of pattern	Spottiswoode and Stevens (2010)

In addition, in some cases, the response to the same colour models is affected by the presence of a parasite (Moksnes and Røskaft 1989; Astie and Reboreda 2005), so colour is not the only cue used in discrimination (Segura et al. 2016). Rather interesting is the ability of South American host species which were tested with a white egg: almost all tested species exhibited fine-tuned recognition ability of white, which is an adaptation towards brood parasitism by shiny cowbirds (Table 22.1).

22.3.2 The Role of Chromatic Versus Achromatic Cues in Egg Discrimination

At present, we can accurately measure the reflectance of birds' eggs and even simulate (on the basis of the sensitivity of bird cones) how reflected radiation is perceived by birds. From this information, the degree of similarity from the perspective of the host's eye can be assessed to determine the level of colour mimicry of parasitic eggs in the nest of the host. The Vorobyev-Osorio model (Vorobyev and Osorio 1998) calculates chromatic and achromatic contrasts between two coloured objects in a visual space that depends on the number of receptor types of the signal receiver in JNDs (just noticeable differences).

Table 22.2 Experiments testing the effect of chromatic and achromatic contrasts on the probability of avian brood parasite eggs being rejected (R) by hosts

	Findings	Source
Common cuckoo (*Cuculus canorus*) hosts		
Chaffinch (*Fringilla coelebs*)	Discrimination gradually increased with increasing differences in chromatic contrast	Avilés et al. (2010)
Great reed warbler (*Acrocephalus arundinaceus*)	Cuckoo eggs were R with poorer chromatic with their eggs, but showed no similar effect for achromatic contrast	Cherry et al. (2007b)
Great reed warbler	R is based on chromatic contrast	Honza et al. (2011)
Blackcap (*Sylvia atricapilla*)	R is based on degree of chromatic matching	Polačiková et al. (2007)
Brown-headed (*Molothrus ater*) hosts		
American robin (*Turdus migratorius*)	Chromatic difference between natural and model eggs predicts R	Croston and Hauber (2014b)

Table 22.2 shows that chromatic contrast is more important in open-nesting hosts, whereas achromatic contrasts have been suggested to play a crucial role in egg discrimination of species nesting in dark nests where colour information is less important (Avilés et al. 2006; Langmore et al. 2009). This explains why achromatic contrasts do not appear to be important cues for the majority of both cuckoo and cowbird hosts.

22.3.3 Ultraviolet Reflectance

In the following studies in which egg appearance was measured using spectrophotometry, the importance of particular wavelengths for egg discrimination was documented in the spotless starling (*Sturnus unicolor*) (Avilés et al. 2006), great reed warbler (*Acrocephalus arundinaceus*) (Cherry et al. 2007b), song thrush (Honza et al. 2007b), blackcap (Polačiková et al. 2007), magpie (*Pica pica*) (Avilés et al. 2004; Soler et al. 2003), and several cowbird hosts (Underwood and Sealy 2008). This strongly suggests that UV vision is used in egg discrimination by birds.

Abernathy and Peer (2015) propose that hosts with brighter UV-reflecting eggs should be more likely to reject UV-blocked eggs than hosts with duller UV-reflecting eggs. Šulc et al. (2016) suggest that such signals may play a more important role when parasitic eggs are non-mimetic rather than mimetic when hosts can use additional cues, such as spotting pattern, to discriminate. Of the total 11 host species (Table 22.3) of hosts that have been tested using own or conspecific UV-blocked eggs, at least two species (blackcap; Honza and Polačiková 2008) and brown thrasher (*Toxostoma rufum*; Abernathy and Peer 2015) appear to use the UV range as the sole cue for discrimination.

Table 22.3 Experimental tests of the role of ultraviolet reflectance in egg discrimination (R) of avian brood parasite eggs by hosts

	Type of manipulation	UV effect on R	Source
Common cuckoo (*Cuculus canorus*) hosts			
Blackcap (*Sylvia atricapilla*)	UV-blocked conspecific egg	Positive	Honza and Polačiková (2008)
Reed warbler (*Acrocephalus scirpaceus*)	Conspecifc egg coated with UV^-	Positive	Šulc et al. (2016)
	Non-mimetic UV^+ egg	Positive	Šulc et al. (2016)
	Non-mimetic UV^- egg	Positive	Šulc et al. (2016)
Brown-headed cowbird (*Molothrus ater*) hosts			
Brown thrasher (*Toxostoma rufum*)	UV^- blocked own egg	Positive	Abernathy and Peer (2015)
American robin (*Turdus migratorius*)	UV^- blocked own egg	Positive	Abernathy and Peer (2015)
Gray catbird (*Dumetella carolinensis*)	UV^- blocked own egg	Positive	Abernathy and Peer (2015)
Northern cardinal (*Cardinalis cardinalis*)	UV^- blocked own egg	No	Abernathy and Peer (2016)
Shiny cowbird (*Molothrus bonariensis*) hosts			
Chalk-browed mockingbird (*Mimus saturninus*)	UV^- plaster white egg	No	De la Colina et al. (2012)
	UV^+ plaster white egg	No	De la Colina et al. (2012)
Great spotted cuckoo (*Cuculus satoratus*) hosts			
Magpie (*Pica pica*)	UV^- blocked real cuckoo egg	No	Avilés et al. (2006)
Oriental cuckoo hosts			
Yellow-bellied prinia (*Prinia flaviventris*)	UV^- blocked conspecific egg	No	Yang et al. (2014)
	UV^- blocked non-mimetic egg	Positive	Yang et al. (2014)
Plain prinia (*Prinia inornata*)	UV-blocked conspecific egg	No	Yang et al. (2014)
	UV^- blocked non-mimetic egg	No	Yang et al. (2014)

22.3.4 Egg Pole

It is well known that maculated bird eggs have more spots at the blunt pole than at the sharp and that the overall surface area of eggshell around the blunt pole is evidently larger than around the sharp pole. Therefore, one should expect that blunt poles have the potential for greater amount of information content of eggshell signals. Polačiková et al. (2007) drew attention to this in the brood parasitism context, showing that colour characteristics of the blunt part of natural conspecific

Table 22.4 Experiments testing the effect of egg size and shape on the probability of avian brood parasite eggs being rejected (R) by hosts

	Findings	Source
Common Cuckoo (*Cuculus canorus*) hosts		
Rufous bush chat (*Cercotrichas galactotes*)	No R of giant eggs, R of both, host sized and cuckoo sized model eggs	Álvarez (2000)
Yellow-browed leaf Warbler (*Phylloscopus humei*)	R decisions are based on the relative size of eggs in the clutch	Marchetti (2000)
Reed warbler (*Acrocephalus scirpaceus*)	Model eggs twice as large as Cuckoo eggs were more likely to be R than normal-sized cuckoo eggs	Davies and Brooke (1988)
Reed warbler	Egg size differences apparently affect the mode and speed, but not R	Stokke et al. (2010)
Marsh warbler (*Acrocephalus palustris*)	R was not dependent on difference in size	Antonov et al. (2006)
Great reed warbler (*Acrocephalus arundinaceus*)	Egg shape influence R, rounder and symmetrical were R more often than more elongated model eggs	Zoelei et al. (2012)
Blackbird (*Turdus merula*)	Size affected ejection but not nest desertion	Soler et al. (2015)
Diederik cuckoo (*Chrysococcyx caprius*) hosts		
Village weaver (*Ploceus cucullatus*)	Test with conspecifics eggs showed no effect of shape and mass on R	Lahti and Lahti (2002)
Brown-headed (*Molothrus ater*) hosts		
American robin (*Turdus migratorius*)	Size was the least important parameter affecting R	Rothstein (1982)
American robin (*Turdus migratorius*)	Shape does not appear to influence egg R	Underwood and Sealy (2006a)
Gray catbird (*Dumetella carolinensis*)	Shape does not appear to influence egg R	Underwood and Sealy (2006a)
Warbling vireo (*Vireo vireo*)	Size did not affect probability R	Underwood and Sealy (2006b)
Yellow warbler (*Setophaga petechia*)	Egg size was not generally used as criterion for R	Guigueno et al. (2014)
Shiny cowbird (*Molothrus bonariensis*) hosts		
Red-crested cardinal (*Paroaria coronata*)	Egg width is an important cue for R	Segura et al. (2016)
Greater honeyguide (*Indicator indicator*) hosts	Commonest hosts did not discriminate against experimental egg that differed from their own in shape and size	Spottiswoode (2013)
Several hosts		

eggs experimentally added to blackcap nests may play a major role in the recognition of parasitic egg. Further studies by Polačiková et al. (2010), Polačiková and Grim (2010), and Zoelei et al. (2012) have revealed that host species rejected eggs manipulated at the blunt pole at significantly higher rates than eggs manipulated at sharp poles, indicating that they perceive critical recognition cues at the blunt pole. Polačiková and Grim (2010) regard the presence of egg recognition cues at the blunt egg pole as a general phenomenon in birds parasitized by interspecific parasites. To confirm this, it is necessary to study the effects of the appearance of the blunt egg pole of real parasitic eggs.

22.3.5 Egg Volume and Shape

Rigorous research studying the effects of the volume and shape of parasitic eggs with respect to rejection started in the early 1980s by Rothstein (1982) and has continued since then (Table 22.4). A variety of experimental methods have been used: conspecific eggs (Lahti and Lahti 2002; Underwood and Sealy 2006a), oversized eggs (Davies and Brooke 1988; Álvarez 2000), model eggs (Marchetti 2000; Antonov et al. 2006), and, in a single study, real parasitic eggs (Segura et al. 2016). With the exception of two studies (Marchetti 2000; Guigueno et al. 2014), results showed that egg size and/or shape were not generally used as a rejection cue.

Concluding Remarks and Future Directions

It is evident that the past decade has brought considerable progress in elucidating the particular cues that birds use during the process of discriminating parasitic eggs. Evaluation of egg colouration has improved considerably with the advent of spectrophotometric techniques allowing objective quantification of colour, including the UV-reflectant range that is invisible to humans. The use of video cameras has proven very useful in studying discriminatory processes, and further applications of this methodology should allow for better designed experiments related to discrimination.

We have a relatively good knowledge of the characteristics of parasitic eggs responsible for recognition, and subsequent rejection, in cuckoo and cowbird hosts. Our review suggests that some aspects of egg appearance could be more important for hosts than others. Individual cues could also interact with each other and play different roles in different circumstances, for example, UV signals could be more important in the recognition of non-mimetic eggs rather than mimetic. As the majority of published studies have focused on individual cues, we encourage the study of combined cues potentially responsible for rejection. This is because selection has shaped egg size, shape, colour, luminance, and patterns, and these cues together could contribute to egg detection and rejection behaviour. Future research should also explore different

(continued)

functions of eggshell components, e.g. shape, size, colour, and spottedness in egg discrimination in less well-known brood parasitic systems.

There is an apparent lack of knowledge on whether birds can recognize the shape of a natural parasitic egg. More studies are therefore needed with natural parasitic eggs to identify whether shape and volume are valid cues for egg recognition. In addition, further research is needed to clarify how hosts use the information content of the eggshell around the blunt pole across all brood parasitism systems, testing potential hosts in natural conditions.

Future work should also focus on improving visual models by incorporating physiologically appropriate, individual specific cone densities/ absorbance spectra, as well as nest site-specific egg, nest lining, and ambient light availability data. In the future, new techniques such as 3D printing should provide opportunities for more extensive experimentation on the potential biological or evolutionary significance of size and shape variation of foreign eggs in rejection decisions.

Acknowledgements We are grateful to Manuel Soler, Daniel Hanley, and Phil Cassey for their helpful comments and suggestions. This work was partly supported by the Grant Agency of the Czech Republic (grant no. 17-12262S to MH) and the National Research Foundation of South Africa (grant 96257 to MIC).

References

Abernathy VE, Peer BD (2015) Mechanisms of egg recognition in brown-headed cowird hosts: the role of ultraviolet reflectance. Anim Behav 109:73–79

Abernathy VE, Peer BD (2016) Reduced ultraviolet reflectance does not affect egg rejection by Northern cardinals (*Cardinalis cardinalis*). Wilson J Ornithol 128:334–342

Aidala Z, Croston R, Schwartz J, Tong L, Hauber ME (2015) The role of egg-nest contrast in the rejection of brood parasitic eggs. J Exp Biol 218:1126–1136

Álvarez F (1999) Attractive non-mimetic stimuli in cuckoo *Cuculus canorus* eggs. Ibis 141:142–144

Álvarez F (2000) Response to common cuckoo *Cuculus canorus* model egg size by a parasitized population of rufous bush chat *Cercotrichas galactotes*. Ibis 142:683–686

Álvarez F, Arias de Reyna L, Segura M (1976) Experimental brood parasitism of the magpie (*Pica pica*). Anim Behav 24:907–916

Antonov A, Stokke BG, Moksnes A, Røskaft E (2006) Egg rejection in marsh warblers (*Acrocephalus palustris*) heavily parasitized by common cuckoos (*Cuculus canorus*). Auk 123:419–430

Antonov A, Stokke BG, Moksnes A, Røskaft E (2008) Getting rid of the cuckoo *Cuculus canorus* egg: why do hosts delay rejection? Behav Ecol 19:100–107

Antonov A, Stokke BG, Moksnes A, Røskaft E (2009) Evidence for discrimination preceding failed rejection attempts in a small cuckoo host. Biol Lett 5:169–171

Astie AA, Reboreda JC (2005) Creamy – bellied thrush defenses against shiny cowbird brood parasitism. Condor 107:788–796

Avilés JM (2008) Egg colour mimicry in the common cuckoo *Cuculus canorus* as revealed by modelling host retinal function. Proc R Soc B 275:2345–2352

Avilés JM, Møller AP (2004) Meadow pipit *(Anthus pratensis)* egg appearance in cuckoo (*Cuculus canorus*) sympatric and allopatric populations. Biol J Linn Soc 79:543–549

Avilés JM, Soler JJ, Soler M, Møller AP (2004) Rejection of parasitic eggs in relation to egg appearance in magpies. Anim Behav 67:951–958

Avilés JM, Soler JJ, Pérez-Contreras T (2006) Dark nests and egg colour in birds: a possible functional role of ultraviolet reflectance in egg detectability. Proc R Soc B 273:2821–2829

Avilés JM, Vikan JR, Fossoy F, Antonov A, Moksnes A, Røskaft E, Stokke BG (2010) Avian colour perception predicts behavioural responses to experimental brood parasitism in chaffinches. J Evol Biol 23:293–301

Briskie JV, Sealy SG, Hobson KA (1992) Behavioral defenses against avian brood parasitism in sympatric and allopatric host populations. Evolution 46:334–340

Brooke M de L, Davies NB, Noble DG (1998) Rapid decline of host defences in response to reduced cuckoo parasitism: behavioural flexibility of reed warblers in a changing world. Proc R Soc B 265:1277–1282

Cassey P, Honza M, Grim T, Hauber ME (2008) The modelling of avian visual perception predicts behavioural rejection responses to foreign egg colours. Biol Lett 4:515–517

Cherry MI, Bennett ATD (2001) Egg colour matching in an African cuckoo, as revealed by ultraviolet-visible reflectance spectrophotometry. Proc Biol Soc 268:565–571

Cherry MI, Bennett ATD, Moskát C (2007a) Do cuckoos choose nests of great reed warblers on the basis of host egg appearance? J Evol Biol 20:1218–1222

Cherry MI, Bennett ATD, Moskát C (2007b) Host intra-clutch variation, cuckoo egg matching and egg rejection by great reed warblers. Naturwissenschaften 94:441–447

Croston R, Hauber ME (2014a) High repeatability of egg rejection in response to experimental brood parasitism in the American robin (*Turdus migratorius*). Behaviour 151:703–718

Croston R, Hauber ME (2014b) Spectral tuning and perceptual differences do not explain the rejection of brood parasitic eggs by American robins (*Turdus migratorius*). Behav Ecol Sociobiol 68:351–362

Cruz A, Wiley JW (1989) The decline of an adaptation in the absence of a presumed selection pressure. Evolution 43:55–62

Cuthill IC, Partridge JC, Bennett ATD, Church SC, Hart NS, Hunt E (2000) Ultraviolet vision in birds. Adv Study Behav 29:159–214

Davies NB (2000) Cuckoos, cowbirds and other cheats. T & AD Poyser, London

Davies NB, Brooke MD (1988) Cuckoos versus reed warblers: adaptations and counteradaptations. Anim Behav 36:262–284

Davies NB, Brooke MD (1989) An experimental study of co-evolution between cuckoo, *Cuculus canorus*, and its hosts. 2. Host egg markings, chick discrimination and general discussion. J Anim Ecol 58:225–236

Davies NB, Brooke M de L, Kacelnik A (1996) Recognition errors and probability of parasitism determine whether reed warblers should accept or reject mimetic cuckoo eggs. Proc R Soc Lond B 263:925–931

De La Colina MA, Pompilio L, Hauber ME, Reboreda JC, Mahler B (2012) Different recognition cues reveal the decision rules used for egg rejection by hosts of a variably mimetic avian brood parasite. Anim Cogn 15:881–889

Endler JA (1990) On the measurement and classification of color in studies of animal color patterns. Biol J Linn Soc 41:315–352

Guigueno MF, Sealy SG (2009) Nest sanitation plays a role in egg burial by yellow warblers. Ethology 115:247–256

Guigueno MF, Sealy SG (2012) Nest sanitation in passerine birds: implications for egg rejection in hosts of brood parasites. J Ornithol 153:35–52

Guigueno MF, Sealy S, Westphal A (2014) Rejection of parasitic eggs in passerine hosts: size matters more for a non-ejecter. Auk 131:583–594

Hanley D, Samas P, Hauber M, Grim T (2015) Who moved my eggs? An experimental test of the egg arrangement hypothesis for the rejection of brood parasitic eggs. Anim Cogn 18:299–305

Hanley D, Šulc M, Brennan PLR, Hauber M, Grim T, Honza M (2016) Dynamic egg color mimicry. Ecol Evol 6:4192–4202

Hanley D, Grim T, Igic B, Samaš P, Lopez AV, Shawkey MD, Hauber ME (2017) Egg discrimination along a gradient of natural variation in eggshell coloration. Proc R Soc B 284:2016–2592

Hart NS (2001) The visual ecology of avian photoreceptors. Proc Retin Eye Res 20:675–703

Hart NS, Partridge JC, Cuthill IC, Bennet ATD (2000) Visual pigments, oil droplets, ocular media and cone photoreceptor distribution in two species of passerine bird: the blue tit (*Parus caeruleus L.*) and the blackbird (*Turdus merula*). J Comp Physiol A 186:375–387

Hauber ME, Sherman PW (2001) Self-referent phenotype matching: theoretical considerations and empirical evidence. Trends Neurosci 24:609–616

Hauber ME, Moskát C, Bán M (2006) Experimental shift in hosts´ acceptance threshold of inacurate-mimic brood parasite eggs. Biol Lett 2:177–189

Hauber ME, Tong LG, Bán M, Croston R, Grim T, Waterhouse GIN, Shawkey MD, Barron AB, Moskat C (2015) The value of artificial stimuli in behavioral research: making the case for egg rejection studies in avian brood parasitism. Ethology 121:521–528

Higuchi H (1989) Responses of the bush warbler *Cettia diphone* to artificial eggs of *Cuculus canorus* in Japan. Ibis 131:94–98

Honza M, Moskát C (2008) Egg rejection behaviour in the great reed warbler (*Acrocephalus arundinaceus*): the effect of egg type. J Ethol 26:38–395

Honza M, Polačiková L (2008) Experimental reduction of ultraviolet wavelengths reflected from parasitic eggs affects rejection behaviour in the blackcap *Sylvia atricapilla*. J Exp Biol 211:2519–2523

Honza M, Požgayová M, Procházka P, Tkadlec E (2007a) Consistency in egg rejection behaviour: responses to repeated brood parasitism in the blackcap (*Sylvia atricapilla*). Ethology 113:344–351

Honza M, Polačiková L, Procházka P (2007b) Ultraviolet and green parts of the colour spectrum affect egg rejection in the song thrush (*Turdus philomelos*). Biol J Linn Soc 92:269–276

Honza M, Procházka P, Morongová K, Čapek M, Jelínek V (2011) Do nest light conditions affect rejection of parasitic eggs? A test of the light environment hypothesis. Ethology 117:539–546

Igic B, Numez W, Voss HU, Croston R, Aidala Z, López AV, van Tatenhove A, Holford ME, Shawkey MD, Hauber ME (2015) Using 3D printed eggs to examine the egg-rejection behaviour of wild birds. PeerJ 3:e965

Jackson WM (1993) Causes of couspecific nest parasitism in the northern masked weaver. Beh Ecol Soc 2:119–126

Lahti D (2015) The limits of artificial stimuli in behavioral research: the Umwelt gamble. Ethology 121:529–537

Lahti DC, Lahti AR (2002) How precise is egg discrimination in weaverbirds? Anim Behav 63:1135–1142

Langmore NE, Stevens M, Maurer G, Kilner RM (2009) Are dark cuckoo eggs cryptic in host nests? Anim Behav 78:461–468

Lawes MJ, Kirkman S (1996) Egg recognition and interspecific brood parasitism rates in red bishops (Aves: *Ploceidae*). Anim Behav 52:553–563

Li D, Ruan Y, Wang Y, Chang AK, Wan D, Zhang Z (2016) Egg-spot matching in common cuckoo parasitism of the oriental warbler: effects of host nest availability and egg rejection. Avian Res 7:21

Lindholm AK, Thomas RJ (2000) Differences between populations of reed warblers in defences against brood parasitism. Behaviour 137:25–42

Marchetti K (2000) Egg rejection in a passerine bird: size does matter. Anim Behav 59:877–883

Martín-Vivaldi M, Soler M, Møller AP (2002) Unrealistically high costs of rejecting artificial model eggs in cuckoo *Cuculus canorus* hosts. J Avian Biol 33:295–301

Mermoz ME, Reboreda JC (1994) Brood parasitism of the shiny cowbird, *Molothrus bonariensis*, on the brown-and-yellow marshbird, *Pseudoleistes virescens*. Condor 96:716–721

Moksnes A, Røskaft E (1989) Adaptations of meadow pipits to parasitism by the common cuckoo. Behav Ecol Sociobiol 24:25–30

Moksnes A, Røskaft E (1993) Rejection of cuckoo (*Cuculus canorus*) eggs by meadow pipits (*Anthus pratensis*). Behav Ecol 4:120–127

Moksnes A, Røskaft E (1995) Egg-morphs and host preferences in the common cuckoo (*Cuculus canorus*) – an analysis of cuckoo and host eggs from European museum collections. J Zool 236:625–648

Moksnes A, Røskaft E, Braa AT (1991) Rejection behaviour by common cuckoo hosts towards artificial brood parasite eggs. Auk 108:348–354

Moreno J, Lobato E, Morales J (2011) Eggshell blue-green colouration fades immediately after oviposition: a cautionary note about measuring natural egg. Ornis Fenn 88:51–56

Moskát C, Hauber M (2007) Conflict between egg recognition and egg rejection decisions in common cuckoo *Cuculus canorus* hosts. Anim Cogn 10:377–386

Moskát C, Szekely T, Kisbenedek T, Karcza Z, Bártol I (2003) The importance of nest cleaning in egg rejection behaviour of great reed warblers *Acrocephalus arundinaceus*. J Avian Biol 34:16–19

Moskát C, Avilés JM, Bán M, Hargitai R, Zölei A (2008a) Experimental support for the use of egg uniformity in parasite discrimination by cuckoo hosts. Behav Ecol Sociobiol 62:1885–1890

Moskát C, Szekely T, Cuthill IC, Kisbenedek T (2008b) Hosts' responses to parasitic eggs: which cues elicit hosts' egg discrimination? Ethology 114:186–194

Moskát C, Bán M, Szekely T, Komdeur J, Lucassen RWG, van Boheemen LA, Hauber ME (2010) Discordancy or template-based recognition? Dissecting the cognitive basis of the ejection of foreign eggs in hosts of avian brood parasites. J Exp Biol 213:1976–1983

Moskát C, Zoelei A, Bán M, Elek Z, Tong L, Geltsch N, Hauber ME (2014) How to spot a stranger's egg? A mimicry-specific discordancy effect in the recognition of parasitic eggs. Ethology 120:616–626

Nakamura H, Kubota S, Suzuki R (1998) Co-evolution between the common cuckoo and its major hosts in Japan. In: Rothstein SI, Robinson SK (eds) Parasitic birds and their hosts. Oxford University Press, Oxford, pp 94–112

Navarro JY, Lahti DC (2014) Light dulls and darkens bird eggs. PLoS One 9:e116112

Ortega CP, Cruz A (1988) Mechanisms of egg acceptance by marsh-dwelling blackbirds. Condor 90:349–358

Payne R, Payne L (1998) Brood parasitism by cowbirds: risks and effects on reproductive success and survival in indigo buntings. Behav Ecol 9:64–73

Peer BD, Rothstein SK, Delaney SK, Fleischer RC (2007) Defence behaviour against brood parasitism is deeply rooted in mainland and island scrub-jays. Anim Behav 73:55–63

Polačiková L, Grim T (2010) Blunt egg pole holds cues for alien egg discrimination: experimental evidence. J Avian Biol 41:111–116

Polačiková L, Honza M, Procházka P, Topercer J, Stokke BG (2007) Colour characteristics of the blunt egg pole: cues for recognition of parasitic eggs as revealed by reflectance spectrophotometry. Anim Behav 74:419–427

Polačiková L, Stokke BG, Procházka P, Honza M, Moksnes A, Røskaft E (2010) The role of blunt egg pole characteristics for recognition of eggs in the song thrush (*Turdus philomelos*). Behaviour 147:465–478

Polačiková L, Takasu F, Stokke BG, Moksnes A, Røskaft E, Cassey P, Hauber ME, Grim T (2013) Egg arrangement in avian clutches covaries with the rejection of foreign eggs. Anim Cogn 16:819–828

Požgayová M, Procházka P, Honza M (2009) Sex-specific defence behaviour against brood parasitism in a host with female-only incubation. Behav Process 81:34–38

Požgayová M, Procházka P, Polačiková L, Honza M (2011) Closer clutch inspection – quicker egg ejection: timing of host responses toward parasitic eggs. Behav Ecol 22:46–51

Prather JW, Cruz A, Weaver PF, Wiley JW (2007) Effects of experimental egg composition on rejection by village weavers (*Ploceus cucullatus*). Wilson J Ornithol 119:703–711

Rensch B (1925) Verhalten von Singvögeln bei Aenderung des Geleges. Ornithol Monatschr 33:169–173

Roncalli G, Ibanéz-Álamo JD, Soler M (2017) Size and material of model parasitic eggs affect the rejection response of Western Bonelli's warbler *Phylloscopus bonelli*. Ibis 159:113–123

Rothstein SI (1974) Mechanisms of avian egg recognition: possible learned and innate factors. Auk 91:796–807

Rothstein SI (1982) Mechanisms of avian egg recognition: which egg parameters elicit responses by rejecter species? Behav Ecol Sociobiol 11:229–239

Ruíz-Raya F, Soler M, Sanchez-Perez LLI, Ibanez-Alamo JD (2015) Could a factor that does not affect egg recognition influence the decision of rejection? PLoS One 10:e0135624

Sackmann P, Reboreda JC (2003) A comparative study of shiny cowbird parasitism of two large hosts, the chalk-browed mockingbird and the rufous-bellied thrush. Condor 105:728–736

Sealy SG (1992) Removal of yellow warbler eggs in association with cowbird parasitism. Condor 94:40–54

Sealy SG (1995) Burial of cowbird eggs by parasitized yellow warblers: an empirical and experimental study. Anim Behav 49:877–889

Sealy SG, Underwood TJ (2012) Egg discrimination by hosts and obligate brood parasites: a historical perspective and new synthesis. Chin Birds 3:274–294

Segura LN, Di Sallo FG, Mahler B, Reboreda JC (2016) Red-crested cardinals use color and width as cues to reject shiny cowbird eggs. Auk 133:308–315

Soler M (1990) Relationship between the great spotted cuckoo *Clamator glandarius* and its corvid hosts in a recently colonized area. Ornis Scand 21:212–223

Soler M (2014) Long-term coevolution between avian brood parasites and their host. Biol Rev 89:688–704

Soler M, Møller AP (1990) Duration of sympatry and coevolution between the great spotted cuckoo and its magpie host. Nature 343:748–750

Soler JJ, Martinez JG, Soler M, Møller AP (1999a) Host sexual selection and cuckoo parasitism: an analysis of nest size in sympatric and allopatric magpie *Pica pica* populations parasitized by the great spotted cuckoo *Clamator glandarius*. Proc R Soc B 266:1765–1771

Soler JJ, Martinez JG, Soler M, Møller AP (1999b) Genetic and geographic variation in rejection behavior of cuckoo eggs by European magpie populations: an experimental test of rejecter-gene flow. Evolution 53:947–956

Soler JJ, Soler M, Møller AP (2000) Host recognition of parasite eggs and the physical appearance of host eggs: the magpie and its brood parasite the great spotted cuckoo. Etología 8:9–16

Soler M, Martín-Vivaldi M, Pérez-Contreras T (2002) Identification of the sex responsible recognition and the method of ejection of parasitic eggs in some potential common cuckoo hosts. Ethology 108:1093–1101

Soler JJ, Avilés JM, Soler M, Møller AP (2003) Evolution of host egg mimicry in a brood parasite, the great spotted cuckoo. Biol J Linn Soc 79:551–563

Soler M, Fernandez-Morante J, Espinosa F, Martín-Vivaldi M (2012) Pecking but accepting the parasitic eggs may not reflect ejection failure: the role of motivation. Ethology 118:662–672

Soler M, Ruiz-Raya F, Roncalli G, Ibáñez-Álamo JD (2015) Nest desertion cannot be considered an egg-rejection mechanism in a medium-sized host: an experimental study with the common blackbird *Turdus merula*. J Avian Biol 46:369–377

Soler M, Ruiz-Raya F, Roncalli G, Ibanez-Alamo JD (2017) Relationships between egg-recognition and egg-ejection in a grasp-ejector species. PLoS One 12:e0166283

Spottiswoode C (2013) A brood parasite selects for its own egg traits. Biol Lett 9:20130573

Spottiswoode C, Stevens M (2010) Visual modelling shows that avian host parents use multiple visual cues in rejecting parasitic eggs. Proc Natl Acad Sci U S A 107:8672–8676

Starling M, Heinsohn R, Cockburn A, Langmore NE (2006) Cryptic gentes revealed in pallid cuckoos *Cuculus pallidus* using reflectance spectrophotometry. Proc R Soc B 273:1929–1934

Stoddard MC, Stevens M (2010) Pattern mimicry of host eggs by the common cuckoo, as seen through a bird's eye. Proc R Soc B 277:1387–1393

Stoddard MC, Kilner RM, Town C (2014) Pattern recognition algorithm reveals how birds evolve individual egg pattern. Nat Commun 5:4117

Stokke BG, Hafstad I, Rudolfsen G, Moksnes A, Møller AP, Røskaft E, Soler M (2008) Predictors of resistance to brood parasitism within and among reed warbler populations. Behav Ecol 19:612–620

Stokke BG, Polačiková L, Dyrcz A, Hafstad I, Moksnes E, Røskaft E (2010) Responses of reed warblers *Acrocephalus scirpaceus* to non-mimetic eggs of different sizes in a nest parasitism experiment. Acta Ornithol 45:98–104

Šulc M, Procházka P, Čapek M, Honza M (2016) Birds use eggshell UV reflectance when recognizing non-mimetic parasitic eggs. Behav Ecol 27:677–684

Swynnerton CFM (1918) Rejections by birds of eggs unlike their own: with remarks on some of the cuckoo problems. Ibis Ser 10 6:127–154

Underwood TJ, Sealy SG (2006a) Influence of shape on egg discrimination in American robins and gray catbirds. Ethology 112:164–173

Underwood TJ, Sealy S (2006b) Parameters of brown-headed cowbird *Molothrus ater* egg discrimination in warbling vireos *Vireo gilvus*. J Avian Biol 37:457–466

Underwood TJ, Sealy SG (2008) UV reflectance of eggs of brown-headed cowbirds (*Molothrus ater*) and accepter and rejecter hosts. J Ornithol 149:313–321

Vikan JR, Stokke BG, Rutila J, Huhta E, Moksnes A, Røskaft E (2010) Evolution of defences against cuckoo (*Cuculus canorus*) parasitism in bramblings (*Fringilla montifringilla*): a comparison of four populations in Fennoscandia. Evol Ecol 24:1141–1157

Vorobyev M, Osorio D (1998) Receptor noise as a determinant of colour thresholds. Proc R Soc Lond B 265:351–358

Vorobyev M, Osorio D, Bennett ATD, Marshall NJ, Cuthill IC (1998) Tetrachromacy, oil droplets and bird plumage colours. J Comp Physiol A 183:621–633

Yang CC, Liu Y, Zeng LJ, Liang W (2014) Egg color variation, but not egg rejection behavior, changes in a cuckoo host breeding in the absence of brood parasitism. Ecol Evol 4:2239–2246

Zoelei A, Hauber M, Geltsch N (2012) Asymmetrical signal content of egg shape as predictor of egg rejection by great reed warblers, hosts of the common cuckoo. Behaviour 149:391–406

Anti-brood Parasite Defences: The Role of Individual and Social Learning

23

Daniela Campobello, Spencer G. Sealy, and Justin A. Welbergen

Abstract

In this chapter, we consider the ways in which learning is involved in the anti-brood parasitism defences that hosts deploy across the nesting cycle. Brood parasitism varies in space and through time, and hosts have accordingly evolved plastic defences that can be tuned to local conditions. Hosts can achieve their defence plasticity by individual and social learning, as well as by experience-independent mechanisms. While these mechanisms can profoundly affect the coevolutionary dynamics between hosts and their brood parasites, our understanding of how they feature across the host nesting cycle is far from complete. Hosts can actively defend themselves against brood parasitism via a variety of behaviours, including nest defence, egg discrimination and chick discrimination. Such anti-brood parasite defences rely on the host's ability to recognise and then defend against the parasitic threat, and there is good evidence that both these components of discrimination can be influenced by learning. To date, most research has focused on the function of learning in nest defence, but the learning mechanisms underlying egg discrimination are much better understood; and despite some notable exceptions, the role of learning in chick discrimination remains largely unexplored. An important challenge now is to understand the observed plasticity of anti-brood parasite defences in the context of environmental heterogeneity and specifically in terms of variation in the presence, detection and reliability of parasitism cues.

D. Campobello (✉)
Department STEBICEF, Section of Animal Biology, Università di Palermo, Palermo, Italy
e-mail: daniela.campobello@unipa.it

S. G. Sealy
Department of Biological Sciences, University of Manitoba, Winnipeg, MB, Canada

J. A. Welbergen
Hawkesbury Institute for the Environment, Western Sydney University, Richmond, NSW, Australia

M. Soler (ed.), *Avian Brood Parasitism*, Fascinating Life Sciences,
https://doi.org/10.1007/978-3-319-73138-4_23

23.1 Introduction

When defences are costly, behavioural plasticity is advantageous if the risk of encountering enemies varies in space or in time (Caro 2005, Chap. 25). Accordingly, when brood parasitism varies spatiotemporally, many host species adjust their defences in accordance with parasitism risk (Davies and Brooke 1988; Hosoi and Rothstein 2000; Forsman and Monkkonen 2001; Davies et al. 2003; Langmore et al. 2009; Welbergen and Davies 2009, 2012; Patten et al. 2011; Kleindorfer et al. 2013). However, it is often not clear how hosts achieve such plastic responses to brood parasitism. Yet, depending on its prevalence and mode of acquisition, behavioural plasticity can either impede or promote genetic evolution (Price et al. 2003), and so it is of fundamental importance to our understanding of brood parasite–host systems as models for coevolution.

Behavioural plasticity can be adaptive because it enables organisms to respond to changing ecological demands and so exploit their environment more efficiently (Johnston 1982; DeWitt and Scheiner 2004; Shettleworth 2010). It depends on the presence, detection and reliability of relevant environmental cues (Getty 1996; Tollrian and Harvell 1999; Campobello et al. 2015), and in brood parasite–host systems, such cues can convey information about the likelihood of brood parasitism, providing the means for potential hosts to 'tune' their defences to local conditions. Such tuning can be an innate or a learnt response or a result from some complex interplay of both (Mery and Burns 2010). Thus, how environmental cues become associated with a behavioural response is key to understanding the mechanisms underpinning behavioural plasticity in brood parasite hosts.

Learning occurs when behaviour is modified by experience (Dukas 1998; Shettleworth 2010) and can be classified as associative, relational or perceptual (Zentall et al. 2014). Associative learning occurs when an individual learns an association between two stimuli; for example, when a host learns to associate an adult brood parasite at the nest with the subsequent appearance of an odd egg in its clutch (e.g. Davies and Brooke 1988; Moksnes and Røskaft 1989; Moksnes et al. 1993). Relational learning occurs when an individual develops the ability to categorise a stimulus as distinct from another; for instance, when a host learns to differentiate the odd egg in its clutch (see also discordancy hypothesis: Rothstein 1975; Marchetti 2000). Finally, perceptual learning occurs when an individual's ability to categorise such stimuli improves with experience; for example, when a host improves its internal template of the appearance of its own eggs (true recognition: Rothstein 1975; Lotem et al. 1995; Lahti and Lahti 2002).

Like any other phenotypic trait (Stearns 1992), whether or not learning evolves is subject to trade-offs between the associated fitness benefits and costs. When costs outweigh benefits, innate, experience-independent mechanisms of response plasticity are favoured by selection; whereas when benefits outweigh costs, learnt plasticity is likely to arise instead (Mery and Burns 2010). Costs of learning include the development and maintenance of the complex neural structures necessary for processing, storing and retrieving information (Dukas 1999; Shettleworth 2010). Other costs include a period of trial-and-error during which individuals exhibit

Table 23.1 Hypothesised contributions to behavioural plasticity of individual learning, social learning and experience-independent processes ('+' = low relative contribution; '++' = medium relative contribution; '+++' = high relative contribution), under low, medium and high levels of environmental heterogeneity (see text)

Levels of environmental heterogeneity	Individual learning	Social learning	Experience-independent processes
Low	+	++	+++
Medium	++	+++	+
High	+++	+	+

suboptimal behaviour; this 'cost of being naïve' can be an important obstacle for the evolution of learnt plasticity as it occurs again in every generation. Against such costs are the benefits from the additional level of plasticity afforded by learning (Johnston 1982; Heyes 1994; Laland et al. 1996; Shettleworth 2010), which can be substantial especially when environmental conditions vary spatially within the home range and/or temporally within the lifetime of an individual (Stephens 1991).

When individual learning is costly and error-prone, individuals may learn socially and use public information to guide their responses instead (Boyd and Richerson 1988; Laland et al. 1996; Doligez et al. 2004; Campobello and Hare 2007). Social learners that learn vicariantly by eavesdropping on the experiences of conspecifics (Heyes 1994) may avoid many of the trial-and-error costs associated with learning individually and so gain an advantage relative to individual learners (Galef and Laland 2005). However, whether individuals should rely on public information depends on the availability of experienced individuals and, importantly, the reliability of the information acquired (Boyd and Richerson 1988; Laland et al. 1996).

In heterogeneous environments where conditions change rapidly through space and time, public information has only local relevance that becomes quickly outdated (e.g. Heyes 1994; Giraldeau et al. 2002; Lefebvre and Giraldeau 1996). In such environments, social learning may be more costly than individual learning, so that the latter would then become the favoured mode of information acquisition and use. On the other hand, under low and moderate levels of environmental heterogeneity, social learning is expected to be favoured over individual learning (Boyd and Richerson 1988) (Table 23.1). Indeed, this is supported by comparative studies that show that individuals are more prone to engage in individual learning in more heterogeneous environments (Giraldeau et al. 2002) and that this provides an advantage relative to social learning under these conditions (Laland et al. 1996; Freeberg et al. 1999; Doligez et al. 2004).

Brood parasite–host systems exhibit fine-scale, spatial (Campobello and Sealy 2009; Welbergen and Davies 2012; Soler et al. 2013) and temporal (Brooke et al. 1998; Lindholm 1999; Campobello 2008; Stokke et al. 2008) autocorrelations of defences with parasitism risk, indicating that brood parasites exert a geographical mosaic of selection on their hosts and vice versa (Stokke et al. 2008). This heterogeneous selective backdrop thus provides rich opportunities for individually and socially learnt defences to evolve. In the next section, we assess the evidence for the role of individual and social learning in anti-brood parasite defences.

23.2 Learning Anti-brood Parasite Defences

Hosts can actively defend themselves against brood parasitism via a variety of behaviours, including mobbing (Duckworth 1991; Welbergen and Davies 2008), egg discrimination (Davies and Brooke 1988) and chick discrimination (Langmore et al. 2003). Such anti-brood parasite defences rely on the host's ability to recognise and then defend against (reject) the parasitic threat, and both these components of discrimination can potentially be influenced by learning. There is ample of evidence to show that hosts use several cues to discriminate against adults, eggs, nestlings and fledglings of brood parasites in accordance with the threats they pose (Duckworth 1991; Uyehara and Narins 1995; Gill et al. 1997, 2008; Welbergen et al. 2001; Gill and Sealy 2004; Honza et al. 2006; Welbergen and Davies 2008, 2009, 2011; Campobello and Sealy 2010; Trnka and Prokop 2012; Li et al. 2015). However, as explained above, such plastic defences do not necessarily imply learning mechanisms, and indeed there is evidence for learnt (e.g. Lotem 1993; Davies and Welbergen 2009; Campobello and Sealy 2011b; Langmore et al. 2012) and experience-independent (e.g. Langmore et al. 2003) discrimination, or both (Welbergen and Davies 2012), from a variety of systems.

Recent work has identified multiple lines of host defence and parasite offence leading to distinct coevolutionary arms races at successive stages of the host nesting cycle, from nest building through the egg and chick/fledgling stages (Brooke and Davies 1988; Langmore et al. 2003; Welbergen and Davies 2009; Campobello 2011a, b; De Mársico et al. 2012) (reviewed in Feeney et al. 2014). Below, we consider the ways in which learning is involved in the defences at each of these stages of the host nesting cycle (Table 23.2).

23.2.1 Learning at the Frontline

Successful brood parasitism critically depends on the ability of the adult parasite to lay its egg in a host's nest, so there is a premium on host adaptations that prevent the parasite from accessing the nest (reviewed in Feeney et al. 2012). Nest defence has been shown to effectively deter a parasite from laying (Welbergen and Davies 2009; Feeney and Langmore 2013); however, to be effective, hosts should be able to recognise and treat an adult brood parasite as a unique threat (Neudorf and Sealy 1992; Welbergen and Davies 2008; Campobello and Sealy 2010; Davies and Welbergen 2009), although not always (e.g. Canestrari et al. 2009).

Several host species, including reed warblers (*Acrocephalus scirpaceus*), yellow warblers (*Setophaga petechia*) and superb fairywrens (*Malurus cyaneus*) appear to learn to recognise and treat brood parasites as different from other nest intruders (Gill and Sealy 2004; Campobello and Sealy 2010; Welbergen and Davies 2008; Feeney and Langmore 2013) (Table 23.3), with some species even differentiating between different brood parasites (Yang et al. 2014). However, it is not yet clear what exactly is learnt, although this likely involves some combination of morphological features (such as head and beak), integrated with certain stimulus elements (such as a yellow

Table 23.2 Learnt anti-brood parasitism defences at the different stages of the host nesting cycle, together with their potential cues (direct, indirect and social), learning mechanisms (associative, relational, perceptual) and sources of information (personal, social)

	Frontline	Egg stage	Chick/fledgling stage
Learnable trait	Nest defence	Egg discrimination	Chick/fledgling discrimination
Potential cues	*Direct cues:*	*Direct cues:*	*Direct cues:*
	(D) adult parasite morphology [1–6]	(D) egg colour[22–25]	(I) gape colour, pattern[36]
	(D) parasite with egg in beak [7,8]	(D) egg pattern[22,23]	(I) begging posture[37]
		(D) egg size[26]	(D) begging calls[37]
		(D) egg shape[27]	(I) head feather appearance[38]
			(D) number of chicks in the nest[38,39]
	Indirect cues:	*Indirect cues:*	*Indirect cues:*
	(D) egg missing from clutch[7,8]	(D) adult parasite sighted on nest[28,29]	(U) adult parasite sighted on nest[28,29]
	(I) parasite sighted in area[9–14]	(D) adult parasite sighted in area[11]	(I) parasite sighted in area[40]
	(D) parasite perch distance [9,15–17]	(U) parasite perch distance[15–17]	(U) parasite perch distance[15–17]
	(I) time of year[9,18]	(I) time of year[30]	(U) time of year[30]
	(I) host density[9]	(U) host density[9]	(U) host density[9]
	Social cues:	*Social cues:*	*Social cues:*
	(D) alarm calls[4,7,19,20]	(U) egg rejection behaviour	(U) chick rejection behaviour
	(D) mobbing behaviours[4,7,19]		
Potential mechanism	(I) Associative[7–14]	(D) Associative[11,29–31]	(I) Associative[29]
	(U) Relational[32,33]	(D) Relational[32,33]	(I) Relational[27,28]
	(D) Perceptual[1,2]	(D) Perceptual[34,45]	(I) Perceptual[26,31]
Potential source of information	(D) Personal experience[21]	(D) Personal experience[22–35]	(D) Personal experience[36,40]
	(D) Social interaction[4,7,19]	(U) Social interaction [4,7,19]	(U) Social interaction [4,7,19]

The evidence base for each is indicated as either direct (D), indirect (I), or unknown (U)
[1]Welbergen and Davies (2011); [2]Trnka et al. (2012); [3]Thorogood and Davies (2016); [4]Davies and Welbergen (2009); [5]Thorogood and Davies (2012); [6]Welbergen and Davies (2008); [7]Campobello and Sealy (2011b); [8]Campobello and Sealy (2011b); [9]Welbergen and Davies (2009); [10]Briskie et al. (1992); [11]Davies et al. (2003); [12]Langmore et al. (2012); [13]Røskaft et al. (2002); [14]Yang et al. (2014); [15]Welbergen and Davies (2012); [16]Lindholm and Thomas (2000); [17]Gill et al. (1997); [18]Campobello and Sealy (2010); [19]Feeney and Langmore (2013); [20]Gill and Sealy (2003, 2004); [21]Hobson and Sealy (1989); [22]Moksnes et al. (2000); [23]Soler et al. (2012); [24]Feeney et al. (2015); [25]Honza et al. 2007; [26]Soler et al. (2013); [27]Zölei et al. (2012); [28]Moksnes and Røskaft (1989); [29]Moksnes et al. (1993); [30]Brooke et al. (1988); [31]Davie and Brooke (1988); [32]Rothstein (1975); [33]Marchetti (2000); [34]Lotem et al. (1995); [35]Lahti and Lahti (2002); [36]Langmore et al. (2011); [37]Colombelli-Negrel et al. (2012); [38]Soler et al. (2014a, b); [39]Langmore et al. (2003); [40]Langmore et al. (2009)

Table 23.3 Selection of studies involving learning in anti-brood parasite defences found at the different stages of the host nesting cycle

Continent	Brood parasite species	Host species	Learning mode	References
Nest defence				
Europe	Common cuckoo (*Cuculus canorus*)	Reed warbler (*Acrocephalus scirpaceus*)	I + S	Lindholm and Thomas (2000), Davies et al. (2003), Welbergen and Davies (2008), Davies and Welbergen (2009), Welbergen and Davies (2009), Campobello and Sealy (2011a), Thorogood and Davies (2012, 2016), Welbergen and Davies (2012)
		Great reed warbler (*Acrocephalus arundinaceus*)	EI/U	Trnka et al. (2013)
		On 14 host species	U	Røskaft et al. (2002)
Asia	Oriental cuckoo (*Cuculus optatus*)	Yellow-bellied prinia (*Prinia flaviventris*)	U	Yang et al. (2014)
North America	Brown-headed cowbird (*Molothrus ater*)	Yellow warbler (*Setophaga petechia*)	I	Hobson and Sealy (1989), Briskie et al. (1992), Gill et al. (1997), Tewksbury et al. (2002), Gill and Sealy (2003, 2004), Campobello and Sealy (2011b)
		Song sparrow (*Melospiza melodia*)	U	Smith et al. (1984)
Australia	Horsfield's bronze-cuckoos (*Chalcites basalis*)	Superb fairy-wren (*Malurus cyaneus*)	S	Feeney and Langmore (2013)
	Shining bronze-cuckoo (*Chalcites lucidus*)	Superb fairy-wren	U	Langmore et al. (2012)
		Yellow-rumped thornbill (*Acanthiza chrysorrhoa*)	U	Medina and Langmore (2016)

(continued)

Table 23.3 (continued)

Continent	Brood parasite species	Host species	Learning mode	References
Egg discrimination				
Europe	Common cuckoo	Reed warbler	I + S	Moksnes et al. (2000), Thorogood and Davies (2016)
		Rufous-tailed scrub robin (*Cercotrichas galactotes*)	U	Soler et al. (2012)
	Great spotted cuckoo (*Clamator glandarius*)	Magpie (*Pica pica*)	U	Soler et al. (1994, 2013)
Asia	Oriental cuckoo	Yellow-bellied prinia	U	Yang et al. (2014)
Africa	Cuckoo finch (*Anomalospiza imberbis*)	Tawny-flanked prinia (*Prinia subflava*)	I	Feeney et al. (2015)
Central America	Shiny cowbird (*Molothrus bonariensis*)	Village weavers (*Ploceus cucullatus*)	U	Robert and Sorci (1999)
Chick/fledgling discrimination				
Chicks:				
Australia	Horsfield's bronze-cuckoo	Superb fairy-wren	I	Langmore et al. (2009), Colombelli-Negrel et al. (2012)
Fledglings:				
Europe	Great spotted cuckoo	Magpie	I	Soler et al. (2014a, b)

Learning mode: *I* individual, *S* social, *EI* experience-independent ('innate'), *U* undetermined

eyering and underpart barring) that are separately important (Welbergen and Davies 2011; Trnka and Prokop 2012; Nováková et al. 2017). In addition to 'what' is learnt, 'how' hosts learn to discriminate against adult brood parasites is currently a topic of heightened interest and debate, but there is evidence that both individual and social learning mechanisms play a role.

Discrimination against adult brood parasite at the nest can potentially be learnt from personal experience. For example, from personal encounters with nest intruders, a host could learn that adult brood parasites can safely be attacked, or it could learn that adult brood parasites are special enemies (e.g. from the subsequent appearance of a foreign egg or chick). However, such learning would tend to provide limited fitness returns because it only allows hosts to enhance their frontline defences for their next breeding attempt. Interestingly, some brood parasites, such as the common cuckoo (*Cuculus canorus*), remove a host egg when parasitizing (Davies and Brooke 1988), whereas others, such as the brown-headed cowbird (*Molothrus ater*), first remove the host egg before parasitizing the clutch on a subsequent visit (Sealy 1992). Thus, contrary to egg removal by cuckoos, egg removal by cowbirds provides an immediate cue for impending parasitism, and it is not surprising then that some hosts, like yellow warblers, appear to use this information to enhance their nest defences specifically against cowbirds(Campobello and Sealy 2011a).

While personal encounters with adult brood parasite at the nest can provide opportunities for learning frontline defences, brood parasites are secretive and parasitic laying is rapid (Chance 1940; Wyllie 1981), meaning there is generally little opportunity for individual learning of direct cues, especially in cuckoo hosts. In addition, some brood parasitic cuckoos resemble birds of prey (Davies and Welbergen 2008; Welbergen and Davies 2011; Gluckman and Mundy 2013), so individual learning also entails a predation risk in these cases. Under such conditions, it may be more beneficial to learn about the risk of parasitism based on indirect cues, such as the presence of brood parasites in the general area or environmental factors that correlate with the risk of parasite encounters (Welbergen and Davies 2009; Patten et al. 2011).

Discrimination against adult brood parasites at the nest may also be learnt by eavesdropping on conspecifics (Davies and Welbergen 2009; Feeney and Langmore 2013), and this has been suggested to be more important than personal experience in the development of frontline defences (Campobello and Sealy 2011b). Some hosts possess specific, functionally referential alarm calls that are elicited almost exclusively by the presence of brood parasites (e.g. Hobson and Sealy 1989; Gill and Sealy 2004; Langmore et al. 2012), whereas others have graded alarm calls that denote a brood parasite in a more probabilistic manner (Welbergen and Davies 2008). In both instances, however, such calls and associated behaviours may allow naïve individuals to learn about the parasitism threat from informed conspecifics and mount frontline defences.

In some of the more intensely studied systems, hosts show evidence of a mix of individual and social learning around their frontline defences. Reed warblers appear to use both personal (Lindholm and Thomas 2000; Welbergen and Davies 2009, 2012) and social information (Davies and Welbergen 2009; Campobello and Sealy 2011b; Thorogood and Davies 2012) to vary their frontline defences strategically against brood parasitism from common cuckoos (Table 23.3). Similarly, superb fairywrens show strong microgeographic variation in nest defences suggestive of individual learning (Langmore et al. 2012), as well as experimental evidence of social learning (Feeney and Langmore 2013) of defences against Horsfield's bronze-cuckoos (*Chalcites basalis*). Depending on conditions, however, hosts may prefer to use social information over personal information (Campobello and Sealy 2011b), or the reverse (Campobello and Sealy 2011a), and here an important challenge is to understand the modes of learning in terms of the costs of information acquisition and reliability of personal versus social cues.

23.2.2 Learning at the Egg Stage

Once a nest is parasitized, the next response at the host's disposal is recognition and rejection of the parasite egg. Egg discrimination appears to be mediated by two learning processes: relational learning, whereby an individual learns to identify the odd egg, or eggs, in its clutch (discordancy hypothesis: Rothstein 1975; Marchetti 2000), and perceptual learning, whereby an individual learns to match the egg in its clutch to an internal template of the appearance of its own eggs (true recognition: Rothstein 1975; Lotem et al. 1995; Lahti and Lahti 2002). In hosts, these types of learning can operate simultaneously (e.g. Moskát et al. 2010; Bán et al. 2013), but which process occurs under what conditions is an issue that to our knowledge has not yet been addressed.

While there is some direct evidence that hosts learn to discriminate against parasitic eggs as a plastic response to parasitism (Lotem et al. 1992), there is overwhelming indirect evidence to this effect. For example, the rapidity of increase in rejection rates recorded in recently parasitized host populations is thought to be incompatible with genetic adaptation and instead must involve learning (Nakamura 1990; Soler et al. 1994, 2013; Robert and Sorci 1999). More frequently, egg discrimination is thought to be a learnt trait because, like frontline defences, it is adjusted according to the perceived risks (e.g. Davies et al. 2003) or costs (e.g. Welbergen et al. 2001) of parasitism, and it has been hypothesised that egg discrimination may also be socially transmitted (Soler 2011). Mechanisms driving egg discrimination are described in more detail in Chap. 24.

23.2.3 Learning at the Chick/Fledgling Stage

Once the parasitic egg hatches, the next defence at the host's disposal is recognition and rejection of the parasite chick. Hosts of non-evicting brood parasites, such as cowbirds, finches and some parasitic cuckoos, could in principle discriminate against chicks based on the same perceptual and relational learning mechanisms as in egg discrimination; however, there is remarkably little direct evidence for this, with the exception perhaps of intraspecific brood parasitic American coots (*Fulica americana*) (Shizuka and Lyon 2010).

In the case of evicting and specialised parasites, such as most species of cuckoo, there is no opportunity for relational learning, because the newly hatched cuckoo chick evicts all the host's eggs and nestlings from the nest. Perceptual learning would be risky under these circumstances because if a host forms a chick recognition template based on its first breeding attempt, and the host happens to be parasitized during that attempt, the host is likely to reject its own chicks in all subsequent attempts (cost of mis-imprinting: Lotem 1993). However, there is little evidence to suggest that the evicting habits of some brood parasites have evolved in response to chick discrimination by hosts (Fraga 1998; Davies 2000). In addition, some studies have shown that when host and chicks of evicting parasites are experimentally forced to be in the same nest, hosts nevertheless fail to discriminate (Martín-Gálvez et al. 2005; Grim et al. 2009), suggesting that learnt chick discrimination does not feature in the evolutionary arms race between evicting brood parasites and these hosts.

At present, there are very few documented cases of learnt discrimination of parasite chicks or fledglings (Table 23.3). Indeed, evidence of any host discrimination at this stage of the nesting cycle is rare (for discussion see Grim 2006), so it is not surprising that we know even less about the underlying learning mechanisms. In superb fairywrens experienced females commit fewer recognition errors than naïve ones (Langmore et al. 2003, 2009) and are more likely to abandon parasite chicks when parasitism risk is higher (Langmore et al. 2001, 2009). In this case, learning has been recorded at the embryonic level, where hosts learn a 'password' from their parents. The parasite's embryo appears to lack these learning abilities, although it is not yet clear what features the hosts use to reject the parasitic chick (Colombelli-Negrel et al. 2012). Great spotted cuckoo (*Clamator glandarius*) fledglings, reared together with foster magpie (*Pica pica*) siblings in mixed broods, were fed less than their foster siblings and less than those fed in cuckoo-only broods (Soler et al. 2014a). However, parasitized magpie hosts were more likely to allofeed cuckoo fledglings from other magpie broods than unparasitized magpies (Soler et al. 2014b). These findings imply that magpies develop a chick recognition template at each breeding attempt, and while such perceptual learning provides some benefit (preferential feeding of own chicks), it also entails costs (allofeeding of cuckoo fledglings). There are other cases of discrimination at the chick/fledgling stage, including the video recording of large-billed gerygone (*Gerygone magnirostris*) evicting chicks of little bronze-cuckoo (*Chalcites minutillus*), but whether this discrimination is experience-independent or learnt remains to be explored (Tokue and Ueda 2010).

As for the earlier-stage defences, chick/fledgling discrimination does not necessarily imply learning (nor indeed 'recognition'—Grim et al. 2003). For example, experimental evidence shows that reed warblers desert cuckoo chicks on the basis on the duration of parental care, as cuckoo chicks require care for longer than warbler chicks (Grim 2007). While it could perhaps be argued that the duration of parental care is analogous to morphological features that form the basis for recognition elsewhere, these findings nevertheless challenge our understanding of the discriminatory mechanisms underlying anti-brood parasite defences.

Concluding Remarks and Future Directions

Recently, a common theme has emerged in the brood parasitism literature whereby brood parasites use mimicry as a general strategy for circumventing discrimination mechanisms of their victims, and in turn, hosts counteract exploitation by improving their enemy recognition at successive stages of the nesting cycle. It is now clear that learning is a ubiquitous process

(continued)

throughout these stages, but while learning can profoundly affect the coevolutionary dynamics between hosts and their brood parasites, it has hitherto received relatively little attention.

To date, most research has focused on the role of learning in nest defence, and there is evidence that both individual and social learning mechanisms play a role in the acquisition of frontline defences. It is generally assumed that public information used for enhancing nest defences comes from conspecifics; however, there is evidence to suggest that heterospecifics too may eavesdrop on nest defence behaviour. It would be interesting to examine whether individuals use information from heterospecifics to enhance their frontline defences. If so, this would have important implications for the resistance of avian assemblages to brood parasitism, as host individuals then also benefit from interspecific interactions. Hosts have a plethora of potential cues at their disposal for mounting plastic defences that are useful in a world where parasitism varies greatly in time and space. Hosts can use direct cues, such as the sight of brood parasite adults, eggs or chicks, or indirect cues, such as environmental factors that correlate with the risk of parasite encounters, to learn about the risk of parasitism and modify their defences accordingly. The challenge then is to understand the observed plasticity of anti-brood parasite defences in the context of environmental heterogeneity, and specifically in terms of variation in the costs of information acquisition and in terms of the presence, detection, and reliability of relevant cues.

Acknowledgements Constructive comments from Manuel Soler and William E. Feeney greatly improved the early drafts of this chapter.

References

Bán M, Moskát C, Barta Z, Hauber ME (2013) Simultaneous viewing of own and parasitic eggs is not required for egg rejection by a cuckoo host. Behav Ecol 24:1014–1021

Boyd R, Richerson PJ (1988) An evolutionary model of social learning: the effects of spatial and temporal variation. In: Zentall TR, Galef BG Jr (eds) Social learning. Lawrence Erlbaum Associates, Hillsdale, pp 29–48

Briskie JV, Sealy SG, Hobson KA (1992) Behavioural defences against avian brood parasitism in sympatric and allopatric populations. Evolution 46:334–340

Brooke ML, Davies NB (1988) Egg mimicry by cuckoos *Cuculus canorus* in relation to discrimination by hosts. Nature 335:630–632

Brooke ML, Davies NB, Noble DG (1998) Rapid decline of host defences in response to reduced cuckoo parasitism: behavioural flexibility of reed warblers in a changing world. Proc R Soc Lond B 265:1277–1282

Campobello D (2008) A comparative approach to the study of avian nest defence: experience and adaptive significance. PhD thesis, University of Manitoba, Canada

Campobello D, Hare JF (2007) Information transfer determined by association of neighbours in European bee-eater, *Merops apiaster,* colonies. Ethol Ecol Evol 19:237–243

Campobello D, Sealy SG (2009) Avian brood parasitism in a Mediterranean region: hosts and habitat preferences of Common Cuckoos. Bird Study 56(3):389–400

Campobello D, Sealy SG (2010) Enemy recognition of reed warbler, *Acrocephalus scirpaceus*: threats and reproductive value act independently in nest defence modulation. Ethology 116:498–508

Campobello D, Sealy SG (2011a) Use of social over personal information enhances nest defence against avian brood parasitism. Behav Ecol 22:422–428

Campobello D, Sealy SG (2011b) Nest defence against avian brood parasites is promoted by egg-removal events in a cowbird-host system. Anim Behav 82:885–891

Campobello D, Hare JF, Sarà M (2015) Social phenotype extended to communities: expanded multilevel social selection analysis reveals fitness consequences of interspecific interactions. Evolution 69:916–925

Canestrari D, Marcos JM, Baglione V (2009) Cooperative breeding in carrion crows reduces the rate of brood parasitism by great spotted cuckoos. Anim Behav 77:1337–1344

Caro TM (2005) Antipredator defences in birds and mammals. University of Chicago Press, Chicago

Chance EP (1940) The truth about the cuckoo. Country Life, London

Colombelli-Negrel D, Hauber ME, Robertson J, Sulloway FJ, Hoi H, Griggio M, Kleindorfer S (2012) Embryonic learning of vocal passwords in superb fairy-wrens reveals intruder cuckoo nestlings. Curr Biol 22:2155–2160

Davies NB (2000) Cuckoos, cowbirds and other cheats. T&AD Poyser, London

Davies NB, Brooke ML (1988) Cuckoos versus reed warblers: adaptations and counteradaptations. Anim Behav 36:262–284

Davies NB, Welbergen JA (2008) Cuckoo–hawk mimicry? An experimental test. Proc R Soc Lond B 275:1817–1822

Davies NB, Welbergen JA (2009) Social transmission of a host defence against cuckoo parasitism. Science 324:1318–1320

Davies NB, Butchart SHM, Burke TA, Chaline N, Stewart IRK (2003) Reed warblers guard against cuckoos and cuckoldry. Anim Behav 65:285–295

De Mársico MC, Gantchoff MG, Reboreda JC (2012) Host–parasite coevolution beyond the nestling stage? Mimicry of host fledglings by the specialist screaming cowbird. Proc R Soc Lond B 279:3401–3408

DeWitt TJ, Scheiner SM (2004) Phenotypic variation from single genotypes: a primer. In: DeWitt TJ, Scheiner SM (eds) Phenotypic plasticity. Oxford University Press, New York, pp 1–9

Doligez B, Pärt T, Danchin E (2004) Prospecting in the collared flycatcher: gathering public information for future breeding habitat selection? Anim Behav 67:457–466

Duckworth JW (1991) Responses of breeding reed warblers, *Acrocephalus scirpaceus*, to mounts of sparrowhawk, *Accipiter nisus*, cuckoo, *Cuculus canorus*, and jay, *Garrulus glandarius*. Ibis 133:68–74

Dukas R (1998) Cognitive ecology: the evolutionary ecology of information processing and decision making. University of Chicago Press, Chicago

Dukas R (1999) Costs of memory: ideas and predictions. J Theor Biol 197:41–50

Feeney WE, Langmore NE (2013) Social learning of a brood parasite by its host. Biol Lett 9:20130443

Feeney WE, Welbergen JA, Langmore NE (2012) The frontline of avian brood parasite–host coevolution. Anim Behav 84:3–12

Feeney WE, Welbergen JA, Langmore NE (2014) Advances in the study of coevolution between avian brood parasites and their hosts. Annu Rev Ecol Evol Syst 45:227–246

Feeney WE, Troscianko J, Langmore NE, Spottiswoode CN (2015) Evidence for aggressive mimicry in an adult brood parasitic bird, and generalized defences in its host. Proc R Soc Lond B 282:20150795

Forsman JT, Monkkonen M (2001) Responses by breeding birds to heterospecific song and mobbing call playbacks under varying predation risk. Anim Behav 62:1067–1073

Fraga RM (1998) Interactions of the parasitic screaming and shiny cowbirds (*Molothrus rufoaxillaris* and *M. bonariensis*) with a shared host, the bay-winged cowbird (*M. badius*). In: Rothstein SI, Robinson SK (eds) Parasitic birds and their hosts. Oxford University Press, Oxford, pp 173–193

Freeberg TM, Duncan SD, Kast TL, Enstrom DA (1999) Cultural influences on female mate choice: an experimental test in cowbirds, *Molothrus ater*. Anim Behav 57:421–426

Galef BG Jr, Laland KN (2005) Social learning in animals: empirical studies and theoretical models. Bio Sci 55:489–499

Getty T (1996) The maintenance of phenotypic plasticity as a signal detection problem. Am Nat 148:378–385

Gill SA, Sealy SG (2003) Tests of two functions of alarm calls given by yellow warblers during nest defence. Can J Zool 81:1685–1690

Gill SA, Sealy SG (2004) Functional reference in an alarm signal given during nest defence: seet calls of yellow warblers denote brood-parasitic brown-headed cowbirds. Behav Ecol Sociobiol 56:71–80

Gill SA, Neudorf DL, Sealy SG (1997) Host responses to cowbirds near the nest: cues for recognition. Anim Behav 53:1287–1293

Gill SA, Neudorf DLH, Sealy SG (2008) Do hosts discriminate between sexually dichromatic male and female brown-headed cowbirds? Ethology 114:548–556

Giraldeau L-A, Valone TJ, Templeton JJ (2002) Potential disadvantages of using socially acquired information. Philos Trans R Soc B 357:1559–1566

Gluckman T-L, Mundy NI (2013) Cuckoos in raptors' clothing: barred plumage illuminates a fundamental principle of Batesian mimicry. Anim Behav 86:1165–1181

Grim T (2006) The evolution of nestling discrimination by hosts of parasitic birds: why is rejection so rare? Evol Ecol Res 8:785–802

Grim T (2007) Experimental evidence for chick discrimination without recognition in a brood parasite host. Proc R Soc Lond B 274:373–381

Grim T, Kleven O, Mikulica O (2003) Nestling discrimination without recognition: a possible defence mechanism for hosts towards cuckoo parasitism? Proc R Soc Lond B 270:S73–S75

Grim T, Rutila J, Cassey P, Hauber ME (2009) Experimentally constrained virulence is costly for common cuckoo chicks. Ethology 115:14–22

Heyes CM (1994) Social learning in animals: categories and mechanisms. Biol Rev 69:207–231

Hobson KA, Sealy SG (1989) Responses of yellow warblers to the threat of cowbird parasitism. Anim Behav 38:510–519

Honza M, Šicha V, Procházka P, Ležalová R (2006) Host nest defense against a color-dimorphic brood parasite: great reed warblers (*Acrocephalus arundinaceus*) versus common cuckoos (*Cuculus canorus*). J Ornithol 147:629–637

Honza M, Polačikovà L, Prochàzka P (2007) Ultraviolet and green parts of the colour spectrum affect egg rejection in the song thrush (*Turdus philomelos*). Biol J Linn Soc 92:269–276

Hosoi SA, Rothstein SI (2000) Nest desertion and cowbird parasitism: evidence for evolved responses and evolutionary lag. Anim Behav 59:823–840

Johnston TD (1982) Selective costs and benefits in the evolution of learning. Adv Study Behav 12:65–106

Kleindorfer S, Evans C, Colombelli-Négrel D, Robertson J, Griggio M, Hoi H (2013) Host response to cuckoo song is predicted by the future risk of brood parasitism. Front Zool 10:1–10

Lahti DC, Lahti AR (2002) How precise is egg discrimination in weaverbirds? Anim Behav 63:1135–1142
Laland KN, Richerson PJ, Boyd R (1996) Developing a theory of animal social learning. In: Heyes CM, Galef BG Jr (eds) Social learning in animals: the roots of culture. Academic Press, San Diego, pp 129–154
Langmore NE, Stevens M, Maurer G, Heinsohn R, Hall ML, Peters A, Kilner RM (2001) Visual mimicry of host nestlings by cuckoos. Proc R Soc Lond B 278:2455–2463
Langmore NE, Hunt S, Kilner RM (2003) Escalation of a coevolutionary arms race through host rejection of brood parasitic young. Nature 422:157–160
Langmore NE, Cockburn A, Russell AF, Kilner RM (2009) Flexible cuckoo chick-rejection rules in the superb fairy-wren. Behav Ecol 20:978–984
Langmore NE, Feeney WE, Crowe-Riddell J, Luan H, Louwrens KM, Cockburn A (2012) Learned recognition of brood parasitic cuckoos in the superb fairy-wren *Malurus cyaneus*. Behav Ecol 23:798–805
Lefebvre L, Giraldeau L-A (1996) Is social learning an adaptive specialization? In: Heyes CM, Galef BG Jr (eds) Social learning in animals. Academic Press, San Diego, pp 107–128
Li D, Wei H, Zhang Z, Liang W, Stokke BG (2015) Oriental reed warbler, *Acrocephalus orientalis* nest defence behaviour towards brood parasites and nest predators. Behaviour 152:1601–1621
Lindholm AK (1999) Brood parasitism by the cuckoo on patchy reed warbler populations in Britain. J Anim Ecol 68:293–309
Lindholm AK, Thomas RJ (2000) Differences between populations of reed warblers in defences against brood parasitism. Behaviour 137:25–42
Lotem A (1993) Learning to recognize nestlings is maladaptive for cuckoo *Cuculus canorus* hosts. Nature 362:743–745
Lotem A, Nakamura H, Zahavi A (1992) Rejection of cuckoo eggs in relation to host age – a possible evolutionary equilibrium. Behav Ecol 3:128–132
Lotem A, Nakamura H, Zahavi A (1995) Constraints on egg discrimination and cuckoo-host co-evolution. Anim Behav 49:1185–1209
Marchetti K (2000) Egg rejection in a passerine bird: size does matter. Anim Behav 59:877–883
Martín-Gálvez D, Soler M, Soler JJ, Martín-Vivald M, Palomino JJ (2005) Food acquisition by common cuckoo chicks in rufous bush robin nests and the advantage of eviction behaviour. Anim Behav 70:1313–1321
Medina I, Langmore NE (2016) Batten down the thatches: front-line defences in an apparently defenceless cuckoo host. Anim Behav 112:195–201
Mery F, Burns JG (2010) Behavioural plasticity: an interaction between evolution and experience. Evol Ecol 24:571–583
Moksnes A, Røskaft E (1989) Adaptations of meadow pipits to parasitism by the common cuckoo. Behav Ecol Sociobiol 24:25–30
Moksnes A, Røskaft E, Korsnes L (1993) Rejection of cuckoo (*Cuculus canorus*) eggs by meadow pipits (*Anthus pratensis*). Behav Ecol 4:120–127
Moksnes A, Røskaft E, Hagen LG, Honza M, Mörk C, Olsen PH (2000) Common cuckoo, *Cuculus* canorus, and host behaviour at reed warbler, *Acrocephalus scirpaceus*, nests. Ibis 142:247–258
Moskát C, Bán M, Székely T, Komdeur J, Lucassen RWG, van Boheemen LA, Hauber ME (2010) Discordancy or template-based recognition? Dissecting the cognitive basis of the rejection of foreign eggs in hosts of avian brood parasites. J Exp Biol 213:1976–1983
Nakamura H (1990) Brood parasitism by the cuckoo *Cuculus canorus* in Japan and the start of new parasitism on the azure-winged magpie *Cyanopica cyana*. Jpn J Ornithol 39:1–18
Neudorf DL, Sealy SG (1992) Reactions of four passerine species to threats of predation and cowbird parasitism: enemy recognition or generalized responses? Behaviour 123:84–105
Nováková N, Veselý P, Fuchs R (2017) Object categorization by wild ranging birds—winter feeder experiments. Behav Process 143:7–12
Patten M, Reinking D, Wolfe D (2011) Hierarchical cues in brood parasite nest selection. J Ornithol 152:521–532

Price TD, Qvarnström A, Irwin DE (2003) The role of phenotypic plasticity in driving genetic evolution. Proc R Soc Lond B 270:1433–1440

Robert M, Sorci G (1999) Rapid increase of host defence against brood parasites in a recently parasitized area: the case of village weavers in Hispaniola. Proc R Soc Lond B 266:941–946

Røskaft E, Moksnes A, Stokke BG, Bicěk V, Moskàt C (2002) Aggression to dummy cuckoos by potential European cuckoo hosts. Behaviour 139:613–628

Rothstein SI (1975) Mechanisms of avian egg-recognition: do birds know their own eggs? Anim Behav 23:268–278

Sealy SG (1992) Removal of yellow warbler eggs in association with cowbird parasitism. Condor 94:40–54

Shettleworth SJ (2010) Cognition, evolution, and behaviour. Oxford University Press, New York

Shizuka D, Lyon BE (2010) Coots use hatch order to learn to recognize and reject conspecific brood parasitic chicks. Nature 463:223–226

Smith JNM, Arcese P, McLean IG (1984) Age, experience, and enemy recognition by wild song sparrows. Behav Ecol Sociobiol 14:101–106

Soler JJ, Martín-Gálvez D, de Neve L, Soler M (2013) Brood parasitism correlates with the strength of spatial autocorrelation of life history and defensive traits in Magpies. Ecology 94 (6):1338–1346

Soler M (2011) Could egg rejection behaviour be transmitted by social learning? Anim Behav 81: e1–e6

Soler M, Martinez J, Soler J, Møller A (1994) Micro-evolutionary change in host response to a brood parasite. Behav Ecol Sociobiol 35:295–301

Soler M, Martín-Vivaldi M, Fernández-Morante J (2012) Conditional response by hosts to parasitic eggs: the extreme caseof the rufous-tailed scrub robin. Anim Behav 84:421–426

Soler M, de Neve L, Roncalli G, Macías-Sánchez E, Ibáñez-Álamo JD, Pérez-Contreras T (2014a) Great spotted cuckoo fledglings are disadvantaged by magpie host parents when reared together with magpie nestlings. Behav Ecol Sociobiol 68:333–342

Soler M, Pérez-Contreras T, Ibáñez-Álamo JD, Roncalli G, Macías-Sánchez E, de Neve L (2014b) Great spotted cuckoo fledglings often receive feedings from other magpie adults than their foster parents: which magpies accept to feed foreign cuckoo fledglings? PLoS One 9:e107412

Stearns SC (1992) The evolution of life histories. Oxford University Press, Oxford

Stephens DW (1991) Change, regularity, and value in the evolution of animal learning. Behav Ecol 2:77–89

Stokke BG, Hafstad I, Rudolfsen G, Moksnes A, Møller AP, Røskaft E, Soler M (2008) Predictors of resistance to brood parasitism within and among reed warbler populations. Behav Ecol 19:612–620

Tewksbury JJ, Martin TE, Hejl SJ, Kuehn MJ, Jenkins JW (2002) Parental care of a cowbird host: caught between the costs of egg-removal and nest predation. Proc R Soc Lond B 269:423–429

Thorogood R, Davies NB (2012) Cuckoos combat socially transmitted defences of reed warbler hosts with a plumage polymorphism. Science 337:578–580

Thorogood R, Davies NB (2016) Combining personal with social information facilitates host defences and explains why cuckoos should be secretive. Sci Rep 6:19872

Tokue K, Ueda K (2010) Mangrove gerygones *Gerygone laevigasert* eject little bronze-cuckoo *Chalcites minutillus* hatchlings from parasitized nests. Ibis 152:835–839

Tollrian R, Harvell CD (1999) The ecology and evolution of inducible defences. Princeton University Press, Princeton, NJ

Trnka A, Prokop P (2012) The effectiveness of hawk mimicry in protecting cuckoos from aggressive hosts. Anim Behav 83:263–268

Trnka A, Požgayová M, Samaš P, Honza M (2013) Repeatability of host female and male aggression towards a brood parasite. Ethology 119:907–917

Uyehara JC, Narins PM (1995) Nest defence by willow flycatchers to brood-parasitic intruders. Condor 97:361–368

Welbergen JA, Davies NB (2008) Reed warblers discriminate cuckoos from sparrowhawks with graded alarm signals that attract mates and neighbours. Anim Behav 76:811–822

Welbergen JA, Davies NB (2009) Strategic variation in mobbing as a front line of defence against brood parasitism. Curr Biol 19:235–240

Welbergen JA, Davies NB (2011) A parasite in wolf's clothing: hawk mimicry reduces mobbing of cuckoos by hosts. Behav Ecol 22:574–579

Welbergen JA, Davies NB (2012) Direct and indirect assessment of parasitism risk by a cuckoo host. Behav Ecol 23:783–789

Welbergen J, Komdeur J, Kats R, Berg M (2001) Egg discrimination in the Australian reed warbler (*Acrocephalus australis*): rejection response toward model and conspecific eggs depending on timing and mode of artificial parasitism. Behav Ecol 12:8–15

Wyllie I (1981) The cuckoo. Batsford, London

Yang C, Wang L, Cheng S-J, Hsu Y-C, Liang W, Møller A (2014) Nest defenses and egg recognition of yellow-bellied prinia against cuckoo parasitism. Naturwissenschaften 101:727–734

Zentall TR, Wasserman EA, Urcuioli PJ (2014) Associative concept learning in animals. J Exp Anal Behav 101(1):130–151

Zölei A, Hauber ME, Geltsch N, Moskát C (2012) Asymmetrical signal content of egg shape as predictor of egg rejection by great reed warblers, hosts of the common cuckoo. Behaviour 149:391–406

Cognitive Decision Rules for Egg Rejection 24

Thomas Manna, Csaba Moskát, and Mark E. Hauber

Abstract

Egg rejection is the best studied behavioral adaptation by hosts to avian brood parasitism. Investigations of the mechanism(s) by which a host accomplishes the task of perceiving and deciding to reject a foreign egg have been a hotbed of debate and discovery for decades. The two most often tested cognitive explanations for this behavior are: (1) the host rejects the egg most dissimilar from the other eggs in the nest (discordancy mechanism) and (2) the host compares each egg to an internal template of the appearance of its own eggs (template recognition mechanism). While many years of published work have purported sole support for the template recognition hypothesis (for instance, hosts can experimentally reject foreign eggs which do not represent a quantitative minority in the clutch), in recent years an increasingly prevalent argument that the two mechanisms are working in tandem has come to light. Furthermore, there is also a steadily building body of work indicating that hosts have plastic discrimination thresholds, such that the extent to which parasitic eggs must be different from a host's own egg before rejection occurs appears to be both socio-ecological context dependent and shaped by earlier experiences through a learning component. Overall, the cognitive architecture of

T. Manna
Department of Psychology, Hunter College and the Graduate Center, City University of New York, New York, NY, USA

C. Moskát
MTA-ELTE-MTM Ecology Research Group, A Joint Research Group of the Hungarian Academy of Sciences, The Biological Institute of the Eötvös Loránd University and the Hungarian Natural History Museum, Budapest, Hungary

M. E. Hauber (✉)
Department of Animal Biology, School of Integrative Biology, University of Illinois, Urbana-Champaign, IL, USA
e-mail: markehauber@gmail.com

M. Soler (ed.), *Avian Brood Parasitism*, Fascinating Life Sciences,
https://doi.org/10.1007/978-3-319-73138-4_24

egg rejection decisions appears to be complex and shaped by the particular coevolutionary histories of hosts and parasites.

24.1 Introduction

The coevolutionary arms race between avian brood parasites and their hosts (Rothstein 1990) can be conceptualized as a war on two fronts distinguished temporally: before and after the parasite lays its egg in the host nest. The first takes place prior to the insertion of the parasite egg and primarily comprises of the host's attempts to prevent nest infiltration by the parasite (Welbergen and Davies 2009; Mclean 1987, see Chap. 17). The second pertains to host defense mechanisms following successful egg deposition by a parasite (egg, chick, and nestling rejection; nest desertion). At this stage, the generally (but not always, e.g., Servedio and Hauber 2006; Krüger 2011; Hauber et al. 2014) most effective way to avoid investing parental care in unrelated offspring is the destruction, marginalization, or removal of the parasitic egg(s) prior to hatching. Accordingly, the identification and rejection of foreign eggs are among the most prevalent and also best studied adaptations to brood parasitism among commonly parasitized species (Rothstein 1974; Brooke and Davies 1988; Feeney et al. 2014).

Whereas several physical and socio-ecological factors may play key roles in egg rejection behavior such as disruptions in the arrangement ("shuffling") of eggs in a clutch (Polačiková et al. 2013, but see Hanley et al. 2015) and the observation of a nearby parasite individual by the host (Bártol et al. 2002), this chapter will focus strictly on the perceptual cognitive rules in assessing foreign eggs in the nest. Rothstein (1975) examined egg rejection in an attempt to assess the cognitive mechanism or decision "rule" that the host individual might use to determine which eggs from a parasitized clutch to eject. He made reference to two candidate rules: the "odd egg out" or discrimination by discordancy mechanism and the template-based "true recognition."

Under the discordancy mechanism, hosts will selectively reject the egg(s) which are most dissimilar in appearance to others in the same clutch. This is perhaps the simpler of the cognitive rules, as the host's discrimination task is not dependent on memory. Discordancy may seem to represent a reasonably successful strategy, as parasitic eggs are usually in the minority of any given parasitized clutch relative to the host's own eggs (Hauber 2001). However, due to incidental variation in the eggshells of the host's own clutch, this strategy may become highly error-prone (Lotem et al. 1995), and the host may accept foreign eggs or even reject its own egg (s). Own-egg rejection under the discordancy mechanism is especially likely in populations with such high parasitic pressure that multiple parasitism may place a host's own egg in the minority of its clutch (Stevens et al. 2013). Nevertheless, the discordancy rule is a relatively parsimonious possible mechanism in part due to its direct nature, and during the early decades of egg rejection research, it was accepted as the sole explanation for the behavior of rejecter species, as it lined up empirically with naturalistic field observation (but see below).

Under the template recognition mechanism, on the other hand, the host can discriminate a foreign egg by comparing its visual features to that of an innate or learned recognition template of the host's own eggs' appearance. This is considered "true recognition" because it requires a cross reference to the host individual's internal knowledge and, thus, a neural encoding of the template (Hauber and Sherman 2001). Template-based recognition can also explain diverse experimental patterns of parasitic egg rejection in cases of single parasitism where discordancy is evoked, but is typically passed over in favor of discordancy because it is less parsimonious (requires neural mechanisms for memory and internal comparisons). Consequently, templates are specifically and singularly evoked to explain patterns of rejection, such as color dissimilarity-based increased rejection of foreign eggs from nests of multiply parasitized hosts where own eggs are in the minority or are absent altogether (Bán et al. 2013).

Fortunately, the logistics of experimentally teasing apart the relative prevalence of discordancy vs template-based foreign egg discrimination mechanisms in a given host population are relatively straightforward. For instance, a series of artificial parasitism experiments which vary in the relative makeup of host vs foreign eggs would have obvious predictions: hosts operating under the discordancy rule should reject the egg(s) which constitute the minority of the clutch even if they are its own eggs and (2) conversely, hosts operating under the template recognition rule should preferentially reject foreign eggs, even when they represent the majority in the clutch. Many such experiments over the past 40 years have been carried out, resulting primarily in steadily mounting support for the template recognition hypothesis (Victoria 1972; Rothstein 1974, 1975; Moksnes 1992; Lotem et al. 1995; Sealy and Bazin 1995; Peer and Sealy 2001; Lahti and Lahti 2002; Lyon 2007; Strausberger and Rothstein 2009; Moskát et al. 2010, 2014b, c, Bán et al. 2013; Lang et al. 2014; Wang et al. 2015).

24.2 Support for Template-Based Recognition

Victoria (1972) carried out 322 egg replacement trials on a captive colony of African village weaverbirds *Ploceus cucullatus*, which, in natural conditions, are commonly parasitized by the dideric cuckoo *Chrysococcyx caprius*. The experiment demonstrated rejection behavior even in the absence of the host's own eggs and concluded that the weaverbirds possessed a relatively precise memory of the appearance of their own eggs, especially since rejection rate increased proportionally to the difference between the experimental egg and the egg it replaced. Following this, Rothstein (1975) demonstrated a similar capability in the gray catbird *Dumetella carolinensis*. Moksnes (1992) carried out an experiment on chaffinch *Fringilla coelebs* and brambling *Fringilla montifringilla* nests in which all but one egg in a host's clutch was replaced with a foreign egg. Under the condition wherein the experimental egg starkly contrasted the host's natural eggshell, 10 out of 11 nests showed rejection of the foreign eggs, some going so far as to reject every egg until only the original host egg remained (these nests were ultimately abandoned by the host). Similarly, Peer and Sealy (2001) tested the same capability in the great-tailed grackle *Quiscalus mexicanus*, by swapping two eggs out of a clutch of three eggs with artificial eggs. In all ten experimental nests, both artificial eggs were ejected even though the remaining natural grackle egg had represented the discordant egg.

Lahti and Lahti (2002) followed up Victoria's 1972 study on *P. cucullatus* with a rigorous field study of rejection rates in response to various levels of mimicry as well as relative proportion of host vs experimental eggs. The authors observed the rejection of the single foreign egg in conditions wherein there were two host eggs, one host egg, and indeed zero host eggs. In fact, host subjects were not statistically better discriminators when they had their own eggs available in the clutch for comparison to the foreign egg. The authors point out that a template-based rule would be particularly adaptive for *P. cucullatus* when compared to discordancy, as with an average clutch size of two eggs, this species' own eggs cannot be expected to maintain a majority in a parasitized clutch. Importantly, even though the study employed eggs with varying levels of mimicry, no distinct threshold of differential from the host's own egg was found to elicit rejection universally, indicating individual variation in egg rejection decision-making in a single population (see evidence for plastic threshold of discrimination below). This study, along with Victoria (1972) and Rothstein (1975) above, was one of many which explicitly exhibited host individuals' ability to discriminate and reject foreign eggs even in the absence of one's own eggs for comparison. Subsequently, the great reed warbler *Acrocephalus arundinaceus*, a host of the common cuckoo *Cuculus canorus*, has also demonstrated this ability with empirical rigor and through the manipulation of either egg maculation (Moskát et al. 2010) or shell background coloration (Bán et al. 2013).

24.3 Evidence for Mixture of Discordancy and Template-Based Rules

None of the above is meant to imply that the discordancy mechanism does not exist in any host population, nor that discordancy and template-based heuristics in egg rejection decision-making are mutually exclusive, indeed there is mounting evidence that they are not (Marchetti 2000; Moskát et al. 2010, 2014c; Stevens et al. 2013; Yang et al. 2014). One of the more compelling cases for the influence of a discordancy mechanism is the phenomenon of many host populations' rejection rates reducing as the relative proportion of foreign eggs in the nest to own eggs increases (Moskát et al. 2009, 2010; Stevens et al. 2013). If template-based recognition were the only rule in place, hosts should be able to distinguish foreign eggs from their own just as easily no matter their propensity in the clutch.

Recently, Yang et al. (2014) examined a population of ashy-throated parrotbills *Paradoxornis alphonsianus*, a host of the common cuckoo. This is an attractive host-brood parasite study system as *P. alphonsianus* individuals lay immaculate white, gray, or blue eggs but only one color per female. The race, or gens, of *C. canorus* which parasitizes *P. alphonsianus* mimics the host's eggshell pattern as is often the case with *C. canorus* (Brooke and Davies 1988), but the authors in previous experiments found no indication that parasite individuals preferentially target hosts which lay the same color eggshell as they do (Yang et al. 2013). In this follow-up experiment, the researchers artificially parasitized nests with conspecific eggs not matching their own phenotype with variation in propensity of host vs foreign eggs. Contrary to most studies detailed so far, this host population required at least one own egg in the nest for reference to reject foreign eggs; nests containing only manipulated eggs were accepted, indicating that rather than cross referencing a memory of its own eggs' appearance; these individuals were employing "online-self phenotype referencing" (Hauber and Sherman 2001), rather than template-based discrimination, in real time when deciding whether or not to eject. In addition, some individuals preferentially rejected their own eggs when they were in the minority, seemingly adhering to a discordancy rule. It is possible that this species' lack of an internal template may be due in part to the immaculate nature of their eggshells, as maculation (spotting) has been theorized to play a critical cognitive role in the internalization of one's own eggshell phenotype (Davies 2000; Stoddard et al. 2014).

Stevens et al. (2013) demonstrated how the discordancy component to the African tawny-flanked prinia *Prinia subflava*'s rejection strategy may be being exploited by the cuckoo finch *Anomalospiza imberbis*, of which it is the most common host. The authors first experimentally demonstrated that overall, hosts were consistently likely to reject a foreign egg across clutches with host/foreign egg ratios of 1:1, 1:2, and 1:3, showing strong evidence of a template-based mechanism in place. However, rejection rates decrease as the relative proportion of foreign eggs in the clutch increases, providing evidence for the influence of a discordancy rule (i.e., as the foreign eggs are becoming less discordant, rejection rates are decreasing). The authors go on to demonstrate that in the wild, individual *A. imberbis* parasite females will very often target the same host female twice, laying two eggs in the same nest. This, it seems, is a novel strategy which exploits the tendency for acceptance rates to increase as parasite egg count increases. However, the reported pattern of decreasing foreign egg rejection rates across increasing multiple parasitism is also consistent with the mechanism that increased overall color and pattern variation of foreign eggs in the clutch induces a cognitive interference for template- or online self-referencing-based discrimination mechanisms in that the foreign eggs represent a broader range of variation in multiple than single parasitism, thereby reducing the efficiency by which recognition threshold-based mechanism can discriminate own vs foreign eggs (Hauber et al. 2006; Bán et al. 2013; see below). To fully assess these alternatives requires artificial egg introductions into host nests where the variability of the foreign eggs is experimentally controlled. Bán et al. (2013) conducted such an experiment partially (Fig. 24.1), but a fully factorial study is still needed.

Earlier, Moskát et al. (2010) had conducted a similar experiment on the great reed warbler, host of the common cuckoo, demonstrating that the discordancy and

Fig. 24.1 An experimental great reed warbler nest from the Bán et al. (2013) study of multiple (artificial) parasitism of great reed warbler clutches, with all different egg colors. In this nest, each egg is dyed with a different highlighter pen to alter the background color, with the original maculation pattern left visible. In this treatment, the host does not have its own egg color available as a referent for egg rejection and thus must use a template-based recognition mechanism

template-based rulesets could be experimentally disentangled and exhibited in a single population. The authors provided the test population with one of three nest setups: (1) with one dyed egg and the rest unaltered, (2) with all but one egg dyed, and (3) with all eggs dyed (Fig. 24.2). In the condition where the unmanipulated egg represented a minority, the host individuals rejected that egg (own egg) above control levels, indicating the influence of the discordancy rule. However, a full 33% of the nests which contained all manipulated eggs with no own eggshell pattern for reference matching showed rejection behavior, a clear sign of a template-based ruleset. From this study it can be concluded that these rulesets can coexist in a single population and even influence individual behavior to each mechanism's extreme (rejecting own minority egg due to discordancy and rejecting parasite eggs in the absence of any own eggs due to template-based matching).

Then Moskát et al. (2014c) investigated the phenomenon of rejection rates decreasing in cases of multiple parasitism from a perspective of variable clutch heterogeneity. The authors reasoned that as the number of foreign eggs in a nest increases, the overall intraclutch variability will increase resulting in a disruption of discordancy-based decision-making. They manipulated several great reed warbler nests with experimental eggs varying in color (and thereby extent of mimicry) and minority/majority status. They discovered that a host's rejection rate of a given experimental egg color was largely contingent on the color of the surrounding eggs in the clutch, and concluded that discordancy-based egg rejection decisions are context dependent. Specifically, hosts seem to take into account the appearance of the entire clutch and the specific extents of mimicry both in the contextual minority and majority with regard to color. They termed this a mimicry-dependent discordancy rule, and its prevalence in other host–parasite systems still remains to be explored fully.

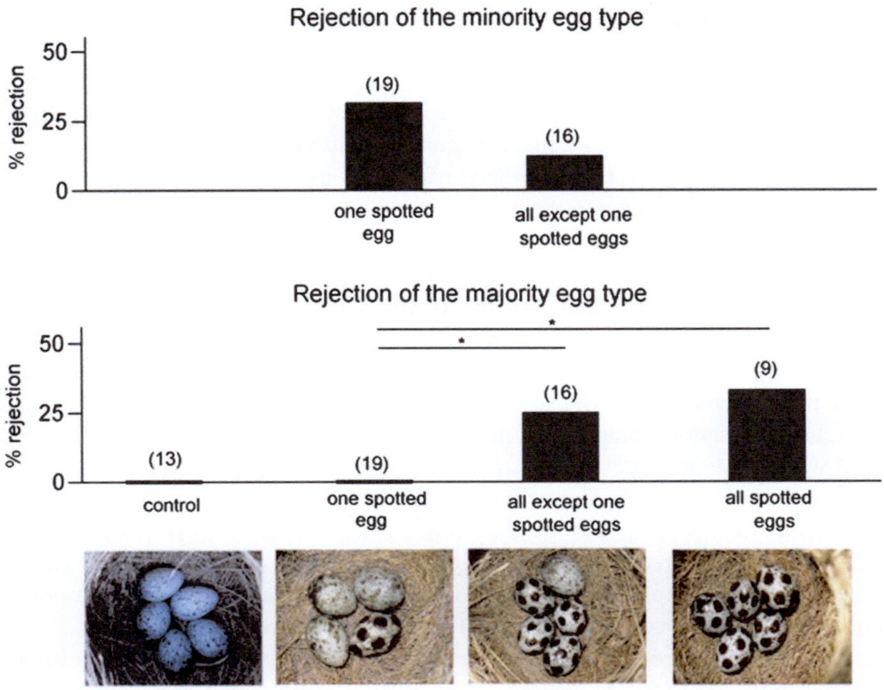

Fig. 24.2 Great reed warblers' responses to experimentally manipulated and non-manipulated own eggs (modified from Moskát et al. 2010). Manipulations were effected by adding artificial spots to the eggshell surface. Significant pairwise differences between rejection rates of the main categories is shown by asterisk (Fisher's exact test, $P < 0.05$). Great reed warblers rejected some of their own eggs when in the minority of the clutch, thereby supporting the discordancy-based recognition mechanism. Great reed warblers rejected spotted eggs even when these were in the majority and the totality of the manipulated clutch, thereby supporting the template-based mechanism

24.4 Evidence for Plasticity of Discrimination Thresholds

Regardless of which egg rejection decision rule is being employed, the cognitive task of a host deciding whether or not to eject an egg is essentially the same. Whether it is comparing the egg to an own egg in the nest, to an internal template of an own egg, or to the majority of the clutch, the host must make a decision (Moskát and Hauber 2007) regarding if the egg is different *enough* in comparison to putative own (majority) eggs to reject. Natural intraclutch variation is common (Cherry et al. 2007) and can itself be variable from individual to individual (Lotem et al. 1995), so what constitutes "different enough" is not conceptually, empirically, or experimentally straightforward.

Lotem et al. (1995) examined rejection rates at a study site of the oriental reed warbler *Acrocephalus orientalis*, a host of the common cuckoo in Japan, in response to both artificial and natural parasitism. Coinciding with many other studies detailed in this chapter, they found that rejection rates were higher for experimental egg types most different in appearance to the hosts' own. Interestingly, the egg types which were rejected less often took longer to reject, perhaps indicating a more effortful, difficult cognitive decision. These eggs, compared to the more often rejected extreme eggs, can be thought of as being closer to the threshold of acceptance, or the minimum amount of difference from template required to be perceived to execute rejection behavior, while the variable rates of rejection can be attributed to this threshold being unique to each host individual. The authors also discovered two important points regarding the mid-season, younger breeders in the host colony: (1) compared to the older females, their intraclutch variation in eggshell appearance was quite high, and (2) their acceptance rate of foreign eggs was also relatively high. The authors reasoned that high intraclutch variation could result in higher collateral, own-egg rejection rates if they were prone to reject odd eggs, as any egg in a highly variable clutch may seem odd and that this accounts for a lower overall rejection rate even of foreign eggs. Importantly, they posited that experience with own eggs during a host individual's first breeding season might constitute a learning component for establishing the template for recognition and adjusting a special case of optimal threshold of acceptance (Reeve 1989) for subsequent breeding seasons.

This host age-dependent foreign egg rejection in the related great reed warblers was recently confirmed by Moskát et al. (2014a) who found that young, naïve breeders are prone to accept experimental parasitic eggs both at the onset and the conclusion of egg laying, whereas older, experienced breeders are increasingly more likely to reject foreign eggs as egg laying progresses. Similar age-dependent patterns and increases in rates of egg rejection were reported for individually known aged magpie *Pica pica* hosts of the great spotted cuckoo *Clamator glandarius* in Spain (Molina-Morales et al. 2014). In turn, Stokke et al. (2007) generated a model to predict changes in a host's threshold of acceptance based on clutch characteristics. In keeping with the conclusions of Lotem et al. (1995), they assumed that experience with one's own eggs during a host individual's first breeding season is critical to the learning process and that longer learning processes (template development and refinement) can be expected in hosts with greater intraclutch variation due again to the increased risk of recognition error and collateral own egg rejection. Finally, working on the North American catbird-cowbird system, Strausberger and Rothstein (2009) hypothesized that brood parasites might hijack this imprinting process by parasitizing first-time breeders, causing the hosts to develop their recognition templates on parasitic eggs and ultimately accept such eggs in the future. Indeed, the authors demonstrated experimentally that hosts of the brown-headed cowbird *Molothrus ater* decrease rates of rejection of foreign eggs in nests where the misimprinting event on a cowbird egg was artificially induced.

It appears that discrimination thresholds are not only adjusted between breeding seasons; mounting evidence indicates that such acute plasticity may be present over various stages in the egg laying cycle as well. Moskát and Hauber (2007) measured foreign egg rejection rates of the great reed warbler within 1 day just before egg laying and during the egg laying cycle. They encountered a dramatic increase in rejection of common cuckoo eggs (28–75%) between pre-laying and after the first egg was laid (in this case, the parasite replaced the host egg). This appears to indicate that the innate template of egg appearance in the host was refined by the laying female's own experience with her egg once it was laid, causing such a dramatic leap in the threshold of acceptance in mere hours. Additionally, rejection rates dropped back down as the laying cycle proceeded and the number (and, presumably, the variability) of host eggs in the nest following parasitism increased, showcasing several major shifts in the acceptance threshold of host individuals in this population over a single egg laying bout. Wang et al. (2015) found similar rejection rate variation in the yellow-bellied prinia *Prinia flaviventris* across several stages of a single egg laying cycle, with further variation based on extent of mimicry.

A host's threshold of foreign egg rejection is not only modulated by experience with its own eggs; it can also demonstrate plasticity as a result of experience rejecting a foreign egg. Honza et al. (2007) parasitized Eurasian blackcap *Sylvia atricapilla* nests by repeating experimental parasitism events on the same individual to gauge how rejection/acceptance response compared to a second attempt. They found not only that acceptors/rejecters were consistent in their decision between trials but that those individuals which rejected during the first trial were *quicker* (showed a shorter latency) to reject during the second trial, indicating acute learning. Hauber et al. (2006) demonstrated perhaps the clearest example of experience-dependent threshold of acceptance modulation for a rejecter species. They demonstrate that the great reed warbler will typically accept a near-mimetic egg with slightly altered maculation and will reject a more drastically altered egg which has had its background color completely masked. This should come as no surprise by now: hosts are more likely to reject a foreign or experimentally manipulated own egg as the less mimetic it becomes (Hauber et al. 2015). However, what the authors discovered was that individuals which had previously rejected the more drastically altered egg would also, in the same egg laying cycle, typically reject the near-mimetic egg which was accepted by the naïve individuals. The authors account for this acute modulation of the individual's entire decision-making heuristic by pointing out the original, extremely manipulated egg acted as a sign of parasitic threat in the area, likely prompting an increase in attention and a narrowing of the thresholds of the various defense mechanisms and cognitive decision rules of the affected host individuals.

Concluding Remarks and Future Directions

Avian brood parasitism is a truly remarkable model system for coevolution, and cognitive adaptations by hosts to reject foreign eggs in response to the pattern and strength of parasitism are clear predictions of such an arms race. So rarely can we find such an intimate, one to one relationship between parasite and host such that responsive and compensatory traits can be so readily attributed to each other's advances. Egg rejection in particular represents a phenomenal driving force in this increasingly complex game state. It has driven the common cuckoo to the seemingly herculean feat of host-specific eggshell mimicry which in turn may have driven host species to develop more sophisticated cognitive faculties and perhaps even something resembling self-knowledge.

There should be no doubt now that the discordancy and template-based rules thought to dictate egg rejection decision-making are not necessarily mutually exclusive on the species, population, or even individual level. Clever experimental designs have disentangled them to show us they were entangled in the first place, and now variable levels of each can be measured in various species of various locations to gain a clearer picture of the ever-morphing shape of egg rejection behavior worldwide and how it is a product of contextual parasite–host dynamics. The final decision to eject an egg or not is reliant on such variables as relative cognitive ability of the host, characteristics of the clutch, experience, and relative parasitism pressure. There are host populations under low pressure where multiple parasitism is rare, and so discordancy mechanisms have a relatively powerful influence. As well, the lack of maculation on their eggshells may account for their poor recognition template, which would in turn strengthen the case for maculation as a critical feature in functional templates of other species. In contrast, populations of hosts exist under high pressure and with a small clutch size, such that the discordancy rule would be highly maladaptive, and as such template recognition is strong enough to reject foreign eggs in the absence of one's own. In these ways, the relative prevalence of egg rejection rulesets can, when measured rigorously, open new frames of reference for brood parasite dynamics on a comparative scale.

Acknowledgments We thank M. Soler for his invitation to contribute this chapter. We also thank Eivin Røskaft, Francisco Ruiz-Raya, and Manolo Soler for their comments on the manuscript. For financial support, we thank the Human Frontier Science Program and the National Science Foundation (to MEH). This study was also supported by the National Research, Development and Innovation Office (NKFIH, NN118194 to CM).

References

Bán M, Moskát C, Barta Z, Hauber ME (2013) Simultaneous viewing of own and parasitic eggs is not required for egg rejection by a cuckoo host. Behav Ecol 24:1014–1021

Bártol I, Karcza Z, Moskát C, Røskaft E, Kisbenedek T (2002) Responses of great reed warblers *Acrocephalus arundinaceus* to experimental brood parasitism: the effects of a cuckoo *Cuculus canorus* dummy and egg mimicry. J Avian Biol 33:420–425

Brooke MDL, Davies NB (1988) Egg mimicry by cuckoos *Cuculus canorus* in relation to discrimination by hosts. Nature 335:630–632

Cherry MI, Bennett ATD, Moskát C (2007) Host intra-clutch variation, cuckoo egg matching and egg rejection by great reed warblers. Naturwissenschaften 94:441–447

Davies NB (2000) Cuckoos, cowbirds and other cheats. T & AD Poyser, London

Feeney WE, Welbergen JA, Langmore NE (2014) Advances in the study of coevolution between avian brood parasites and their hosts. Annu Rev Ecol Evol Syst 45:227–246

Hanley D, Samas P, Hauber ME, Grim T (2015) Who moved my eggs? An experimental test of the egg arrangement hypothesis for the rejection of brood parasitic eggs. Anim Cogn 18:299–305

Hauber ME (2001) Site selection and repeatability in brown-headed cowbird (*Molothrus ater*) parasitism of eastern phoebe (*Sayornis phoebe*) nests. Can J Zool 79:1518–1523

Hauber ME, Sherman PW (2001) Self-referent phenotype matching: theoretical considerations and empirical evidence. Trends Neurosci 24:609–616

Hauber ME, Moskát C, Bán M (2006) Experimental shift in hosts' acceptance threshold of inaccurate-mimic brood parasite eggs. Biol Lett 2:177–180

Hauber ME, Samaš P, Anderson MG, Rutila J, Low J, Cassey P, Grim T (2014) Life-history theory predicts host behavioural responses to experimental brood parasitism. Ethol Ecol Evol 26:349–364

Hauber ME, Tong L, Bán M, Croston R, Grim T, Waterhouse GIN, Shawkey MD, Barron AD, Moskát C (2015) The value of artificial stimuli in behavioral research: making the case for egg rejection studies in avian brood parasitism. Ethology 121:521–528

Honza M, Požgayová M, Procházka P, Tkadlec E (2007) Consistency in egg rejection behaviour: responses to repeated brood parasitism in the blackcap (*Sylvia atricapilla*). Ethology 113:344–351

Krüger O (2011) Brood parasitism selects for no defence in a cuckoo host. Proc Biol Sci 278:2777–2783

Lahti DC, Lahti AR (2002) How precise is egg discrimination in weaverbirds? Anim Behav 63:1135–1142

Lang AK, Bollinger EK, Peer BD (2014) Effect of parasite-to-host egg ratio on egg rejection by a brown-headed cowbird host. Auk 131:694–701

Lotem A, Nakamura H, Zahavi A (1995) Constraints on egg discrimination and cuckoo-host co-evolution. Anim Behav 49:1185–1209

Lyon B (2007) Mechanism of egg recognition in defenses against conspecific brood parasitism: American coots (*Fulica americana*) know their own eggs. Behav Ecol Sociobiol 61:455–463

Marchetti K (2000) Egg rejection in a passerine bird: size does matter. Anim Behav 59:877–883

McLean IG (1987) Response to a dangerous enemy: should a brood parasite be mobbed? Ethology 75:235–245

Moksnes A (1992) Egg recognition in chaffinches and bramblings. Anim Behav 44:993–995

Molina-Morales M, Martínez JG, Martín-Gálvez D, Dawson D, Burke T, Aviles JM (2014) Cuckoo hosts shift from accepting to rejecting parasitic eggs across their lifetime. Evolution 68:3020–3029

Moskát C, Hauber ME (2007) Conflict between egg recognition and egg rejection decisions in common cuckoo (*Cuculus canorus*) hosts. Anim Cogn 10:377–386

Moskát C, Hauber ME, Avilés JM, Bán M, Hargitai R, Honza M (2009) Increased host tolerance of multiple cuckoo eggs leads to higher fledging success of the brood parasite. Anim Behav 77:1281–1290

Moskát C, Bán M, Székely T, Komdeur J, Lucassen RWG, van Boheemen LA, Hauber ME (2010) Discordancy or template-based recognition? Dissecting the cognitive basis of the rejection of foreign eggs in hosts of avian brood parasites. J Exp Biol 213:1976–1983

Moskát C, Bán M, Hauber ME (2014a) Naïve hosts of avian brood parasites accept foreign eggs, whereas older hosts fine-tune foreign egg discrimination during laying. Front Zool 11:45

Moskát C, Hauber ME, Elek Z, Gommers M, Bán M, Groenewoud F, Versluijs TSL, Hoetz CWA, Komdeur J (2014b) Foreign egg retention by avian hosts in repeated brood parasitism: why do rejecters accept? Behav Ecol Sociobiol 68:403–413

Moskát C, Zölei A, Bán M, Elek Z, Tong L, Geltsch N, Hauber ME (2014c) How to spot a stranger's egg? A mimicry-specific discordancy effect in the recognition of parasitic eggs. Ethology 120:616–626

Peer BD, Sealy SG (2001) Mechanism of egg recognition in the great-tailed grackle (*Quiscalus mexicanus*). Bird Behav 14:71–73

Polačiková L, Takasu F, Stokke BG, Moksnes A, Røskaft E, Cassey P, Hauber ME, Grim T (2013) Egg arrangement in avian clutches covaries with the rejection of foreign eggs. Anim Cogn 16:819–828

Reeve HK (1989) The evolution of conspecific acceptance thresholds. Am Nat 133:407–435

Rothstein SI (1974) Mechanisms of avian egg recognition: possible learned and innate factors. Auk 91:796–807

Rothstein SI (1975) Mechanisms of avian egg-recognition: do birds know their own eggs? Anim Behav 23:268–278

Rothstein SI (1990) A model system for coevolution: avian brood parasitism. Annu Rev Ecol Syst 21:481–508

Sealy SG, Bazin RC (1995) Low frequency of observed cowbird parasitism on eastern kingbirds: host rejection, effective nest defense, or parasite avoidance? Behav Ecol 6:140–145

Servedio MR, Hauber ME (2006) To eject or to abandon? Life history traits of hosts and parasites interact to influence the fitness payoffs of alternative anti-parasite strategies. J Evol Biol 19:1585–1594

Stevens M, Troscianko J, Spottiswoode CN (2013) Repeated targeting of the same hosts by a brood parasite compromises host egg rejection. Nat Commun 4:2475

Stoddard MC, Kilner RM, Town C (2014) Pattern recognition algorithm reveals how birds evolve individual egg pattern signatures. Nat Commun 5:4117

Stokke BG, Takasu F, Moksnes A, Røskaft E (2007) The importance of clutch characteristics and learning for antiparasite adaptations in hosts of avian brood parasites. Evolution 61:2212–2228

Strausberger BM, Rothstein SI (2009) Parasitic cowbirds may defeat host defense by causing rejecters to misimprint on cowbird eggs. Behav Ecol 20:691–699

Victoria JK (1972) Clutch characteristics and egg discriminative ability of the African village weaverbird *Ploceus cucullatus*. Ibis 114:367–376

Wang L, Yang C, Møller AP, Liang W, Lu X (2015) Multiple mechanisms of egg recognition in a cuckoo host. Behav Ecol Sociobiol 69:1761–1767

Welbergen JA, Davies NB (2009) Strategic variation in mobbing as a front line of defense against brood parasitism. Curr Biol 19:235–240

Yang C, Cai Y, Liang W (2013) Eggs mimicry of common cuckoo (*Cuculus canorus*) utilizing ashy-throated parrotbill (*Paradoxornis alphonsianus*) host. Chin Birds 4:51–56

Yang C, Møller AP, Røskaft E, Moksnes A, Liang W, Stokke BG (2014) Reject the odd egg: egg recognition mechanisms in parrotbills. Behav Ecol 25:1320–1324

Phenotypic Plasticity in Egg Rejection: Evidence and Evolutionary Consequences 25

Francisco Ruiz-Raya and Manuel Soler

Abstract

Rejection of parasitic eggs is the most common and effective defence used by hosts to mitigate the fitness costs imposed by avian brood parasites. Although egg rejection importantly relies on the cognitive abilities of parasitized individuals, both theoretical models and experimental studies have found that some hosts are able to modify their response according to the current conditions of parasitism, which reflects the existence of phenotypic plasticity in host defences. In environments in which the risk of parasitism is variable, plastic responses can be favoured by natural selection as they will allow hosts to avoid potential rejection costs under low risk of parasitism. In this chapter, we review the current evidence of plastic responses in egg rejection and discuss both the evolution and the long-term consequences of phenotypic plasticity for brood parasite–host coevolution. In addition, we suggest addressing the study of egg rejection as a complex process affected by multiple components and governed by decision-making and host motivation, which has important implications for host responses. Despite its apparent benefits, phenotypic plasticity is scarce among host species. Thus, the evolution of phenotypic plasticity in brood parasite–host systems deserves special attention as the maintenance or the loss of plastic responses involves important evolutionary consequences, affecting the long-term outcome of the interaction between brood parasites and their hosts. We conclude this chapter with some suggestions to deal with phenotypic plasticity in future egg-rejection studies.

F. Ruiz-Raya (✉)
Departamento de Zoología, Facultad de Ciencias, Universidad de Granada, Granada, Spain
e-mail: fraruiz@correo.ugr.es

M. Soler
Departamento de Zoología, Facultad de Ciencias, Universidad de Granada, Granada, Spain

Grupo Coevolución, Unidad Asociada al CSIC, Universidad de Granada, Granada, Spain

M. Soler (ed.), *Avian Brood Parasitism*, Fascinating Life Sciences,
https://doi.org/10.1007/978-3-319-73138-4_25

25.1 Introduction

Antagonistic interactions between avian brood parasites and their hosts usually result in a coevolutionary arms race in which adaptations and counteradaptations evolve on both sides (Davies 2000). The high fitness costs imposed by brood parasitism select for host defences at different stages of the breeding cycle: hosts attack brood parasites that approach their nests, reject the parasitic egg, remove the young parasitic chick or refuse to feed parasitic nestlings or fledglings (Soler 2014). However, recognition and rejection of parasitic eggs is the most common and effective strategy used by hosts against brood parasitism (Rothstein 1990; Davies 2000), which has led some brood parasites to evolve sophisticated egg mimicry and host species to fine-tune their abilities to recognize parasitic eggs (Brooke and Davies 1988).

Both theoretical models (Rothstein 1990; Takasu 1998; Robert et al. 1999; Servedio and Lande 2003) and empirical evidence (Martín-Gálvez et al. 2006, 2007) have pointed to the importance of the genetic component of hosts' egg recognition abilities. In many host species, such abilities are maintained even after generations of allopatry with brood parasites (Lahti 2006; Hale and Briskie 2007; Peer et al. 2011; Soler 2014; Ruiz-Raya et al. 2016), while, in other cases, evolutionary change has led to a decline in the egg-rejection behaviour of some hosts after long period of isolation from parasites (Kuehn et al. 2014). There are occasions, however, when the decline of parasitism rate is accompanied by a decrease in rejection rates too rapid to reflect genetic change in populations (Soler et al. 2012a; Thorogood and Davies 2013). These findings suggest that egg rejection does not rely exclusively on genetically determined cognitive abilities, but hosts are able to modify their rejection decisions under different environmental contexts, such as the risk of parasitism (Stokke et al. 2005; Moskát and Hauber 2007; Soler et al. 2012b). Given the costs linked to the expression of antiparasitic behaviours, natural selection should favour the emergence of flexible host defences in circumstances in which the risk of parasitism is low. For example, highly mimetic parasitic eggs can lead hosts to eject their own eggs by mistake (recognition costs) or to break some eggs of their own during the ejection process (ejection costs) (Davies 2000). Furthermore, cuckoo-hawk mimicry (Davies and Welbergen 2008; Welbergen and Davies 2011) makes it dangerous to approach an intruder that might be a potentially lethal enemy. Thus, when the probability of being parasitized is low, phenotypic plasticity in host defences would allow individuals to avoid such costs.

Phenotypic plasticity can be defined as the property of a single genotype to produce different phenotypes in response to distinct environmental conditions (Pigliucci 2001), which has been evidenced for a large array of traits including morphological, physiological, life history and behavioural traits (West-Eberhard 2003; Pigliucci 2005; Nussey et al. 2007). This relationship between phenotype and environment is usually represented by "reaction norms" (Fig. 25.1), whose slope reflects the degree of plasticity of genotypes (Schlichting and Pigliucci 1998; Pigliucci 2005). Furthermore, in cases in which there is variation within the population for the slope of the reaction norm (non-parallel reaction norms), it can be said

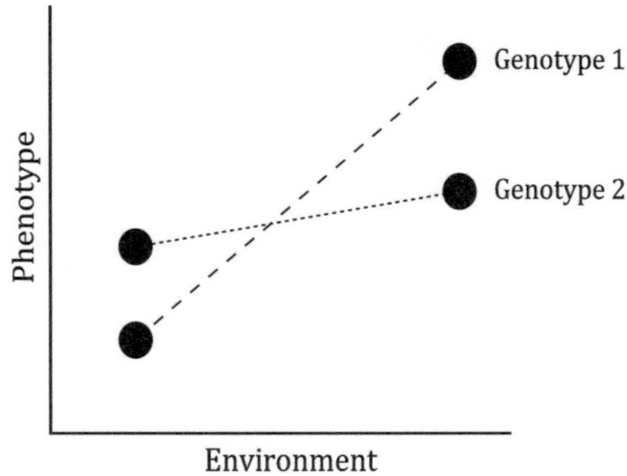

Fig. 25.1 Genotypic reaction norms used to illustrate phenotypic plasticity. Dashed lines represent the reaction norms for two single genotypes exposed to two different environments. Different slopes show the different degree of plasticity between genotype 1 (more plastic) and genotype 2 (less plastic). Based on Fig. 1 in Pigliucci (2005)

that population shows a G × E effect (i.e. genotype × environment) (Pigliucci 2005). This approach is also used to study between-individual variations in animal behaviour through the "behavioural reaction norms", which are characterized not only by the individual slopes of reaction norms (plasticity) but also from individual differences in the elevation of reaction norms (personality) (Dingemanse et al. 2010; Dingemanse and Wolf 2013). In highly variable environments, individuals can obtain fitness benefits from their ability to respond plastically when compared with those that do not, so phenotypic plasticity can become adaptive (Ghalambor et al. 2007). Behavioural plasticity has been suggested to be particularly important in variable environments since the development and expression of animal behaviour may be immediate, reversible and especially sensitive to environmental changes (West-Eberhard 2003; Fordyce 2006; Snell-Rood 2013).

The strength of interactions between brood parasites and their hosts shows a remarkable spatiotemporal heterogeneity, and some hosts have been found able to adjust their defences accordingly (see below). Such flexible defences are important from an evolutionary perspective as they can determine the result of ecological interactions (Fordyce 2006). The existence of flexible defences in hosts has been documented regarding nest vigilance (Davies et al. 2003; Feeney and Langmore 2015), mobbing to adult parasites and nest defence (Lindholm and Thomas 2000; Welbergen and Davies 2009, 2012; Langmore et al. 2012; Thorogood and Davies 2013; Kuehn et al. 2016), egg rejection (see Table 25.1) and chick rejection (Langmore et al. 2009a). Throughout this chapter, we review the current evidences of flexible defences in hosts of obligate brood parasites by focusing on the egg-rejection behaviour and the host decision-making. In addition, we discuss both the evolutionary origin and the main consequences of phenotypic plasticity on the coevolutionary history between brood parasites and their hosts.

Table 25.1 Studies showing potential evidence of plastic host responses towards parasitic eggs

Evidence of plastic response	Brood parasite	Host species	Reference
Responses to spatial heterogeneity	*Cuculus canorus*	*Acrocephalus scirpaceus*	Lindholm and Thomas (2000)
	Cuculus canorus	*Acrocephalus scirpaceus*	Lindholm (2000)
	Cuculus canorus	*Acrocephalus scirpaceus*	Stokke et al. (2008)
	Molothrus bonariensis	*Ploceus cucullatus*	Cruz et al. (2008)
Responses to temporal heterogeneity	*Clamator glandarius*	*Pica pica*	Soler et al. (1994)
	Clamator glandarius	*Pica pica*	Soler et al. (1998)
	Cuculus canorus	*Acrocephalus scirpaceus*	Brooke et al. (1998)
	Cuculus canorus	*Cyanopica cyana*	Nakamura et al. (1998)
	Molothrus bonariensis	*Ploceus cucullatus*	Robert and Sorci (1999)
	Cuculus canorus	*Cercotrichas galactotes*	Soler et al. (2012a)
	Cuculus canorus	*Acrocephalus scirpaceus*	Thorogood and Davies (2013)
Seasonal changes in egg rejection	*Molothrus ater*	*Dendroica petechia*	Burgham and Picman (1989)
	Cuculus canorus	*Cercotrichas galactotes*	Álvarez (1996)
	Cuculus canorus	*Acrocephalus scirpaceus*	Brooke et al. (1998)
	Cuculus canorus	*Acrocephalus scirpaceus*	Lindholm (2000)
	Cuculus canorus	*Cercotrichas galactotes*	Soler et al. (2012a)
	Cuculus canorus	*Acrocephalus scirpaceus*	Thorogood and Davies (2013)
	Molothrus ater	*Turdus migratorius*	Lang et al. (2014)
Response to the parasite presence	*Cuculus canorus*	*Acrocephalus scirpaceus*	Davies and Brooke (1988)
	Cuculus canorus	*Anthus pratensis*	Moksnes and Røskaft (1989)
	Cuculus canorus	*Anthus pratensis*	Moksnes et al. (1993)
	Cuculus canorus	*Acrocephalus scirpaceus*	Moksnes et al. (2000)
	Cuculus canorus	*Acrocephalus arundinaceus*	Bartol et al. (2002)

(continued)

Table 25.1 (continued)

Evidence of plastic response	Brood parasite	Host species	Reference
Lack of consistency in egg rejection	*Cuculus canorus*	*Cercotrichas galactotes*	Álvarez (1996)
	Cuculus canorus	*Cercotrichas galactotes*	Soler et al. (2000)
	Cuculus canorus[a]	*Sylvia atricapilla*	Honza et al. (2007)
	Molothrus ater	*Quiscalus quiscula*	Peer and Rothstein (2010)
	Cuculus canorus	*Acrocephalus scirpaceus*	Guigueno and Sealy (2012)
Response to parasite retaliation	*Clamator glandarius*	*Pica pica*	Soler et al. (1995)
	Clamator glandarius	*Pica pica*	Soler et al. (1999b)
Recognition without rejection	*Cuculus canorus*	*Acrocephalus scirpaceus*	Lindholm (2000)
	Cuculus canorus	*Acrocephalus arundinaceus*	Moskát and Hauber (2007)
	Cuculus canorus	*Cercotrichas galactotes*	Soler et al. (2012b)
	Cuculus canorus[a]	*Turdus merula*	Ruiz-Raya et al. (2015)
	Cuculus canorus[a]	*Turdus merula*	Soler et al. (2017)

[a]Potential host of the common cuckoo

25.2 Evidence of Phenotypic Plasticity in Egg Rejection

The existence of a conditional component in egg-rejection behaviour has been predicted by several theoretical models, which widely describe the importance of both the risk of parasitism and potential rejection costs on host responses (Davies et al. 1996; Rodríguez-Gironés and Lotem 1999; Holen and Johnstone 2006; Stokke et al. 2007). In addition to theoretical predictions, many empirical studies (see below) have suggested that phenotypic plasticity is behind the variations in rejection rates found among both host populations and individuals, especially those cases where there is a variable risk of parasitism.

Ecological interactions between brood parasites and their hosts show spatial heterogeneity resulting in the emergence of both "coevolutionary hot spots" (population with intense interactions between parasites and hosts) and "cold spots" (populations showing scarce or absent interactions) (Thompson 2005; Møller and

Soler 2012). As a result, host populations vary in the extent of local adaptations against brood parasitism, leading to differences in egg-rejection behaviour among populations (Davies and Brooke 1989; Soler and Møller 1990; Briskie et al. 1992; Soler et al. 1999a). Although such differences may reflect genotypic differences (Martín-Gálvez et al. 2007), studies on metapopulation dynamics have revealed the importance of phenotypic plasticity in explaining variations in rejection rates among host populations. For example, the presence of common cuckoo *Cuculus canorus* (cuckoo hereafter) parasitism has been proven to be the main predictor to explain the spatial variation in host resistance existing among different populations of reed warbler *Acrocephalus scirpaceus* across Europe (Stokke et al. 2008). In British populations of this species that are not genetically isolated, individuals from unparasitized populations reject eggs at lower rates and show a less aggressive response towards cuckoos than those from parasitized populations (Lindholm and Thomas 2000). Most importantly, reed warblers from both parasitized and unparasitized populations were able to recognize (individuals from all populations pecked the model eggs) and eject experimental model eggs, but they differ in their tendency to reject them, confirming that phenotypic plasticity was responsible for the differences in host responses between nearby populations (Lindholm 2000). Similarly, some hosts of the shiny cowbird *Molothrus bonariensis*, such as the village weavers *Ploceus cucullatus* on Hispaniola, show higher rejection rates in those areas where the parasites are present (Cruz et al. 2008), confirming that differences between host population can be explained by the spatial distribution of brood parasites.

In addition to geographic variations, risk of parasitism also varies across a temporal scale as the result of changes in the density of parasite populations over the time (reviewed in Møller and Soler 2012). In some cases, when parasitism pressure increases in a certain area, or brood parasites expand their geographical ranges, some hosts respond by increasing their egg-rejection rate. In Guadix, southern Spain, the increase in parasitism rate by great spotted cuckoos *Clamator glandarius* over a decade was followed by a marked increase of egg rejection by common magpies *Pica pica* (Soler et al. 1994, 1998). A similar increase in host defences was also found in one of the major hosts of the common cuckoo in Japan, the azure-winged magpie *Cyanopica cyana*, which was also related to a strong increase in cuckoo parasitism (Nakamura et al. 1998). In other cases, parasite populations experience significant declines over a few years as a result of various ecological factors, and, in such circumstances, host defences can show a rapid decrease within host populations as parasitism rate decreases. The most extreme response to such decline in parasitism was that found by Soler et al. (2012a) in Los Palacios, southern Spain, where rufous-tailed scrub robins *Cercotrichas galactotes* drastically reduced their egg-rejection rate from about 64.7% to 0% within 10 years following the cuckoos' disappearance from the area. These results are similar to those previously found on Wicken Fen, England, where a decline in cuckoo parasitism from 16% to 2–6% over a 12-year period resulted in a significant decline in egg-rejection behaviour by reed warblers from 75% to 25% during the same period of time (Brooke et al. 1998). Interestingly, this trend has been maintained in this English population during the subsequent 15 years, and it may reflect a phenomenon common to other reed warbler populations across Europe (Thorogood and Davies 2013).

Although changes in host defences as those described above could also be due to rapid evolutionary change, the fact that hosts are able to modify their rejection behaviour in response to changes in parasitism pressure within the same breeding season supports the idea that these variations are due to phenotypic plasticity. For instance, rufous-tailed scrub robins show a drastic decline in rejection rate when cuckoos leave the breeding area to go to their winter quarters (Álvarez 1996; Soler et al. 2012a). Other host species, such as reed warblers, have been also found to weaken their rejection response as the breeding season progresses (Brooke et al. 1998; Lindholm 2000; Thorogood and Davies 2013); although the opposite effect, an increase in rejection rates later in the breeding season, has also been described in some host species of the brown-headed cowbird *Molothrus ater* (Lang et al. 2014), which could also be explained by phenotypic plasticity. Of course, this seasonal decline in rejection rates could be the consequence of factors linked to life history traits instead of changes in parasitism pressure, such as inexperienced individuals breeding later (Lotem et al. 1992) or lower chances of renesting at the end of the breeding season (Burgham and Picman 1989). However, rufous-tailed scrub robins did not show differences in their rejection rates between young and experienced individuals (Soler et al. 2000), which suggests that phenotypic plasticity is behind the within-season changes in the egg-rejection behaviour found in some host species.

Flexible responses to brood parasitism are also observable within populations as individuals may respond differently to parasitism and show variations in their individual level of defence, which could be considered valuable evidence of phenotypic plasticity. Several studies have shown that the presentation of a female cuckoo model increases the likelihood of egg rejection in several host species, such as reed warblers (Davies and Brooke 1988), meadow pipits *Anthus pratensis* (Moksnes and Røskaft 1989; Moksnes et al. 1993) and great reed warblers *Acrocephalus arundinaceus* (Bartol et al. 2002) (but see Lindholm 2000; Soler et al. 2012a), meaning that individuals exposed to greater risk of parasitism are more likely to reject parasitic eggs, at least in the mentioned species. In fact, naturally parasitized reed warblers reject the parasitic egg more frequently when they discover the cuckoo near the nest (Moksnes et al. 2000). In other cases, individuals show lack of consistency in their rejection behaviour, and they vary their response to experimental parasitism as a consequence of the balance between cost and benefits of egg rejection (Soler et al. 2000). In response to repeated parasitism, individual yellow warblers *Setophaga petechia* may either accept or reject the parasitic egg after recognizing it (Guigueno and Sealy 2012). This individual lack of consistency in subsequent parasitism events has been documented even in rarely parasitized species, such as the common grackle *Quiscalus quiscula* (Peer and Rothstein 2010).

One of the most striking examples of conditional host behaviours is that exhibited in response to the mafia tactics used by some parasites, which have been particularly studied in the great spotted cuckoo-magpie system. After parasitism, great spotted cuckoos systematically punish rejecter magpies by predating those nests where the parasitic egg has disappeared, drastically reducing the reproductive success of hosts and, therefore, the advantage of egg rejection (Soler et al. 1995). The benefit obtained by the great spotted cuckoo is a greater probability of acceptance of the parasite egg since punished magpies change their behaviour from rejection to acceptance in subsequent parasitism events (Soler et al. 1999b). Thus, magpie response to the great spotted cuckoos' mafia tactics reveals the ability of hosts to adjust their response based on previous experience. These changes in the host response are more frequent in areas with a high density of brood parasites in which hosts are likely to be parasitized by several cuckoos, which means that magpies can plastically adjust their response according to the risk of suffering a second retaliation (Soler et al. 1999b; Chap. 15).

Host defences are often costly since rejection behaviour may involve the desertion of the complete clutch, recognition errors during egg ejection, accidental breaking of some own eggs or even retaliation by the parasite. As mentioned earlier, flexible defences can be favoured by natural selection in order to avoid such costs under certain ecological conditions, such as a low risk of parasitism. The use of video cameras in recognition experiments has made possible the detailed study of host responses to parasitic eggs, and it has shown that hosts can recognize more eggs than they finally reject, thus confirming the complex nature of the egg-rejection process. Antonov et al. (2009) found that eastern olivaceous warblers *Hippolais pallida* pecked the experimental egg very often, but they did not always eject it because of physical restrictions that made ejection harder, such as the impossibility shown by olivaceous warblers to puncture strong egg shells. But acceptance of previously recognized eggs is not always the outcome of ejection failure, as was revealed by the fact that some host species such as rufous-tailed scrub robins, which can easily eject parasitic eggs, frequently peck the experimental egg but they finally accept it (Soler et al. 2012b). According to the authors, acceptance may occur if hosts do not have the necessary motivation to assume the potential costs associated with egg ejection, perhaps due to a low risk of parasitism in the area. Thus, an increase in potential costs of egg ejection would lead to higher acceptance rates, even if hosts have previously recognized the parasitic egg. Recent studies have experimentally demonstrated that hosts can decide to accept foreign eggs due to physical constraints imposed by certain characteristics of the parasitic egg that do not affect recognition but hinder egg ejection (Ruiz-Raya et al. 2015; Soler et al. 2017).

25.3 Why Have Flexible Defences Evolved in Hosts?

Evidence of flexible behaviours described in the previous section reveals that hosts can adjust their response to the current conditions of brood parasitism. But what ecological conditions make plastic responses adaptive for hosts? Why have plastic responses been described in just a few host species? Behaviour, like any phenotypic trait, is either favoured or penalized by natural selection as a result of a fitness trade-off between associated costs and benefits under certain ecological contexts. However, ecological conditions are often highly variable, and no behavioural trait can be considered consistently optimal, so behavioural plasticity will allow individuals to track rapid environmental changes and respond appropriately. According to Mery and Burns (2010), the evolution of behavioural plasticity requires four conditions: (1) environmental heterogeneity, (2) reliable cues, (3) that benefits of plasticity outweigh the costs, and (4) the existence of genetic basis to plasticity. Throughout this section, we will discuss the importance of these conditions for the evolution of plastic responses by hosts.

25.3.1 Environmental Heterogeneity

Environmental changes force both individuals and populations to rapidly respond and adapt to the current ecological context. This response may occur through genetic changes within populations resulting from microevolutionary processes guided by natural selection. However, environmental heterogeneity sometimes involves variation on such a fine spatiotemporal scale that individuals respond too rapidly to represent genetic changes within populations. Brood parasite populations are usually variable in space and time, so hosts are likely to experience wide fluctuations in parasitism rates. This variation would make phenotypic plasticity adaptive for hosts and would explain the presence of plastic responses in some host populations. For example, populations of common cuckoo parasitizing reed warblers are usually restricted to small patches in wetland, which makes them prone to local extinction; as a consequence, parasitism rates frequently vary between reed warbler populations and years (Lindholm 1999). On the other hand, after dispersion, young reed warblers probably occupy territories where the rate of parasitism will be different from that suffered by their parents. But variation in risk of parasitism may also result from annual cuckoo movements among host populations, which might be an adaptive behaviour allowing cuckoos to increase the probability of finding naïve hosts with less effective defences (Lotem et al. 1995; Langmore et al. 2009a, 2012). Recently, the existence of spatiotemporal variation in host use due to habitats constraints has been revealed in the great spotted cuckoo, indicating that parasitism pressure differs among environments for host species (Baglione et al. 2017). Given this temporal and spatial heterogeneity in parasitism rates, hosts would benefit from the ability to assess changes in the local risk of parasitism and adjust their defences accordingly, leading to the emergence of behavioural plasticity in some host populations.

25.3.2 Reliable Cues of Parasitism

The ability to rapidly respond to variations in parasitism conditions implies that hosts effectively assess the risk of parasitism in their area. Thus, the evolution of plastic defences requires the use of reliable cues that make possible an accurate monitoring of changes in the risk of parasitism. But how do hosts assess the risk of parasitism to which they are subjected? Hosts might assess the risk of parasitism from direct encounters with parasites. Several studies have found that some hosts increased their rejection rate after observing the parasite near the nest (Davies and Brooke 1988;

Moksnes et al. 2000; Bartol et al. 2002); however, this response has been documented in a few species, while, in other cases, the presence of the parasite does not seem to be enough to modify the host behaviour (Lindholm 2000; Soler et al. 2012a). Furthermore, some brood parasites such as the cuckoo show secretive habits that make encounters with parasites unlikely (Davies 2000), as well as cuckoo-hawk mimicry makes any inspection approach potentially lethal to hosts (Davies and Welbergen 2008; Welbergen and Davies 2011). Another possibility is that hosts are able to assess the activity of parasites in their own territories and adjust their behaviour based on such information. Brooke et al. (1998) evaluated the effect of proximity to naturally parasitized nests, where parasite activity will be easier to detect, on the rejection rate of reed warblers. Although they did not find greater rejection rates in nests located near parasitized nests, reed warblers from a small unparasitized population just 11 km away did not show rejection behaviour, which suggests that hosts were able to track the cuckoo activity. Given the scarce and potentially risky nature of direct encounters with parasites, it has been suggested that some hosts may use both direct and indirect cues to assess specific risk of parasitism. Welbergen and Davies (2012) found that nest defence by reed warblers declined with distance to potential cuckoo perches (indirect cue), places from where cuckoos usually locate the host nests and whose distance to the host nest is related to the risk of parasitism (Øien et al. 1996). Interestingly, this effect was found both in parasitized and unparasitized sites, suggesting that direct cues of parasitism, such as the cuckoo presence, can be modulated by the use of indirect cues (Welbergen and Davies 2012).

The existence of reliable cues of parasitism is therefore a necessary condition for the evolution of plastic defences, which could explain the lower degree of plastic responses shown by hosts of other brood parasites. As was suggested by Thorogood and Davies (2013), the degree of specialization between the common cuckoo and its hosts is so high that parasite activity may be a more reliable cue for cuckoo's hosts than for hosts of more generalist parasites. For instance, parasite activity alone might not be a reliable indicator of the risk of parasitism for any particular host species of the brown-headed cowbird given the more generalist nature of this brood parasite, which could difficult the evolution of plastic responses in these host species. It is clear (Table 25.1) that documented cases of plasticity in brown-headed cowbird hosts are much scarcer (only three species) than in common cuckoo hosts (seven species). This should not be surprising considering that most cowbird hosts are acceptors and phenotypic plasticity would be adaptive later, once egg-rejection ability has evolved and, mainly, when the risk of parasitism is highly variable (Soler 2014).

25.3.3 Benefits and Costs of Plastic Responses

Evolution of behavioural plasticity in host responses is necessarily subject to a trade-off between costs and benefits of plasticity, which explains the current variation in the occurrence of flexible defences among species, populations or individuals. As suggested above, benefits of phenotypic plasticity become noticeable in changing environments, so environmental variation is considered the main selective force for phenotypic plasticity (Moran 1992; Ghalambor et al. 2007; Hendry 2016). Regarding behaviour, we can differentiate two forms of plasticity. On the one hand, activational behavioural plasticity, or "innate" plasticity, allows individual to express a particular behaviour in response to the external context, so that, different environmental conditions will result in the expression of different behaviours (Snell-Rood 2013). Developmental behavioural plasticity, on the other hand, refers to the expression of different behavioural phenotypes in different contexts as a consequence of different developmental trajectories, which includes modifications of behaviour as a result of experience (learning) (Mery and Burns 2010; Snell-Rood 2013). The characteristics of the environmental variation will determine the relative benefits of each of these two types of behavioural plasticity. When individuals have to cope with variations within their lifetime, reversible activational plasticity will be favoured. In contrast, in those situations where environmental variations occur between generations or exceed the dispersal capacity of individuals but remain relatively constant within generations, learning costs would be minimized and irreversible plasticity would be favoured (Mery and Burns 2010; Snell-Rood 2013). Some host species may experience significant variations in the rate of parasitism throughout their lifetime, so evolution could favour the appearance of flexible responses (reversible) in such species. On the other hand, when the parasitism pressure remains stable over time, the level of defence could be maintained in the population according to such invariable risk of parasitism (Zölei et al. 2015).

As we mentioned, behavioural plasticity has been proved to be advantageous for hosts when facing environments with variable risk of parasitism. So why do most host species lack this capacity? Given that no trait is infinitely or ideally plastic, the most likely response is that there are restrictions both for the evolution of plasticity and for its maintenance. Studies on the evolution of phenotypic plasticity have suggested the existence of two main types of constraints on the evolution and maintenance of plasticity: *costs*, which lead to lower fitness when a feature is produced through plasticity rather than constitutively, and *limits*, referring to the impossibility of reaching the optimal trait value (DeWitt et al. 1998; Murren et al. 2015). Specifically, costs linked to activational behavioural plasticity may be related to the maintenance of the sensory and regulatory mechanisms to detect environmental changes and a further development of possible motor responses (Snell-Rood 2013). Unlike activational behavioural plasticity, which is usually an immediate response to environmental variations, selection of developmental behavioural plasticity, in particular learning, is usually linked to a trial-and-error process. This involves a period of suboptimal behaviour that is usually costly since it requires a considerable investment in time and energy, as well as making errors, what is known as "costs of being naïve" (Mery and Burns 2010; Snell-Rood 2013). In addition, learning has been also shown to involve important constitutive costs since it requires a remarkable neurobiological and morphological reorganization (Mery and Burns 2010). Therefore, if environmental conditions fail to favour phenotypic plasticity, the costs of plasticity per se could lead to the loss of plastic responses.

Individual experience can enhance the host ability to respond against brood parasitism and results in more effective defences. In some cases, experience is the result of the interaction with other individuals, which can also modify the behaviour and lead to plastic responses. It has been suggested that social transmission of defences (social learning) represents an additional factor in the rapid acquisition of defences by some hosts. For instance, social learning is especially important in the transmission of mobbing behaviour towards cuckoos in reed warblers, which acquire information by observing conspecific from adjacent territories (Davies and Welbergen 2009; Campobello and Sealy 2011). These cues will allow hosts to track fine-scale variations in parasitism risk and respond accordingly in future encounters with parasites. In the case of egg rejection, a similar mechanism of social transmission is unlikely to evolve due to the difficulty for individuals to witness an egg-rejection event on the part of their conspecifics. However, other mechanisms of social transmission have been suggested to explain the extraordinarily rapid increase in rejection rates observed in some host species. Such mechanisms might be based on communicative systems, and they would be especially important in species with remarkable cognitive abilities such as the common magpie or the azure-winged magpie (Soler 2011); however, further studies are needed to prove the existence of such mechanisms. Unlike individual learning, social transmission of defences would allow hosts to track environmental changes in risk of parasitism while avoiding the potential costs associated with trial-and-error processes (Mery and Burns 2010).

25.3.4 Genetic Basis of Plasticity

Since phenotypic plasticity is a property of genotypes and it is subject to evolution, an underlying genetic basis would be expected in those host species showing plastic responses. From a general point of view, three models have been proposed to address the genetic basis of plasticity (Scheiner 1993): (1) *overdominance*, which states that plasticity is an inverse function of heterozygosity; (2) *pleiotropy*, which states that plasticity comes from the differential expression of genes in different environments; and (3) *epistasis*, which predicts that plasticity results from the interaction between genes that determine the magnitude of the response to environmental factors with genes that determine the mean expression of a trait. Empirical studies reveal that, while there is little evidence for the overdominance model, both pleiotropic and epistatic effects have been proved important in any plastic response, just as is probably the case with most of the phenotypic complex traits (Scheiner 1993; Pigliucci 2005). Therefore, current evidence suggests that specific properties of genotypes could favour the evolution of plastic responses also in hosts of avian brood parasites.

25.4 The Egg-Rejection Process: Decision-Making and the Role of Motivation

The lack of egg rejection exhibited by some host populations has been considered one of the most puzzling issues in the study of brood parasitism (Stokke et al. 2005), especially considering the high fitness costs linked to rearing the chick of many brood parasites. Over the last few decades, several hypotheses have been proposed to explain the acceptance of parasitic eggs within the framework of the coevolutionary arms race between brood parasites and hosts. Sometimes, lack of rejection might reflect an initial stage in the arms race between brood parasites and hosts resulting from a recent parasitism, so hosts would not have had enough time to evolve egg-rejection defences [evolutionary-lag hypothesis (Rothstein 1990; Davies 2000)]. Another possibility is that acceptance is adaptive under certain circumstances, which would occur whenever costs of maintaining egg rejection are higher than costs of parasitism [evolutionary-equilibrium hypothesis (Lotem et al. 1992, 1995; Lotem and Nakamura 1998)]. According to the metapopulation dynamics theory, the presence of acceptors within specific populations might also be due to gene flow of "acceptor alleles" from non-parasitized populations (Martínez et al. 1999; Soler et al. 1999a), as a possible consequence of the spatial structure of habitats (Røskaft et al. 2002, 2006). In other cases, the evolution of cryptic eggs by the parasite has prevented hosts to evolve recognition abilities, as occurs in some Australian cuckoos (Brooker et al. 1990; Langmore et al. 2009b). In addition, as some traits of parasite eggs such as the eggshell thickness can make ejection difficult, some species could accept due to the impossibility to puncture the parasitic egg, which would imply *rejection failures* instead of *recognition failures* (Antonov et al. 2009). On the other hand, egg acceptance could also be due to successful frontline defences blocking the evolution of egg-rejection behaviours (Britton et al. 2007). But there is one last possibility to be considered: in some cases, the absence of egg rejection might reflect *acceptance decisions*, which means that some hosts choose to accept the parasitic egg even after recognition (Soler et al. 2012b, 2017; Ruiz-Raya et al. 2015). Under this framework, host decision-making would play a central role in the egg-rejection process, being affected by the interaction of multiple elements such as clutch characteristics, recognition abilities of host, the presence of conditional stimuli and the host genotype (Stokke et al. 2005). The study of cognitive phenotypes in a "judgement and decision-making" framework has been suggested as an important point to gain a better understanding of the processes guiding animal decisions in behavioural ecology studies (Mendelson et al. 2016), which also includes egg-rejection studies in brood parasite–host systems (Ruiz-Raya and Soler 2018). Egg rejection can therefore be understood as a complex and potentially plastic process in which different stages are differentiated, namely, judgement, decision and action itself.

25.4.1 Judgement

Judgements allow hosts to arrive at an understanding of the environment, which is necessary to carry out a specific response towards the parasitic egg. Thus, egg rejection requires that hosts successfully assess information and realize that their nests have been parasitized; that is, recognition of the parasitic egg must occur. As suggested by Ruiz-Raya and Soler (2018), egg-rejection studies require the use of a unified and consistent terminology in which terms such as egg recognition deserve to be properly differentiated from other cognitive abilities affecting judgement, such as discrimination, categorization or evaluation. According to the authors, *egg discrimination* refers to the cognitive process by which hosts can distinguish two or more different stimuli from the parasitized clutch, and it therefore relies on the intensity of the stimulus and lead to the signal detection needed to initiate the decision-making process (Rodríguez-Gironés and Lotem 1999). Otherwise, *egg recognition* can be considered as the process leading to the host response, which implies that host has been able to identify the odd egg as a parasitic egg. Certain characteristics of the cognitive stimulus, such as the mimicry of the parasitic egg or the degree of intraclutch variation are determining factors that affect the host judgement. Thus, highly mimetic parasitic eggs will hinder egg discrimination, whereas high intraclutch variation will increase the likelihood of committing recognition errors (Stokke et al. 2005). On the other hand, cognitive abilities of hosts can also significantly affect egg recognition. Such cognitive abilities will be determined by the host genotype, and, at the population level, egg rejection will be partly affected by the frequencies of different genotypes present in that population. In accordance with the existing parasitism pressure and rejection costs, natural selection will favour or penalize certain genotypes and, along with metapopulation processes such as gene flow or drift, will determine the frequencies of "rejecters" and "acceptors" within populations. When the cognitive stimulus (i.e. the signal) is weak and/or the host's cognitive abilities are not fine enough, the egg-rejection process can be wrecked in the recognition stage, and the outcome of the egg-rejection process will be egg acceptance.

25.4.2 Decision

Once the parasitic egg is recognized, hosts must choose among two different options before carrying out the action itself: they have to decide between acceptance and rejection of the parasitic egg. But how do individuals determine which strategy is optimal? During the decision stage, some hosts are able to integrate information from the judgement with that derived from conditional stimuli in a process that will determine their "tendency to rejection" or motivation. *Motivation* is revealed as a crucial concept in the psychology of decision-making, and it can be understood as the host's tendency to assume the potential costs of egg rejection under certain parasitism conditions (i.e. trade-offs between costs of egg rejection and risk of parasitism) (Soler et al. 2012b). As previously seen, some hosts vary their response to parasitism according to the current risk of parasitism, their experience or the stage of the breeding season, which leads some of them to accept the parasitic egg. Of course, judgement and decision are deeply linked in the decision-making process since they both depend on the accuracy with which hosts evaluate the available information about parasitism. The combination of prior and new information will allow hosts to improve the evaluation of their current situation, a process known as "Bayesian updating" (Piersma and Gils 2011). As will be seen in the following section, the assessment of these potential costs is continuously updated, even in the last stage of the egg-rejection process (action itself), which can also significantly influence decision-making as it could result in new judgements and therefore new decisions.

25.4.3 Action

Action is the last stage of the egg-rejection process, and it can be defined as the specific behaviour resulting from the host decision-making. The egg-rejection process can also be interrupted in this last stage since some physical characteristics of parasitic eggs make ejection particularly hard. For example, unusual strong shells can make ejection difficult to small hosts (puncture ejectors) and can force the acceptance of previously recognized eggs (Antonov et al. 2009). During the action stage, hosts can assess the potential costs of ejection by gathering information on the physical characteristics of the parasite egg that may hinder ejection. For instance, Soler et al. (2012b) suggested that some hosts use weak touches to the parasitic egg as a mechanism to assess the shell strength and therefore the potential costs of egg puncture ejection. These behaviours contribute to the Bayesian updating process and allow hosts to add new information to the prior information on parasitism conditions, which might lead to new judgements and therefore modify the outcome of the decision-making process (Fig. 25.2). In this context, a low risk of parasitism would involve insufficient motivation to assume potential ejection costs, resulting in acceptance decisions. However, when the conditions of parasitism make the host's motivation higher, hosts could increase the strength and frequency of their pecking in order to puncture the parasitic egg, assuming the possibility of breaking one of their own eggs in the attempt. At this point, if puncture ejection is not possible, higher motivation would be required to desert the nest as it is a much more costly strategy [see Fig. 6 in Soler et al. (2012b)]. In grasp ejectors, who eject the parasitic egg by grasping it with the beak and taking it out of the nest, the action stage can be affected by other egg traits such as egg size or mass. In fact, it has been recently shown how such traits can lead to acceptance of previously recognized eggs, or a delay in rejection decisions, when host motivation is not enough (Ruiz-Raya et al. 2015; Soler et al. 2017).

25.5 Phenotypic Plasticity and Brood Parasite–Host Coevolution

The existence of plastic responses makes it possible for individuals to rapidly respond to changes in ecological conditions affecting their environment. Importantly, these ecological conditions include phenotypes of other individuals with which they interact, so phenotypic plasticity has important consequences on both

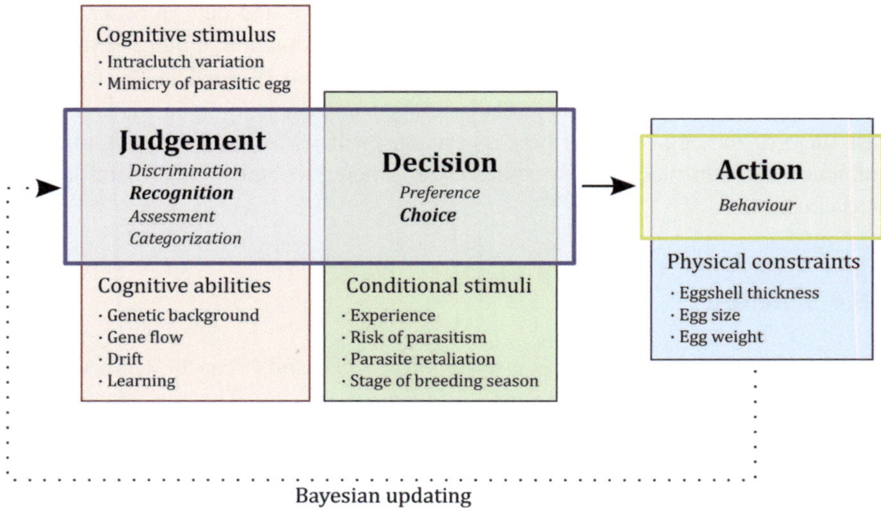

Fig. 25.2 The egg-rejection process and its three main stages: judgement, decision and action. Both judgement and decision are involved in the host decision-making (purple box), which will result in the host behaviour (action, yellow box). Each stage of the egg-rejection process can be affected independently, which will determine the outcome of the process. Host can add new information regarding parasitism conditions during any phase of the egg-rejection process, even from the action stage (i.e. after decision), which may result in an updated decision-making process leading to new decisions (Bayesian updating). In italics, important terms related to each stage of the egg-rejection process are highlighted (terms based on Ruiz-Raya and Soler 2018)

the strength of ecological interactions and the evolutionary trajectories of the species involved (Agrawal 2001; Fordyce 2006). In brood parasite–host interactions, the existence of fine-tuned plastic responses (i.e. close to the optimal phenotype) would allow some hosts to successfully respond to changes in parasitism pressure within populations. When a host is able to rapidly reach optimal behaviour after increases in parasitism rates, parasite fitness will be affected, and the potential outcome might be the host switching or the local extinction of the brood parasite. Phenotypic plasticity would allow certain hosts species, such as reed warbler, to retain their rejection abilities after periods of low parasitism pressure, even when the expression of their defences is reduced (Lindholm and Thomas 2000; Stokke et al. 2008; Thorogood and Davies 2013). This will make it difficult for a future reutilization of the host population by brood parasites since hosts will be able to rapidly respond to changes in the parasitism pressure.

However, although plastic responses have been documented in several host species, phenotypic plasticity in egg rejection seems to be the exception rather than the rule. In most host species, rejection behaviour is a fixed trait retained in the absence of brood parasitism (even after speciation events; see references above), and continuous coevolutionary cycles have been shown to be absent in some brood parasite–host systems (Soler et al. 1998; Rothstein 2001; Peer et al. 2007). In fact, recent evidences have emphasized that the absence of coevolutionary cycles is the most frequent situation in host–brood parasite systems (Soler 2014). So what role does phenotypic plasticity play in most host species? Phenotypic plasticity is crucial to tolerate and then adapt to new environmental conditions, allowing populations to move more easily to another adaptive peak (Price et al. 2003; Ghalambor et al. 2007). Therefore, plastic responses would be especially important in the early stages of brood parasitism (Soler 2014), allowing hosts to benefit from the ability to adjust their behaviour to the new selection pressure. As long as the risk of parasitism is variable and unpredictable, hosts that show plastic responses will have higher fitness than those that do not, and phenotypic plasticity will be adaptive. By contrast, if the pressure or virulence of parasitism increases and is maintained over time, plastic responses may cease to be adaptive because of the costs of phenotypic plasticity itself, which would favour the evolution of the canalized phenotype (Pigliucci et al. 2006; Soler and Soler 2017), and the rejection behaviour would be fixed. This loss of plasticity can be understood as an alteration in the reaction norm (i.e. a flat reaction norm) derived from selection operating only in the new environment (Pigliucci et al. 2006; Ghalambor et al. 2007). In the absence of parasitism, the evolution of relaxed fixed host defences as response to reduced risk of parasitism will lead to coevolutionary cycles (Nuismer and Thompson 2006). However, egg-rejection behaviour will be maintained unless recognition errors and rejection costs are high, and recent studies have shown that the occurrence of such errors is low among host species (Stokke et al. 2016). Therefore, the maintenance of egg-rejection abilities shown by many species over long periods of time indicates that costs associated to the maintenance of such antiparasitic defence are insignificant, which has important implications in the evolutionary trajectory of brood parasites and their hosts: coevolutionary cycles are replaced by successful resistance as the main outcome of brood parasite–host coevolution (Rothstein 2001; Soler 2014). In fact, when the coevolutionary trajectories of brood parasite–host systems are reviewed, it was found that first, fixed responses and absence of coevolutionary cycles are frequent in host-brood parasite systems and, second, the few species showing phenotypic plasticity in their responses are usually involved in coevolutionary cycles, probably reflecting initial stages of parasitism (Soler 2014). According to Soler (2014), coevolutionary cycles are likely an intermediate phase of the interactions between brood parasites and their hosts, whose outcome will be the extinction or the acquisition of successful resistance by the host, forcing parasites to specialize in less suitable host species.

In some cases, phenotypic plasticity results in the emergence of reciprocal phenotypic changes among players in ecological interactions, which has important evolutionary implications (Agrawal 2001). In brood parasite–host systems, mathematical models suggest that emergence of some parasites' strategies such as mafia behaviour might be promoted by the host's plastic responses (Chakra et al. 2014). When retaliation occurrence is moderate, plastic responses will be beneficial for hosts and, at the same time, brood parasites will benefit from the existence of such plastic responses to force the acceptance of the egg parasites by punishing rejecter individuals. As the mafia strategy expands in the population, hosts would benefit from unconditional acceptance of parasitic eggs, leading the parasite population back to a non-mafia strategy. Thus, the occurrence of the mafia strategy within parasite population will oscillate in time, and plastic host responses would be crucial for its evolution. The magpies' ability to express plastic defences is a determinant of the evolution of plastic virulence in parasites (e.g. retaliatory behaviours), while the existence of such plastic virulence will favour the maintenance of plastic defences in hosts (Soler and Soler 2017, Chap. 15).

Concluding Remarks and Future Directions

In this chapter, we have seen how some host species are able to modify their egg-rejection behaviour according to the perceived risk of parasitism, which reveals the existence of plastic defences against brood parasitism. Such plastic responses are favoured by natural selection under highly variable risk of parasitism since it allows hosts to avoid the expression of costly defences when the risk of parasitism is low. Future work should address the mechanisms used by hosts to accurately assess the risk of parasitism in their sites, which is crucial for the evolution of plastic defences. In addition, further studies should focus on the costs and limitations linked to the host plastic responses, which will determine the maintenance or loss of plasticity under low parasitism pressure. From a general perspective, studies on brood parasite–host coevolution should consider the potential effects of phenotypic plasticity on both the maintenance of host defences over the time and the long-term outcome of ecological interactions. Recent evidence that some potential host species may recognize more eggs than they eventually reject indicates that egg rejection can be viewed as a complex process in which multiple components interact and where decision-making and host motivation play a central role. Interestingly, the existence of phenotypic plasticity remains untested in many potential host species, so egg recognition experiments in different populations and/or under different risk of parasitism need to be carried out in most host species. These results, when available, will open a new avenue of research in which comparative studies would allow to answer crucial questions about the evolution of egg rejection. More work is also required to clarify the possible role of social learning on the defence acquisition by hosts and its importance on egg rejection. Finally, in view of the significant decline experienced by some parasites during the last years, phenotypic plasticity will be crucial to understand how host populations adapt to such changing environment.

Acknowledgements We greatly thank Naomi Langmore and Brian Peer, who provided useful comments which significantly improved this chapter.

References

Agrawal AA (2001) Phenotypic plasticity in the interactions and evolution of species. Science 294:321–326

Álvarez F (1996) Model cuckoo *Cuculus canorus* eggs accepted by rufous bush chats *Cercotrichas galactotes* during the parasite's absence from the breeding area. Ibis 138:340–342

Antonov A, Stokke BG, Moksnes A, Røskaft E (2009) Evidence for egg discrimination preceding failed rejection attempts in a small cuckoo host. Biol Lett 5:169–171

Baglione V, Bolopo D, Canestrari D, Martínez JG, Roldán M, Vila M, Soler M (2017) Spatiotemporal variation of host use in a brood parasite: the role of the environment. Behav Ecol 28:49–58

Bartol I, Karcza Z, Moskát C, Røskaft E, Kisbenedek T (2002) Responses of great reed warblers *Acrocephalus arundinaceus* to experimental brood parasitism: the effects of a cuckoo *Cuculus canorus* dummy and egg mimicry. J Avian Biol 33:420–425

Briskie JV, Sealy SG, Hobso KA (1992) Behavioral defenses against avian brood parasitism in sympatric and allopatric host populations. Evolution 46:334–340

Britton NF, Planqué R, Franks NR (2007) Evolution of defence portfolios in exploiter-victim systems. Bull Math Biol 69:957–988

Brooke MDL, Davies NB (1988) Egg mimicry by cuckoos *Cuculus canorus* in relation to discrimination by hosts. Nature 335:630–632

Brooke MDL, Davies NB, Noble DG (1998) Rapid decline of host defences in response to reduced cuckoo parasitism: behavioural flexibility of reed warblers in a changing world. Proc R Soc B 265:1277–1282

Brooker LC, Brooker MG, Brooker AMH (1990) An alternative population/genetics model for the evolution of egg mimesis and egg crypsis in cuckoos. J Theor Biol 146:123–143

Burgham MCJ, Picman J (1989) Effect of brown-headed cowbirds on the evolution of yellow warbler anti-parasite strategies. Anim Behav 38:298–308

Campobello D, Sealy SG (2011) Use of social over personal information enhances nest defense against avian brood parasitism. Behav Ecol 22:422–428

Chakra MA, Hilbe C, Traulsen A (2014) Plastic behaviors in hosts promote the emergence of retaliatory parasites. Sci Rep 4:4251. https://doi.org/10.1038/srep04251

Cruz A, Prather JW, Wiley JW, Weaver PF (2008) Egg rejection behavior in a population exposed to parasitism: village weavers on Hispaniola. Behav Ecol 19:398–403

Davies NB (2000) Cuckoos, cowbirds and other cheats. T. & A. D. Poyser, London

Davies NB, Brooke MDL (1988) Cuckoos versus reed warblers: adaptations and counteradaptations. Anim Behav 36:262–284

Davies NB, Brooke MDL (1989) An experimental study of co-evolution between the cuckoo, *Cuculus canorus*, and its hosts. I. Host egg discrimination. J Anim Ecol 58:207–224

Davies NB, Welbergen JA (2008) Cuckoo-hawk mimicry? An experimental test. Proc R Soc B 275:1817–1822

Davies NB, Welbergen JA (2009) Social transmission of a host defense against cuckoo parasitism. Science 324:1318–1320

Davies NB, Brooke MDL, Kacelnik A (1996) Recognition errors and probability of parasitism determine whether reed warblers should accept or reject mimetic cuckoo eggs. Proc R Soc B 263:925–931

Davies NB, Butchart SHM, Burke T, Chaline N, Stewart IR (2003) Reed warblers guard against cuckoos and cuckoldry. Anim Behav 65:285–295

DeWitt TJ, Sih A, Wilson DS (1998) Cost and limits of phenotypic plasticity. Trends Ecol Evol 13:77–81

Dingemanse NJ, Wolf M (2013) Between-individual differences in behavioural plasticity within populations: causes and consequences. Anim Behav 85:1031–1039

Dingemanse NJ, Kazem AJN, Réale D, Wright J (2010) Behavioural reaction norms: animal personality meets individual plasticity. Trends Ecol Evol 25:81–89

Feeney WE, Langmore NE (2015) Superb fairy-wrens (*Malurus cyaneus*) increase vigilance near their nest with the perceived risk of brood parasitism. Auk 132:359–364

Fordyce JA (2006) The evolutionary consequences of ecological interactions mediated through phenotypic plasticity. J Exp Biol 209:2377–2383

Ghalambor CK, McKay JK, Carroll SP, Reznick DN (2007) Adaptive versus non-adaptive phenotypic plasticity and the potential for contemporary adaptation in new environments. Funct Ecol 21:394–407

Guigueno MF, Sealy SG (2012) Increased investigation of manipulated clutches suggests egg recognition without rejection in a brown-headed cowbird (*Molothrus ater*) host, the yellow warbler (*Setophaga petechia*). Auk 129:17–25

Hale K, Briskie JV (2007) Response of introduced european birds in New Zealand to experimental brood parasitism. J Avian Biol 38:198–204

Hendry AP (2016) Key questions on the role of phenotypic plasticity in eco-evolutionary dynamics. J Hered 107:25–41

Holen ØH, Johnstone RA (2006) Context-dependent discrimination and the evolution of mimicry. Am Nat 167:377–389

Honza M, Požgayová M, Procházka P, Tkadlec E (2007) Consistency in egg rejection behaviour: responses to repeated brood parasitism in the blackcap (*Sylvia atricapilla*). Ethology 113:344–351

Kuehn MJ, Peer BD, Rothstein SI (2014) Variation in host response to brood parasitism reflects evolutionary differences and not phenotypic plasticity. Anim Behav 88:21–28

Kuehn MJ, Peer BD, McCleery RA, Rothstein SI (2016) Yellow warbler defenses are retained in the absence of brood parasitism but enhanced by experience with cowbirds. Behav Ecol 27:279–286

Lahti DC (2006) Persistence of egg recognition in the absence of cuckoo brood parasitism: pattern and mechanism. Evolution 60:157–168

Lang AK, Bollinger EK, Peer BD (2014) Effect of parasite-to-host egg ratio on egg rejection by a Brown-headed cowbird host. Auk 131:694–701

Langmore NE, Cockburn A, Russell AF, Kilner RM (2009a) Flexible cuckoo chick-rejection rules in the superb fairy-wren. Behav Ecol 20:978–984

Langmore NE, Stevens M, Maurer G, Kilner RM (2009b) Are dark cuckoo eggs cryptic in host nests? Anim Behav 78:461–468

Langmore NE, Feeney WE, Crowe-Riddell J, Luan H, Louwrens KM, Cockburn A (2012) Learned recognition of brood parasitic cuckoos in the superb fairy-wren *Malurus cyaneus*. Behav Ecol 23:799–805

Lindholm AK (1999) Brood parasitism by the cuckoo on patchy reed warbler populations in Britain. J Anim Ecol 68:293–309

Lindholm AK (2000) Tests of phenotypic plasticity in reed warbler defences against cuckoo parasitism. Behaviour 137:43–60

Lindholm AK, Thomas RJ (2000) Differences between populations of reed warblers in defences against brood parasitism. Behaviour 137:25–42

Lotem A, Nakamura H (1998) Evolutionary equilibria in avian brood parasitism. In: Rothstein S, Robinson S (eds) Parasitic birds and their hosts: studies in coevolution. Oxford University Press, New York, pp 223–235

Lotem A, Nakamura H, Zahavi A (1992) Rejection of cuckoo eggs in relation to host age: a possible evolutionary equilibrium. Behav Ecol 3:128–132

Lotem A, Nakamura H, Zahavi A (1995) Constraints on egg discrimination and cuckoo-host co-evolution. Anim Behav 49:1185–1209

Martínez JG, Soler JJ, Soler M, Møller AP, Burke T (1999) Comparative population structure and gene flow of a brood parasite, the great spotted cuckoo (*Clamator glandarius*), and its primary host, the magpie (*Pica pica*). Evolution 53:269–278

Martín-Gálvez D, Soler JJ, Martínez JG, Krupa AP, Richard M, Soler M, Møller AP, Burke T (2006) A quantitative trait locus for recognition of foreign eggs in the host of a brood parasite. J Evol Biol 19:543–550

Martín-Gálvez D, Soler JJ, Martínez JG, Krupa AP, Soler M, Burke T (2007) Cuckoo parasitism and productivity in different magpie subpopulations predict frequencies of the 457bp allele: a mosaic of coevolution at a small geographic scale. Evolution 61:2340–2348

Mendelson TC, Fitzpatrick CL, Hauber ME, Pence CH, Rodríguez RL, Safran RJ, Stern CA, Stevens JR (2016) Cognitive phenotypes and the evolution of animal decisions. Trends Ecol Evol 31:850–859

Mery F, Burns JG (2010) Behavioural plasticity: an interaction between evolution and experience. Evol Ecol 24:571–583

Moksnes A, Røskaft E (1989) Adaptations of meadow pipits to parasitism by the common cuckoo. Behav Ecol Sociobiol 24:25–30

Moksnes A, Røskaft E, Korsnes L (1993) Rejection of cuckoo (*Cuculus canorus*) eggs by meadow pipits (*Anthus pratensis*). Behav Ecol 4:120–127

Moksnes A, Røskaft E, Hagen LG, Honza M, Mork C, Olsen PH (2000) Common cuckoo *Cuculus canorus* and host behaviour at reed warbler *Acrocephalus scirpaceus* nests. Ibis (Lond 1859) 142:247–258

Møller AP, Soler JJ (2012) A coevolutionary framework based on temporal and spatial ecology of host-parasite interactions: a missing link in studies of brood parasitism. Chin Birds 3:259–273

Moran NA (1992) The evolutionary maintenance of alternative phenotypes. Am Nat 139:971–989

Moskát C, Hauber ME (2007) Conflict between egg recognition and egg rejection decisions in common cuckoo (*Cuculus canorus*) hosts. Anim Cogn 10:377–386

Murren CJ, Auld JR, Callahan H, Ghalambor CK, Handelsman CA, Heskel MA, Kingsolver JG, Maclean HJ, Masel J, Maughan H, Pfennig DW, Relyea RA, Seiter S, Snell-Rood E, Steiner UK, Schlichting CD (2015) Constraints on the evolution of phenotypic plasticity: limits and costs of phenotype and plasticity. Heredity 115:293–301

Nakamura H, Kubota S, Suzuki R (1998) Coevolution between the common cuckoo and its major hosts in Japan. In: Rothstein SI, Robinson SK (eds) Parasitic birds and their hosts: studies in coevolution. Oxford University Press, New York, pp 94–112

Nuismer SL, Thompson JN (2006) Coevolutionary alternation in antagonistic interactions. Evolution 60:2207–2217

Nussey DH, Wilson AJ, Brommer JE (2007) The evolutionary ecology of individual phenotypic plasticity in wild populations. J Evol Biol 20:831–844

Øien IJ, Honza M, Moksnes A, Roskaft E (1996) The risk of parasitism in relation to the distance from reed warbler nests to cuckoo perches. J Anim Ecol 65:147–153

Peer BD, Rothstein SI (2010) Phenotypic plasticity in common grackles (*Quiscalus quiscula*) in response to repeated brood parasitism. Auk 127:293–299

Peer BD, Rothstein SI, Delaney KS, Fleischer RC (2007) Defence behaviour against brood parasitism is deeply rooted in mainland and island scrub-jays. Anim Behav 73:55–63

Peer BD, Kuehn MJ, Rothstein SI, Fleischer RC (2011) Persistence of host defence behaviour in the absence of avian brood parasitism. Biol Lett 7:670–673

Piersma T, van Gils JA (2011) The flexible phenotype: a body-centred integration of ecology, physiology, and behaviour. Oxford University Press, Oxford

Pigliucci M (2001) Phenotypic plasticity: beyond nature and nurture. Johns Hopkins University Press, Baltimore

Pigliucci M (2005) Evolution of phenotypic plasticity: where are we going now? Trends Ecol Evol 20:481–486

Pigliucci M, Murren CJ, Schlichting CD (2006) Phenotypic plasticity and evolution by genetic assimilation. J Exp Biol 209:2362–2367

Price TD, Qvarnström A, Irwin DE (2003) The role of phenotypic plasticity in driving genetic evolution. Proc R Soc B 270:1433–1440

Robert M, Sorci G (1999) Rapid increase of host defence against brood parasites in a recently parasitized area: the case of village weavers in Hispaniola. Proc R Soc B 266:941–946

Robert M, Sorci G, Møller A, Hochberg M, Pomiankowski A, Pagel M (1999) Retaliatory cuckoos and the evolution of host resistance to brood parasites. Anim Behav 58:817–824

Rodríguez-Gironés MA, Lotem A (1999) How to detect a cuckoo egg: a signal-detection theory model for recognition and learning. Am Nat 153:633–648

Røskaft E, Moksnes A, Stokke BG, Moskát C, Honza M (2002) The spatial habitat structure of host populations explains the pattern of rejection behavior in hosts and parasitic adaptations in cuckoos. Behav Ecol 13:163–168

Røskaft E, Takasu F, Moksnes A, Stokke BG (2006) Importance of spatial habitat structure on establishment of host defenses against brood parasitism. Behav Ecol 17:700–708

Rothstein SI (1990) A model system for coevolution: avian brood parasitism. Annu Rev Ecol Syst 21:481–508

Rothstein SI (2001) Relic behaviours, coevolution and the retention versus loss of host defences after episodes of avian brood parasitism. Anim Behav 61:95–107

Ruiz-Raya F, Soler M (2018) Rejection of parasitic eggs: an updated terminology for a complex process. J Avian Biol. https://doi.org/10.1111/jav.01484

Ruiz-Raya F, Soler M, Sánchez-Pérez LL, Ibáñez-Álamo JD (2015) Could a factor that does not affect egg recognition influence the decision of rejection? PLoS One 10:1–10. https://doi.org/10.1371/journal.pone.0135624

Ruiz-Raya F, Soler M, Roncalli G, Abaurrea T, Ibáñez-Álamo JD (2016) Egg rejection in blackbirds *Turdus merula*: a by-product of conspecific parasitism or successful resistance against interspecific brood parasites? Front Zool 13:16. https://doi.org/10.1186/s12983-016-0148-y

Scheiner SM (1993) Genetics and evolution of phenotypic plasticity. Annu Rev Ecol Syst 24:35–68

Schlichting C, Pigliucci M (1998) Phenotypic evolution: a reaction norm perspective. Sinauer Associates, Sunderland, MA

Servedio MR, Lande R (2003) Coevolution of an avian host and its parasitic cuckoo. Evolution 57:1164–1175

Snell-Rood E (2013) An overview of the evolutionary causes and consequences of behavioural plasticity. Anim Behav 85:1004–1011

Soler M (2011) Could egg rejection behaviour be transmitted by social learning? Anim Behav 81: e1–e6

Soler M (2014) Long-term coevolution between avian brood parasites and their hosts. Biol Rev 89:688–704

Soler M, Møller AP (1990) Duration of sympatry and coevolution between the great spotted cuckoo and its magpie host. Nature 343:748–750

Soler JJ, Soler M (2017) Evolutionary change: facultative virulence by brood parasites and tolerance and plastic resistance by hosts. Anim Behav 125:101–107

Soler M, Soler JJ, Martínez JG, Møller AP (1994) Micro-evolutionary change in host response to a brood parasite. Behav Ecol Sociobiol 35:295–301

Soler M, Soler JJ, Martínez JG, Møller AP (1995) Magpie host manipulation by great spotted cuckoos: evidence for an avian mafia. Evolution 49:770–775

Soler M, Soler JJ, Martínez JG, Pérez-Contreras T, Møller AP (1998) Micro-evolutionary change and population dynamics of a brood parasite and its primary host: the intermittent arms race hypothesis. Oecologia 117:381–390

Soler JJ, Martínez JG, Soler M, Møller AP (1999a) Genetic and geographic variation in rejection behavior of cuckoo eggs by European magpie populations: an experimental test of rejecter-gene flow. Evolution 53:947–956

Soler JJ, Sorci G, Soler M, Møller AP (1999b) Change in host rejection behavior mediated by the predatory behavior of its brood parasite. Behav Ecol 10:275–280

Soler M, Palomino JJ, Martín-Vivaldi M, Soler JJ (2000) Lack of consistency in the response of rufous-tailed scrub robins *Cercotrichas galactotes* towards parasitic common cuckoo eggs. Ibis 142:151–154

Soler M, Martín-Vivaldi M, Fernández-Morante J (2012a) Conditional response by hosts to parasitic eggs: the extreme case of the rufous-tailed scrub robin. Anim Behav 84:421–426

Soler M, Fernández-Morante J, Espinosa F, Martín-Vivaldi M (2012b) Pecking but accepting the parasitic eggs may not reflect ejection failure: the role of motivation. Ethology 118:662–672

Soler M, Ruiz-Raya F, Roncalli G, Ibáñez-Álamo JD (2017) Relationships between egg-recognition and egg-ejection in a grasp-ejector species. PLoS One 12:e0166283
Stokke BG, Moksnes A, Røskaft E (2005) The enigma of imperfect adaptations in hosts of avian brood parasites. Ornithol Sci 4:17–29
Stokke BG, Takasu F, Moksnes A, Røskaft E (2007) The importance of clutch characteristics and learning for antiparasite adaptations in hosts of avian brood parasites. Evolution 61:2212–2228
Stokke BG, Hafstad I, Rudolfsen G, Moksnes A, Møller AP, Røskaft E, Soler M (2008) Predictors of resistance to brood parasitism within and among reed warbler populations. Behav Ecol 19:612–620
Stokke BG, Røskaft E, Moksnes A, Møller AP, Antonov A, FossøY F, Liang W, López-Iborra G, Moskát C, Shykoff JA, Soler M, Vikan JR, Yang C, Takasu F (2016) Disappearance of eggs from nonparasitized nests of brood parasite hosts: the evolutionary equilibrium hypothesis revisited. Biol J Linn Soc 118:215–225
Takasu F (1998) Modelling the arms race in avian brood parasitism. Evol Ecol 12:969–987
Thompson JN (2005) The geographic mosaic of coevolution. University of Chicago Press, Chicago
Thorogood R, Davies NB (2013) Reed warbler hosts fine-tune their defenses to track three decades of cuckoo decline. Evolution 67:3545–3555
Welbergen JA, Davies NB (2009) Strategic variation in mobbing as a front line of defense against brood parasitism. Curr Biol 19:235–240
Welbergen JA, Davies NB (2011) A parasite in wolf's clothing: hawk mimicry reduces mobbing of cuckoos by hosts. Behav Ecol 22:574–579
Welbergen JA, Davies NB (2012) Direct and indirect assessment of parasitism risk by a cuckoo host. Behav Ecol 23:783–789
West-Eberhard MJ (2003) Developmental plasticity and evolution. Oxford University Press, Oxford
Zölei A, Bán M, Moskát C (2015) No change in common cuckoo *Cuculus canorus* parasitism and great reed warblers' *Acrocephalus arundinaceus* egg rejection after seven decades. J Avian Biol 46:570–576

Part VIII

Coevolutionary Interactions at the Nestling and Fledgling Stages

The Evolution of Nest Sharing and Nest Mate Killing Strategies in Brood Parasites

26

Csaba Moskát, Mark E. Hauber, and Matthew I. M. Louder

Abstract

Chicks of avian brood parasites either cohabit with host nestlings (e.g., cowbirds, non-evictor cuckoos) or eventually eliminate their potential competitors and are raised as the sole inhabitants of the foster parents' nest. This latter phenomenon, termed as direct killing, involves the young brood parasitic chick evicting all other eggs or hatchlings from the nest (e.g., evictor cuckoos) or killing their nest mates with their bill hook (e.g., honeyguides) and monopolize all food resources delivered by host parents. We overview the types of variability in parasite chick strategies toward nest mates and also discuss the potential reasons why competition vs. eviction behaviors have evolved in different parasitic lineages. We explore the possibility that one of the key factors for the evolution of nest mate acceptance versus direct killing was the virulence of the parasitic chick in relation to host nestling (competitive ability) and the ability to manipulate provisioning by the host parents. For extensive life history studies of the conflict between unrelated hosts and brood parasitic nestlings, we suggest that additional avenues for

C. Moskát (✉)
MTA-ELTE-MTM Ecology Research Group, A Joint Research Group of the Hungarian Academy of Sciences, The Biological Institute of the Eötvös Loránd University, Budapest, Hungary
e-mail: moskat.csaba@nhmus.hu

M. E. Hauber
Department of Animal Biology, School of Integrative Biology, University of Illinois, Urbana-Champaign, IL, USA

M. I. M. Louder
Department of Animal Biology, School of Integrative Biology, University of Illinois, Urbana-Champaign, IL, USA

Department of Psychology, Hunter College and the Graduate Center, The City University of New York, New York, NY, USA

Department of Biology, East Carolina University, Greenville, NC, USA

M. Soler (ed.), *Avian Brood Parasitism*, Fascinating Life Sciences,
https://doi.org/10.1007/978-3-319-73138-4_26

future research should explore and understand why and how evolutionary adaptations of brood parasitic nestlings have evolved.

26.1 Introduction

The eviction of all nest contents by common cuckoo (*Cuculus canorus*) chicks is an astonishing phenomenon, known since Aristotle (Schulze-Hagen et al. 2009). Modern studies in ornithology, ecology, and evolutionary biology have revealed the variability of the extent of nest mate eviction strategies within and across diverse avian brood parasite lineages (Davies 2000). One of the central questions in brood parasitism studies remains how the brood parasitic nestling can survive in the nest of a foreign bird species, which may show resistance toward accepting and provision suboptimally the foreign offspring. We provide an overview of the two main variants of nestling strategy in avian brood parasitism (direct killing vs. nest mate acceptance) and explore when, how, and why the parasitic chick cohabits with host nest mates or eliminates such competitors from the nest.

26.2 Chick–Chick Conflicts in the Bird Nest: Sibling Competition and Siblicide

Many bird species adjust clutch or brood size according to food availability [reviews in Mock and Parker (1997) and Ricklefs (2002)], but nestling behavioral strategies can further enhance an individual's survival when food is limited, through begging competition (Leonard et al. 2000) or sibling negotiation (Roulin et al. 2000). As food availability changes during the breeding season, and a chick's need for a particular prey-size shifts during ontogeny (Drummond 2002), competition among nestlings can lead to the survival of only those chicks that possess a competitive edge. For example, in raptors and herons, the earlier-hatched chicks with larger bodies show more aggressive behaviors, whereas in a handful of passerine species, the younger chicks may beg more than do older ones (e.g., Cotton et al. 1999). Alternatively, begging competition could be disadvantageous for those individuals having an underdeveloped ability to compete (later hatched and smaller chicks), so asynchronous hatching may help those nestlings that hatch earlier to survive (Price and Ydenberg 1995; Hauber 2003a).

Many eggs and nestlings perish as the consequence of the active killing by their nest mates even in nonparasitized avian broods. However, there is no single mechanism achieving this process, termed as siblicide or cainism as it is often called in large raptors (Simmons 1988). Siblicide can often be regarded as the result of falling due to scrambling in the overcrowded nest (competition for space) and/or competition for food in large nestlings that are close to fledging (e.g., in cattle egrets *Bubulcus ibis*; Fujioka 1985; Ploger and Mock 1986). If siblicide happens early in the nestling cycle, when the weaker junior chick dies from the injuries of intensive pecking by the senior chick (e.g., in the black eagle *Aquila verreauxii*, Gargett 1978), the remaining chick will receive all food delivered by the parents. In these situations, hatching asynchrony may help the evolution of siblicide. Interestingly, unlike in some insects and amphibians, avian siblicide rarely leads to cannibalism (Mock et al. 1990). Gonzalez-Voyer et al. (2007) used a comparative analysis to identify the feeding method used by parents (directly to the nestlings or indirectly to the nest bottom), clutch size, and the length of time chicks stay in the nest as contributing factors in the evolution of siblicide. Indirect feeding, long nestling period, and small clutch size positively correlated with brood mate aggressive competition, while feeding rate and size of food parcels were not correlated with aggression.

26.3 The Case of Brood Parasites: Sharing the Nest or Killing Host Nestlings

Adult brood parasites may choose hosts with better parental abilities (Soler et al. 1995; Hauber 2001; Louder et al. 2015) but thereafter have no influence over how their young are fed. At the nest, host parents may choose which chick to feed depending on aspects of its phenotype, including begging intensity and body size (Payne and Payne 2002), and so nestling competition is especially fierce when brood parasites are raised along with the host's chicks (e.g., *Molothrus* cowbirds in the Americas, *Vidua* finches in Africa, *Scythrops* and *Clamator* cuckoos in the New and Old Worlds, respectively). These parasites' adaptations, such as asynchronously earlier hatching, more rapid growth, and elevated begging rates, enable brood parasites to outcompete host nestlings for food (Hauber 2003a; see Chaps. 27 and 30).

Many brood parasites in the cuckoo family Cuculidae evolved strategies that eliminate their nest mates, enabling the parasite to monopolize all food provisioned by the host parents. In the cuckoos *Cuculus*, glossy cuckoos *Chrysococcyx*, and *Cacomantis* genera, as well as the *Cercococcyx*, *Pachycoccyx*, *Surniculus* spp., and two species of koels in Australasia (*Eudynamys cyanocephala* and *E. taitensis*), the so-called eviction behavior has evolved (Davies 2000), where the young parasitic chick ejects all nest contents, eggs or young nestlings, from the nest. If a young parasitic chick gets into contact with an egg or young, it will carry them on its back up the side of a nest and push them over the nest rim (Wyllie 1981; Fig. 26.1). Once the host eggs or nestlings are displaced from the nest cup, the host parents typically lose interest in their offspring and provide care solely to the young cuckoo inside the nest. In open-cup nesters, eviction behavior works with almost 100% efficiency (Anderson et al. 2009), whereas in cavity nesters with steeper nest cups, eviction success can be as low as 50% (Grim et al. 2009a; see Chap. 16). In general, the eviction of a host egg is probably easier for a young cuckoo chick than the eviction of a host nestling, as the nestlings may catch the nest material with their claws and so make the eviction trial more difficult (Fig. 26.1a, b). This behavior is likely absent from other species of koels, *Clamator* and *Scythrops*, because the eggs of their hosts are much larger and more difficult to lift (Wyllie 1981). Altogether at least three independent origins are known where nest mate killing behavior has evolved among all avian brood parasitic lineages (Table 26.1).

A video-recording study revealed that young common cuckoos evicted the first egg about 40 hours after hatching and displaced all nest contents within 1–2 days

Fig. 26.1 A young common cuckoo chick evicts a host egg from a great reed warbler nest (**a**) (Photo credit: Nikoletta Geltsch). Eviction appears more difficult when the young cuckoo chick evicts great reed warbler host nestlings (**b**) (Photo credit: Csaba Moskát)

Table 26.1 Evolutionarily independent origins of nest mate killing vs. nest mate sharing strategies in brood parasites

Order	Family	Representative genera	Nest mate killing	Nest mate sharing	References
Anseriformes	Anatidae	*Heteronetta*	No	Yes	Lyon and Eadie (2004)
Cuculiformes	Cuculidae (Cuculinae: Cuculini)	*Cuculus, Scynthrops*	Yes	Yes	Davies (2000) and Payne (2005)
Cuculiformes	Cuculidae (Cuculinae: Phaenicophaeni)	*Clamator*	No	Yes	Soler and de Neve (2013)
Cuculiformes	Cuculidae (Neomorphinae)	*Tapera*	Yes	No	Morton and Farabaugh (1979)
Passeriformes	Icteridae	*Molothrus*	No	Yes	Hauber (2003a)
Passeriformes	Viduidae	*Vidua*	No	Yes	Payne and Payne (2002)
Piciformes	Indicatoridae	*Indicator*	Yes	No	Spottiswoode and Koorevaar (2012)

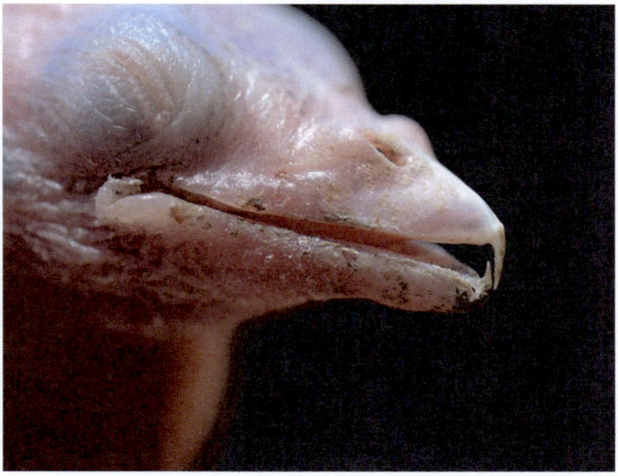

Fig. 26.2 The African brood parasite, the greater honeyguide (*Indicator indicator*), has a hook on its bill and uses it for aggressive attacks toward host eggs and nestlings (Spottiswoode and Koorevaar 2012) (Photo credit: Claire Spottiswoode)

(Honza et al. 2007). The eviction behavior typically exists for about 4 days (Heinroth and Heinroth 1966) or even later (up to 7 days; Grim et al. 2009a). In nests of the great reed warbler (*Acrocephalus arundinaceus*), about 95% of common cuckoos' ejections were carried out during the hosts' egg stage, and in only 5% of cases were the host hatchlings subject to cuckoos' eviction (Honza et al. 2007). This behavior seems to be risky for an evictor chick; in about 2% of parasitized nests of reed warblers (*Acrocephalus scirpaceus*), the cuckoo chick also fell out of the nest and died (Wyllie 1981). Although parents often feed altricial fledglings around their nests, they do not feed evicted nestlings outside the nest structure, leaving them to starve and die (see Chap. 16).

Although research on brood parasites and their hosts has boomed in the last three decades, how this highly specialized and invariable phenomenon eviction behavior has repeatedly evolved in unrelated parasitic lineages remains a mystery. In Australia, the Horsfield's bronze-cuckoo (*Chalcites basalis*) and the shining bronze-cuckoo (*C. lucidus*) evicted eggs and chicks similarly to the common cuckoo, with 33–48 hours after hatching for eggs and up to 96 hours for chicks (Brooker and Brooker 1989). Furthermore, nestlings in the lineages of honeyguides (Indicatoridae) in Africa and of the striped cuckoo (*Tapera naevia*) in Central and South America evolved modified bill hooks and, upon hatching, actively peck and pierce their host nest mates to death (Morton and Farabaugh 1979; Spottiswoode and Koorevaar 2012; Fig. 26.2). Another avian brood parasite in South America, the pavonine cuckoo (*Dromococcyx pavoninus*), also removes host eggs and kills nest mates with its strong bill (Sánchez-Martínez et al. 2017).

26.4 Did Eviction Behavior Originate from Siblicide?

Recently Wang and Kimball (2012) set out to derive the evolutionary origins of eviction behavior from siblicide. This study mapped the presence of this behavior onto phylogenies and suggested that eviction behavior of Cuculidae originated from siblicidal clades of birds (Pelecaniformes, Sphenisciformes, and Gruiformes). This study also concluded that nest mate killing behavior originated from Accipitridae and Strigiformes, which clade also contains the siblicidal hornbill (*Bucorvus leadbeateri*), and that in non-evictor cuckoos, the killing behavior could not be manifested due to physiological, ecological, and other evolutionary reasons.

As eviction behavior is controlled by instinct in cuckoos, it is programmed rigorously to prevent nestling competition in the host nest. Fulmer and Hauber (2016) developed an extension of the 2012 model focusing on when siblicide between chicks, and brood reduction due to brood parasitism is linked with ecological and social, including kin selected, constraints. This latter model integrates genetic relatedness and parental resource availability (energetic costs) when considering the evolution of evictor brood parasitism as evolved from an ancestral siblicidal strategy. The authors concluded that costs of siblicidal brood reduction must exceed the costs of parental care for heterospecific mixed broods to permit the evolution of interspecific brood parasitism.

26.5 Trade-Offs in Virulence for Nestling Brood Parasites

The degree by which avian brood parasites reduce the fecundity of their hosts, termed virulence, is highly variable (Hauber 2003b; Kilner 2005; Medina and Langmore 2016; Soler and Soler 2017). Competition for food and space inside the nest directly contributes to the survival of parasitic nestlings. Consequently, reductions in the host's clutch can lower the competition for parental resources from the viewpoint of brood parasites. Adult brood parasites often reduce the number of competitors for their offspring by removing or pecking one or more of the host eggs to reduce chick-chick competition, as it frequently happens in otherwise non-evictor *Molothrus* cowbirds (Peer 2006; Tuero et al. 2012) and *Clamator* cuckoos (Soler et al. 1997, 1999). However, because excessive egg removal by adult parasites can induce nest abandonment for their hosts (Hill and Sealy 1994; Moskát et al. 2011), selection may favor virulence caused by nestlings through competition and direct killing. Upon hatching, avian brood parasites vary markedly in the virulence imposed on hosts, from direct killing of all host's offspring to relatively benign coexistence within the nest. If nest mate killing minimizes competition for resources, why then are all parasites not extremely virulent?

Surprisingly, host nest mates can provide direct, fitness relevant benefits to parasites (see Chap. 27). Specifically, an interesting mathematical model was developed by Gloag et al. (2012), comparing the trade-off of nest mate acceptance and nest mate killing strategies of avian brood parasites. They suggest that this trade-off is influenced by the ability of brood parasite chick to increase the provisioning rate of the host parents (see, e.g., Kilner et al. 2004) and the competitive ability both of the host and parasitic chicks. If competitive ability of the parasite chick is low, nest mate killing is the best strategy for the brood parasite in order to maximize its food intake. If competitive ability of the parasite chick is high, nest mate acceptance might be a better strategy (Fig. 26.3). This model also revealed that brood size of the hosts affects

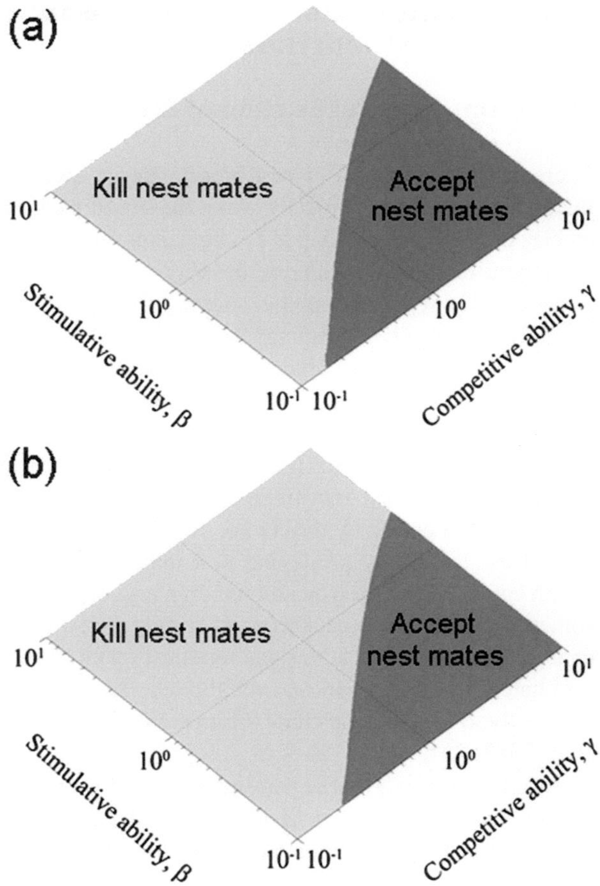

Fig. 26.3 The adaptiveness of two nest mate strategies (nest mate acceptance vs. killing) of brood parasite chicks, shown by the results of the provisioning trade-off model [adapted from Gloag et al. (2012)]. Figure (**a**) shows a case for a singleton brood parasite and two host chicks; Figure (**b**) shows the case when the number of host nestlings is doubled. Stimulative ability expresses how the presence of the singleton brood parasite affects the provisioning rate of adult hosts, and competitive ability is expressed for the brood parasitic chick in relation to host chicks [see for more details in Gloag et al. (2012)]

the adaptiveness of the nest mate killing strategy; it seems to be more advantageous when host's brood size is larger. A field experiment of this model on the non-evictor shiny cowbird (*Molothrus bonariensis*), a generalist brood parasite parasitizing more than 200 host species, strongly supported this model (Gloag et al. 2012).

Adult brood parasites may help to increase the virulence of their offspring. For example, egg destruction during laying by adult great spotted cuckoos (*Clamator glandarius*) is adaptive because it improved hatching and fledging success of the brood parasite (Soler et al. 1997). These adaptations were also supported in brown-headed cowbirds (Wood and Bollinger 1997; Hoover 2003; Kilner et al. 2004) and in shiny cowbirds, at least in large host nests, in which parasitic nestlings are most successful when the clutch is reduced (Astie and Reboreda 2009; Fiorini et al. 2009). Furthermore, non-evictor brood parasitic nestlings can gain a competitive edge from a shorter incubation period (see Chap. 30).

26.6 Can the Evictor Brood Parasite Survive When Sharing the Nest with Host Nestlings?

26.6.1 Unsuccessful Eviction of Brood Parasite Chicks

There are rare reports of evictor brood parasite chicks that failed to evict their nest mates (Cramp 1985), typically when the brood parasite hatches later than the host chicks, or when the nest cup is too high and/or deep (Grim et al. 2009a), resulting in a competitive situation with host nestlings (Wyllie 1981). For example, Petrescu and Béres (1997) reported on the successful cohabitation of common cuckoo and fieldfare *Turdus pilaris* nestlings. Molnár (1939) reported that late-hatching common cuckoo chicks were not able to evict host eggs/nestlings from great reed warbler nests and that some late-hatching cuckoos died in the nests of this host. As the cuckoo embryo has a one-day advantage in incubation when it is laid (due to "internal incubation," Birkhead et al. 2011), the cuckoos have a relatively wide time window in which to lay their eggs to ensure first-hatching of their offspring. The optimal laying period by the cuckoo for successful parasitism of great reed warblers starts from the empty pre-laying nest until the onset of incubation, i.e., when the fourth egg is laid in the typically five-egg host clutch (Geltsch et al. 2016). However, sometimes even those cuckoo eggs laid in early incubation will hatch before host eggs, and cuckoos that hatch synchronously with hosts still have the chance to evict host hatchlings from the nest, even though nest mate eviction behavior is physiological costly to the parasitic chick (Grim et al. 2009a, b; Anderson et al. 2009; Hargitai et al. 2012).

Eviction behavior by the parasite is most effective when the brood parasitic chick hatches the earliest in the nest (Payne and Payne 1998; Honza et al. 2007), and difficulties can arise when hatching later. Synchronous hatching of the brood parasite together with hosts causes the necessity of chick eviction, instead of eggs, which seems to be no problem for some parasitic chicks (Honza et al. 2007; Gill 1998; Geltsch et al. 2016; Sánchez-Martínez et al. 2017). A few days delay in hatching of the brood parasite in relation to hosts' present a more difficult situation for the brood parasitic chick, but in several cases successful eviction might be expected, as it was observed in splendid fairy-wren nests (*Malurus splendens*), the hosts of Horsfield's bronze-cuckoos in Australia (e.g., Payne and Payne 1998). When the brood parasite hatches several days later after the hosts had hatched, the chance for a successful eviction is low. However, there are reports when the late-hatched parasitic chick was able to evict its nest mates. For example, a striped cuckoo successfully evicted hosts that were 7 days older (Sánchez-Martínez et al. 2017).

26.6.2 The Case of the Common Redstart and Other Cavity Nesters

Common cuckoos prefer to parasitize hosts that build open nests. However, there are many, and regionally common, records of common cuckoos that parasitized cavity-nesting species (Makatsch 1955; Grim and Samas 2016). Attempts with species using wide-entrance hollows proved to be more successful (e.g., in the black redstart, *Phoenicurus ochruros*, Makatsch 1955), and common cuckoos regularly parasitize the common redstart (*Phoenicurus phoenicurus*), a hollow-nester (Rutila et al. 2002; Samas et al. 2016; Chap. 16). Redstarts are able to rear cuckoo nestlings with similar weight as regular hosts from open-cup nests (Grim and Samas 2016). The chance for successful eviction by a cuckoo chick is limited in redstarts' breeding holes (Samas et al. 2016), so cuckoo chicks in this host often experience cohabitation with host nestlings (Rutila et al. 2002). Despite this constraint, about 50% of cuckoo nestlings are able to fledge successfully both when raised alone or together with host nestlings (Rutila et al. 2002; Thomson et al. 2016; Samas et al. 2016), and an experimental study casually linked the reduced rates of survival to the cuckoo on a per capita basis to mixed compared to singleton experimental broods (Grim et al. 2009a, b).

26.6.3 Mixed Brood Experiments of Host and Brood Parasite Nestlings Reveal the Consequences of Non-eviction

A handful of studies have experimentally forced the chick of an evictor brood parasite to share a nest with the hosts' own offspring. These studies simulated the rare cases when evictions fail and show four main disadvantages the brood parasitic chick could face in the absence of eviction behavior.

1. *Parasitic chick loss*. It is unclear whether a common cuckoo chick naturally has a high competitive ability relative to host chicks. Mixed brood experiments of host and cuckoo nestlings revealed some details on this puzzle. When two great reed warbler chicks were forced to share the nest with a young cuckoo in similar of the age when eviction instinct has been terminated (after 5 days from hatching), many nestlings fell out of the nest cup, indicating severe competition between host and parasitic chicks (Moskát and Hauber 2010; Fig. 26.4). Typically the cuckoo wins (remains in the nest), but sometimes even a cuckoo can be the chick that is jostled out of the nest. Interestingly, no chicks fell from the nest in either of the two types of control treatments, when a single cuckoo or three host chicks were in the nest alone, without the presence of the other species. Such competition and nest mate loss are likely related to a problem of space in the nest, as sometimes we observed the loss of one host chick when too many host nestlings (i.e., six, instead of the typical four to five host nestlings) shared the same nest cup naturally (C. Moskát unpublished observations).
2. *Changes in prey size*. Hosts typically feed common cuckoo chicks with larger items than their own nestlings (Grim and Honza 1997; Kilner et al. 1999). This seems to be the consequence of the larger body size of the brood parasitic chick compared with similar-aged host chicks, e.g., the well-known *Acrocephalus* hosts. Our experimental broods, comprised of two great reed warblers and one common cuckoo, also revealed small differences in prey size compared to single species clutches, with a pattern of the cuckoo chicks receiving the larger prey items than their host chick nest mates (Hauber and Moskát 2008). However, variability in prey

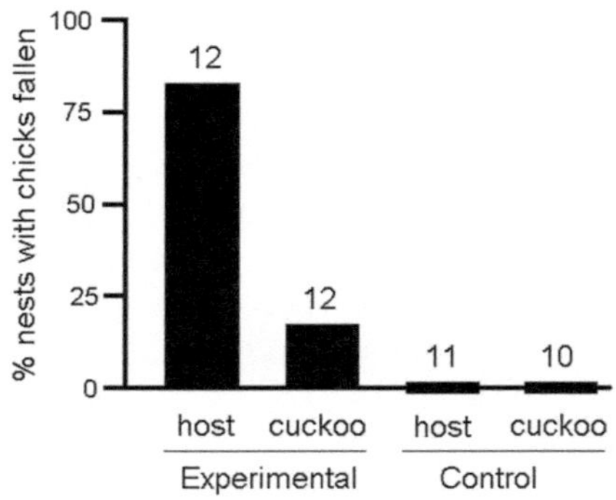

Fig. 26.4 The frequency of nests containing great reed warblers and common cuckoos with fallen chicks in experimental mixed and control broods; mixed broods contained two hosts and one cuckoo nestlings, while control nests had either three hosts or one cuckoo nestlings (Moskát and Hauber 2010)

size may depend on the host species. For example, the rufous-tailed scrub robin (*Cercotrichas galactotes*) did not differentiate food size between chicks in mixed broods of hosts and common cuckoos (Martín-Gálvez et al. 2005).

3. *Slower chick development*. The presence of host chicks in addition to a cuckoo chick in a nest also slows down the development of the cuckoo chick. However, cuckoos typically stay longer in nests than host chicks (e.g., great reed warbler chicks fledge at 11–12 days, but cuckoos fledge at 18–21 days, Anderson et al. 2009), and after the fledging of host chicks, the remaining cuckoo rapidly gains weight. Common cuckoos in mixed broods show plasticity in weight gain rates and fledged with the same weight and time as singleton cuckoos (Geltsch et al. 2012).
4. *Parent-offspring conflict*. Moskát and Hauber (2010) reported that when great reed warbler host chicks fledged successfully from experimental mixed broods, their parents preferred to feed the common cuckoo chick that stayed still in the nest. If host fledglings sought to solicit more parental feedings, they had to return to the nest from about 3–10 m distance, stay on the rim, and be fed while competing with the aggressively begging cuckoo nestling still in the nest. Host fledglings returned to the nest in three out of the four cases where hosts successfully fledged. However, this behavior of host fledglings was never observed in unparasized nests. Simultaneously rearing both nestling and fledglings of host and brood parasitic chicks is also a conflict situation for the foster parents of non-evictor brood parasites (see Chap. 30).

26.7 To Kill or Not to Kill: What Are the Deciding Factors?

Nest mate eviction behavior, including its attempts and completion, of brood parasitic chicks is physiologically costly per se (e.g., reduced growth, Anderson et al. 2009; fledging delay, Grim et al. 2009b; oxidative damage, Hargitai et al. 2012) and, by definition, can be advantageous for brood parasitic chicks only when the garnered benefits outweigh the costs. However, the cost-benefit trade-offs of this behavior seem to be also shared between closely related taxa, so that the evolution of eviction behavior is more frequent in closely related sets of taxa, including all 12 recognized *Cuculus* cuckoo species. The evolutionarily recent evictor *Chrysococcyx* genus has 12 species, again all evictors (Friedmann 1964). In contrast, eviction has not evolved in *Clamator* cuckoos, an evolutionarily ancient and independent parasitic genus (Sorenson and Payne 2005) with only four species. Soler and Soler (2017) show that this difference in diversity of parasitic cuckoos is probably caused by the lower virulence of *Clamator* species, as they do not evict their nest mates.

Below we overview the main factors which may contribute to the evolution of the chick-killing strategy in brood parasites or, alternatively, help the evolution of the nest mate acceptance strategy.

26.7.1 The Importance of Host Size

The competitive ability of the brood parasite determines what strategy, nest mate killing and nest mate acceptance, leads to the maximum food intake. Consequently, nest mate acceptance is expected in species parasitizing larger hosts and killing toward species parasitizing smaller hosts. However, if size difference is large enough, the parasitic nestlings lose their superior competitive ability (Rivers et al. 2010; Gloag et al. 2012; Bolopo et al. 2015). For generalist brood parasites, the optimum may change by different host species ("the generalist's dilemma"; Gloag et al. 2012); therefore an ability to adjust their behavior to the current and most common hosts (Kilner 2005), otherwise the evolution of host- or host-type-specific lineages, would be selected more strongly (Mahler et al. 2007; Gloag et al. 2012). The host's egg and chick mass, as well as nest size and shape, also influence the parasite's ability (and its costs) to successfully compete and/or evict the nest mates (Hauber 2003a; Grim et al. 2009a; Hargitai et al. 2012).

26.7.2 The Importance of Breeding Strategy

The breeding strategy hypothesis suggest that for clutch size to be adaptive in a bird species, it should be as large as the parents are able to provision the young (Lack 1947). In brood parasitism by non-evictor species, the parasitic chick(s) cohabits with host nest mates but may still have extraordinary needs in terms of additional feedings, for example, and so the effective brood size is greater than what is predicted by the clutch size. In turn, in evictor brood parasites effective brood size is reduced because the parasite eliminates nest mates. A relevant hypothesis related to host breeding strategy proposes to explain why parasitic young do or do not kill their nest mates (Soler 2001, 2002; Soler and de Neve 2013). In some bird species, parents feed their nestlings inequitably, favoring the larger and more intensively begging nestlings, and consequently some of their own young often starve when facing food shortages (brood reducers). In contrast, other bird species feed their young equitably or even preferentially feed the younger and smaller chicks (Cotton et al. 1999), and as a result, all or almost all of their nestlings are fledged (clutch-size adjusters; Soler 2001, 2002). Under ecologically harsh circumstances, e.g., when food is limited, clutch-size adjusters may change to the brood reducer strategy (Soler 2001). Given the critical role of parental decision rules in provisioning chicks with different begging strategies in the nest, young of brood parasites should therefore be adapted to the provisioning strategy of their particular (suite of) host species (Soler 2002). Specifically, in a trade-off effort by the parasite, its chick should be optimizing the positive and negative potential consequences of being alone in the nest or competing with nest mates. For example, unlike when nests contain a single egg, when parasitic young are alone in the nest, hosts rarely abandon their brood, whereas when host nest mates have a strong competitive ability and clutch-size adjuster parents distribute food evenly among chicks, the parasite may lose out from securing sufficient provisioning (Soler 2002; Soler and de Neve 2013). Therefore, when the brood parasite is a generalist, utilizing several host species at a time with different breeding strategies, or the availability of the preferred host species is limited, the eviction strategy seems to be a generally useful adaptation, as seen in the common cuckoo (Soler 2001). Finally, this hypothesis also provides a plausible explanation why there is almost no overlap in host species (breeding strategy) among the use of nest mate evictors and nest mate tolerant brood parasitic species, such as seen in the common cuckoos and great spotted cuckoos across their areas of breeding distribution, including the lack of shared host species use in the regions of their sympatry (e.g., in Europe; Soler 2002).

26.7.3 Multiple Parasitism

Multiple parasitism has been reported to be frequent in several non-evictor brood parasite–host systems (e.g., Soler and Soler 2000; Hoover 2003; Gloag et al. 2012, 2014; Payne 1977; Spottiswoode 2013; de Marsico and Reboreda 2008). It also occurs in evictor brood parasite–host systems (e.g., Moskát and Honza 2002; Zölei et al. 2015). In common cuckoos, typically different females parasitize the same nests in multiple parasitism (Moskát and Honza 2002; Moskát et al. 2009). Evictor brood parasitic chicks evict all host and brood parasitic eggs or hatchlings from the nest, avoiding the occurrence of extraordinary large clutch sizes. Thus, nest mate killing behavior of evictor brood parasites, e.g., in the common cuckoo, seems to be the key factor in the persistence of the brood parasite population in the context of frequent multiple parasitism, either having a positive effect (limited host defense, Moskát et al. 2009) or a negative effect (the waste of all but one brood parasitic eggs in the same nest) on brood parasites (Takasu and Moskát 2011). Multiple parasitism in cuckoo hosts could be a generally relevant strategy in areas where both hosts and cuckoos have high densities, not only in the unique population of great reed warblers breeding in Central Hungary, where about 24–52% of the parasitized nests proved to be multiply parasitized (Zölei et al. 2015). For instance, in the oriental reed warbler (*Acrocephalus orientalis*), about 10% of the parasitized clutches contained two cuckoo eggs (Li et al. 2016). Another example is the European robin (*Erithacus rubecula*) in Hungary (Varga 1994), and multiple parasitism was also detected in late populations of the red-backed shrike (*Lanius collurio*) in Germany (Rey 1895; Makatsch 1955). However, in many common cuckoo populations, this multiple parasitism-related regulatory mechanism could not be manifested because of the low (probably often historically decreasing) levels of parasitism rates [see for an overview of parasitism rates in Davies (2000)]. Interestingly, multiple parasitism could be responsible not only for the necessity of eviction behavior (see above) but also for the lack of direct killing behavior in the great spotted cuckoo (Soler and de Neve 2013). Unlike in the common cuckoo, where multiple parasitism occurs through laying by different female cuckoos into the same host nest (Moskát et al. 2009), female great spotted cuckoos repeatedly lay eggs in the same magpie nests (Martínez et al. 1998b). Therefore, the eviction of nest contests by great spotted cuckoos would include killing kin offspring and be disfavored through kin selection (Martínez et al. 1998a).

Concluding Remarks and Future Directions

We reviewed how one of the most effective ways to monopolize resources is to eliminate their nest mates from the nest. Direct killing behavior of brood parasite chicks likely originated from siblicidal behaviors, although many details and the socio-ecological evolutionary drivers remain unknown. Although there are examples when an otherwise evictor brood parasite chick could survive in the host nest following failed eviction attempts, subtle changes in the socio-ecological circumstances may destabilize this relationship. One such factor may be multiple parasitism, which increases the need of parasitic nestlings for parental provisioning and reduces personal space availability for optimal positioning and growth in the nest. Recent studies pointed to the importance of host size, as larger brood parasitic chicks may have an advantage to survive when host nestlings are smaller, although exceptions also exist, depending on hosts' food delivery strategy. In turn, breeding strategy can also be regarded as an important factor related to the evolution of nest mate killing behavior. Finally, drawing parallels between avian and microbial parasitism, conceptual and empirical studies both have addressed the factors that might affect the virulence of the parasite chick in host nests, and we saw above how they can be limited by ecological and physiological constraints. Although some flexibility may be expected to be adaptive in nest mate killing and acceptance strategies, the failure of nest mate killing in evictor brood parasites is typically only caused by environmental or reduced own physiological constraints. For example, the common cuckoo chicks often cannot evict eggs or nestlings when they hatch in nests built in cavities; subsequently, they pay a survival, growth, and fledgling delay costs for cohabitation.

Future studies may contribute to this knowledge in detail, particularly by exploring the effects of facultative virulence on brood parasites and nest mate acceptance on their hosts in different lineages. We propose that this new

(continued)

research should take into consideration chicks' size and weight, the generalist vs. specialist strategies of brood parasites, as well as the ecological potential for multiple parasitism. For example, comparative studies controlling for phylogenetic relationships in both brood parasites and host species are much needed to test the breeding strategy hypothesis and its specific predictions.

Acknowledgments We thank Manolo Soler for his invitation to contribute this chapter. We also thank Ross Gloag, Jarkko Rutila, and Manolo Soler for useful comments on an earlier version of the manuscript. For financial support, we thank the Hungarian National Research, Development and Innovation Office (NKFIH, NN118194 to CM). This study was also supported by the Human Frontier Science Program and the National Science Foundation (#1456524 and #1456612 to MEH and MIML).

References

Anderson MG, Moskát C, Bán M, Grim T, Cassey P, Hauber ME (2009) Egg eviction imposes a recoverable cost of virulence in chicks of a brood parasite. PLoS One 4(11):e7725

Astie AA, Reboreda JC (2009) Function of egg punctures by shiny cowbirds in parasitized and nonparasitized creamy-bellied thrush nests. J Field Ornithol 80:336–343

Birkhead TR, Hemmings N, Spottiswoode CN, Mikulica O, Moskát C, Bán M, Schulze-Hagen K (2011) Internal incubation and early hatching in brood parasitic birds. Proc R Soc B 278:1019–1024

Bolopo D, Canestrari D, Roldán M, Baglione V, Soler M (2015) High begging intensity of great spotted cuckoo nestlings favours larger-size crow nest mates. Behav Ecol Sociobiol 69:873–882

Brooker MG, Brooker LC (1989) The comparative breeding behaviour of two sympatric cuckoos, Horsfield's bronze-cuckoo *Chrysococcyx basalis* and the shining bronze-cuckoo *C. lucidus* in Western Australia: a new model for the evolution of egg morphology and host specificity in avian brood parasites. Ibis 131:528–547

Cotton PA, Wright J, Kacelnik A (1999) Chick begging strategies in relation to brood hierarchies and hatching asynchrony. Am Nat 153:412–420

Cramp S (ed) (1985) Handbook of the birds of Europe, the Middle East and North Africa, vol 4. Oxford University Press, Oxford

Davies NB (2000) Cuckoos, cowbirds and other cheats. T&AD Poyser, London

Drummond H (2002) Begging versus aggression in avian broodmate competition. In: Wright J, Leonard ML (eds) The evolution of nestling begging: competition, cooperation and communication. Kluwer Academic, Dordrecht, pp 337–360

Fiorini VD, Tuero DT, Reboreda JC (2009) Shiny cowbirds synchronize parasitism with host laying and puncture host eggs according to host characteristics. Anim Behav 77:561–568

Friedmann H (1964) Evolutionary trends in the avian genus *Clamator*. Smithson Misc Coll 146:1–127

Fujioka M (1985) Sibling competition and siblicide in asynchronously-hatching broods of the cattle egret, *Bubulcus ibis*. Anim Behav 33:1228–1242

Fulmer AG, Hauber ME (2016) A siblicidal origin for avian brood parasitism? J Ornithol 157:219–227

Gargett V (1978) Sibling aggression in the black eagle in the Matopos, Rhodesia. Ostrich 49:57–63

Geltsch N, Hauber ME, Anderson MG, Bán M, Moskát C (2012) Competition with a host nestling for parental provisioning imposes recoverable costs on parasitic cuckoo chick's growth. Behav Process 90:378–383

Geltsch N, Bán M, Hauber ME, Moskát C (2016) When should common cuckoos *Cuculus canorus* lay their eggs in host nests? Bird Study 63:46–51

Gill BJ (1998) Behavior and ecology of the shining cuckoo *Chrysococcyx lucidus*. In: Rothstein SI, Robinson SK (eds) Parasitic birds and their hosts: studies in coevolution. Oxford University Press, Oxford, pp 143–151

Gloag R, Tuero DT, Fiorini VD, Reboreda JC, Kacelnik A (2012) The economics of nestmate-killing in avian brood parasites: a provisions trade-off. Behav Ecol 23:132–140

Gloag R, Keller LA, Langmore NE (2014) Cryptic cuckoo eggs hide from competing cuckoos. Proc R Soc B 281:20141014

Gonzalez-Voyer A, Székely T, Drummond H (2007) Why do some siblings attack each other? Comparative analysis of aggression in avian broods. Evolution 61:1946–1955

Grim T, Honza M (1997) Differences in parental care of reed warbler (*Acrocephalus scirpaceus*) to its own nestling and parasitic cuckoo (*Cuculus canorus*) chicks. Folia Zool 46:135–142

Grim T, Samas P (2016) Growth performance of nestling cuckoos *Cuculus canorus* in cavity nesting hosts. Acta Ornithol 51:175–188

Grim T, Rutila J, Cassey P, Hauber ME (2009a) The cost of virulence: an experimental study of egg eviction by brood parasitic chicks. Behav Ecol 20:1138–1146

Grim T, Rutila J, Cassey P, Hauber ME (2009b) Experimentally constrained virulence is costly for common cuckoo chicks. Ethology 115:14–22

Hargitai R, Costantini D, Moskát C, Bán M, Muriel J, Hauber ME (2012) Variation in plasma oxidative status and testosterone level in relation to egg-eviction effort and age of brood-parasitic common cuckoo nestlings. Condor 114:782–791

Hauber ME (2001) Site selection and repeatability in brown-headed cowbird (*Molothrus ater*) parasitism of eastern Phoebe (*Sayornis phoebe*) nests. Can J Zool 79:1518–1523

Hauber ME (2003a) Hatching asynchrony, nestling competition, and the cost of interspecific brood parasitism. Behav Ecol 14:224–235

Hauber ME (2003b) Interspecific brood parasitism and the evolution of host clutch sizes. Evol Ecol Res 5:559–570

Hauber ME, Moskát C (2008) Shared parental care is costly for nestlings of common cuckoos and their great reed warbler hosts. Behav Ecol 19:79–86

Heinroth O, Heinroth M (1966) Die Vögel Mitteleuropas. I. Band. Edition Leipzig, Leipzig

Hill DP, Sealy SG (1994) Desertion of nests parasitized by cowbirds: have clay-coloured sparrows evolved an anti-parasite defence? Anim Behav 48:1063–1070

Honza M, Voslajerová K, Moskát C (2007) Eviction behaviour of the common cuckoo *Cuculus canorus* chicks. J Avian Biol 38:385–389

Hoover J (2003) Multiple effects of brood parasitism reduce the reproductive success of prothonotary warblers, *Protonotaria citrea*. Anim Behav 65:923–934

Kilner RM (2005) The evolution of virulence in brood parasites. Ornithol Sci 4:55–64

Kilner RM, Noble DG, Davies NB (1999) Signals of need in parent-offspring communication and their exploitation by the common cuckoo. Nature 397:667–672

Kilner RM, Madden JR, Hauber ME (2004) Brood parasitic cowbird nestlings use host young to procure resources. Science 305:877–879

Lack D (1947) The significance of clutch size. Ibis 89:302–352

Leonard ML, Horn AG, Gozna A, Ramen S (2000) Brood size and begging intensity in nestling birds. Behav Ecol 11:196–201

Li D, Zhang Z, Grim T, Liang W, Stokke BG (2016) Explaining variation in brood parasitism rates between potential host species with similar habitat requirements. Evol Ecol 30:905–923

Louder MI, Schelsky WM, Albores AN, Hoover JP (2015) A generalist brood parasite modifies use of a host in response to reproductive success. Proc R Soc B 282:20151615

Lyon BE, Eadie JM (2004) An obligate brood parasite trapped in the intraspecific arms race of its hosts. Nature 432:390–393

Mahler B, Confalonieri VA, Lovette IJ, Reboreda JC (2007) Partial host fidelity in nest selection by the shiny cowbird (*Molothrus bonariensis*), a highly generalist avian brood parasite. J Evol Biol 20:1918–1923

Makatsch W (1955) Der Brutparasitismus in der Vogelwelt. Neumann Verlag, Radebul und Berlin

de Marsico MC, Reboreda JC (2008) Egg-laying behavior in screaming cowbirds: why does a specialist brood parasite waste so many eggs? Condor 110:143–153

Martínez JG, Burke T, Dawson D, Soler JJ, Soler M, Møller AP (1998a) Microsatellite typing reveals mating patterns in the brood parasitic great spotted cuckoo (*Clamator glandarius*). Mol Ecol 7:289–297

Martínez JG, Soler JJ, Soler M, Burke T (1998b) Spatial patterns of egg laying and multiple parasitism in a brood parasite: a non-territorial system in the great spotted cuckoo (*Clamator glandarius*). Oecologia 117:286–294

Martín-Gálvez D, Soler M, Soler JJ, Martin-Vivaldi M, Palomino JJ (2005) Food acquisition by common cuckoo chicks in rufous bush robin nests and the advantage of eviction behaviour. Anim Behav 70:1313–1321

Medina I, Langmore NE (2016) The evolution of host specialization in avian brood parasites. Ecol Lett 19:110–118

Mock DM, Parker GA (1997) The evolution of sibling rivalry. Oxford University Press, Oxford

Mock DW, Drummond H, Stinson CH (1990) Avian siblicide. Am Sci 78:438–449

Molnár B (1939) About the cuckoo. Observations and experiments on the eviction instinct of the cuckoo chick. Published by the Author, Szarvas, p 24

Morton ES, Farabaugh SM (1979) Infanticide and other adaptations of the nestling striped cuckoo *Tapera naevia*. Ibis 121:212–213

Moskát C, Hauber ME (2010) Chick loss from mixed broods reflects severe nestmate competition between an evictor brood parasite and its hosts. Behav Process 83:311–314

Moskát C, Honza M (2002) European cuckoo *Cuculus canorus* parasitism and host's rejection behaviour in a heavily parasitized great reed warbler *Acrocephalus arundinaceus* population. Ibis 144:614–622

Moskát C, Hauber ME, Avilés JM, Bán M, Hargitai R, Honza M (2009) Increased host tolerance of multiple cuckoo eggs leads to higher fledging success of the brood parasite. Anim Behav 77:1281–1290

Moskát C, Rosendaal EC, Boers M, Zölei A, Bán M, Komdeur J (2011) Post-ejection nest-desertion of common cuckoo hosts: a second defense mechanism or avoiding reduced reproductive success? Behav Ecol Sociobiol 65:1045–1053

Payne LL (1977) Clutch size, egg size and the consequences of single versus multiple parasitism in parasitic finches. Ecology 58:500–513

Payne RB (2005) The cuckoos. Oxford University Press, Oxford

Payne RB, Payne LL (1998) Nestling eviction and vocal begging behaviors in the Australian glossy cuckoos *Chrysococcyx basalis* and *C. lucidus*. In: Rothstein SI, Robinson SK (eds) Parasitic birds and their hosts. Oxford University Press, Oxford, pp 152–169

Payne RB, Payne LL (2002) Begging for parental care from another species: behavioural specialization and generalization in brood parasitic finches. In: Wright J, Leonard ML (eds) The evolution of begging: competition, cooperation competition, cooperation and communication. Kluwer Academic, Dordrecht, pp 429–450

Peer BD (2006) Egg destruction and egg removal by avian brood parasites: adaptiveness and consequences. Auk 123:16–22

Petrescu A, Béres I (1997) The cuckoo (*Cuculus canorus* L.) is parasitising a nest of fieldfare (*Turdus pilaris* L.) (Aves) in Maramures (Romania). Trav Mus Natl Hist Nat Grigore Antipa 37:135–139

Ploger BJ, Mock DW (1986) Role of sibling aggression in distribution of food to nestling cattle egrets, *Bubulcus ibis*. Auk 103:768–776

Price K, Ydenberg R (1995) Begging and provisioning in broods of asynchronously-hatched yellow-headed blackbird nestlings. Behav Ecol Sociobiol 37:201–208

Rey O (1895) Beobachtungen über den Kuckuck bei Leipzig aus dem Jahre 1894. J Ornithol 43:30–43

Ricklefs RE (2002) Sibling competition and the evolution of brood size and development rate in birds. In: Wright J, Leonard ML (eds) The evolution of begging. Kluwer Academic, Dordrecht, pp 283–301

Rivers JW, Loughin TM, Rothstein SI (2010) Brown-headed cowbird nestlings influence nestmate begging, but not parental feeding, in hosts of three distinct sizes. Anim Behav 79:107–116

Roulin A, Kölliker M, Richner H (2000) Barn owl (*Tyto alba*) siblings vocally negotiate resources. Proc R Soc B 267:459–463

Rutila J, Latja R, Koskela K (2002) The common cuckoo *Cuculus canorus* and its cavity nesting host, the redstart *Phoenicurus phoenicurus*: a peculiar cuckoo-host system? J Avian Biol 33:414–419

Samas P, Rutila J, Grim T (2016) The common redstart as a suitable model to study cuckoo-host coevolution in a unique ecological context. BMC Evol Biol 16:255

Sánchez-Martínez MA, David S, Londono GA, Robinson SK (2017) Brood parasitism by the enigmatic and rare pavonine cuckoo in Amazonian Peru. Auk 134:330–339

Schulze-Hagen K, Stokke BG, Birkhead TR (2009) Reproductive biology of the European cuckoo *Cuculus canorus*: early insights, persistent errors and the acquisition of knowledge. J Ornithol 150:1–16

Simmons R (1988) Offspring quality and the evolution of cainism. Ibis 130:339–357

Soler M (2001) Begging behaviour of nestlings and food delivery by parents: the importance of breeding strategy. Acta Ethol 4:59–63

Soler M (2002) Breeding strategy and begging intensity: influences on food delivery by parents and host selection by parasitic cuckoos. In: Wright J, Leonard ML (eds) The evolution of begging: competition, cooperation competition, cooperation and communication. Kluwer Academic, Dordrecht, pp 413–427

Soler M, de Neve L (2013) Brood mate eviction or brood mate acceptance by brood parasitic nestlings? An experimental study with the non-evictor great spotted cuckoo and its magpie host. Behav Ecol Sociobiol 67:601–607

Soler JJ, Soler M (2000) Brood-parasite interactions between great spotted cuckoos and magpies: a model system for studying coevolutionary relationships. Oecologia 125:309–320

Soler JJ, Soler M (2017) Evolutionary change: facultative virulence by brood parasites and tolerance and plastic resistance by hosts. Anim Behav 125:101–107

Soler JJ, Soler M, Møller AP, Martínez JG (1995) Does the great spotted cuckoo choose magpie hosts according to their parenting ability. Behav Ecol Sociobiol 36:201–206

Soler M, Soler JJ, Martínez JG (1997) Great spotted cuckoos improve their reproductive success by damaging magpie host eggs. Anim Behav 54:1227–1233

Soler M, Soler JJ, Pérez-Contreras T (1999) The cost of host egg damage caused by a brood parasite: experiments on great spotted cuckoos (*Clamator glandarius*) and magpies (*Pica pica*). Behav Ecol Sociobiol 46:381–386

Sorenson MD, Payne RB (2005) Molecular systematics: cuckoo phylogeny inferred from mitochondrial DNA sequences. In: Payne RB (ed) Bird families of the world: cuckoos. Oxford University Press, Oxford, pp 68–94

Spottiswoode CN (2013) A brood parasite selects for its own egg traits. Biol Lett 9:20130573

Spottiswoode CN, Koorevaar J (2012) A stab in the dark: chick killing by brood parasitic honeyguides. Biol Lett 8:241–244

Takasu F, Moskát C (2011) Modeling the consequence of increased host tolerance toward avian brood parasitism. Popul Ecol 53:187–193

Thomson RL, Tolvanen J, Forsman JT (2016) Cuckoo parasitism in a cavity nesting host: near absent egg-rejection in a northern redstart population under heavy apparent (but low effective) brood parasitism. J Avian Biol 47:363–370

Tuero DT, Fiorini VD, Reboreda JC (2012) Do shiny cowbird females adjust egg pecking behavior according to the level of competition their chick face in host nests? Behav Process 89:137–142

Varga F (1994) Cuckoo observations around the source of the river Zagyva. Published by the Author, Salgótarján

Wang N, Kimball RT (2012) Nestmate killing by obligate brood parasitic chicks: is this linked to obligate siblicidal behavior? J Ornithol 153:825–831

Wood DR, Bollinger EK (1997) Egg removal by brown-headed cowbirds: a field test of the host incubation efficiency hypothesis. Condor 99:851–857

Wyllie I (1981) The cuckoo. Batsford, London

Zölei A, Bán M, Moskát C (2015) No change in common cuckoo *Cuculus canorus* parasitism and great reed warblers' *Acrocephalus arundinaceus* egg rejection after seven decades. J Avian Biol 46:570–576

Begging Behaviour, Food Delivery and Food Acquisition in Nests with Brood Parasitic Nestlings

27

Manuel Soler

Abstract

The existence of a brood parasitic nestling in a host nest implies an intrusion in the parent–offspring communication system, which will have important implications in both food delivery by parents and food acquisition by nestlings. The aim of this chapter is to review such implications taking into account two issues that are crucial in both food delivery and food acquisition in nonparasitized nests: breeding strategy of the species (brood reducer or clutch-size adjuster) and difference in size among nestlings in a brood. First, I review mechanisms allowing brood parasitic nestlings to secure parental provisioning from unrelated caregivers, such as exaggerated begging, mimicry of host begging calls, emitting a begging call that stimulates a wide range of hosts, tuning the begging call in a way that optimizes food provisioning, mimicking the begging calls of a brood, integrating visual and vocal nestling displays and procuring host assistance at the nest. Second, I review evidence showing that exaggerated begging behaviour exhibited by brood parasitic nestlings influences begging behaviour of their nestmates and food distribution decisions by foster parents. In addition, I present novel data testing the "integration of signals" hypothesis.

27.1 Introduction

When a passerine bird arrives with food at its nest, it finds a group of chicks exhibiting brightly coloured gapes and producing repetitive loud calls. These begging signals are part of a communication system by which nestlings solicit food from their parents, which regulate their effort and resource allocation according to the

M. Soler (✉)
Departamento de Zoología, Facultad de Ciencias, Universidad de Granada, Granada, Spain

Grupo Coevolución, Unidad Asociada al CSIC, Universidad de Granada, Granada, Spain
e-mail: msoler@ugr.es

M. Soler (ed.), *Avian Brood Parasitism*, Fascinating Life Sciences,
https://doi.org/10.1007/978-3-319-73138-4_27

conspicuousness and intensity of the begging displays used by their offspring (Wright and Leonard 2002). However, if its nest has been parasitized by the common cuckoo *Cuculus canorus* (the most famous brood parasite species), what it finds is a monstrous chick with an enormous gape of a bright orange-red colour calling at a very high rate.

Obligate avian brood parasites, by laying their eggs in the nests of other species (the hosts; Davies 2000, Chap. 1), parasitize the parental care of other breeding species (Roldán and Soler 2011), which means that brood parasitic nestlings are not related to either their nestmates or their foster parents. Thus, the existence of a brood parasitic nestling in a host nest implies an intrusion in the parent–offspring communication system, which will have important implications for both food delivery by parents and food acquisition by nestlings.

The aim of this chapter is to review such implications. However, first, it is necessary to introduce one issue that has been proposed (Soler 2001, 2002) to be crucial for food delivery by parents and food acquisition by nestlings in both nonparasitized and parasitized nests: the breeding strategy of the species, which is also related with nestling brood parasitic strategy (i.e. nestmate killers or non-killers).

27.2 Implications of the Breeding Strategy on Food Delivery and Food Acquisition

Two main breeding strategies have been described for parent birds: brood reducers, in which some chicks starve [when females lay an optimistic clutch size and later, during the nestling period, adjust brood size to food supply provoking starvation of the smaller nestling(s)] and clutch-size adjusters (clutch adjusters afterwards), in which all hatched nestlings typically survive to fledge [when females directly adjust clutch size (Lack 1947)]. This suggests that parents adopting these different strategies follow different provisioning rules. In clutch adjusters, parents tend to distribute food equitably among their nestlings, preferentially feeding young that are smaller than their nestmates or in poorer condition, while in brood reducers parents selectively feed larger nestlings independently of begging intensity (Soler 2001, 2002). Both breeding strategies cannot be clearly separated given that most clutch adjusters might be facultative brood reducers when food is limited. However, this does not contradict this idea; it merely means that in the case of food shortage, parents can avoid total brood loss by allocating more food to larger nestlings. This occurs even though the junior nestlings beg more intensively, which implies that parents are able to make active decisions regarding food allocation and preferentially feed nestlings with the best future prospects (Soler 2002; Caro et al. 2016; Soler et al. 2017).

The fact that different potential host species use different rules when distributing food among their nestlings is important for brood parasites, which should preferentially choose hosts of only one strategy, and be adapted to the breeding strategy of the host species, as it occurs in the two European brood parasites, the great spotted

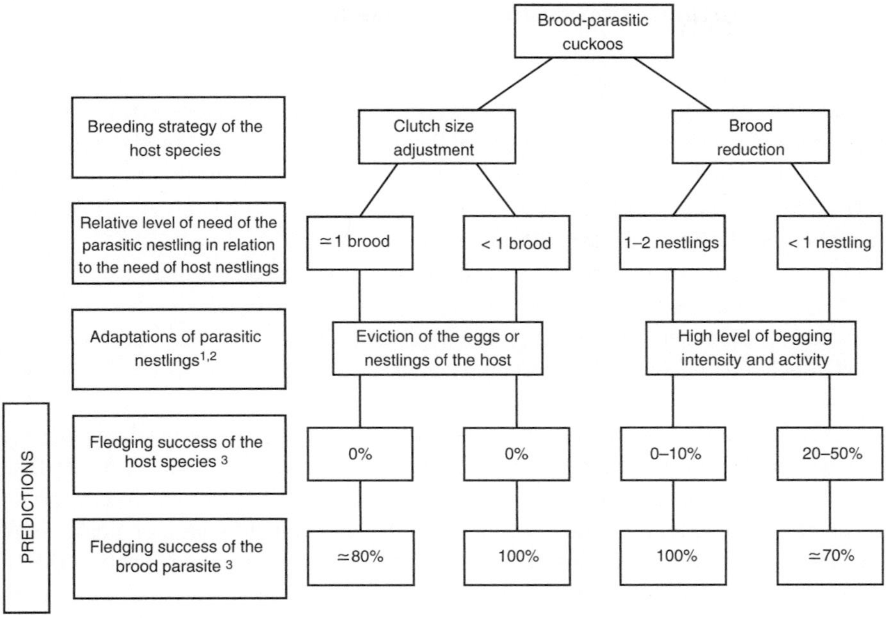

Fig. 27.1 Model showing how the breeding strategy of the host species and the relative level of need of the parasitic nestling in relation to the need of the host nestlings can influence host selection and the adaptations developed by different species of brood parasitic cuckoos. Some predictions are also specified. (1) In all cases a short incubation period will be advantageous for cuckoo nestlings. (2) A very high level of begging intensity should also be found in brood parasitic cuckoo nestlings parasitizing host species that adjust clutch size, mainly when the relative level of need of the cuckoo nestling is approximately that of the entire host brood, because parasitic nestlings need their hosts to deliver food at maximum capacity. (3) Fledging success is the percentage of eggs hatched that result in nestlings fledged. Numbers here are approximate without considering nest predation. Figure redrawn from Fig. 1 in Soler (2002)

cuckoo (*Clamator glandarius*) and the common cuckoo (Soler 2002). According to the breeding strategies' model (Fig. 27.1), another relevant point directly affecting both host selection and the tactics of the brood parasitic nestling is the level of need of the parasitic nestling in relation to need of the host brood. Therefore, both the breeding strategy of hosts and the relative need (relative body size) of the parasite nestling in comparison to host nestlings should affect the success of both brood parasite and host nestlings (Fig. 27.1). This model explains things such as absence of overlapping in host selection between the common and great spotted cuckoos in Europe and why the much bigger common cuckoo nestling is not successful in starving host nestlings to death when they are experimentally obliged to share the nest (Martín-Gálvez et al. 2005; Grim et al. 2009).

27.2.1 Importance of Nestlings' Relative Size

The fact that parents in most species (i.e. brood reducers and in clutch adjusters when food is limited) selectively feed larger nestlings (see above) implies that the relative size of nestlings in a brood will be one of the main factor affecting food distribution. It is well known that the largest nestling in the brood receives most of the food from provisioning parents (e.g. Bengtsson and Ryden 1983; Teather 1992; Price and Ydenberg 1995; Smiseth et al. 2003).

In non-killing brood parasites, parasitic nestlings often have to compete with host nestlings for the food delivered by foster parents. In these cases, the size difference between parasitic and host nestlings has important implications for both intra-brood food allocation by foster parents and the fate of both parasitic and host chicks in their competition for food (Table 27.1).

Usually, brood parasitic nestlings that share the nest with host nestlings enjoy a size advantage, but what happens when host nestlings are significantly larger than brood parasitic nestlings? In brood parasitic New World cowbirds (*Molothrus* spp.), it has been shown that parasitic nestlings usually outcompete young of smaller-size species [reviewed in Robinson et al. (1995), Lorenzana and Sealy (1999); Table 27.1], but when cowbirds parasitize larger-size hosts, they are not able to overcome the size advantage of host nestlings (Dearborn 1998; Lichtenstein and Sealy 1998; Lichtenstein 2001; Dearborn and Lichtenstein 2002; Rivers 2007) because of hosts' increased competitive ability. In nests parasitized by both shiny cowbirds (*M. bonariensis*) and brown-headed cowbirds (*M. ater*), it has been reported that the larger nestlings receive the majority of the food provisioned by adults independently of being the host or the parasite (Dearborn and Lichtenstein 2002; Rivers 2007).

However, even though non-killing brood parasitic nestlings of larger size than their host nestmates are much better competitors, they potentially face the problem that small host parents are simply unable to provision them at a sufficiently high rate. On the other hand, in nests of larger host species, usually some host nestlings survive to fledge, which reduces number of feeds received by the parasitic nestling. Host size, relative to the size of the brood parasite, can thus both increase and decrease the food supplied to the parasitic chicks. This might explain the lack of relationship between host mass and growth reported in brown-headed cowbird nestlings in 20 different host species (Kilpatrick 2002).

Most brood parasitic cuckoos use as hosts species smaller than themselves. The only case studied until now in which one cuckoo parasitizes a host species larger than itself is the great spotted cuckoo parasitizing its secondary host, the carrion crow (*Corvus corone*), which is about double the size of the parasite. It was found that great spotted cuckoo nestlings were fed at a similar rate than host chicks because their exaggerated begging behaviour allowed them to overcome their size disadvantage (Bolopo et al. 2015). Brown-headed cowbirds also usually parasitize smaller host species, and when parasitizing larger ones, parasitic nestlings are less efficient in obtaining food from their foster parents than their larger host nestmates (Dearborn and Lichtenstein 2002; Rivers 2007). Thus, as it occurs in brood reducers, size

Table 27.1 Overview of several results concerning non-evictor brood parasite–host systems in relationship to relative size of host and parasite species

Brood parasite	Host	Difference in size	Do hosts bring more food to parasitized nests?	Which nestling is preferentially fed in mixed broods?	Influence on host begging	Reference
Clamator glandarius	*Corvus corone*	P < H		H	Decrease	Bolopo et al. (2015)
Clamator Glandarius	*Pica pica*	P = H[a]	No	P[b]		Soler et al. (1995)
Clamator Glandarius	*Pica pica*	P = H		H[c]		Soler et al. (2018)
Molothrus ater	*Agelaius phoeniceus*	P = H	No	H	No effect[d]	Glassey and Forbes (2003)
Molothrus Ater	*Agelaius Phoeniceus*	P = H	No	No difference	No effect	Rivers (2007) and Rivers et al. (2010)
Molothrus Ater	*Agelaius Phoeniceus*	P = H	Yes			Grayson et al. (2013)
Molothrus Ater	*Agelaius Phoeniceus*	P = H	No[e]			Yasukawa et al. (2016)
Molothrus Ater	*Dendroica petechia*	P > H		P		Lichtenstein and Sealy (1998)
Molothrus Ater	*Dendroica Petechia*	P > H		P		Lichtenstein and Dearborn (2004)
Molothrus Ater	*Melospiza melodia*	P > H		No effect[f]		Zanette et al. (2005)
Molothrus Ater	*Melospiza Melodia*	P > H		No effect[g]	Increase	Pagnucco et al. (2008)
Molothrus Ater	*Passerina cyanea*	P > H	Yes		Increase	Dearborn (1999)
Molothrus Ater	*Passerina Cyanea*	P > H	Yes	P	Increase	Dearborn et al. (1998)
Molothrus Ater	*Protonotaria citrea*	P > H	Yes			Hoover and Reetz (2006)

(continued)

Table 27.1 (continued)

Brood parasite	Host	Difference in size	Do hosts bring more food to parasitized nests?	Which nestling is preferentially fed in mixed broods?	Influence on host begging	Reference
Molothrus Ater	*Sayornis phoebe*	P > H	Yes[8]			Hauber and Montenegro (2002)
Molothrus Ater	*Sayornis Phoebe*	P > H		P		Kilner et al. (2004)
Molothrus Ater	*Spizella pusilla*	P > H	No	P	Increase	Rivers (2007) and Rivers et al. (2010)
Molothrus Ater	*Toxostoma rufum*	P < H	No	H	Decrease	Rivers (2007) and Rivers et al. (2010)
Molothrus Ater	*Vireo bellii*	P > H		No difference		Rivers et al. (2014)
Molothrus bonariensis	*Mimus saturninus*	P < H		H		Gloag et al. (2012)
Molothrus Bonariensis	*Turdus rufiventris*	P < H		H		Lichtenstein (2001)
Molothrus Bonariensis	*Turdus Rufiventris*	P < H		H		Lichtenstein and Dearborn (2004)
Molothrus Bonariensis	*Troglodytes aedon*	P > H		P		Gloag et al. (2012)
Molothrus Rufoaxillaris	*Agelaioides badius*	P > H	Yes			Ursino et al. (2011)

P parasite, *H* host

[a]In natural conditions (size P > H because earlier hatching)

[b]In naturally parasitized nests, parasitic nestlings are larger than host nestlings because they hatch earlier

[c]In experimental conditions (size P = H)

[d]Continued begging by the parasite induced host nestlings to beg after food was allocated more often than in unparasitized nests

[e]Female redwings departed to forage and returned with food faster in response to redwing begging than to cowbird begging

[f]Host nestlings have the same mass in parasitized and unparasitized nests

[g]Hosts were fed at a similar rate in parasitized than in unparasitized nests

[h]Nest attendance rates were positively related with the proportion of cowbirds reared in a brood

differences between parasitic and host nestlings are crucial for determining which nestling will be feed in a parasitized nest (Table 27.1).

27.2.2 Nestmate Killers Versus Non-evictors

Brood parasitic nestlings are expected to behave selfishly because they are not genetically related with their foster parents or nestmates (Davies 2000). Given that sharing the nest with host nestlings is costly because it implies to share feeds as well, it would be expected that brood parasite nestlings would evolve mechanisms to eliminate host offspring, which would allow them to monopolize all feedings provisioned by foster parents. This is in fact the case; nestlings of many brood parasitic species destroy all host offspring either by evicting them from the nest or by killing them using specialized bill structures (Davies 2000; Kilner 2005, see Chap. 26). However, in other species, parasitic nestlings do not attempt to destroy host offspring and instead share the nest with them (Davies 2000). In these non-evictor species, parasitic nestlings engage in costly competition with host nestlings, which usually leads to starvation of host nestlings as a consequence of different adaptations that make the parasitic nestlings more efficient in getting food than host nestlings (see Chap. 30). However, this is not always the case, and sometimes host nestlings outcompete parasitic nestlings for food (Table 27.1), mainly when the former are larger than the latter (see Sect. 27.2.1).

Thus, the key question is why not all brood parasitic species have evolved nestmate killing strategies. Various adaptive explanations have been proposed [reviewed in Kilner (2005) and Wang and Kimball (2012), Chap. 26]. Although it has been suggested that none of them can fully account for the phylogenetic distribution of nestmate killing (Wang and Kimball 2012, Chap. 26), here I suggest that the model based upon the breeding strategies of the host complemented with the evolutionary lag hypothesis applied to more recently evolved brood parasites (i.e. *Molothrus* spp.) can be quite generalizable. In short, the breeding strategies' model suggests that, considering that evicting or killing host offspring is a costly strategy (Chap. 26), this strategy would only be adaptive when eviction costs outweigh the cost of competition for food delivered by foster parents (Soler 2002). When brood parasitic nestlings are reared in the nests of brood-reducing species, such costly strategy is not necessary (at least when the parasite is the largest chick in the brood), given that brood-reducing host species preferentially feed the larger chicks in the nest, the parasite would enjoy this advantage and be able to monopolize the food delivered at the nest outcompeting host nestmates (Soler 2002). In contrast, parasites of clutch adjusters, which parents tend to distribute food equitably among their nestlings, would be less likely to survive. Therefore, nestmate killing would be more likely to be selected for in species that parasitize clutch adjusters.

Especially when the relative level of need of the parasitic nestling is equivalent to that of a complete brood of host nestlings, nestmate destruction will be a compulsory strategy (Soler 2002, Fig. 27.1). This is the case of common cuckoo nestlings, which are more than double larger than host nestlings. When they are experimentally forced to share the nest with host nestlings, they suffer from a decrease in food received, growth and survival, while the growth and survival of their host nestmates is little affected (Martín-Gálvez et al. 2005; Hauber and Moskát 2008; Grim et al. 2009).

Several papers have supported the breeding strategies' model in some cuckoo hosts (see references above), but what happens with non-evictor cowbirds? Likely they parasitize both clutch adjusters and brood reducers. Why then eviction or nestmate killing strategies have not evolved in this group? There are two main explanations: first, that when cowbirds parasitize small host species, size difference is not as large as in the common cuckoo–host systems, i.e. the relative level of need of the cowbird nestling is smaller than the need of a host brood, and thus the evolution of offspring killing strategy is not vital. Second, the evolutionary origin of parasitic cowbirds is much recent than that of cuckoos (Davies 2000). Thus, the lack of eviction or nestmate killing strategies in cowbirds could also be explained by evolutionary lag (Chap. 26).

27.3 Food Acquisition

Brood parasitic nestlings face the challenge of acquiring food from unrelated host foster parents, which requires them to manipulate the parent–nestling communication system in their own benefit. In the case of nestmate killers, the greatest challenge is to get adult foster parents to bring an appropriate amount of food to the nest, while in non-evictors, the greatest challenge is to ensure they receive the necessary quantity of food during intra-brood competition. Many potential morphological, physiological and behavioural adaptations have been proposed as contributors to this endeavour (e.g. early hatching, host-attuned growth rates, nestmate killing behaviour and host chick mimicry; see Chap. 30); however, the most important factor driving manipulation of host parental behaviour to their benefit is begging behaviour used by parasitic nestlings. Begging displays of brood parasitic nestlings not only include vocalizations, posturing and movements more exaggerated than those of host nestlings (see Sect. 27.3.1) but sometimes additional morphological and behavioural characteristics that contribute to increase begging efficiency, for instance, brightly coloured gapes (Álvarez 2004), palatal papillae (Soler et al. 1995), gape-coloured patch on the wings (Tanaka and Ueda 2005) and wing shaking (Grim 2008).

Parents usually feed the first nestling that starts to beg after arrival at the nest. Thus, another potential adaptation of parasitic nestlings would be to evolve a more rapid begging response to cues of parental arrival than their host nestmates, which has been found in brown-headed cowbirds (Rivers 2007) and great spotted cuckoos (Soler et al. 2018). This rapid response could incur in erroneous begging (i.e. in the absence of parents), which can increase the risk of nest predation (Haskell 2002; Ibáñez-Álamo et al. 2012). This is the case of brown-headed cowbird nestlings (Hauber 2003; Rivers 2009); however, in an experiment under laboratory conditions, it has been demonstrated that nestlings of both parasitic great spotted cuckoo and magpie hosts rarely showed erroneous begging when food was abundant, but under conditions of restricted food, magpie nestlings increased erroneous begging while cuckoo nestlings did not (Soler et al. 2018).

How nestling parasites can solve the problem of securing parental provisioning from unrelated caregivers? Several non-exclusive mechanisms have been suggested, which will be commented upon in turn.

27.3.1 Exaggerated Begging

This is the most commonly presented strategy. In general, brood parasitic nestlings consistently show more exaggerated levels of begging than host nestlings (Dearborn 1998; Lichtenstein and Sealy 1998; Kilner and Davies 1999; Kilner et al. 1999a; Grim and Honza 2001; Rivers 2007), which often are more effective in stimulating provisioning than those elicited by host's own nestlings (Davies et al. 1998; Dearborn 1998; Kilner et al. 1999a; Rivers 2007; Gloag and Kacelnik 2013). This pattern is expected since parasitic nestlings are not constrained by inclusive fitness costs; i.e. exaggerated begging enhances the competitive ability of the parasite at the expense of their host nestmates with which brood parasitic nestlings are not genetically related (Briskie et al. 1994). This hypothesis has been recently supported by the finding that begging intensity of brown-headed cowbird nestlings is greater in a population with low rate of multiparasitism compared to another population in which parasitized nests are frequently occupied by related parasitic nestlings (Rivers and Peer 2016).

Higher intensity of begging in brood parasitic nestlings than in host nestlings is not an absolute rule. Recently, it has been reported that the begging displays of brood parasitic brown-headed cowbird nestlings are not more exaggerated than those of the closely related red-winged blackbird (*Agelaius phoeniceus*) (Rivers et al. 2013). Similarly, it was reported that great spotted cuckoo nestlings only rarely were able to survive to fledge in the nests of jackdaws (*Corvus monedula*), a sporadically used host that exhibits more intense and exaggerated begging displays than parasitic cuckoo nestlings. This was explained as a consequence of this brood-reducing species breeding in safe hole nests allowing the evolution of more intense begging (Soler 2002). In a series of cross-fostering experiments, it was demonstrated, first, that cuckoo nestlings experimentally introduced in jackdaw nests were outcompeted by the jackdaw nestlings in all cases ($n = 4$) and, second, that all jackdaw nestlings experimentally introduced in magpie (the main host of the great spotted cuckoo) nests survived, while most magpie nestlings starved ($n = 8$) (Soler 2002). These experiments clearly demonstrated that jackdaw nestlings with their exaggerated begging behaviour are able to outcompete both parasitic great spotted cuckoo and their magpie host nestlings.

Although exaggerated begging and the absence of indirect fitness costs in brood parasitic nestlings have provoked doubts about the honesty of their begging displays (Redondo and Zuñiga 2002; Rivers 2007), strong evidence is accumulating showing that begging behaviour by brood parasites honestly reflects their short-term need and body condition (Soler et al. 2012; Rivers et al. 2013), as theoretical models predict in non-brood parasites (Royle et al. 2002).

27.3.2 Mimicry of Host Begging Calls

In some brood parasitic species, it has been found that the parasitic nestling mimics the begging calls of its host's nestlings, not only in non-evicting species, where it is possible to learn to call appropriately by mimicking the calls of their nestmates, but also in some evicting species (Davies 2000). Similarity between the begging calls of parasitic nestlings and their hosts has been reported in 53% of parasitic species whose begging calls have been described (Chap. 28), which implies that such similarity is a more widespread phenomenon than previously appreciated (see Chap. 28 for more detailed information on this topic).

27.3.3 Emitting a Begging Call That Stimulates a Wide Range of Hosts

Begging calls adapted to the environment (i.e. host nests in which they are developing) are expected to have evolved in specialized brood parasites (Chap. 30). However, in generalist brood parasites, in which a female lays her eggs in the nests of multiple host species, the most appropriate strategy for these nestlings would be to use "generalist" begging calls that would act as an effective stimulus of provisioning from multiple potential hosts species (Davies 2000; Payne and Payne 2002). This hypothesis has been experimentally tested by broadcasting begging calls of the generalist brood parasitic shiny cowbird and of its house wren (*Troglodytes aedon*) host at unparasitized house wren broods. In agreement with the hypothesis, it was found that cowbird begging calls were more effective eliciting parental provisioning than host calls and, furthermore, that they triggered an increase in parental provisioning even in nests of a non-host species (Gloag and Kacelnik 2013).

27.3.4 Tuning the Begging Call in a Way That Optimizes Food Provisioning

This mechanism suggests that newly hatched brood parasitic nestlings can adjust their begging calls according to provisioning rewards obtained from foster parents (Davies 2000). Strong support for this mechanism has been found in generalist brood parasites (West and King 1998; Langmore et al. 2008) and in a specialist brood parasite (at the female level), the common cuckoo (Butchart et al. 2003; Madden and Davies 2006). The strongest support was provided by cross-fostering common cuckoo eggs between host species. In this study, it was found that difference in the structure of begging calls of common cuckoo nestlings when reared by different hosts was due to a trial-and-error mechanism of learning, which allows cuckoo nestlings to use the begging call most effective in exploiting foster parents (Madden and Davies 2006). Interestingly, in the great spotted cuckoo, a specialist brood parasite, it has been found that reported differences in begging calls between parasitic nestlings reared in nests of their two main host species are the consequence of phenotypic plasticity allowing parasitic nestlings to learn efficient begging displays when exploiting foster parents of different species (Roldán et al. 2013). Traditionally, it is thought that nestlings of specialist brood parasites do not modify the structure of their begging calls when parasitizing different hosts (Payne and Payne 1998). However, differences found in the great spotted cuckoo, though small, suggest that adjusting the begging to optimize food provisioning by foster parents may be a more frequent mechanism than previously suspected, even in specialist brood parasites.

27.3.5 Mimicking the Begging Calls of a Brood

Davies et al. (1998) found that 1-week-old common cuckoo nestling mimics the begging calls of a brood of four reed warbler (*Acrocephalus scirpaceus*) young. This highly attractive finding is broadly cited, frequently extending this possibility to other brood parasite–host systems. What we know is that nestling cuckoos use begging calls that mimic a brood when they are 6–8 days old; but, first, this case has not been found in any other brood parasite species since then, and, second, during the first days of the nestling period, the cuckoo nestling does not produce begging calls at all. Thus, mimicking the begging calls of a brood of reed warblers by the cuckoo nestling when 7 days old could be merely a coincidence. Likely, greater caution would be needed when citing this finding out of this context.

27.3.6 Integrating Visual and Vocal Nestling Displays

Kilner et al. (1999a) demonstrated that the high rate of begging calls by the cuckoo chick in reed warbler nests compensates for a much smaller gape area than that of a whole host brood ("integration of signals" hypothesis).

Considering that cuckoo females lay their eggs in the nests of 15 different host species (Moksnes and Røskaft 1995), the integration of signals hypothesis raises the question of whether cuckoos will have to tune into different offspring–parent communication systems in the nests of different hosts, particularly when taking into account that such species differ considerably in size (which determines gape area) and begging call rate. Different host species should integrate visual and vocal signals of their chicks in different ways (Kilner et al. 1999b). Therefore, cuckoo chicks have to tune into different offspring–parent communication systems in nests of different host species, implying that the rate of cuckoo begging calls should differ between hosts (Kilner et al. 1999b; Kilner and Davies 1999). This prediction was partially supported in Butchart et al.'s (2003) study, in which they found that there are important differences between cuckoo host races in the relationship between call rate and age (Butchart et al. 2003). Here I present a test of this important prediction of the integration of signals hypothesis based on an analysis of the differences in the vocalizations of cuckoo and host nestlings in the nests of four host species based on unpublished results by M. Soler, M. Martín-Vivaldi, E. Røskaft, C. Moskát and J.J. Palomino. The two host species previously studied by Butchart et al. (2003) are the reed warbler (data obtained from Fig. 27.2 in Kilner and Davies 1999) and the great reed warbler (*Acrocephalus arundinaceus*), and the two never previously studied in this respect are the rufous-tailed scrub robin (*Cercotrichas galactotes*) and the robin (*Erithacus rubecula*).

27.3.6.1 Methods

The rufous-tailed scrub robin was studied in Los Palacios (southern Spain 37°9′N, 2°14′W) in 1994–1996. A total of 212 nests were found of which 34 were parasitized by the common cuckoo (16%). This host species is a small (20–27 g) passerine that is considered a regular cuckoo host (Álvarez 1994). The robin, another regular cuckoo host (Moksnes and Røskaft 1995) of small size (16–22 g), was studied in Sierra Nevada (southern Spain 37°10′N, 3°25′W) during 1998 and 1999. Parasitism rate was 13.5%; the common cuckoo parasitized 5 out of 37 nests. The great reed warbler (24–40 g) was studied in Kiskunlacháza and Kunszentmiklós (Central Hungary 47°02′–47°10′N, 19°02′–19°15′E) during 2000. Here the parasitism rate was up to 65.6% (Moskát and Honza 2000). These three host species are considerably bigger than the reed warbler (12 g).

Begging calls were recorded with a Sony CCD-TRO5E PAL video camera recorder placed about 50 cm from the nest (rufous-tailed scrub robin nests) or with a Sennheiser MKE 2–3 tie clip microphone attached to the side of the nest and connected to the same video camera placed at 6–8 m from the nest (robin nests). The great reed warblers were recorded with Sony TR-707 or Sony TR-2000 video cameras and HI 8 (8 mm) video tape, mounted on stands placed about 1 m from the nests. Begging calls were digitized, and sonograms were produced by using the software Sound-Edit 2.0.5. A total of four (8–9 days old), two (7 and 13 days old, respectively) and six (7–10 days old) cuckoo chicks were recorded from rufous-tailed scrub robin, robin and great reed warbler nests, respectively. Calling rates of whole broods were recorded in eight (8–10 days old) rufous-tailed scrub robin, four (4–7 days old) robin and six (7–8 days old) read warbler nests.

To analyse whether the structure of cuckoo begging calls mimics those of its host species, three people blind on the precedence of samples were asked to classify all sonograms from cuckoo nestlings into three groups (because we had calls of cuckoos reared in three host species) according to similarities in call structure. We expected that if cuckoo begging calls mimic those of their foster species, those sonograms of chicks reared by the same host species would be grouped together.

We scored the number of calls produced during the first 6 s of each nest visit by parents. For each brood and for each single cuckoo, we scored the number of calls in two to four nest visits of parents separated by 10–30 min during 2-h recordings. We used the mean number of calls per 6 s for the four recordings for each brood and for

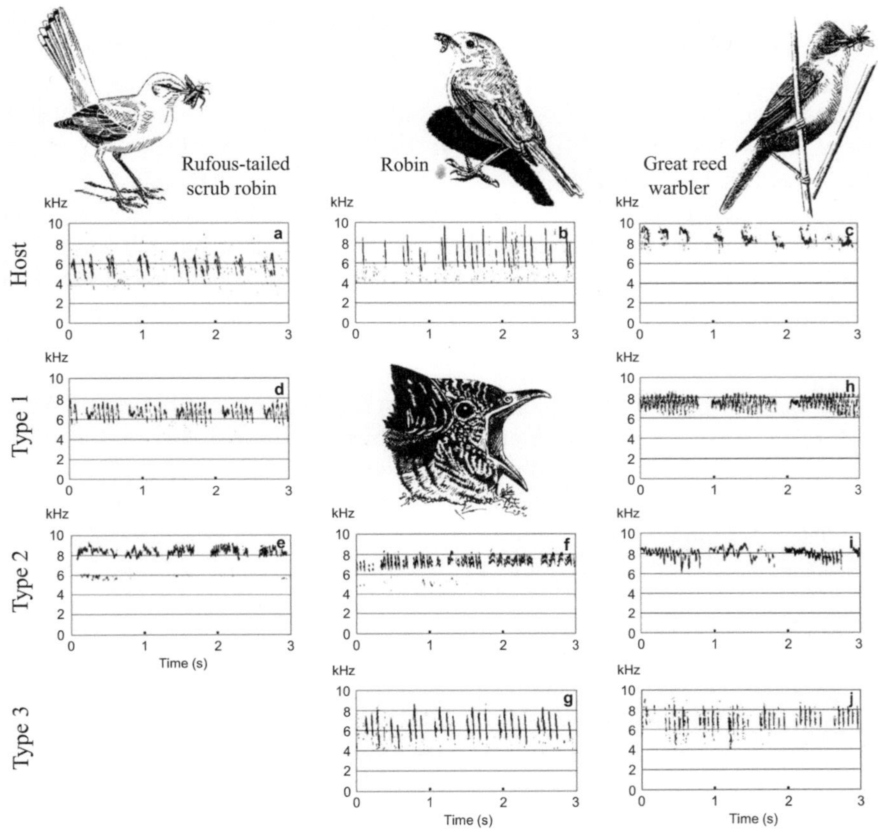

Fig. 27.2 Sonograms (3 s) of the begging calls of (**a**) a brood of three rufous-tailed scrub robin chicks, (**b**) a brood of five robin chicks, (**c**) a brood of five great reed warbler chicks and (**d**–**j**) single cuckoo chicks reared in rufous-tailed scrub robin nests (**d**, **e**), robin nests (**f**, **g**) and great reed warbler nests (**h**, **i**, **j**). Figure based on unpublished data by M. Soler, M. Martín-Vivaldi, E. Røskaft, C. Moskát and J.J. Palomino

each single cuckoo in the analyses. Repeatability of calling rates of the same cuckoo chick at different times was high ($R_{11,\ 32} = 0.95$, $P < 0.0001$).

27.3.6.2 Results

Based on the appearance of sonograms in Fig. 27.2, it seems clear that begging calls of cuckoo chicks do not resemble those of the host chicks. There are differences in the structure of call notes, but this difference is not related to each cuckoo host race. In fact, there are three different versions of the cuckoo chick begging calls, all three of which were shown by cuckoo nestlings reared in different host species and different European study areas. The sonogram type 1 was found in cuckoos reared in two rufous-tailed scrub robin nests and three great reed warbler nests; type 2 with a more irregular pattern was found in cuckoos reared in two rufous-tailed scrub robin nests, one robin nest and two great reed warbler nests; and type 3 was found in cuckoos reared in one robin nest and one great reed warbler nest (Fig. 27.2).

We found that cuckoo chicks in nests of these three host species have similar begging rates (Fig. 27.2). Sonograms are furthermore similar to those of cuckoo chicks in reed warbler nests (Davies et al. 1998). Number of calls per 6 s (mean calling rate ± s.d.) produced by cuckoo nestlings was similar in the nests of rufous-tailed scrub robins (70.1 ± 12.8), robins (70.0 ± 18.8), great reed warblers (78.6 ± 16.4) and reed warblers (68.4 ± 8.8; Kruskal–Wallis, H3 = 3.21, $P = 0.4$). The repeatability of cuckoo chick calling rates within host species was 0 (R3, 18 = 0, $P = 0.48$) indicating more variance in calling rates among cuckoo chicks reared by the same species than among cuckoo chicks reared by different host species.

Calling rates of whole broods of the four host species (rufous-tailed scrub robins, 39.8 ± 7.8; robin, 39.3 ± 6.6; great reed warbler, 24.4 ± 6.3; and reed warbler, 34.5 ± 10.0) were significantly different (Kruskal-Wallis, H3 = 9.36, $P = 0.025$), mainly as a consequence of calling rates of great reed warbler chicks being lower than that of the other three species.

27.3.6.3 Discussion

Butchart et al. (2003), by measuring several call properties (maximum, minimum and peak frequencies; bandwidth; and note duration), found no differences in the structure of cuckoo begging call notes among host species.

By classifying cuckoo call types by their sonogram, we were able to show that there is variability in call structure within each type of host parasitized, which is greater than the variation shown among the host species in whose nests the cuckoo chicks were found. The three types of sonograms identified in this study vary considerably in their structure. We have found these three types of structure of common cuckoo nestling begging calls almost equally distributed among the three cuckoo races.

With respect to calling rate of nestling cuckoos, we have found that, despite the fact that the four host species differ considerably in body size (see above) and in calling rates, vocalizations of cuckoo chicks do not differ significantly among these four cuckoo host races in this parameter at a given age. Results obtained by Butchart et al. (2003) are similar in the sense that in their group of four cuckoo host races, they also found that great reed warbler cuckoos presented the higher calling rate and reed warbler cuckoos showed the lower one.

Thus, these data do not support the integration of signals hypothesis given that vocalizations of cuckoo chicks do not differ among host species. However, more studies with larger sample sizes and including more host species are needed. These findings showing no differences of cuckoo begging calls are consistent with the hypothesis that begging rates of cuckoo chicks are always exaggerated regardless of host size or host species (Soler and Soler 1999).

27.3.7 Procuring Host Assistance at the Nest

Kilner et al. (2004) found that brown-headed cowbird nestlings grow faster and survive better when they share the nest with eastern phoebe (*Sayornis phoebe*) host nestlings than when raised alone. They suggested that these results were the consequence of cowbird nestlings obtaining assistance from host nestlings in stimulating foster parents to increase provision to the brood. They demonstrated that the parasitic nestling was able to monopolize the extra food delivered because of its greater competitive ability. Later Kilner (2005), given that comparative studies show that cowbirds have lower mortality rate (Kilner 2003) and high growth rate (Kilner et al. 2004) when raised with host young than when raised alone, proposed this advantage of brown-headed cowbird nestlings in mixed broods to be extendible to other brood parasite–host systems. However, the prediction that a brood parasitic nestling would receive more food and/or grow faster when sharing the nest with host nestlings than when alone has not been supported in other studies.

The first experimental tests of this prediction were made using the evicting common cuckoo, and, contrary to the prediction, it was found that the presence of host chicks in the nest caused a decrease in both food received (Martín-Gálvez et al. 2005), growth rate and survival (Hauber and Moskát 2008; Grim et al. 2009). However, this is not surprising because the common cuckoo is an evictor species. Thus, it is expected that common cuckoo nestlings will develop better alone than sharing the nest with host nestlings (Soler 2002, see Sect. 27.2.2).

What happens in non-evictor brood parasites? The above-mentioned prediction of the begging assistance hypothesis was not supported in the great spotted cuckoo–magpie host system. Rather, nestling great spotted cuckoos showed a tendency for the contrary: they grow better when raised alone than when raised together with one or two magpie nestlings (Soler and de Neve 2013).

The prediction of the begging assistance hypothesis has only been supported in a study (Gloag et al. 2012) in which nestling shiny cowbirds were experimentally introduced in the nests of house wren (*Troglodytes aedon*), a smaller-size host. However, in the same study, when cowbird nestlings were introduced in the nests of the larger-size chalk-browed mockingbirds (*Mimus saturninus*) hosts, they obtained less feeds and had a lower mass growth when reared with host nestlings than when reared alone (Gloag et al. 2012). The same was found to occur when brown-headed cowbird nestlings share the nest with the larger-size brown thrashers (*Toxostoma rufum*) host nestlings (Rivers et al. 2010). Even more, in this case the larger-size host nestlings are the ones which benefit from the intense begging of the parasitic nestling stimulating parents to increase nest provisioning (Rivers et al. 2010), a result that has also been reported when great spotted cuckoos parasitize their larger-size carrion crow hosts (Bolopo et al. 2015).

All these results considered together indicate that although elevated begging emanating from nests shared by brood parasite and host nestlings increases host provisioning (Gloag et al. 2012), this does not mean that the parasitic nestling would be able to monopolize the extra food delivered. In fact, previously reported results clearly show that the larger nestlings are the ones who successfully compete for food, regardless whether they are hosts or parasites (see Sect. 27.2.1).

27.4 Can the Presence of a Parasitic Nestling Lead to Changes in the Behaviour of Host Nestlings?

This question has been addressed only rarely, but, in most cases, it has been found that the presence of one parasitic nestling in the nest provokes changes in the behaviour of its host nestmates, more frequently an increase in their begging intensity (Table 27.1). This effect was also demonstrated in a comparative study, which showed that loudness of begging in different host species increased with parasitism pressure (Boncoraglio et al. 2009, but see Rivers et al. 2010). Only in one host species, the red-winged blackbird, no change has been found (Glassey and Forbes 2003; Rivers et al. 2010); however, this host is atypical since in this species nestlings beg so exaggeratedly as the brood parasite nestling (Rivers et al. 2013). In three host species, it has been found that the presence of the parasitic nestling leads to an increase in the intensity of begging, while in two host species, the effect has been the opposite (Table 27.1). Interestingly, a decrease of the intensity of begging occurs in species that are larger than the parasite, while an increase occurs in species that are smaller than the parasite (Table 27.1). This means that the effect of the presence of one parasitic nestling on the begging behaviour of host nestlings is the consequence of host nestlings being hungry when they are smaller than the parasite but not when they are larger than the parasite (see Sect. 27.2.1).

It has been demonstrated that nestlings are able to adjust their begging behaviour according to their nestmate size and begging intensity (Leonard and Horn 1998; Rivers 2007) and that they are able to learn to adjust their begging behaviour according to provisioning rewards (Kedar et al. 2000, Sect. 27.3.4). Thus, the changes in begging behaviour described above are expected to be widespread among host species but surely are not the result of an evolved response to brood parasitism but just a response to the size and/or competitive ability of nestmates. This means that begging signals in parasitized nests are the consequence of a dynamic interaction between the brood parasite and host nestlings (Hauber and Kilner 2007; Pagnucco et al. 2008). Even more, it has been suggested that host nestlings would be selected to mimic exaggerated begging signals of the brood parasite when both share the nest (Hauber and Kilner 2007).

27.5 Food Delivery

It is generally accepted that nestling begging influences food distribution decisions by parents and that brood parasitic nestlings exhibit more exaggerated levels of begging than host nestlings (see Sect. 27.3.1). However, an increased provisioning to parasitized broods compared to unparasitized ones has been supported in only some cases (Table 27.1). Others, however, have failed to find a clear effect of the presence of a brood parasite on host parental provisioning (Table 27.1). For example, it has been reported that foraging rates do not differ significantly between parasitized and unparasitized broods (Soler et al. 1995; Glassey and Forbes 2003; Pagnucco et al. 2008; Rivers et al. 2010; Grim et al. 2017) or that host nestlings achieved similar mass in parasitized and unparasitized nests (Zanette et al. 2005; Skipper 2008). In those cases in which provisioning rate is higher in parasitized than in unparasitized nests, the crucial question is who benefit from this increase, the parasite or the hosts. According to current empirical and experimental available information, the response is that generally nestlings that receive the majority of food brought to the nest are larger ones, regardless whether they are hosts or parasites (see Sect. 27.2.1).

A relevant question is which nestling, the host or the parasite, is preferentially fed in mixed broods. Traditionally, it was thought that it should be the parasite because of their more intense begging and other adaptations (Davies 2000). However, curiously, the host has been reported to be favoured in seven studies and the parasite in six studies (Table 27.1). But, if we take into account the size difference between host and parasite nestlings, much clearer results appear: in brood parasite–host systems in which the parasite is smaller than the host, the latter is favoured in five studies, while the former is not favoured in any study, while when the parasite is larger than the host nestling, the parasite is favoured in five studies, while the host is not favoured in any study (Table 27.1). These clear results highlight the fact that hosts favour the larger nestling in the nests regardless whether it is the host or the parasite as I have emphasized above (Sect. 27.2.1).

27.6 Evidence of Discrimination Based on Food Delivery by Hosts

Nestling recognition and rejection is a potential host defence that may be highly efficient against brood parasites. Rejection of the parasitic nestling can occur by directly ejecting it, deserting the parasitized nest, or by refusing to feed it (see Chap. 29). Although in several studies, it has been shown that brood parasitic or heterospecific nestlings cross-fostered to nests of potential host species starve (see Grim 2006 and Chap. 29), only four of these cases of presumed discrimination can be considered the consequence of selective feeding of own host nestlings and/or of starving the introduced brood parasite nestling to death, two involving hosts of the evicting common cuckoo and two involving hosts of non-evicting brood parasites.

Reed warblers (*Acrocephalus scirpaceus*), a small host species of the common cuckoo, sometimes become insensitive to parasite demands and desert the nest provoking the starvation of the cuckoo nestling (Grim et al. 2003). However, this behaviour was considered a discrimination mechanism not involving recognition but the response of foster parents when the duration of present care exceeds the optimal nestling period (Grim 2007).

The blackbird (*Turdus merula*) is a potential host species that is parasitized only rarely. When nestling cuckoos are experimentally introduced in their nests in the absence of host nestlings, the cuckoo's growth is very poor, and they survive for only 1 or 2 days, starving to death in all cases, while lone blackbird nestlings always survived until fledging (Grim et al. 2011).

Most shiny cowbird parasitic nestlings starve to death when parasitizing nests of a larger host species, the rufous-bellied thrush (*Turdus rufiventris*), because foster parents frequently favour their own nestlings even removing prey items from cowbird mouths passing them to host nestlings (Lichtenstein 2001).

There are two pieces of evidence showing parental discrimination by magpies against great spotted cuckoos. First, in an experimental study in which size-matched broods of different composition (only magpie nestlings, only great spotted cuckoo nestlings or mixed broods) were created (Soler and de Neve 2012), great spotted cuckoo nestlings sometimes starve even in the absence of size asymmetries, while in none of the nests a magpie chick starved. These results could be the consequence of magpies being reluctant to feed cuckoo nestlings, which would imply the existence of discrimination ability by magpies. However, the possibility of cuckoo nestlings receiving lower-quality food items (Soler 2008) or being sensitive to some particular component of the diet (e.g. cereal grains) cannot be dismissed (Soler and de Neve 2012). Second, magpies sometimes ignore active begging of great spotted cuckoo nestlings while waking up one of their own nestlings to feed it (Soler et al. 2017).

Concluding Remarks and Future Directions

Although the breeding strategies' model has received considerable support, additional research on this topic is clearly warranted. A comparative study of hosts selection by brood parasitic cuckoos controlling by phylogeny and considering breeding strategy as a continuous variable (i.e. number of nestlings that on average starve in each potential host species) would allow to test the predictions that each cuckoo species should be adapted to brood reducers or clutch adjusters and that overlapping in host selection should be minimal.

In spite of brood parasitic nestlings exhibiting more exaggerated levels of begging than host nestlings, the majority of food brought to the nest by foster parents is provided to larger nestlings regardless whether they are hosts or parasites. Few investigations have quantified food distribution in brood parasite–host systems in which the parasite is smaller than the host. Future research efforts should therefore focus on these systems.

Additional work would be needed to determine whether the finding that common cuckoo nestlings trick their reed warbler foster parents by mimicking the begging calls of a complete brood can be generalized or not. Sonograms should be obtained in parasitized reed warbler nests from cuckoo nestlings at different ages.

The analysis of the differences in the vocalizations of cuckoo and host nestlings in the nests of four host species does not support the integration of

(continued)

signals hypothesis given that vocalizations of cuckoo chicks do not differ among host species (Fig. 27.2). Similar studies using other host species would contribute to elucidate whether this hypothesis can be considered to be broadly supported or not.

Sonograms of cuckoo nestlings reared in the nests of four different host species (Fig. 27.2) show three types of sonograms of common cuckoo nestling begging calls, which are almost equally distributed among the three cuckoo races. This is a fascinating finding that deserves additional work.

In general, more information is needed not only about begging behaviour of cuckoo and host nestlings but also about cuckoo growth and survival in different current and old host species. Also, cross-fostering experiments introducing one brood parasitic nestling in the nest of different host species could provide new information about adaptations and counter-adaptations involved in brood parasitic nestling–foster parent relationships that could illuminate the topic.

Acknowledgements I thank the reviewers Tomas Grim, Rebecca Kilner and James River for their constructive and helpful comments and José Ángel Soler Ortiz and Manuel Martín-Vivaldi for making Figs. 27.1 and 27.2, respectively. In addition, I thank M. Martín-Vivaldi, E. Røskaft, C. Moskát and J.J. Palomino for allowing me to include in this chapter our unpublished results testing the integration of signals hypothesis.

References

Álvarez F (1994) A gens of cuckoo *Cuculus canorus* parasitizing rufous bush chat *Cercotrichas galactotes*. J Avian Biol 25:239–243

Álvarez F (2004) The conspicuous gape of the nestling common cuckoo *Cuculus canorus* as a supernormal stimulus for rufous bush chat *Cercotrichas galactotes* hosts. Ardea 92:63–68

Bengtsson H, Ryden O (1983) Parental feeding rate in relation to begging behavior in asynchronously hatched broods of the great tit *Parus major*. Behav Ecol Sociobiol 12:243–251

Bolopo D, Canestrari D, Roldán M, Baglione V, Soler M (2015) High begging intensity of great spotted cuckoo nestlings favours larger-size crow nest mates. Behav Ecol Sociobiol 69:873–882

Boncoraglio G, Saino N, Garamszegi LZ (2009) Begging and cowbirds: brood parasites make hosts scream louder. Behav Ecol 20:215–221

Briskie JV, Naugler CT, Leech SM (1994) Begging intensity of nestling birds varies with sibling relatedness. Proc R Soc B 258:73–78

Butchart SHM, Kilner RM, Fuisz T, Davies NB (2003) Differences in the nestling calls of hosts and host-races of the common cuckoo, *Cuculus canorus*. Anim Behav 65:345–354

Caro SM, Griffin AS, Hinde CA, West SA (2016) Unpredictable environments lead to the evolution of parental neglect in birds. Nat Commun 7:10985

Davies NB (2000) Cuckoos, cowbirds and other cheats. Poyser, London

Davies NB, Kilner RM, Noble DG (1998) Nestling cuckoos, *Cuculus canorus*, exploit hosts with begging calls that mimic a brood. Proc R Soc B 265:673–678

Dearborn DC (1998) Begging behavior and food acquisition by brown-headed cowbird nestlings. Behav Ecol Sociobiol 43:259–270

Dearborn DC (1999) Brown headed cowbird nestling vocalizations and risk of nest predation. Auk 116:448–457

Dearborn DC, Lichtenstein G (2002) Begging behaviour and host exploitation in parasitic cowbirds. In: Wright J, Leonard ML (eds) The evolution of begging: competition, cooperation and communication. Kluwer Academic, Dordrecht, pp 361–388

Dearborn DC, Anders AD, Thompson FR III, Faaborg J (1998) Effects of cowbird parasitism on parental provisioning and nestling food acquisition and growth. Condor 100:325–334

Glassey B, Forbes S (2003) Why brown-headed cowbirds do not influence red-winged blackbird parent behaviour. Anim Behav 65:1235–1246

Gloag R, Kacelnik A (2013) Host manipulation via begging call structure in the brood-parasitic shiny cowbird. Anim Behav 86:101–109

Gloag R, Tuero DT, Fiorini VD, Reboreda JC, Kacelnik A (2012) The economics of nestmate killing in avian brood parasites: a provisions trade-off. Behav Ecol 23:132–140

Grayson P, Glassey B, Forbes S (2013) Does brood parasitism induce paternal care in a polygynous host? Ethology 119:439–495

Grim T (2006) The evolution of nestling discrimination by hosts of parasitic birds: why is rejection so rare? Evol Ecol Res 8:785–802

Grim T (2007) Experimental evidence for chick discrimination without recognition in a brood parasite host. Proc R Soc B 274:373–381

Grim T (2008) Wing-shaking and wing-patch as nestling begging strategies: their importance and evolutionary origins. J Ethol 26:9–15

Grim T, Honza M (2001) Does supernormal stimulus influence parental behaviour of the cuckoo's host? Behav Ecol Sociobiol 49:322–329

Grim T, Kleven O, Mikulica O (2003) Nestling discrimination without recognition: a possible defence mechanism for hosts towards cuckoo parasitism? Proc R Soc B 270:S73–S75

Grim T, Rutila J, Cassey P, Hauber ME (2009) Experimentally constrained virulence is costly for common cuckoo chicks. Ethology 115:14–22

Grim T, Samaš P, Moskát C, Kleven O, Honza M, Moksnes A, Røskaft E, Stokke BG (2011) Constraints on host choice: why do parasitic birds rarely exploit some common potential hosts? J Anim Ecol 80:508–518

Grim T, Tyller Z, Samaš P (2017) Unusual diet of brood parasite nestlings and its fitness consequences. Auk 134:732–750

Haskell DG (2002) Begging behavior and nest predation. In: Wright J, Leonard ML (eds) The evolution of begging: competition, cooperation and communication. Kluwer Academic, Dordrecht, pp 163–172

Hauber ME (2003) Lower begging responsiveness of host versus parasitic Brown-headed cowbird (*Molothrus ater*) nestlings is related to species identity but not to early social experience. J Comp Psychol 117:24–30

Hauber ME, Kilner RM (2007) Coevolution, communication, and host chick mimicry in parasitic finches: who mimics whom? Behav Ecol Sociobiol 61:497–503

Hauber ME, Montenegro K (2002) What are the costs of raising a brood parasite? Comparing host parental care at parasitized and non-parasitized broods. Etología 10:1–9

Hauber ME, Moskat C (2008) Shared parental care is costly for nestlings of common cuckoos and their great reed warbler hosts. Behav Ecol 19:79–86

Hoover JP, Reetz MJ (2006) Brood parasitism increases provisioning rate and reduces offspring recruitment and adult return rates in a cowbird host. Oecologia 149:165–173

Ibáñez-Álamo JD, Arco L, Soler M (2012) Experimental evidence for a predation cost of begging using active nests and real chicks. J Ornithol 153:801–807

Kedar H, Rodríguez-Gironés MA, Yedvab S, Lotem A, Winkler DW (2000) Learning modifies offspring signaling of need to their parents. Proc R Soc B 267:1723–1727

Kilner RM (2003) How selfish is a cowbird nestling? Anim Behav 66:569–576

Kilner RM (2005) The evolution of virulence in brood parasites. Ornithol Sci 4:55–64

Kilner RM, Davies NB (1999) How selfish is a cuckoo chick? Anim Behav 58:797–808

Kilner RM, Davies NB, Noble DG (1999a) Signals of need in parent-offspring communication and their exploitation by the common cuckoo. Nature 397:667–672

Kilner RM, Noble DG, Davies NB (1999b) How should cuckoo chicks signal in different host nests? Trends Ecol Evol 14:322

Kilner RM, Madden JR, Hauber ME (2004) Brood parasitic cowbird nestlings use host young to procure resources. Science 305:877–879

Kilpatrick AM (2002) Variation in growth of brown-headed cowbird (*Molothrus ater*) nestlings and energetic impacts on their host parents. Can J Zool 80:145–153

Lack D (1947) The significance of clutch size. Ibis 89:302–352

Langmore NE, Maurer G, Adcock GJ, Kilner RM (2008) Socially acquired host-specific mimicry and the evolution of host races in Horsfield's bronze-cuckoo *Chalcites basalis*. Evolution 62:1689–1699

Leonard ML, Horn AG (1998) Need and nestmates affect begging in tree swallows. Behav Ecol Sociobiol 42:431–436

Lichtenstein G (2001) Low success of shiny cowbird chicks parasitizing rufous-bellied thrushes: chick–chick competition or parental discrimination? Anim Behav 61:401–413

Lichtenstein G, Dearborn DC (2004) Begging and short-term need in cowbird nestlings: how different are brood parasites? Behav Ecol Sociobiol 56:352–359

Lichtenstein G, Sealy SG (1998) Nestling competition, rather than supernormal stimulus, explains the success of parasitic brown-headed cowbird chicks in yellow warbler nests. Proc R Soc B 265:249–254

Lorenzana JC, Sealy SG (1999) A meta-analysis of the impact of parasitism by the brown-headed cowbird on its hosts. Stud Avian Biol 18:241–253

Madden JR, Davies NB (2006) A host-race difference in begging calls of nestling cuckoos *Cuculus canorus* develop through experience and increases host provisioning. Proc R Soc B 273:2343–2345

Martín-Gálvez D, Soler M, Soler JJ, Martín-Vivaldi M, Palomino JJ (2005) Food acquisition by common cuckoo chicks in rufous bush robin nests and the advantage of eviction behaviour. Anim Behav 70:1313–1321

Moksnes A, Røskaft E (1995) Egg-morphs and host preference in the common cuckoo (*Cuculus canorus*): an analysis of cuckoo and host eggs from European museum collections. J Zool Lond 236:625–648

Moskát C, Honza M (2000) Effect of nest and nest site characteristics on the risk of cuckoo *Cuculus canorus* parasitism in the great reed warbler *Acrocephalus arundinaceus*. Ecography 23:335–341

Pagnucco K, Zanette L, Clinchy M, Leonard ML (2008) Sheep in wolf's clothing: host nestling vocalizations resemble their cowbird competitor's. Proc R Soc B 275:1061–1065

Payne RB, Payne LL (1998) Nestling eviction and vocal begging behaviors in the Australian glossy cuckoos *Chrysococcyx basalis and C. lucidus*. In: Rothstein SI, Robinson SK (eds) Parasitic birds and their hosts. Oxford University Press, Oxford, pp 152–169

Payne RB, Payne LL (2002) Begging for parental care from another species: specialization and generalization in brood-parasitic finches. In: Wright J, Leonard ML (eds) The evolution of begging: competition, cooperation and communication. Kluwer Academic, Dordrecht, pp 429–449

Price K, Ydenberg R (1995) Begging and provisioning in broods of asynchronously- hatched yellow-headed blackbird nestlings. Behav Ecol Sociobiol 37:201–208

Redondo T, Zuñiga JM (2002) Dishonest begging and host manipulation by *Clamator* cuckoos. In: Wright J, Leonard ML (eds) The evolution of begging: competition, cooperation and communication. Kluwer Academic, Dordrecht, pp 389–412

Rivers JW (2007) Nestmate size, but not short-term need, influences begging behavior of a generalist brood parasite. Behav Ecol 18:222–230

Rivers JW (2009) Parent-absent begging in the Brown-headed cowbird (*Molothrus ater*): the role of short-term need and nestmate size. Behav Ecol Sociobiol 63:707–717

Rivers JW, Peer BD (2016) Relatedness constrains virulence in an obligate avian brood parasite. Ornithol Sci 15:191–201

Rivers JW, Loughin TM, Rothstein SI (2010) Brown-headed cowbird nestlings influence nestmate begging, but not parental feeding, in hosts of three distinct sizes. Anim Behav 79:107–116

Rivers JW, Blundell MA, Loughin TM, Peer BD, Rothstein SI (2013) The exaggerated begging behavior of an obligate avian brood parasite is shared with a nonparasitic close relative. Anim Behav 86:529–536

Rivers JW, Blundell MA, Rothstein SI (2014) Mismatched begging displays between foreign and host offspring reduced brood parasite fitness. Behav Ecol 25:785–793

Robinson SK, Rothstein SI, Brittingham MC, Petit LJ, Grzybowski JA (1995) Ecology and behaviour of cowbirds and their impact on host populations. In: Martin TE, Finch DM (eds) Ecology and management of neotropical migratory birds. Oxford University Press, Oxford, pp 428–460

Roldán M, Soler M (2011) Parental care parasitism: how unrelated offspring attain acceptance by foster parents? Behav Ecol 22:679–691

Roldán M, Martín-Gálvez D, Rodríguez J, Soler M (2013) Breeding biology and fledgling survival in a carrion crow *Corvus corone* population of southern Spain: a comparison of group and pair breeder. Acta Ornithol 48:221–235

Royle NJ, Hartley IR, Parker GA (2002) Begging for control: when are offspring solicitation behaviours honest? Trends Ecol Evol 17:434–440

Skipper BR (2008) Influence of parental provisioning and Brown-headed Cowbird (*Molothrus ater*) parasitism on Bobolink (*Dolichonyx oryzivorus*) nestling mass. Doctoral dissertation, University of Nebraska at Kearney

Smiseth PT, Bu RJ, Eikenæs AK, Amundsen T (2003) Food limitation in asynchronous bluethroat broods: effects on food distribution, nestling begging, and parental provisioning rules. Behav Ecol 14:793–801

Soler M (2001) Begging behaviour of nestlings and food delivery by parents: the importance of breeding strategy. Acta Ethol 4:59–63

Soler M (2002) Breeding strategy and begging intensity: influences on food delivery by parents and host selection by parasitic cuckoos. In: Wright J, Leonard ML (eds) The evolution of begging: competition, cooperation and communication. Kluwer Academic, Dordrecht, pp 413–427

Soler M (2008) Do hosts of interspecific brood parasites feed parasitic chicks with lower quality prey? Anim Behav 76:1761–1763

Soler M, de Neve L (2012) Great spotted cuckoo nestlings but not magpie nestlings starve in experimental age-matched broods. Ethology 118:1036–1044

Soler M, de Neve L (2013) Brood mate eviction or brood mate acceptance by brood parasitic nestlings? An experimental study with the non-evictor great spotted cuckoo and its magpie host. Behav Ecol Sociobiol 67:601–607

Soler M, Soler JJ (1999) The cuckoo chick tricks their reed warbler foster parents, but what about other host species? Trends Ecol Evol 14:296–297

Soler M, Martínez JG, Soler JJ, Møller AP (1995) Preferential allocation of food by magpie *Pica pica* to great spotted cuckoo *Clamator glandarius* chicks. Behav Ecol Sociobiol 37:7–13

Soler M, de Neve L, Roldán M, Macías-Sánchez E, Martín-Gálvez D (2012) Do great spotted cuckoo nestlings beg dishonestly? Anim Behav 83:163–169

Soler M, Macías-Sánchez E, Martín-Gálvez D, de Neve L (2017) Complex feeding behaviour by magpies in nests with great spotted cuckoo nestlings. J Avian Biol. https://doi.org/10.1111/jav.01473

Soler M, de Neve L, Macías-Sánchez E, Pérez-Contreras T (2018) Despite a higher vigilance, great spotted cuckoos incur less erroneous begging when hungry than their magpie host nestlings. Behav Ecol Sociobiol (in press)

Tanaka KD, Ueda K (2005) Horsfield's hawk-cuckoo nestlings simulate multiple gapes for begging. Science 308:653

Teather KL (1992) An experimental study of competition for food between male and female nestlings of the red-winged blackbird. Behav Ecol Sociobiol 31:81–87

Ursino CA, De Marsico MC, Sued M, Farall A, Reboreda JC (2011) Brood parasitism disproportionately increases nest provisioning and helper recruitment in a cooperatively breeding bird. Behav Ecol Sociobiol 65:2279–2286

Wang N, Kimball R (2012) Nestmate killing by obligate brood parasitic chicks: is this linked to obligate siblicidal behavior? J Ornithol 153:825–831

West MJ, King AP (1998) Female visual displays affect the development of male song in the Cowbird. Nature 334:244–246

Wright J, Leonard M (2002) The evolution of begging: competition, cooperation and communication. Kluwer Academic, Dordrecht

Yasukawa K, Berrios HK, Johannes AW (2016) A test of the nestling discrimination hypothesis for parasitism of red-winged blackbirds (*Agelaius phoeniceus*) by brown-headed cowbirds (*Molothrus ater*). Wilson J Ornithol 128:441–445

Zanette L, MacDougall-Shakleton E, Clinchy M, Smith JNM (2005) Brown-headed cowbirds skew host offspring sex ratios. Ecology 86:815–820

Begging Call Mimicry by Brood Parasite Nestlings: Adaptation, Manipulation and Development

28

Gabriel A. Jamie and Rebecca M. Kilner

Abstract

Begging calls provide nestling brood parasites with a powerful and flexible tool for avoiding rejection, altering parental provisioning and competing with host nestmates. Despite much research into the topic, no synthesis of parasite vocal strategies for host manipulation has yet been made. In this chapter, we begin by reviewing the literature on reported similarity between the begging calls of avian brood parasites and their hosts. We show that such similarity is a more widespread phenomenon than previously appreciated. Secondly, we examine the selection pressures that drive the evolution of begging call mimicry by avian brood parasites, assess their importance and illustrate them with empirical examples. Finally, we propose a theoretical framework to explain variation in the ways that brood parasite begging calls develop. We suggest that the mode of development can be predicted from a consideration of the accuracy of genetic cues (as mediated by parasite specialisation levels) and the benefits to the young parasite of using environmental cues to modulate their begging call (as influenced by levels of discrimination shown by host parents). Perhaps the main contribution of this chapter is to highlight how little we know about brood parasitic begging calls. This points the way for future work on this topic.

28.1 Introduction

Studies of avian brood parasitism have revealed a multitude of strategies employed by parasite parents and offspring to dupe host parents and allow parasites to integrate into host families (Rothstein 1990; Davies 2011; Feeney et al. 2014). Considerable attention has been given to the visual trickery of hosts by brood parasites through

G. A. Jamie (✉) · R. M. Kilner
Department of Zoology, University of Cambridge, Cambridge, UK
e-mail: gaj29@cam.ac.uk

M. Soler (ed.), *Avian Brood Parasitism*, Fascinating Life Sciences,
https://doi.org/10.1007/978-3-319-73138-4_28

mimicry of host egg and chick appearance (Langmore and Spottiswoode 2012). By contrast, despite much research in the area, no equivalent examination of parasite vocal strategies for host manipulation has yet been undertaken. Such an analysis not only allows us to organise interesting natural history into a predictive theoretical framework but also provides valuable insights into the evolution of host–parasite interactions that are not evident from studying visual mimicry alone.

A key distinction between vocal and visual strategies for host manipulation is that vocal behaviour has much greater potential for plasticity. Processes, such as learning, which underpin plasticity, can generate vocal similarity between parasite and host within a matter of days rather than requiring successive generations of genetic evolution (see Chap. 27). This sets the stage for vocal similarities to arise at a pace that outstrips visual ones with potentially important evolutionary consequences (Price et al. 2003; West-Eberhard 2003; Whitman and Agrawal 2009; Verzijden et al. 2012). For example, plasticity in begging calls could allow a parasite nestling to persist in a new host environment in a single generation, exposing it to novel selection pressures and altering the course of its genetic evolution (Pfennig et al. 2010). Conversely, such plasticity could also allow the offspring of a generalist brood parasite to be successful in the nests of a range of host species without exhibiting genetic specialisation to any one host in particular.

Furthermore, the plastic nature of begging calls means that they can be used by parasites to manipulate host parents into provisioning them. Parasites can tune into the host's offspring–parent communication channel and update their displays over the course of development depending on their condition and need (Kilner et al. 1999; Davies 2011).

Finally, although begging call development can be highly plastic, calls can also develop without any influence of the host environment. Whilst some nestling brood parasites plastically develop host-specific begging calls, other species beg with calls that are unchanged by the host environment in which the parasitic nestling develops. This raises the question of why variation in begging call development exists. We examine the selection pressures underpinning mimetic vocal begging and use this to develop an adaptive framework, illustrated with empirical examples, that allows us to understand variation in begging call development for different brood parasites.

28.2 Begging Call Similarity Between Avian Brood Parasites and Their Hosts: A Survey of the Literature

We begin by surveying the literature on begging call similarity between all avian brood parasites and their hosts (Table 28.1). This provides the empirical foundation for the conceptual sections that follow.

We collated published, and some unpublished, information on begging calls of all interspecific avian brood parasites. For each species, we noted whether similarity has been reported between its begging calls and those of its host(s), how that similarity was assessed ("subjectively" by human comparison or "quantitatively" by analytic comparison of sonograms) and whether that similarity is to an individual host chick

Table 28.1 Begging call information for those avian brood parasites whose begging calls have been described

English name	Scientific name	Parasite begging calls			Parasite specialisation level (from Erritzøe et al. 2012)		Hosts known to show discrimination against parasite nestlings? (see Grim 2006; and Chap. 29)
		Reported similarity to host's?	Nature of similarity	How similarity assessed	Species-level	Individual-level	
Cuckoos	*Cuculidae*						
American striped cuckoo	*Tapera naevia*	Yes (Morton and Farabaugh 1979)	To individual host chick (Morton and Farabaugh 1979)	Subjectively (Morton and Farabaugh 1979)	Generalist	Unknown	Unknown
Great spotted cuckoo	*Clamator glandarius*	Yes (Redondo and Arias de Reyna 1988; Mundy 1973), however, no evidence of similarity in a subsequent, more rigorous, study (Roldán et al. 2013)	To individual host chick (Redondo and Arias de Reyna 1988; Mundy 1973)	Subjectively (Redondo and Arias de Reyna 1988; Mundy 1973) and quantitatively (Roldán et al. 2013)	Specialist	Specialist	Yes (Soler et al. 1995)
Levaillant's cuckoo	*Clamator levaillantii*	Yes (Jubb 1952; Mundy 1973; Steyn 1973)	To individual host chick (Steyn 1973) and adult (Steyn 1973; Mundy 1973; Jubb 1952)	Subjectively (Jubb 1952; Mundy 1973; Steyn 1973)	Generalist	Unknown	Unknown
Jacobin cuckoo	*Clamator jacobinus*	Yes (Erritzøe et al. 2012; Jamie and de Silva Wijeyeratne 2014; A. Ridley unpublished; Fry et al. 2002)	To individual host chick (Erritzøe et al. 2012; A. Ridley unpublished; Fry et al. 2002) and	Subjectively (Erritzøe et al. 2012; Jamie and de Silva Wijeyeratne 2014; A. Ridley	Generalist	Unknown	Unknown

(continued)

Table 28.1 (continued)

English name	Scientific name	Parasite begging calls			Parasite specialisation level (from Erritzøe et al. 2012)		Hosts known to show discrimination against parasite nestlings? (see Grim 2006; and Chap. 29)
		Reported similarity to host's?	Nature of similarity	How similarity assessed	Species-level	Individual-level	
			adult (Jamie and Wijeyeratne 2014)	unpublished; Fry et al. 2002)			
Yellow-billed cuckoo	*Coccyzus americanus*	No (Payne 2005; Potter 1980; Erritzøe et al. 2012)			Generalist	Unknown	Unknown
Black-billed cuckoo	*Coccyzus erythropthalmus*	No (Erritzøe et al. 2012; Payne 2005; Spencer 1943)			Generalist	Unknown	Unknown
Thick-billed cuckoo	*Pachycoccyx audeberti*	Yes (Erritzøe et al. 2012; Fry et al. 2002; Vernon 1984)	To individual host chick (Erritzøe et al. 2012; Fry et al. 2002; Vernon 1984)		Regional specialist	Probably specialist	Unknown
Common koel	*Eudynamys scolopaceus*	No (Payne 2005; Gosper 1997)			Generalist	Unknown	Yes (Dewar 1907)
Long-tailed koel	*Urodynamis taitensis*	Yes (McLean and Waas 1987; Ranjard et al. 2010)	To individual host chick (McLean and Waas 1987; Ranjard et al. 2010)	Quantitatively (Ranjard et al. 2010) and subjectively (McLean and Waas 1987)	Generalist	Unknown	Unknown
Channel-billed cuckoo	*Scythrops novaehollandiae*	Yes (Courtney 1967; Lord 1956)	To individual host chick (Courtney 1967; Lord 1956)	Subjectively (Courtney 1967; Lord 1956)	Generalist	Unknown	Unknown

Diederik Cuckoo	*Chrysococcyx caprius*	No—but does vary nonmimetically between host races (Reed 1968)			Generalist	Gentes	Unknown
Klaas's cuckoo	*Chrysococcyx klaas*	No (Payne 2005; Erritzøe et al. 2012; Skead 1995)			Generalist	Unknown	Unknown
Horsfield's bronze cuckoo	*Chalcites basalis*	Yes (Langmore et al. 2008)	To individual host chick (Langmore et al. 2008)	Quantitatively (Langmore et al. 2008)	Generalist	Generalist	Chick rejection (Langmore et al. 2003)
Shining bronze cuckoo	*Chalcites lucidus*	Yes (Anderson et al. 2009; Ranjard et al. 2010; McLean and Waas 1987)	To individual host chick (Anderson et al. 2009; Ranjard et al. 2010; McLean and Waas 1987)	Quantitatively (Anderson et al. 2009; Ranjard et al. 2010) and subjectively (McLean and Waas 1987)	Specialist	Unknown	Chick rejection (Langmore et al. 2003, 2011; Sato et al. 2015)
Little bronze cuckoo	*Chalcites minutillus*	Yes (N. Langmore unpublished)	To individual host chick (N. Langmore unpublished)	Subjectively (N. Langmore unpublished)	Specialist	Specialist	Chick rejection (Sato et al. 2010; Tokue and Ueda 2010)
Black-eared cuckoo	*Misocalius osculans*	Yes (Erritzøe et al. 2012)	To individual host chick (Erritzøe et al. 2012)	Subjectively (Erritzøe et al. 2012)	Generalist	Unknown	Unknown
Pallid cuckoo	*Cacomantis pallidus*	Yes (Courtney 1967; Serventy and Whittell 1962)	To individual host chick (Courtney 1967; Serventy and Whittell 1962)	Subjectively (Courtney 1967; Serventy and Whittell 1962)	Generalist	Unknown	Unknown
Fan-tailed cuckoo	*Cacomantis flabelliformis*	No (Erritzøe et al. 2012; Clunie 1973)			Generalist	Unknown	Unknown

(continued)

Table 28.1 (continued)

English name	Scientific name	Parasite begging calls			Parasite specialisation level (from Erritzøe et al. 2012)		Hosts known to show discrimination against parasite nestlings? (see Grim 2006; and Chap. 29)
		Reported similarity to host's?	Nature of similarity	How similarity assessed	Species-level	Individual-level	
Rusty-breasted cuckoo	*Cacomantis sepulcralis*	No (Grim 2008)			Generalist	Unknown	Unknown
Common hawk-cuckoo	*Hierococcyx varius*	Yes (Ali and Whistler 1936)	To individual host chick (Ali and Whistler 1936)	Subjectively (Ali and Whistler 1936)	Generalist	Unknown	Unknown
Black cuckoo	*Cuculus clamosus*	No (Payne 2005; Skead 1946)			Generalist	Unknown	Unknown
Red-chested cuckoo	*Cuculus solitarius*	No (Payne 2005; Salewski and Grafe 1999)			Generalist	Unknown	Unknown
Indian cuckoo	*Cuculus micropterus*	Yes (Erritzøe et al. 2012; Payne 2005)	To individual host chick (Erritzøe et al. 2012; Payne 2005)	Subjectively (Erritzøe et al. 2012; Payne 2005)	Generalist	Unknown	Unknown
African cuckoo	*Cuculus gularis*	No (Fry et al. 2002)			Generalist	Unknown	Unknown
Common cuckoo	*Cuculus canorus*	Yes for some hosts (Butchart et al. 2003; Davies et al. 1998; Erritzøe et al. 2012; Madden and Davies 2006; Payne 2005)	To brood of host chicks (Davies et al. 1998)	Quantitatively (Butchart et al. 2003; Davies et al. 1998; Madden and Davies 2006)	Generalist	Gentes	Yes, in one population but not others (Davies et al. 1998; Grim et al. 2003)

Honeyguides	*Indicatoridae*						
Eastern green-backed honeyguide	*Prodotiscus zambesiae*	No (Vernon 1987; Short and Horne 2001)			Generalist	Unknown	Unknown
Lesser honeyguide	*Indicator minor*	No (Short and Horne 2001)			Generalist	Unknown	Unknown
Scaly-throated honeyguide	*Indicator variegatus*	No (Short and Horne 2001)					Unknown
Greater honeyguide	*Indicator indicator*	Yes (Jubb 1966)	To brood of host chicks (Jubb 1966)	Subjectively (Jubb 1966)	Generalist	Generalist	Unknown
Cowbirds	*Icteridae*						
Screaming cowbird	*Molothrus rufoaxillaris*	Yes (De Mársico et al. 2012)	To individual host chick (De Mársico et al. 2012)	Quantitatively (De Mársico et al. 2012)	Specialist	Specialist	Chick rejection (De Mársico et al. 2012)
Brown-headed cowbird	*Molothrus ater*	No (Dearborn 1998; Rivers 2006)		Quantitatively (Rivers 2006)	Generalist	Generalist	No evidence of discrimination (Dearborn 1998; Rivers et al. 2010)
Shiny cowbird	*Molothrus bonariensis*	No (Gloag and Kacelnik 2013; Tuero et al. 2015)		Quantitatively (Gloag and Kacelnik 2013; Tuero et al. 2015)	Generalist	Generalist	Yes, by at least some hosts (Dearborn and Lichtenstein 2002; Lichtenstein 2001)
Indigobirds/ Whydahs	*Viduidae*						
Village indigobird	*Vidua chalybeata*	No (Payne and Payne 2002)		Quantitatively (Payne and Payne 2002)	Specialist	Specialist	Yes (Payne et al. 2001)

(continued)

Table 28.1 (continued)

English name	Scientific name	Parasite begging calls			Parasite specialisation level (from Erritzøe et al. 2012)		Hosts known to show discrimination against parasite nestlings? (see Grim 2006; and Chap. 29)
		Reported similarity to host's?	Nature of similarity	How similarity assessed	Species-level	Individual-level	
Pin-tailed whydah	*Vidua macroura*	Yes (G. Jamie unpublished)	To individual host chick (G. Jamie unpublished)	Quantitatively (G. Jamie unpublished)	Prob. regional specialist	Specialist	Yes (Schuetz 2005)
Broad-tailed Paradise whydah	*Vidua obtusa*	Yes (G. Jamie unpublished)	To individual host chick (G. Jamie unpublished)	Quantitatively (G. Jamie unpublished)	Specialist	Specialist	Unknown
Cuckoo finch	*Anomalospiza imberbis*	No (C. Spottiswoode unpublished)		Subjectively (C. Spottiswoode unpublished)	Generalist	Specialist	Yes (Spottiswoode et al. 2012)

or to a brood of host chicks. Additionally, we noted the level of specialisation shown by each parasite species and whether any of its hosts are known to exhibit chick discrimination or rejection. These last two parameters are included because of their importance in predicting how begging calls are likely to develop in the parasite (see Sect. 28.3).

We found that there are many gaps in knowledge of brood parasite begging calls: indeed the begging calls of 64 of the 100 or so species of brood-parasitic birds remain entirely unknown. The largest gaps in knowledge are among the *Vidua* finches, the Asian and African cuckoos and the honeyguides. Importantly, most publications describing vocal similarity between parasite and host calls are largely anecdotal. Only 11 species of avian brood parasite have had their begging calls quantitatively compared to those of their host(s) (Table 28.1). Most studies instead have small sample sizes and assess similarity subjectively. This is problematic, as evidenced by a recent quantitative study which found no evidence for begging call mimicry of hosts by great spotted cuckoos (Roldán et al. 2013), despite earlier works based on small sample sizes and subjective assessment of similarity suggesting evidence of mimicry (Mundy 1973; Redondo and Arias de Reyna 1988). Therefore, there is still work to be done validating claims of vocal similarity between parasitic and host young and providing detail about the development and information content of parasite begging calls.

Of those parasitic species whose begging calls have been described, similarity between host chick begging calls and parasitic nestling vocalisations has been reported from at least 53% (Table 28.1). Of the 11 species whose calls have been quantitatively compared to host calls, similarity has been reported from 7 (Table 28.1). These findings run counter to the prevailing view in the literature. For example, in his monograph *The Cuckoos*, Payne states that the “Begging calls of most cuckoos differ from the begging calls of their hosts” (Payne 2005, p. 148). However, our literature survey reveals that, of the 25 parasitic cuckoo species for which begging calls have been described, vocal similarity between the begging calls of cuckoos and their hosts has been reported from at least 60% (Table 28.1).

The literature survey also shows that vocal similarity between avian brood parasites and their hosts has been reported from six of the seven independent transitions to parasitic lifestyles in birds (in all three transitions in cuckoos and in each of the cowbird, finch and honeyguide transitions). The only transition lacking reports of vocal similarity is that of the black-headed duck (*Heteronetta atricapilla*) which is highly precocial and produces no begging calls (Lyon and Eadie 2013). Thus, vocal similarity between avian brood parasites and their hosts is potentially a more widespread phenomenon than currently appreciated.

28.3 What Selection Pressures Underpin the Evolution of Vocal Mimicry in Brood Parasites?

Having documented the extent of begging call similarity between avian brood parasites and their hosts, we now examine the processes that might drive these similarities. Five hypotheses have been suggested. Two of them consider the evolution of vocal mimicry to be independent of the evolutionary interactions between brood parasites and their hosts. They suggest that similarity could be the consequence of (1) phylogenetic inertia or (2) shared ecology in the host nests (Grim 2005). However, neither is likely to be responsible for avian parasite-host begging call similarity. Phylogenetic inertia is unlikely due to the large evolutionary distances between most avian brood parasites and their hosts. Likewise although ecological factors, such as predation, have been shown to have some influence on call parameters like peak frequency and amplitude (Briskie et al. 1999), there is no evidence they can explain the majority of variation in nestling begging call structure between species. For example, closely related nestling Estrildid finch species occupying the same habitat in southern Zambia, and presumably subject to very similar predation pressures, have highly divergent begging call structures (G. Jamie unpublished). Therefore, whilst shared ecology may account for minor similarities in peak frequency between parasite and host, it is unlikely that it can explain major convergences in overall call structure.

We confine our attention to the remaining three hypotheses in which vocal mimicry is considered to have evolved as a direct consequence of the evolutionary interactions between brood parasites and their hosts. These are (3) rejection of foreign chicks by hosts either through chick ejection or nest abandonment (Langmore et al. 2008), (4) the need for parasites to tune into parent-offspring communication systems in order to manipulate host parents to supply them with enough food (Davies et al. 1998; Kilner et al. 1999) and (5) competition with nestmates for access to parental investment (Hauber and Kilner 2007; Pagnucco et al. 2008).

In relation to hypothesis 3, chick rejection has evolved in the hosts of several avian brood parasites (Table 28.1). These parasites exhibit begging calls that closely match those of their hosts (Langmore et al. 2008; De Mársico et al. 2012; Ranjard et al. 2010). Therefore, it appears vocal mimicry of hosts is essential for survival in the nests of hosts exhibiting chick rejection behaviour.

Even in systems where hosts do not reject parasitic chicks outright, some parasites have evolved begging calls that closely match those of their hosts. In these cases, vocal mimicry may have evolved to tune into parent-offspring communication rules to manipulate host parents into feed them adequately (hypothesis 4). For example, nestling *Vidua* finches visually mimic the appearance of host chicks. Rather than being rejected by host parents, nonmimetic chicks are instead fed less than chicks which look like their own offspring (Schuetz 2005). Pin-tailed (*Vidua macroura*) and broad-tailed paradise whydahs (*V. obtusa*) nestlings mimic the begging calls as well as the visual appearance of their respective hosts common waxbill (*Estrilda astrild*) and orange-winged pytilia (*Pytilia afra*) (G. Jamie unpublished). Given the lack of chick rejection by these hosts, it suggests that begging call mimicry is important for soliciting sufficient levels of investment from host parents rather than to prevent removal of the parasite from the nest by host parents (Schuetz 2005; Jamie unpublished).

In common cuckoos, the large size of cuckoo chicks relative to host chicks means that cuckoo chicks must tap into parent-offspring communication systems to manipulate parental provisioning (Davies et al. 1998). Here, mimicry of the whole brood is necessary for the cuckoo to signal its hunger state to the host parent and compensate for the deficient visual stimulus it provides—only one gape rather than several (Kilner et al. 1999). Some avian brood parasites, such as shiny cowbirds (*Molothrus bonariensis*), can also vocally manipulate host behaviour without using mimetic begging calls (Gloag and Kacelnik 2013; Tuero et al. 2015). Such situations arise where host parents are responsive to certain generic characteristics of a hungry chick rather than requiring species-specific calls.

Hypothesis 5 suggests that the evolution of vocal mimicry in nestling avian brood parasites is driven by competition with nestmates for access to parental investment (Hauber and Kilner 2007; Pagnucco et al. 2008). This hypothesis applies to non-evictor brood parasite species (see Chap. 28) such as Viduidae finches, cowbirds, *Clamator* cuckoos, channel-billed cuckoo (*Scythrops novaehollandiae*) and Asian koel (*Eudynamys scolopaceus*). Most of these have been reported to have begging calls similar to those of their hosts (Table 28.1). The traditional explanation for the evolution of trait similarity between parasites and hosts is that the parasite has converged on the characteristics of the host offspring (reviewed in Hauber and Kilner 2007). However, it is also possible that some similarity between host and parasite begging calls arises through host adaptations to resemble the parasitic chick. For example, one study experimentally parasitised song sparrow (*Melospiza melodia*) nests with brown-headed cowbird (*Molothrus ater*) chicks. Host nestlings in parasitised nests plastically altered aspects of their begging call (higher frequency, louder) so that they more closely resembled aspects of the parasite chick's begging calls (Pagnucco et al. 2008). This process of the host converging on the parasite call to compete more effectively for parental care could also explain some of the vocal similarities between host and parasite nestlings in other systems where parasites are raised alongside host young (Hauber and Kilner 2007).

28.4 The Development of Begging Calls: An Adaptive Framework

In this final section, we present a framework to explain variation in begging call development of different brood parasite species. Our aim is to outline a verbal model that predicts when it is adaptive for begging call structure in nestling parasites to be inflexible versus phenotypically plastic. We explain the mode of begging call development by focusing on the parasite's level of specialisation and the benefits to parasitic offspring of modulating their begging calls in response to environmental cues. These benefits depend on the levels of discrimination or rejection shown by host parents against odd-sounding chicks.

Theoretical analyses can be used to predict when selection will favour plasticity versus genetically controlled development. One approach is to treat genetic and environmental cues that might influence development as competing sources of information and identify the conditions in which one source of information is superior to the other in yielding adaptive development (Leimar 2009). An organism's genotype is a statistical record of the selection pressures experienced by previous generations. Thus genes are good cues when environments are stable from generation to generation—the past events that selected those genes are good predictors of selection pressures an individual will experience during their life (Leimar 2009).

We caricature the development of begging calls as either inflexible and insensitive to the particular host environment in which the brood parasitic chick is raised or plastic and flexibly modulated by the host environment. In reality, there might be a gradient between these extremes, with the development of different call parameters being affected to a greater or lesser extent by environmental cues. During plastic development, brood parasite chicks could modulate their begging development in response to parental provisioning behaviour. Non-evictor species could instead (or as well) alter begging in response to hearing calls from host nestmates. Although there is certainly evidence of plasticity in brood parasite begging call development, the detailed mechanisms underpinning plasticity remain to be identified (Langmore et al. 2008).

28.4.1 The Adaptive Value of Genetic Cues: Specialists vs. Generalists

For brood-parasitic nestlings, the accuracy of genetic cues in predicting the environment in which they develop depends on the level of host specialisation exhibited by that parasite species. The more specialised the parasite on a given host species, the more likely it is that selection pressures experienced by the nestling's ancestors will match the selection pressures the nestling experiences after hatching. Therefore, to predict the adaptive value of genetic cues in influencing begging call development, we need to know the degree of host specialisation for that brood parasite species. Three categories of host specialisation can be distinguished:

1. Specialists at the Species Level

   These species parasitise only one host species or a few closely related hosts, sharing very similar nest and nestling traits. Examples include most *Vidua* finches (Payne and Payne 2002), the screaming cowbird (*Molothrus rufoaxillaris*) (De Mársico et al. 2012), the little bronze cuckoo (*Chalcites minutillus*) and the shining bronze cuckoo (*C. lucidus*) (Ranjard et al. 2010). For nestlings of these species, genetic cues can accurately predict the host nest in which the nestling will hatch and the selection pressures it will consequently face. Here we should expect begging call structure to be primarily determined genetically, with minor modifications depending on the condition of the parasite nestling.

2. Specialists at the Individual Level but Generalists at the Species Level

   In these species, an individual parasite targets only one host species in their lifetime, but other members of that same brood parasitic species might specialise on different host species. Examples of this include the common cuckoo (Moksnes and Røskaft 1995) and the cuckoo finch (*Anomalospiza imberbis*) (Spottiswoode and Stevens 2010, 2011, 2012). Crucially, however, host specialisation is often confined to the female line, giving rise to female host races or "gentes" (Moksnes and Røskaft 1995; Gibbs et al. 2000; Spottiswoode and Stevens 2011, 2012), whilst males mate promiscuously across the races and exhibit no host specialisation at all (Marchetti et al. 1998; Gibbs et al. 2000). This complicates predictions of genetic cue accuracy, particularly given that sex is determined chromosomally in birds, and females are the heterogametic sex (females have ZW chromosomes, males have ZZ). For female brood-parasitic nestlings, genetic cues associated with the W chromosome are exclusively inherited from the mother and so can accurately predict the host nest in which the nestling will hatch and the selection pressures it will consequently face. But this is not true for male nestlings. Infidelity in the male line with respect to past use of host species means that the Z chromosomes potentially carry mixed messages about previous host use. Two possibilities then emerge. The first is that male and female nestling parasites use different strategies to develop an adaptive begging call, with females being reliant on accurate genetic cues and males perhaps making more use of environmental cues to direct the development of their begging calls. The second possibility involves maternal effects operating in the egg before hatching. If gene products associated with the maternal W chromosome are deposited in the egg at laying, then both sons and daughters could use these inherited cues to develop an adaptive begging call (see Madden and Davies 2006).

3. Generalists at the Individual Level

   In these species, an individual parasite will lay her eggs in the nests of multiple host species during her life. Examples of this include the Horsfield's bronze cuckoo (*Chalcites basalis*) (Langmore et al. 2008), the shiny cowbird (*Molothrus bonariensis*) (Jaramillo and Burke 1999) and the brown-headed cowbird (Friedmann and Kiff 1985). The challenge for the brood-parasitic nestling here is to develop an adaptive begging call when genetic cues provide little or no information about host identity. One obvious solution is to use environmental cues of host identity to develop host-specific plastic begging calls (Langmore et al. 2008). Another solution involves deploying a genetically fixed bet-hedging begging call: one that is effective enough to secure adequate care from any host species the parasitic nestling might be raised by and whose structure is unchanged by the host environment (Gloag and Kacelnik 2013).

This brief overview shows that we can predict the degree to which genetic cues might guide the development of nestling begging calls from the degree of host specialisation by the brood parasite. But it also tells us that this is not sufficient to predict the mode of begging call development in every brood-parasitic species. As outlined in the previous paragraph, when individual brood parasites are generalists, for example, parasite offspring have multiple solutions available to solicit host parental investment. Therefore, to successfully predict the mode of begging call development, we need to know the scale of fitness benefits the parasitic chick stands to gain if it uses environmental cues to modify its calls. Are these gains substantial, preventing its otherwise certain death or are the benefits relatively trivial?

28.4.2 The Benefits Gained from Using Environmental Cues to Direct Begging Call Development

The costs and benefits of using environmental cues to direct begging call development depend on the levels of discrimination or rejection that hosts use against foreign chicks. At one end of the continuum are species of nestling brood parasite whose hosts can recognise cuckoo chicks as alien and reject them, either by flinging them from the nest (Sato et al. 2010; Tokue and Ueda 2010) or by abandoning them to starve to death (Langmore et al. 2003; Soler and de Neve 2012). For these parasitic offspring, close mimicry of host young can prevent rejection and therefore death. Here there is an extremely high fitness benefit to be gained from sounding like host young. If genetic cues are too inaccurate to achieve this, then the payoffs of environmentally induced begging call mimicry are high.

The hosts of other species of brood-parasitic offspring, however, do not exhibit chick rejection, showing chick discrimination instead. In these hosts, the relationship between rates of parental provisioning in response to nestling begging displays varies depending on whether the nestling is a parasite or a host, but the parasite is never actively rejected or abandoned. For these species, the key function of brood-parasitic nestling calls is to secure adequate provisioning to survive to independence. This is particularly important when brood parasitic chicks kill host offspring and so must solicit care single-handedly from their hosts. Here the fitness benefits of environmentally induced call mimicry depend on how host parents use nestling begging calls to refine their provisioning behaviour. For some host species, it appears that only a relatively small fraction of their own offspring's call structure is used to regulate provisioning at the nest (e.g. Madden and Davies 2006). This means that the brood-parasitic nestling can gain high fitness benefits by environmentally modifying its call to a small degree. It needs only attune its begging call to match to these structural components of the host begging call to secure adequate provisioning. By contrast, some avian parents are particularly sensitive to features of begging calls that even their own offspring do not produce (Gloag and Kacelnik 2013). Brood parasites that can produce these sorts of calls need not mimic host nestlings at all to secure care successfully. Furthermore, their call is likely to be effective in a range of host species.

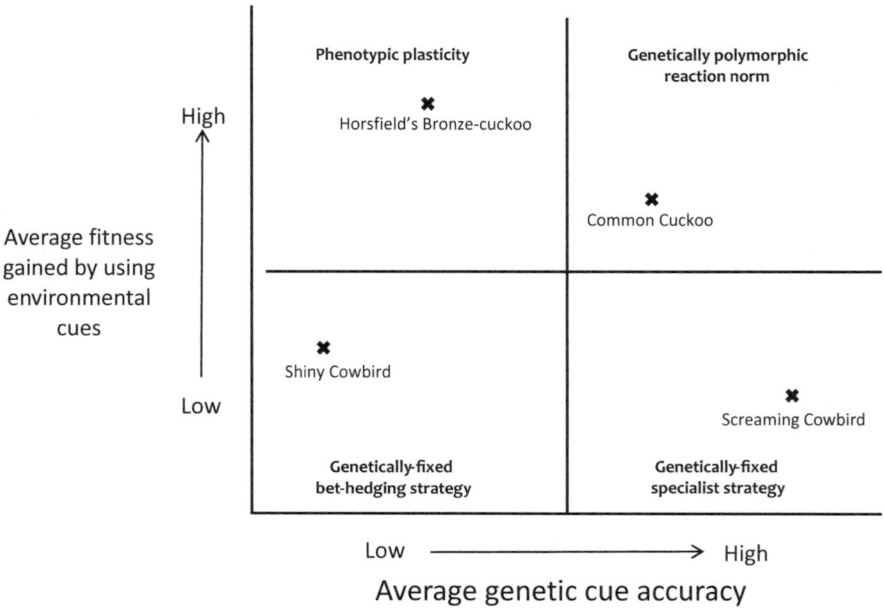

Fig. 28.1 An adaptive framework to explain variation in the mode of development of brood-parasite begging calls

28.4.3 Predicting the Mode of Begging Call Development: An Adaptive Framework

We have now described two orthogonal axes for predicting the development of nestling begging calls, which are very similar to the axes described by Leimar (Leimar 2009) in predicting the adaptive value of developmental mechanisms in general. On one axis is the accuracy of genetic cues in predicting the host environment and therefore the adaptive value of any calls produced in that environment. On the other is the fitness benefit to be gained by the parasitic nestling from environmental induction of begging call structure as mediated by the discriminatory behaviour of the host parents (Fig. 28.1).

We can now divide Fig. 28.1 into four arbitrary quadrants, within each of which we predict a particular mode of begging call development. These predictions are as follows: When genetic cue accuracy is low, because individual brood parasites are generalists, but the benefit gained from using environmental cues to modulate begging call structure is high, then we expect to see brood parasitic begging calls exhibiting phenotypic plasticity. When genetic cue accuracy is low and the benefit gained from using environmental cues to modulate begging call structure is also low, then here we expect to see a genetically fixed bet-hedging begging call, attuned to no host species in particular but nevertheless effective at securing care from many different hosts. When genetic cue accuracy is high, because individual brood parasites specialise on a particular host species, and the benefit of environmentally modulating call structure is low, then here we expect to see a genetically fixed begging call that is insensitive to the host nest environment. Finally, when genetic cue accuracy is high and the benefits of environmentally modulating call structure are also high, then here we expect to see genetically polymorphic norms of reaction. This means that individuals can modulate their begging calls to suit the host environment in which they are raised, but that there are genetic differences among chicks from different host races in the extent of call modulation in response to a common environment.

28.4.4 Testing the Adaptive Framework: Four Case Studies

1. Phenotypic Plasticity

   We now test these ideas (Fig. 28.1) with four case studies where there have been sufficient observational and experimental work to consider them within this framework. We start with the Horsfield's bronze cuckoo, *Chalcites basalis*. Individual females of this species are generalists, and no genetically distinct host races have been identified (Joseph et al. 2002; Langmore et al. 2008; Langmore and Kilner 2009). However, the majority of hosts used are fairywrens (*Malurus* spp.), and thornbills (*Acanthiza* spp.) are secondary hosts (Brooker and Brooker 1989). The default expectation for a Horsfield's bronze cuckoo nestling is therefore that it will hatch in a fairy-wren nest. The cost to the Horsfield's bronze cuckoo chick of making the wrong begging call is potentially very high indeed: if it begs like a thornbill nestling in a fairy-wren nest, then it will be abandoned by its hosts to die (Langmore et al. 2008).

   According to our verbal model (Fig. 28.1), we should expect the Horsfield's bronze cuckoo to exhibit a phenotypically plastic begging call: there is a low to moderate chance that genetic cues will accurately predict the host species a chick is to be raised by and a very high fitness gain from environmental modulation of the begging call if necessary. This is indeed the case. Using cross-fostering experiments, Langmore et al. (2008) showed that the structure of the nestling cuckoo's calls is modified following parasitism by experience with their foster parents. Specifically, they found that Horsfield's bronze cuckoo chicks innately express begging calls that match those of their primary host, the superb fairywren (*Malurus cyaneus*). However, if the chick finds itself in a buff-rumped thornbill (*Acanthiza pusilla*) nest, the cuckoo starts to produce highly variable begging calls. It then relies on "social shaping" (a form of instrumental conditioning by which human parents teach toddlers to form words from their babbles) via interactions with host parents to modify its calls and mimic those produced by a buff-rumped thornbill (Langmore et al. 2008). Here the call repertoire is reduced to those that are most effective at eliciting feeding from host parents.

   Hosts may evolve counterstrategies to limit parasitic chicks' ability to develop mimetic begging calls. Superb fairywrens have been shown to call to their eggs during incubation. After hatching, host nestlings are able to produce elements from their mother's incubation call whereas parasitic Horsfield's bronze cuckoo nestlings are not (Colombelli-Negrel et al. 2012). This parent-specific "password", learned embryonically by host young, might help host parents detect cuckoo nestlings, although there is no direct evidence that the "password" alone is sufficient to prevent chick rejection. The suggestion is that cuckoo nestlings fail to learn the incubation call because they have a shorter incubation period than host young and are therefore exposed to the incubation call for fewer days (Colombelli-Negrel et al. 2012). [An alternative interpretation is that the "password" plays no role in preventing chick rejection, and the cuckoo is detected by hosts using other cues (Langmore et al. 2009)].

2. Genetically Fixed Bet-Hedging Call

   Female shiny cowbirds are also generalist brood parasites, targeting more host species than perhaps any other brood parasite (Jaramillo and Burke 1999). Genetic cues alone are therefore unlikely to predict the host species that will raise the brood parasitic chick. Consistent with our predictions (Fig. 28.1), shiny cowbirds seemingly have a bet-hedging begging call that is effective at securing care from diverse avian parents (Gloag and Kacelnik 2013). The rate of begging has been shown to vary between host environments (after controlling for parasite chick condition and need) but call structure seems genetically fixed (Tuero et al. 2015). Additionally, when shiny cowbird chicks were cross-fostered into the nests of baywing (*Agelaioides badius*), their calls did not develop to resemble those made by host young (De Mársico et al. 2012). Similarly, in brown-headed cowbirds, there is no evidence that begging call structure varies between host environments although the average time spent begging was found to vary between host environments depending on the physical size of nestmates (Rivers 2006). However, further experimental work is needed in this species to examine how begging calls are modulated in response to the provisioning rules of different host species (Rivers 2006).

3. Genetically Fixed Mimetic Begging Call

   Unlike the shiny cowbird, the screaming cowbird is an ultra-specialist parasitising the baywing almost exclusively (De Mársico et al. 2012). Genetic cues in this brood parasite are thus remarkably accurate in predicting the host species that will raise the brood-parasitic nestling. Environmental cues are therefore redundant in this regard and might even be a more costly way of acquiring the appropriate begging call, given that plasticity requires accurate and repeated sampling of environmental cues to be accurate (Frankenhuis and Panchanathan 2011). The benefits of environmentally induced begging call development in this species are therefore likely to be very low. According to our model (Fig. 28.1), with high genetic cue accuracy but little fitness to be gained from environmentally induced begging calls, we should expect to see genetically fixed begging calls. Cross-fostering experiments apparently support this prediction. When screaming cowbird nestlings were cross-fostered to be raised by chalk-browed mockingbirds (*Mimus saturninus*), they retained their characteristic begging call, suggesting a strong genetic influence to call development (De Mársico et al. 2012).

4. Genetically Polymorphic Reaction Norm

Our final case study comes from the common cuckoo. In this species, individual females tend to specialise in parasitising a single host species, whereas males can mate promiscuously across the female host races (Marchetti et al. 1998; Gibbs et al. 2000; Fossøy et al. 2011, 2016). Genetic cues inherited from the mother are therefore highly accurate in predicting the host that will rear the cuckoo chick, whereas genetic cues inherited from the father are less accurate—giving a moderately high level of genetic cue accuracy on average. The fitness gained from modifying the cuckoo nestling's begging call in response to environmental cues is also relatively high. Common cuckoo nestlings are large in relation to nestlings of their hosts; they evict host young from the nest and so single-handedly face the challenge of eliciting sustained and elevated provisioning rates with their begging call (Kilner and Davies 1999; Kilner et al. 1999). Their calls also differ among the different host species they target (Butchart et al. 2003), suggesting that calling is specifically attuned to the different host species to elicit adequate levels of care.

In short, it seems that common cuckoo nestlings stand to gain moderately high fitness from the environmental modification of their begging calls to suit different hosts because this allows them to procure care more effectively. We should therefore expect that them to exhibit a genetically polymorphic reaction norm (Fig. 28.1): this means we might see modulation of the begging call according to the host environment but that host races should still exhibit some differences in their calls even when raised by the same host species. This is exactly what was found in a cross-fostering study carried out on the common cuckoo. Here, cuckoos from eggs laid in reed warbler nests were transferred to dunnock (*Prunella modularis*) nests and developed begging calls more similar to cuckoos that were naturally found in dunnock nests. Nevertheless, they still retained some signature of their reed warbler host origin (Madden and Davies 2006). Thus, reed warbler-cuckoo chicks modulated their begging call in a dunnock environment yet did not converge completely on the calls produced by Dunnock-cuckoo nestlings—just as expected with a genetically polymorphic reaction norm.

Concluding Remarks and Future Directions

Perhaps the main contribution of this chapter is to highlight how little we know about brood parasitic begging calls and to point the way for future work on this topic. To date, brood-parasitic nestling begging calls have been described in a small number of avian brood parasites, and many of these reports are based on studies with small sample sizes and without quantitative comparison of sonograms and without consideration of how the birds themselves might hear these begging calls (Table 28.1). Much more natural history remains to be described. Importantly, though, the function of brood parasitic chick calling cannot be discerned from sonograms alone. Field experiments and comparative analyses are needed to determine how selection influences the

(continued)

development of the nestling begging call, as there are multiple reasons for hosts and brood parasites to share similar begging calls. Finally, we have highlighted a completely new area of research on brood parasites, by showing how they lend themselves ideally to adaptive analyses of behavioural development. We set out new theory predicting the mode of begging call development, which can be tested in future work by means of cross-fostering experiments. Interesting avenues arising from this work will be to determine whether any brood parasites learn their begging calls through interactions with nestmates and to discover precisely how some host parents train brood-parasitic nestlings to beg like host chicks.

Acknowledgements We would like to thank Nick Davies, Cecilia de Mársico, Tomáš Grim, Manuel Soler, Claire Spottiswoode and Rose Thorogood for comments on earlier versions of this manuscript that have greatly helped to improve its contents. We also thank Naomi Langmore, Amanda Ridley and Claire Spottiswoode for making available unpublished information on little bronze cuckoo, Jacobin cuckoo and cuckoo finch begging calls, respectively. Our research is funded by a Research Project Grant from the Leverhulme Trust.

References

Ali SA, Whistler H (1936) The ornithology of Travancore and Cochin. Part VI. J Bombay Nat Hist Soc 39:3–35

Anderson MG, Ross HA, Brunton DH, Hauber M (2009) Begging call matching between a specialist brood parasite and its host: a comparative approach to detect coevolution. Biol J Linn Soc 98:208–216

Briskie JV, Martin PR, Martin TE (1999) Nest predation and the evolution of nestling begging calls. Proc R Soc B 266:2153–2159

Brooker M, Brooker L (1989) The comparative breeding behaviour of two sympatric cuckoos, Horsfield's Bronze-Cuckoo *Chrysococcyx basalis* and the Shining Bronze-Cuckoo *C. lucidus*, in Western Australia: a new model for the evolution of egg morphology and host specificity in avian brood parasites. Ibis 131:528–547

Butchart SHM, Kilner RM, Fuisz T, Davies NB (2003) Differences in the nestling begging calls of hosts and host-races of the common cuckoo, *Cuculus canorus*. Anim Behav 65:345–354

Clunie F (1973) Fan-tailed cuckoo parasitises Fiji Warbler. Notornis 20:168

Colombelli-Negrel D, Hauber ME, Robertson J, Sulloway FJ, Hoi H, Griggio M, Kleindorfer S (2012) Embryonic learning of vocal passwords in superb fairy-wrens reveals intruder cuckoo nestlings. Curr Biol 22:2155–2160

Courtney J (1967) The juvenile food-begging call of some fledgling cuckoos – vocal mimicry or vocal duplication by natural selection. Emu 67:154–157

Davies NB (2011) Cuckoo adaptations: trickery and tuning. J Zool 284:1–14

Davies NB, Kilner RM, Noble DG (1998) Nestling cuckoos, *Cuculus canorus*, exploit hosts with begging calls that mimic a brood. Proc R Soc B 265:673–678

De Mársico MC, Gantchoff MG, Reboreda JC (2012) Host-parasite coevolution beyond the nestling stage? Mimicry of host fledglings by the specialist screaming cowbird. Proc R Soc B 279:3401–3408

Dearborn DC (1998) Begging behavior and food acquisition by brown-headed cowbird nestlings. Behav Ecol Sociobiol 43:259–270

Dearborn DC, Lichtenstein G (2002) Begging behaviour and host exploitation in parasitic cowbirds. In: Wright J, Leonard ML (eds) The evolution of begging. Kluwer Academic, Dordrecht

Dewar D (1907) An inquiry into the parasitic habits of the Indian koel. J Bombay Nat Hist Soc 17:765–782

Erritzøe J, Mann CF, Brammer FP, Fuller RA (2012) Cuckoos of the world. Christopher Helm, London

Feeney WE, Welbergen JA, Langmore NE (2014) Advances in the study of coevolution between avian brood parasites and their hosts. Annu Rev Ecol Evol Syst 45:227–246

Fossøy F, Antonov A, Moksnes A, Roskaft E, Vikan JR, Moller AP, Shykoff JA, Stokke BG (2011) Genetic differentiation among sympatric cuckoo host races: males matter. Proc R Soc B 278:1639–1645

Fossøy F, Sorenson MD, Liang W, Ekrem T, Moksnes A, Moller AP, Rutila J, Roskaft E, Takasu F, Yang C, Stokke BG (2016) Ancient origin and maternal inheritance of blue cuckoo eggs. Nat Commun 7:10272

Frankenhuis WE, Panchanathan K (2011) Balancing sampling and specialization: an adaptationist model of incremental development. Proc R Soc B 278:3558–3565

Friedmann H, Kiff L (1985) The parasitic cowbirds and their hosts. In: Proceedings of the Western Foundation of Vertebrate Zoology, vol 2, pp 225–304

Fry CH, Keith S, Urban EK (2002) The birds of Africa, parrots to woodpeckers, vol 3. Christopher Helm, London

Gibbs HL, Sorenson MD, Marchetti K, Brooke ML, Davies NB, Nakamura H (2000) Genetic evidence for female host-specific races of the common cuckoo. Nature 407:183–186

Gloag R, Kacelnik A (2013) Host manipulation via begging call structure in the brood-parasitic shiny cowbird. Anim Behav 86:101–109

Gosper D (1997) Aspects of breeding of the common koel *Eudynamys scolopacea* and one of its biological hosts, the Magpie-lark *Grallina cyanoleuca*. Aust Bird Watch 17:11–19

Grim T (2005) Mimicry vs. similarity: which resemblances between brood parasites and their hosts are mimetic and which are not? Biol J Linn Soc 84:69–78

Grim T (2006) The evolution of nestling discrimination by hosts of parasitic birds: why is rejection so rare? Evol Ecol Res 8:785–802

Grim T (2008) Begging behavior of fledgling rusty-breasted cuckoo (*Cacomantis sepulcralis*). Wilson J Ornithol 120:887–890

Grim T, Kleven O, Mikulica O (2003) Nestling discrimination without recognition: a possible defence mechanism for hosts towards cuckoo parasitism? Proc R Soc B 270:S73–S75

Hauber ME, Kilner RM (2007) Coevolution, communication, and host chick mimicry in parasitic finches: who mimics whom? Behav Ecol Sociobiol 61:497–503

Jamie GA, de Silva WG (2014) Similarity of the calls of juvenile pied cuckoo *Clamator jacobinus* and its Sri Lankan host species, yellow-billed babbler *Turdoides affinis*. Forktail 30:133–134

Jaramillo A, Burke P (1999) New world blackbirds: the icteridae. Christopher Helm, London

Joseph L, Wilke T, Alpers D (2002) Reconciling genetic expectations from host specificity with historical population dynamics in an avian brood parasite, Horsfield's Bronze-Cuckoo *Chalcites basalis* of Australia. Mol Ecol 11:829–837

Jubb R (1952) Some notes on birds of Southern Rhodesia. Ostrich 23:162–164

Jubb R (1966) Red-billed hoopoe and a greater honey-guide. Bokmakierie 18:66–67

Kilner RM, Davies NB (1999) How selfish is a cuckoo chick? Anim Behav 58:797–808

Kilner RM, Noble DG, Davies NB (1999) Signals of need in parent-offspring communication and their exploitation by the common cuckoo. Nature 397:667–672

Langmore NE, Kilner RM (2009) Why do Horsfield's bronze-cuckoo *Chalcites basalis* eggs mimic those of their hosts? Behav Ecol Sociobiol 63:1127–1131

Langmore NE, Spottiswoode CN (2012) Visual trickery in avian brood parasites. In: Hughes DP, Brodeur J, Thomas F (eds) Host manipulation by parasites. Oxford University Press, Oxford

Langmore NE, Hunt S, Kilner RM (2003) Escalation of a coevolutionary arms race through host rejection of brood parasitic young. Nature 422:157–160

Langmore NE, Maurer G, Adcock GJ, Kilner RM (2008) Socially acquired host-specific mimicry and the evolution of host races in Horsfield's bronze-cuckoo *Chalcites basalis*. Evolution 62:1689–1699

Langmore NE, Cockburn A, Russell AF, Kilner RM (2009) Flexible cuckoo chick-rejection rules in the superb fairy-wren. Behav Ecol 20(5):978–984

Langmore NE, Stevens M, Maurer G, Heinsohn R, Hall ML, Peters A, Kilner RM (2011) Visual mimicry of host nestlings by cuckoos. Proc R Soc B 278:2455–2463

Leimar O (2009) Environmental and genetic cues in the evolution of phenotypic polymorphism. Evol Ecol 23:125–135

Lichtenstein G (2001) Low success of shiny cowbird chicks parasitizing rufous-bellied thrushes: chick-chick competition or parental discrimination? Anim Behav 61:401–413

Lord E (1956) The birds of the Murphy's Creek district, southern Queensland. Emu 56:100–128

Lyon BE, Eadie JM (2013) Patterns of host use by a precocial obligate brood parasite, the black-headed duck: ecological and evolutionary considerations. Chin Birds 4:71–85

Madden JR, Davies NB (2006) A host-race difference in begging calls of nestling cuckoos *Cuculus canorus* develops through experience and increases host provisioning. Proc R Soc B 273:2343–2351

Marchetti K, Nakamura H, Gibbs HL (1998) Host-race formation in the common cuckoo. Science 282:471–472

McLean I, Waas JR (1987) Do cuckoo chicks mimic the begging calls of their hosts? Anim Behav 35:1896–1907

Moksnes A, Røskaft E (1995) Egg-morphs and host preferences in the common cuckoo (*Cuculus canorus*): an analysis of cuckoo and host eggs from European museum collections. J Zool 236:625–648

Morton ES, Farabaugh SM (1979) Infanticide and other adaptations of the nestling striped cuckoo *Tapera naevia*. Ibis 121:212–213

Mundy P (1973) Vocal mimicry of their hosts by nestlings of the great spotted cuckoo and striped crested cuckoo. Ibis 115:602–604

Pagnucco K, Zanette L, Clinchy M, Leonard ML (2008) Sheep in wolf's clothing: host nestling vocalizations resemble their cowbird competitor's. Proc R Soc B 275:1061–1065

Payne RB (2005) The cuckoos. Oxford University Press, Oxford

Payne RB, Payne LL (2002) Begging for parental care from another species: specialization and generalization in brood-parasitic finches. In: Horn AG, Leonard ML (eds) The evolution of begging: competition, cooperation and communication. Kluwer Academic, Dordrecht

Payne RB, Woods JL, Payne LL (2001) Parental care in estrildid finches: experimental tests of a model of *Vidua* brood parasitism. Anim Behav 62:473–483

Pfennig DW, Wund MA, Snell-Rood EC, Cruickshank T, Schlichting CD, Moczek AP (2010) Phenotypic plasticity's impacts on diversification and speciation. Trends Ecol Evol 25:459–467

Potter EF (1980) Notes on nesting yellow-billed cuckoos. J Field Ornithol 51:17–29

Price TD, Qvarnstrom A, Irwin DE (2003) The role of phenotypic plasticity in driving genetic evolution. Proc R Soc B 270:1433–1440

Ranjard L, Anderson MG, Rayner MJ, Payne RB, McLean I, Briskie JV, Ross HA, Brunton DH, Woolley SMN, Hauber ME (2010) Bioacoustic distances between the begging calls of brood parasites and their host species: a comparison of metrics and techniques. Behav Ecol Sociobiol 64:1915–1926

Redondo T, Arias de Reyna L (1988) Vocal mimicry of hosts by great spotted cuckoo *Clamator glandarius*: further evidence. Ibis 130:540–544

Reed RA (1968) Studies of the diderik cuckoo *Chrysococcyx caprius* in the Transvaal. Ibis 110:321–331

Rivers JW (2006) Nest mate size, but not short-term need, influences begging behavior of a generalist brood parasite. Behav Ecol 18:222–230

Rivers JW, Loughin TM, Rothstein SI (2010) Brown-headed cowbird nestlings influence nestmate begging, but not parental feeding, in hosts of three distinct sizes. Anim Behav 79:107–116

Roldán M, Soler M, Márquez R, Soler JJ (2013) The vocal begging display of great spotted cuckoo *Clamator glandarius* nestlings in nests of its two main host species: genetic differences or developmental plasticity? Ibis 155:867–876

Rothstein SI (1990) A model system for coevolution: avian brood parasitism. Annu Rev Ecol Syst 21:481–508

Salewski V, Grafe TU (1999) New tape recordings of three West African birds. Malimbus 21:117–121

Sato NJ, Tokue K, Noske RA, Mikami OK, Ueda K (2010) Evicting cuckoo nestlings from the nest: a new anti-parasitism behaviour. Biol Lett 6:67–69

Sato NJ, Tanaka KD, Okahisa Y, Yamamichi M, Kuehn R, Gula R, Ueda K, Theuerkauf J (2015) Nestling polymorphism in a cuckoo-host system. Curr Biol 25:R1164–R1165

Schuetz JG (2005) Reduced growth but not survival of chicks with altered gape patterns: implications for the evolution of nestling similarity in a parasitic finch. Anim Behav 70:839–848

Serventy D, Whittell H (1962) Birds of Western Australia. Paterson Brokensha Pty, Perth

Short LL, Horne JFM (2001) Toucans, barbets and honeyguides. Oxford University Press, Oxford

Skead CJ (1946) Record of a young black cuckoo (*Surniculoides clamosus*). Ostrich 17:359–360

Skead CJ (1995) Life-history notes on East Cape bird species (1940–1990), vols 1 & 2. Algoa Regional Services Council, Port Elizabeth

Soler M, de Neve L (2012) Great spotted cuckoo nestlings but not magpie nestlings starve in experimental age-matched broods. Ethology 118:1036–1044

Soler M, Soler JJ, Martinez JG, Møller AP (1995) Chick recognition and acceptance: a weakness in magpies exploited by the parasitic great spotted cuckoo. Behav Ecol Sociobiol 37:243–248

Spencer OR (1943) Nesting habits of the black-billed cuckoo. Wilson Bull 55:11–22

Spottiswoode CN, Stevens M (2010) Visual modeling shows that avian host parents use multiple visual cues in rejecting parasitic eggs. Proc Natl Acad Sci U S A 107:8672–8676

Spottiswoode CN, Stevens M (2011) How to evade a coevolving brood parasite: egg discrimination versus egg variability as host defences. Proc R Soc B 278:3566–3573

Spottiswoode CN, Stevens M (2012) Host-parasite arms races and rapid changes in bird egg appearance. Am Nat 179:633–648

Spottiswoode CN, Kilner RM, Davies NB (2012) Brood parasitism. In: Royle NJ, Smiseth PT, Kölliker M (eds) The evolution of parental care. Oxford University Press, Oxford

Steyn P (1973) Some notes on the breeding biology of the striped cuckoo. Ostrich 44:163–169

Tokue K, Ueda K (2010) Mangrove gerygones *Gerygone laevigaster* eject little bronze-cuckoo *Chalcites minutillus* hatchlings from parasitized nests. Ibis 152:835–839

Tuero DT, Gloag R, Reboreda JC (2015) Nest environment modulates begging behavior of a generalist brood parasite. Behav Ecol 27:204–210

Vernon CJ (1984) The breeding biology of the thick-billed cuckoo. In: Proceedings of the fifth Pan-African ornithological congress, pp 825–840

Vernon CJ (1987) On the eastern green-backed honeyguide. Honeyguide 33:6–12

Verzijden MN, ten Cate C, Servedio MR, Kozak GM, Boughman JW, Svensson EI (2012) The impact of learning on sexual selection and speciation. Trends Ecol Evol 27:511–519

West-Eberhard M (2003) Developmental plasticity and evolution. Oxford University Press, New York

Whitman DW, Agrawal AA (2009) What is phenotypic plasticity and why is it important? In: Whitman D, Ananthakrishnan T (eds) Phenotypic plasticity of insects. Science, Enfield, NH

Host Defences Against Brood Parasite Nestlings: Theoretical Expectations and Empirical Evidence

29

Tomáš Grim

Abstract

In the past, chick discrimination was assumed to be non-existent without virtually any research invested to check the reality. Models of brood parasitism considered the benefits of chick rejection small and costs too high; consequently, the nestling stage was long ignored in studies of host–parasite coevolution. Remarkably, the majority of recent studies that addressed parasite chick biology did find evidence for host behaviours that alleviate the costs of parasitism during nestling stage. Most of the hosts that (apparently) discriminate against parasite chicks are acceptors of natural parasite eggs; this pattern is in line with the rarer enemy model. The main impetus for future work is therefore not naively assuming but empirically checking the (non)-existence of chick discrimination to show how common is chick discrimination in reality. This will allow to elucidate mechanisms of chick discrimination, both those that specifically evolved as a response to past parasitism pressure and those stemming from non-specific general host life history traits, and factors that facilitate or constrain their evolution.

29.1 Introduction

Life cycles of both brood parasites and hosts develop through a series of stages, from egg laying, through incubation, hatching, nestling and fledgling periods to independence during non-breeding periods. At each stage the interactions involve costs and benefits for both parties and, consequently, affect overall population and evolutionary dynamics of both parasites and hosts (Dawkins and Krebs 1979). Despite the large and increasing research effort into brood parasite–host systems, we are still far

T. Grim (✉)
Department of Zoology and Laboratory of Ornithology, Palacký University, Olomouc, Czech Republic
e-mail: tg.cuckoo@gmail.com

M. Soler (ed.), *Avian Brood Parasitism*, Fascinating Life Sciences,
https://doi.org/10.1007/978-3-319-73138-4_29

from a coherent understanding of the complete host–parasite life cycles. This is because of an imbalance in research efforts across all life stages: a great majority of studies have addressed, and still addresses, the incubation stage (egg rejection and mimicry); few studies have focused on the egg laying and hatching processes and nestling and fledgling periods.

Here, I focus on the nestling period (Soler 2009), which has been typically neglected in comparison with the egg stage (Grim 2007a). Recent advances in this tiny field of study suggest that host–parasite interactions during the nestling stage may have fundamental effects on parasite–host coevolution.

29.2 Theoretical Explanations

Most historical discussions of chick discrimination, i.e. differential behavioural response of hosts towards own vs. foreign chicks (e.g. Langmore et al. 2003; Grim 2007b; Shizuka and Lyon 2010), tried to explain its absence (or, at best, rarity), taking the lack of such host adaptation for granted. Today, all previous explanations for the absence are invalid because good empirical evidence for various modes of chick discrimination has accumulated during the last ca. 15 years (Table 29.1). All previous explanations for the rarity of chick discrimination were reviewed and refuted with theoretical arguments or empirical evidence (Grim 2006). To avoid unnecessary duplication, I am not going to repeat the arguments here. If anything, post-2006 empirical data provide additional support for rejecting traditional explanations. Here I give just one example.

It has been argued that evicting parasites, like common cuckoos (*Cuculus canorus*), eliminate comparative material (host own chicks) which prevents host chick recognition; this hypothesis was in line (at the time of publication) with chick discrimination being reported mainly from non-evicting parasite–host systems. However, there are at least six different cognitive systems that might be employed as a basis for chick discrimination without material for comparison (see section "Simultaneous Comparison Constraint" in Grim 2006). All those cognitive mechanisms are known in birds and work in other social and behavioural contexts; this rejects the notion that discrimination is impaired by the lack of comparative cues. I see no theoretical reason why chick discrimination should be an exception. Indeed, empirical data support this view: the best evidence for chick discrimination comes from hosts of evicting parasites (Langmore et al. 2003; Fig. 29.1a). In fact, it was even more often reported from evicting than from non-evicting parasites (Table 29.1).

The "material for comparison" idea stemmed from findings of early egg discrimination studies—but there is ample evidence that even foreign egg discrimination is not limited by the absence of own eggs for comparison (Bán et al. 2013). The hypothesis of missing comparative material probably reflected a notion that cuckoos evict immediately after hatching; this notion is wrong because cuckoos start to evict only when they are 1 or 2 days old (Honza et al. 2007; Grim et al. 2009b). Therefore hosts do have material for comparison, either eggs (why should a host not assess a contrast between, e.g. own egg colour and parasite egg skin colour?) or their own

Table 29.1 Overview of host–parasite systems with reported cases of (apparent) chick discrimination or mimicry

Host	Parasite	Natural egg rejection rate (%)	Evictor parasite	References
Anecdotal observations				
Acrocephalus arundinaceus	*Cuculus canorus*	57[a]	+	Honza et al. (2010)
Alauda arvensis	*Cuculus canorus*	0[b]	+	Hegemann and Voesten (2011)
Phoenicurus phoenicurus	*Cuculus canorus*	0	+	Grim and Rutila (Chap. 16)
Turdoides spp.	*Clamator jacobinus*	0[c]	–	Jourdain (1925) and Gaston (1976)
Phenotypic similarity				
Acanthiza chrysorrhoa	*Chalcites lucidus*	0[d]	+	Langmore et al. (2011)
Corvus splendens	*Eudynamys scolopacea*	0	–	Dewar (1907)
Estrildidae spp.	Viduinae spp.	0[e]	–	Nicolai (1964, 1974), Payne et al. (2001) and Schuetz (2005a, b)
Gerygone igata	*Chalcites lucidus*	0	+	McLean and Waas (1987), McLean and Rhodes (1991), Gill (1998), McLean and Maloney (1998) and Thorogood et al. (2017)
Malurus coronatus	*Chalcites basalis*	0[f]	+	Langmore et al. (2011)
Mohoua spp.	*Urodynamis taitensis*	67[g]	+	McLean and Waas (1987)
Ploceidae spp.	*Chrysococcyx caprius*	0[c]	+	Reed (1968)

(continued)

Table 29.1 (continued)

Host	Parasite	Natural egg rejection rate (%)	Evictor parasite	References
Turdoides jardineii	*Clamator levaillantii*	0[c]	–	Mundy (1973) and Steyn (1973)
Tyrannidae spp.	*Dromococcyx pavoninus*	0[h]	+[h]	Sánchez-Martínez et al. (2017)
Direct evidence				
Acrocephalus scirpaceus	*Cuculus canorus*	38	+	Grim et al. (2003) and Grim (2007b)
Agelaioides badius	*Molothrus rufoaxillaris*	0	–	Fraga (1998) and De Mársico et al. (2012)
Anumbius annumbi	*Molothrus bonariensis*	0[i]	–	Delhey et al. (2011)
Estrilda astrild	*Vidua macroura*	0[j]	–	Schuetz (2005a,b)
Eumyias thalassinus	*Cuculus canorus*	0	+	Yang et al. (2013)
Gerygone flavolateralis	*Chalcites lucidus*	0[k]	+	Sato et al. (2015)
Gerygone laevigaster	*Chalcites minutillus*	0	+	Tokue and Ueda (2010)
Gerygone magnirostris	*Chalcites minutillus*	0	+	Sato et al. (2010b)
Hirundo daurica	*Chrysococcyx maculatus*	0	+	Yang et al. (2015)
Malurus cyaneus	*Chalcites basalis*	0	+	Langmore et al. (2003, 2008, 2009, 2011) and Colombelli-Négrel et al. (2012, 2014)

Passer cinnamomeus	Various cuckoo spp. (potentially)	0	+	Huo et al. (2018)
Pica pica	*Clamator glandarius*	5[l]	–	Redondo (1993), Soler et al. (1995), Soler and De Neve (2012), Roldán et al. (2013) and Soler et al. (2013a, 2017)
Prunella modularis	*Cuculus canorus*	0	+	Davies and Brooke (1989)
Turdus merula	*Cuculus canorus*	?[m]	+	Grim et al. (2011)
Turdus rufiventris	*Molothrus bonariensis*	0	–	Lichtenstein (2001)

Also cases based on mere similarity (cf. Grim 2005, 2013) are included to highlight any promising study systems for future research. If rejection rates of natural egg parasitism were not reported or cited in the study given in "References" column, I retrieved data from another sources (see footnotes). If only results from experimental parasitism with model eggs were known, I assumed that acceptance of dissimilar non-mimetic models implies acceptance of more similar natural parasite eggs (Grim 2005). I have excluded *Psarocolius* and *Cacicus* spp. vs *Scaphidura oryzivora* system (Table 1 in Grim 2006) because subsequent studies did not verify original findings (see Davies 2000, p. 186)

[a]Trnka et al. (2012)

[b]Antonov et al. (2010)

[c]Parasite and host eggs very similar (Davies 2000; Payne 2005), making rejection of natural parasite eggs unlikely

[d]Brooker and Brooker (1989)

[e]Includes ca. 15 host–parasite species pairs where parasite chick mimicry seem to have evolved independently (Davies 2000)

[f]No evidence of natural cuckoo egg rejection but experimental data not available (M. L. Hall, pers. comm.)

[g]Briskie (2003). Only artificial models, sample sizes small ($n = 3$)

[h]Hosts included two *Leptopogon* spp. and one *Mionectes* sp. No naturally laid parasite eggs were ejected. Observed desertion rates of parasitized nests (5 out of 9) did not differ statistically (Fisher's exact test: $p = 0.64$) from rates expected (3 out of 9) for this sample of nests ($n = 9$) from background desertion rates at control nests (21 out of 63). The pavonine cuckoo does not show evicting behaviour but kills host progeny via bill hook; thus, it is raised alone just like chicks of evicting cuckoo species are

[i]Ca. 10% cowbird eggs disappeared but host own eggs disappeared at the same rate (10%; hosts remove their own eggs punctured by cowbirds). Experiments in different populations did not reveal any egg rejection by hosts (Delhey et al. 2011)

[j]Payne et al. (2001)

[k]Supplemental Information from Sato et al. (2015) suggests no specific rejection of parasite eggs

[l]M. Soler, pers. comm.

[m]Blackbirds reject roughly half of non-mimetic and mimetic models and even conspecific eggs (Samas et al. 2014); responses to natural cuckoo eggs were not tested so far and are hard to predict because blackbirds do not reject foreign eggs based on absolute cue dissimilarity discrimination thresholds but show colour categorization (Hanley et al. 2017). Cuckoo chicks cross-fostered into blackbird nests die soon, even after evicting host progeny; this is unlikely to be explained by unsuitable diet (Grim et al. 2011) and calls for more research

Fig. 29.1 (**a**) Deserted Horsfield's bronze cuckoo (*Chalcites basalis*) chick (3–4 days old) in a superb fairywren (*Malurus cyaneus*) nest in Australia (nest and chick moved from original position to facilitate photography). (**b**) Deserted common cuckoo (*Cuculus canorus*) chick (14 days old) in a reed warbler (*Acrocephalus scirpaceus*) nest in the Czech Republic. (**c**) Deserted common cuckoo chick (17 days old) in a common redstart (*Phoenicurus phoenicurus*) nest with a deserted host chick (same age) after other host chicks fledged in Finland. Photo credits: (**a**) N. Langmore, (**b**) O. Mikulica, (**c**) T. Grim

hatched chicks. At least in some nests, the cuckoo and host chicks coexist for several days, in some hosts regularly, and even up to fledging (Chap. 16). This provides as much time (or much longer, in host–parasite mixed broods) as is needed for host responses to cuckoo eggs (Hanley et al. 2016).

Further, I am going to discuss, chronologically, only potentially viable explanations (for unviable ones see Grim 2006).

29.2.1 Misimprinting Model

Lotem (1993) assumed that hosts imprint on both their eggs and chicks for the lifetime. Given his model assumptions, he showed that in hosts of evicting parasites, costs of misimprinting on parasite chicks are higher than benefits of correct imprinting on own chicks. Due to its elegant simplicity, the model was generally accepted as a good explanation for why hosts do not reject foreign chicks.

Theoretically, the generality of misimprinting model is limited because it may apply only under special conditions (Planqué et al. 2002; Grim 2006; Britton et al. 2007). The model did not take into account rejection costs and errors and was based on poorly supported assumptions. Specifically, misimprinting model assumed, without any empirical evidence, that *chick* recognition is only learned and innate recognition is improbable. Today, there is good empirical evidence for innate recognition in various biological contexts (egg, adult, enemy, etc.; Grim 2006), including chick recognition (Langmore et al. 2003, 2009). Misimprinting model also assumed, with limited empirical evidence, that *egg* recognition is based on imprinting. However, the majority of studies do not support this contention (Soler et al. 2013b and references therein). Recognition, even innate one, is not strictly necessary for chick discrimination (Grim et al. 2003; Schuetz 2005a). At the time of publication, the model was apparently in line with empirical data because chick discrimination seemed to be more prevalent in non-evicting parasites; today, the opposite is true: most cases of chick discrimination come from evicting parasites (Table 29.1).

I suspect that the misimprinting model, although admirable for its parsimony, might be partly responsible for low research effort on chick discrimination: it took the absence of chick discrimination for granted and by providing an explanation for the *absence* of the phenomenon it apparently made chick discrimination a "closed case". This view is in line with the fact that its publication did not elicit any empirical tests of chick discrimination. Today, both theoretical reasons and empirical data make it clear that the misimprinting model should no longer be considered a viable *general* explanation for the lack or (apparent?) rarity of chick discrimination. In contrast, alternative models (rarer enemy, strategy blocking) fit empirical data well.

29.2.2 Rarer Enemy Model

The rarer enemy model (Grim 2006) is an extension of the classic verbal model, the rare enemy effect (Dawkins 1982). Although parasite eggs are rare enemies, parasite chicks are even rarer enemies. This is primarily because host interactions with chicks are pre-empted by host removing potential parasite chicks before they hatch (egg rejection) and also via parasite egg infertility and nest failure due to predation or inclement weather; even egg acceptors face parasite chicks less often than eggs (this point has not been considered in any other theoretical model). All these factors additively decrease *effective* parasitism rate during the chick stage. The core of the rarer enemy idea is that these factors prevent parasitism rate at the chick stage to reach the threshold that selection pressures must overcome in order to positively select for chick discrimination (Fig. 1 in Grim 2006). The view that it is hosts themselves that create low parasitism rate at the chick stage (via egg rejection) fundamentally differs from alternative explanations. Although the model is "only" verbal, each step of the suggested coevolutionary dynamics is supported by published mathematical models (Fig. 1a–f in Grim 2006).

The rarer enemy model explicitly predicted that chick discrimination should evolve mostly in hosts that are forced to accept parasite eggs from any reason (egg mimicry, crypsis, shared diet effects on egg colour, phylogenetic and physical constraints, etc.; Grim 2002, 2005). Novel evidence accumulated after the model was published (Grim 2006) consistently supported the model (Grim 2011): the most persuasive evidence for chick discrimination came from pure acceptors of natural parasitism (Table 29.1). Still, the model predicted coexistence of imperfect egg and imperfect chick discrimination (Grim 2006). An example of this might be provided by reed warblers (*Acrocephalus scirpaceus*) from a population where hosts always accept cuckoo eggs unless they directly witness the cuckoo during laying (Moksnes et al. 2000) and sometimes desert cuckoo chicks (Grim et al. 2003; Fig. 29.1b). Indeed, this population may experience a strong parasitism pressure only recently (Igic et al. 2012).

The idea can be extended to other stages of the host–parasite arms race. For example, also successful nest defence pre-empts selection pressure on both egg and chick rejection: e.g. "sitting on the nest" by yellow warblers (*Setophaga petechia*) blocks parasite access to the nest (Hobson and Sealy 1989) and might explain why these hosts accept foreign eggs. Testing this hypothesis will require data on *natural* host–parasite interactions during laying (Moksnes et al. 2000; see Chap. 18). Dummy experiments are useful for other purposes, e.g. enemy recognition studies (Hobson and Sealy 1989), but useless in this respect because they cannot determine if the parasite female is deterred by hosts or not. Further, a trade-off between earlier and later defences can work not only at the host species/population level (as discussed so far) but at individual level too. Here, the rarer enemy effect predicts that females that are egg rejecters should be more likely chick acceptors, whereas egg acceptors should be more chick discriminating; note that female is the sex responsible for discrimination of both eggs and chicks in most birds (Davies 2000; Langmore et al. 2003).

29.2.3 Strategy Blocking Model

The mathematical model (Britton et al. 2007; see also Planqué et al. 2002 for a less general version of the model) is based on an idea that adaptiveness of a particular strategy is context-dependent, i.e. its fitness is affected by the frequency of other strategies in "defence portfolios". Britton et al. (2007) showed that a strategy might be adaptive on its own but can get prevented ("blocked") by another ("blocking") strategy. For example, the model suggested that a population of naive hosts (i.e. non-defending acceptors of both egg and chick parasites) can be invaded by

Table 29.2 Differences between rarer enemy and strategy blocking models. See the original studies for rationales

Prediction	Rarer enemy (Grim 2006)	Strategy blocking (Britton et al. 2007)
Egg rejection is blocked by chick rejection	No	Yes
Chick rejection can coexist with egg rejection	Yes	No
Patterns of host defences under natural conditions	Only acceptors of natural (but not necessarily experimental) parasite eggs should show chick discrimination	No explicit prediction

egg rejecter strategy but such a population of egg rejecters would not be later invaded by either chick rejecters or all rejecters (i.e. rejecters of both eggs and chicks). The crucial and contra-intuitive insight is that a mixed strategy of rejecting both eggs and chicks is maladaptive and cannot invade a population of either egg rejecters or chick rejecters (at least given the model structure).

Strategy blocking is similar to rarer enemy effect but there are important differences between the two (Table 29.2). Essentially, the two models can be seen as a one-sided (rarer enemy) vs. double-sided (strategy blocking) trade-off between strategies within a defence portfolio. The two models are clearly different from other models both conceptually and in their predictions (e.g. Lotem 1993; Redondo 1993; Lawes and Marthews 2003; Sato et al. 2010a). Future modelling and empirical studies are needed to clarify which conditions best explain patterns of chick discrimination observed in nature.

29.2.4 Egg Dilution Model

The final published theoretical explanation of conditions necessary for the evolution of chick discrimination takes into account multiple parasitism. Sato et al. (2010a) suggested that hosts in populations where multiple parasitism is common may adaptively accept a first-laid parasite egg (even though they can discriminate the foreign egg) and delay the rejection of the parasite into the chick period. The rationale is that not rejecting the first parasite egg increases the total clutch size and the second-laying parasite female might remove the egg of the first female, effectively helping hosts with rejecting the first parasite. However, the assumptions of the model are unlikely to occur empirically in most brood parasite–host systems, severally limiting its applicability and making it an unlikely general explanation for chick discrimination patterns (Grim 2017).

29.3 Empirical Evidence

Australian hosts, superb fairywrens (*Malurus cyaneus*), desert parasite chicks (Langmore et al. 2003) and fine-tune their innate chick rejection via learning (Langmore et al. 2009). This has resulted in an evolution of admirable level of both acoustic (Langmore et al. 2008) and visual mimicry (Langmore et al. 2011). The last study found beautiful similarity between chicks of several host–parasite pairs, suggesting that chick discrimination may be much more widespread than thought previously (Table 29.1).

Reed warblers desert some nests where the brood, be it a cuckoo (Grim et al. 2003) or own chicks (in nests with experimentally prolonged nestling period), needs longer care than host own brood under natural conditions (Grim 2007b). Such preprogrammed parental care represents "discrimination without recognition" because warblers do not recognize their vs. foreign chicks (Davies 2000) yet are able to get rid off of the parasite. As argued by Soler et al. (2013a), deserting cuckoo nestlings is unlikely a by-product of host fledgling process: altricial species decrease feeding to force nestlings to leave the nest but this "forcing" behaviour does not result in chick death.

Recently, chick discrimination by ejection (i.e. host parent grasped a cuckoo chick and removed it from its nest) was video-recorded in three gerygone species (Table 29.1). This provides direct evidence that physical constraints are impotent as an explanation for acceptance of foreign chicks (as already supported by indirect evidence: Grim 2006). Additionally, there is an older observational and experimental evidence of foreign chick discrimination from another gerygone species (McLean and Rhodes 1991). A closely related thornbill (*Acanthiza*) host is parasitized by apparently mimetic cuckoo chicks (Table 29.1). This phylogenetic distribution (Nyári and Joseph 2012) suggests that chick discrimination may represent an ancestral trait in the Acanthizidae; clearly, other related species in this clade deserve more attention. The same holds for various *Malurus* fairywrens (Langmore et al. 2003, 2011; Colombelli-Négrel et al. 2016).

Some of these cases (Table 29.1) unambiguously represent chick rejection, while others may be more parsimoniously explained as by-products of host general life history traits, namely, indigestible food, nest design, etc. (Grim et al. 2011). More important than the origin of (apparent) chick discrimination is its effects on parasite–host coevolutionary dynamics: for example, fosterer non-willingness to feed a chick with phenotype different from host chicks effectively results in chick discrimination (lowered parasite fitness, improved host fitness: this is certain) no matter what was the evolutionary origin of that behaviour (host–parasite coevolution, parent–offspring conflict, sibling competition, exploitation of preexisting preferences, etc.: this is uncertain). Thus, even general life history traits (Grim et al. 2011) might contribute to host avoidance or diminishing of misplaced investment into alien chicks (Soler 2008; Yang et al. 2013; Grim et al. 2017).

Despite increasing empirical evidence, there is completely untapped cornucopia of potentially relevant study systems. These include the many introduced host populations that were typically released from parasite pressures (reviewed in Grim and Stokke 2016). Such efforts will allow to study phylogenetic patterns of host responses to foreign chicks (Wang and Kimball 2012), their covariation with other host traits (Trnka and Grim 2014), and determining conditions which favour or constrain the evolution of chick discrimination (Britton et al. 2007). Empirical studies are fundamental, including metareplication, i.e. repeating the same study design across phylogeny, space and time (Grim et al. 2011; Grim and Stokke 2016). For example, Soler et al. (2013a) replicated the study of Grim (2007b) and found no evidence for the preprogrammed parental care in their model species. Empirical studies at nestling stage should be integrated with studies of all other stages of parasite and host life cycles: ignoring a life cycle stage can produce wrong conclusions, e.g. highly biased fitness estimates (Chap. 16). Comprehensive "all stages" approach was attempted rarely (Grim et al. 2011; Li et al. 2016; Chap. 16) and never really satisfactorily so far.

29.4 Constraints on Our Understanding of Host Defences

Understanding of chick discrimination has been hindered by both inappropriate methods in empirical research and prejudices in theoretical considerations. For the development of this field, it is essential that these errors are well understood and avoided in the future.

29.4.1 Non-experimental Data

An observation that hosts accept *natural* parasite chicks does not mean the host does not possess chick rejection abilities. Just like egg mimicry forces egg rejecters to accept (Igic et al. 2012), chick mimicry forces chick rejecters to accept (Langmore et al. 2011). Observations are inconclusive and cannot be used as evidence for the absence of chick discrimination—this flawed way of reasoning was standard in the past. The sole way to discover whether a host shows chick discrimination is experimental, either cross-fostering of various chicks with divergent phenotypes (Grim and Samaš 2016) or manipulation of host and parasite phenotypes (Redondo 1993).

An observation that parasite chicks are similar to host ones does not necessarily imply that chicks are mimetic. Host progeny (eggs, nestlings, fledglings) may resemble parasite progeny from various reasons other than mimicry (Grim 2005). Recent common descent of host and parasite will likely create a superficial—and flawed—impression of mimicry and indirectly of discrimination. This highlights the crucial importance of experiments, especially in the cases of mere "Phenotypic similarity" (Table 29.1). For example, for half a century at least (Voipio 1953), the rufous morph of the common cuckoo female was assumed to mimic falcons (*Falco* spp.). However, a closer examination of plumage patterns and cues known to be used by hosts for enemy recognition casts doubts on the hypothesis. Indeed, experimental data reject it (Trnka et al. 2015). This case warns us to infer mimicry from mere similarity, i.e. without experimental manipulations (Grim 2013).

Parasite chicks can die very soon after hatching and may easily go undetected (Delhey et al. 2011). Parasite chicks can also perish shortly before (Grim et al. 2003; Fig. 29.1c) or even after fledging (De Mársico et al. 2012); such cases might easily by misinterpreted as "parasite successful" because avian ecologists often use artificial cut-off points to score nest fates before fledging.

29.4.2 Preferential Feeding of Super-Chicks

If we adopt a view that parents *always* prefer larger, higher quality or more vigorously begging chicks (host exploitation hypothesis: Redondo 1993; see also Tanaka and Ueda 2005), then cuckoo chick discrimination will seem impossible—a parasite chick is almost always larger immediately after the emergence from the egg and begs more than host chicks do (Redondo 1993). What are the consequences of such constrained assumptions? Consistent parental/fosterer preference for a larger chick inevitably leads to ever-increasing discrepancy between the size of the largest chick (fed more and more) and the smaller chick (fed less and less); such a positive feedback leads to brood reduction (starvation or even death of smaller chicks). Even without any empirical data, this scenario seems unlikely to apply generally. Indeed, only some birds prefer to feed larger chicks, creating size asymmetries and leading to brood reduction (brood reducers *sensu* Soler 2002), whereas others prefer an egalitarian distribution of food, creating evenly sized broods and avoiding brood reduction (clutch adjusters *sensu* Soler 2002). The common cuckoo has specialized in parasitizing "clutch adjusters" species, and therefore cuckoo chicks need to be alone in the host nest, which makes egg eviction a compulsory strategy (Soler 2002; Grim et al. 2009a; Chap. 16).

29.4.3 Potential Cues for Chick Discrimination

Altricial chicks show low variation overall, and parents recognize them via their signals of need and quality (begging), creating a "blind alley" situation for the evolution of chick discrimination (Redondo 1993). I have refuted this argument previously (Grim 2006). Since then, a large body of evidence accumulated, showing that cues for potential chick discrimination are indeed more varied than realized previously (Table 29.3) and are present for the whole time after the parasite emerges from the egg, from hatchlings, through nestlings, to fledglings (for the last stage, see Chap. 30 and Tyller et al. 2018). Even for conspecific chicks, there is evidence of individual-level cues and recognition based on such fine-scale cues, in both parasitic and non-parasitic taxa (Kilner 2006; Shizuka and Lyon 2010, 2011; Levréro et al. 2009). For example, juvenile gapes, the skin, and feathers reflect UV-part of the light spectrum at least in some passerine nestlings and fledglings, and parents use these signals to adjust their provisioning (Tanner and Richner 2008 and references therein).

Table 29.3 Diversity of potential chick discrimination cues, exemplified by the common cuckoo

Cue	Host	Parasite	Reference
Brood size	Several chicks	Single chick	Anderson and Hauber (2007)
Body size	Normal	Larger	Wyllie (1981)
Colour			
Skin	Normal	(Different)	Wyllie (1981)
Gape	Normal	(Different)	Noble et al. (1999)
Natal down	(Present)	Absent	Harrison (2002)
Plumage	Normal	Different	Voipio (1953)
Sound			
Begging calls	Normal	(Different)	Butchart et al. (2003)
Host absent vocalizations	Absent	Present	Šicha et al. (2007)
Behaviour			
Eviction	Absent	Present	Anderson et al. (2009)
Wing-shaking	Symmetrical	Asymmetrical	Grim (2008)
Smell			
Skin smell	Normal	?	No studies yet
Repulsive secretions	Absent	Present	Trnka et al. (2016)
Demands			
Length of care	Normal	Protracted	Grim (2007b)
Intensity of care	Normal	Elevated	Grim et al. (2003)

"Normal" refers to non-parasitized host nests. Parentheses denote traits when the cuckoo chick is different from host chicks in some hosts but not in others

29.4.4 Adaptiveness of Late Chick Discrimination

What is the benefit of late chick desertion for the host that has already lost its own progeny? Previous mistaken investment does not imply that additional investment is adaptive. What matters is future benefits, regardless of previous ("sunk") costs (see the concept of "Concorde fallacy": Dawkins and Carlisle 1976). This applies not only to late chick discrimination (Grim et al. 2003) but also to even later fledgling discrimination (De Mársico et al. 2012; Soler et al. 2014). Mistakenly invested costs are continuously increasing every single day the host cares for foreign nestlings (including conspecific nestlings: Shizuka and Lyon 2010; Samas et al. 2014). Therefore every single day the host avoids such care adds to the host's fitness.

Concluding Remarks and Future Directions

To move the field forward, it is now essential to focus on discovering more chick discrimination systems and then build theoretical models that could be informed by *quantitative* estimates of *empirically* determined costs and benefits. Both observational and experimental approaches will be necessary, including manipulation of chick phenotypic traits, cross-fostering of parasites

(continued)

and both conspecific and heterospecific chicks as controls. Laboratory studies might be important too but should be validated by experiments under natural conditions to confirm that laboratory results are not biased due to artificial environment in captivity.

Acknowledgements I am grateful to E. Røskaft, A. Moksnes and B. G. Stokke for their hospitality during my stay at NTNU in Trondheim in 2002 during which I conceived the rarer enemy model and to Ø. Holen for inviting me to give a seminar at the University of Oslo in October 2002 where I first presented the concept (thus, the model was developed independently from conclusions of Langmore et al. 2003). I am indebted to R. Planqué, N. F. Britton and N. R. Franks for their work on our joint manuscript (during 2003–2005) which integrated their idea of defence portfolios with my concept of rarer enemy; after being rejected six times, we decided to publish two separate papers (Grim 2006; Britton et al. 2007). For the comments on the draft, I am grateful to M. C. De Mársico and R. Planqué. The manuscript benefited from comments by D. Martín-Gálvez, R. Kilner and M. Soler. The Human Frontier Science Program (RGY69/2007 and RGY83/2012) and the Czech Science Foundation (P506/12/2404) supported my research.

References

Anderson MG, Hauber ME (2007) A recognition-free mechanism for reliable rejection of brood parasites. Trends Ecol Evol 22:283–286

Anderson MG, Moskát C, Bán M, Grim T, Cassey P, Hauber ME (2009) Egg eviction imposes a recoverable cost of virulence in chicks of a brood parasite. PLoS One 4:e7725

Antonov A, Stokke BG, Ranke PS, Fossøy F, Moksnes A, Røskaft E (2010) Absence of egg discrimination in a suitable cuckoo *Cuculus canorus* host breeding away from trees. J Avian Biol 41:501–504

Bán M, Moskát C, Barta Z, Hauber ME (2013) Simultaneous viewing of own and parasitic eggs is not required for egg rejection by a cuckoo host. Behav Ecol 24:1014–1021

Briskie JV (2003) Frequency of egg rejection by potential hosts of the New Zealand cuckoos. Condor 105:719–727

Britton NF, Planqué R, Franks NR (2007) Evolution of defence portfolios in exploiter–victim systems. Bull Math Biol 69:957–988

Brooker MG, Brooker LC (1989) The comparative breeding behaviour of two sympatric cuckoos, Horsfield's bronze-cuckoo *Chrysococcyx basalis* and the shining bronze-cuckoo *C. lucidus*, in Western Australia: a new model for the evolution of egg morphology and host specificity in avian brood parasites. Ibis 131:528–547

Butchart SHM, Kilner RM, Fuisz T, Davies NB (2003) Differences in the nestling begging calls of hosts and host-races of the common cuckoo, *Cuculus canorus*. Anim Behav 65:345–354

Colombelli-Négrel D, Hauber ME, Robertson J, Sulloway FJ, Hoi H, Griggio M, Kleindorfer S (2012) Embryonic learning of vocal passwords in superb fairywrens reveals intruder cuckoo nestlings. Curr Biol 22:2155–2160

Colombelli-Négrel D, Hauber ME, Kleindorfer S (2014) Prenatal learning in an Australian songbird: habituation and individual discrimination in superb fairy-wren embryos. Proc R Soc Lond B 281:20141154

Colombelli-Négrel D, Webster MS, Dowling JL, Hauber ME, Kleindorfer S (2016) Vocal imitation of mother's calls by begging red-backed fairywren nestlings increases parental provisioning. Auk 133:273–285

Davies NB (2000) Cuckoos, cowbirds and other cheats, 1st edn. T. & A. D. Poyser, London
Davies NB, Brooke ML (1989) An experimental study of co-evolution between the cuckoo, *Cuculus canorus*, and its hosts. II. Host egg markings, chick discrimination and general discussion. J Anim Ecol 58:225–236
Dawkins R (1982) The extended phenotype, 1st edn. Oxford University Press, Oxford
Dawkins R, Carlisle TR (1976) Parental investment, mate desertion and a fallacy. Nature 262:131–133
Dawkins R, Krebs JR (1979) Arms races between and within species. Proc R Soc Lond B 205:489–511
De Mársico MC, Gantchoff MG, Reboreda JC (2012) Host–parasite coevolution beyond the nestling stage? Mimicry of host fledglings by the specialist screaming cowbird. Proc R Soc Lond B 279:3401–3408
Delhey K, Carrizo M, Verniere L, Mahler B, Peters A (2011) Rejection of brood-parasitic shiny cowbird *Molothrus bonariensis* nestlings by the firewood-gatherer *Anumbius annumbi*? J Avian Biol 42:463–467
Dewar D (1907) An inquiry into the parasitic habits of the Indian koel. J Bombay Nat Hist Soc 17:765–782
Fraga RM (1998) Interactions of the parasitic screaming and shiny cowbirds (*Molothrus rufoaxillaris* and *M. bonariensis*) with a shared host, the bay-winged cowbird (*M. badius*). In: Rothstein SI, Robinson SK (eds) Parasitic birds and their hosts, 1st edn. Oxford University Press, New York, pp 173–193
Gaston AJ (1976) Brood parasitism by the pied crested cuckoo *Clamator jacobinus*. J Anim Ecol 45:331–345
Gill BJ (1998) Behavior end ecology of the shining cuckoo, *Chrysococcyx lucidus*. In: Rothstein SI, Robinson SK (eds) Parasitic birds and their hosts, 1st edn. Oxford University Press, New York, pp 143–151
Grim T (2002) Why is mimicry in cuckoo eggs sometimes so poor? J Avian Biol 33:302–305
Grim T (2005) Mimicry vs. similarity: which resemblances between brood parasites and their hosts are mimetic and which are not? Biol J Linn Soc 84:69–78
Grim T (2006) The evolution of nestling discrimination by hosts of parasitic birds: why is rejection so rare? Evol Ecol Res 8:785–802
Grim T (2007a) Equal rights for chick brood parasites. Ann Zool Fenn 44:1–7
Grim T (2007b) Experimental evidence for chick discrimination without recognition in a brood parasite host. Proc R Soc Lond B 274:373–381
Grim T (2008) Wing-shaking and wing-patch as nestling begging strategies: their importance and evolutionary origins. J Ethol 26:9–15
Grim T (2011) Ejecting chick cheats: a changing paradigm? Front Zool 8:14
Grim T (2013) Mimicry, signalling, and co-evolution. Ethology 119:270–277
Grim T (2017) Chick discrimination versus adaptive parasitic egg acceptance: the egg dilution effect hypothesis revisited. Ornithol Sci 16:163–170
Grim T, Samaš P (2016) Growth performance of nestling cuckoos *Cuculus canorus* in cavity nesting hosts. Acta Ornithol 51:175–188
Grim T, Stokke BG (2016) In the light of introduction: importance of introduced populations for the study of brood parasite-host coevolution. In: Weis JS, Sol D (eds) Biological invasions and animal behaviour, 1st edn. Cambridge University Press, Cambridge, pp 133–157
Grim T, Kleven O, Mikulica O (2003) Nestling discrimination without recognition: a possible defence mechanism for hosts towards cuckoo parasitism? Proc R Soc Lond B 270:S73–S75
Grim T, Rutila J, Cassey P, Hauber ME (2009a) Experimentally constrained virulence is costly for common cuckoo chicks. Ethology 115:14–22
Grim T, Rutila J, Cassey P, Hauber ME (2009b) The cost of virulence: an experimental study of egg eviction by brood parasitic chicks. Behav Ecol 20:1138–1146

Grim T, Samaš P, Moskát C, Kleven O, Honza M, Moksnes A, Røskaft E, Stokke BG (2011) Constraints on host choice: why do parasitic birds rarely exploit some common potential hosts? J Anim Ecol 80:508–518

Grim T, Tyller Z, Samaš P (2017) Unusual diet of brood parasite nestlings and its fitness consequences. Auk 134:732–750

Hanley D, Šulc M, Brennan PLR, Hauber ME, Grim T, Honza M (2016) Dynamic egg color mimicry. Ecol Evol 6:4192–4202

Hanley D, Grim T, Igic B, Samaš P, López AV, Shawkey MD, Hauber ME (2017) Egg discrimination along a gradient of natural variation in eggshell coloration. Proc R Soc Lond B 284:20162592

Harrison C (2002) Bird nests, eggs and nestlings of Britain and Europe, 1st edn. Harper, London

Hegemann A, Voesten R (2011) Can skylarks *Alauda arvensis* discriminate a parasite nestling? Possible case of nestling cuckoo *Cuculus canorus* ejection by its host parents. Ardea 99:117–120

Hobson KA, Sealy SG (1989) Responses of yellow warblers to the threat of cowbird parasitism. Anim Behav 38:510–519

Honza M, Vošlajerová K, Moskát C (2007) Eviction behaviour of the common cuckoo *Cuculus canorus* chicks. J Avian Biol 38:385–389

Honza M, Procházka P, Šicha V, Požgayová M (2010) Nest defence in a cuckoo host: great reed warblers risk themselves equally for their own and parasitic chicks. Behaviour 147:741–756

Huo J, Yang C, Su T, Liang W, Møller AP (2018) Russet sparrows spot alien chicks from their nests. Avian Res 9:12

Igic B, Cassey P, Grim T, Greenwood DR, Moskát C, Rutila J, Hauber ME (2012) A shared chemical basis of avian host–parasite egg colour mimicry. Proc R Soc Lond B 279:1068–1076

Jourdain FCR (1925) A study of parasitism in the cuckoos. Proc Zool Soc Lond 44:639–667

Kilner RM (2006) Function and evolution of color in young birds. In: Hill G, McGraw K (eds) Bird coloration, 1st edn. Harvard University Press, London, pp 201–232

Langmore NE, Hunt S, Kilner RM (2003) Escalation of a coevolutionary arms race through host rejection of brood parasitic young. Nature 422:157–160

Langmore NE, Maurer G, Adcock GJ, Kilner RM (2008) Socially acquired host-specific mimicry and the evolution of host races in Horsfield's bronze-cuckoo *Chalcites basalis*. Evolution 62:1689–1699

Langmore NE, Cockburn A, Russell AF, Kilner RM (2009) Flexible cuckoo chick-rejection rules in the superb fairy-wren. Behav Ecol 20:978–984

Langmore NE, Stevens M, Maurer G, Heinsohn R, Hall ML, Peters A, Kilner RM (2011) Visual mimicry of host nestlings by cuckoos. Proc R Soc Lond B 278:2455–2463

Lawes MJ, Marthews TR (2003) When will rejection of parasite nestlings by hosts of nonevicting avian brood parasites be favored? A misimprinting-equilibrium model. Behav Ecol 14:757–770

Levréro F, Durand L, Vignal C, Blanc A, Mathevon N (2009) Begging calls support offspring individual identity and recognition by zebra finch parents. C R Biol 332:579–589

Li D, Zhang Z, Grim T, Liang W, Stokke BG (2016) Explaining variation in brood parasitism rates between potential host species with similar habitat requirements. Evol Ecol 30:905–923

Lichtenstein G (2001) Low success of shiny cowbird chicks parasitizing rufous-bellied thrushes: chick–chick competition or parental discrimination? Anim Behav 61:401–413

Lotem A (1993) Learning to recognize nestlings is maladaptive for cuckoo *Cuculus canorus* hosts. Nature 362:743–745

McLean IG, Maloney RF (1998) Brood parasitism, recognition and response: the options. In: Rothstein SI, Robinson SK (eds) Parasitic birds and their hosts, 1st edn. Oxford University Press, New York, pp 255–272

McLean IG, Rhodes G (1991) Enemy recognition and response in birds. Curr Ornithol 6:173–211

McLean IG, Waas JR (1987) Do cuckoo chicks mimic the begging calls of their hosts? Anim Behav 35:1896–1898

Moksnes A, Røskaft E, Hagen LG, Honza M, Mørk C, Olsen PH (2000) Common cuckoo *Cuculus canorus* and host behaviour at reed warbler *Acrocephalus scirpaceus* nests. Ibis 142:247–258
Mundy PJ (1973) Vocal mimicry of their hosts by nestlings of the great spotted cuckoo and striped crested cuckoo. Ibis 115:602–604
Nicolai J (1964) Der Brutparasitismus der Viduinae als ethologisches Problem. Z Tierpsychol 21:129–204
Nicolai J (1974) Mimicry in parasitic birds. Sci Am 231:92–98
Noble DG, Davies NB, Hartley IR, McRae SB (1999) The red gape of the nestling cuckoo (*Cuculus canorus*) is not a supernormal stimulus for three common hosts. Behaviour 136:759–777
Nyári ÁS, Joseph L (2012) Evolution in Australasian mangrove forests: multilocus phylogenetic analysis of the gerygone warblers (Aves: Acanthizidae). PLoS One 7:e31840
Payne RB (2005) The cuckoos, 1st edn. Oxford University Press, Oxford
Payne RB, Woods JL, Payne LL (2001) Parental care in estrildid finches: experimental tests of a model of *Vidua* brood parasitism. Anim Behav 62:473–483
Planqué R, Britton NF, Franks NR, Peletier MA (2002) The adaptiveness of defence strategies against cuckoo parasitism. Bull Math Biol 64:1045–1068
Redondo T (1993) Exploitation of host mechanisms for parental care by avian brood parasites. Etología 3:235–297
Reed RA (1968) Studies of the diederik cuckoo *Chrysococcyx caprius* in the Transvaal. Ibis 110:321–331
Roldán M, Soler M, Márquez R, Soler JJ (2013) The vocal begging display of great spotted cuckoo *Clamator glandarius* nestlings in nests of its two main host species: genetic differences or developmental plasticity? Ibis 155:867–876
Samas P, Hauber ME, Cassey P, Grim T (2014) Host responses to interspecific brood parasitism: a by-product of adaptations to conspecific parasitism? Front Zool 11:34
Sánchez-Martínez MA, David S, Londoño GA, Robinson SK (2017) Brood parasitism by the enigmatic and rare pavonine cuckoo in Amazonian Peru. Auk 134:330–339
Sato NJ, Mikami OK, Ueda K (2010a) Egg dilution effect hypothesis: a condition under which parasitic nestling ejection behaviour will evolve. Ornithol Sci 9:115–121
Sato NJ, Tokue K, Noske RA, Mikami OK, Ueda K (2010b) Evicting cuckoo nestlings from the nest: a new anti-parasitism behaviour. Biol Lett 6:67–69
Sato NJ, Tanaka KD, Okahisa Y, Yamamichi M, Kuehn R, Gula R, Ueda K, Theuerkauf J (2015) Nestling polymorphism in a cuckoo-host system. Curr Biol 25:R1151–R1165
Schuetz JG (2005a) Low survival of parasite chicks may result from their imperfect adaptation to hosts rather than expression of defenses against parasitism. Evolution 59:2017–2024
Schuetz JG (2005b) Reduced growth but not survival of chicks with altered gape patterns: implications for the evolution of nestling similarity in a parasitic finch. Anim Behav 70:839–848
Shizuka D, Lyon BE (2010) Coots use hatch order to learn to recognize and reject conspecific brood parasitic chicks. Nature 463:223–226
Shizuka D, Lyon BE (2011) Hosts improve the reliability of chick recognition by delaying the hatching of brood parasitic eggs. Curr Biol 21:515–519
Šicha V, Procházka P, Honza M (2007) Hopeless solicitation? Host-absent vocalization in the common cuckoo has no effect on feeding rate of reed warblers. J Ethol 25:147–152
Soler M (2002) Breeding strategy and begging intensity: influences on food delivery by parents and host selection by parasitic cuckoos. In: Wright J, Leonard ML (eds) The evolution of begging, 1st edn. Kluwer, Dordrecht, pp 413–427
Soler M (2008) Do hosts of interspecific brood parasites feed parasitic chicks with lower-quality prey? Anim Behav 76:1761–1763
Soler M (2009) Co-evolutionary arms race between brood parasites and their hosts at the nestling stage. J Avian Biol 40:237–240
Soler M, De Neve L (2012) Great spotted cuckoo nestlings but not magpie nestlings starve in experimental age-matched broods. Ethology 118:1036–1044
Soler M, Soler JJ, Martinez JG, Møller AP (1995) Chick recognition and acceptance – a weakness in magpies exploited by the parasitic great spotted cuckoo. Behav Ecol Sociobiol 37:243–248

Soler M, Pérez-Contreras T, De Neve L (2013a) Magpies do not desert after prolonging the parental care period: an experimental study. Behav Ecol 24:1292–1298

Soler M, Ruiz-Castellano C, Carra LG, Ontanilla J, Martín-Galvez D (2013b) Do first-time breeding females imprint on their own eggs? Proc R Soc Lond B 280:20122518

Soler M, De Neve L, Roncalli G, Marcías-Sánchez E, Ibáñez-Álamo JD, Pérez-Contreras T (2014) Great spotted cuckoo fledglings are disadvantaged by magpie host parents when reared together with magpie nestlings. Behav Ecol Sociobiol 68:333–342

Soler M, Macías-Sánchez E, Martín-Gálvez D, De Neve L (2017) Complex feeding behaviour by magpies in nests with great spotted cuckoo nestlings. J Avian Biol 48:1406–1413

Steyn P (1973) Some notes on the breeding biology of the striped cuckoo. Ostrich 44:163–169

Tanaka KD, Ueda K (2005) Signal exploitation by parasitic young in birds: a new categorization of manipulative signals. Ornithol Sci 4:49–54

Tanner M, Richner H (2008) Ultraviolet reflectance of plumage for parent–offspring communication in the great tit (*Parus major*). Behav Ecol 19:369–373

Thorogood R, Kilner RM, Rasmussen JL (2017) Grey gerygone hosts are not egg rejecters, but shining bronze-cuckoos lay cryptic eggs. Auk 134:340–349

Tokue K, Ueda K (2010) Mangrove gerygones *Gerygone laevigaster* eject little bronze-cuckoo *Chalcites minutillus* hatchlings from parasitized nests. Ibis 152:835–839

Trnka A, Grim T (2014) Testing for correlations between behaviours in a cuckoo host: why do host defences not covary? Anim Behav 92:185–193

Trnka A, Požgayová M, Procházka P, Prokop P, Honza M (2012) Breeding success of a brood parasite is associated with social mating status of its host. Behav Ecol Sociobiol 66:1187–1194

Trnka A, Trnka M, Grim T (2015) Do rufous common cuckoo females indeed mimic a predator? An experimental test. Biol J Linn Soc 116:134–143

Trnka A, Požgayová M, Procházka P, Capek M, Honza M (2016) Chemical defence in avian brood parasites: production and function of repulsive secretions in common cuckoo chicks. J Avian Biol 47:288–293

Tyller Z, Kysučan M, Grim T (2018) Post-fledging behavior of the common cuckoo (*Cuculus canorus*) attended by the chaffinch (*Fringilla coelebs*): a comprehensive approach to study the least known stage of brood parasite-host coevolution. Wilson J Ornithol 130

Voipio P (1953) The hepaticus variety and the juvenile types of the cuckoo. Ornis Fenn 30:97–117

Wang N, Kimball R (2012) Nestmate killing by obligate brood parasitic chicks: is this linked to obligate siblicidal behavior? J Ornithol 153:825–831

Wyllie I (1981) The cuckoo, 1st edn. Batsford, London

Yang C, Stokke BG, Antonov A, Cai Y, Shi S, Moksnes A, Røskaft E, Møller AP, Liang W, Grim T (2013) Host selection in parasitic birds: are open-cup nesting insectivorous passerines always suitable cuckoo hosts? J Avian Biol 44:216–220

Yang C, Wang L, Chen M, Liang W, Møller AP (2015) Nestling recognition in red-rumped and barn swallows. Behav Ecol Sociobiol 69:1821–1826

Parasite Adaptations During the Nestling and Fledgling Stages

30

María C. De Mársico, Vanina D. Fiorini, Diego T. Tuero, Ros Gloag, Cynthia A. Ursino, and Juan C. Reboreda

Abstract

Young of altricial brood parasites are fully dependent on their foster parents for a considerable period of time before being able to survive on their own. Therefore, they are expected to exhibit traits that allow them to avoid host defences and manipulate host's parental behaviour to their favour. Many morphological, physiological and behavioural traits have been proposed as adaptations for brood parasitism based on their apparent selective advantage for parasitic chicks. In this chapter, we describe and discuss these putative adaptations to examine, in the light of available evidence, whether or not those traits have evolved specifically to increase parasites' fitness. We show that whereas some adaptations are well-supported and indisputable, other traits require a closer scrutiny taking into account the parasite's evolutionary history before concluding that they are true adaptations for brood parasitism. We propose some future directions for research on key adaptations to parasitism and coevolutionary interactions between parasites and their hosts during the nestling and fledgling stages.

30.1 Introduction

Successful parasitism requires that young parasites generate the appropriate signals to elicit provisioning from their foster parents and deal with competition within the brood. It has long been recognized that parasitic nestlings possess many behavioural

M. C. De Mársico (✉) · V. D. Fiorini · D. T. Tuero · C. A. Ursino · J. C. Reboreda
Facultad de Ciencias Exactas y Naturales, Departamento de Ecología, Genética y Evolución & IEGEBA-CONICET, Universidad de Buenos Aires, Buenos Aires, Argentina
e-mail: de_marsico@ege.fcen.uba.ar

R. Gloag
Behaviour and Genetics of Social Insects Lab, School of Life and Environmental Sciences, Faculty of Science, University of Sydney, Sydney, NSW, Australia

M. Soler (ed.), *Avian Brood Parasitism*, Fascinating Life Sciences,
https://doi.org/10.1007/978-3-319-73138-4_30

and physiological traits that allow them to secure parental care and manipulate host's parental behaviour (Payne 1977; Rothstein 1990). These include nestmate-killing behaviours, early hatching, rapid growth rates and/or exuberant begging displays (Payne 1977; Rothstein 1990; Redondo 1993; Davies 2000). Despite their intuitive selective advantage, however, these putative adaptations should be tested taking into account the parasite's evolutionary history before concluding that they have evolved specifically for brood parasitism (Mermoz and Ornelas 2004; Birkhead et al. 2011).

In theory, the fitness costs imposed on hosts by parasitic nestlings that either kill or outcompete host young may set the stage for a coevolutionary arms race, where hosts evolve defences against parasitic nestlings and parasites evolve counter-adaptations that select for improved host defences and so on (Rothstein 1990). However, it was assumed until recently that nestling rejection was unlikely to evolve when hosts have to learn the appearance of their own young during the first breeding attempt because the costs of incurring recognition or rejection errors would outweigh the benefits of this defence (Lotem 1993; Lawes and Marthews 2003; Grim 2006a). Today, new evidences for the evolution of host adaptations against parasitic chicks and reciprocal counter-adaptations in parasite populations (i.e. chick mimicry) are driving a paradigm shift (Langmore et al. 2003; Grim 2007, 2011; Soler 2009; Sato et al. 2010; De Mársico et al. 2012; Feeney et al. 2014; Chap. 29).

Davies (2011) has coined the terms 'trickery' and 'tuning' to distinguish between parasite's traits that have evolved as a result of coevolutionary interactions with its host (trickery) and traits that have not coevolved with host defences but which enhance parasite's success (tuning). For instance, mimicking host-specific cues to avoid host discrimination would be trickery (see also Grim 2005), whereas adjusting begging displays to better exploit host's sensory preferences would be tuning (Davies 2011). The trickery-tuning distinction may become diffuse when failure to tune into host's life history ultimately results in parasite's death or when parasite tuning might also lead to reciprocal adaptations (Davies 2011). However, it provides a useful framework to pinpoint co-evolved adaptations in parasites and their hosts, which has historically been a major focus of research on avian brood parasitism (Rothstein 1990).

Trickery and tuning adaptations similar to those observed at the nestling stage can also appear during the post-fledging period since fledgling parasites depend on their hosts for several weeks before attaining nutritional independence. Juveniles of obligate brood parasites also face another major challenge: at some point, they must abandon their foster parents and meet their conspecifics to continue their life cycle. Obviously, they are adapted to do so, or those species would not exist! However, how juvenile parasites recognize their own kind remains poorly understood (Göth and Hauber 2004). Indeed, very little is known about parasites' biology after they fledge from host nests. The paucity in research during this critical life period arises from the difficulties associated with tracking mobile juveniles. Despite this, recent insights into the social interactions and dispersal patterns of parasitic juveniles suggest potential adaptations at this last stage of the nesting cycle (Soler and Soler 1999; Hauber et al. 2000, 2001; Hauber 2002; Soler et al. 2014a; Louder et al. 2015).

Here we review the main proposed adaptations for brood parasitism at the nestling and fledgling stages (Tables 30.1 and 30.2). We describe how young

Table 30.1 Overview of proposed adaptations for brood parasitism at the nestling stage

Challenges to solve	Proposed adaptations	Study species	Adaptive role supported?	References
Deal with intra-brood competition	Early hatching	Common cuckoo (*Cuculus canorus*), African cuckoo (*C. gularis*) and honeyguides (*Indicator indicator*)	No	Birkhead et al. (2011) Honza et al. (2001, 2015)
		Cowbirds (*Molothrus* spp.)	No	McMaster and Sealy (1998) Mermoz and Ornelas (2004)
		Great spotted cuckoo (*Clamator glandarius*)	No	Soler (1990)
	'Nestmate-killing' behaviour	Old World cuckoos (Family Cuculidae) Honeyguides (Family Indicatoridae) Striped cuckoo (*Tapera naevia*)	Yes	Kilner (2005) Hauber and Moskát (2008) Grim et al. (2009) Spottiswoode and Koorevaar (2012) Wang and Kimball (2012)
Grow optimally in host nests	Rapid growth rates	Cowbirds	No	Mermoz and Ornelas (2004) Remeš (2010)
		Great spotted cuckoo	No	Soler and Soler (1991)
	Increased digestive efficiency	Great spotted cuckoo	Partially	Soler et al. (2014b)
Stimulate parental care	Exaggerated begging displays/Tuning into host's sensory preferences/ Displaying conspicuous traits (e.g. wing patches, palatal papillae)	Common cuckoo	Partially	Kilner et al. (1999) Madden and Davies (2006)
		Great spotted cuckoo (*Clamator glandarius*)	Partially	Soler et al. (1999) Soler et al. (1995a)
		Cowbirds	Mixed evidence	Lichtenstein (2001) Dearborn and

(continued)

Table 30.1 (continued)

Challenges to solve	Proposed adaptations	Study species	Adaptive role supported?	References
				Lichtenstein (2002) Rivers et al. (2010, 2013) Gloag and Kacelnik (2013)
		Horsfield's hawk cuckoo (*Cuculus fugax*)	Partially	Tanaka and Ueda (2005) Grim (2008)
Avoid host defenses	Chick mimicry	Horsfield's bronze cuckoo (*Chalcites basalis*)	Presumably yes	Langmore et al. (2003, 2011)
		Little bronze cuckoos (*Chrysococcyx minutillus*)	Partially	Sato et al. (2010), Tokue and Ueda (2010)
		Whydahs and indigobirds (*Vidua* spp.)	Mixed evidence	Payne et al. (2001) Schuetz (2005a, b) Hauber and Kilner (2007)

Selected references provide evidence either supporting or not the hypothesized adaptation. 'Partially' supported means that empirical data suggest a fitness benefit of the proposed adaptation, but comparative phylogenetic studies are lacking. 'Mixed evidence' means conflicting results about the putative adaptive role

parasites can solve the problem of securing parental provisioning from hatching to independence and discuss, in the light of available evidence, whether those seemingly adaptive traits have evolved specifically as trickery or tuning adaptations for brood parasitism. Finally, we suggest future directions for research on brood parasitism at the nestling and fledgling stages.

30.2 Proposed Parasite Adaptations During the Nestling Stage

30.2.1 Early Hatching

Hatching earlier than host chicks can be key to the survival of parasitic young, because it ensures that nestling parasites have the size advantage needed to either eject host nestmates or compete with them for food. The notion that early hatching represents an adaptation to brood parasitism stems from long-standing observations

Table 30.2 Overview of proposed adaptations for brood parasitism at the fledgling stage

Challenges to solve	Proposed adaptations	Study species	Adaptive role supported?	References
Securing parental care out of the nest	Exaggerated begging signals/tuning into host's sensorial preferences	Brown-headed cowbird (*Molothrus ater*)	Partially	Woodward (1983) Sealy and Lorenzana (1997) Rasmussen and Sealy (2006)
		Great spotted cuckoo (*Clamator glandarius*)	No	Soler et al. (2014a)
	Joined dispersal	Great spotted cuckoo	Partially	Soler et al. (1995b, 2014a)
	Change to better caregivers	Great spotted cuckoo	Partially	Soler et al. (2014a, c)
Avoiding host discrimination	Host fledgling mimicry	Screaming cowbird (*M. rufoaxillaris*)	Yes	Fraga (1998) De Mársico et al. (2012)
Conspecific recognition	'Password' for species recognition	Brown-headed cowbird	Partially	King and West (1977) Graham and Middleton (1988) Hauber et al. (2001)
		Great spotted cuckoo	No	Soler and Soler (1999)
		Indigobirds (*Vidua funerea* and *V. purpurascens*)	Presumably yes	Payne et al. (2000) DaCosta and Sorenson (2014)
	Self-referencing	Brown-headed cowbird	Partially	Hauber et al. (2000)
	Facilitation by adult parasites	Brown-headed cowbird	Mixed evidence	Hahn and Fleischer (1995) Hauber (2002) Louder et al. (2015)

(continued)

Table 30.2 (continued)

Challenges to solve	Proposed adaptations	Study species	Adaptive role supported?	References
		Great spotted cuckoo	Partially	Soler and Soler (1999) Soler et al. (1995b)

Selected references provide evidence either supporting the hypothesized adaptation or not. 'Partially' supported means that empirical data suggest a fitness benefit of the proposed adaptation, but comparative studies are lacking. 'Mixed evidence' means that different studies showed conflicting results

that parasite eggs hatch sooner than would be allometrically expected for their size (Hamilton and Orians 1965; Payne 1977; Briskie and Sealy 1990). Early hatching might be achieved via maternal traits (e.g. internal incubation; Chap. 20) or via the rapid development of parasite embryos themselves.

Researchers have looked to characteristics of parasite's eggshells for evidence of rapid embryo growth, with equivocal results. The eggs of brown-headed cowbirds (*Molothrus ater*) are more porous that those of two closely related non-parasitic icterids, consistent with cowbird eggs achieving the high rates of gas flux needed for rapid embryo growth (Jaeckle et al. 2012). However, a phylogenetic analysis of cowbirds and their relatives found no evidence that these parasites had shorter-than-expected incubation times (Mermoz and Ornelas 2004). In common cuckoos (*Cuculus canorus*), eggshells are *less* porous than those of non-parasites, which would not favour rapid growth but might instead help them conserve the energetic reserves needed for hatching (Portugal et al. 2014). Indeed, hatching from parasite eggs, which are typically thicker shelled than those of non-parasites, may require its own adaptations, such as stronger hatching muscles (Honza et al. 2001, 2015).

30.2.2 Host-Attuned Growth Rates

Host species broadly differ in life history traits, provisioning effort, and predation pressures, all of which can affect the survival and growth of parasitic chicks. Given that growth patterns may have important consequences for fitness, the question arises of whether parasites are attuned to grow optimally in host nests.

Two major drivers of growth rates in birds are nest predation risk (Martin et al. 2011; Mainwaring and Hartley 2012) and within-brood competition (Royle et al. 1999). Parasitic chicks may benefit from growing as fast as possible if this reduces their exposure to predation (Remeš and Martin 2002; Remeš and Matysioková 2016) and/or increases their competitive ability (Royle et al. 1999). However, empirical data do not support this expectation. Rather, growth rates vary within parasite species depending on host characteristics and brood size. For instance, brown-headed cowbird nestlings grow faster when parasitizing host species with shorter nestling periods (Kilpatrick 2002; Remeš 2010) and shiny cowbird (*M. bonariensis*) nestlings grow more rapidly in nests of small- and medium-sized hosts (De Mársico et al. 2010; Gloag et al. 2012; but see Tuero et al. 2013). In addition, a comparative study indicates that cowbirds have not evolved faster growth rates than their non-parasitic relatives (Mermoz and Ornelas 2004). Similarly, common cuckoo nestlings show considerable variation in growth rate among host species, probably due to differences in provisioning rates or host quality (Kleven et al. 1999; Butchart et al. 2003; Grim 2006b). Common cuckoo nestlings also showed decreased growth rates when they were forced to compete with host nestmates in mixed broods (Hauber and Moskát 2008; Grim et al. 2009; Geltsch et al. 2012),which further indicate that growth dynamics of parasite nestlings can be constrained by the rearing environment.

Parasitic nestlings could achieve rapid growth rates via increased digestive efficiency relative to host nestlings. Soler et al. (2014b, 2017) examined this putative adaptation by analysing the digestive performance of hand-fed great spotted cuckoo (*Clamator glandarius*) nestlings and those of its primary host, the magpie (*Pica pica*). Great spotted cuckoo nestlings usually grow faster than magpies and ingest more food under similar conditions (Soler and Soler 1991; Soler et al. 1995a, 2014b). Nevertheless, controlled biochemical analyses of nestlings' faeces failed to support the hypothesis of better nutrient assimilation in cuckoos, suggesting that faster growth rates result from parasite's higher competitive ability rather than increased digestive efficiency (Soler et al. 2017).

30.2.3 Nestmate-killing Behaviour

Young of obligate brood parasites are expected to behave selfishly since they are not constrained by kin selection (Hamilton 1964). A striking example is the nestmate-killing behaviour found in most Old World cuckoos (Jenner 1788; Honza et al. 2007), honeyguides (*Indicator* spp.; Spottiswoode and Koorevaar 2012) and the striped cuckoo (*Tapera naevia*) (Morton and Farabaugh 1979). Nestmate-killing behaviours typically result in parasite nestlings becoming the sole occupants of the nest, hence its most obvious benefit is to monopolize parental care (Kilner 2005; Hauber and Moskát 2008; Grim et al. 2009). Various adaptive explanations have been proposed for nestmate-killing in brood parasites (reviewed in Kilner 2005; see also Soler 2002), but none of them can fully account for the phylogenetic distribution of this trait (Wang and Kimball 2012; see also Grim 2006c).

What are the evolutionary origins of nestmate-killing in obligate brood parasites? Is it linked to obligate siblicide? Wang and Kimball (2012) examined this link. They found that clades of parasite species exhibiting nestmate-killing behaviours were nested within clades containing obligate siblicidal species and, conversely, clades with 'nestmate-acceptor' parasites did not feature siblicidal behaviour (Wang and Kimball 2012). The exception is the genus *Clamator*, which do not exhibit nestmate-killing but occurs within a larger clade containing nestmate-killing parasites

(Wang and Kimball 2012). The evolution of virulence in obligate brood parasites may have resulted from a predisposition for nestmate killing in certain lineages combined with ecological and evolutionary constraints (Wang and Kimball 2012; Fulmer and Hauber 2016). Nevertheless, the adaptiveness of nestmate-killing likely depends on multiple factors including host's food allocation strategies and the competitive ability of parasitic chicks (Soler 2002; Kilner et al. 2004; Rivers 2007; Gloag et al. 2012; Soler and de Neve 2013).

30.2.4 Exaggerated Begging Displays

Nestling birds communicate their need through begging displays that comprise visual (posturing and coloured gapes) and vocal signals (begging calls). Begging signals have been hypothesized to carry honest information about offspring need because they convey direct (i.e. energy expenditure, predation risk) and indirect fitness costs (i.e. competition with full- or half-sibs) to nestlings (Godfray 1995; Kilner and Johnstone 1997). Hence, it is expected that nestlings will beg at intensities that reflect their hunger level (i.e. short-term need) and body condition (i.e. long-term need) (Budden and Wright 2001).

Early ornithologists have noticed that parasitic nestlings beg louder than host young (e.g. Friedmann 1929; Fig. 30.1). These observations found empirical support from studies showing exaggerated begging displays in different parasite species (Gochfeld 1979; Kilner et al. 1999; Soler et al. 1999; Rivers 2007). This exaggeration involves not only acoustic signals but also visual ones, such as body stretching, conspicuous gape colours (Álvarez 2004), wing-shaking (Grim 2008) or wing patches that simulate extra gapes in the nest (Tanaka et al. 2005; Tanaka and Ueda 2005). The exaggeration of begging signals is expected since parasitic chicks are only constrained by the direct costs of begging (Godfray 1995). This has raised the question of whether parasite's begging displays are honest signals of need or not. There is growing evidence for a positive relationship between hunger level and begging intensity of parasitic chicks, supporting the view that despite their exaggeration begging displays convey reliable information to host parents, as predicted by models of honest signalling (Kilner and Davies 1999; Lichtenstein 2001; Hauber and Ramsey 2003; Tanaka and Ueda 2005; Soler et al. 2012; but see Rivers 2007).

The ubiquity of intense begging behaviours among brood parasites has led to the assumption that this is an adaptation for brood parasitism. However, few studies have tested this hypothesis taking evolutionary history into account. Rivers et al. (2013) quantified nestling begging intensity in the brown-headed cowbird and a close non-parasitic relative, the red-winged blackbird (*Agelaius phoeniceus*) under similar conditions. Contrary to the expectation, they found that begging intensity of nestling cowbirds was similar or lower than that of blackbirds (Rivers et al. 2013). By contrast, Lichtenstein (2001) showed that screaming cowbird (*Molothrus rufoaxillaris*) nestlings beg more intensely than those of its phylogenetically related host, the baywing (*Agelaioides badius*), after controlling for short-term need. Given

Fig. 30.1 Shiny cowbird (*Molothrus bonariensis*) chick exhibiting its begging display in a nest of a common host, the chalk-browed mockingbird (*Mimus saturninus*), which contains two host chicks. Photo credit: Vanina D. Fiorini

such mixed evidence, it is too early to either conclude or dismiss that increased begging intensity is a tuning adaptation of parasite young.

30.2.5 Host Chick Mimicry

Active host defences against parasitism at the chick stage are either less common or less commonly documented than those at the egg stage (see Chap. 29). Direct evidence of host parents rejecting cuckoo nestlings comes from hosts of the Australasian bronze cuckoos (*Chalcites* spp.), which either drag newly hatched cuckoo nestlings out of the nest (*Gerygone* sp.; Sato et al. 2010; Tokue and Ueda 2010) or abandon lone cuckoo chicks (superb fairy wrens, *Malurus cyaneus*; Langmore et al. 2003). Bronze cuckoos are thus expected to have evolved counter-adaptations to evade host rejection. Consistent with visual mimicry as one such counter-defence, bronze cuckoo nestlings are more similar in skin colour, flange colour and feather colour to their respective hosts than they are to congener cuckoos (Langmore et al. 2011; Fig. 30.2). However, experimental evidence is needed to confirm that hosts use visual cues to recognize and reject parasite chicks and thus that the evolution of visual similarities with hosts is driven by host defences (Grim 2005). Indeed, Langmore et al. (2003) found that superb fairy wrens reject shining bronze cuckoo chicks (*C. lucidus*) more consistently than Horsfield's bronze cuckoo chicks (*C. basalis*), despite the two being similarly close visual matches for the fairy wren's young (Langmore et al. 2003, 2011). Thus visual mimicry in this case must be, at best, only one of several cues used to identify cuckoos. Horsfield's bronze cuckoos may benefit from mimicry of additional cues used by fairy wrens to make rejection decisions, such as rictal flange colour or begging call structure (Langmore et al. 2011; Colombelli-Négrel et al. 2012).

Vidua finches closely resemble their hosts in the colours and patterns of their gapes, but whether similarity in this case derives from counter-defences by parasites remains unresolved (Payne et al. 2001). In experimental manipulations, red-billed

Fig. 30.2 Little bronze cuckoo (*Chrysococcyx minutillus*) chick (top) showing striking visual similarity to the chicks of its host, the large-billed gerygone (*Gerygone magnirostris*). Photo credit: Hee-Jin Noh

firefinches (*Lagonosticta senegala*) were more likely to abandon the nestlings of non-parasitic estrildid finches, which look unlike their own young, than they were the nestlings of the parasitic village indigobird (*V. chalybeata*), which closely resemble their own young (Payne et al. 2001). However, common waxbills (*Estrilda astrild*) parasitized by another gape-mimicking *Vidua* (*V. macroura*) showed no active defence against parasitic nestlings in the field (Schuetz 2005a) nor were they more likely to abandon waxbill young with modified gape colour than those with unmanipulated gapes (Schuetz 2005b). Other explanations for the visual similarity of *Vidua* sp. and their host nestlings do not require an assumption of active host defence at the nestling stage. For example, gape mimicry may be driven by increased exploitation of host's sensory biases during provisioning, or a coevolutionary arm's race in which intrabrood conflict has led host young to converge on parasite morphology (Hauber and Kilner 2007).

30.3 Proposed Parasite Adaptations During the Fledgling Stage

The post-fledging period represents a critical yet understudied life stage. In many species, fledglings are flightless by the time they leave the nest, and post-fledging parental care becomes crucial to enhance their survival (Naef-Daenzer and Grüebler 2016). Therefore, it is expected that fledgling brood parasites deploy tactics to secure parental care (Table 30.2).

30.3.1 Tuning Adaptations

Fledglings of the brown-headed cowbird exhibit loud and persistent begging calls even after being able to forage for themselves (Woodward 1983). Most notably, they can engage individuals other than their foster parents to provision them (reviewed in Sealy and Lorenzana 1997). These so-called 'auxiliary feedings' of parasitic fledglings have been reported for another 12 parasite species and involved provisioning adults of either the same or a different species than the original hosts, suggesting that some parasites may be adapted to 'tune' into a broad range of parent-offspring communication systems (Sealy and Lorenzana 1997).

Auxiliary feedings seem to occur more frequently among parasitic than non-parasitic fledglings (Shy 1982; Sealy and Lorenzana 1997) and could enhance the survival of parasitic fledglings when hosts cease to feed them prematurely or provide inadequate amounts of food (Sealy and Lorenzana 1997). Yet, whether parasitic fledglings are adapted to look for auxiliary feedings or exploit them opportunistically is not clear. So far, the only parasite species believed to regularly use this tactic are the pallid cuckoo (*Cuculus pallidus*; Sealy and Lorenzana 1997) and the great spotted cuckoo (Soler et al. 1995b, 2014c). The latter may abandon their natal group and join other conspecific fledglings to find new caregivers (Soler et al. 1995b, 2014a, c). Looking for auxiliary feedings could be an adaptive response of great spotted cuckoo fledglings that fail to tune into host's sensory preferences and become undernourished when reared in mixed broods (Soler et al. 2014a, c). Consistent with this idea, cuckoo fledglings were more likely to elicit alloparental feedings from adult magpies that have reared only cuckoo chicks than from those that have cared for mixed broods (Soler et al. 2014c).

30.3.2 Trickery Adaptations

Just as in the nestling stage, fledgling parasites may outcompete host young if they are able to monopolize parental feedings after leaving the nest. Rasmussen and Sealy (2006) compiled 102 reports of post-fledging parental feedings in hosts of the brown-headed cowbird that cared for mixed broods and found that hosts fed only the cowbird fledglings in 95% of these cases. Although most of these reports were anecdotal and likely subjected to biases, it is possible that reproductive losses caused by cowbird parasitism just before or after fledging are higher than previously thought.

If high enough, post-fledging fitness costs could drive the evolution of host defences against fledgling parasites, which may potentially lead to a coevolutionary arms race at this last stage of the nesting cycle. The most compelling evidence of coevolution at the fledgling stage comes from the host specialist screaming cowbird and its primary host, the baywing. Fledgling screaming cowbirds match closely the plumage coloration and begging calls of baywing fledglings (Fraga 1998; De Mársico et al. 2012). This resemblance seems to be a case of true mimicry evolved in response to host discrimination because baywing parents provide prolonged parental care to screaming cowbird fledglings but refuse to feed non-mimetic fledglings of the closely related shiny cowbird (Fraga 1998; De Mársico et al. 2012). Why baywings have evolved rejection of parasitic fledglings instead of parasitic eggs or chicks is intriguing. Delaying discrimination could be adaptive if it makes recognition of alien young easier and allows hosts to save energy for future reproduction. Similar trickery adaptations could occur in other parasite species (e.g. the Jacobin cuckoo, *Clamator jacobinus*) (reviewed in Grim 2006a; Feeney et al. 2014), but those reports require confirmation.

30.3.3 Adaptations for Conspecific Recognition

Juveniles of obligate brood parasites must also solve the problem of meeting their conspecifics after having been reared by heterospecific hosts (Table 30.2). In birds, conspecific recognition templates are shaped through social experience with individuals (e.g. parents or siblings) from whom developing young can reliably learn species-specific phenotypes (ten Cate and Voss 1999; Slagsvold and Hansen 2001). So, how do young parasites avoid sexually misimprinting and develop the appropriate species-recognition templates?

Hauber et al. (2001) suggested that brood parasites may use ‘passwords’ for species recognition. Young parasites would be able to innately recognize certain phenotypic traits that reliable signals conspecific identity, which in turn would trigger learning of additional cues for species recognition from the password-giver’s phenotype (Soler and Soler 1999; Hauber et al. 2001). In the brown-headed cowbird, the ‘chatter’ call of adult males and females has been proposed as such a password (Hauber et al. 2001). A similar mechanism could be at play in parasitic whydahs and indigobirds (*Vidua* spp.). Male indigobirds exhibit ‘mimicry songs’ that are common to conspecific males that share the same host species and ‘non-mimicry songs’ that are common to conspecific males from the same local neighbourhood but vary over larger geographic scales (Payne et al. 1998; DaCosta and Sorenson 2014). Young males that disperse outside their local ‘dialect neighbourhood’ may use mimicry songs to identify conspecifics, from which they can subsequently learn new repertoires of non-mimicry songs (DaCosta and Sorenson 2014).

A second mechanism for species recognition could be self-referencing, which is developing a conspecific recognition template based on the individual’s own phenotype (the ‘armpit effect’; Dawkins 1982). This idea has found some support in a choice experiment performed with brown-headed cowbird juveniles that were dyed to manipulate their phenotype (Hauber et al. 2000). Juveniles approached more quickly and associated preferentially with cowbird females coloured like themselves (Hauber et al. 2000), indicating that they could use salient features of their own phenotype to identify conspecifics in the absence of, or in addition to, other recognition cues.

Another possibility is that adult parasites themselves facilitate conspecific recognition by providing opportunities for social learning of species-specific cues to juveniles. This idea arose from studies showing social interactions between adult females and juveniles in the brown-headed cowbird (Woodward 1983; Hahn and Fleischer 1995; Hauber 2002) and the great spotted cuckoo (Soler and Soler 1999). Such interactions could facilitate an early socialization if parasitic juveniles follow conspecific females to foraging flocks or communal roosting sites. However, a recent study based on intensive monitoring of radio-tagged cowbirds failed to find temporal correlation in the departures of females and juveniles out of the juveniles' natal area (Louder et al. 2015). The authors suggest that cowbird juveniles may use their solitary excursions outside the natal area to locate conspecifics and gradually segregate from hosts before sexual misimprinting can occur (Louder et al. 2015).

Adult great spotted cuckoos had been observed approaching parasitic fledglings (Soler and Soler 1999). These observations indirectly support the facilitation hypothesis, but the role of adult parasites in the socialization of juvenile cuckoos remains to be determined. Contrary to non-parasitic birds, adult and juvenile cuckoos migrate separately, with adults leaving the breeding areas earlier than juveniles (Soler et al. 1994). This suggests that cuckoo juveniles are able to locate wintering grounds without the assistance of conspecific adults (Soler et al. 1994).

Concluding Remarks and Future Directions

Testing of putative adaptations for brood parasitism requires comparison of parasites not to their hosts but to their non-parasitic relatives. Comparative studies that take into account the parasite's evolutionary history would help to understand if seemingly beneficial traits (e.g. early hatching) are true adaptations or 'pre-adaptations'. Some traits such as growth patterns and begging behaviour can vary widely within parasite species according to environmental factors. To study whether such flexibility reflects ecological constraints or adaptive phenotypic plasticity would be a fruitful subject for future research. Also, additional studies that quantify the fitness costs and benefits of nestmate-killing *versus* nestmate acceptance may help to better understand the evolution of these strategies.

The unexpected discovery of chick rejection by some hosts has led to renewed focus on host–parasite coevolution during the nestling and fledging stages. Some recent studies suggest the evolution of visual or vocal mimicry in parasitic chicks, but more research is needed into how hosts integrate perceptual cues in rejection decisions. This will help to determine whether host

(continued)

resemblance in parasite chicks is a trickery adaptation or not. Understanding how phylogenetically widespread is chick rejection behaviour and how parasitism drives selection on host young would further shed light on the matter.

The post-fledgling stage is perhaps the most neglected topic in the study of brood parasitism. As new technologies to track juvenile parasites in the wild become increasingly available, we can expect to increase our knowledge of the brood parasitic life cycle beyond the nestling stage. Further experimental studies on social interactions during the post-fledgling stage and the development of species-recognition templates in young parasites may uncover additional key adaptations to brood parasitism.

Acknowledgements We thank Rebecca Kilner, Tomáš Grim and Manuel Soler for helpful comments on the manuscript. MCDM, VF, DT and JCR are research fellows of Consejo Nacional de Investigaciones Científicas y Técnicas (CONICET). CAU was supported by a CONICET Fellowship. RG was supported by a University of Sydney Fellowship.

References

Álvarez F (2004) The conspicuous gape of the nestling common cuckoo *Cuculus canorus* as a supernormal stimulus for rufous bush chat *Cercotrichas galactotes* hosts. Ardea 92:63–68

Birkhead TR, Hemmings N, Spottiswoode CN, Mikulica O, Moskát C, Bán M, Schulze-Hagen K (2011) Internal incubation and early hatching in brood parasitic birds. Proc R Soc Lond B 278:1019–1024

Briskie JV, Sealy SG (1990) Evolution of short incubation periods in the parasitic cowbirds *Molothrus* spp. Auk 107:789–794

Budden AE, Wright J (2001) Begging in nestling birds. Curr Ornithol 16:83–118

Butchart SHM, Kilner RM, Fuisz T, Davies NB (2003) Differences in the nestling begging calls of hosts and host-races of the common cuckoo, *Cuculus canorus*. Anim Behav 65:345–354

Colombelli-Négrel D, Hauber ME, Robertson J, Sulloway FJ, Hoi H, Griggio M, Kleindorfer S (2012) Embryonic learning of vocal passwords in superb fairy-wrens reveals intruder cuckoo nestlings. Curr Biol 22:2155–2160

DaCosta JM, Sorenson MD (2014) An experimental test of host song mimicry as a species recognition cue among male brood parasitic indigobirds (*Vidua* spp.). Auk 131:549–558

Davies NB (2000) Cuckoos, cowbirds and other cheats. T & AD Poyser, London

Davies NB (2011) Cuckoo adaptations: trickery and tuning. J Zool 284:1–14

Dawkins R (1982) The extended phenotype. The gene as the unit of selection. Oxford University Press, New York

De Mársico MC, Mahler B, Reboreda JC (2010) Reproductive success and nestling growth of the baywing parasitized by screaming and shiny cowbirds. Wilson J Ornithol 122:417–431

De Mársico MC, Gantchoff MG, Reboreda JC (2012) Host–parasite coevolution beyond the nestling stage? Mimicry of host fledglings by the specialist screaming cowbird. Proc R Soc Lond B 279:3401–3408

Dearborn DC, Lichtenstein G (2002) Begging behaviour and host exploitation in parasitic cowbirds. In: Wright W, Leonard ML (eds) The evolution of begging. Competition, cooperation and communication. Kluwer Academic, Dordrecht, pp 361–387

Feeney WE, Welbergen JA, Langmore NE (2014) Advances in the study of coevolution between avian brood parasites and their hosts. Annu Rev Ecol Evol Syst 45:227–246

Fraga RM (1998) Interactions of the parasitic screaming and shiny cowbirds (*Molothrus rufoaxillaris* and *M. bonariensis*) with a shared host, the bay-winged cowbird (*M. badius*). In: Rothstein SI, Robinson SK (eds) Parasitic birds and their hosts. Studies on coevolution. Oxford University Press, New York, pp 173–193

Friedmann H (1929) The cowbirds: a study in the biology of social parasitism. CC Thomas, Springfield, IL

Fulmer AG, Hauber ME (2016) A siblicidal origin for avian brood parasitism? J Ornithol 157:219–227

Geltsch N, Hauber ME, Anderson MG, Bán M, Moskát C (2012) Competition with a host nestling for parental provisioning imposes recoverable costs on parasitic cuckoo chick's growth. Behav Process 90:378–383

Gloag R, Kacelnik A (2013) Host manipulation via begging call structure in the brood-parasitic shiny cowbird. Anim Behav 86:101–109

Gloag R, Tuero DT, Fiorini VD, Reboreda JC, Kacelnik A (2012) The economics of nestmate killing in avian brood parasites: a provisions trade-off. Behav Ecol 23:132–140

Gochfeld M (1979) Begging by nestling shiny cowbirds: adaptive or maladaptive. Living Bird 17:41–50

Godfray HC (1995) Signaling of need between parents and young: parent-offspring conflict and sibling rivalry. Am Nat 146:1–24

Göth A, Hauber ME (2004) Ecological approaches to species recognition in birds through studies of model and non-model species. Ann Zool Fenn 41:823–842

Graham DS, Middleton AL (1988) Conspecific recognition by juvenile brown-headed cowbirds. Bird Behav 8:14–22

Grim T (2005) Mimicry vs. similarity: which resemblances between brood parasites and their hosts are mimetic and which are not? Biol J Linn Soc 84:69–78

Grim T (2006a) The evolution of nestling discrimination by hosts of parasitic birds: why is rejection so rare? Evol Ecol Res 8:785–802

Grim T (2006b) Cuckoo growth performance in parasitized and unused hosts: not only host size matters. Behav Ecol Sociobiol 60:716–723

Grim T (2006c) Low virulence of brood parasitic chicks: adaptation or constraint? Ornithol Sci 5:237–242

Grim T (2007) Equal rights for chick brood parasites. Ann Zool Fenn 44:1–7

Grim T (2008) Wing-shaking and wing-patch as nestling begging strategies: their importance and evolutionary origins. J Ethol 26:9–15

Grim T (2011) Ejecting chick cheats: a changing paradigm? Front Zool 8:14–19

Grim T, Rutila J, Cassey P, Hauber ME (2009) Experimentally constrained virulence is costly for common cuckoo chicks. Ethology 115:14–22

Hahn DC, Fleischer RC (1995) DNA fingerprint similarity between female and juvenile brown-headed cowbirds trapped together. Anim Behav 49:1577–1580

Hamilton WD (1964) The genetical evolution of social behaviour. J Theor Biol 7:1–16

Hamilton WJ, Orians G (1965) Evolution of brood parasitism in altricial birds. Condor 67:361–382

Hauber ME (2002) First contact: a role for adult-offspring social association in the species recognition system of brood parasites. Ann Zool Fenn 39:291–305

Hauber ME, Kilner RM (2007) Coevolution, communication, and host chick mimicry in parasitic finches: who mimics whom? Behav Ecol Sociobiol 61:497–503

Hauber ME, Moskát C (2008) Shared parental care is costly for nestlings of common cuckoos and their great reed warbler hosts. Behav Ecol 19:79–86

Hauber ME, Ramsey CK (2003) Honesty in host-parasite communication signals: the case for begging by fledgling brown-headed cowbirds *Molothrus ater*. J Avian Biol 34:339–344

Hauber ME, Sherman PW, Paprika D (2000) Self-referent phenotype matching in a brood parasite: the armpit effect in brown-headed cowbirds (*Molothrus ater*). Anim Cogn 3:113–117

Hauber ME, Russo SA, Sherman PW (2001) A password for species recognition in a brood-parasitic bird. Proc R Soc Lond B 268:1041–1048

Honza M, Picman J, Grim T, Novak V, Capek M, Mrlik V (2001) How to hatch from an egg of great structural strength: a study of the common cuckoo. J Avian Biol 32:249–255

Honza M, Vošlajerová K, Moskát C (2007) Eviction behaviour of the common cuckoo *Cuculus canorus* chicks. J Avian Biol 38:385–389

Honza M, Feikusova K, Prochazaka P, Picman J (2015) How to hatch from the common cuckoo (*Cuculus canorus*) egg: implications of strong eggshells for the hatching muscle (musculus complexus). J Ornithol 156:679–685

Jaeckle WB, Kiefer M, Childs B, Harper RG, Rivers JW, Peer BD (2012) Comparison of eggshell porosity and estimated gas flux between the brown-headed cowbird and two common hosts. J Avian Biol 43:486–490

Jenner E (1788) Observations on the natural history of the cuckoo. Philos Trans R Soc Lond 78:219–237

Kilner RM (2005) The evolution of virulence in brood parasites. Ornithol Sci 4:55–64

Kilner RM, Davies NB (1999) How selfish is a cuckoo chick? Anim Behav 58:797–808

Kilner RM, Johnstone R (1997) Begging the question: are offspring solicitation behaviours signals of need? Trends Ecol Evol 12:11–15

Kilner RM, Noble DG, Davies NB (1999) Signals of need in parent–offspring communication and their exploitation by the common cuckoo. Nature 397:667–672

Kilner RM, Madden JR, Hauber ME (2004) Brood parasitic cowbirds use host young to procure food. Science 305:877–879

Kilpatrick AM (2002) Variation in growth of brown-headed cowbird (*Molothrus ater*) nestlings and energetic impacts on their host parents. Can J Zool 80:145–153

King AP, West MJ (1977) Species identification in the North American cowbird: appropriate responses to abnormal song. Science 195:1002–1004

Kleven O, Moksnes A, Røskaft E, Honza M (1999) Host species affects the growth rate of cuckoo (*Cuculus canorus*) chicks. Behav Ecol Sociobiol 47:41–46

Langmore NE, Hunt S, Kilner RM (2003) Escalation of a coevolutionary arms race through host rejection of brood parasitic young. Nature 422:157–160

Langmore NE, Stevens M, Maurer G, Heinsohn R, Hall ML, Peters A, Kilner RM (2011) Visual mimicry of host nestlings by cuckoos. Proc R Soc Lond B 278:2455–2463

Lawes MJ, Marthews TR (2003) When will rejection of parasite nest-lings by hosts of nonevicting avian brood parasites be favored? A misimprinting-equilibrium model. Behav Ecol 14:757–770

Lichtenstein G (2001) Selfish begging by screaming cowbirds, a mimetic brood parasite of the bay-winged cowbird. Anim Behav 61:1151–1158

Lotem A (1993) Learning to recognize nestlings is maladaptive for cuckoo *Cuculus canorus* hosts. Nature 362:743–745

Louder MIM, Ward MP, Schelsky WM, Hauber ME, Hoover JP (2015) Out on their own: a test of adult-assisted dispersal in fledgling brood parasites reveals solitary departures from hosts. Anim Behav 110:29–37

Madden JR, Davies NB (2006) A host-race difference in begging calls of nestling cuckoos *Cuculus canorus* develops through experience and increases host provisioning. Proc R Soc Lond B 273:2343–2351

Mainwaring MC, Hartley IR (2012) Causes and consequences of differential growth in birds: a behavioural perspective. Adv Study Behav 44:225–277

Martin TE, Lloyd P, Bosque C, Barton DC, Biancucci AL, Cheng YR, Ton R (2011) Growth rate variation among passerine species in tropical and temperate sites: an antagonistic interaction between parental food provisioning and nest predation risk. Evolution 65:1607–1622

McMaster DG, Sealy SG (1998) Short incubation periods of brown-headed cowbirds: how do cowbird eggs hatch before yellow warbler eggs? Condor 100:102–111

Mermoz ME, Ornelas JF (2004) Phylogenetic analysis of life-history adaptations in parasitic cowbirds. Behav Ecol 15:109–119

Morton ES, Farabaugh SM (1979) Infanticide and other adaptations of the nestling striped cuckoo *Tapera naevia*. Ibis 121:212–213

Naef-Daenzer B, Grüebler MU (2016) Post-fledging survival of altricial birds: ecological determinants and adaptation. J Field Ornithol 87:227–250
Payne RB (1977) The ecology of brood parasitism in birds. Annu Rev Ecol Syst 8:1–28
Payne RB, Payne LL, Woods JL (1998) Song learning in brood-parasitic indigobirds *Vidua chalybeata*: song mimicry of the host species. Anim Behav 55:1537–1553
Payne RB, Payne LL, Woods JL, Sorenson MD (2000) Imprinting and the origin of parasite–host species associations in brood-parasitic indigobirds, *Vidua chalybeata*. Anim Behav 59:69–81
Payne RB, Woods JL, Payne LL (2001) Parental care in estrildid finches: experimental tests of a model of *Vidua* brood parasitism. Anim Behav 62:473–483
Portugal SJ, Hauber ME, Maurer G, Stokke BG, Grim T, Cassey P (2014) Rapid development of brood-parasitic cuckoo embryos cannot be explained by increased gas exchange through the eggshell. J Zool 293:219–226
Rasmussen JL, Sealy SG (2006) Hosts feeding only brown-headed cowbird fledglings: where are the host fledglings? J Field Ornithol 77:269–279
Redondo T (1993) Exploitation of host mechanisms for parental care by avian brood parasites. Etología 3:235–297
Remeš V (2010) Explaining postnatal growth plasticity in a generalist brood parasite. Naturwissenschaften 97:331–335
Remeš V, Martin TE (2002) Environmental influences on the evolution of growth and developmental rates in passerines. Evolution 56:2505–2518
Remeš V, Matysioková B (2016) Survival to independence in relation to pre-fledging development and latitude in songbirds across the globe. J Avian Biol 47:610–618
Rivers JW (2007) Nest mate size, but not short-term need, influences begging behavior of a generalist brood parasite. Behav Ecol 18:222–230
Rivers JW, Loughin TM, Rothstein SI (2010) Brown-headed cowbird nestlings influence nestmate begging, but not parental feeding, in hosts of three distinct sizes. Anim Behav 79:107–116
Rivers JW, Blundell MA, Loughin TM, Peer BD, Rothstein SI (2013) The exaggerated begging behaviour of an obligate avian brood parasite is shared with a nonparasitic close relative. Anim Behav 86:529–536
Rothstein SI (1990) A model system for coevolution: avian brood parasitism. Annu Rev Ecol Evol Syst 21:481–508
Royle NJ, Hartley IR, Owens IP, Parker GA (1999) Sibling competition and the evolution of growth rates in birds. Proc R Soc Lond B 266:923–932
Sato NJ, Tokue K, Noske RA, Mikami OK, Ueda K (2010) Evicting cuckoo nestlings from the nest: a new anti-parasitism behaviour. Biol Lett 6:67–69
Schuetz JG (2005a) Low survival of parasite chicks may result from their imperfect adaptation to hosts rather than expression of defences against parasitism. Evolution 59:2017–2024
Schuetz JG (2005b) Reduced growth but not survival of chicks with altered gape patterns: implications for the evolution of nestling similarity in a parasitic finch. Anim Behav 70:839–848
Sealy SG, Lorenzana JC (1997) Feeding of nestling and fledgling brood parasites by individuals other than the foster parents: a review. Can J Zool 75:1739–1752
Shy MM (1982) Interspecific feeding among birds: a review. J Field Ornithol 53:370–393
Slagsvold T, Hansen BT (2001) Sexual imprinting and the origin of obligate brood parasitism in birds. Am Nat 158:354–367
Soler M (1990) Relationships between the great spotted cuckoo *Clamator glandarius* and its corvid hosts in a recently colonized area. Ornis Scand 21:212–223
Soler M (2002) Breeding strategy and begging intensity: influences on food delivery by parents and host selection by parasitic cuckoos. In: Wright W, Leonard ML (eds) The evolution of begging. Competition, cooperation and communication. Kluwer Academic, Dordrecht, pp 413–427
Soler M (2009) Co-evolutionary arms race between brood parasites and their hosts at the nestling stage. J Avian Biol 40:237–240

Soler M, de Neve L (2013) Brood mate eviction or brood mate acceptance by brood parasitic nestlings? An experimental study with the non-evictor great spotted cuckoo and its magpie host. Behav Ecol Sociobiol 67:601–607

Soler M, Soler JJ (1991) Growth and development of great spotted cuckoos and their magpie host. Condor 1:49–54

Soler M, Soler JJ (1999) Innate versus learned recognition of conspecifics in great spotted cuckoos *Clamator glandarius*. Anim Cogn 2:97–102

Soler M, Palomino JJ, Martínez JG, Soler JJ (1994) Activity, survival, independence and migration of fledgling great spotted cuckoos. Condor 96:802–805

Soler M, Martínez JG, Soler JJ, Møller AP (1995a) Preferential allocation of food by magpies *Pica pica* to great spotted cuckoo *Clamator glandarius* chicks. Behav Ecol Sociobiol 37:7–13

Soler M, Palomino JJ, Martínez JG, Soler JJ (1995b) Communal parental care by monogamous magpie hosts of fledgling great spotted cuckoos. Condor 97:804–810

Soler M, Soler JJ, Martínez JG, Moreno J (1999) Begging behaviour and its energetic cost in great spotted cuckoo and magpie host chicks. Can J Zool 77:1794–1800

Soler M, De Neve L, Roldán M, Macías-Sánchez E, Martín-Gálvez D (2012) Do great spotted cuckoo nestlings beg dishonestly? Anim Behav 83:163–169

Soler M, De Neve L, Roncalli G, Macías-Sánchez E, Ibáñez-Álamo JD, Pérez-Contreras T (2014a) Great spotted cuckoo fledglings are disadvantaged by magpie host parents when reared together with magpie nestlings. Behav Ecol Sociobiol 68:333–342

Soler M, De Neve L, Pérez-Contreras T, Rubio LA (2014b) Comparison of digestive efficiency in the parasitic great spotted cuckoo and its magpie host nestlings. Biol J Linn Soc 111:280–289

Soler M, Pérez-Contreras T, Ibáñez-Álamo JD, Roncalli G, Macías-Sánchez E, De Neve L (2014c) Great spotted cuckoo fledglings often receive feedings from other magpie adults than their foster parents: which magpies accept to feed foreign cuckoo fledglings? PLoS One 9:e107412

Soler M, Rubio LA, Pérez-Contreras T, Ontanilla J, De Neve L (2017) Nestling great spotted cuckoo intestinal digestibility is less efficient than that of magpie host nestlings. Biol J Linn Soc 122:675–680

Spottiswoode CN, Koorevaar J (2012) A stab in the dark: chick killing by brood parasitic honeyguides. Biol Lett 8:241–244

Tanaka KD, Ueda K (2005) Horsfield's hawk-cuckoo nestlings simulate multiple gapes for begging. Science 308:653–653

Tanaka KD, Morimoto G, Ueda K (2005) Yellow wing-patch of a nestling Horsfield's hawk cuckoo *Cuculus fugax* induces miscognition by hosts: mimicking a gape? J Avian Biol 36:461–464

ten Cate C, Voss DR (1999) Sexual imprinting and evolutionary processes in birds: a reassessment. Adv Study Behav 28:2–31

Tokue K, Ueda K (2010) Mangrove Gerygones *Gerygone laevigaster* eject Little Bronze-cuckoo *Chalcites minutillus* hatchlings from parasitized nests. Ibis 152:835–839

Tuero DT, Fiorini VD, Mahler B, Reboreda JC (2013) Do sex ratio and development differ in sexually size-dimorphic shiny cowbirds (*Molothrus bonariensis*) parasitizing smaller and larger hosts? Biol J Linn Soc 110:442–448

Wang N, Kimball R (2012) Nestmate killing by obligate brood parasitic chicks: is this linked to obligate siblicidal behavior? J Ornithol 153:825–831

Woodward PW (1983) Behavioral ecology of fledgling brown-headed cowbirds and their hosts. Condor 85:151–163

MIX
Papier aus verantwortungsvollen Quellen
Paper from responsible sources
FSC® C105338

If you have any concerns about our products,
you can contact us on
ProductSafety@springernature.com

In case Publisher is established outside the EU,
the EU authorized representative is:
Springer Nature Customer Service Center GmbH
Europaplatz 3, 69115 Heidelberg, Germany

Printed by Libri Plureos GmbH
in Hamburg, Germany